Duct Acoustics

Using a hands-on approach, this self-contained toolkit covers topics ranging from the foundations of duct acoustics to the acoustic design of duct systems, through practical modeling, optimization and measurement techniques. Discover in-depth analyses of one- and three-dimensional models of sound generation, propagation and radiation, as techniques for assembling acoustic models of duct systems from simpler components are described. Identify the weaknesses of mathematical models in use and improve them by measurement when needed. Cope with challenges in acoustic design, and improve understanding of the underlying physics, by using the tools described. An essential reference for engineers and researchers who work on the acoustics of fluid machinery ductworks.

Erkan Dokumacı is Professor Emeritus of Mechanical Engineering at Dokuz Eylül University (İzmir, Turkey). He is currently on the Editorial Advisory Board for the *Journal of Sound and Vibration* and is the author of software used for acoustic design of mufflers and silencers, as well as many scientific papers on duct acoustics.

Duct Acoustics

Fundamentals and Applications to Mufflers and Silencers

ERKAN DOKUMACI
Dokuz Eylül University

CAMBRIDGE
UNIVERSITY PRESS

University Printing House, Cambridge CB2 8BS, United Kingdom

One Liberty Plaza, 20th Floor, New York, NY 10006, USA

477 Williamstown Road, Port Melbourne, VIC 3207, Australia

314–321, 3rd Floor, Plot 3, Splendor Forum, Jasola District Centre, New Delhi – 110025, India

79 Anson Road, #06–04/06, Singapore 079906

Cambridge University Press is part of the University of Cambridge.

It furthers the University's mission by disseminating knowledge in the pursuit of education, learning, and research at the highest international levels of excellence.

www.cambridge.org
Information on this title: www.cambridge.org/9781108840750
DOI: 10.1017/9781108887656

First published 2021

A catalogue record for this publication is available from the British Library.

Library of Congress Cataloging-in-Publication Data
Names: Dokumaci, Erkan, 1942– author.
Title: Duct acoustics : fundamentals and applications to mufflers and silencers / Erkan Dokumaci, Dokuz Eylül University.
Description: Cambridge, UK ; New York, NY : Cambridge University Press, 2021. | Includes bibliographical references and index.
Identifiers: LCCN 2020042217 (print) | LCCN 2020042218 (ebook) | ISBN 9781108840750 (hardback) | ISBN 9781108887656 (epub)
Subjects: LCSH: Gas tubing–Soundproofing. | Air ducts–Acoustic properties–Mathematical models. | Fluids–Acoustic properties–Mathematical models. | Hydraulic machinery–Noise. | Gas-machines–Noise. | Acoustical engineering.
Classification: LCC TP757 .D65 2021 (print) | LCC TP757 (ebook) | DDC 620.2/3–dc23
LC record available at https://lccn.loc.gov/2020042217
LC ebook record available at https://lccn.loc.gov/2020042218

ISBN 978-1-108-84075-0 Hardback

To Figen and Pınar

Contents

Preface

Duct-borne noise is a major contributor to the environmental noise: sound waves generated by fluid machinery action on the working fluid are transmitted along the connecting ductwork and cause a nuisance when radiated from an inlet or outlet terminal open to a noise sensitive external environment. It is traditional to control transmission of duct-borne sound energy by mounting silencers, mufflers and resonators on the connecting ductwork. These components themselves consist of ducts (or pipes or tubes) of various types and sizes, which are arranged in multifarious configurations so that the transmission of the incident sound energy is controlled by absorption, dissipation, reflection or interference. Engineering of ductworks with such noise control devices and/or other components, such as after-treatment devices and turbochargers, whose primary purpose is not noise control, to achieve a given reduction of duct-borne noise reduction is a major area of application of duct acoustics, the science of sound waves in spaces confined by guiding surfaces. The scope of applications of duct acoustics is large and includes the intake and exhaust ductworks of fluid machines of all types and sizes in surface and air transport vehicles, heating, ventilation and air conditioning, gas transport, refrigeration, power generation and many other industries.

In this book, acoustic design of fluid machinery ductworks is described from first principles. Because of the focus of the book, the discussion is limited to the linear branch of the subject. This may seem to be a serious limitation, but in reality acoustic wave motions encountered in practical duct systems remain mostly in the linear range, which is fortunate, since it allows for the modeling of complex ductworks and inclusion of various effects, which would otherwise be impracticable for engineering analyses.

The objective of the book is to assist readers to acquire an understanding of the principles and mathematical models that form the foundations of duct acoustics and to develop analytical skills for acoustic modeling and analysis of practical ductworks including mufflers and silencers. It gives in-depth analyses of one- and three-dimensional models of sound propagation in ducts and certain basic duct arrangements and uses a network approach to extend the analysis to duct systems which consist of an assembly of these components. Network methods have been successful over the years in meeting the needs of engineers and researchers for quick and insightful approaches to acoustic design of complex duct systems at low frequencies.

They are used in this book also for acoustic modeling of duct systems in three dimensions.

In view of its objectives, the presentation focuses, as much as possible, on solutions that are transparent to the physics of the systems considered and to straightforward computations. Exact or approximate analytical solutions serve this purpose incomparably better than numerical methods. The success of numerical methods largely depends on the availability of supporting tools, such as solid modeling and pre-processing codes, large system solvers and post-processing codes. Such capabilities are usually provided in software written by teams of specialists and it makes no sense here to attempt to describe implementations of specific numerical methods in all details. However, recourse to a numerical method is inevitable for the development of three-dimensional models of ducts with irregular geometry or solutions of basic problems that are not tractable to analysis. In Chapter 2, we describe how solutions obtained by using the finite element method and the boundary element method for ductwork components can be used in acoustic models of ductwork assemblies. In Chapter 6 we discuss the application of the finite element method for the calculation of transverse modes in straight and curved ducts. In Chapter 9, acoustic reflection and radiation from an open end of a duct is considered essentially in the theoretical formalism of the boundary element method.

The design adopted for the book covers the acoustic aspects of the design process as a source–path–receiver problem and encompasses ductworks with subsonic low Mach number mean flows, ranging in size from toy engines to large power generation plants. Concept generation, source characterization and prototype development are usually the major pain-points in acoustic design of duct systems. These challenges are interlinked through tasks such as acoustic modeling, computation, optimization and measurement, to which the chapters of the book are devoted. This content should prove useful for engineers and researchers with interest in acoustic design of fluid machinery ducting components and networks to read and keep as a reference text. The book should also be useful as supplementary reading in courses on engineering acoustics and noise control, since duct acoustics is a topic which is normally taught in these courses.

The subject matter is presented in 13 chapters and 3 appendices. It is largely self-contained and, in general, demands knowledge and skills in engineering mathematics and physics at senior undergraduate level to proceed through most of the sections. The NIST database "Digital Library of Mathematical Functions" may be consulted whenever a primer is needed for the properties of functions occurring in solutions.

The first chapter describes the basic analytic concepts and operations which are invoked throughout the book. They are collected in this preliminary chapter as a primer and also to avoid interruption of the continuity of discussions on the principal subjects.

The second chapter describes the block diagram-based network approach for construction of acoustic models of duct systems from the simpler components (which are described in later chapters). This topic is considered early in the book because it describes the format used in later chapters in mathematical representation of acoustic models of ductwork components and their assemblies.

Chapter 3 introduces the general analytic theory of one-dimensional sound propagation in ducts and presents acoustic models for uniform, non-uniform and inhomogeneous ducts with hard or finite impedance walls and parallel sheared mean flow. Some coupled duct configurations such as area changes and junctions occur recurrently in practical systems. One-dimensional models of these configurations are given in Chapter 4.

Chambers and resonators are used as noise control devices in almost all industrial duct systems. In Chapter 5, transmission loss is defined and, using the acoustic models and the assembly techniques described in previous chapters, transmission loss characteristics of various chamber and resonator types are demonstrated. Also discussed is the calculation of the shell noise and mean pressure loss (or back pressure), which may impose trade-offs on effective use of these devices in duct-borne noise control.

Chapter 6 introduces the three-dimensional analytic theory of sound propagation in ducts and presents acoustic models of hard-walled and lined uniform ducts. Also discussed are the effects of gradual cross section non-uniformity, the circular curvature of the duct axis, and sheared and vortical mean flows. Chapter 7 describes acoustic three-dimensional models of several coupled duct configurations. The acoustic models described in Chapters 6 and 7 extend the one-dimensional acoustic elements described in Chapters 3 and 4 to three dimensions. In Chapters 3 to 7, the fluid is assumed to be inviscid. Effects of the viscosity and thermal conductivity of the fluid are considered in Chapter 8.

In most duct systems, propagation of duct-borne sound terminates with a duct which opens to an exterior environment. Chapter 9 describes modeling of open ends of ducts and the acoustic field radiated from an open end. This enables acoustic model of a duct system to be extended from the source to the receiver. Then, to complete the source–path–receiver model, it remains to derive an acoustic model for the source process. This step is taken in Chapter 10, where we describe analytical actuator-disk models for the basic acoustic source mechanisms, namely, non-steady mass and heat injection and force application, applications of which are demonstrated on internal combustion engines, turbomachinery and combustion chambers.

Chapter 11 describes calculation of the sound pressure level at a point in the acoustic field of an in-duct source. Insertion loss of a silencer is shown to be represented approximately by source-independent parameters under certain conditions. The discussion encompasses multi-modal sound propagation and radiation and the ASHRAE method of silencer sizing in ventilation and air distribution systems.

Chapter 12 describes the contemporary measurement methods in duct acoustics. Acoustic measurements are necessary in order to validate theoretical models and also to develop acoustic models when theoretical approaches tend to be inadequate or impossible. The multiple wall-mounted microphone method is introduced from first principles and its applications to the measurement of the characteristics of acoustic sources and of passive system elements are described.

Acoustic design of ductworks such as fluid machinery intake and exhaust systems usually requires large number of iterations for concept validation and prototype development. The network approach is ideally suited for this purpose, but systematic search and optimization methods are indispensable for quick and efficient progress. The last chapter, Chapter 13, discusses the acceleration of iterative design calculations and handling uncertainties about model parameters. We also present an approach which brings an inverse perspective to the conventional target based acoustic design calculations.

The chapters are organized so that they are developed one after the other; however, the book need not necessarily be studied in this order, although skimming through all chapters is recommended on first reading. Some topics have been included in appendixes for reference. Duct acoustics builds on the fundamental conservation laws of fluid dynamics and Appendix A gives derivations of these laws in integral and differential forms. Appendix B describes the basic bulk-acoustic properties of rigid-frame fibrous materials, which are frequently used in ducts for sound absorption. Appendix C gives a review of the existing lumped-parameter models of impedance of compact apertures. Knowledge of the impedance of apertures is needed when designing noise control components involving perforated ducts, baffle plates and absorptive liner shields.

The symbols used in the book are defined when first used. When an expression is taken from a different section, the original notation is preserved, unless stated otherwise. Equations and figures are numbered consecutively within the chapters and appendices. References are quoted in the main text in square brackets and listed at the end of each chapter. Citations aim to direct the reader to a source publication or a proof of a non-obvious result or to further reading, rather than giving a comprehensive coverage of the literature, which is, in a subject like duct acoustics, not humanly possible anyway. The presentation is influenced by the writings of classical and contemporary authors on acoustics, and works and personalities on the subject in various research centers have been a source of inspiration all the way.

1 Some Preliminaries

1.1 Introduction to the Linear Theory of Sound Wave Motion

What we generally call sound wave motion is a form of unsteady motion of a fluid and is governed by the fundamental laws of conservation of mass, momentum and energy and the thermodynamic state equations pertinent to the fluid in question. These equations are classical and are given in Appendix A, both in integral and differential forms, together with their derivations. There are many books treating these equations in more detail and giving their solutions for specific fluid flow problems. In this book, we are concerned about their applications in duct acoustics. The dilemma here is that, since these equations are non-linear, they can be solved only numerically even for the simplest of duct systems in industry and this is very expensive in terms of computational resources and time. On the other hand, it is a heuristic fact that in most noise control problems, including fluid machinery duct-borne noise, it is adequately accurate to assume that sound wave motion remains within linear limits. This is a very convenient situation that is worth taking advantage of, because linear differential equations are very much easier to solve than non-linear ones and linearization of non-linear equations is a mathematically well-understood process having its roots in Taylor series expansions of continuous functions. But, passing from the basic non-linear equations to those governing linear sound wave motions is not trivial and involves some mathematical and conceptual difficulties. In particular, we are confronted with certain subtle issues when dealing with turbulent flows in this manner. In this section, we discuss how we go about these problems to develop one-dimensional and three-dimensional linear theories of sound wave motion that are used in this book for modeling of sound transmission in ducts and duct systems.

1.1.1 Linearization Hypothesis

In many problems of unsteady fluid flow, it is permissible to assume that properties of flow at a given point $\mathbf{x}$ at time t may be represented by an asymptotic expansion of the form $q = q_o + q' + q'' + \cdots$, where $q = q(t, \mathbf{x}) \in p, \rho, \mathbf{v}, T, c, \gamma, \ldots$ denotes any property of the fluid, for example, pressure (p), density (ρ), particle velocity ($\mathbf{v}$), absolute temperature (T), speed of sound (c), the ratio of specific heat coefficients (γ) and so on. The same symbol with subscript "o," like $q_o = q_o(\mathbf{x})$, denotes the steady (time-independent) part of a property and the corresponding part of the flow is called mean flow. The same symbol with a prime, like $q' = q'(t, \mathbf{x})$, denotes a fluctuating part of zero mean that is much small compared to q_o, that is, $q'/q_o = O(\varepsilon)$, except for the particle velocity which scales as $|\mathbf{v}'|/c_o = O(\varepsilon)$, where $\varepsilon << 1$.[1] Similarly, two primes, like in $q'' = q''(t, \mathbf{x})$, denotes a fluctuating part of zero mean that is much small compared to q', that is, $q''/q_o = O(\varepsilon^2)$, except for the particle velocity which scales as $|\mathbf{v}''|/c_o = O(\varepsilon^2)$, and so on. In this book, we allow $|\mathbf{v}_o|$ to be well in the subsonic range, that is, substantially smaller than c_o.

The fundamental non-linear equations of unsteady fluid motion are linearized by substituting in them $q = q_o + q' + q'' + \cdots$ and then invoking two hypotheses:

(i) mean flow satisfies the fundamental non-linear equations;
(ii) linear sound wave motion is given by the $O(\varepsilon)$ terms.

The latter hypothesis is largely based on our experience with the amplitudes of sound waves we measure in everyday life and engineering applications. For example, the threshold of pain of human hearing is about $p'_{rms} = 200$ Pa, which is only 0.2% of the atmospheric pressure, p_o, where the subscript "*rms*" denotes the root-mean-square value (see Section 1.2.4). Very close to a jet engine or inside the exhaust ports of an internal combustion engine p'_{rms}/p_o is only about 2–3%, which is about as large as it can be expected to get in many engineering applications.

The $O(\varepsilon)$ terms in expansions $q = q_o + q' + q'' + \cdots$ are usually referred to as the acoustic terms. As will be seen in later chapters, acoustic pressure and density in a perfect gas are related as $p'/p_o = \gamma_o \rho'/\rho_o$ under isentropic conditions. Consequently, acoustic density is also small to $O(\varepsilon)$. On the other hand, the particle velocity $\mathbf{v}' = \mathbf{e}v'$ of longitudinal wave motion in direction of the unit vector $\mathbf{e}$ scales with the acoustic pressure as $v' \sim p'/\rho_o c_o$ [1]. Therefore, $v'/c_o \sim p'/p_o$, showing that v'/c_o is also small to $O(\varepsilon)$. These orders of magnitudes set the stage for the linearization of the fundamental non-linear equations of unsteady fluid motion.

1.1.2 Partitioning Turbulent Fluctuations

A complication arises, however, if the flow is turbulent, because then time-dependent fluctuating structures of multiple scales (called eddies) are also convected with the mean flow and it becomes necessary to separate these from acoustic fluctuations. Turbulent fluctuations may be assumed to be of constant density, but, at the present time, the partitioning problem has no clear-cut solution. For this reason, here we take a pragmatic approach and look into the extent to which linearization may work without detailed modeling of turbulent fluctuations.

Turbulence is known as the most extensively studied, but the least understood, topic of fluid dynamics. However, it is a well-established empirical fact that flow in a duct turns from an orderly sheared form (called laminar) to a chaotic (turbulent) one at about $\mathrm{Re} = \rho_o U_o D/\mu_o \approx 2000$, where Re is the Reynolds number, D and U_o denote,

[1] The rate of growth of a function is called its order and denoted by the letter O (which is also called the Landau symbol). For example, if $f(x)$ is of $O(g(x))$, this means that f does not grow faster than g.

respectively, a characteristic dimension of the duct section and a characteristic velocity of the flow, and μ_o denotes the shear viscosity of the fluid. Based on Lighthill's theory of aerodynamically generated sound, we know that turbulence can generate sound waves and attenuate sound waves. Acoustic power radiated, W_V, say, from a compact region, V, of subsonic turbulence may be estimated qualitatively from Lighthill's theory, but this theory involves rather advanced ideas which could not be treated in this book fully. So, it suffices to state that, if $M = U_c/c_o << 1$, that is, the flow is subsonic in low Mach number range, W_V scales as $W_V{\sim}\rho_o\left(U_c^8/c_o^5\right)(V/L)$, where U_c and L denote the appropriate convection velocity and length scales of the turbulence [2]. This is the famous eight-power law for sound power radiated from free turbulence. Taking the volume V as L^3, this law can be written as $W_V{\sim}\rho_o U_c^8 L^2/c_o^5$. A quantity which is useful for further insight is the quotient of W_V and the mechanical power supplied to the fluid, since this quotient may be considered as the acoustic efficiency of turbulence [3]. Thus, correlating the mechanical power supplied to the fluid with $\left(\rho_o U_c^2/2\right)L^2 U_c$, we obtain $W_V/\rho_o U_c^3 L^2{\sim}M^5$, which shows that the acoustic efficiency of turbulence is of $O\left(M^5\right)$. For subsonic low Mach number flows $(M^2 << 1)$, this is small compared to the efficiency of the usual ducted primary fluid machinery sources. For example, it may similarly be shown that the corresponding acoustic efficiency of pressure loading on rotating blades as a sound source is of $O\left(M^3\right)$ (see Section 10.4), and that due to the unsteady mass injection is of $O(M)$(see Section 10.2), for the characteristic flow velocities and length scales involved in these processes [3].

Thus, the effect of turbulence as an acoustic source may be neglected in subsonic low Mach number flows dominated by primary fluid machinery sources. However, even if turbulence fluctuations are of secondary importance as such, they cause scattering of incident acoustic waves and absorb energy from the acoustic waves. Both of these mechanisms lead to attenuation of sound waves. Based on Lighthill's theory, Tack and Lambert give the following semi-empirical upper-bound estimate for the fraction of the incident acoustic energy scattered by isotropic turbulence per unit duct length, namely, $0.0032\, M_o^2(\omega L/c_o)^6/L$, where M_o denotes the Mach number of the free stream velocity, L denotes the size of eddies (which can be no larger than the duct section size), and ω denotes the radian frequency of the incident sound wave [4]. Thus, the effect of scattering may be neglected at relatively low frequencies and subsonic low Mach numbers. But absorption tends to be important at lower frequencies. It is attributed to the irreversible straining of eddies by the sound wave and is mainly confined in a duct to the turbulent boundary layer where the particle velocity gradients are large. The analysis of the phenomenon is difficult because it requires complete understanding of the turbulent boundary layer, which in turn requires accurate solutions of the non-linear equations of unsteady fluid motion including effects of viscosity and thermal conductivity and all length scales of turbulent and acoustic fluctuations. For this reason, we rely on the progress achieved by simplified models at subsonic low Mach numbers. The simplest model assumes that the sound field in a duct is not influenced by the presence of turbulence and vice versa. Then the problem reduces to the study of sound waves convected with a mean flow that is characterized by the velocity profile. This model, which is usually referred to as the quasi-laminar model, is adopted throughout this book. It is adequately accurate at relatively high frequencies, but underestimates attenuation at low frequencies. Howe, taking into account the frequency dependence of the straining of turbulence by sound waves, proposed a semi-empirical model which correlates satisfactorily with measurements also at low frequencies [5]. This model is described in Section 3.9.3.4, where it is also shown how it can be integrated with the quasi-laminar model.

1.1.3 Linearization of Inviscid Fluid Flow

Thus, unsteady motions of fluids may be linearized assuming a laminar mean flow or a quasi-laminar one with a subsonic low Mach number, if the mean flow is turbulent. The process is described here for inviscid fluids and differential forms of the fluid dynamic equations with no source terms; however, the corresponding equations with source and viscothermal terms, and their integral forms are linearized in exactly the same way.

From Equations (A.21) and (A.23), the conservation of mass and momentum in a source-free region of a duct containing an inviscid fluid can be expressed in differential forms as:

$$\frac{\partial \rho}{\partial t} + \mathbf{v}\cdot\nabla\rho + \rho\nabla\cdot\mathbf{v} = 0 \tag{1.1}$$

$$\rho\left(\frac{\partial \mathbf{v}}{\partial t} + \mathbf{v}\cdot\nabla\mathbf{v}\right) + \nabla p = 0, \tag{1.2}$$

respectively, where ∇ denotes the gradient operator and $\nabla\cdot$ denotes the divergence operator. To obtain the linearized forms of these equations, we substitute $p = p_o + p' + \cdots$, $\rho = \rho_o + \rho' + \cdots$ and $\mathbf{v} = \mathbf{v}_o + \mathbf{v}' + \cdots$. Equation (1.1) then becomes

$$\left(\frac{\partial \rho_o}{\partial t} + \mathbf{v}_o\cdot\nabla\rho_o + \rho_o\nabla\cdot\mathbf{v}_o\right) + \left(\frac{\partial \rho'}{\partial t} + \mathbf{v}_o\cdot\nabla\rho' + \rho_o\nabla\cdot\mathbf{v}' + \mathbf{v}'\cdot\nabla\rho_o + \rho'\nabla\cdot\mathbf{v}_o\right)$$
$$+ \left(\frac{\partial \rho''}{\partial t} + \mathbf{v}_o\cdot\nabla\rho'' + \mathbf{v}'\cdot\nabla\rho' + \mathbf{v}''\cdot\nabla\rho_o + \rho_o\nabla\cdot\mathbf{v}'' + \rho'\nabla\cdot\mathbf{v}' + \rho''\nabla\cdot\mathbf{v}_o\right) + O(\varepsilon^3) = 0 \tag{1.3}$$

In view of the first linearization hypothesis stated in Section 1.1.1, the first bracket vanishes identically, giving the following continuity equation for the mean flow, namely,

$$\nabla\cdot(\rho_o\mathbf{v}_o) = 0 \tag{1.4}$$

since mean flow is time independent.

The terms in the second bracket in Equation (1.3) contain only single-primed variables. These terms represent the $O(\varepsilon)$ terms. This may not be immediately evident from stipulations like $\rho'/\rho_o = O(\varepsilon)$ and $|\mathbf{v}'|/c_o = O(\varepsilon)$, which we have made in Section 1.1.1 about the order of magnitudes of the acoustic (single primed) variables. To see that the consideration of single-primed variables as $O(\varepsilon)$ terms is compatible with these stipulations, observe that we can always write identities like $\mathrm{L}q' = q_o\mathrm{L}(q'/q_o) + (q'/q_o)\mathrm{L}q_o \sim O(\varepsilon)$, for thermodynamic properties $q \in p, \rho, \ldots,$ where L denotes a linear partial differential operator, for example, $\mathrm{L} = \mathbf{v}_o.\nabla$, and $\nabla\cdot\mathbf{v}' = c_o\nabla\cdot(\mathbf{v}'/c_o) + (\mathbf{v}'/c_o)\cdot\nabla c_o \sim O(\varepsilon)$ for the particle velocity.

It can similarly be seen that double primed quantities represent $O(\varepsilon^2)$ terms. Then, the third bracket in Equation (1.3), which contains only double-primed quantities and products of single-primed quantities, is of $O(\varepsilon^2)$ and it is neglected by the linearization hypothesis as being small compared to the first-order terms in the second bracket. Thus, the linearized form of Equation (1.1) is given by the vanishing of the second bracket in Equation (1.2), that is,

$$\frac{\partial \rho'}{\partial t} + \mathbf{v}_o\cdot\nabla\rho' + \rho_o\nabla\cdot\mathbf{v}' + \mathbf{v}'\cdot\nabla\rho_o + \rho'\nabla\cdot\mathbf{v}_o = 0. \tag{1.5}$$

This is called the linearized (or acoustic) continuity equation.

Equation (1.2) is linearized similarly (here, for brevity, we give the terms up to $O(\varepsilon^2)$ only):

$$\left(\rho_o\frac{\partial \mathbf{v}_o}{\partial t} + \rho_o\mathbf{v}_o\cdot\nabla\mathbf{v}_o + \nabla p_o\right) + \left(\rho_o\left(\frac{\partial \mathbf{v}'}{\partial t} + \mathbf{v}_o\cdot\nabla\mathbf{v}' + \mathbf{v}'\cdot\nabla\mathbf{v}_o\right)\right.$$
$$\left. + \rho'\mathbf{v}_o\cdot\nabla\mathbf{v}_o + \rho'\frac{\partial \mathbf{v}_o}{\partial t} + \nabla p'\right) + O(\varepsilon^2) = 0 \tag{1.6}$$

Again, the first bracket vanishes identically, giving the momentum equation for the mean flow:

$$\rho_o\mathbf{v}_o\cdot\nabla\mathbf{v}_o + \nabla p_o = 0 \tag{1.7}$$

and the linearized form of Equation (1.2) follows as

$$\rho_o\left(\frac{\partial \mathbf{v}'}{\partial t} + \mathbf{v}_o\cdot\nabla\mathbf{v}' + \mathbf{v}'\cdot\nabla\mathbf{v}_o\right) + \rho'\mathbf{v}_o\cdot\nabla\mathbf{v}_o + \nabla p' = 0, \tag{1.8}$$

which is called linearized (or acoustic) momentum equation.

Sound wave motion in a duct containing an inviscid fluid is thus governed by Equations (1.5) and (1.8). Much of the material contained in this book is concerned with the solutions of these equations for the acoustic variables p', ρ' and $\mathbf{v}'$, given the fluid type and the mean flow properties. Therefore, one more scalar equation is required for closing the solution of the acoustic variables. For inviscid fluids, this equation comes from the conservation of energy for the unsteady motion. As shown in Appendix A.3.3, in a region of an inviscid fluid free of sources, the energy equation reduces to

$$\frac{\partial s}{\partial t} + \mathbf{v}\cdot\nabla s = 0, \tag{1.9}$$

where s denotes the specific entropy of the fluid. This means that the specific entropy of a fluid particle remains constant (but that constant can be different for different particles), which in turn means that the flow is reversible, that is, adiabatic (no heat transfer is involved) and reversible. Since the specific entropy is not involved in the continuity and momentum equations, we use the thermodynamic state equation for the specific entropy, Equation (A.15), to write Equation (1.9) in terms of the pressure and particle velocity as:

$$\frac{\partial p}{\partial t} + \mathbf{v}\cdot\nabla p = c^2\left(\frac{\partial \rho}{\partial t} + \mathbf{v}\cdot\nabla\rho\right), \tag{1.10}$$

where the speed of sound, c, may be decomposed as $c = c_o + c' + \cdots$, like other fluid properties. Upon linearization of Equation (1.10), we obtain for the acoustic fluctuations:

$$\frac{\partial p'}{\partial t} + \mathbf{v}_o\cdot\nabla p' + \mathbf{v}'\cdot\nabla p_o = c_o^2\left(\frac{\partial \rho'}{\partial t} + \mathbf{v}_o\cdot\nabla\rho' + \mathbf{v}'\cdot\nabla\rho_o + 2c'\mathbf{M}_o\cdot\nabla\rho_o\right), \tag{1.11}$$

where $\mathbf{M}_o = \mathbf{v}_o/c_o$ denotes the local Mach number of the mean flow velocity. The corresponding equation for the mean flow is

$$\mathbf{v}_o\cdot\nabla p_o = c_o^2\mathbf{v}_o\cdot\nabla\rho_o. \tag{1.12}$$

A form of mean flow which satisfies this is $\nabla p_o = c_o^2\nabla\rho_o$, which implies that the mean specific entropy, s_o, is constant everywhere. This is called homentropic flow. Then Equation (1.11) may be simplified as

$$\frac{\partial p'}{\partial t} + \mathbf{v}_o\cdot\nabla p' = c_o^2\left(\frac{\partial \rho'}{\partial t} + \mathbf{v}_o\cdot\nabla\rho' + 2c'\mathbf{M}_o\cdot\nabla\rho_o\right). \tag{1.13}$$

For subsonic low Mach number mean flows, the term involving c' may be neglected as being small compared to the remaining terms in Equations (1.11) and (1.13). On this premise, Equation (1.13) is usually implemented tacitly as

$$\frac{\partial p'}{\partial t} + \mathbf{v}_o\cdot\nabla p' = c_o^2\left(\frac{\partial \rho'}{\partial t} + \mathbf{v}_o\cdot\nabla\rho'\right). \tag{1.14}$$

On the other hand, in many applications, the mean flow gradients are small and the assumptions $p_o \approx$ constant and $\rho_o \approx$ constant (or $T_o \approx$ constant) may be adequately accurate for practical purposes. Then Equation (1.12) is also satisfied and, since in view of the state postulate of thermodynamics (Appendix A.2) we may also assume that $c_o \approx$ constant, Equation (1.11) becomes simply

$$p' = c_o^2\rho'. \tag{1.15}$$

Thus, the acoustic continuity and momentum equations, Equations (1.5) and (1.8), may be closed by Equation (1.15) or other derivatives of Equation (1.11) for the determination of p', ρ' and $\mathbf{v}'$. Solutions of these equations are sometimes considered by stipulating a mean flow which is not completely compatible with the corresponding non-linear inviscid mean flow equations. For example, the mean flow is often assumed to be sheared with a non-uniform axial velocity profile, although the viscosity of the fluid is neglected in the acoustic equations. In this case, we may discard the incompatible mean flow equations, because the acoustic equations still provide an adequately accurate mathematical model (for example, a quasi-laminar model) for studying the effect of refraction in shear layers.

1.1.4 Evolution of Non-Linear Waves

For further justification of linearization and insight into the effect of the neglected $O(\varepsilon^n)$, $n \geq 2$, terms, it is useful to review briefly some implications of the fluid dynamic equations for one-dimensional flow. Putting $\mathbf{v} = \mathbf{e}v$ and $\nabla = \mathbf{e}\partial/\partial x$, Equations (1.1) and (1.2) become, $\partial\rho/\partial t + \partial(\rho v)/\partial x = 0$ and $\rho(\partial v/\partial t + v\partial v/\partial x) + \partial p/\partial x = 0$, respectively. Also, from Equation (A.15a), we have $\mathrm{d}p = c^2\mathrm{d}\rho$ for isentropic motion. It was shown by Riemann in 1860 that these equations are tantamount to the partial differential equation $\partial p/\partial t + (c+v)\partial p/\partial x = 0$ for flow in the positive x direction [1, 6]. This result shows that spatial forms of pressure waveforms are convected with velocity $c + v$ and that, to an observer moving with velocity $c + v$, a pressure waveform appears to be stationary in shape. This may be seen by applying the Galilean transformation $y = x - (c+v)t$ and $\tau = t$, giving $\partial p/\partial\tau = 0$, which implies the general solution to be a function of the form $f(x - (c+v)t)$. But, $c + v$ is a function of the pressure. Therefore, to a fixed observer, each point of a pressure waveform moves with different velocity and, consequently, the shape of a pressure waveform at time $t = 0$ is continuously distorted as it is convected with the flow. Then, if the speed of sound is an increasing function of the pressure (which is usually the case), points of the waveform with higher pressure move faster than those at lower pressure and portions of the waveform where the pressure is increasing become steeper with time, eventually causing the waveform to become multi-valued, which signifies occurrence of a shock, a moving plane of discontinuity in fluid properties, giving rise to finite amplitude waves. Here, we are concerned about the conditions for the occurrence of shock formation.

As in the linearization procedure, we introduce the decomposition $p = p_o + p'$ and so on for other variables, but now we neglect the mean flow $(v_o = 0)$ and allow the acoustic wave $p'(x,t) = p'(x - (c+v)t)$ to have finite amplitudes. Typically, let us assume that, initially, p' is a traveling sinusoidal acoustic pressure waveform of radian frequency ω and amplitude $\hat{p}$, such as $p' = \hat{p}\sin\omega(t - x/c)$. It may then be shown that for a perfect gas, the earliest value of x for which shock formation occurs is $x_{\max} \approx (p_o/\hat{p})(c_o/\omega)(2\gamma_o/(1+\gamma_o)$[1]. For example, at exhaust conditions of internal combustion engines, $x_{\max} = 1.2 \times 10^7/\hat{p}f$ meters, where f denotes the frequency in Hz. At 1000 Hz, this is $12000/\hat{p}$ meters, or 600 meters for exhaust noise pressure amplitudes of about 5% of the atmospheric pressure. Evolution of waveform distortions until the shock may be estimated from the Fubini-Ghiron solution $p' = \sum_{n=1}^{\infty}\hat{p}_n \sin(n\omega(t - x/c_o))$, where $\hat{p}_n/\hat{p} = 2\mathrm{J}_n(nx/x_{\max})$ and J_n denotes a Bessel function of order n. Amplitude of the fundamental harmonic $(n = 1)$ decreases with $x/x_{\max}$, but this is compensated with the growth of an infinite number of higher-order harmonic amplitudes.

In general, therefore, the distances needed for harmonic acoustic perturbations with amplitudes initially in the linear range to travel freely before they may be subjected to substantial distortion are unrealistically long compared to the actual distances involved in practical ductworks and frequencies of interest. Furthermore, the viscous and thermal losses, which are neglected in the foregoing considerations, tend to act in opposition to the non-linear wave distortion process [1].

1.2 Representation of Acoustic Waves in the Frequency Domain

In the previous section, the linear acoustic equations are given in the time domain (time appears explicitly). If we are interested in the evolution of the wave motion immediately after the source (for example, a fluid machine) is switched on, it is necessary to obtain solutions of these equations as a function of time for the appropriate initial conditions on the acoustic variables. However, transient wave motions which the initial conditions generate vanish rapidly due to damping and the source settles into a steady-state operation after a few cycles. In most practical applications of duct acoustics, it suffices for acoustic design purposes to consider the acoustic wave motion at such steady operating points of the source, as the damping, which is unavoidably present in real systems, is usually sufficient for preventing growth of resonances that may be excited during the transient stage. Then the initial conditions are not required, and the Fourier transform may be used for removing the explicit dependence of the acoustic equations on time. This approach is known as the transformation to the frequency domain and is used throughout the present book. In this section, we review the basic Fourier transform based frequency domain concepts which are invoked throughout the book.

1.2.1 Fourier Transform

The Fourier transform [7], is an integral operation which maps a function given in time domain, $h = h(t)$, say, to the function

$$h(\omega) = \int_{-\infty}^{\infty} h(t)\,\mathrm{e}^{\mathrm{i}\omega t}\mathrm{d}t, \tag{1.16}$$

where, i denotes the unit imaginary number, $\mathrm{i} = \sqrt{-1}$, and ω denotes the Fourier transform variable which has the unit of radians per second, that is, $\omega = 2\pi f$, where f denotes frequency in Hertz (Hz). Thus, Equation (1.16) represents a mapping from time domain to frequency domain. It should be noted that we use (here and throughout the book) the same symbol for both time and frequency domain representations of a function. The domain distinction is usually clear from the argument of the function or the context of the discussion. The reader should be warned that, although Equation (1.16) is a commonly used definition of the Fourier transform, there is no fixed convention regarding the sign of exponent and the value of a constant non-dimensional multiplication factor in the integrand. This is worthwhile to keep in mind when comparing transformed equations given by different authors.

If $h(\omega)$ is the Fourier transform of $h(t)$, then the Fourier transform of $\breve{h}(t)$ is $\breve{h}(-\omega)$, where an inverted over-arc denotes the complex conjugate. Consequently, for real time-functions: $h(-\omega) = \breve{h}(\omega)$. Thus, the magnitude of the Fourier transform of an acoustic variable is symmetrical about the origin of the frequency axis, that is, $|h(-\omega)| = |h(\omega)|$. Physically, however, only the positive frequencies are meaningful. Negative frequencies occur as a mathematical artifact of the theory.

The Fourier transform of the partial time-derivatives in the linearized acoustic equations are taken by using the following property of the transformation: If $h(\omega)$ is the Fourier transform of $h(t)$, then

$$\int_{-\infty}^{\infty} \frac{\partial h}{\partial t} \mathrm{e}^{\mathrm{i}\omega t} \mathrm{d}t = \int_{-\infty}^{\infty} \left(\lim_{\Delta t \to 0} \frac{h(t+\Delta t)\mathrm{e}^{\mathrm{i}\omega t} - h(t)\mathrm{e}^{\mathrm{i}\omega t}}{\Delta t} \right) \mathrm{d}t = -\mathrm{i}\omega h(\omega), \tag{1.17}$$

which follows after taking the limit operation out of the integral and noting that the Fourier transform of $h(t+\tau)$ is $\mathrm{e}^{-\mathrm{i}\omega\tau} h(\omega)$. Thus, the acoustic equations given in the time domain may be transformed to the frequency domain simply by replacing the operator $\partial/\partial t$ by $-\mathrm{i}\omega$.

Fourier transforms of acoustic state variables $q' \in p', \rho', \mathbf{v}', \ldots$ are determined by solving the Fourier transformed forms of the acoustic continuity and momentum equations. The corresponding solutions in time domain can be recovered by using the formula

$$h(t) = \frac{1}{2\pi} \int_{-\infty}^{\infty} h(\omega) \mathrm{e}^{-\mathrm{i}\omega t} \mathrm{d}\omega, \tag{1.18}$$

which is called the inverse Fourier transform. This integral shows that, each point of $h(\omega)$ contributes nothing to $h(t)$, but indicates the relative weighting of each frequency component. So, the larger $|h(\omega)|$ is, the larger will be the contribution to $h(t)$ in a narrow frequency band $(\omega - \Delta\omega/2, \omega + \Delta\omega/2)$, which can be calculated by integrating the area under $h(\omega)$ in this band.

The Fourier transform and the inverse Fourier transform can be computed efficiently by using the numerical algorithm called the fast Fourier transform [7]. However, the inverse Fourier transformation is seldom necessary in practical applications, since the frequency domain solutions contain more useful information for acoustic design than their counterparts in the time domain. Indeed, not only is the acoustic response of flow duct systems frequency selective, but so is human hearing [1].

1.2.2 Periodic Functions

The above considerations hold for any continuous function of time. If $h(t)$ is periodic of fundamental period $T = 2\pi/\varpi$, where T is the smallest number which satisfies $h(t) = h(t+T)$, it can be represented by the exponential Fourier series

$$h(t) = \sum_{n=-\infty}^{\infty} h_n \mathrm{e}^{-\mathrm{i}n\varpi t} \tag{1.19}$$

with

$$h_n = \frac{1}{T} \int_{-T/2}^{T/2} h(t) \mathrm{e}^{\mathrm{i}n\varpi t} \mathrm{d}t. \tag{1.20}$$

It can be shown that $h_n = h_T(n\varpi)/T$, where $h_T(\omega)$ denotes the Fourier transform of the truncated (windowed) part of $h(t)$ in the interval $-T/2 \leq t \leq T/2$. For real periodic functions: $h_{-n} = \breve{h}_n$, since $h_T(-\omega) = \breve{h}_T(\omega)$ for real functions. Thus, a periodic function has all its amplitude components h_n at discrete frequencies, which are integer multiples of the fundamental frequency ϖ, and each of these has a definite contribution. The Fourier transform of Equation (1.19) is

$$h(\omega) = 2\pi \sum_{n=-\infty}^{\infty} h_n \delta(\omega - n\varpi), \tag{1.21}$$

since the Fourier transform of $\mathrm{e}^{-\mathrm{i}n\varpi t}$ is $\delta(\omega - n\varpi)$, where $\delta(x)$ denotes a unit impulse function at $x = 0$. Thus, the area (weight) of each impulse is 2π times the value of its corresponding coefficient in the complex Fourier series.

The simplest form of Equation (1.19) is a single frequency function $h(t) = h\mathrm{e}^{-\mathrm{i}\varpi t} + \breve{h}\mathrm{e}^{\mathrm{i}\varpi t} = \mathrm{Re}\{2h\mathrm{e}^{-\mathrm{i}\varpi t}\} = \mathrm{Re}\{\hat{h}\mathrm{e}^{-\mathrm{i}\varpi t}\}$, where ϖ denotes the radian frequency, $\hat{h}$ is called the complex amplitude and Re denotes the real part of a complex number. So, if the acoustic variables are assumed a priori to be single frequency functions of the form $\hat{h}\mathrm{e}^{-\mathrm{i}\varpi t}$, it suffices to take the real parts of the corresponding solutions of the acoustic equations. In fact, if we assume a priori that the time-dependence of the variables in acoustic equations is of the form $\mathrm{e}^{-\mathrm{i}\varpi t}$, we get, after the common factor $\mathrm{e}^{-\mathrm{i}\varpi t}$ is cancelled out, relationships between the complex amplitudes of the acoustic variables, which are in exactly of the same form as the Fourier transformed equations. For this reason, the transformation of the acoustic equations from time domain to frequency domain is often effected in the literature informally by assuming a priori that the acoustic variables have $\mathrm{e}^{-\mathrm{i}\varpi t}$ time-dependence. Then, the resulting acoustic equations yield relations between the complex amplitudes of the acoustic variables, but these equations may also be understood as the Fourier transformed acoustic equations, if the complex amplitudes are replaced by the corresponding Fourier transforms of the acoustic variables. Again, caution is needed when comparing results of different authors, because some authors prefer to assume $\mathrm{e}^{\mathrm{i}\varpi t}$ time-dependence.

Conversely, acoustic variables determined in frequency domain may be transformed back to time domain on single frequency basis, if they are known a priori to encompass only discrete frequencies. Then, for example, if $h(\omega)$ is a variable determined in frequency domain, the contribution of a typical single frequency component of frequency ϖ to $h(\omega)$ is $2\pi h(\varpi)\delta(\omega-\varpi)$. The inverse Fourier transform of this is $h\mathrm{e}^{-\mathrm{i}\varpi t}$, where $h=h(\varpi)$. But, in the Fourier transform $h(\omega)$, each single frequency component, such as $h\mathrm{e}^{-\mathrm{i}\varpi t}$, occurs with its complex conjugate $\breve{h}\mathrm{e}^{\mathrm{i}\varpi t}$, since both positive and negative frequencies are included and $h(-\omega)=\breve{h}(\omega)$ for real functions. Therefore, a single frequency component of frequency ϖ of $h(\omega)$ contributes to $h(t)$ by $h\mathrm{e}^{-\mathrm{i}\varpi t}+\breve{h}\mathrm{e}^{\mathrm{i}\varpi t}$, that is, by $2\mathrm{Re}\{h\mathrm{e}^{-\mathrm{i}\varpi t}\}$. It should be noted that, the root-mean-square value of the single frequency function $2\mathrm{Re}\{h\mathrm{e}^{-\mathrm{i}\varpi t}\}$ is $\sqrt{2}|h|$, which corresponds to a peak amplitude of $2|h|$, and that only positive frequencies are relevant.

1.2.3 Impulse Sampling

An impulse-sampled function, $h^{(s)}(t)$, of the continuous function $h(t)$ is defined as

$$h^{(s)}(t)=h(t)\sum_{k=-\infty}^{\infty}\delta(t-kT_s), \tag{1.22}$$

where T_s denotes the sampling period. Fourier transform of this sampled function is

$$h^{(s)}(\omega)=f_s\sum_{k=-\infty}^{\infty}h(\omega-k\omega_s), \tag{1.23}$$

where $\omega_s=2\pi f_s$ and $f_s=1/T_s$ is called sampling frequency. This result shows that, apart from the scaling factor f_s, the Fourier transform of an impulse-sampled function is a periodic repetition, of period T_s, of the Fourier transform of the continuous function, $h(\omega)$. Therefore, if we evaluate $h^{(s)}(\omega)$, we can see $h(\omega)$ in the frequency range $-\omega_s\leq\omega\leq\omega_s$, provided that

$$f_s\geq 2f_{\max}, \tag{1.24}$$

where $f_{\max}$ denotes the maximum frequency of interest in $h(\omega)$. This is known as the Nyquist sampling criterion.

If the Nyquist condition is not satisfied, the periodic repetitions of $h(\omega)$ will overlap and $h(\omega)$ itself will not be visible exactly at some frequencies lower than $f_{\max}$. In practice, overlap (also called aliasing) is unavoidable, because we must necessarily use functions in a finite time window and time-limited functions are not band limited in the frequency domain. On the other hand, a function in a finite time window is represented mathematically by the product of $h(t)$ with a rectangular gate function having the value of unity in the window and zero elsewhere. The Fourier transform of the product of two time functions is given by the convolution of the Fourier transforms of these functions. For this reason, when we take the Fourier transform of an impulse-sampled function $h^{(s)}(t)$ limited to a finite time window,

$h^{(s)}(\omega)$ convolves with the Fourier transform of the rectangular gate function. The latter is characterized by a main lobe and infinite number of side lobes, the amplitudes of which attenuate at the rate of 6 dB/octave, the amplitude of the first side lobe being 13.3 dB lower than that of the main lobe. Because of the convolution process, the side lobes can interfere with the computed Fourier transform even in frequency ranges of the Nyquist criterion. This interference, which is known as leakage, may be of some concern, as it can result in an inaccurate representation of the Fourier transform of the actual signal. Leakage is usually combated by using different window functions than the rectangular gate function, which aim to make the main lobe narrower and to generate side lobes which attenuate at large rates with frequency.

An efficient numerical procedure known as the fast Fourier transform is widely used for numerical calculation of the Fourier transform of continuous functions [7]. This is executed by impulse sampling the function at $N = 2^m$ equidistant points at times $t = 0, T_s, 2T_s, \ldots, (N-1)T_s$, in a window of length $T = (N-1)T_s$. The procedure yields the values of the Fourier transform at equidistant frequencies:

$$f = 0, \quad \mp\frac{f_s}{N}, \quad \mp 2\frac{f_s}{N}, \ldots, \quad \mp\left(\frac{N}{2}-1\right)\frac{f_s}{N}, \quad \mp\frac{f_s}{2},$$

where $f = \omega/2\pi$. Thus, the Fourier transform is sampled at frequency intervals $\Delta f = f_s/N$ or, if N is large enough, $\Delta f = 1/T$. Accordingly, the time interval should be selected so as to obtain a desired frequency resolution. The fast Fourier transform is invertible; that is, if the Fourier transform of a function is known and is sampled at the foregoing $2N+1$ frequency points, the inverse fast Fourier transform yields the values of the function at times $t = 0, T_s, 2T_s, \ldots, (N-1)T_s$.

1.2.4 Power Spectral Density

Root-mean-square value of an acoustic variable is defined by the formula

$$h_{rms} = \sqrt{\frac{1}{T}\int_0^T h^2(t)\mathrm{d}t} = \sqrt{\langle h^2(t)\rangle}, \tag{1.25}$$

where T is called averaging time and the second equality defines a shorthand notation frequently used to denote a time average. If $h(t)$ is a random function, this definition tacitly assumes that it is stationary and ergodic [7]. Since acoustic variables have, by definition, zero mean parts, the root-mean-square value can be interpreted as temporal standard deviation. The following Parseval's formula links the root-mean-square value of a variable with its Fourier transform:

$$\int_0^T h^2(t)\mathrm{d}t = \frac{1}{2\pi}\int_{-\infty}^{\infty} |h_T(\omega)|^2 \mathrm{d}\omega, \tag{1.26}$$

where $h_T(\omega)$ denotes the Fourier transform of the truncated (windowed) part of the function in the interval $0 \leq t \leq T$. Therefore, h_{rms} can be expressed as

$$h_{rms}^2 = \frac{1}{2\pi} \int_{-\infty}^{\infty} S_h(\omega) d\omega, \tag{1.27}$$

where

$$S_h(\omega) = \frac{|h_T(\omega)|^2}{T} \tag{1.28}$$

is called power spectral density of $h(t)$. In contrast to $h(\omega)$, $S_h(\omega)$ cannot be inverted back to $h(t)$, since it contains no phase information.

The integral in Equation (1.27) is often evaluated by splitting it into contiguous frequency bands. Then, for real time signals, it can be expressed as

$$h_{rms}^2 = \sum_{n=1}^{\infty} h_{rms}^2(\bar{\omega}_n), \tag{1.29}$$

where

$$h_{rms}^2(\bar{\omega}_n) = \int_{\omega_n}^{\omega_{n+1}} |h(\omega)|^2 S_x(\omega) d\omega \tag{1.30}$$

and $\omega_n \leq \omega \leq \omega_{n+1}$ denotes the width of band n and $\bar{\omega}_n$ is called the center frequency. In practice, usually two types of frequency bands are used, namely, constant and proportional. The bandwidth of constant bands does not vary with frequency and the center frequency may be defined by the arithmetic mean $\bar{\omega}_n = (\omega_n + \omega_{n+1})/2$. The most common proportional bands are known as 1/N octave bands. The bandwidths of 1/N octave bands are defined as $\omega_{n+1}/\omega_n = 2^{1/N}$ and the center frequency of band n is given by the geometric mean $\bar{\omega}_n = \sqrt{\omega_n \omega_{n+1}}$. Thus, a given frequency band can be covered by 1/N octave bands contiguously by selecting the center frequencies appropriately.

The power spectral density of a periodic function consists of impulse functions located at frequencies $n\varpi$, $n = 0, \mp 1, \mp 2, \ldots$:

$$S_h(\omega) = 2\pi \sum_{n=-\infty}^{\infty} |h_n|^2 \delta(\omega - n\varpi). \tag{1.31}$$

Then, the root-mean-square value of a periodic signal is given simply by

$$h_{rms}^2 = \sum_{n=-\infty}^{\infty} |h_n|^2, \tag{1.32}$$

which is Parseval's formula for periodic functions.

1.3 Representation of Waves in the Wavenumber Domain

1.3.1 Spatial Fourier Transform

Many important problems of duct acoustics are concerned about sound propagation in ducts which are axially uniform and homogeneous. In such cases, the linear acoustic equations can also be Fourier transformed with respect to the duct axis. Let $h(\omega, x_1, x_2, x_3)$ denote the Fourier transform of function $h(t, x_1, x_2, x_3)$, which is also function of the Cartesian coordinates x_1, x_2, x_3, where x_1 denotes the duct axis. The spatial Fourier transform of h with respect to x_1 is defined similarly as the temporal Fourier transform:

$$h(\omega, \kappa, x_2, x_3) = \int_{-\infty}^{\infty} h(\omega, x_1, x_2, x_3)\,\mathrm{e}^{-\mathrm{i}\kappa x_1}\,\mathrm{d}x_1, \tag{1.33}$$

where the transform variable κ is called the axial wavenumber. As an example, consider the application of this transform to the temporal Fourier transform of Equation (1.5) for a uniform and homogeneous duct with $\rho_o =$ constant, $\mathbf{v}_o = \mathbf{e}_1 v_{o1} + \mathbf{e}_2 v_{o2} + \mathbf{e}_3 v_{o3}$, where v_{o1}, v_{o2}, v_{o3} denote components of $\mathbf{v}_o$ in x_1, x_2, x_3 directions, respectively. Assume for simplicity that v_{o1}, v_{o2}, v_{o3} are constants. Then, upon applying the spatial counterparts of Equations (1.16) and (1.17), Equation (1.5) transforms to:

$$\mathrm{i}(-\omega + \kappa v_{o1})\rho' + v_{o2}\frac{\partial \rho'}{\partial x_2} + v_{o3}\frac{\partial \rho'}{\partial x_3} + \rho_o\left(\mathrm{i}\kappa v_1' + \frac{\partial v_2'}{\partial x_2} + \frac{\partial v_3'}{\partial x_3}\right) = 0. \tag{1.34}$$

Inspection of this result shows that acoustic equations can be transformed to the wavenumber domain simply by replacing the operator $\partial/\partial x_1$ by $\mathrm{i}\kappa$. Likewise, the spatial Fourier transform may also be taken by assuming a priori that acoustic variables have $\mathrm{e}^{\mathrm{i}\kappa x_1}$ dependence (or $\mathrm{e}^{-\mathrm{i}\kappa x_1}$ dependence, if time dependence is assumed to be $\mathrm{e}^{\mathrm{i}\varpi t}$).

Spatial Fourier transformed acoustic equations will have one less spatial derivative than the only temporal Fourier transformed equations, which can be convenient when getting solutions. As will be seen in later chapters, solutions of the acoustic equations for uniform ducts are feasible for ω and κ satisfying $\Delta(\omega, \kappa) = 0$. This equation is called the dispersion equation and its actual form depends on the duct and its acoustic boundary conditions. Roots of the dispersion equation give the axial wavenumbers as functions of ω, and each wavenumber corresponds to a wave mode in the duct. The time-domain characteristics of wave modes can be determined by taking first the inverse spatial Fourier transform

$$h(\omega, x_1, x_2, x_3) = \frac{1}{2\pi}\int_{-\infty}^{\infty} h(\omega, \kappa, x_2, x_3)\,\mathrm{e}^{\mathrm{i}\kappa x_1}\,\mathrm{d}\kappa \tag{1.35}$$

and then the inverse temporal Fourier transform.

1.3.2 Briggs' Criterion

By studying the properties of the temporal and spatial Fourier transforms, Briggs has shown that the information about the directions of wave propagation can be obtained without explicit evaluation of inverse Fourier transforms [8]. First, the wavenumbers are computed for a fixed (real) ω and then for $\omega + \mathrm{i}\chi$, where χ is an arbitrarily large positive number. Briggs' criterion states that: if an axial wavenumber κ originally (for ω) having a positive imaginary part, acquires a negative imaginary part for $\omega + \mathrm{i}\chi$ as $\chi \to +\infty$, then the corresponding wave mode propagates in the $+x_1$ direction. Conversely, if an axial wavenumber κ originally (for ω) having a negative imaginary part, acquires a positive imaginary part for $\omega + \mathrm{i}\chi$, $\chi \to +\infty$, then the corresponding wave mode propagates in the $-x_1$ direction.

For real wavenumbers, Briggs' criterion can be stated as: if $\mathrm{Re}\{V_g\} > 0$, then propagation is in the $+x_1$ direction, and $\mathrm{Re}\{V_g\} < 0$ implies propagation in $-x_1$ direction. Here,

$$V_g = \frac{\mathrm{d}\omega}{\mathrm{d}\kappa} \tag{1.36}$$

and is called the group velocity.

If a complex axial wavenumber propagating in the $+x$ direction has a positive imaginary part, the corresponding wave mode decays in the direction of propagation (since $\mathrm{e}^{\mathrm{i}\kappa x_1}$ dependence is being assumed); but, if it has a negative imaginary part, it amplifies in the direction of propagation. Similarly, if a complex axial wavenumber propagating in the $-x_1$ direction has a negative imaginary part, it amplifies in the direction of propagation; but, if it has positive imaginary part, it decays in the direction of propagation.

A wave mode which decays in the direction of propagation is stable and called evanescent. Amplification of a wave mode in the direction of propagation is known as convective instability, which means that the wave amplitudes increase with distance traveled but remain finite at fixed locations. Instability can also be of the absolute type; that is, wave amplitudes blow up at every point in time at every location. A necessary condition for absolute instability is the occurrence of an axial wavenumber of multiplicity two [8].

1.4 Intensity and Power of Sound Waves

The term acoustic energy is usually understood as part of the fluid's total energy associated with the presence of a sound wave. In an inviscid (ideal) fluid, this is purely mechanical energy and consists of the sum of kinetic and potential (compression strain) energies associated with the sound wave motion [1, 6]. But in a Newtonian fluid with non-negligible viscosity, the presence of a sound wave is also associated with some thermal energy dissipated due to friction. In practice, however, the physically relevant metric for sound energy is the mechanical power it transmits when sound waves reach a reception point such as a microphone or human ear. In this book, we understand acoustic energy as this mechanical part of the fluid's total energy associated with the presence of a sound wave.

Consider a control volume V of surface Σ, which contains no internal energy sources and is free of external fields and shaft work. The integral form of the energy equation for the unsteady motion of a Newtonian fluid in this control volume is given by Equation (A.11a). Denoting the rate of viscous work done by the fluid in the control volume on the surroundings by $\dot{W}_V$, and the rate of heat loss from the control volume to the surroundings by conduction by $\dot{Q}_c$, Equation (A.11a) can be expressed as

$$-\frac{\partial}{\partial t}\int_V \rho e \mathrm{d}V = \int_\Sigma h^o \rho \mathbf{v}\cdot\mathbf{n}\mathrm{d}\Sigma + \dot{Q}_c + \dot{W}_V \tag{1.37}$$

Here, the term which is of interest to us is the time-average of the integral on the right-hand side, that is, $\int_\Sigma \langle h^o(\rho\mathbf{v})\cdot\mathbf{n}\rangle \mathrm{d}\Sigma$, where angled brackets denote time-averaging. This integral gives the total time-averaged power of the fluid crossing normal to Σ. We assume that the momentum density, $\rho\mathbf{v}$, and the specific stagnation enthalpy, h^o, may be represented, as the intensive properties of the fluid, by asymptotic expansions $\rho\mathbf{v} = (\rho\mathbf{v})_o + (\rho\mathbf{v})' + \cdots$ and $h^o = (h^o)_o + (h^o)' + \cdots$, respectively, and hypothesize that the fluctuating parts of $O(\varepsilon)$ are solely due to the presence of sound waves (Section 1.1.1). This is a crucial assumption, because it means that all acoustic ramifications of turbulence fluctuations and losses incurring from flow-acoustic interactions are neglected in the subsequent analysis. Then the total time-averaged power associated with the presence of sound waves may be expressed as [9]:

$$W^o = \int_\Sigma \langle (h^o)'(\rho\mathbf{v})'\cdot\mathbf{n}\rangle \mathrm{d}\Sigma \tag{1.38}$$

because terms of $O(\varepsilon)$ have zero mean by definition and vanish on time-averaging. Upon evaluating the momentum density and the specific stagnation enthalpy to $O(\varepsilon)$ and using the linearized state equation $h' = T_o s' + p'/\rho_o$ (see Equation (A.16)), the foregoing equation may be expanded as:

$$W^o = \int_\Sigma \big\langle (T_o s' + \mathbf{v}_o\cdot\mathbf{v}' + p'/\rho_o)\ (\rho_o\mathbf{v}' + \mathbf{v}_o\rho')\cdot\mathbf{n}\big\rangle \mathrm{d}\Sigma \tag{1.39}$$

But this is not fully mechanical power; some part of it is thermal power dissipated by friction due to the viscosity of the fluid. The contributor of this part in the foregoing integral is the fluctuating entropy term, because an entropy increase corresponds to a change in the amount of heat energy per unit volume, which must equal the loss of mechanical energy per unit volume per unit time from the sound wave due to fluid viscosity. Then, discarding this term, the acoustic power (the mechanical part of the time-averaged power associated with the presence of sound waves) crossing normal to Σ can be expressed as

$$W = \int_{\Sigma} \langle \mathbf{N}''\cdot\mathbf{n} \rangle \mathrm{d}\Sigma, \tag{1.40}$$

where

$$\mathbf{N}'' = (\rho_o \mathbf{v}' + \mathbf{v}_o \rho') \left(\mathbf{v}_o\cdot\mathbf{v}' + \frac{p'}{\rho_o} \right) \tag{1.41}$$

denotes the instantaneous local flux (rate of transport) of acoustic energy, which is called acoustic intensity. Thus, the time-averaged acoustic power transmitted in a duct normal to the duct sections can be calculated by using the formula

$$W = \int_{S} \langle \mathbf{N}''\cdot\mathbf{n} \rangle \mathrm{d}S, \tag{1.42}$$

where S denotes the duct cross-sectional area. It should be noted that, the double prime syntax is used to denote the part of a fluctuating quantity that is of $O(\varepsilon^2)$. The true acoustic intensity is given by $\mathbf{N} = \mathbf{N}' + \mathbf{N}'' + O(\varepsilon^3)$; however, only the $O(\varepsilon^2)$ term is relevant for the calculation of the time-averaged acoustic power, since $\langle \mathbf{N} \rangle = \langle \mathbf{N}'' \rangle$.

In the literature, it is traditional to deduce expressions for the acoustic intensity from an energy continuity equation which is derived as a corollary to the acoustic continuity, momentum and energy equations and is of the general form

$$\frac{\partial \chi''}{\partial t} + \nabla\cdot\mathbf{N}'' = \psi'', \tag{1.43}$$

where χ'' represents the acoustic energy density and ψ'' is called the production term, the double prime syntax still denoting quantities small to the second order. This equation derives its rational from the fact that, upon application of the divergence theorem, Equation (A.1), for a fixed control volume of volume V of surface Σ, it turns into a statement of conservation of acoustic energy: $\mathbf{N}''$ crossing normal to the surface Σ, plus the rate of change of χ'' in V, is balanced by ψ'' in V. A discussion of the ramifications of this approach is given by Morfey, who also presents a general formulation including fluid viscosity and thermal conductivity and non-uniform mean flow [10]. This formulation partitions the density fluctuations that are non-acoustic, by hypothesizing that the effect of viscosity and thermal conductivity on density fluctuations is negligible. Then, the acoustic momentum density fluctuations $(\rho_o \mathbf{v}' + \mathbf{v}_o \rho')$ are evaluated by using Equation (1.15) for ρ'. Apart from this difference, the expression of Morfey's theory for $\mathbf{N}''$ is the same as Equation (1.41). The acoustic fields considered in this book do not contain non-acoustic density fluctuations by definition (Section 1.1.1), and Equation (1.41) is applicable.

An expression for χ'' corresponding to $\mathbf{N}''$ may be derived from Equations (1.5) and (1.8). Letting $\mathbf{m}' = \mathbf{v}_o \rho' + \rho_o \mathbf{v}'$, the former becomes $\partial\rho'/\partial t + \nabla\cdot\mathbf{m}' = 0$. Assuming homentropic mean flow (Section 1.1.2), Equation (1.8) can be expressed as $\partial\mathbf{v}'/\partial t + \nabla E' = s'(p_o/c_{vo})\nabla(1/\rho_o)$, where $E' = \mathbf{v}_o\cdot\mathbf{v}' + p'/\rho_o$, $c_v \approx c_{vo}$ denotes the specific heat coefficient at constant volume and s' denotes specific entropy fluctuations, which enters as we use the linearized form of Equation (A.15a). Upon adding the dot product of the modified form of Equation (1.8) with $\mathbf{m}'$, to E' times the modified form of Equation (1.5), Equation (1.43) is obtained with $\mathbf{N}''$ given by Equation (1.41), $\psi'' = -\mathbf{m}'\cdot(s'p_o/c_{vo})\nabla(1/\rho_o)$ and

$$\chi'' = \frac{1}{2}\rho_o\mathbf{v}'\cdot\mathbf{v}' + \rho'\mathbf{v}_o\cdot\mathbf{v}' + \int\frac{p'\mathrm{d}\rho'}{\rho_o}. \tag{1.44}$$

The third term on the right-hand side of this equation represents the specific potential (compression strain) energy of the fluid associated with the presence of acoustic waves [1]. The specific kinetic energy of the fluid is given by $(\rho_o + \rho')(\mathbf{v}_o + \mathbf{v}')^2/2$, the $O(\varepsilon^2)$ term of which is $\rho_o\mathbf{v}'\cdot\mathbf{v}'/2 + \rho'\mathbf{v}_o\cdot\mathbf{v}'$. Therefore, χ'' gives the specific acoustic energy (sum of specific kinetic and potential energies) associated with the presence of acoustic waves. If Equation (1.15) is used, the specific potential energy becomes $p'^2/2\rho_o c_o^2$ and Equation (1.44) reduces to the form given by Morfey [10], where all terms involving the specific entropy (and also the viscosity and thermal conductivity, which are neglected in the foregoing derivation) are also collected in the production term.

Equation (1.42) may be expressed explicitly as

$$W = \int_S \left\langle p'\mathbf{v}'\cdot\mathbf{n} + \rho_o\mathbf{v}_o\cdot\mathbf{v}'\mathbf{v}'\cdot\mathbf{n} + \frac{\mathbf{v}_o\cdot\mathbf{n}}{\rho_o}p'\rho' + \mathbf{v}_o\cdot\mathbf{n}\mathbf{v}_o\cdot\mathbf{v}'\rho'\right\rangle \mathrm{d}S. \tag{1.45}$$

Since this expression is in the time domain, it is desirable to have an equivalent expression which may be evaluated directly by using the solutions obtained in the frequency domain. The simplest way to derive this expression is to assume that the acoustic variables are single frequency functions of the form $h(t) = 2\mathrm{Re}\{h\mathrm{e}^{-\mathrm{i}\omega t}\}$ (see Section 1.2.2). Then, Equation (1.45) may be written as

$$\begin{aligned} W = 4\int_S \frac{1}{T}\int_o^T \Bigg(& \mathrm{Re}\{p'\mathrm{e}^{-\mathrm{i}\omega t}\}\mathrm{Re}\{\mathbf{v}'\mathrm{e}^{-\mathrm{i}\omega t}\}\cdot\mathbf{n} + \rho_o\mathbf{v}_o\cdot\mathrm{Re}\{\mathbf{v}'\mathrm{e}^{-\mathrm{i}\omega t}\}\mathrm{Re}\{\mathbf{v}'\mathrm{e}^{-\mathrm{i}\omega t}\}\cdot\mathbf{n} \\ & + \frac{\mathbf{v}_o\cdot\mathbf{n}}{\rho_o}\mathrm{Re}\{p'\mathrm{e}^{-\mathrm{i}\omega t}\}\mathrm{Re}\{\rho'\mathrm{e}^{-\mathrm{i}\omega t}\} + \mathbf{v}_o\cdot\mathbf{n}\mathbf{v}_o\cdot\mathrm{Re}\{\mathbf{v}'\mathrm{e}^{-\mathrm{i}\omega t}\}\mathrm{Re}\{\rho'\mathrm{e}^{-\mathrm{i}\omega t}\}\Bigg)\mathrm{d}t\mathrm{d}S, \end{aligned} \tag{1.46}$$

where $T = 2\pi/\omega$. Upon integration we obtain:[2]

$$W = 2\int_S \left(\mathrm{Re}\{\breve{p}'\mathbf{v}'\}\cdot\mathbf{n} + \rho_o\mathbf{v}_o\cdot\mathrm{Re}\{\breve{\mathbf{v}}'\mathbf{v}'\}\cdot\mathbf{n} + \frac{\mathbf{v}_o\cdot\mathbf{n}}{\rho_o}\mathrm{Re}\{\breve{p}'\rho'\} + \mathbf{v}_o\cdot\mathbf{n}\mathbf{v}_o\cdot\mathrm{Re}\{\breve{\mathbf{v}}'\rho'\}\right)\mathrm{d}S \tag{1.47}$$

or

$$W = 2\mathrm{Re}\int_S \left(\breve{p}'\mathbf{v}'\cdot\mathbf{n} + \rho_o\mathbf{v}_o\cdot\breve{\mathbf{v}}'\mathbf{v}'\cdot\mathbf{n} + \frac{\mathbf{v}_o\cdot\mathbf{n}}{\rho_o}\breve{p}'\rho' + \mathbf{v}_o\cdot\mathbf{n}\mathbf{v}_o\cdot\breve{\mathbf{v}}'\rho'\right)\mathrm{d}S \tag{1.48}$$

[2] Recall that $\mathrm{Re}\{ab\} = \mathrm{Re}\{a\}\mathrm{Re}\{b\} - \mathrm{Im}\{a\}\mathrm{Im}\{b\}$.

where an inverted over-arc denotes the complex conjugate. Counterparts of this expression are often given in the literature in terms of the complex amplitudes of the acoustic variables. Equation (1.48) can be converted to complex amplitudes, by replacing $h \in p', \rho', \ldots$ by $\hat{h}/2$, where a circumflex denotes the complex amplitude (Section 1.2.2).

Three-dimensional acoustic fields in ducts consist of a linear combination of, theoretically, an infinite number of wave modes, and acoustic variables may be defined as vectors with components representing the contributions of these modes. For operations with the vector variables, it is convenient to write Equation (1.48) in a form that implies matrix operations:

$$W = 2\mathrm{Re}\int_S \left(\left(\breve{p}'\right)^{\mathrm{T}} \mathbf{v}'\cdot\mathbf{n} + \rho_o \mathbf{v}_o \cdot \left(\breve{\mathbf{v}}'\right)^{\mathrm{T}} \mathbf{v}'\cdot\mathbf{n} + \frac{\mathbf{v}_o\cdot\mathbf{n}}{\rho_o}\left(\breve{p}'\right)^{\mathrm{T}} \rho' + \mathbf{v}_o\cdot\mathbf{n}\mathbf{v}_o\cdot\left(\breve{\mathbf{v}}'\right)^{\mathrm{T}} \rho' \right) \mathrm{d}S, \tag{1.49}$$

where the superscript "T" denotes matrix transposition. Note that this form is not compulsory and in each of the four terms in the large brackets, either one of the two acoustic variables may be complex conjugated and transposed, provided that the transposed variable is moved to the leading position.

1.5 Introduction to the Linear System View of Duct Acoustics

Acoustic models based on the linear theory of acoustic wave motion inherit all generic mathematical properties of linear dynamic systems. In the time domain, a single-input-single-output linear system is associated with the mathematical operation $y(t) = \mathrm{L}x(t)$, where $x = x(t)$ and $y = y(t)$ are, respectively, the input (excitation) and output (response) of the system and L denotes a linear mathematical operator (involving differentiation and/or integration). If the input and output are vectors of variables, the system is called multi-input-multi-output system. In this case, the mathematical operation is written in matrix format as $\mathbf{y}(t) = \mathbf{L}\mathbf{x}(t)$, where $\mathbf{x} = \mathbf{x}(t)$ and $\mathbf{y} = \mathbf{y}(t)$ are, respectively, the input and output vectors of the system and $\mathbf{L}$ denotes a matrix of linear mathematical operations. Most properties of single-input-single-output systems can be extended to multi-input-multi-output systems by changing to matrix notation.

An arbitrary input can be represented by the expansion:

$$x(t) = \int_{-\infty}^{\infty} x(\tau)\delta(t-\tau)\mathrm{d}\tau. \tag{1.50}$$

Here, $\delta(t)$ denotes an impulse function of unit strength placed at time $t = 0$, which is defined formally by the property:

$$\int_{-\infty}^{\infty} g(t)\delta(t)\mathrm{d}t = g(0), \tag{1.51}$$

where $g = g(t)$ denotes an arbitrary function. Therefore, the foregoing expansion actually models the input $x(t)$ as the sum of an infinite number of impulses having strengths equal to the value of the input function at the time of their action.

Let the response of a linear system to a unit impulse input at time $t = \tau$ be denoted by $h(t, \tau) = \mathrm{L}\delta(t - \tau)$. Then,

$$y(t) = \mathrm{L}\left\{ \int_{-\infty}^{\infty} x(\tau)\delta(t-\tau)\mathrm{d}\tau \right\} = \int_{-\infty}^{\infty} x(\tau)h(t,\tau)\mathrm{d}\tau. \tag{1.52}$$

In words, the system output is given by the sum of the outputs of the system to its individual responses to the constituent pulses of the actual input. This is called superposition property of linear systems.

A linear system is called time-invariant if $h(t, \tau) = h(t - \tau)$, that is, if the response of the system to a unit impulse at time $t = \tau$ is same as its response to the unit impulse at $t = 0$ translated to time $t = \tau$. Then, Equation (1.52) reduces to the convolution

$$y(t) = \int_{-\infty}^{\infty} x(\tau)h(t-\tau)\mathrm{d}\tau, \tag{1.53}$$

which is also known as Duhamel's integral. This relationship acquires a more practical form when transformed to frequency domain. Taking the Fourier transform of the both sides of Equation (1.53), we obtain:

$$y(\omega) = \int_{-\infty}^{\infty} \left(\int_{-\infty}^{\infty} h(t-\tau)\mathrm{e}^{\mathrm{i}\omega t}\mathrm{d}t \right) x(\tau)\mathrm{d}\tau = h(\omega) \int_{-\infty}^{\infty} \mathrm{e}^{\mathrm{i}\varpi\tau} x(\tau)\mathrm{d}\tau, \tag{1.54}$$

that is,

$$y(\omega) = h(\omega)x(\omega), \tag{1.55}$$

where $h(\omega)$ is called the frequency response function of the system. Since Equations (1.53) and (1.55) are Fourier transforms of each other, they carry the same information; however, the latter is more widely used, because $h(\omega)$ is relatively simpler to obtain than the unit impulse response $h(t)$. The term frequency response function is usually abbreviated as FRF, but it should be noted that this terminology is normally employed when the input and output of the system are scalar variables. For a multi-input-multi-input system, Equation (1.55) becomes a matrix equation of the form:

$$\mathbf{y}(\omega) = \mathbf{h}(\omega)\mathbf{x}(\omega), \tag{1.56}$$

where $\mathbf{h}(\omega)$ may be called the frequency response matrix. In this book, however, we will usually refer to it as a wave transfer matrix, since the input and output are associated with a wave motion.

Frequency response functions and matrices are extensively used in duct acoustics as descriptors of generation, propagation and radiation of duct-borne sound waves. They are derived from solutions of the governing acoustic equations in frequency domain without being concerned about the actual input. But if we are interested in the output of the system, for example, the sound pressure radiated from the exhaust tailpipe of an engine, the knowledge of the actual input gains importance. It is clear from Equation (1.55) that we can determine the Fourier transform of the output only if we know the Fourier transform of the input.

In some applications, the input may be known not as a time function or its Fourier transform, but as root-mean-square values in frequency bands which cover the frequency range of interest contiguously. Then we can only estimate the root-mean-square values of the output in the same bands. Taking the modulus of Equation (1.55):

$$|y(\omega)|^2 = |h(\omega)|^2|x(\omega)|^2. \tag{1.57}$$

Therefore, from Equation (1.22), the power spectral density of the input and output signals are related as

$$S_y(\omega) = |h(\omega)|^2 S_x(\omega). \tag{1.58}$$

Thus,

$$\int_{-\infty}^{\infty} S_y(\omega)\mathrm{d}\omega = \int_{-\infty}^{\omega} |h(\omega)|^2 S_x(\omega)\mathrm{d}\omega. \tag{1.59}$$

Assuming real time signals and splitting the integral over the bands on the positive frequency axis,

$$\sum_{i=1}^{\infty}\int_{\omega_i}^{\omega_{i+1}} S_y(\omega)\mathrm{d}\omega = \sum_{i=1}^{\infty}\int_{\omega_i}^{\omega_{i+1}} |h(\omega)|^2 S_x(\omega)\mathrm{d}\omega, \tag{1.60}$$

where $\omega_i \leq \omega \leq \omega_{i+1}$ denotes the width of band i. Thus, the root-mean-square value of the output in a typical frequency band is given by

$$y_{rms}^2(\bar{\omega}_i) = \int_{\omega_i}^{\omega_{i+1}} |h(\omega)|^2 S_x(\omega)\mathrm{d}\omega, \tag{1.61}$$

where $\bar{\omega}_i$ denotes the center frequency of band i. If the magnitude of the frequency response function remains sufficiently uniform over a band, the root-mean-square values of the input and output signals may be related in constant or proportional frequency bands (Section 1.2.4) as:

$$y_{rms}(\bar{\omega}_i) \approx |h(\bar{\omega}_i)|x_{rms}(\bar{\omega}_i). \tag{1.62}$$

Obviously, the narrower the bandwidth, the more accurate this description will be.

References

[1] A.D. Pierce, *Acoustics: An Introduction to Its Physical Principles and Applications*, (New York: McGraw-Hill, 1981).

[2] W.K. Blake, *Mechanics of Flow-Induced Sound and Vibration: General Concepts and Elementary Sources,* vol. 1, (London: Academic Press, 2017).
[3] P.O.A.L. Davies, M. Heckl and G.L. Koopman, in G. Bianchi (ed.), *Noise Generation and Control in Mechanical Engineering*, (New York: Springer-Verlag Wien GMBH, 1982).
[4] D.H. Tack and R.F. Lambert, Influence of shear flow on sound attenuation in a lined duct, *J. Acoust. Soc. Am.* **38** (1965) 655–666.
[5] M.S. Howe, *Acoustics of Fluid–Structure Interactions*, (Cambridge: Cambridge University Press, 1998).
[6] M.J. Lighthill, *Waves in Fluids*, (Cambridge: Cambridge University Press, 1979).
[7] W.H. Press, S.A. Teukolsky, W.T. Vetterling and B.P. Flannery, *Numerical Recipes in C: The Art of Scientific Computing*, (Cambridge: Cambridge University Press, 1992).
[8] R.J. Briggs, *Electron-Stream Interactions in Plasmas*, (Cambridge MA: MIT Press,1964).
[9] E. Dokumaci, On the effect of viscosity and thermal conductivity on sound power transmitted in uniform circular ducts, *J. Sound Vib.* **363** (2016), 560–570.
[10] C.L. Morfey, Acoustic energy in non-uniform flows, *J. Sound Vib.* **14** (1971), 159–170.

2 Introduction to Acoustic Block Diagrams

2.1 Introduction

Practical duct systems, which are integral parts of fluid machinery, consist of various types of ducts that are connected in various ways. Mathematic models of sound transmission in these systems are derived from the acoustic equations governing sound wave motion in the constituent ducts and the conditions of continuity and compatibility of acoustic variables at their connections. Such calculations are most efficiently carried out by using a network technique. This consists of regarding the overall ductwork as an assembly of simpler components, such as ducts and some coupled duct arrangements. Mathematical models describing sound transmission in these components are called acoustic elements.

Network methods are quite popular in duct acoustics. The early methods were based on electro-acoustic analogies in the frequency domain and capitalized on the schematics of electric circuits [1]. The analogy is usually established by considering acoustic pressure as analogous to electrical voltage and acoustic volume velocity (product of acoustic particle velocity and duct cross-sectional area) to electrical current, and is implemented by representing the damping, mass and compliance associated with the acoustic wave motion in a duct by analogous electrical resistance, inductance and capacitance, respectively. In this way, duct systems can be represented by an equivalent electrical circuit and analyzed by invoking Kirchhoff laws or the loop method.

The electro-acoustic analogy may also be implemented by using network techniques such as linear graphs [2], and bond graphs [3]. However, lumped acoustic models tacitly assume plane-wave propagation in ducts and neglect the presence of mean flow. Furthermore, they can be accurate at very low frequencies because they must correspond to ducts of compact lengths (Section 3.5.4). This means that duct length must be much less than $\lambda/2\pi$, where $\lambda = c_o/f$ denotes the wavelength of the plane wave and f denotes the frequency. Thus, for example, for a 10 cm long duct, lumped acoustic models may be used up to a frequency of about 50 Hz, this threshold being inversely proportional to the duct length. Then, an adequately accurate acoustic model of a 10 cm long duct valid up to about 1000 Hz at room temperature ($c_o \approx 100\pi$ m/s) requires that the duct is divided into about 20 lumped acoustic elements. Thus, analysis of duct systems by using lumped acoustic element networks can quickly become computationally inefficient as frequency or duct length increases.

Low-frequency sound wave transmission, however, is of considerable interest in fluid machinery ductworks, because it is more difficult to control than high-frequency noise and resolving the acoustic problems at low frequencies usually solves the acoustic design problem. This has motivated development of network methods that utilize continuous acoustic elements based on the plane-wave theory, because such elements can be of adequate accuracy in a much wider low-frequency range than lumped elements.[1] The basic form of these methods is usually referred to in the literature as the transfer matrix method and is normally applicable to ductworks consisting of sequential acoustic elements, but recursive techniques may be used for extension to specific non-sequential situations [4]. Several variants exist depending on the variables used in the definition of the plane acoustic field in a duct. A review of the work before 1997 is given by Glav and Åbom, who also propose a linear graph scheme for assembling arbitrary duct systems from plane-wave duct elements [5]. More recent similar plane-wave algorithms proposed for specific systems are described in references [6–7].

Plane wave acoustic elements may become inaccurate for relatively large duct sizes, even though low frequencies may still be of interest. For example, in large engine installations, power plants and ventilation systems, ducts of diameter 0.5 m to 1 m or more are not uncommon and the frequency range of interest normally exceeds the plane-wave range (see Footnote 1). Furthermore, in some applications, acoustic analysis may have to be taken well over the plane-wave range in order to predict compliance with environmental noise limits. However, no published work exists on network methods which allow use of three-dimensional acoustic elements. Consequently, the finite element method or the boundary element method holds supreme in such situations, but largely lacks the virtues of network methods mentioned above.

In this book we use a network approach to guide the reader through the physics and mathematics of duct acoustics. In this approach, acoustic models of duct systems are represented by a network of blocks, which is called a block diagram, where every block represents an acoustic element. An important advantage of the block diagrams used in our discussions is that their topology is independent of the dimensionality of the blocks. A given block diagram can be analyzed either by using one-dimensional or three-dimensional acoustic elements for the blocks.

Block diagrams are constructed by exploding the hardware of a given duct system into ducts and some basic duct arrangements. This is done on the understanding that adequately accurate acoustic elements of these components, such as those derived and discussed in this book, are available to the analyst. Thus, a network of blocks contains information about both the hardware of the system considered and its acoustic model.

[1] Precise statements about the frequency range of plane-wave propagation in a duct can only be made when the problem is solved in three-dimensions (Section 6.5). For example, for a hollow circular hard-walled duct of diameter D with negligible mean flow, the exact criterion may be stated as $\lambda > 1.706D$, if the first axisymmetric mode is excited.

This chapter explains the general concepts and mathematics of blocks and networks of blocks. Acoustic elements are categorized (Section 2.2), characterized as mathematical entities in the frequency domain (Section 2.3) and analysis of assembled blocks is demonstrated by examples (Section 2.4). In view of the focus of the book on the physics of acoustic elements, formulations given in later chapters are mostly based on analytical solutions. Acoustic elements can also be determined by measurement (Sections 12.3–12.5), the finite element method or the boundary element method. The use of the latter numerical methods, which are widely implemented by using commercial software, are described in Section 2.5. Finally, in Section 2.6 we discuss the development of codes based on the acoustic block diagram approach described in this chapter.

2.2 Classification of Acoustic Models of Ducts

In this book we often use the term "element" in referring to acoustic models of ducts or subassemblies of ducts. Acoustic elements may be classified by the number and type of free terminals they have for connection to the free terminals of other acoustic elements. A free terminal is also called a port and is understood to be a duct section normal to the duct axis.

2.2.1 Classification by Number of Ports

According to the number of ports, acoustic elements may be grouped as

- One-port elements
- Two-port elements
- Multi-port elements

In general, an n-port element is defined as an acoustic element that has $n \in 1, 2, \ldots$ ports for connection to other elements. For visualizing an acoustic element, we use a rectangle (evoking a building block), with the ports shown on the sides of the rectangle by little circles (as shown in Figures 2.1–2.4). The default sign convention

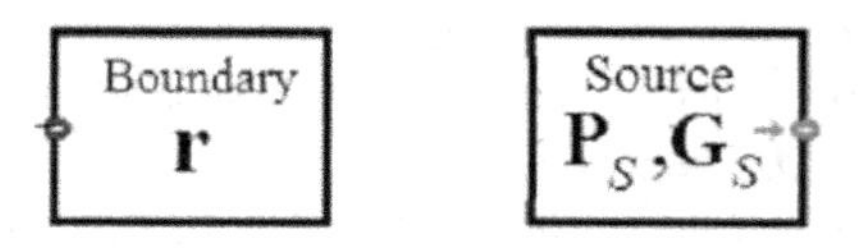

Figure 2.1 Blocks representing boundary and source type one-port elements.

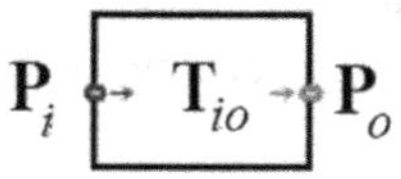

$\mathbf{P}_i$ $\mathbf{T}_{io}$ $\mathbf{P}_o$

Figure 2.2 Block representation of an acoustic two-port element with outlet-to-inlet causality.

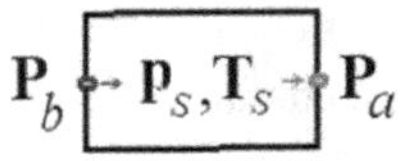

P_b p_s, T_s P_a

Figure 2.3 A two-port source block.

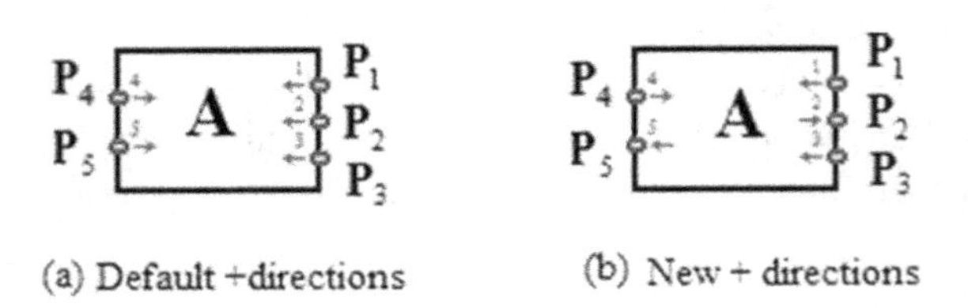

(a) Default +directions (b) New + directions

Figure 2.4 An acoustic multi-port block with different positive acoustic flow directions at some ports.

used for the positive direction of acoustic flow at ports in the formulation of an acoustic element is indicated in the corresponding block by arrows besides the ports. A port is called an inlet port if the arrow direction is into the port and an outlet port if the arrow direction is out of the port. The sign of the mean flow velocity at a port is also determined by the sign convention used for the positive direction of acoustic flow, but it is important to note that the mean flow need not necessarily be in the positive direction of acoustic flow (that is, the inlet and outlet ports are not determined by the direction of mean flow). In acoustic calculations, we must keep track of the mean flow paths in acoustic elements and determine the sign of the mean flow velocity according to the sign convention chosen for the positive acoustic flow directions (see Section 5.7).

2.2.2 Classification by Type of Port

Ports of different elements may be connected, if the acoustic fields in the neighboring ducts match at the interface. When there is a discontinuity between the adjacent acoustic fields, the matching conditions require special considerations. We defer such problems to the formulation stage of acoustic elements involving duct arrangements with discontinuities (Chapters 4 and 7). Then, continuity of acoustic pressure, p', and the component of the acoustic particle velocity in the direction of the duct axis, v', are sufficient for the continuity of neighboring in-duct acoustic fields (the latter ensures continuity of acoustic mass flux). On this premise, each port of an acoustic element is associated with the values of acoustic variables p' and v'. According to how p' and v' are defined at the ports, it suffices to classify acoustic elements in this book as:

- One-dimensional elements
- Modal elements

2.2.2.1 One-Dimensional Elements

Traditionally, one-dimensionality of sound wave motion is stipulated by a priori plane-wave assumption, that is, that sound waves propagate with planar wavefronts. Then p' and v' are understood to be uniformly distributed at port sections. More generally, however, ports of one-dimensional elements may be defined with variables $\overline{p'}$ and $\overline{v'}$, where an overbar denotes averaging over the cross-sectional area of a port (Chapter 3). Then, in the frequency domain, $\overline{p'}$ and $\overline{v'}$ are functions of only frequency and the position on the duct axis.

There is an element of arbitrariness in using $\overline{p'}$ and $\overline{v'}$ as port variables, because, continuity of any two variables that are independent linear combinations of $\overline{p'}$ and $\overline{v'}$ can also ensure continuity of $\overline{p'}$ and $\overline{v'}$. In fact, it transpires that the use of the variables p^+ and p^- defined by the following transformation usually provides better insight to the physics of one-dimensional wave motion (Chapter 3):

$$\begin{bmatrix} p^+ \\ p^- \end{bmatrix} = \frac{1}{2}\begin{bmatrix} 1 & \rho_o c_o \\ 1 & -\rho_o c_o \end{bmatrix}\begin{bmatrix} \overline{p'} \\ \overline{v'} \end{bmatrix}, \tag{2.1}$$

where ρ_o and c_o denote the fluid density and the speed of sound and p^+ and p^- are called acoustic pressure wave components. The physical significance of p^+ and p^- lies in the fact that they may be associated with one-dimensional waves that travel in the positive and negative directions of the duct axis, respectively. In fact, the use of pressure wave components on plane waves is common in the literature, although they are often referred to as incident and reflected plane waves. However, at this point we may consider Equation (2.1) just as a mathematical transformation and postpone physical interpretations until Section 3.4 and later.

One-dimensional acoustic elements may be formulated either by using the fundamental variables $\overline{p'}$ and $\overline{v'}$ or the pressure wave components p^+ and p^- as port variables. This will be clear from the analyses; however, in this book we generally prefer the use of pressure wave components eventually, as they lead to canonical relations (Chapters 3 and 4). Then, each port of a one-dimensional acoustic element is associated with the vector

$$\mathbf{P} = \begin{bmatrix} p^+ \\ p^- \end{bmatrix} \tag{2.2}$$

and transformation of this to the fundamental variables $\overline{p'}$ and $\overline{v'}$ is clear from Equation (2.1).

2.2.2.2 Modal Elements

One-dimensional models are justified, because sound waves propagate in ducts with approximately plane wavefronts at low frequencies. In general, however, p' and v' are functions of position on a duct section as well as the position on the duct axis and the frequency. Analysis of acoustic equations in three dimensions indicate that p' and v' may be conceived as linear superposition of infinite number of wave modes as $p' = p'_1 + p'_2 + \cdots$, $v' = v'_1 + v'_2 + \cdots$, where p'_j and v'_j denote the contribution of mode $j = 1, 2, \ldots, \infty$, which are, in general, non-uniformly distributed over duct sections and decay with distance along the duct axis due to the presence of duct walls and the damping present in the medium. They are ordered in the order of increasing decay rates. Since modes with large decay rates disappear quickly, p' and v' at a port are dominated by a number of modes with smaller decay rates. This allows calculations with such infinite mode sums to be carried out by taking a sufficiently large but finite number of modes. The number of modes taken into account is called modality.

In general, p'_j and v'_j are functions of the spatial coordinates; however, for several duct models of practical importance, the transverse coordinates separate naturally from the axial coordinate. Then the port variables may be defined in the frequency domain by the vector

$$\mathbf{P} = \begin{bmatrix} \mathbf{P}^+ \\ \mathbf{P}^- \end{bmatrix}, \tag{2.3}$$

with

$$\mathbf{P}^{\mp} = \begin{bmatrix} p_1^{\mp} \\ p_2^{\mp} \\ \vdots \\ p_m^{\mp} \end{bmatrix}, \tag{2.4}$$

where m denotes the modality and p_j^+ and $p_j^-, j = 1, 2, \ldots, m$, which are called modal acoustic pressure wave components, are functions of only the axial coordinate and may be related to p'_j and v'_j by transformations akin to Equation (2.1). The superscripts "$\mp$" in these definitions have similar meanings as for one-dimensional acoustic pressure wave components; that is, they denote waves traveling in positive and negative directions of the duct axis.

In this book, acoustic two-port or multi-port models with ports defined by modal pressure wave components as in Equation (2.3) are called modal elements. Typically, straight uniform ducts can be represented by modal two-port elements (Section 6.4). Consequently, any two-port or multi-port duct system with ports on straight uniform ducts can also be represented by a modal element.

Each port of a modal element may have a different modality; however, such diversity is not really necessary in practical applications and we may assume, without loss of generality, that all ports of a modal element are of identical modality. Then, modality may also be considered as a property of modal acoustic elements.

In general, the higher the frequency, the larger should be the modality. For insight on the nature of this mechanism, it is convenient to recall briefly the acoustic modes in hard-walled lossless straight ducts with an inviscid fluid (Section 6.5). Decaying modes still occur in this ideal case and are commonly referred to as evanescent wave modes. As frequency increases, an initially evanescent mode turns into a mode with zero decay rate, which is then called propagating or cut-on mode. The frequency at which this transition occurs is called cut-on frequency. The cut-on frequency is zero for the first mode, which means that it is cut-on at all frequencies, and increases with the order of modes, so that higher-order modes cut-on progressively as frequency increases. At least all cut-on modes must be included in the modality. Evanescent modes are generated at discontinuities, such as area changes, and a sufficient number of them in increasing decay rates should also be included in the modality of ports that are close to a discontinuity.

2.3 Mathematical Models of Acoustic Elements

Acoustic elements are formulated by using default positive directions for acoustic flow at their ports. These directions are indicated in the block representation of elements by arrows besides the ports (see Figures 2.1–2.4). Compatibility of the arrow directions at connected nodes is monitored when assembling block diagrams (Section 2.4).

2.3.1 One-Port Elements

One-port elements are used to apply constraints to port variables. They are connected to ports where the wave propagation is subject to external constraints.

One-port elements may be source type or boundary type. Block representations of these two genres are shown in Figure 2.1. The boundary type one-port is called a boundary element. It has only one port and is described mathematically by a reflection matrix, $\mathbf{r}$, which is defined as

$$[\mathbf{r} \quad -\mathbf{I}]\mathbf{P} = 0 \text{ or } \mathbf{P}^- = \mathbf{r}\mathbf{P}^+, \tag{2.5}$$

where $\mathbf{P}$ denotes the port variables vector, Equation (2.2) or (2.3), and $\mathbf{I}$ denotes a unit matrix of conforming size. Note that no positive acoustic flow direction is specified, because the port inherits this from the connected port. Since the port variables vector is in general of size $(2m) \times 1$, the reflection matrix is of size $m \times m$, where m denotes the modality.

Sound propagation in a duct system terminates at the inlet port of an acoustic boundary in the way defined by the reflection matrix. Ideal acoustic boundary types are represented by the following reflection matrices:

$$\text{rigidly closed} : \mathbf{r} = \mathbf{I} \text{ or } \mathbf{P}^+ = \mathbf{P}^-, \text{implying } v' = 0, \tag{2.6}$$

$$\text{pressure release} : \mathbf{r} = -\mathbf{I} \text{ or } \mathbf{P}^+ = -\mathbf{P}^-, \text{implying } p' = 0, \tag{2.7}$$

$$\text{anechoic} : \mathbf{r} = \mathbf{0}, \text{implying } \mathbf{P}^- = 0. \tag{2.8}$$

For one-dimensional elements $\mathbf{P}^{\mp} = p^{\mp}$ and Equation (2.5) reduce to

$$r = \frac{p^-}{p^+}, \tag{2.9}$$

where r is called the plane-wave reflection coefficient, and Equations (2.6)–(2.8) reduce to the scalar relationships $r = 1$, $r = -1$ and $r = 0$, respectively. It is important to note that the reflection matrix or coefficient at an open end of a duct is not included explicitly in the above list. This is because there is no such simple way of simulating the acoustic wave reflection process at open duct terminations and the subject is treated separately in Chapter 9. Nevertheless, the pressure release condition may be used as a first approximation at low frequencies (see Figure 9.7). At this point, it is important to add that, at an open termination, sound waves are radiated to the environment outside the termination and, hence, the radiation characteristics gain importance, in addition to reflection, as attributes of a boundary type one-port. However, we postpone further considerations of this issue to Chapter 9.

The source type one-port is called the one-port source. It has only an outlet port (arrow outward from the element) and activates an acoustic wave motion at the port to which it is connected. In general, an acoustic source constrains the variables vector of the port at which it is applied as

$$\mathbf{P}_S = \mathbf{G}_S\mathbf{P}, \tag{2.10}$$

where $\mathbf{P}$ denotes the vector of port variables of the activated port, $\mathbf{P}_S$ is the generalized source pressure strength vector and $\mathbf{G}_S$ is the non-dimensional source matrix. In general, $\mathbf{P}_S$ is of size $q \times 1$, $\mathbf{G}_S$ is of size $q \times 2q$ and $\mathbf{P}$ is of size $(2q) \times 1$, where q denotes the number of modes generated by the source process. If the source process is assumed to generate only one-dimensional sound waves, then $q = 1$, $\mathbf{G}_S = [\,G_{S1} \quad G_{S2}\,]$, $\mathbf{P}_S = P_S$ and Equation (2.10) simplifies to $P_S = G_{S1}p^+ + G_{S2}p^-$. In view of Equation (2.1), this can be expressed as

$$P_S = \frac{G_{S1} + G_{S2}}{2}\left(\rho_o c_o \frac{G_{S2} - G_{S2}}{G_{S1} + G_{S2}} v' + p'\right). \tag{2.11}$$

It is convenient to re-scale P_S as $p_S = 2P_S/(G_{S1} + G_{S2})$ so that this equation is written in the traditional form

$$p_S = Z_S v' + p', \tag{2.12}$$

where p_S and $Z_S = \rho_o c_o(G_{S2} - G_{S2})/(G_{S1} + G_{S2})$ are called the source pressure strength and source impedance, respectively.

Mathematical models of primary ducted acoustic source mechanisms in fluid machinery are discussed in Chapter 10. An acoustic model of a duct system must have at least one such source, because no source means no duct-borne sound. However, we are often interested in the acoustics of parts of ductworks that are located well downstream of the primary source process. Such parts must be considered as driven by an acoustic source that is equivalent to the actual source process, in that it produces the same acoustic field, in the part considered, as the actual source. The primary source process may consist of single or multiple one-port sources, but the equivalent source usually reduces to a single one-port source. The mathematical characteristics of the latter can sometimes be predicted (see Section 10.3), but they are usually determined by measurement (see Section 12.4).

2.3.2 Two-Port Elements

An acoustic element having two ports, each of which is associated with port variables vector $\mathbf{P}$ defined by Equation (2.2) or (2.3), is called a two-port element. Two-ports may be of passive type or active type. The latter is called a two-port source. A two-port element is understood to be of the passive type, unless the genre is specified explicitly.

Block representation of a (passive) two-port element is shown in Figure 2.2. In this book, the sign convention used in mathematical formulation of two-port elements for positive acoustic flow directions is such that of the two ports, one is the inlet port and the other is the outlet port of the element. In the frequency domain, the block in Figure 2.2 is tantamount to the mathematical relationship

$$\mathbf{P}_{in} = \mathbf{T}_{io}\mathbf{P}_{out}, \tag{2.13}$$

where the matrix $\mathbf{T}_{io}$ is called the wave transfer matrix with outlet-to-inlet causality, which means that $\mathbf{P}_{in}$ is determined from $\mathbf{T}_{io}$, given $\mathbf{P}_{out}$. In linear system parlance, $\mathbf{T}_{io}$ represents the frequency response matrix of a multiple-input-multiple-output system (see Section 1.5), but following the common acoustic network terminology, we refer to it as the wave transfer matrix, where the adjective wave is used to indicate that it relates the modal or one-dimensional acoustic pressure wave components. Obviously, the wave transfer matrix with inlet-to-outlet causality is simply $\mathbf{T}_{oi} = \mathbf{T}_{io}^{-1}$. We will not generally spell out the type of causality of a wave transfer matrix, as this will be evident from the subscripts used.

A one-dimensional two-port element is characterized by a 2×2 wave transfer matrix:

$$\mathbf{T}_{io} = \begin{bmatrix} T_{11} & T_{12} \\ T_{21} & T_{22} \end{bmatrix} \tag{2.14}$$

and a multi-modal two-port element is characterized by a $2m \times 2m$ wave transfer matrix:

$$\mathbf{T}_{io} = \begin{bmatrix} \mathbf{T}_{11} & \mathbf{T}_{12} \\ \mathbf{T}_{21} & \mathbf{T}_{22} \end{bmatrix}, \tag{2.15}$$

where the submatrices are of size $m \times m$.

A two-port source element also has two ports; however, these are activated ports and are not to be understood as inlet and outlet ports, though arrows are still used to indicate the default positive acoustic flow directions (Figure 2.3). A two-port source may be characterized mathematically as

$$\mathbf{P}_S = \begin{bmatrix} \mathbf{G}_{Sa} & \mathbf{G}_{Sb} \end{bmatrix} \begin{bmatrix} \mathbf{P}_a \\ \mathbf{P}_b \end{bmatrix}, \tag{2.16}$$

where the subscripts "a" and "b" denote the element ports. Equation (2.16) is usually expressed in the form

$$\mathbf{p}_S = \mathbf{T}_S\mathbf{P}_a + \mathbf{P}_b, \tag{2.17}$$

where $\mathbf{T}_S = (\mathbf{G}_{Sb})^{-1}\mathbf{G}_{Sa}$ is called the wave transfer matrix of the two-port source and $\mathbf{p}_S = (\mathbf{G}_{Sb})^{-1}\mathbf{P}_S$ is called its pressure strength vector. Mathematical two-port models of ducted source mechanisms in fluid machinery are discussed in Chapter 10.

2.3.3 Multi-Port Elements

Acoustic elements with more than two ports are called multi-ports. Multi-port elements may be characterized mathematically in the frequency domain by an equation of the form

$$\begin{bmatrix} \mathbf{A}_{11} & \mathbf{A}_{12} & \cdots & \mathbf{A}_{1n} \\ \mathbf{A}_{21} & \mathbf{A}_{22} & \cdots & \mathbf{A}_{2n} \\ \vdots & \vdots & \ddots & \vdots \\ \mathbf{A}_{k1} & \mathbf{A}_{k2} & \cdots & \mathbf{A}_{kn} \end{bmatrix} \begin{bmatrix} \mathbf{P}_1 \\ \mathbf{P}_2 \\ \vdots \\ \mathbf{P}_n \end{bmatrix} = \mathbf{0}. \tag{2.18}$$

Since this equation is not in wave transfer equation format, it is called, for lack of better terminology, a wave transmission equation and the coefficients matrix is referred to as the wave transmission matrix. We will denote the latter briefly as $\mathbf{A}_{1,2,\ldots,n} = [\mathbf{A}_1 \quad \mathbf{A}_2 \quad \ldots \quad \mathbf{A}_n]$, where the subscripts denote the numbers of the ports, $\mathbf{A}_j$ denotes the jth super column of the wave transmission matrix corresponding to port number $j = 1, 2, \ldots, n$, and n denotes the number of ports. For one-dimensional elements, the pressure wave components vectors in Equation (2.18) are defined as

$$\mathbf{P}_j = \begin{bmatrix} p^+ \\ p^- \end{bmatrix}_j \tag{2.19}$$

and for modal elements as

$$\mathbf{P}_j = \begin{bmatrix} \mathbf{P}^+ \\ \mathbf{P}^- \end{bmatrix}_j, \tag{2.20}$$

where

$$\mathbf{P}_j^{\mp} = \begin{bmatrix} p_1^{\mp} \\ p_2^{\mp} \\ \vdots \\ p_m^{\mp} \end{bmatrix}_j. \tag{2.21}$$

The submatrices $\mathbf{A}_{ij}$, $i = 1, 2, \ldots, k$, are of size $(\beta m) \times 2m$, where m denotes modality of the element, $\beta = 1$ or $\beta = 2$ depending on the multi-port type and k denotes the number of $\mathbf{A}_{ij}$ in one column, which is determined again by the element type. Both β and k will be obvious from the mathematical formulations of acoustic elements in later chapters. Two-port elements are encompassed in Equation (2.18) with $k = 1$, $n = 2$, $\beta = 2$. In this case, Equation (2.18) reduces to $\mathbf{A}_{11}\mathbf{P}_1 + \mathbf{A}_{12}\mathbf{P}_2 = \mathbf{0}$, which is equivalent to Equation (2.13) with the wave transfer matrix given by $\mathbf{T}_{12} = -\mathbf{A}_{11}^{-1}\mathbf{A}_{12}$.

The notion of inlet and outlet ports is not well-defined for multi-port elements. For this reason, they are formulated by using temporary sign conventions for positive acoustic flow directions at their ports. For example, Figure 2.4a shows a five-port element with tentative positive acoustic flow directions, from which all ports may be understood as inlet ports (arrows point into the block). Mathematically, this block represents Equation (2.18) with $n = 5$. The sign convention used for positive acoustic flow direction at a port of a multi-port element may be changed by multiplying the submatrices on the super column of the wave transmission matrix in Equation (2.18) corresponding to the port in question by a unit skew matrix of size $2m \times 2m$. For example, if the sign convention at port $j = J$ is to be changed, then the submatrices $\mathbf{A}_{iJ}$, $i = 1, 2, \ldots, k$, are post-multiplied by the matrix

$$\mathbf{S} = \begin{bmatrix} \mathbf{0} & \mathbf{I} \\ \mathbf{I} & \mathbf{0} \end{bmatrix}, \tag{2.22}$$

where $\mathbf{I}$ denotes a unit matrix of size $m \times m$. If the temporary sign conventions at some ports are to be changed as shown in Figure 2.4b, the relevant super-columns of matrix $\mathbf{A}$ should be post-multiplied by matrix $\mathbf{S}$. Thus, the block in Figure 2.4b represents the wave transmission equation

$$\begin{bmatrix} \mathbf{A}_{11} & \mathbf{A}_{12}\mathbf{S} & \ldots & \mathbf{A}_{15}\mathbf{S} \\ \mathbf{A}_{21} & \mathbf{A}_{22}\mathbf{S} & \ldots & \mathbf{A}_{25}\mathbf{S} \\ \vdots & \vdots & \ddots & \vdots \\ \mathbf{A}_{k1} & \mathbf{A}_{k2}\mathbf{S} & \ldots & \mathbf{A}_{k5}\mathbf{S} \end{bmatrix} \begin{bmatrix} \mathbf{P}_1 \\ \mathbf{P}_2 \\ \vdots \\ \mathbf{P}_5 \end{bmatrix} = \mathbf{0}. \tag{2.23}$$

In this block, ports 2 and 5 may now be conceived as outlet ports.

2.4 Assembly of Blocks

Block diagram networks of duct systems are constructed by exploding the hardware of a given duct system into ducts and some basic coupled duct arrangements, on the understanding that acoustic models of these components are available. These components are then replaced by their acoustic blocks and these blocks are assembled (de-exploded) by connecting their ports according to the corresponding hardware connectivity. Therefore, before assembling blocks, we must check if they have the same modality and inspect the compatibility of the default positive acoustic flow directions at the ports to be connected and change the incompatible directions, if necessary, as described in Section 2.3.3. A useful strategy is to keep the arrow directions of one-port and two-port elements fixed, since the ports of these elements are physically meaningful as inlet and outlet ports, and change the arrow directions only at ports of multi-port elements.

An assembled network of blocks is signified by lines joining ports of different blocks (for example Figure 2.5d). It is understood by definition that a port may not be connected to more than one port. The resulting diagram of connected blocks is called a block diagram. The acoustic wave transfer equations of the blocks of a block diagram can be manipulated so that the port variables at all ports can be computed in a systematical manner. In the following, this is demonstrated separately for block diagrams consisting of only two-port elements and those also including multi-port elements.

2.4.1 Assembly of Two-Ports

Consider the expansion chamber shown in Figure 2.5a. This relatively simple duct system is chosen for the clarity of the discussion. No generality is lost by this, because all duct systems are modeled and analyzed similarly. The chamber is assumed to be driven by an equivalent one-port source from port 1 (the actual source may be, for example, an engine exhaust process far upstream of this plane) and it has a boundary at port 6 (for example, the open end of the tailpipe of an exhaust system).

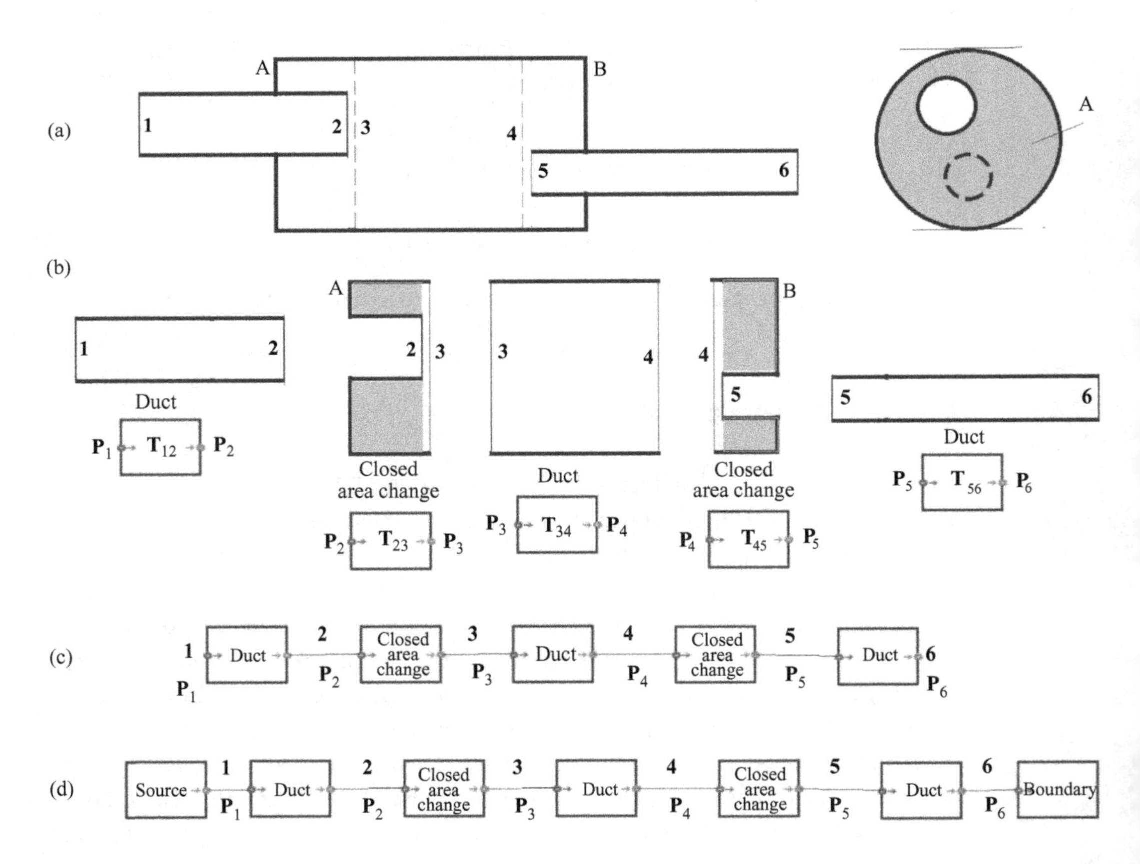

Figure 2.5 (a) Expansion chamber with extended inlet and outlet ducts, (b) exploded view of the chamber components and the corresponding acoustic blocks, (c) block diagram model of the chamber without source and boundary elements, (d) block diagram with source and boundary elements.

Consider exploding the chamber into components. As indicated in Figure 2.5a and b, one possibility is to explode it into three ducts, ducts 1–2, 3–4 and 5–6. Let us assume that we have appropriate acoustic models (of the same modality) for these ducts (Chapter 3 or 6). The corresponding blocks are shown in Figure 2.5b with their default port arrow directions and wave transfer matrices ($\mathbf{T}_{12}$, $\mathbf{T}_{34}$ and $\mathbf{T}_{56}$). But these ducts are not connected in unison; section areas of the connecting ports (2 with 3 and 4 with 5) are discontinuous. Therefore, we need intermediate acoustic elements in order to match the acoustic fields in the neighboring ducts. Acoustic elements modeling sound wave transmission at such sudden axial area changes in one- and three-dimensions are derived in Sections 4.3 and 7.4, respectively. In this example, the area changes are of the closed type, which means that the side-branch ducts (the annular ducts indicated in Figure 2.5b with gray filling), which are closed by end-caps A and B, are integrated into the formulation, so that the ports 2 and 3 and ports 4 and 5 can be connected by using intermediate two-port elements. The corresponding blocks, assumed to be of the same modality as the duct blocks, are also shown in Figure 2.5b with their default port arrow directions and wave transfer matrices ($\mathbf{T}_{23}$ and $\mathbf{T}_{45}$). Inspection of the arrow directions in the blocks in Figure 2.5b shows that the default sign conventions used for positive acoustic flow directions at the connecting ports are compatible. Thus, assembling the blocks we obtain a block diagram model of the chamber shown in Figure 2.5c. This diagram does not include the source and the boundary condition yet, but as we will see in Chapter 5, such block diagrams can still provide useful information for acoustic design of duct systems. The block diagram in Figure 2.5c is completed by applying the source at port 1 and the boundary condition at port 6, as shown in Figure 2.5d.

We may now embark on the calculation of the port variables vectors. The objective is to calculate all port variables vectors, given the one-port source equation at port 1, the boundary condition at port 6 and the wave transfer matrices of all two-ports. To this end, we first write down all equations of the acoustic elements in Figure 2.5d. Two-port equations are (Section 2.3.2):

$$\mathbf{P}_1 = \mathbf{T}_{12}\mathbf{P}_2 \tag{2.24}$$

$$\mathbf{P}_2 = \mathbf{T}_{23}\mathbf{P}_3 \tag{2.25}$$

$$\mathbf{P}_3 = \mathbf{T}_{34}\mathbf{P}_4 \tag{2.26}$$

$$\mathbf{P}_4 = \mathbf{T}_{45}\mathbf{P}_5 \tag{2.27}$$

$$\mathbf{P}_5 = \mathbf{T}_{56}\mathbf{P}_6. \tag{2.28}$$

Note that these equations have output-to-input causality. One-port equations are (Section 2.3.1):

$$\mathbf{P}_S = \mathbf{G}_S\mathbf{P}_1 \tag{2.29}$$

$$\begin{bmatrix}\mathbf{r}_6 & -\mathbf{I}\end{bmatrix}\mathbf{P}_6 = 0. \tag{2.30}$$

From the two-port equations, it follows that

$$\mathbf{P}_1 = \mathbf{T}_{12}\mathbf{T}_{23}\mathbf{T}_{34}\mathbf{T}_{45}\mathbf{T}_{56}\mathbf{P}_6 = \mathbf{T}_{16}\mathbf{P}_6, \tag{2.31}$$

where the matrix multiplications are compatible, since the blocks have the same modality. Therefore, $[\mathbf{r}_6 \quad -\mathbf{I}]\mathbf{T}_{61}\mathbf{P}_1 = 0$ and combining this with the one-port source equation gives

$$\mathbf{P}_1 = \begin{bmatrix} [\mathbf{r}_6 \quad -\mathbf{I}]\mathbf{T}_{61} \\ \mathbf{G}_S \end{bmatrix}^{-1} \begin{bmatrix} \mathbf{0} \\ \mathbf{P}_S \end{bmatrix}. \tag{2.32}$$

All remaining port variables vectors can now be calculated chain-wise from the two-port equations. An important feature of Equation (2.32) is that, apart from the source and boundary characteristics, it is dependent only on the wave transfer matrix of the acoustic path between source port and the boundary port, $\mathbf{T}_{61}$. This result is general and can be applied to all block diagrams with an equivalent source and a single termination. For this reason, we will usually skip the step of adding the source and terminal boundary blocks to block diagrams.

Application of Equation (2.32) is often hindered in practice by lack of knowledge about characteristics of the one-port source. In the absence of this data, the best we can do is to compute the port variables vectors in terms of the variables vector of another port. For this purpose, the most important result of the foregoing analysis is Equation (2.31). This is deducible directly from the block diagram of the chamber in Figure 2.5c and shows that the wave transfer matrix of multiple two-port blocks connected in series is given by the product of the wave transfer matrices of the constituent two-port blocks in the order of the series, provided that they have output-to-input causality. For example, if $\mathbf{P}_6$ is taken as the reference vector, then the port variable vectors can be determined chain-wise from the two-port equations. In fact, $\mathbf{P}_6$ (or, generally, the port vector at the boundary) is the best choice, since the acoustic boundary condition at port 6, $\mathbf{r}_6$, can be computed separately. (See Section 2.3.1 or Chapter 9 for open terminations.) Therefore, in view of Equation (2.30), all port variables vectors can be computed in terms of $\mathbf{P}_6^+$ (or p_6^+ in the one-dimensional case). Many calculations on acoustic design of duct systems are carried out on this basis and examples are shown in several places in this book.

It should be noted that when the wave transfer matrix $\mathbf{T}_{25} = \mathbf{T}_{23}\mathbf{T}_{34}\mathbf{T}_{45}$ is known, the chamber itself (with side branches) may be considered as an acoustic two-port element.

A two-port source, Equation (2.17), can be combined with passive acoustic two-ports on either side of the source by chain multiplication. For example, if $\mathbf{P}_a = \mathbf{T}_{a2}\mathbf{P}_2$, then $\mathbf{P}_b = -\mathbf{T}_S\mathbf{T}_{a2}\mathbf{P}_2 + \mathbf{p}_S$, and if $\mathbf{P}_1 = \mathbf{T}_{1b}\mathbf{P}_b$, then $\mathbf{P}_1 = \mathbf{T}_{1b}(-\mathbf{T}_S\mathbf{T}_{a2}\mathbf{P}_2 + \mathbf{p}_S)$.

2.4.2 Assembly of Multi-Ports

Duct systems that are exploded into components, including some with multi-port blocks, can be assembled systematically by using the global matrix technique. This essentially consists of collating the wave transfer equations of all components in a single matrix. The blocks should still be of the same modality and the positive acoustic

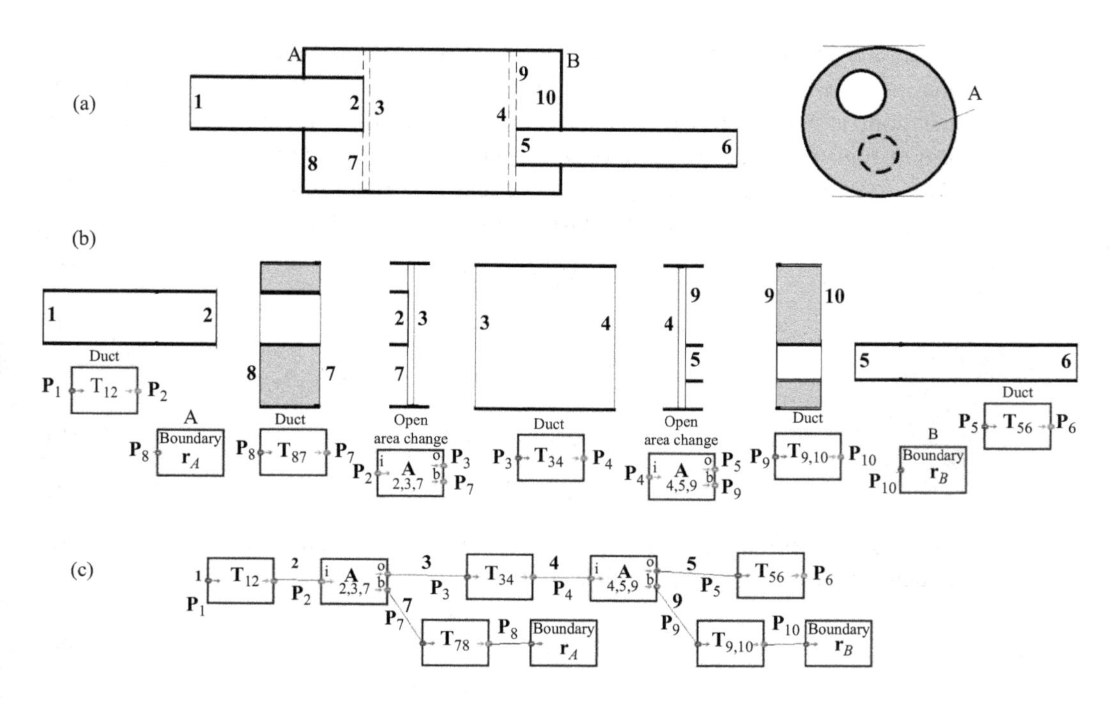

Figure 2.6 (a) Expansion chamber with extended inlet and outlet ducts, (b) exploded view of the chamber and the corresponding acoustic blocks, (c) block diagram model of the chamber without source and outlet boundary elements.

flow directions at the ports to be connected should be compatible. Application of this approach is explained here by two examples.

The first is, again, an expansion chamber. But as depicted in Figure 2.6a and b, it is now exploded into five ducts, namely, ducts 1–2, 3–4, 5–6, 7–8 and 9–10. Note that the side-branch ducts 7–8 and 9–10 are annular ducts, which are indicated by a gray filling. They are terminated at ports 8 and 10 by boundaries (end-caps A and B), which are assumed to have reflection matrices $\mathbf{r}_A$ and $\mathbf{r}_B$, respectively. For example, $\mathbf{r}_A = \mathbf{r}_B = \mathbf{I}$ or $r_A = r_B = 1$ in the one-dimensional case, if the end-caps are rigid. But, in this case, port 3 interfaces with ports 2 and 7, and port 4 interfaces with ports 5 and 9. Since the cross-sectional areas of these ports are not in unison, we must now use an intermediate acoustic element to match the acoustic variables at the connecting ports. Here the appropriate intermediate acoustic elements are the open area-changes discussed in Sections 4.3 or 7.2 and we use three-port elements, because side-branch ducts 7–8 and 9–10 are disjointed from the area discontinuities. The blocks corresponding to these three-ports are shown in Figure 2.6b with their default port arrow directions and wave transmission matrices, $\mathbf{A}_{2,3,7}$ and $\mathbf{A}_{4,5,9}$.

Inspection of the arrow directions in blocks in Figure 2.5b shows that the default sign conventions used for positive acoustic flow directions at the connecting ports are compatible, except for duct 8–7. This incompatibility can be rectified by inverting the causality of the wave transfer matrix of duct 8–7. Hence, upon joining the ports, we obtain a block diagram model of the chamber as shown in Figure 2.6c. This diagram may now be completed by applying the one-port source at port 1 and the boundary condition at port 6, similarly to Figure 2.5d. But, this is not really necessary, because use of Equation (2.32) requires only $\mathbf{T}_{61}$ and this can be determined from the block diagram in Figure 2.6c. Thus, we again begin by writing down all equations of the acoustic elements in Figure 2.6c. Two-port equations are (Section 2.3.2):

$$\mathbf{P}_1 = \mathbf{T}_{12}\mathbf{P}_2 \tag{2.33}$$

$$\mathbf{P}_7 = \mathbf{T}_{78}\mathbf{P}_8 \tag{2.34}$$

$$\mathbf{P}_3 = \mathbf{T}_{34}\mathbf{P}_4 \tag{2.35}$$

$$\mathbf{P}_9 = \mathbf{T}_{9,10}\mathbf{P}_{10} \tag{2.36}$$

$$\mathbf{P}_5 = \mathbf{T}_{56}\mathbf{P}_6. \tag{2.37}$$

Note that, in these equations, two-port wave transfer matrices are of size $2m \times 2m$, where m denotes the modality of acoustic elements. Three-port equations are (Section 2.3.3):

$$\begin{bmatrix} \mathbf{A}_2 & \mathbf{A}_3 & \mathbf{A}_7 \end{bmatrix} \begin{bmatrix} \mathbf{P}_2 \\ \mathbf{P}_3 \\ \mathbf{P}_7 \end{bmatrix} = 0 \tag{2.38}$$

$$\begin{bmatrix} \mathbf{A}_4 & \mathbf{A}_5 & \mathbf{A}_9 \end{bmatrix} \begin{bmatrix} \mathbf{P}_4 \\ \mathbf{P}_5 \\ \mathbf{P}_9 \end{bmatrix} = 0, \tag{2.39}$$

where the matrices $\mathbf{A}_j$, j = 2,3,7,4,5,9 are of size $(3m) \times 2m$. One-port boundary equations are (Section 2.3.1):

$$\begin{bmatrix} \mathbf{r}_A & -\mathbf{I} \end{bmatrix}\mathbf{P}_8 = \mathbf{R}_8\mathbf{P}_8 = 0 \tag{2.40}$$

$$\begin{bmatrix} \mathbf{r}_B & -\mathbf{I} \end{bmatrix}\mathbf{P}_{10} = \mathbf{R}_{10}\mathbf{P}_{10} = 0, \tag{2.41}$$

where $\mathbf{I}$ denotes an $m \times m$ unit matrix and the matrices $\mathbf{R}_8, \mathbf{R}_{10}$ are of size $m \times 2m$. The foregoing equations may be collected in single homogeneous matrix equation as:

$$\begin{bmatrix} \mathbf{I} & -\mathbf{T}_{12} & \mathbf{0} & \mathbf{0} & \mathbf{0} & \mathbf{0} & \mathbf{0} & \mathbf{0} & \mathbf{0} & \mathbf{0} \\ \mathbf{0} & \mathbf{0} & \mathbf{0} & \mathbf{0} & \mathbf{0} & \mathbf{0} & \mathbf{I} & -\mathbf{T}_{78} & \mathbf{0} & \mathbf{0} \\ \mathbf{0} & \mathbf{0} & \mathbf{I} & -\mathbf{T}_{34} & \mathbf{0} & \mathbf{0} & \mathbf{0} & \mathbf{0} & \mathbf{0} & \mathbf{0} \\ \mathbf{0} & \mathbf{0} & \mathbf{0} & \mathbf{0} & \mathbf{0} & \mathbf{0} & \mathbf{0} & \mathbf{0} & \mathbf{I} & -\mathbf{T}_{9,10} \\ \mathbf{0} & \mathbf{0} & \mathbf{0} & \mathbf{0} & \mathbf{I} & -\mathbf{T}_{56} & \mathbf{0} & \mathbf{0} & \mathbf{0} & \mathbf{0} \\ \mathbf{0} & \mathbf{A}_2 & \mathbf{A}_3 & \mathbf{0} & \mathbf{0} & \mathbf{0} & \mathbf{A}_7 & \mathbf{0} & \mathbf{0} & \mathbf{0} \\ \mathbf{0} & \mathbf{0} & \mathbf{0} & \mathbf{A}_4 & \mathbf{A}_5 & \mathbf{0} & \mathbf{0} & \mathbf{0} & \mathbf{A}_9 & \mathbf{0} \\ \mathbf{0} & \mathbf{0} & \mathbf{0} & \mathbf{0} & \mathbf{0} & \mathbf{0} & \mathbf{0} & \mathbf{R}_8 & \mathbf{0} & \mathbf{0} \\ \mathbf{0} & \mathbf{0} & \mathbf{0} & \mathbf{0} & \mathbf{0} & \mathbf{0} & \mathbf{0} & \mathbf{0} & \mathbf{0} & \mathbf{R}_{10} \end{bmatrix} \begin{bmatrix} \mathbf{P}_1 \\ \mathbf{P}_2 \\ \mathbf{P}_3 \\ \mathbf{P}_4 \\ \mathbf{P}_5 \\ \mathbf{P}_6 \\ \mathbf{P}_7 \\ \mathbf{P}_8 \\ \mathbf{P}_9 \\ \mathbf{P}_{10} \end{bmatrix} = \mathbf{0}, \tag{2.42}$$

where $\mathbf{I}$ now denotes a $(2m) \times 2m$ unit matrix. The coefficients matrix of this set of homogeneous equations is called the global matrix of the system. But, note that this equation may also be considered as the wave transmission equation of a 10-port element.

Counting the number of equations, we see that Equation (2.42) comprises $18m$ scalar equations, whilst each of port variables vectors $\mathbf{P}_j$, $j = 1, 2, \ldots, 10$, contain $2m$ variables. Thus, Equation (2.42) constitutes a system of $18m$ homogeneous equations in $20m$ unknowns. Therefore, any of the vectors $\mathbf{P}_j$ can be expressed in terms of any vector $\mathbf{P}_k$, $k \neq j$. Consider, for example, the reduction of Equation (2.42) to the two-port $\mathbf{P}_1 = \mathbf{T}_{16}\mathbf{P}_6$, so that Equation (2.32) may be used. To this end, we first move the vector $\mathbf{P}_6$ to the right-hand side of Equation (2.42). Let $\mathbf{C}$ denote the new coefficients matrix on the left-hand side. This is obtained from the global matrix by deleting the columns corresponding to the vector $\mathbf{P}_6$. Next, the modified form of Equation (2.42) is multiplied from left by the inverse of matrix $\mathbf{C}$ to obtain

$$\begin{bmatrix} \mathbf{P}_1 \\ \mathbf{P}_2 \\ \mathbf{P}_3 \\ \mathbf{P}_4 \\ \mathbf{P}_5 \\ \mathbf{P}_7 \\ \mathbf{P}_8 \\ \mathbf{P}_9 \\ \mathbf{P}_{10} \end{bmatrix} = \mathbf{C}^{-1} \begin{bmatrix} \mathbf{0} \\ \mathbf{0} \\ \mathbf{0} \\ \mathbf{0} \\ -\mathbf{T}_{56} \\ \mathbf{0} \\ \mathbf{0} \\ \mathbf{0} \\ \mathbf{0} \end{bmatrix} \mathbf{P}_6. \tag{2.43}$$

Inversion of $\mathbf{C}$ is possible because it is a square non-singular matrix. Thus, all port variables can be calculated from Equation (2.43) in terms of $\mathbf{P}_6$ and the transfer matrix $\mathbf{T}_{16}$ (of size $2m \times 2m$) is extracted from the first super row, that is, $\mathbf{P}_1 = \mathbf{T}_{16}\mathbf{P}_6$. We may now use Equation (2.32) to compute $\mathbf{P}_1$ and, hence, all port variables vectors.

The second example shown in Figure 2.7a is usually referred to as straight-through resonator or muffler. It is essentially an expansion chamber with the inlet and outlet ducts bridged by a perforated duct. The annular space between the casing and the perforated pipe may be packed with sound absorbent material.

Consider exploding the muffler into components. Inspection of the topology of the muffler shows that a difficulty arises because ducts 3-5 and 2-7 are coupled through the perforations and cannot be exploded separately. Therefore, we need an acoustic element that models sound transmission in ducts coupled by perforations (Sections 4.7–4.8 and Section 7.7 are devoted to the formulation of such acoustic elements in one- and three-dimensions). The particular arrangement of this example is known as two-duct (see Figure 4.10a) and can be represented by an acoustic multi-port element with four ports, namely, ports 2, 3, 5, 7. Thus, we may now explode the muffler into this two-duct and the ducts 1-2, 4-3, 5-6 and 7-8, as shown in Figure 2.7b, where the corresponding blocks are also shown. Ducts 4-3 and 5-6 are annular ducts and connect with duct 3–5 in unison. Hence, upon joining the ports, we obtain a block diagram

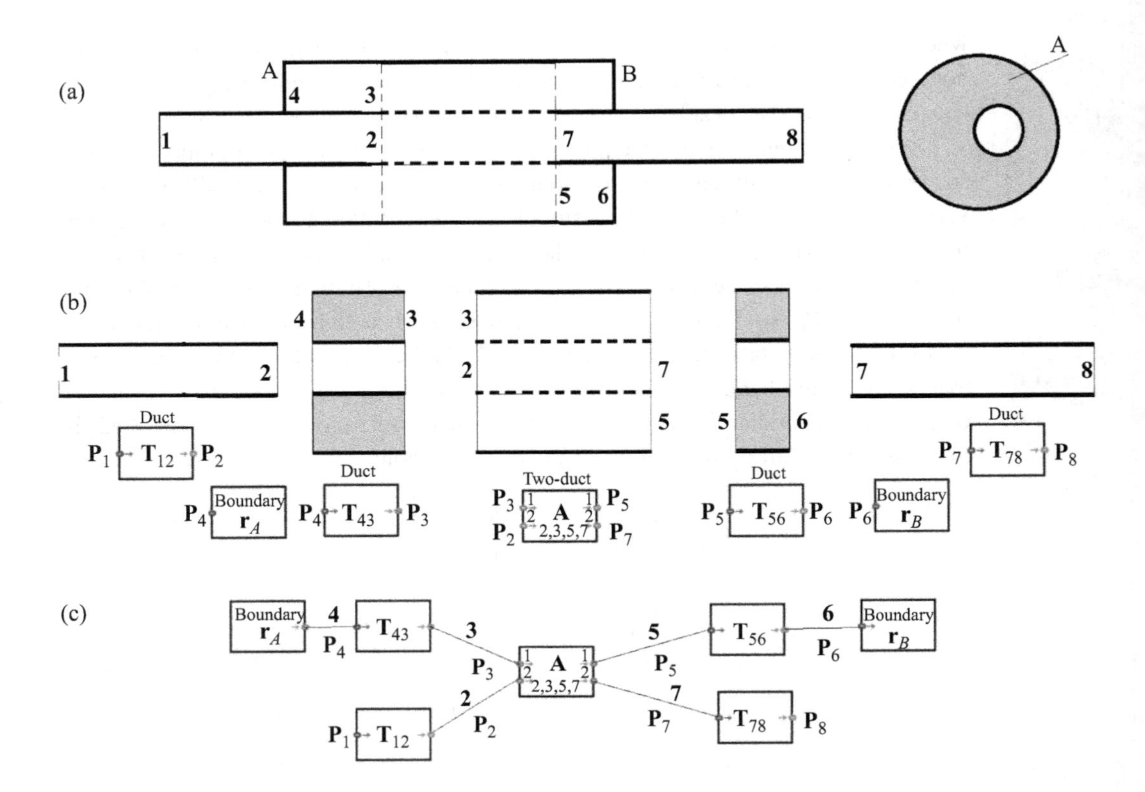

Figure 2.7 (a) Straight-through muffler, (b) exploded view of the muffler and the corresponding acoustic blocks, (c) block diagram model of the muffler without source and outlet boundary elements.

model of the muffler shown in Figure 2.7c. This diagram may now be completed by applying the one-port source at port 1 and the boundary condition at port 8; however, we may again skip this step because we can use Equation (2.32) with $\mathbf{T}_{61}$ replaced by $\mathbf{T}_{81}$ and this can be determined from the block diagram in Figure 2.7c, as described next. Again, we begin by listing the two-port equations:

$$\mathbf{P}_1 = \mathbf{T}_{12}\mathbf{P}_2 \tag{2.44}$$

$$\mathbf{P}_4 = \mathbf{T}_{43}\mathbf{P}_3 \tag{2.45}$$

$$\mathbf{P}_5 = \mathbf{T}_{56}\mathbf{P}_6 \tag{2.46}$$

$$\mathbf{P}_7 = \mathbf{T}_{78}\mathbf{P}_8. \tag{2.47}$$

One-port boundary equations are (Section 2.3.1):

$$[\mathbf{r}_A \quad -\mathbf{I}]\mathbf{P}_4 = \mathbf{R}_4\mathbf{P}_4 = 0 \tag{2.48}$$

$$[\mathbf{r}_B \quad -\mathbf{I}]\mathbf{P}_6 = \mathbf{R}_6\mathbf{P}_6 = 0. \tag{2.49}$$

The coupled perforated duct section is represented by the four-port equation (for example, Section 4.7.1):

$$\begin{bmatrix} \mathbf{P}_3 \\ \mathbf{P}_2 \end{bmatrix} = \begin{bmatrix} \mathbf{S}_{11} & \mathbf{S}_{12} \\ \mathbf{S}_{21} & \mathbf{S}_{22} \end{bmatrix} \begin{bmatrix} \mathbf{P}_5 \\ \mathbf{P}_7 \end{bmatrix}. \tag{2.50}$$

Collecting the foregoing relations, the global matrix of the system is obtained as

$$\begin{bmatrix} \mathbf{I} & -\mathbf{T}_{12} & \mathbf{0} & \mathbf{0} & \mathbf{0} & \mathbf{0} & \mathbf{0} & \mathbf{0} \\ \mathbf{0} & \mathbf{0} & \mathbf{I} & \mathbf{0} & -\mathbf{S}_{11} & \mathbf{0} & -\mathbf{S}_{12} & \mathbf{0} \\ \mathbf{0} & \mathbf{I} & \mathbf{0} & \mathbf{0} & -\mathbf{S}_{21} & \mathbf{0} & -\mathbf{S}_{22} & \mathbf{0} \\ \mathbf{0} & \mathbf{0} & \mathbf{I} & -\mathbf{T}_{34} & \mathbf{0} & \mathbf{0} & \mathbf{0} & \mathbf{0} \\ \mathbf{0} & \mathbf{0} & \mathbf{0} & \mathbf{0} & \mathbf{I} & -\mathbf{T}_{56} & \mathbf{0} & \mathbf{0} \\ \mathbf{0} & \mathbf{0} & \mathbf{0} & \mathbf{0} & \mathbf{0} & \mathbf{I} & -\mathbf{T}_{78} & \mathbf{0} \\ \mathbf{0} & \mathbf{0} & \mathbf{0} & \mathbf{r}_4 & \mathbf{0} & \mathbf{0} & \mathbf{0} & \mathbf{0} \\ \mathbf{0} & \mathbf{0} & \mathbf{0} & \mathbf{0} & \mathbf{0} & \mathbf{r}_6 & \mathbf{0} & \mathbf{0} \end{bmatrix} \begin{bmatrix} \mathbf{P}_1 \\ \mathbf{P}_2 \\ \mathbf{P}_3 \\ \mathbf{P}_4 \\ \mathbf{P}_5 \\ \mathbf{P}_6 \\ \mathbf{P}_7 \\ \mathbf{P}_8 \end{bmatrix} = \mathbf{0}. \tag{2.51}$$

This is a linear homogeneous system of $14m$ algebraic equations in $16m$ scalar unknowns and any vector $\mathbf{P}_j, j \in 1, 2, \ldots, 8$, can be expressed in terms of $\mathbf{P}_k$, $k \neq j$. For example, the transfer matrix $\mathbf{T}_{18}$ can be determined by passing $\mathbf{P}_8$ to the right of Equation (2.51) and solving the resulting system of $14m$ equations, as explained in the previous example.

2.4.3 Optimization of Global Matrix Size

A disadvantage of the global matrix approach is that the size of the global matrix increases with the number of elements in a block diagram. For this reason, it is advantageous to develop strategies for reducing the size of the global matrix. One useful strategy is to assemble all two-port elements that are connected in series into a single two-port wave transfer matrix before the formation of the global matrix (Section 2.4.1). Another effective strategy is to formulate the multi-port elements with movable ports, since then ducts can be connected to the ports without having to use additional duct elements in the block diagram. Movable ports are introduced relatively easily into multi-port formulations. Consider a generic n-port element with wave transmission equation given by Equation (2.18). Making a port, port $j (\leq n)$ say, movable means that $\mathbf{P}_j$ is referred to a new port, j' say, through the two-port equation $\mathbf{P}_j = \mathbf{T}_{j,j'}\mathbf{P}_{j'}$, where $\mathbf{T}_{j,j'}$ denotes the wave transfer matrix of the duct along which the port is moved. Therefore, the wave transmission equation of the n-port with port j made movable may be obtained from Equation (2.18) by post-multiplying the submatrices $\mathbf{A}_{ij}$, $i = 1, 2, \ldots, k$, by $\mathbf{T}_{j,j'}$. For example, referring to the five-port block in Figure 2.4a, if we make port 2 movable, then the wave transmission equation of the element becomes

$$\begin{bmatrix} \mathbf{A}_{11} & \mathbf{A}_{12}\mathbf{T}_{2,2'} & \cdots & \mathbf{A}_{15} \\ \mathbf{A}_{21} & \mathbf{A}_{22}\mathbf{T}_{2,2'} & \cdots & \mathbf{A}_{25} \\ \vdots & \vdots & \ddots & \vdots \\ \mathbf{A}_{k1} & \mathbf{A}_{k2}\mathbf{T}_{2,2'} & \cdots & \mathbf{A}_{k5} \end{bmatrix} \begin{bmatrix} \mathbf{P}_1 \\ \mathbf{P}_{2'} \\ \vdots \\ \mathbf{P}_n \end{bmatrix} = \mathbf{0}. \tag{2.52}$$

Application of these strategies is demonstrated in Figure 2.8 for a folded duct resonator. A causal block diagram of the resonator is shown in Figure 2.8a. Here,

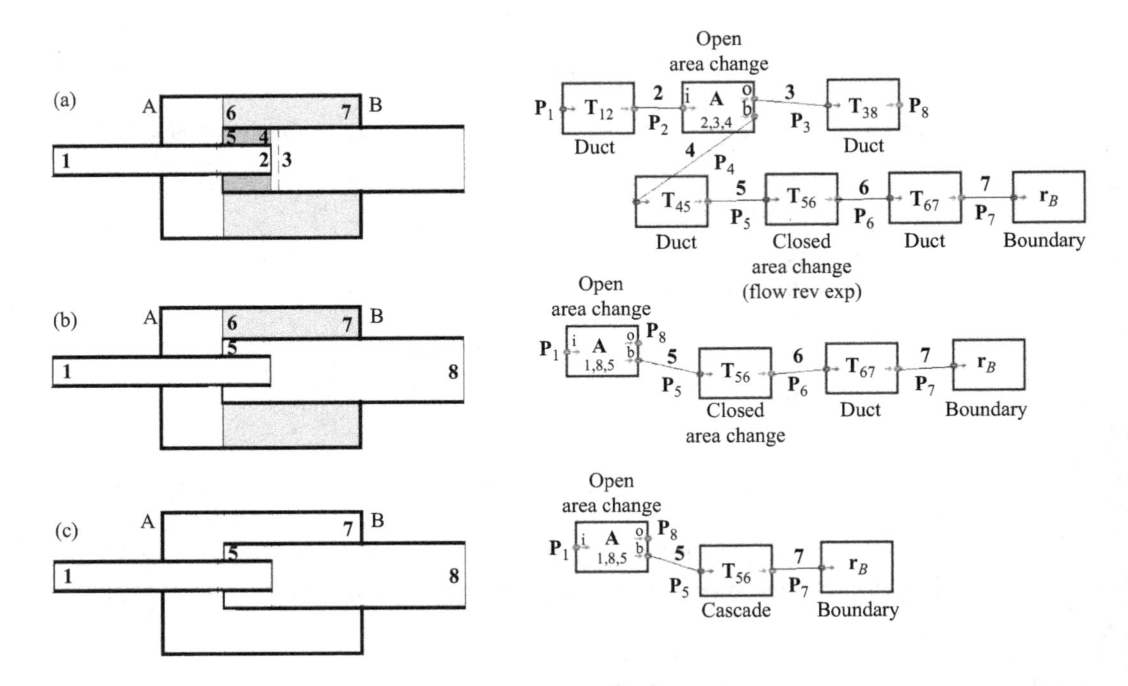

Figure 2.8 Simplification of an acoustic model of a folded duct resonator by using three-ports with extendable port lengths and assembly of series blocks.

the resonator is exploded into ducts 1-2, 3-8, 5-4, 6-7, a closed area change of flow-reversing type 5-6, which is a two-port as it includes the annular duct terminating with end-cap A in its formulation, and an open area change 2-3-4, which is a three-port element. This block diagram yields a global matrix of size $14m \times 16m$, where m denotes the modality of acoustic elements. But, suppose that, the open area-change element 2-3-4 was formulated with movable ports. Then, duct 1-2 can be integrated at port 2, duct 3-8 can be integrated at port 3 and duct 4-5 can be integrated at port 4, as depicted in Figure 2.8b. These extensions do not change the size of the three-port, which has now the ports 1-8-5, but eliminates three ducts from the block diagram in Figure 2.8a. The resulting three-port now covers a substantially larger domain. The corresponding block diagram (which is shown in Figure 2.15b), gives a global matrix of size $8m \times 10m$. The size of the global matrix can be reduced further if the global matrix is formed after the two two-ports in series in Figure 2.8b are assembled into a single two-port element. The system block diagram then reduces to that shown in Figure 2.8c, which yields a global matrix of size $6m \times 8m$.

2.4.4 Contraction of Assembled Modal Two-Ports

Many practical duct systems are represented by block diagrams that are equivalent to a two-port model. This means that however complex the block diagram may be, it has only one inlet port and one outlet port, all remaining ports of the blocks being either connected or closed. Then, the global matrix can be reduced to a two-port form, but, since blocks of a block diagram have the same modality, m, the reduction may give a model with port variables vectors that have insignificant modes included at the frequencies of interest. For example, referring to the expansion chamber in Figure 2.5 or 2.6, let the diameters of the inlet and outlet ducts, ducts 1-2 and 5-6, respectively, be such that, at the frequencies of interest, the acoustic fields in these ducts are dominated by plane waves. But we may still want to take $m > 1$ in order to account of the three-dimensional effects in the chamber and at the cross-sectional area discontinuities. Then, the two-port equation $\mathbf{P}_1 = \mathbf{T}_{16}\mathbf{P}_6$ comes out with modality $m > 1$, although $m = 1$ is adequately accurate for ducts 1-2 and 5-6. So, we may consider eliminating the insignificant modes (usually called evanescent modes) from the inlet and outlet port variables vectors.

In general, a port variables vector of a modal two-port model, which is of the form of Equation (2.3), may be partitioned, at a given frequency, as

$$\mathbf{P} = \begin{bmatrix} \mathbf{P}_b \\ \mathbf{P}_e^+ \\ \mathbf{P}_e^- \end{bmatrix} \tag{2.53}$$

where

$$\mathbf{P}_b = \begin{bmatrix} \mathbf{P}_b^+ \\ \mathbf{P}_b^- \end{bmatrix}, \quad \mathbf{P}_b^\mp = \begin{bmatrix} p_1^\mp \\ p_2^\mp \\ \vdots \\ p_b^\mp \end{bmatrix}, \tag{2.54}$$

and

$$\mathbf{P}_e^\mp = \begin{bmatrix} p_{b+1}^\mp \\ p_{b+2}^\mp \\ \vdots \\ p_m^\mp \end{bmatrix}. \tag{2.54a}$$

Here, the modal pressure wave components in vectors $\mathbf{P}_e^\pm$ are all associated with evanescent modes (or modes with large decaying rates) and $\mathbf{P}_b^\pm$ include at least all propagating modes (or modes with smaller decaying rates). Upon introducing these subvectors, Equation (2.13) may be expressed in partitioned form as

$$\begin{bmatrix} \mathbf{P}_b \\ \mathbf{P}_e^+ \\ \mathbf{P}_e^- \end{bmatrix}_{in} = \begin{bmatrix} \mathbf{B}_{11} & \mathbf{B}_{12} & \mathbf{B}_{13} \\ \mathbf{B}_{21} & \mathbf{B}_{22} & \mathbf{B}_{23} \\ \mathbf{B}_{31} & \mathbf{B}_{32} & \mathbf{B}_{33} \end{bmatrix} \begin{bmatrix} \mathbf{P}_b \\ \mathbf{P}_e^+ \\ \mathbf{P}_e^- \end{bmatrix}_{out}. \tag{2.55}$$

Let the inlet port be sufficiently away from any upstream discontinuity and the outlet port sufficiently away from any discontinuity downstream of it, so that no evanescent modes can reach it. Then we may take $\left(\mathbf{P}_e^+\right)_{in} = 0$, $\left(\mathbf{P}_e^-\right)_{out} = 0$ and eliminating $\left(\mathbf{P}_e^+\right)_{out}$ from the first two equations in Equation (2.55) yields the two-port equation

$$(\mathbf{P}_b)_{in} = \mathbf{T}_{io}(\mathbf{P}_b)_{out}, \tag{2.56}$$

where the wave transfer matrix is given by

$$\mathbf{T}_{io} = \mathbf{B}_{11} - \mathbf{B}_{12}\mathbf{B}_{22}^{-1}\mathbf{B}_{21}. \tag{2.57}$$

Thus, the size of the original modal two-port is reduced from $2m \times 2m$ to $2b \times 2b$.

This reduction is called contraction. In particular, if $b = 1$, the size of the wave transfer matrix reduces to that of an one-dimensional two-port and the element can be connected to other one-dimensional or plane-wave elements, however, it is still classified as multi-modal element, because three-dimensional effects are included in its internal acoustic fields. This is an important capability; because it allows study of three-dimensional acoustic effects in duct systems by using one-dimensional block diagram models. (See Section 7.8 for applications.)

2.5 Acoustic Elements Based on Numerical Methods

In subsequent chapters, because of this book's focus on solutions that are transparent to the physics of the systems considered, formulations of acoustic models in one- and three-dimensions are based as much as possible on exact or approximate analytical solutions of the acoustic equations of continuity, momentum and energy. However, in cases where this is not possible or satisfactory, it is necessary to use a numerical method. With present-day digital computers, calculations involved in numerical methods are not, in general, of great issue; however, casting of numerical solutions in a form suitable for network methods may require additional manipulations. This may be a decisive factor in the choice of a numerical method for one-dimensional duct models in the frequency domain, because numerical performances of major methods for linear differential equations in one independent variable are more or less similar. In Chapter 3, we describe a numerical approach which can yield the solutions directly in the wave transfer matrix format of one-dimensional two-port duct elements (Equation (2.14)).

Numerical determination of modal acoustic elements is also relatively straightforward in principle, but not always easy in implementation, when the dependency of the acoustic field on the transverse and axial coordinates can be separated analytically. In this case, the transverse functions are governed by a partial differential equation in two-dimensions and there is an abundance of numerical methods which may be used to obtain solutions. Apart from the finite element and the boundary element methods, which are discussed presently in more detail, we may mention the finite difference method and the spectral methods, which may be used effectively for common duct section shapes. In the finite difference method, derivatives are approximated over local grids and, hence, accuracy is limited to the extent to which the grid can be refined without being affected by the round-off errors and the precision with which the acoustic boundary conditions on the duct walls can be applied.

Spectral methods [8], on the other hand, are based on expansion of the exact transverse functions, φ, say, into a linear combination of linearly independent global basis functions, β_k, as $\varphi(\bullet) \approx \sum_{k=0}^{\infty} a_k \beta_k(\bullet)$, where $(\bullet)$ denotes the transverse coordinates and a_k denote undetermined constants. In the classical Galerkin method, the basis functions are selected so that each satisfies the acoustic boundary conditions on the duct walls and the coefficients a_k are determined so that the error incurred, called residual, from the substitution of this expansion in the governing differential equation is orthogonal to the basis functions. When the differential equation has a weak (variational) form, this procedure becomes equivalent to the Rayleigh–Ritz method for the same basis functions. However, the Rayleigh–Ritz method can be applied with less strict continuity conditions on the basis functions. Other well-known spectral methods are the collocation method and the tau method. In the collocation method, which is often referred to as the pseudo-spectral method, a_k are selected so that the boundary conditions are satisfied and the residual vanishes at as many discrete points on a duct section as possible. In the tau method, a_k are also selected so that the boundary conditions are satisfied, but the residual is made orthogonal to as many basis functions as possible. Note that in the tau and collocation methods, the basis functions are not required to satisfy the boundary conditions individually. They are usually constructed from trigonometric functions for periodic circumferential dependency of a transverse function and the Chebyshev or Legendre polynomials for the radial dependency.

If the acoustic field in a duct does not separate into transverse and axial functions, the governing acoustic equations have to be solved numerically over the entire domain of the duct. In this case the finite element method and the boundary element method are more commonly used because of their versatility in dealing with arbitrary geometry, although the availability of general-purpose commercial software for the implementation of these methods also has a role in this situation. Indeed, the success of these methods largely depends on the availability of efficient solid modeling and preprocessing (automatic meshing) codes, large system solvers, post-processing codes and advanced computer programming techniques (parallel processing). Nowadays engineers and researchers rely on software written by teams of specialists (usually commercial) in implementation of these numerical methods and it makes no sense to attempt to describe them here in detail. In this section we discuss the underlying principles of the finite element and the boundary element methods only in sufficient detail to show how the results produced by these methods may be transformed to the modal wave transfer equation format adopted in this book.

2.5.1 The Finite Element Method

Application of the finite element method involves two major steps. In the first step, the volume of a duct, V of surface Σ, say, is divided into a large number, N, of contiguous polyhedrals (typically, tetrahedrons) of volumes V^e, $e = 1, 2, \ldots, N$, called finite elements, which are considered to be connected to each other at points called nodes (typically, vertices of tetrahedrons). Such division of volume V also divides the boundary Σ into a large number, M, say, of contiguous surfaces of area Σ^s, $s = 1, 2, \ldots, M$. Division of volume V into finite elements may involve an approximation of the actual geometry, if the surfaces Σ^s of the elements do not fit Σ exactly.

In the second step, the acoustic fields in finite elements are approximated by using a spectral method, usually the Galerkin method. For simplicity of presentation, and as a prelude to later applications for calculation of transverse duct modes, here we will demonstrate the process for a homogeneous duct with negligible mean flow; however, it can also be applied similarly without these restrictions. In this case, acoustic continuity and momentum equations, Equations (1.5) and (1.8), simplify to $\partial \rho'/\partial t + \rho_o \nabla \cdot \mathbf{v}' = 0$ and $\rho_o \partial \mathbf{v}'/\partial t + \nabla p' = 0$, respectively, where the acoustic pressure and density are related by Equation (1.15). Upon eliminating the particle velocity, the foregoing equations yield, in the frequency domain, the classical three-dimensional Helmholtz equation for the acoustic pressure

$$\nabla^2 p' + k_o^2 p' = 0, \tag{2.58}$$

where $k_o = \omega/c_o$ and c_o is the speed of sound. For any p' that is not an exact solution, $\nabla^2 p' + k_o^2 p' = R$, where R is called the residual. The global Galerkin method requires that $\int_V wR\mathrm{d}V = 0$, where w denotes a basis function of the expansion representing p' and, obviously, it is necessary that second derivatives of w exist. However, using the identity $\nabla \cdot w\nabla p' - \nabla w \cdot \nabla p' = w\nabla \cdot \nabla p'$ and invoking the divergence theorem, Equation (A.1), the foregoing residual statement may be expressed as

$$\int_V (-\nabla w \cdot \nabla p' + k_o^2 w p')\mathrm{d}V + \int_\Sigma w\mathbf{n} \cdot \nabla p' \mathrm{d}\Sigma = 0, \tag{2.59}$$

where $\mathbf{n}$ denotes the outward unit normal vector of Σ. This equation is called the weak (variational) form of Equation (2.58) [9]. Since the derivatives on w and p' are now equally divided, the basis functions need to be continuous only up to the first derivative. A spectral method based on the weak form (if it exists) of a partial differential equation is usually distinguished as the Rayleigh–Ritz method.

The global nature of approximation implied in Equation (2.59) can be diverted to local approximations by splitting the domain and boundary integrals over the finite elements as

$$\sum_{e=1}^{N} \int_{V^e} (-\nabla w \cdot \nabla p' + k_o^2 w p')\mathrm{d}V + \sum_{s=1}^{M} \int_{\Sigma^s} w v_n' \mathrm{d}\Sigma = 0, \tag{2.60}$$

where, in writing the boundary integral, we have substituted $\mathbf{n} \cdot \nabla p' = \mathrm{i}\rho_o \omega v_n'$ from the acoustic momentum equation and $v_n' = \mathbf{n} \cdot \mathbf{v}'$ denotes the normal velocity of the boundary. The standard procedure now is the evaluation of the integrands by taking

$$p'(x_1, x_2, x_3) \approx p^e(x_1, x_2, x_3) = \sum_{i=1}^{n^e} \psi_i^e(x_1, x_2, x_3) p_i^e = \mathbf{\Psi}^e \mathbf{p}^e \tag{2.61}$$

and $w = \psi_i^e$, $i = 1, 2, \ldots, n^e$. Here, n^e denotes the number of nodes of a finite element, ψ_i^e are called shape functions, which are defined in a global rectangular coordinate system, (x_1, x_2, x_3), say, and $p_i^e = p^e((x_1)_i, (x_2)_i, (x_3)_i)$ denotes the acoustic pressure at node $i \in 1, 2, \ldots, n^e$ with coordinates $(x_1)_i, (x_2)_i, (x_3)_i$. The shape functions are usually determined by polynomial interpolation and satisfy the condition $\psi_j^e((x_1)_i, (x_2)_i, (x_3)_i) = 1$ if $i = j$, else 0. They need not be of high order, since they are localized to finite element domains.

The second equality in Equation (2.61) defines the approximation in question in matrix notation with $\mathbf{p}^e = \{p_1^e \quad p_2^e \quad \ldots \quad p_{n^e}^e\}$ and $\mathbf{\Psi}^e = [\psi_1^e \quad \psi_2^e \quad \ldots \quad \psi_{n^e}^e]$. Similarly, we write $v_n' = \mathbf{\Psi}^s \mathbf{v}^s$ with $\mathbf{v}^s = \{v_1^s \quad v_2^s \quad \ldots \quad v_{n^s}^s\}$ and $\mathbf{\Psi}^s = [\psi_1^s \quad \psi_2^s \quad \ldots \quad \psi_{n^s}^s]$, where v_j^s denotes the value of v_n' at node $j \in 1, 2, \ldots, n^s$ on Σ^s, n^s being the number of nodes of Σ^s, and the shape functions ψ_j^s are the shape function of the element whose surface include Σ^s. After these substitutions, Equation (2.60) may be expressed in matrix notation as

$$\sum_{e=1}^{N} \left(\mathbf{K}^e - k_o^2 \mathbf{M}^e\right) \mathbf{p}^e + \mathrm{i}\rho_o \omega \sum_{s=1}^{M} \mathbf{F}^s \mathbf{v}^s = 0, \tag{2.62}$$

where

$$\mathbf{K}^e = \int_{V^e} \nabla(\mathbf{\Psi}^e)^T \cdot \nabla \mathbf{\Psi}^e \mathrm{d}V \tag{2.63}$$

$$\mathbf{M}^e = \int_{V^e} (\mathbf{\Psi}^e)^T \mathbf{\Psi}^e \,\mathrm{d}V \tag{2.64}$$

$$\mathbf{F}^s = \int_{\Sigma^s} (\mathbf{\Psi}^e)^T \mathbf{\Psi}^s \mathrm{d}\Sigma. \tag{2.65}$$

The sums in Equation (2.62) may be assembled by collating all equations on the global vectors $\mathbf{p}'$ and $\mathbf{v}_\Sigma'$, which are defined as

$$\mathbf{p}' = \begin{bmatrix} p_1' \\ p_2' \\ \vdots \\ p_n' \end{bmatrix} = \begin{bmatrix} \mathbf{p}_\Sigma' \\ \mathbf{p}_V' \end{bmatrix}, \mathbf{v}_\Sigma' = \begin{bmatrix} (v_n')_1 \\ (v_n')_2 \\ \vdots \\ (v_n')_m \end{bmatrix}, \tag{2.66}$$

where n denotes the total number of nodes of the finite element mesh, m denotes the number of nodes on Σ, and $\mathbf{p}_V'$ and $\mathbf{p}_\Sigma'$ contain, respectively, the terms of $\mathbf{p}'$ at the interior nodes and at the boundary nodes. The vector $\mathbf{p}'$ includes all distinct elements of finite element vectors $\mathbf{p}^e$, $e = 1, 2, \ldots, N$, and the vector $\mathbf{v}_\Sigma'$ includes all distinct terms of $\mathbf{v}^s$, $s = 1, 2, \ldots, M$. Upon fitting Equation (2.62) to these global vectors, we obtain

$$\begin{bmatrix} \mathbf{Z}_{11} & \mathbf{Z}_{12} \\ \mathbf{Z}_{21} & \mathbf{Z}_{22} \end{bmatrix} \begin{bmatrix} \mathbf{p}_\Sigma' \\ \mathbf{p}_V' \end{bmatrix} = \begin{bmatrix} \mathbf{F}\mathbf{v}_\Sigma' \\ \mathbf{0} \end{bmatrix}, \tag{2.67}$$

where the partitioned matrix represents the square symmetric matrix $\mathbf{Z} = \mathbf{K} - \omega^2 \mathbf{M}$, where the matrices $\mathbf{K}$ and $\mathbf{M}$ are derived from the finite element matrices $\mathbf{K}^e$ and $\mathbf{M}^e$, $e = 1, 2, \ldots, N$, and the matrix $\mathbf{F}$ is derived from the matrices $\mathbf{F}^s$, $s = 1, 2, \ldots, M$.

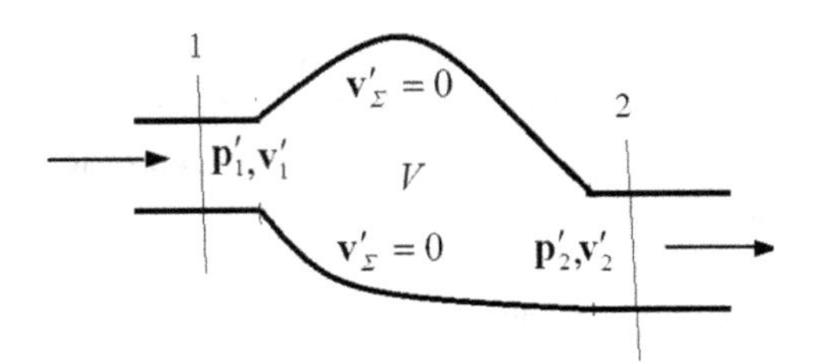

Figure 2.9 A non-uniform curved duct with uniform duct extensions.

Equation (2.67) represents the normal output format of the finite element method. It may be condensed to the form

$$\mathbf{D}\mathbf{p}'_\Sigma = \mathbf{v}'_\Sigma, \tag{2.68}$$

where $\mathbf{D} = \mathbf{F}^{-1}\left(\mathbf{Z}_{11} - \mathbf{Z}_{12}\mathbf{Z}_{22}^{-1}\mathbf{Z}_{21}\right)$.

Now consider an arbitrary two-port duct system (Figure 2.9). Equation (2.68) may be partitioned further as

$$\begin{bmatrix} \mathbf{D}_{11} & \mathbf{D}_{12} & \mathbf{D}_{1W} \\ \mathbf{D}_{21} & \mathbf{D}_{22} & \mathbf{D}_{2W} \\ \mathbf{D}_{W1} & \mathbf{D}_{W2} & \mathbf{D}_{WW} \end{bmatrix} \begin{bmatrix} \mathbf{p}'_1 \\ \mathbf{p}'_2 \\ \mathbf{p}'_W \end{bmatrix} = \begin{bmatrix} \mathbf{v}'_1 \\ \mathbf{v}'_2 \\ \mathbf{v}'_W \end{bmatrix}, \tag{2.69}$$

where the subscript "W" denotes the duct walls and the subscripts "1" and "2" denote the port sections ($\Sigma = 1 \cup 2 \cup W$). The elements of $\mathbf{p}'_j$ and $\mathbf{v}'_j$ correspond, one-to-one, to nodes on ports $j \in 1, 2$, and elements of $\mathbf{p}'_W$ and $\mathbf{v}'_W$ correspond to nodes on the remaining surfaces of Σ. In general, the latter may be connected as $\mathbf{p}'_W = \mathbf{Z}_W \mathbf{v}'_W$, where $\mathbf{Z}_W$ denotes a square diagonal matrix of acoustic impedances at nodes on W. Hence, eliminating $\mathbf{p}'_W$, we obtain:

$$\begin{bmatrix} \mathbf{D}_{11} + \mathbf{D}_{1W}\left(\mathbf{Z}_W^{-1} - \mathbf{D}_{WW}\right)^{-1}\mathbf{D}_{W1} & \mathbf{D}_{12} + \mathbf{D}_{1W}\left(\mathbf{Z}_W^{-1} - \mathbf{D}_{WW}\right)^{-1}\mathbf{D}_{W2} \\ \mathbf{D}_{21} + \mathbf{D}_{2W}\left(\mathbf{Z}_W^{-1} - \mathbf{D}_{WW}\right)^{-1}\mathbf{D}_{W1} & \mathbf{D}_{22} + \mathbf{D}_{2W}\left(\mathbf{Z}_W^{-1} - \mathbf{D}_{WW}\right)^{-1}\mathbf{D}_{W2} \end{bmatrix} \begin{bmatrix} \mathbf{p}'_1 \\ \mathbf{p}'_2 \end{bmatrix} = \begin{bmatrix} \mathbf{v}'_1 \\ \mathbf{v}'_2 \end{bmatrix}, \tag{2.70}$$

which we write briefly as

$$\begin{bmatrix} \mathbf{E}_{11} & \mathbf{E}_{12} \\ \mathbf{E}_{21} & \mathbf{E}_{22} \end{bmatrix} \begin{bmatrix} \mathbf{p}'_1 \\ \mathbf{p}'_2 \end{bmatrix} = \begin{bmatrix} \mathbf{v}'_1 \\ \mathbf{v}'_2 \end{bmatrix}. \tag{2.71}$$

This is, in fact, the finite element model of an acoustic two-port element. It can be transformed to modal two-port model if its ports are on ducts with modal acoustic elements. Typically, as implied in Figure 2.9, if the ports are on uniform straight ducts, we may use Equation (6.66) to express the nodal acoustic pressure vectors in Equation (2.70) as $\mathbf{p}'_1 = \mathbf{\Phi}_1\left(\mathbf{P}_1^+ + \mathbf{P}_1^-\right)$ and $\mathbf{p}'_2 = \mathbf{\Phi}_2\left(\mathbf{P}_2^+ + \mathbf{P}_2^-\right)$, where

$$\mathbf{\Phi}_j = \begin{bmatrix} \varphi_1(\bullet_1) & \varphi_2(\bullet_1) & \cdots & \varphi_m(\bullet_1) \\ \varphi_1(\bullet_2) & \varphi_2(\bullet_2) & & \varphi_m(\bullet_2) \\ & & & \\ \varphi_1(\bullet_N) & \varphi_2(\bullet_N) & & \varphi_m(\bullet_N) \end{bmatrix}_j, \quad j = 1, 2 \tag{2.72}$$

$$\mathbf{P}_j^{\mp} = \begin{bmatrix} p_1^{\mp} \\ p_2^{\mp} \\ \vdots \\ p_m^{\mp} \end{bmatrix}_j, \quad j = 1, 2. \tag{2.73}$$

Here, $\varphi_\mu(\bullet_i)$ denotes value of the transverse eigenfunction of mode μ at node i and m denotes the assumed modality. The nodal normal acoustic velocity vectors in Equation (2.70) are expressed from Equation (6.67) similarly as $\rho_1 c_1 \mathbf{v}_1' = \mathbf{\Phi}_1\left(\mathbf{A}_1^+\mathbf{P}_1^+ + \mathbf{A}_1^-\mathbf{P}_1^-\right)$ and $\rho_2 c_2 \mathbf{v}_2' = \mathbf{\Phi}_2\left(\mathbf{A}_2^+\mathbf{P}_2^+ + \mathbf{A}_2^-\mathbf{P}_2^-\right)$, where $\mathbf{A}_j^{\mp}$ and $\rho_j c_j$ denote the modal admittance matrix defined by Equation (6.64) and the characteristic impedance at port $j \in 1, 2$, respectively. Substituting these modal expansions in Equation (2.71) and rearranging gives the required modal two-port element transfer equation

$$\begin{bmatrix} \mathbf{P}_1^+ \\ \mathbf{P}_1^- \end{bmatrix} = -\begin{bmatrix} \mathbf{E}_{11}\mathbf{\Phi}_1 - \mathbf{\Phi}_1\dfrac{\mathbf{A}_1^+}{\rho_1 c_1} & \mathbf{E}_{11}\mathbf{\Phi}_1 - \mathbf{\Phi}_1\dfrac{\mathbf{A}_1^-}{\rho_1 c_1} \\ \mathbf{E}_{21}\mathbf{\Phi}_1 & \mathbf{E}_{21}\mathbf{\Phi}_1 \end{bmatrix}^{-1}$$

$$\times \begin{bmatrix} \mathbf{E}_{12}\mathbf{\Phi}_2 & \mathbf{E}_{12}\mathbf{\Phi}_2 \\ \mathbf{E}_{22}\mathbf{\Phi}_2 - \mathbf{\Phi}_2\dfrac{\mathbf{A}_2^+}{\rho_2 c_2} & \mathbf{E}_{22}\mathbf{\Phi}_2 - \mathbf{\Phi}_2\dfrac{\mathbf{A}_2^-}{\rho_2 c_2} \end{bmatrix} \begin{bmatrix} \mathbf{P}_2^+ \\ \mathbf{P}_2^- \end{bmatrix}. \tag{2.74}$$

This may be considered as a contracted element, since its modality pertains only to the inlet and outlet sections, the interior being modeled to the accuracy of the finite element mesh.

2.5.2 The Boundary Element Method

Consider again a duct system of volume V with surface Σ. In the boundary element method, only Σ is divided into a large number of contiguous polygonal (typically, triangular) surfaces Σ^e, called boundary elements, which are connected to each other at points called nodes (typically, vertices of triangles). An integral equation formulation of the acoustic equations is used for derivation of the acoustic equations of boundary elements. Here, again, we will demonstrate the process for the Helmholtz equation, Equation (2.58). This is a self-adjoint linear differential equation, meaning that we can always find a vector bilinear form $\mathbf{J}(u, v)$ which satisfies the identity $v\mathrm{L}u - u\mathrm{L}v = \nabla\cdot\mathbf{J}(u, v)$, where $\mathrm{L} = \nabla^2 + k_o^2$ is the Helmholtz operator and u and v denote arbitrary scalar point functions [10]. Upon integrating this identity over the volume V and applying the divergence theorem, we obtain the Green identity $\int_V (v\mathrm{L}u - u\mathrm{L}v)\mathrm{d}V = \int_\Sigma \mathbf{n}\cdot\mathbf{J}(u, v)\mathrm{d}\Sigma$, where Σ denotes the boundary of V and $\mathbf{n}$ denotes its outward unit normal vector. It is easy to show that, for the Helmholtz operator L, the vector bilinear form is $\mathbf{J}(u, v) = v\nabla u - u\nabla v$. Thus, application of the Green identity for V with $u = p'$ and $v = G$ gives

$$\int_V (G\mathrm{L}p' - p'\mathrm{L}G)\mathrm{d}V = \int_\Sigma \mathbf{n}\cdot(G\nabla p' - p'\nabla G)\mathrm{d}\Sigma. \tag{2.75}$$

Let $\mathbf{x} \in V$ and $\mathbf{y} \in \Sigma$ denote any two points. It is convenient to take $G = G(R) = \mathrm{e}^{\mathrm{i}k_o R}/4\pi R$, where $R = |\mathbf{x} - \mathbf{y}|$, since this is a particular solution of $LG(R) = -\delta(\mathbf{y} - \mathbf{x})$, where $\delta(\mathbf{y} - \mathbf{x})$ denotes a Dirac delta function placed at point $\mathbf{x}$. By definition of the Dirac delta function $\int_V p'(\mathbf{y})\delta(\mathbf{y} - \mathbf{x})\mathrm{d}V(\mathbf{y}) = p'(\mathbf{x})$ and, consequently, Equation (2.75) reduces to:

$$p'(\mathbf{x}) = \int_{\Sigma} \mathbf{n}\cdot(G(R)\nabla p' - p'\nabla G(R))\mathrm{d}\Sigma, \quad \mathbf{x} \in V. \tag{2.76}$$

But, this integral formula is not applicable for points $\mathbf{x} \in \Sigma$, because $G(\mathbf{x}, \mathbf{y})$ becomes singular when $\mathbf{x} = \mathbf{y}$. The following classical technique is used for the calculation of acoustic pressure at points on Σ: let surface Σ be distorted in the neighborhood of point $\mathbf{x} \in \Sigma$ by an elementary spherical surface of radius ε centered at $\mathbf{x}$, so that this point now lies in the interior of the modified Σ. Upon evaluating the surface integrals in Equation (2.73) for this distorted boundary and letting $\varepsilon \to 0$, Equation (2.76) becomes [11]:

$$a(\mathbf{x})p'(\mathbf{x}) = \int_{\Sigma} G(R)\frac{\partial p'}{\partial n}\mathrm{d}\Sigma - \int_{\Sigma} p'\mathbf{n}\cdot\nabla G(R)\mathrm{d}\Sigma, \quad \mathbf{x} \in \Sigma, \tag{2.77}$$

where $a(\mathbf{x})$ denotes the inner solid angle, Ω, of surface Σ at point $\mathbf{x} \in \Sigma$ divided by 4π, for example, if $\mathbf{x} \in \Sigma$ and Σ is a smooth surface, then $a(\mathbf{x}) = 1/2$. Noting that $\mathbf{n}\cdot\nabla G(R) = -(\partial G/\partial R)\cos\gamma$, where γ denotes the angle between the unit normal vector $\mathbf{n}$ and the vector $\mathbf{x} - \mathbf{y}$, Equations (2.76) and (2.77) can be combined as

$$a(\mathbf{x})p'(\mathbf{x}) = -\mathrm{i}\omega\rho_o \int_{\Sigma} G(R)v_n'\mathrm{d}\Sigma + \int_{\Sigma} \frac{\partial G}{\partial R}p'\cos\gamma\mathrm{d}\Sigma \quad \begin{cases} a(\mathbf{x}) = 1 & \text{if} \quad \mathbf{x} \in V \\ a(\mathbf{x}) = \dfrac{\Omega(\mathbf{x})}{4\pi} & \text{if} \quad \mathbf{x} \in \Sigma \end{cases} \tag{2.78}$$

where we have used the acoustic momentum equation to express the normal derivative of the acoustic pressure in terms of the component of the particle velocity normal to Σ, that is, $\partial p'/\partial n = -\mathrm{i}\omega\rho_o v_n'$. Equation (2.78) for $\mathbf{x} \in \Sigma$ is known as the interior surface Helmholtz integral equation. For $\mathbf{x} \in V$, it is called the interior Helmholtz integral formula.

The boundary element method essentially consists of expressing the surface integrals in Equation (2.78) as sums of integrals over the areas of boundary elements on Σ and approximating the distributions of p' and v_n' as in Equation (2.61). For the formulation of acoustic models of ducts, we use the interior surface Helmholtz integral equation. Thus, writing Equation (2.78) with $\mathbf{x} \in \Sigma$ for every node of the boundary element mesh, we obtain the following system of n equations:

$$\frac{\Omega_i}{4\pi}p_i' = -\mathrm{i}\omega\rho_o \sum_{e=1}^{N} \int_{\Sigma^e} (G(R_i)\boldsymbol{\Psi}^e\mathrm{d}\Sigma)\mathbf{v}^e + \sum_{e=1}^{N} \int_{\Sigma^e} \left(\frac{\partial G}{\partial R}(R_i)\boldsymbol{\Psi}^e\cos\gamma_i\mathrm{d}\Sigma\right)\mathbf{p}^e, i = 1, 2, \ldots, n. \tag{2.79}$$

Here, n denotes the number nodes of the boundary mesh, N denotes the number of elements of the boundary mesh and subscript "i" denotes value at a node; $\mathbf{p}^e$ and $\mathbf{\Psi}^e$ have the same meaning defined for the finite element method in the previous section, except that elements are now boundary elements, and $\mathbf{v}^e = \{ v_1^e \quad v_2^e \quad \ldots \quad v_{ne}^e \}$, where v_j^e denotes v_n' at node j. It should be noted that in order to circumvent the impact of singularity in the integrands, the integrals in Equation (2.79) are usually evaluated by using a suitable Gauss quadrature. Therefore, R_i and γ_i may be understood as values for node i and a Gauss point (summation of integrands over Gauss points is not shown).

Defining the global vectors $\mathbf{p}_\Sigma'$ and $\mathbf{v}_\Sigma'$ as

$$\mathbf{p}_\Sigma' = \begin{bmatrix} p_1' \\ p_2' \\ \vdots \\ p_n' \end{bmatrix}, \mathbf{v}_\Sigma' = \begin{bmatrix} (v_n')_1 \\ (v_n')_2 \\ \vdots \\ (v_n')_n \end{bmatrix}, \tag{2.80}$$

the system in Equation (2.79) can be collated to the matrix equation

$$\mathbf{\Omega p}_\Sigma' = \mathbf{S}(k_o)\mathbf{v}_\Sigma' + \mathbf{Q}(k_o)\mathbf{p}_\Sigma', \tag{2.81}$$

where $\mathbf{\Omega}$ is a diagonal matrix with non-zero elements equal to $\Omega_i/4\pi$, $i = 1, 2, \ldots, n$ and $\mathbf{S}$ and $\mathbf{Q}$ are non-symmetric square matrices. This equation is in the form of Equation (2.68); that is, it can be written as $\mathbf{Dp}_\Sigma' = \mathbf{v}_\Sigma'$, where $\mathbf{D} = \mathbf{S}^{-1}(\mathbf{\Omega} + \mathbf{Q})$. Consequently, it can be transformed to the modal pressure wave components similarly, provided that the ports of boundary element model are on uniform straight ducts.

The above described boundary element formulation may be applied to duct acoustics problems defined by a linear operator L that has a vector bilinear form $\mathbf{J}(u, v)$. However, when mean flow is present, the versatility of the method is limited and the finite element method is usually the preferred option.

Solution of the interior acoustic pressure field is not required for the formulation of a modal element with either the finite element method or the boundary element method. When the finite element method is used, internal acoustic pressures can be calculated at the mesh nodes directly from the second set of Equation (2.67), that is, $\mathbf{p}_V' = -\mathbf{Z}_{22}^{-1}\mathbf{Z}_{21}\mathbf{p}_\Sigma'$. In the boundary element method, acoustic pressure at any given interior point can be calculated by using the interior Helmholtz integral formula, after Equation (2.81) has been solved.

2.6 Programming Considerations

Writing a code based on the block diagram approach described in this chapter for acoustic modeling and analysis of arbitrary duct systems essentially entails the development of an acoustic element database and an algorithm for the construction of the global matrix from a given block diagram. Acoustic parameters relevant to acoustic design can then be calculated, as described in later chapters of the book, by straightforward programming.

An acoustic element database consists of a suite of subroutines. Each subroutine asks for the relevant input data and outputs the corresponding acoustic element (for example, a wave transfer matrix, if it is a two-port). This database is supported by auxiliary databases containing subroutines for the calculation of the thermodynamic properties of fluids, such as density, speed of sound and specific heat coefficients, equivalent fluid properties of sound absorptive materials and impedances of perforates.

Depending on the intended functions of the code, input data may be arranged in several categories. For the simplicity of the discussion, here we consider fluid machinery running at a given operation point. Then, it suffices to input the frequency, modality, the fluid type and the data specific to the acoustic element for each block of a block diagram. Since the first three are common to all blocks, they can be input as global variables. Depending on the scope of the code and preferences of the programmer, some other parameters may also be made global variables; however, here it suffices to assume that all remaining data are contained in data files tied to the blocks.

Acoustic element data vary from element to element and may be arranged in different ways by different programmers, but the following data must be provided for every element:

- Name of the block (usually the name of the subroutine that calculates it or a pointer to that subroutine).
- Pointer for location of element data file (file structure depends on the preferences of the programmer).
- Code for β and k of the wave transmission matrix (Section 2.3.3).
- Port numbers (distinct integers, except for connected ports which must have common numbers).
- Flags indicating change of the default positive acoustic flow direction at ports (for example, 0: not changed, 1: changed).
- Duct extension lengths at ports (relevant for multi-ports).
- Mean temperature, pressure and velocity as required in the mathematical model.
- Other geometrical and physical data as required by the mathematical model.

The system description may be manual or graphical. In the manual method, the user draws a valid block diagram of the duct system on paper. The code prompts the user to select a block from the list of block names in the acoustic element database and opens a form with controls for the required data input when a selection is made. The aim of the first phase of input is to teach the block diagram to the computer. In this phase, the user inputs the port numbers of the selected block, the code for β and k of the wave transfer or transmission matrix of the block and a pointer for location of the data file and saves the input. In the second phase, the block datasheets are re-opened and all remaining data are input and saved (this phase may of course be integrated partially or fully with the first phase). The important point here is that the port numbers must be assigned by the user so that connected ports have common numbers and rest of the ports have distinct numbers.

In the graphical method, the user opens a graphical interface, where the block diagram can be drawn just as it is drawn on paper. In the first step, the user creates blocks on this interface, again by selecting block names from a list, however, pointers for data-file locations, tentative port numbers and the codes for β and k are assigned automatically by the computer program. This process is repeated for all blocks. Port numbers are tentative, because blocks are not joined yet. In the next step, whilst the user joins the ports to be connected by drawing lines from one port to the other, the program modifies the tentative port numbers so that the connected ports have common port numbers, and records the changes made to the default positive acoustic flow directions at the ports (for compatibility of the connections) on the data file. The rest of the element data are input manually by using the appropriate controls on the graphical user interface.

Obviously, the graphical method is more attractive from a practical point of view, but the design of an efficient user-friendly graphical user interface requires considerable programming effort. Two points are worth paying attention to in designing a graphical interface. Firstly, it should be possible to manage basic acoustic element subassemblies as easily as the basic elements. Ability to quickly create, import, change or delete subassemblies of a system model provides flexibility for component and concept changes and speeds up the acoustic design process. Also, it should be possible to develop subassemblies as separate models that can be activated or deactivated with ease as parts of a complete system when necessary. The second point is concerned about the limited drawing area available on usual computer screens. A large variety of practical flow duct configurations can be constructed from a relatively small database of basic building blocks such as ducts and junctions, but the block diagram can become quickly untidy and confused to read easily. One solution to this problem is to design the graphical interface so that the user can develop subassemblies on different layers, with one of the layers reserved for construction of the complete system model. With the capability to link the subassemblies drawn in sublayers to the main layer as template or specific model, the acoustic model of a duct system can always be represented by a simple block diagram, however complex it may be.

With either the manual method or graphical method, an algorithm is required for sorting the port numbers so that the global matrix can be initialized and the appropriate wave transfer matrix or wave transmission equation of the blocks are placed in allocated areas as they are computed. Here, we give an APL implementation of such an algorithm (the symbol ⍝ means comment; lines in APL unicode font are executed) [12]:

```
G←m GlobalMatrix E;E1;N;I;P;J;K;F;K1;G;name;A;U;i;j;k
⍝ Output:
⍝ G = global matrix of input block diagram
⍝ E1 = echo of an extended form of input matrix E, showing the order port variables
   vectors in the calculated global matrix. Block equations are placed in the global
   matrix in the order they are given in matrix E.
⍝ Input:
⍝ m = modality of block diagram
⍝ E is a heterogeneous N×3 matrix, which must be extracted in another section of the
   code from the element data files such that:
⍝ First column E[;1] contains block names included in the block diagram
⍝ Second column E[;2] contains block codes for β and k
```

⍝ Third column `E[;3]` contains port numbers of the assembled blocks
⍝ Order of blocks in `E` may be arbitrary.
⍝ The following codes are used to specify β and k (this scheme must be extended if not exhaustive for the blocks in the working acoustic element database):
⍝ code $= 1$: $\beta = 2$ and $k = n/2$
⍝ code $= 2$: $\beta = 1$ and $k = n$
⍝ where $n =$ number of ports
⍝ For example, for the block diagram in Figure 2.6:

$$⍝\ E = \begin{bmatrix} \text{duct} & 1 & (1\ \ 2) \\ \text{duct} & 1 & (7\ \ 8) \\ \text{duct} & 1 & (3\ \ 4) \\ \text{duct} & 1 & (9\ \ 10) \\ \text{duct} & 1 & (5\ \ 6) \\ \text{openexp} & 2 & (2\ \ 3\ \ 7) \\ \text{openexp} & 2 & (4\ \ 5\ \ 9) \\ \text{bound} & 2 & 8 \\ \text{bound} & 2 & 10 \end{bmatrix}$$

⍝ Compute number of blocks:

```
N←1↑ ρ E
```

⍝ Calculate β and k and catenate them as columns 4 and 5 to matrix `E` to form matrix `E1`:

```
E1←E,[2](N,2)ρ 0
:for I :in ιN
      :if E[I;2]=1
            E1[I;4]←2 ◇ E1[I;5]←(ρ,⊃E[I;3])÷2
      :elseif E[I;2]=2
            E1[I;4]←1 ◇ E1[I;5]←ρ,⊃E[I;3]
      :endif
:endfor
```

⍝ Enlist port numbers in `E[;3]` in vector `P`:

```
P←∊ E[;3]
```

⍝ Discard non-distinct port numbers from `P`; put the result in vector `J` (in preparation for the next step):

```
J←,P[1]
:for I :in 1↓ ιρ P
      :if ~P[I]∊J
            J←J,P[I]
      :endif
:endfor
```

⍝ Renumber port numbers in vector *P* (without changing port connectivity) so that the new port numbers in vector *K* are one-to-one with the terms in *P* and form a complete set of integers starting with 1.

⍝ The new port numbers also stand for the indices of the super columns of the global matrix; see Equation (2.42):

```
K←(ρP)ρ0 ◇ I←0
:while 0 ∊ K
      I←I+1
      K←K+I×P∊ J[I]
:endwhile
```

⍝ Put the number of ports of blocks in *1×N* vector *J*:

```
J←θ
:for I :in ιN
      J←J,ρ,⊃E[I;3]
 :endfor
```

⍝ Distribute elements of *K*, the new port numbers, to ports of blocks in one-to-one correspondence with the *1×N* heterogeneous vector *E[;3]*, and store in *1×N* heterogeneous vector *F*:

```
F←θ ◇ K1←K
:for I :in ιN
      F←F,⊂J[I]↑ K1
      K1←(J[I])⌽ K1
:endfor
```

⍝ Reshape *F* and catenate to matrix *E1* as sixth column:

```
E1←E1,[2](N,1)ρ F
```

⍝ For example, the final form of matrix *E1* for the block diagram in Figure 2.6 is:

⍝ $$E1 = \begin{bmatrix} \text{duct} & 1 & (1\ \ 2) & 2 & 1 & (1\ \ 2) \\ \text{duct} & 1 & (7\ \ 8) & 2 & 1 & (3\ \ 4) \\ \text{duct} & 1 & (3\ \ 4) & 2 & 1 & (5\ \ 6) \\ \text{duct} & 1 & (9\ \ 10) & 2 & 1 & (7\ \ 8) \\ \text{duct} & 1 & (5\ \ 6) & 2 & 1 & (9\ \ 10) \\ \text{openexp} & 2 & (2\ \ 3\ \ 7) & 1 & 3 & (2\ \ 5\ \ 3) \\ \text{openexp} & 2 & (4\ \ 5\ \ 9) & 1 & 3 & (6\ \ 9\ \ 7) \\ \text{bound} & 2 & 8 & 1 & 1 & 4 \\ \text{bound} & 2 & 10 & 1 & 1 & 8 \end{bmatrix}$$

⍝ Referring to Figure 2.6, the algorithm will form the global matrix with the port variables vectors taken in the following order:

⍝ $\mathbf{P}_1\ \ \mathbf{P}_2\ \ \mathbf{P}_7\ \ \mathbf{P}_8\ \ \mathbf{P}_3\ \ \mathbf{P}_4\ \ \mathbf{P}_9\ \ \mathbf{P}_{10}\ \ \mathbf{P}_5\ \ \mathbf{P}_6$

⍝ Note that this is not same as the order of these vectors in Equation (2.42). However, the block equations are placed in the global matrix in the same order as in Equation (2.42).

⍝ Calculate the total number of scalar acoustic equations to be derived from the block diagram:

```
i←m×+/(⊃E1[;4])×⊃E1[;5]
```

⍝ Calculate the total number scalar port variables:

```
j←2×m×-/K
```

⍝ Initialize the global matrix. Note that it is a complex matrix:

```
G←(2,i,j)ρ 0
```

⍝ Initialize a real unit matrix of size *m* in complex form:

```
U←(2,m,m)ρ 0
:for I :in ιm
        U[1;I;I]←1
:endfor
```

⍝ Place block matrices in the global matrix (recall that two-port transfer matrices are placed in the global matrix in the wave transmission matrix format). This step requires that subroutines for calculation of wave transfer matrices or wave transmission equations of the blocks are available in the code. For the clarity of the algorithm, assume that subroutines named `openexp` and `bound` (see matrix `E` above) are available in the database; other blocks may be included similarly. It is not necessary to consider every block individually; they can be covered in generic groups as one-ports, two-ports, three-ports, multi-ports with variable number of ports, etc. Note that here the subroutines are assumed to have been written such that they can be called by specifying the port numbers of the blocks. This requires that the code scans the element data files to find the matching port numbers and fetch the element data for calculations. However, the subroutines may also be called in different ways according to the preferences of programmers.

```
K←E1[;4]×E1[;5]
:for I :in ιN
     i←,⊃E1[I;6] ◇ J←m×+/(I-1)↑K ◇ F←K[I]
    :if (,⊃E[I;1])≡'duct'
         name←'duct ',⍕,⊃ E[I;3]
         G[;J+ιF×m;(2×m×+/i[1]-1)+ι 2×m]←G[;J+ιF×m;(2×m×
         +/i[1]-1)+ι2×m]+U
         G[;J+ιF×m;(2×m×+/i[2]-1)+ι2×m]←G[;J+ιF×m;(2×m×
         +/i[2]-1)+ι2×m]+(-⍎name)[;ιF×m;ι2×m]
    :elseif (,⊃ E[I;1])≡'openexp'
         name←'openexp ',⍕,⊃ E[I;3]
```

```
            G[;J+⍳F×m;(2×m×+/i[1]-1)+⍳ 2×m]←G[;J+⍳ F×m;(2×m×
            +/i[1]-1)+⍳2×m]+(⍉name)[;⍳ F×m;⍳2×m]
            G[;J+⍳F×m;(2×m×+/i[2]-1)+⍳ 2×m]←G[;J+⍳ F×m;(2×m×
            +/i[2]-1)+⍳2×m]+(⍉name)[;⍳ F×m;⍳2×m]
            G[;J+⍳F×m;(2×m×+/i[3]-1)+⍳ 2×m]←G[;J+⍳ F×m;(2×m×
            +/i[3]-1)+⍳2×m]+(⍉name)[;⍳ F×m;⍳2×m]
        :elseif (,⊃ E[I;1])≡'bound'
            name←'bound ',⍕,⊃ E[I;3]
            G[;J+⍳F×m;(2×m×+/i[1]-1)+⍳ 2×m]←G[;J+⍳ F×m;(2×m×
            +/i[1]-1)+⍳2×m]+(⍉name)
        :endif
:endfor
```

⍝ Output results in a heterogeneous vector as

```
G←(E1) (G)
```

⍝ This algorithm may be tested by replacing the subroutine calls by arbitrary numerical matrices having the same shapes as defined in the subroutine calls.

If the block diagram has only one inlet and only one outlet port, all other ports being either connected or closed, the global matrix can be reduced to a two-port wave transfer matrix, as described in Section 2.4.2. The following program uses output of GlobalMatrix to make this reduction.

```
T←GlobalToTwoPort G;E1;I;P;J;K;T;V;C
```

⍝ This algorithm uses the subroutines `CxMatInv A` (computes the inverse of a complex square matrix `A`) and `A CxMatTimes B` (computes product two complex matrices `A` and `B`). They should be adopted from standard matrix computations packages.

⍝ `G` is the output of `GlobalMatrix`. Thus:

```
E1←⊃ G[1] ⋄  G←⊃ G[2]
```

⍝ Find indices of unconnected ports:

```
P←∊ E1[;6]
J←⍬
:for I :in ⍳⍴P
    :if 1=+/P∊P[I]
        J←J,P[I]
    :endif
:endfor
```

⍝ Check if two-port:

```
:if 2≠⍴J
    'Global matrix does not reduce to a two-port' ⋄ →0
:else
```

⍝ Reduce global matrix to wave transfer matrix with *J[2]* to *J[1]* causality

```
        V←-G[;;(2×m×J[2]-1)+⍳2×m]
        C←(~(⍳(⍴G)[3])∊(2×m×J[2]-1)+⍳2×m)/[3]G
        T←(CxMatInv C) CxMatTimes V
:endif
```

⍝ Get the port numbers and output the wave transfer matrix

```
K←(P∊J)/∊E1[;3]
T←(⌽K) (T)
```

References

[1] L.L. Beranek, *Acoustics*, (New York: Acoustical Society of America, 1996).

[2] J.L. Shearer, A.T. Murphy and H.H. Richardson, *Introduction to System Dynamics*, (Reading: Addison-Westley, 1971).

[3] D. Karnopp, Lumped parameter models of acoustic filters using normal modes and bond graphs, *J. Sound Vib.* **42** (1975), 437–446.

[4] J.F. Dowling and K.S. Peat, An algorithm for the efficient acoustic analysis of silencers of any general geometry, *Applied Acoustic* **65** (2004), 211–227.

[5] R. Glav and M. Åbom, A general formalism for analyzing acoustic 2-port networks, *J. Sound Vib.* **202** (1997), 739–747.

[6] S.N. Panigrahi and M.L. Munjal, Plane wave propagation in generalized multiply connected acoustic filters, *J. Acoust. Soc. of Am.* **118** (2005), 2860–2868.

[7] C-N. Wang, C-H. Wu and T-D. Wu, A network approach for analysis of silencers with/without absorbent material, *Applied Acoustics* **70** (2009), 208–214

[8] D.A. Kopriva, *Implementing Spectral Methods for Partial Differential Equations*, (Dordrecht: Springer Scientific Computation Series, 2009).

[9] J.N. Reddy, *An Introduction to the Finite Element Method*, (New York: McGraw-Hill, Inc., 1993).

[10] R. Courant and D. Hilbert, *Methods of Mathematical Physics,* vol.1, (New York: Interscience, 1953).

[11] B.B. Baker and E.T. Copson, *The Mathematical Theory of Huygen's Principle*, (Oxford: Clarendon Press, 1950).

[12] APL+Win Tutor, APLNow LLC, 2019 (http://tutorial.apl2000.com:9005/)

3 Transmission of Low-Frequency Sound Waves in Ducts

3.1 Introduction

This chapter introduces the theory of one-dimensional sound propagation in ducts and, based on this theory, presents solutions in the two-port wave transfer matrix format of Equation (2.14). The one-dimensional theory holds with adequate accuracy at relatively low frequencies and for subsonic low Mach number mean flows. The actual frequency and Mach number limits depend on the type of the duct. At this point, it suffices to remark that the sound field in a duct consists of a linear combination of infinite number of modes, but the number of modes that propagate without substantial attenuation depends on the frequency. If the frequency is low enough, the acoustic field is dominated only by one mode, which is called fundamental mode or, more generally, the least attenuated mode. This frequency range may be stated to be very roughly $\lambda >> a$, where λ denotes the wavelength and a denotes a characteristic dimension of the duct cross section. Precise statements about the frequency range of plane-wave propagation in a duct can only be made when the problem is solved in three-dimensions (Chapter 6). For example, for a hollow circular hard-walled duct of radius a with negligible mean flow, the actual criterion transpires as $\lambda > 3.4126a$, if the first axisymmetric mode is excited. Recalling that wavelength is defined as $\lambda = c_o/f$, where c_o denotes the speed of sound and f denotes the frequency, this criterion may be expressed as $k_oa < 1.8412$, where $k_o = 2\pi f/c_o$ and the non-dimensional product k_oa is called the Helmholtz number based on duct radius. The term low frequency is generally understood in an acoustic sense, signifying that the Helmholtz number based on a characteristic duct size is less than about unity.

Thus, one-dimensional theory can at best predict the fundamental mode in a duct. But, it is attractive due to its relative simplicity and is important because in many practical applications the frequencies of interest are in the low Helmholtz number range.

Some features of the ducts considered in this chapter are depicted schematically in Figure 3.1. Although this figure shows a uniform rectangular duct for simplicity, the subsequent analysis is applicable for non-uniform ducts of arbitrary cross-sectional shape. The duct geometry is defined in an orthogonal Cartesian frame with axes x_1, x_2, x_3, or the parallel central axes x, y, z, where $x \| x_1$ denotes the duct axis. The unit vectors in positive directions of x_1, x_2, x_3 axes are denoted by $\mathbf{e}_1, \mathbf{e}_2, \mathbf{e}_3$, respectively.

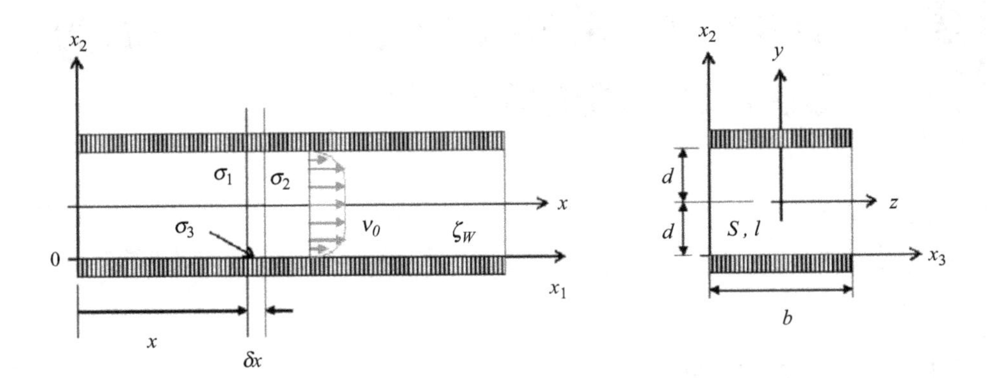

Figure 3.1 An inhomogeneous non-uniform duct with sheared parallel mean flow; duct cross-sectional area, mean-flow velocity, pressure and density and the wall impedance may vary along the duct axis.

The duct contains an inviscid compressible fluid, a typical property of which is denoted as $q = q(t, x_1, x_2, x_3)$, where t denotes the time. Typical fluid properties are pressure, density, particle velocity, absolute temperature and speed of sound, which are denoted, respectively, as $q \in p, \rho, \mathbf{v}, T, c, \ldots$ throughout the book. Presently, we make no stipulations about the duct boundaries, however, we keep in mind the existence of duct walls.

The classical form of the one-dimensional theory of sound propagation in ducts is generally known as the plane-wave theory, which is ubiquitous in the literature on acoustics. The plane-wave theory assumes a priori that $q = q(t, x)$, that is, that all fluid properties are distributed uniformly over a duct section. In the following exposition of the one-dimensional theory, this restriction is partially relaxed in order to include the effects of refraction, boundary layers and finite impedance walls in the analysis.

3.2 One-Dimensional Theory of Sound Propagation in Ducts

3.2.1 Unsteady Flow Equations

In reference to Figure 3.1, consider an elementary source-free inertial control volume $\delta V = S\delta x$ of surface $\sigma = \sigma_1 \cup \sigma_2 \cup \sigma_3$, where σ_1 and σ_2 denote surfaces of area S and $S + \delta S$ normal to the duct axis at sections x and $x + \delta x$, respectively, and σ_3 denotes the surface of δV matching the boundary of $S = S(x)$. At this point, it should be made clear that the term boundary is understood here either as the physical surfaces of the duct walls or as the border of a thin shear layer over them. The model to be used for the acoustic boundary condition will determine the option, as explained in later sections, when specific duct models are considered.

We begin with the integral forms of the conservation equations for mass and momentum in the control volume. From Appendix A.3, these are, respectively,

$$\frac{\partial}{\partial t}\int_{\delta V}\rho \mathrm{d}V+\int_{\sigma}\rho\mathbf{v}\cdot\mathbf{n}\mathrm{d}\sigma=0 \tag{3.1}$$

$$\frac{\partial}{\partial t}\int_{\delta V}\rho\,\mathbf{v}\mathrm{d}V+\int_{\sigma}\rho\mathbf{v}\ \mathbf{v}\cdot\mathbf{n}\mathrm{d}\sigma+\int_{\sigma}p\mathbf{n}\mathrm{d}\sigma=0, \tag{3.2}$$

where $\mathbf{n}$ denotes the outward unit normal vector of σ. Let $v(t,x_1,x_2,x_3)=\mathbf{e}_1\cdot\mathbf{v}$ denote axial component of the particle velocity and $u(t,x_1,x_2,x_3)=\{\mathbf{n}\cdot\mathbf{v}|x_2,x_3\in\sigma_3\}$ denote the component of the particle velocity normal to the duct boundary, where $\mathbf{e}$ denotes the unit vector in the positive direction of the duct axis, x.[1] Splitting the surface integral over surfaces $\sigma_1,\sigma_2,\sigma_3$, Equation (3.1) and the axial component of Equation (3.2) can be expressed as

$$\frac{\partial}{\partial t}\int_{\delta V}\rho\mathrm{d}V-\int_{\sigma_1}\rho v\mathrm{d}\sigma+\int_{\sigma_2}\rho v\mathrm{d}\sigma+\int_{\sigma_3}\rho u\mathrm{d}\sigma=0 \tag{3.3}$$

$$\frac{\partial}{\partial t}\int_{\delta V}\rho\,v\mathrm{d}V-\int_{\sigma_1}\rho\,v^2\mathrm{d}\sigma+\int_{\sigma_2}\rho v^2\mathrm{d}\sigma+\int_{\sigma_3}\rho vu\mathrm{d}\sigma-\int_{\sigma_1}p\mathrm{d}\sigma+\int_{\sigma_2}p\mathrm{d}\sigma+\int_{\sigma_3}\mathbf{e}\cdot\mathbf{n}p\mathrm{d}\sigma=0, \tag{3.4}$$

respectively. Applying the divergence theorem, Equation (A.1), in one dimension, the surface integrals on σ_1 and σ_2 are transformed to a volume integral (over the control volume) and Equations (3.3) and (3.4) can be written, respectively, as

$$\int_x^{x+\delta x}\frac{\partial}{\partial t}\left(\int_S\rho\mathrm{d}S\right)\mathrm{d}x+\int_x^{x+\delta x}\frac{\partial}{\partial x}\left(\int_S\rho v\mathrm{d}S\right)\mathrm{d}x+\int_x^{x+\delta x}\oint_\ell\rho uJ\mathrm{d}\ell\mathrm{d}x=0 \tag{3.5}$$

$$\begin{aligned}&\int_x^{x+\delta x}\frac{\partial}{\partial t}\left(\int_S\rho v\mathrm{d}S\right)\mathrm{d}x+\int_x^{x+\delta x}\frac{\partial}{\partial x}\left(\int_S\rho v^2\mathrm{d}S\right)\mathrm{d}x+\int_x^{x+\delta x}\frac{\partial}{\partial x}\left(\int_S p\mathrm{d}S\right)\mathrm{d}x\\&+\int_x^{x+\delta x}\int_\ell\rho vuJ\mathrm{d}\ell\mathrm{d}x+\int_x^{x+\delta x}\oint_\ell\mathbf{e}\cdot\mathbf{n}pJ\mathrm{d}\ell\mathrm{d}x=0.\end{aligned} \tag{3.6}$$

Here, ℓ denotes a perimeter curve of S. To understand J, assume that the duct surface is a surface of revolution, as is the case for most practical ducts. Then, $\ell=\ell(x)$ and let $\chi=\chi(x)$ form with ℓ a pair of orthogonal curvilinear coordinates over the duct boundary surface with unit metrical coefficients, that is, $\delta s_\ell=\delta\ell$ and $\delta s_\chi=\delta\chi$, where δs_ℓ and δs_χ denote, respectively, the elementary curve lengths along the ℓ and χ

[1] For ducts having an axis in the form of a space curve, x may be understood as the arc length of this curve and $\mathbf{e}$ as its unit tangent vector, the coordinates x_2 and x_3 being in directions of the normal and binormal vectors of the duct axis. However, one-dimensional theory is not sensitive to the duct curvature, except for its effect on the duct length. A pseudo-one-dimensional circularly curved duct element is described in Section 6.10.3.

coordinates and an elemental surface area is given by $\delta\ell\delta\chi$. Then, $\delta s_\chi = (\mathrm{d}\chi/\mathrm{d}x)\delta x$, which gives the definition of J as $J = \mathrm{d}\chi/\mathrm{d}x$. Obviously, if the duct is uniform, then $J = 1$. If the lateral surface of the duct is developed by full rotation of a continuous curve $r = r(x)$ about the duct axis, then $\delta\chi = \sqrt{(\delta x)^2 + (\delta r)^2}$, that is, $J = \sqrt{1 + (\mathrm{d}r/\mathrm{d}x)^2}$. For example, if σ_3 is a conical surface of cone angle ϕ_0, then $r = x\tan(\phi_0)$ and, hence, $J = \sec(\phi_0)$. This result may also be used with a rectangular duct, the depth or width of which varies with duct axis as $r = r(x)$, but the integration must be split into the depth and width surfaces. For example, if width is constant and depth varies as $r = r(x)$, then $J = 1$ on the depth surfaces and $J = \sqrt{1 + (\mathrm{d}r/\mathrm{d}x)^2}$ on the width surfaces.

We now introduce the following ordinary spatial averages (denoted by an overbar):

$$\bar{q}(t,x) = \frac{1}{S}\int_S q(t,x_1,x_2,x_3)\mathrm{d}S, \qquad q \in p, \rho, v \tag{3.7}$$

and the spatial Favre averages (denoted by two overbars):

$$\bar{\bar{q}}(t,x) = \frac{1}{S\bar{\rho}(t,x)}\int_S \rho(t,x_1,x_2,x_3)q(t,x,x_2,x_3)\mathrm{d}S, \qquad q \in v, v^2. \tag{3.8}$$

Also, we note that the last term on the left-hand side of Equation (3.6) represents the projection of the pressure force on an elementary surface $J\mathrm{d}\ell\mathrm{d}x$ of σ_3 on to a plane normal to the duct axis and, therefore, it can be expressed as

$$\int_x^{x+\delta x}\oint_\ell \mathbf{e}\cdot\mathbf{n}pJ\mathrm{d}\ell\mathrm{d}x = -\int_x^{x+\delta x}\left(\frac{\mathrm{d}S}{\mathrm{d}x}\right)p(t,x,\ell)\mathrm{d}x, \tag{3.9}$$

where the minus sign accounts for the correct direction of the axial component of the pressure force on σ_3. Hence, Equations (3.5) and (3.6) yield, respectively, the differential forms:

$$S\left(\frac{\partial\bar{\rho}}{\partial t} + \bar{\bar{v}}\frac{\partial\bar{\rho}}{\partial x}\right) + \bar{\rho}\frac{\partial}{\partial x}(S\bar{\bar{v}}) = -\oint_\ell \rho uJ\mathrm{d}\ell \tag{3.10}$$

$$S\frac{\partial}{\partial t}(\bar{\rho}\bar{\bar{v}}) + \frac{\partial}{\partial x}\left(S\bar{\rho}\overline{\overline{v^2}}\right) + S\frac{\partial\bar{p}}{\partial x} = -\int_\ell \rho vuJ\mathrm{d}\ell + (p(t,x,\ell) - \bar{p}(t,x))\frac{\mathrm{d}S}{\mathrm{d}x}. \tag{3.11}$$

Using Equation (3.10) to eliminate the time derivative of the fluid density, Equation (3.11) is expressed in the preferred form here:

$$S\bar{\rho}\left(\frac{\partial\bar{\bar{v}}}{\partial t} + \bar{\bar{v}}\frac{\partial\bar{\bar{v}}}{\partial x}\right) + \frac{\partial}{\partial x}\left(S\bar{\rho}\left(\overline{\overline{v^2}} - (\bar{\bar{v}})^2\right)\right) + S\frac{\partial\bar{p}}{\partial x} = \oint_\ell \rho u(\bar{\bar{v}} - v)J\mathrm{d}\ell + (p(t,x,\ell) - \bar{p}(t,x))\frac{\mathrm{d}S}{\mathrm{d}x}. \tag{3.12}$$

3.2.2 Equations Governing Acoustic Wave Motion

Equations (3.10) and (3.12) are linearized (see Section 1.1) by invoking the decompositions $q = q_o + q'$, $q \in p, \bar{p}, \rho, \bar{\rho}, v, \bar{v}, \bar{\bar{v}}, u$, where subscript "$o$" denotes the mean part and a prime denotes the (fluctuating) acoustic part (of zero temporal mean) of a property. The ordinary averages of the mean and acoustic parts of a property over a section of the duct obey the rules $\bar{q}_o = (\bar{q})_o$ and $\bar{q}' = (\bar{q})'$. We assume that $\bar{\rho}_o$ and $\bar{p}_o$ are functions of the duct axis only, $v_o = v_o(x, x_2, x_3)$ and $u = u_o(x, \ell) + u'(t, x, \ell)$, where $g = g(\ell)$ means that g is a function of the points on ℓ. For simplicity of notation, we will often drop the overbar notation in denoting the averages of mean-flow properties over a section of the duct, except for the mean-flow velocity, when there is no risk of confusion. For example, $\bar{\rho}_o = \rho_o(x)$, $\bar{p}_o = p_o(x)$ and so on, except for $\bar{v}_o$.

The spatial Favre-averaged particle velocity, $\bar{\bar{v}}$, can be expanded to first order as

$$\bar{\bar{v}} = \frac{1}{S\bar{\rho}(t,x)} \int_S (\rho_o(x) v'(t, x, x_2, x_3) + \rho_o(x) v_o(x, x_2, x_3) + \rho'(t, x, x_2, x_3) v_o(x, x_2, x_3)) \mathrm{d}S. \tag{3.13}$$

Upon evaluating the last integrand in the sense of spatial Favre average and the remaining integrands as ordinary averages, this can be written as

$$\bar{\bar{v}} = \frac{\rho_o \overline{v'} + \overline{\rho'}\, \overline{\overline{v_o}} + \rho_o \overline{v_o}}{\rho_o + \overline{\rho'}} = \overline{v_o} + \overline{v'} + \frac{\overline{\rho'}}{\rho_o} \left(\overline{\overline{v_o}} - \overline{v_o} \right), \tag{3.14}$$

where small terms of the second order are neglected and

$$\overline{\overline{v_o}} = \frac{1}{S\overline{\rho'}(t,x)} \int_S \rho'(t, x, x_2, x_3) v_o(x, x_2, x_3) \mathrm{d}S. \tag{3.15}$$

The spatial Favre average $\overline{\overline{v^2}}$ is expanded to first order similarly:

$$\overline{\overline{v^2}} = \frac{\rho_o \overline{v_o^2} + 2\rho_o \overline{v_o v'} + \overline{\rho'}\, \overline{\overline{v_o^2}}}{\rho_o + \overline{\rho'}} = \overline{v_o^2} + 2\overline{v_o v'} + \frac{\overline{\rho'}}{\rho_o} \left(\overline{\overline{v_o^2}} - \overline{v_o^2} \right). \tag{3.16}$$

We will neglect the terms involving the difference of Favre and ordinary averages in the foregoing equations. This approximation is exact, irrespective of the mean-flow velocity distribution, if the density fluctuations are uniformly distributed over a duct section. This condition, however, is perturbed to some extent by effects such as refraction in mean-flow shear layers, compliance of the duct walls and non-uniformity of the cross sections. We assume, therefore, that such effects remain within ranges in which the corresponding acoustic fluctuations do not deviate substantially from their mean values over a duct section, that is, remain sufficiently planar. This condition imposes certain parameter restrictions for some duct models. We will discuss these restrictions later when specific duct models are considered.

Hence, we implement Equation (3.14) as $\overline{\overline{v}} \approx \overline{v_o} + \overline{v'}$ and Equation (3.16) similarly reduce to $\overline{\overline{v^2}} \approx \overline{v_o^2} + 2\overline{v_o v'}$, where we may take $\overline{v_o v'} \approx \overline{v_o}\,\overline{v'}$, since fluctuations are assumed to remain approximately planar. Then, the continuity equation resulting from linearization of Equation (3.10) for acoustic fluctuations can be expressed as

$$S\left(\frac{\partial \overline{\rho'}}{\partial t} + \overline{v_o}\frac{\partial \overline{\rho'}}{\partial x} + \overline{v'}\frac{\mathrm{d}\overline{\rho_o}}{\mathrm{d}x}\right) + \overline{\rho_o}\frac{\partial}{\partial x}(S\overline{v'}) + \overline{\rho'}\frac{\mathrm{d}}{\mathrm{d}x}(S\overline{v_o}) = -\oint_{\ell} \rho_o u' J \mathrm{d}\ell - \oint_{\ell} \rho' u_o J \mathrm{d}\ell. \tag{3.17}$$

The corresponding continuity equation for the mean flow is

$$\frac{\mathrm{d}}{\mathrm{d}x}(S\overline{\rho_o}\,\overline{v_o}) = -\oint_{\ell} \rho_o u_o J \mathrm{d}\ell. \tag{3.18}$$

On the linearization of the momentum equation, we first note that, the fluctuating part of the last term on the right-hand side of Equation (3.12) is $(p'(t,x,\ell) - \overline{p'})\mathrm{d}S/\mathrm{d}x$. We neglect this term as small compared to the remaining terms, not only because we have restricted the analysis to sufficiently planar acoustic fluctuations, but also because of the fact that use of sufficiently gradual cross sectional area taper, $\mathrm{d}S/\mathrm{d}x$, is usually dictated in practice in order to prevent flow separation and to reduce accompanying flow losses. Then, the linearized momentum equation can be expressed as

$$\begin{aligned} S\overline{\rho_o}\left(\frac{\partial \overline{v'}}{\partial t} + \overline{v_o}\frac{\partial \overline{v'}}{\partial x} + \overline{v'}\frac{\mathrm{d}\overline{v_o}}{\mathrm{d}x}\right) + S\overline{v_o}\frac{\mathrm{d}\overline{v_o}}{\mathrm{d}x}\overline{\rho'} + \frac{\partial}{\partial x}\left(S\beta(\overline{v_o})^2\overline{\rho'}\right) + S\frac{\partial \overline{p'}}{\partial x} \\ = \overline{v_o}\oint_{\ell} \rho_o \varphi_{\mathrm{SF}} u' J \mathrm{d}\ell + \overline{v_o}\oint_{\ell} \rho' \varphi_{\mathrm{SF}} u_o J \mathrm{d}\ell, \end{aligned} \tag{3.19}$$

where

$$\beta = \frac{\overline{v_o^2}}{(\overline{v_o})^2} - 1 \tag{3.20}$$

$$\varphi_{\mathrm{SF}} = 1 - \frac{w_o}{\overline{v_o}} \tag{3.21}$$

and $w_o = v_o(x,\ell)$ denotes the mean slip velocity at the duct boundary. It is assumed that w_o is constant on ℓ (for given x). In general w_o may vary on ℓ, but such variations are difficult, if not impossible, to quantify experimentally. This justifies the assumption of a peripherally uniform slip flow velocity.

The corresponding momentum equation for the mean flow is

$$S\overline{\rho_o}\,\overline{v_o}\frac{\mathrm{d}\overline{v_o}}{\mathrm{d}x} + \frac{\partial}{\partial x}\left(S\overline{\rho_o}\beta(\overline{v_o})^2\right) + S\frac{\mathrm{d}\overline{p_o}}{\mathrm{d}x} = \overline{v_o}\oint_{\ell} \rho_o \varphi_{\mathrm{SF}} u_o J \mathrm{d}\ell + (p_o(x,\ell) - \overline{p_o}(x))\frac{\mathrm{d}S}{\mathrm{d}x}. \tag{3.22}$$

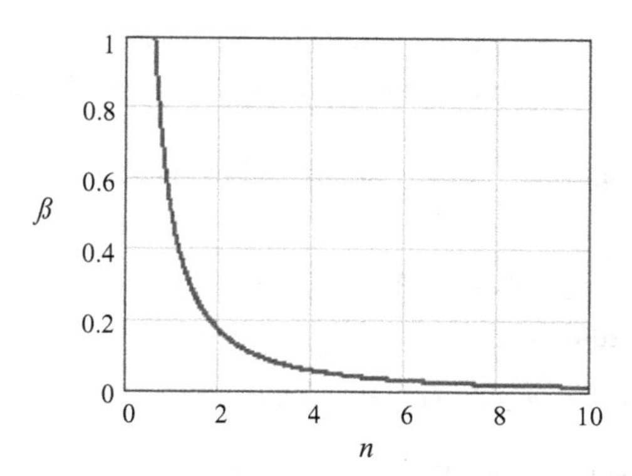

Figure 3.2 "1/n" power law turbulent mean-flow velocity profile β characteristics.

Two characteristic parameters in Equation (3.19) are β and φ_{SF}. The latter is called the slip flow parameter. It enters the equation only when the duct walls have finite impedance ($u' \neq 0$) or are permeable ($u_o \neq 0$). Its typical values are $\varphi_{SF} = 0$ and $\varphi_{SF} = 1$, which correspond, respectively, to uniform (full-slip) mean flow and parallel sheared mean flow with no-slip.

The parameter β is called the mean-flow velocity profile factor. Typical values of β for a circular duct are $\beta = 0$ and $\beta = 1/3$, which correspond to uniform and parabolic mean-flow velocity profiles, respectively. Figure 3.2 shows β as a function of n for the "1/n" power law $v_o(r) \propto (1 - r/a)^{1/n}$, $0 \leq r \leq a$, where a denotes the duct radius. It is seen that the value of β is less than 0.037 for $n \geq 5$, which is typical of a fully developed turbulent mean-flow velocity profile. If the mean flow is assumed to be confined to a core area $S_{\text{core}}(< S)$, then $\beta = S/S_{\text{core}}$.

The linearized continuity and momentum equations, Equations (3.17) and (3.19), provide two equations for the determination of $\overline{p'}$, $\overline{v'}$ and $\overline{\rho'}$. The third equation for their closure comes from the energy equation for unsteady motion inviscid fluids. In Section 1.1.2, we saw that the latter equation is tantamount to the statement that the fluid motion is isentropic. Application of this condition for closure of Equations (3.17) and (3.19) for determination of $\overline{p'}$, $\overline{v'}$ and $\overline{\rho'}$ is discussed in subsequent sections.

3.3 Solution of Linearized Acoustic Equations

This section describes a general analytical method for the derivation of the wave transfer matrices of ducts based on the one-dimensional theory. To this end, it is convenient to introduce the following variables:

(i) The entropy wave component ε', which is defined as:

$$\varepsilon' = c_o^2 \overline{\rho'} - \overline{p'}, \tag{3.23}$$

ε' takes its name from the fact that the linearized form of Equation (A.15a) may be expressed as $\varepsilon' = -\rho_o T_o(\gamma_o - 1)s'$. So, ε' scales with the entropy fluctuations (of zero mean).

(ii) The pressure wave components $p^{\mp}$, which are defined by the transformation

$$\begin{bmatrix} \overline{p'} \\ v' \end{bmatrix} = \begin{bmatrix} 1 & 1 \\ 1/\rho_o c_o & -1/\rho_o c_o \end{bmatrix} \begin{bmatrix} p^+ \\ p^- \end{bmatrix}, \tag{3.24}$$

where c_o denotes the speed of sound.

It is also convenient to introduce the following definitions:

- Homogeneous duct: p_o = constant and ρ_o = constant (and, hence, c_o = constant).
- Uniform duct: all parameters are independent of the axial coordinate x.

Obviously, a uniform duct is also homogeneous.

3.3.1 Homogeneous Ducts ($\varepsilon' = 0$)

In this case the linearized energy equation, Equation (1.11), reduces to Equation (1.15), that is, $p' = c_o^2 \rho'$, which implies, on taking cross sectional averages, $\overline{p'} = c_o^2 \overline{\rho'}$ or $\varepsilon' = 0$. On substituting this relationship, Equations (3.17) and (3.19) can be expressed in frequency domain in the state-space form:

$$\frac{\partial}{\partial x}\begin{bmatrix} \overline{p'} \\ v' \end{bmatrix} = \begin{bmatrix} F_{11} & F_{12} \\ F_{21} & F_{22} \end{bmatrix} \begin{bmatrix} \overline{p'} \\ v' \end{bmatrix}, \tag{3.25}$$

where the coefficients matrix is called the matrizant (or matricant) of this system of equations [1–2]. We will generally prefer to deal with the solution of the foregoing equation after it is transformed to the pressure wave components $p^{\mp}$ defined in Equation (3.24). Applying the latter transformation, Equation (3.25) yields the state-space equation:

$$\frac{\partial}{\partial x}\begin{bmatrix} p^+ \\ p^- \end{bmatrix} = \begin{bmatrix} H_{11} & H_{12} \\ H_{21} & H_{22} \end{bmatrix} \begin{bmatrix} p^+ \\ p^- \end{bmatrix}. \tag{3.26}$$

This is written briefly as

$$\frac{\partial \mathbf{P}}{\partial x} = \mathbf{HP}, \tag{3.27}$$

where

$$\mathbf{P} = \begin{bmatrix} p^+ \\ p^- \end{bmatrix} \tag{3.28}$$

is called pressure wave components vector and the new matrizant is given by

$$\mathbf{H} = \frac{1}{2}\begin{bmatrix} F_{11} + F_{22} + \rho_o c_o F_{21} + \dfrac{F_{12}}{\rho_o c_o} + \dfrac{\mathrm{d}}{\mathrm{d}x}\ln \rho_o c_o & F_{11} - F_{22} + \rho_o c_o F_{21} - \dfrac{F_{12}}{\rho_o c_o} - \dfrac{\mathrm{d}}{\mathrm{d}x}\ln \rho_o c_o \\ F_{11} - F_{22} - \rho_o c_o F_{21} + \dfrac{F_{12}}{\rho_o c_o} - \dfrac{\mathrm{d}}{\mathrm{d}x}\ln \rho_o c_o & F_{11} + F_{22} - \rho_o c_o F_{21} - \dfrac{F_{12}}{\rho_o c_o} + \dfrac{\mathrm{d}}{\mathrm{d}x}\ln \rho_o c_o \end{bmatrix}. \tag{3.29}$$

We will consider the specific forms of the elements of this matrizant in subsequent sections in a sequence of increasing complexity. Here, we discuss some general properties, which will be invoked later for specific cases.

The general solution of Equation (3.26) can be expressed in the form [1–2]:

$$\begin{bmatrix} p^+(x) \\ p^-(x) \end{bmatrix} = \begin{bmatrix} T_{11} & T_{12} \\ T_{21} & T_{22} \end{bmatrix} \begin{bmatrix} p^+(0) \\ p^-(0) \end{bmatrix} \tag{3.30}$$

or, briefly, as

$$\mathbf{P}(x) = \mathbf{T}(x)\mathbf{P}(0), \tag{3.31}$$

where $\mathbf{T}(x)$ denotes the square matrix in Equation (3.30). This is a useful result, because, if the origin $x = 0$ of a duct of length L is taken at its outlet and the section $x = L$ is considered to be its inlet, Equation (3.31) becomes analogous to Equation (2.13) and the matrix $\mathbf{T}(L)$ is recognized as the wave transfer matrix of the duct (with inlet-to-outlet causality). It is proved in the treatise on the theory of linear differential equations that the solution of Equation (3.27) with a matrizant of any size can be expressed in the form of Equation (3.31) with

$$\mathbf{T}(x) = \mathbf{I} + \int_0^x \mathbf{H}(\xi)\mathrm{d}\xi + \int_0^x \mathbf{H}(\xi) \int_0^\xi \mathbf{H}(\zeta)\mathrm{d}\zeta\mathrm{d}\xi + \cdots, \tag{3.32}$$

where $\mathbf{I}$ denotes a unit matrix of conforming size (in this case 2×2) and the infinite series on the right is called the Peano series [2].

3.3.1.1 Uniform Ducts

If the duct is uniform, elements of the matrizant $\mathbf{H}$ are not dependent on x and the sum of the Peano series converges to [2]:

$$\mathbf{T} = \mathrm{e}^{\mathbf{H}x}, \tag{3.33}$$

where the matrix exponential is given by the collineatory transformation:[2]

$$\mathbf{T} = \begin{bmatrix} \varphi_{11} & \varphi_{12} \\ \varphi_{21} & \varphi_{22} \end{bmatrix} \begin{bmatrix} \mathrm{e}^{\mathrm{i}\kappa_1 x} & 0 \\ 0 & \mathrm{e}^{\mathrm{i}\kappa_2 x} \end{bmatrix} \begin{bmatrix} \varphi_{11} & \varphi_{12} \\ \varphi_{21} & \varphi_{22} \end{bmatrix}^{-1}. \tag{3.34}$$

Here, κ_j, $j = 1, 2$, are called axial wavenumbers of the duct and the corresponding vectors $\varphi_j = \{\varphi_{1j} \ \varphi_{2j}\}$ are called wave modes. The axial wavenumbers are given by the roots of the equation $|\mathbf{H} - \mathrm{i}\kappa\mathbf{I}| = 0$, which is called the dispersion equation of the duct. The wave modes are determined by the solution of the homogeneous system of equations $[\mathbf{H} - \mathrm{i}\kappa_j\mathbf{I}]\varphi_j = 0$, where $\mathrm{i}\kappa_j$ and φ_j are called eigenvalues and eigenvectors

[2] Proof: by definition, $\mathrm{e}^{\mathbf{A}} = \mathbf{I} + \mathbf{A} + \mathbf{A}^2/2! + \cdots$. If matrix $\mathbf{A}$ is diagonalizable then $\mathbf{A} = \mathbf{\Phi}\mathbf{\Lambda}\mathbf{\Phi}^{-1}$, where $\mathbf{\Lambda}$ is a diagonal matrix of eigenvalues of $\mathbf{A}$ and $\mathbf{\Phi}$ is a matrix having the corresponding right eigenvectors of $\mathbf{A}$ as its columns. Therefore, $\mathrm{e}^{\mathbf{A}} = \mathbf{I} + \mathbf{\Phi}\mathbf{\Lambda}\mathbf{\Phi}^{-1} + (\mathbf{\Phi}\mathbf{\Lambda}\mathbf{\Phi}^{-1})^2/2! + \cdots = \mathbf{\Phi}\mathrm{e}^{\Lambda}\mathbf{\Phi}^{-1}$.

of matrix **H**. The eigenvectors (also called modes) are unique except for a constant factor. But Equation (3.34) is unique because these constant factors cancel out on evaluation of **T**.

An important corollary to Equation (3.34) is that the pressure wave components can always be represented by the modal transformation:

$$\begin{bmatrix} p^+(x) \\ p^-(x) \end{bmatrix} = \begin{bmatrix} \varphi_{11} & \varphi_{12} \\ \varphi_{21} & \varphi_{22} \end{bmatrix} \begin{bmatrix} \pi^+(x) \\ \pi^-(x) \end{bmatrix}, \tag{3.35}$$

where $\pi^{\mp}$ are called the principal pressure wave components. The transmission of the principal pressure wave components along the duct is characterized by a diagonal wave transfer relationship as

$$\begin{bmatrix} \pi^+(x) \\ \pi^-(x) \end{bmatrix} = \begin{bmatrix} \mathrm{e}^{\mathrm{i}\kappa_1 x} & 0 \\ 0 & \mathrm{e}^{\mathrm{i}\kappa_2 x} \end{bmatrix} \begin{bmatrix} \pi^+(0) \\ \pi^-(0) \end{bmatrix}. \tag{3.36}$$

3.3.1.2 Homogeneous Non-Uniform Ducts

In this case, the matrizant **H** is a function of x and the Peano series in Equation (3.32) can be evaluated only numerically. Alternatively, the following formula may be used for the calculation of the wave transfer matrix:

$$\mathbf{T}(x) = \begin{bmatrix} 1 & 0 \\ 0 & H_{12}(\omega, x) \end{bmatrix}^{-1} \begin{bmatrix} y_2 & y_1 \\ \dfrac{\mathrm{d}y_2}{\mathrm{d}x} & \dfrac{\mathrm{d}y_1}{\mathrm{d}x} \end{bmatrix} \begin{bmatrix} 1 & 0 \\ 0 & H_{12}(\omega, 0) \end{bmatrix} \mathrm{e}^{\int_0^x H_{11}(\omega,\xi)\mathrm{d}\xi}, \tag{3.37}$$

where $y_1 = y_1(x)$ and $y_2 = y_2(x)$ are two independent solutions of the linear differential equation

$$\frac{\mathrm{d}^2 y}{\mathrm{d}x^2} + \left(H_{11} - H_{22} - \frac{\mathrm{d}}{\mathrm{d}x} \ln H_{12} \right) \frac{\mathrm{d}y}{\mathrm{d}x} - H_{12} H_{21} y = 0 \tag{3.38}$$

for arbitrary but linearly independent boundary conditions, typically, for example, $y_1 = 0$, $\mathrm{d}y_1/\mathrm{d}x = 1$ and $y_2 = 1$ $\mathrm{d}y_2/\mathrm{d}x = 0$ at $x = 0$. Equation (3.37) can be proved by noting from Equation (3.31) that $\partial \mathbf{P}/\partial x = (\partial \mathbf{T}/\partial x)\mathbf{T}^{-1}\mathbf{P} = \mathbf{HP}$ and showing that, for y_1 and y_2 satisfying Equation (3.38), the formula in Equation (3.37) satisfies the relationship $(\partial \mathbf{T}/\partial x)\mathbf{T}^{-1} = \mathbf{H}$ identically. A direct derivation of Equation (3.37) is given in reference [3].

3.3.2 Inhomogeneous Ducts ($\varepsilon' \neq 0$)

Closure of Equations (3.17) and (3.19) for the determination of $\overline{p'}$, $\overline{v'}$ and $\overline{\rho'}$ for this case is described in Section 3.10.1 for a perfect gas. Equations (3.17) and (3.19) are complemented by Equation (3.164), and the frequency domain forms of the three equations can be collated in the state-space form (see Equation (3.166)):

$$\frac{\partial}{\partial x}\begin{bmatrix} \overline{p'} \\ \overline{v'} \\ \varepsilon' \end{bmatrix} = \begin{bmatrix} F_{11} & F_{12} & F_{13} \\ F_{21} & F_{22} & F_{23} \\ F_{31} & F_{32} & F_{33} \end{bmatrix} \begin{bmatrix} \overline{p'} \\ \overline{v'} \\ \varepsilon' \end{bmatrix}. \tag{3.39}$$

Using Equations (3.23) and (3.24), this is transformed to the state-space form:

$$\frac{\partial}{\partial x}\begin{bmatrix} \mathbf{P} \\ \varepsilon' \end{bmatrix} = \begin{bmatrix} H_{11} & H_{12} & H_{13} \\ H_{21} & H_{22} & H_{23} \\ H_{31} & H_{32} & H_{33} \end{bmatrix} \begin{bmatrix} \mathbf{P} \\ \varepsilon' \end{bmatrix}, \tag{3.40}$$

where the vector $\mathbf{P}$ is given by Equation (3.28) and H_{ij}, $i,j = 1,2,3$, denote the elements of the system matrizant $\mathbf{H}$, which is now a 3×3 matrix. Solution of this system of equations can be expressed in the form:

$$\begin{bmatrix} \mathbf{P}(x) \\ \varepsilon'(x) \end{bmatrix} = \begin{bmatrix} T_{11} & T_{12} & T_{13} \\ T_{21} & T_{22} & T_{23} \\ T_{31} & T_{32} & T_{33} \end{bmatrix} \begin{bmatrix} \mathbf{P}(0) \\ \varepsilon'(0) \end{bmatrix}, \tag{3.41}$$

where the 3×3 matrix is the wave transfer matrix $\mathbf{T}$ of the duct. This is still given by the Peano series in Equation (3.32) or, if the matrizant is not dependent on x, by Equation (3.33). In the latter case, we may make use of the fact that the collineatory transformation in Equation (3.34) is valid for a matrizant of any size (see Footnote 2). It can be expressed generally as $\mathbf{T} = \mathbf{\Phi\Lambda\Phi}^{-1}$, where $\mathbf{\Phi}$ denotes the modal matrix whose columns are the eigenvectors $\boldsymbol{\varphi}_j = \{ \varphi_{1j} \quad \varphi_{2j} \quad \cdots \}$ of the matrizant $\mathbf{H}$ and $\mathbf{\Lambda}$ denotes a diagonal square matrix with diagonal elements $\mathrm{e}^{\mathrm{i}\kappa_j x}$, $j = 1, 2, \ldots$, where $\mathrm{i}\kappa_j$ are the eigenvalues of $\mathbf{H}$. Thus, Equation (3.34) now applies with κ_j, $j = 1,2,3$, and $\varphi_j = \{ \varphi_{1j} \quad \varphi_{2j} \quad \varphi_{3j} \}$.

The first two equations in Equation (3.41) may be written as

$$\begin{bmatrix} p^+(x) \\ p^-(x) \end{bmatrix} = \begin{bmatrix} T_{11} & T_{12} \\ T_{21} & T_{22} \end{bmatrix} \begin{bmatrix} p^+(0) \\ p^-(0) \end{bmatrix} + \begin{bmatrix} T_{13} \\ T_{23} \end{bmatrix} \varepsilon'(0). \tag{3.42}$$

Therefore, if $\varepsilon'(0) = 0$, the pressure wave components are still related as in Equation (3.30) by a 2×2 wave transfer matrix. The entropy wave is then given by $\varepsilon'(x) = [T_{31} \quad T_{32}]\mathbf{P}(0)$ and it has no effect on the pressure wave components. This reduction is always possible, unless the duct is driven by fluctuating entropy sources (for example, a combustion process) at the origin.

3.3.3 Numerical Matrizant Method

In general, if the matrizant of a duct is an arbitrary function of the duct axis, x, its wave transfer matrix can be determined only numerically. The Runge–Kutta method may be used for this purpose, however, a very efficient alternative numerical procedure which can produce potentially exact results, is to divide a duct of length L length into N axial segments $x_{k-1} \leq x \leq x_k$, k=1,2,..., N, $x_0 = 0$, $x_N = L$, and to approximate the matrizants of segments $\mathbf{H}_k(\omega, x)$ by $\mathbf{H}_k(\omega, \xi_k)$, where $x = \xi_k$ denotes the axial coordinate of the center of segment k, $x_{k-1} \leq x \leq x_k$. Convergence of this approximation to the exact representation of the duct as the number of segments is increased is ensured by the matrizant theory [2].

Since segments are approximated by some uniform ducts, the wave transfer matrix of segment k can be calculated as $\mathbf{T}_{k-1,k} = \mathrm{e}^{\mathbf{H}_k(\omega,\xi_k)L_k}$ with $L_k = x_{k-1} - x_k < 0$, if the origin is taken as x_k; or as $\mathbf{T}_{k,k-1} = \mathrm{e}^{\mathbf{H}_k(\omega,\xi_k)L_k}$ with $L_k = x_k - x_{k-1} > 0$, if the origin is taken as x_{k-1}, $k = 1, 2, \ldots, N$. Then, for example, if the latter scheme is adopted, the wave transfer matrix of the whole duct can be calculated from

$$\mathbf{T}_{N,1} = \mathbf{T}_{N,N-1}\mathbf{T}_{N-1,N-2} \ldots \mathbf{T}_{21}\mathbf{T}_{10}. \tag{3.43}$$

The number of segments required for satisfactory accuracy may be determined relatively easily by inspecting the convergence of this solution.

This approach is called numerical segmentation, in order to differentiate it from the physical segmentation method, in which the actual duct is replaced by a stepwise uniform duct and the wave transfer matrices of the segments are computed from the matrizants of uniform ducts. But physical segmentation introduces discontinuity between the uniform segments and Equation (3.43) may not be used now to compute the overall wave transfer matrix of the duct, because it is necessary to match the acoustic fields in the segments at their interfaces. For example, if the actual duct is non-uniform and is modeled as a stepwise uniform duct, the neighboring uniform segments must be matched by using the axial closed area-change elements (see Section 4.2). Then Equation (3.43) becomes $\mathbf{T}_{N,1} = \mathbf{T}^{(0)}_{N,N-1}$ $\mathbf{A}_{N-1,N-2} \ldots \mathbf{T}^{(0)}_{21}\mathbf{A}_{10}\mathbf{T}^{(0)}_{10}$, where the superscript "(0)" denotes the wave transfer matrices of uniform duct segments and $\mathbf{A}_{k,k-1}$ denotes the wave transfer matrices of the closed area-change elements joining segments k and $k-1$. It should be noted that $\mathbf{T}^{(0)}_{k,k-1}$ are distinct from $\mathbf{T}_{k,k-1}$, since the former matrizants do not include axial gradient terms.

The following APL function implements the numerical matrizant method with equal length segments.

```
T←NumericSegmentation name;A;S;Lk;H;R;k;N;L;a;b;Z
```

⍝ This function computes the wave transfer matrix, *T*, of a duct with 2×2 matrizant that is a function of the duct axis, x, by using the segmentation method described above. The right argument (name) must be an alphanumeric expression which may be used to execute the element subroutine that calculates its matrizant for given x. For example, if the full call for this is `H←ports HOTPIPE x`, the right argument is input as `name← 'ports HOTPIPE'`. Then the expression `⍎name, 'x'` calculates the matrizant at section x of the duct element called `HOTPIPE` with specified ports in the block diagram. We assume for the simplicity of the discussion that element subroutines are written such that they may be called only by specifying their port numbers (see Section 2.6). If they are written differently, the right argument must be modified accordingly. The segmentation method is applied by dividing the duct into *N* segments of equal length. The total duct length *L* and *N* must be included in the element data. So, the first step of the program is to fetch *L* and *N* from the element data file. This part of the program depends on file system commands. At this point, we assume that this is done and the local variables *L* and *N* contain the fetched data.

⍝ Compute segment length:

```
Lk←L÷N
```

⍝ Set initial wave transfer matrix to real unit matrix:

```
T←2 2 2⍴ 0   ⋄   T[1;1;1]←T[1;2;2]←0
```

⍝ Compute wave transfer matrices of segments, assuming that they are uniform with properties equal to their values at the middle of segments:

```
:for k :in ⍳N
```

⍝ Compute matrizant $\mathbf{H}_k = \begin{bmatrix} H_{11} & H_{12} \\ H_{21} & H_{22} \end{bmatrix}_k$ of segment $x_{k-1} \leq x \leq x_k$:

```
H←⍎name,' Lk×k-0.5'
```

⍝ Compute roots of $|\mathbf{H}_k - \lambda \mathbf{I}| = \lambda^2 - (H_{11} + H_{11})\lambda + H_{11}H_{22} - H_{12}H_{21} = 0$ where $\lambda = \mathrm{i}\kappa$, see Equation (3.34):

```
a←-H[;1;1]+H[;2;2]
b←(H[;1;1] ZTIMES H[;2;2])-H[;1;2] ZTIMES H[;2;1]
```

⍝ Use the formula for the roots of the quadratic $\lambda^2 + a\lambda + b = 0$:

```
Z←ZSQRT (a ZTIMES a)-4×b
R←2 1⍴¯0.5×Z+a ⋄ R←R,0.5×Z-a
```

⍝ Roots are $\lambda_1 = R[;1]$ and $\lambda_2 = R[;2]$. Let first root always have positive imaginary part:

```
:if R[2;1]<0
    R←⌽ R
:endif
```

⍝ Compute elements of the wave transfer matrix of principal pressure wave components; see Equation (3.36):

```
A←2 2 2⍴ 0
A[;1;1]←ZEXP -Lk×R[;1]
A[;2;2]←ZEXP -Lk×R[;2]
```

⍝ where *ZEXP* Z computes e^Z for a complex number Z. The minus sign is because the wave transfer matrix of segment $x_{k-1} \leq x \leq x_k$, $\mathbf{T}_{k-1,k} = e^{\mathbf{H}_k(\omega,\xi_k)L_k}$ is computed with origin taken at x_k and, therefore, the wave transfer matrix of the whole duct is computed from $\mathbf{T}_{1,N} = \mathbf{T}_{01}\mathbf{T}_{12}\ldots\mathbf{T}_{N-1,N}$.

⍝ Compute eigenvectors corresponding to $\lambda_1 = R[;1]$ and $\lambda_2 = R[;2]$:

```
S[;1;1]←S[;2;2]←1 0
S[;2;1]←H[;2;1] ZDIV H[;2;2]-R[;1]
S[;1;2]←H[;1;2] ZDIV H[;1;1]-R[;2]
```

⍝ where `a ZDIV b` computes the quotient `a/b` of two complex numbers.

⍝ Compute inverse of matrix `S`:

```
b←(S[;1;1] ZTIMES S[;2;2])-S[;1;2] ZTIMES S[;2;1]
```

⍝ where `a ZTIMES b` computes the product `ab` of two complex numbers:

```
a[;1;1]←S[;2;2] ZDIV b
a[;2;2]←S[;1;1] ZDIV b
a[;2;1]←-S[;2;1] ZDIV b
a[;1;2]←-S[;1;2] ZDIV b
```

⍝ Cascade segment with the previous one:

```
T←T CXMATTIMES a CXMATTIMES A CXMATTIMES S
```

⍝ where `A CXMATTIMES B` computes the product `AB` of two complex matrices.

```
:endfor
```

3.4 Time-Averaged Power of One-Dimensional Acoustic Waves

The time-averaged acoustic power passing through normal to a duct cross section may be computed in the frequency domain by Equation (1.48). Since in the one-dimensional theory the acoustic variables are represented by their averages over a duct section, Equation (1.48) may be expressed as:

$$W = 2\mathrm{Re}\int_S \left(\overline{\breve{p}'}\,\overline{v'} + \rho_o v_o \overline{\breve{v}'}\,\overline{v'} + \frac{v_o}{\rho_o}\overline{\breve{p}'}\,\overline{\rho'} + v_o^2 \overline{\breve{v}'}\,\overline{\rho'} \right) \mathrm{d}S \tag{3.44}$$

where S denotes the duct cross section and an inverted over-arc denotes the complex conjugate. Integrating this and using Equations (3.23) and (3.24), we obtain

$$W = \frac{2S}{\rho_o c_o}\left(1 + 2\overline{M_o} + (1+\beta)\left(\overline{M_o}\right)^2\right)\ |p^+|^2 - \left(1 - 2\overline{M_o} + (1+\beta)\left(\overline{M_o}\right)^2\right)\ |p^-|^2 + \mathrm{Re}\left\{\overline{M_o}\left((1+\overline{M_o})\breve{p} + (1-\overline{M_o})\breve{p}\right)\varepsilon'\right\} \tag{3.45}$$

or, for homogeneous ducts,

$$W = \frac{2S}{\rho_o c_o}\left(1 + 2\overline{M_o} + (1+\beta)\left(\overline{M_o}\right)^2\right)\ |p^+|^2 - \left(1 - 2\overline{M_o} + (1+\beta)\left(\overline{M_o}\right)^2\right)\ |p^-|^2, \tag{3.46}$$

where β represents the effect of refraction in sheared mean flow and $\overline{M_o} = \overline{M_o} = \overline{v_o}/c_o$ and an overbar denotes averaging over the duct cross-section (see Equation (4.7)). This shows that when the entropy wave is neglected, the time-averaged acoustic power transmitted in the duct can be decomposed as $W = W^+ + W^-$, where

$$W^{\mp} = \frac{\mp 2S}{\rho_o c_o}\left(1\mp 2\overline{M_o} + (1+\beta)\left(\overline{M_o}\right)^2\right)\ |p^{+}|^2 \tag{3.47}$$

and $W^{\mp}$ denote the time-averaged power transmitted in $\mp x$ directions, respectively. When β is neglected, this reduces to the commonly quoted expression in the literature, namely,

$$W^{\mp} = \frac{\mp 2S}{\rho_o c_o}\left(1\mp\overline{M_o}\right)^2\ |p^{+}|^2. \tag{3.48}$$

These expressions hold for a single frequency component in the Fourier spectrum of the acoustic pressure. They are often given in the literature in terms of the amplitudes of sinusoidal pressure wave components, $\hat{p}^{\mp} = 2|p^{\mp}|$, where an over-hat denotes the amplitude of a sinusoidal quantity. They may also be expressed in terms of the root-mean-square (*rms*) values of the pressure wave components, that is, $p^{\mp}_{rms} = \sqrt{2}|p^{\mp}|$. For periodic acoustic disturbances of fundamental period T, Equation (3.47) gives

$$W^{\mp} = \frac{\mp 2S}{\rho_o c_o}\left(1\mp 2\overline{M_o} + (1+\beta)\left(\overline{M_o}\right)^2\right)\sum_{n=-\infty}^{\infty}\left|p^{\mp}(n\varpi)\right|^2, \tag{3.49}$$

where $\varpi = 2\pi/T$. For more general acoustic disturbances, such as stationary ergodic random disturbances or transient disturbances, a still further generalization is

$$W^{\mp} = \frac{\mp 2S}{\rho_o c_o}\left(1\mp 2\overline{M_o} + (1+\beta)\left(\overline{M_o}\right)^2\right)\int_{-\infty}^{\infty} S_{p^{\mp}}(\omega)\mathrm{d}\omega, \tag{3.50}$$

where $S_{p^{\mp}}(\omega)$ denote the power spectral density functions of $p^{\mp}(\omega)$. This is computed from the Fourier transforms as $S_{p^{\mp}}(\omega) = \left|p^{\mp}(\omega)\right|^2/T$, where T denotes excitation duration or an averaging time. The integrals are usually evaluated numerically by covering the frequency range of interest by contiguous frequency bands and assuming that the power spectral density remains constant in these bands (see Section 1.2.4).

3.5 Hard-Walled Uniform Ducts

Ducts with rigid and impervious walls are usually referred to briefly as hard-walled ducts. We consider hard-walled uniform ducts first, because they are mathematically the simplest and the most useful for this reason. In this section, we also capitalize on this relative mathematical simplicity to introduce some important concepts on the physics of one-dimensional sound waves.

In one-dimensional theory, the hard-wall condition is applied by putting $u' = u_o = 0$ in Equations (3.17) and (3.19), and also in the corresponding mean-flow equations, Equations (3.18) and (3.22). For a uniform duct with impermeable walls (not necessarily hard), the mean-flow equations are satisfied identically and need not be considered any further.

Three-dimensional sound propagation in a uniform hard-walled duct with a uniform mean flow is governed by the convected wave equation, Equation (6.47). This predicts that the fundamental mode of sound wave propagation is planar and that only this mode propagates in the duct, if the wavelength is large compared to the dimensions of the duct section (for precise criteria, see Section 6.5). For this case, therefore, the one-dimensional theory is exact and reduces to the classical plane-wave theory. If the mean flow is sheared, however, three-dimensional sound propagation in the duct is governed by the Pridmore-Brown equation (see Section 6.8.1). This predicts that the planar nature of the fundamental mode is perturbed due to refraction of sound waves in shear layers. The effect of refraction increases with the shear gradients and the frequency. Exact solutions of the Pridmore-Brown equation for hard-walled circular and two-dimensional rectangular ducts show that in the fundamental mode [4–5], deviations of the acoustic properties from their mean over a duct section remain less than few percent if the local mean-flow velocity is in the subsonic low Mach number range, typically, less than 0.3 or so. The one-dimensional theory is adequately accurate for these ranges of the parameters. (For example, see Section 3.9.3.1.)

3.5.1 Wave Transfer Matrix

For hard-walled uniform ducts, the linearized continuity and momentum equations, Equation (3.17) and (3.19), simplify to

$$\frac{\partial\overline{\rho'}}{\partial t}+\overline{v_o}\frac{\partial\overline{\rho'}}{\partial x}+\rho_o\frac{\partial\overline{v'}}{\partial x}=0 \tag{3.51}$$

$$\rho_o\left(\frac{\partial\overline{v'}}{\partial t}+\overline{v_o}\frac{\partial\overline{v'}}{\partial x}\right)+\beta(\overline{v_o})^2\frac{\partial\overline{\rho'}}{\partial x}+\frac{\partial\overline{p'}}{\partial x}=0. \tag{3.52}$$

Substituting $\overline{p'}=c_o^2\overline{\rho'}$, the foregoing equations can be recast in frequency domain as

$$\left(\bar{M}_o\frac{\partial}{\partial x}-\mathrm{i}k_o\right)\overline{p'}+\rho_o c_o\frac{\partial\overline{v'}}{\partial x}=0 \tag{3.53}$$

$$\alpha^2\frac{\partial\overline{p'}}{\partial x}+\rho_o c_o\left(\bar{M}_o\frac{\partial}{\partial x}-\mathrm{i}k_o\right)\overline{v'}=0, \tag{3.54}$$

where $\bar{M}_o=\overline{v_o}/c_o$ denotes the Mach number of the cross section-averaged mean-flow velocity, $k_o=\omega/c_o$ denotes the wavenumber and

$$\alpha=\sqrt{1+\beta\bar{M}_o^2}. \tag{3.55}$$

Equations (3.53) and (3.54) can be expressed in the state-space form

$$\frac{\partial}{\partial x}\begin{bmatrix}\overline{p'}\\ \overline{v'}\end{bmatrix}=\frac{\mathrm{i}k_o}{\alpha^2-\bar{M}_o^2}\begin{bmatrix}-\bar{M}_o & \rho_o c_o\\ \dfrac{\alpha^2}{\rho_o c_o} & -\bar{M}_o\end{bmatrix}\begin{bmatrix}\overline{p'}\\ \overline{v'}\end{bmatrix}. \tag{3.56}$$

This is transformed to the pressure wave components using Equation (3.29):

$$\frac{\partial}{\partial x}\begin{bmatrix} p^+ \\ p^- \end{bmatrix} = \frac{1}{2}\frac{ik_o}{\alpha^2 - \bar{M}_o^2}\begin{bmatrix} -2\bar{M}_o + \alpha^2 + 1 & \alpha^2 - 1 \\ -\alpha^2 + 1 & -2\bar{M}_o - \alpha^2 - 1 \end{bmatrix}\begin{bmatrix} p^+ \\ p^- \end{bmatrix}. \quad (3.57)$$

The eigenvalues and eigenvectors of the matrizant of this equation are found from the solution of the algebraic eigenvalue problem:

$$\frac{1}{2}\frac{ik_o}{\alpha^2 - \bar{M}_o^2}\begin{bmatrix} -2\bar{M}_o + \alpha^2 + 1 & \alpha^2 - 1 \\ -\alpha^2 + 1 & -2\bar{M}_o - \alpha^2 - 1 \end{bmatrix}\begin{bmatrix} \varphi_1 \\ \varphi_2 \end{bmatrix} = i\kappa\begin{bmatrix} 1 & 0 \\ 0 & 1 \end{bmatrix}\begin{bmatrix} \varphi_1 \\ \varphi_2 \end{bmatrix}. \quad (3.58)$$

Hence, denoting the axial wavenumbers corresponding to the eigenvalues by $i\kappa^+$ and $i\kappa^-$, we find:

$$\kappa^{\mp} = \frac{\mp k_o}{\alpha \mp \bar{M}_o} \quad (3.59)$$

$$\frac{\varphi_{11}}{\varphi_{12}} = \frac{\varphi_{22}}{\varphi_{21}} = \frac{1+\alpha}{1-\alpha} \quad (3.60)$$

Hence, from Equation (3.34), the wave transfer matrix of the duct can be expressed as:

$$\mathbf{T} = \begin{bmatrix} 1+\alpha & 1-\alpha \\ 1-\alpha & 1+\alpha \end{bmatrix}\begin{bmatrix} e^{ik_oK^+x} & 0 \\ 0 & e^{ik_oK^-x} \end{bmatrix}\begin{bmatrix} 1+\alpha & 1-\alpha \\ 1-\alpha & 1+\alpha \end{bmatrix}^{-1}, \quad (3.61)$$

where $\kappa^{\mp} = k_oK^{\mp}$ and

$$K^{\mp} = \frac{\mp 1}{\alpha \mp \bar{M}_o} \quad (3.62)$$

are called propagation constants. Since the values of β are much smaller than about 0.3 for real mean-flow velocity profiles (see Figure 3.2), we may take $\alpha \approx 1$ without substantial loss of accuracy. Consequently, the pressure wave components are not substantially different than the principal pressure wave components. However, we will keep α in the analysis for the sake of generality and also to remind us that the effect of mean-flow velocity profile is included in the analysis.

3.5.2 Traveling Waves and Direction of Propagation

From Equation (3.36), the principal pressure wave components are given in frequency domain by

$$\begin{bmatrix} \pi^+(x) \\ \pi^-(x) \end{bmatrix} = \begin{bmatrix} e^{ik_oK^+x} & 0 \\ 0 & e^{ik_oK^-x} \end{bmatrix}\begin{bmatrix} \pi^+(0) \\ \pi^-(0) \end{bmatrix}. \quad (3.63)$$

This equation can be transformed to time domain by taking inverse Fourier transforms. The definition of the inverse Fourier transform in Equation (1.16) implies the shifting property $h(t - t_o) \leftrightarrow h(\omega)e^{i\omega t_o}$, where $\rightarrow$ denotes Fourier transformation and $\leftarrow$ denotes inverse Fourier transformation (see Section 1.2). Hence the principal pressure wave components can be expressed in the time domain as

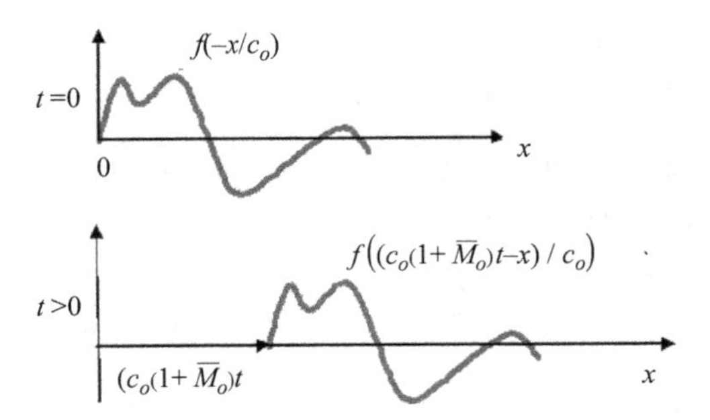

Figure 3.3 A waveform traveling in the +x direction with velocity $c_o(1+\bar{M}_o)$, $\alpha = 1$.

$$\pi^+(t,x) = \pi^+\left(t - \frac{x}{c_o(\alpha + \bar{M}_o)}\right) \tag{3.64}$$

$$\pi^-(t,x) = \pi^-\left(t + \frac{x}{c_o(\alpha - \bar{M}_o)}\right), \tag{3.65}$$

where the velocities $c_o(\alpha \mp \bar{M}_o) = c_o/K^{\mp}$ are known as the phase velocities (also called Doppler velocities). For insight to the physics of this result, let $\xi = (\alpha + \bar{M}_o)t - x/c_o$ and $\eta = (\alpha - \bar{M}_o)t + x/c_o$. Then, the functions on the right-hand sides of Equations (3.64) and (3.65) can be represented as $f(\xi)$ and $g(\eta)$, respectively. At time $t = 0$, the function $f(\xi)$ is $f(-x/c_o)$. At any time $t > 0$, it becomes $f((c_o(\alpha + \bar{M}_o)t - x)/c_o)$, but it still has the same shape as at $t = 0$, except that it is displaced (moved, traveled) in the $+x$ direction by a distance equal to $c_o(\alpha + \bar{M}_o)t$. This is illustrated in Figure 3.3 for $\alpha = 1$. Consequently, to an observer fixed to the x-axis, the function $f(\xi)$ travels with velocity $c_o(\alpha + \bar{M}_o)$ in the $+x$ direction without changing its shape. It is concluded similarly that, to the same observer, the function $g(\eta)$ travels without changing its shape, with velocity $c_o(\alpha - \bar{M}_o)$, in the $-x$ direction. If this kind of longitudinal wave motion is periodic temporally (in time), it is also periodic spatially. To see this, let the temporal fundamental period be denoted by T. Then, for example, from Equations (3.64), $\pi^+(t,x) = \pi^+(t+T,x) = \pi^+(t + T - x/c_o(\alpha + \bar{M}_o))$ or, upon rearranging the argument, $\pi^+(t,x) = \pi^+(t - (x+\lambda)/c_o(\alpha + \bar{M}_o)) = \pi^+(t, x+\lambda)$ where $\lambda = Tc_o(\alpha + \bar{M}_o)$ represents the spatial period of the wave. It is called wavelength. An observer fixed to the x-axis measures the frequency of this wave as $f = 1/T$ in Hz units or as $\omega = 2\pi f$ in rad/s units. But an observer who is moving with a velocity of Mach number $\bar{M}_o$ measures the frequency of the same wave as $\tilde{f} = 1/\tilde{T}$, where $\tilde{T} = \lambda/c_o$, since the wavelength is still the same.

Thus, the principal pressure wave components $\pi^{\mp}$ correspond to wave modes propagating in $\mp x$ directions, respectively. This explains why $K^{\mp}$ are called propagation constants. The superscripts in the notation $\pi^{\mp}$ indicate the direction of propagation of the principal wave modes.

As depicted by Equation (3.35) in frequency domain, $p^{\mp}$ consist of a superimposition of the principal pressure components $\pi^{\mp}$ and, therefore, their time-domain forms are made up of wave modes propagating both in $+x$ and $-x$ directions. But an exception occurs if $\alpha = 1$, as we then have $p^{\mp} = \pi^{\mp}$. Since $\alpha \approx 1$ in the subsonic low Mach number range for practical forms of mean-flow velocity profiles, we may assume, for all practical purposes, that the pressure wave components $p^{\mp}$ represent wave modes propagating in $\mp x$ directions, respectively.

The direction of propagation of the principal wave modes may also be deduced by using the Briggs criterion (see Section 1.3.2). For this purpose, we first note that the propagation constants $K^{\mp}$ are real and related to the axial wavenumbers as $\kappa^{\mp}=k_o K^{\mp}$. Then, from Equation (1.31), the group velocities are $V_g^{\mp} = c_o/K^{\mp}$ and the application of Briggs' criterion immediately shows that the wave mode corresponding to K^+ propagates in $+x$ direction (since $V_g^+ > 0$) and the wave mode corresponding to K_e^- propagates in $-x$ direction (since $V_g^- < 0$). It should be noted that Briggs' criterion predicts the direction of propagation of $\pi^{\mp}$, not of $p^{\mp}$, unless $\alpha = 1$.

3.5.3 Reflection Coefficient and Standing Waves

Two parameters which are frequently used in the description of frequency-domain characteristics of one-dimensional acoustic fields in ducts are reflection coefficient and acoustic impedance. The latter is defined as

$$z(\omega, x) = \frac{\overline{p'}(\omega, x)}{\overline{v'}(\omega, x)}, \tag{3.66}$$

which is a generalization of the classical plane-wave definition of acoustic impedance, in that cross section averages of the acoustic variables are used. Reflection coefficient, on the other hand, is defined as the quotient:

$$r(\omega, x) = \frac{p^-(\omega, x)}{p^+(\omega, x)} \tag{3.67}$$

and, hence, it is related to the acoustic impedance as

$$z(\omega, x) = \rho_o c_o \frac{1 - r(\omega, x)}{1 + r(\omega, x)}. \tag{3.68}$$

If an infinitely long initially quiescent uniform duct is excited from one end by a rigid plane piston vibrating at low frequency, then only the p^+ wave exists in the duct. But if the duct is of finite length, some fraction of the p^+ wave, which is determined by the reflection coefficient of the other end of the duct, is reflected back at the other end as a p^- wave. The latter is superimposed on the existing p^+ wave to form a steady wave motion pattern along the duct. This steady wave is called a standing wave. Axial variations of acoustic pressure in a standing wave can be computed from $\overline{p'} = p^+ + p^-$. For example, assuming $\alpha = 1$ for simplicity, Equation (3.63) gives $p^{\mp}(x) = p^{\mp}(0)\mathrm{e}^{\mathrm{i}k_o K^{\mp} x}$ and the acoustic pressure field in the duct can be expressed in the frequency domain as

$$\overline{p'}(x) = p^+(0)\mathrm{e}^{\mathrm{i}k_o K^+ x}\left(1 + |r(0)|\mathrm{e}^{\mathrm{i}\theta + \mathrm{i}k_o(K^- - K^+)x}\right), \tag{3.69}$$

where we have introduced the polar notation $r(0) = |r(0)|e^{i\theta}$. Hence,

$$\frac{|\overline{p'}(x)|^2}{|p^+(0)|^2} = 1 + |r(0)|^2 + 2|r(0)|\cos(\theta + k_o(K^- - K^+)x). \tag{3.70}$$

The maximum acoustic pressure magnitude is given by

$$\left|\frac{\overline{p'}(x_n)}{p^+(0)}\right| = 1 + |r(0)|, \tag{3.71}$$

where $x_n = (\theta - 2n\pi)/k_o(K^- - K^+)$, $n = 0, 1, 2, \ldots$ Similarly, the minimum acoustic pressure magnitude is

$$\left|\frac{\overline{p'}(x_n)}{p^+(0)}\right| = 1 - |r(0)|, \tag{3.72}$$

where $x_n = (\theta + 2n\pi)/k_o(K^- - K^+)$, $n = 0, 1, 2, \ldots$. Therefore, the quotient

$$\frac{|\overline{p'}(x)|_{\min}}{|\overline{p'}(x)|_{\max}} = \frac{1 - |r(0)|}{1 + |r(0)|} \tag{3.73}$$

depends only on the value of the reflection coefficient r at the origin.

This formula is the basis of the classical standing-wave ratio method of measurement of the reflection coefficient and, hence, acoustic impedance, of a duct termination at $x = 0$. The duct is excited from the other end, $x = L$ (where $|L|$ denotes the duct length), by a loudspeaker and the maximum and minimum sound pressure amplitudes and their axial locations are measured at desired frequencies by traversing a microphone along the duct axis. The magnitude of the reflection coefficient is then calculated from Equation (3.73), and the argument of the reflection coefficient from the measured axial distance between two successive pressure minima (or maxima).

The same method is also used for measurement of the reflection coefficient of sound control materials. In this case, one end of a hard-walled duct is blocked ($\bar{M}_o = 0$ and $\alpha = 1$) by the material to be measured and the duct is excited from the other end. Since the theory assumes plane-wave propagation, the size of the measurement duct must be selected so that only the fundamental mode propagates in the duct at the frequencies of interest (see Section 12.2.3 for modern implementation of the method).

When the acoustic boundary conditions at both ends of a duct are fixed, standing waves can exist in the duct only at certain discrete frequencies, which are called axial natural frequencies.[3] Simple expressions exist for natural frequencies of ducts having the ideal acoustic boundary conditions discussed in Section 2.3. For example, if both ends of the duct are pressure release boundaries, then $\bar{p}'(0) = \bar{p}'(L) = 0$, which imply that $p^+(0)\left(e^{ik_oK^+L} - e^{ik_oK^-L}\right) = 0$, or $\sin\left(k_oL/\left(1 - \bar{M}_o^2\right)\right) = 0$, where L denotes the duct length. Thus, in this case the natural frequencies are given by

[3] These frequencies are often referred to in the literature as axial resonance frequencies. We prefer to reserve the term resonance frequency for frequencies at which the magnitude of the forced response of a linear system attains a maximum value (see Section 5.3.1).

Table 3.1 Natural frequencies and mode of ducts with ideal boundary conditions

Reference	Boundary condition	Natural frequency	Pressure node positions
open–open	$r(0) = -1$ $r(L) = -1$	$f_n = \frac{nc_o(1 - \bar{M}_o^2)}{2L}$ $\varphi_n = \sin(n\pi x/L)$ $n = 1, 2, 3, \ldots$	$x_i = \frac{iL}{n}$ $i = 0, 1, 2, \ldots, n$
closed–open	$r(0) = +1$ $r(L) = -1$	$f_n = \frac{nc_o}{4L}$ $\varphi_n = \cos(n\pi x/2L)$ $n = 1, 3, 5, \ldots$	$x_i = \frac{iL}{2n}$ $i = 1, 3, 5, \ldots, 2n - 1$
closed–closed	$r(0) = +1$ $r(L) = +1$	$f_n = \frac{nc_o}{2L}$ $\varphi = \cos(n\pi x/L)$ $n = 1, 2, 3, \ldots$	$x_i = \frac{iL}{2n}$ $i = 1, 3, 5, \ldots, 2n - 1$

$$f_n = \frac{nc_o(1 - \bar{M}_o^2)}{2L}, \quad n = 1, 2, \ldots \tag{3.74}$$

The corresponding acoustic pressure distribution along the duct axis (the natural standing wave pattern) is given by $\varphi_n(x) = \sin(n\pi x/L)$, except for a constant factor. Thus, acoustic pressure vanishes at points $x_i = iL/n,\ i = 0, 1, 2, \ldots, n$, on the duct axis. Such points are called pressure nodes of standing waves. These results and the corresponding ones for other combinations of ideal boundary conditions are summarized in Table 3.1. Note that ducts having a rigidly closed end are assumed to carry no mean flow and the anechoic duct case is not included, since it does not sustain standing waves.

3.5.4 Lumped Acoustic Elements

Consider a duct without mean flow ($\bar{M}_o = 0$) of length L with its outlet at origin $x = 0$ and inlet at $x = L$ such that $k_oL << 1$ for the frequencies of interest. Retaining only the first order terms in k_oL in Taylor series expansions of the exponentials (that is, $\mathrm{e}^{\mp \mathrm{i}k_oL} \approx 1 \mp \mathrm{i}k_oL$), Equation (3.61) gives $p^+(L) = (1 + \mathrm{i}k_oL)p^+(0)$ and $p^-(L) = (1 - \mathrm{i}k_oL)p^-(0)$. Upon adding and subtracting these equations, the acoustic pressure and particle velocity in the duct can be expressed as:

$$\overline{p'}(0) - \overline{p'}(L) = \rho_o L\left[-\mathrm{i}\omega \overline{v'}(0)\right] \tag{3.75}$$

$$\frac{1}{-\mathrm{i}\omega}\left[\overline{v'}(0) - \overline{v'}(L)\right] = \frac{L}{\rho_o c_o^2}\overline{p'}(0). \tag{3.76}$$

These equations can be transformed back to the time domain by $-\mathrm{i}\omega q' \to \mathrm{d}q'/\mathrm{d}t$ and $(-\mathrm{i}\omega)^{-1}q' \to \int q'\mathrm{d}t)$ substitution. Consequently, from Newton's second law of motion, Equation (3.75) implies that the fluid in the duct behaves like an ideal mass.

Similarly, from Hooke's law of elasticity, Equation (3.76) implies that the fluid in the duct behaves like an ideal spring. The ratio of the inertia force to the spring force can be expressed, in the frequency domain, as either $1 - \overline{p'}(L)/\overline{p'}(0)$ or $-k_o^2L^2/\left(1 - \overline{v'}(L)/\overline{v'}(0)\right)$. Therefore, if $\overline{p'}(L) \approx \overline{p'}(0)$, the fluid in the duct acts mainly like an ideal spring. This will be approximately valid for a duct of a relatively large volume. On the other hand, if $\overline{v'}(L) \approx \overline{v'}(0)$, the fluid in the duct mainly acts like an ideal mass. This condition can be more or less true in a relatively short and narrow duct.

Thus, a lumped parameter acoustic model of a ductwork may be formed by representing ducts by ideal masses and ideal springs. This model can be assembled by using the electric circuit analogy or a network method such as linear graphs (see Section 2.1), but the mass and spring coefficient of ducts must be defined consistently with the physics of the analogous system. The analogy (mechanical or electrical) is usually based on taking the acoustic pressure as the across variable (force or voltage) and the acoustic volume velocity, which is defined as $Q' = \overline{v'}S$, where S denotes the duct cross-sectional area, as the through variable (velocity or current) [6]. The latter choice follows, because acoustic mass flux $\rho_o Q'$ is a conserved quantity at duct junctions (recall that lumped elements neglect mean flow). Accordingly, to derive expressions for the analogous mass and spring coefficient of a duct of cross-sectional area S, we recast Equations (3.75) and (3.76) as:

$$\overline{p'}(0) - \overline{p'}(L) = \frac{\rho_o L}{S}(-\mathrm{i}\omega Q'(0)) \tag{3.77}$$

$$\frac{1}{-\mathrm{i}\omega}[Q'(0) - Q'(L)] = \frac{SL}{\rho_o c_o^2}\overline{p'}(0). \tag{3.78}$$

Hence, the analogous lumped mass (or capacitance) and spring coefficient (or 1/inductance) of a duct are given by $\rho_o L/S$ and $\rho_o c_o^2/SL$, respectively.

Since this analogy is derived from the plane-wave theory and is restricted further by the condition $k_o L << 1$, it can be expected to be adequately accurate only at sufficiently low frequencies in the plane-wave range. Applications of lumped acoustic elements are shown in Sections 5.3.5 and 5.4.8.

3.6 Hard-Walled Homogeneous Ducts with Non-Uniform Cross Section

In this section, we continue with the analysis of hard-walled homogeneous ducts, but now allow the cross-sectional area of the duct to vary along its axis. In this case, the continuity equation for the mean flow, Equation (3.18), implies that $S(x)\overline{v_o}(x) =$ constant. Hence, given the cross-sectional area taper function $S = S(x)$, the average mean-flow velocity can be determined from the proportionality $\overline{v_o}(x) \propto 1/S(x)$. The momentum equation for the mean flow reduces to

$$S\rho_o\overline{v_o}\frac{\mathrm{d}\overline{v_o}}{\mathrm{d}x} + \rho_o\frac{\mathrm{d}}{\mathrm{d}x}\left(S\beta(\overline{v_o})^2\right) = 0. \tag{3.79}$$

3.6.1 Wave Transfer Matrix

The linearized continuity and momentum equations, Equations (3.17) and (3.19), simplify to, respectively,

$$S\left(\frac{\partial \overline{\rho'}}{\partial t}+\overline{v_o}\frac{\partial \overline{\rho'}}{\partial x}\right)+\rho_o\frac{\partial}{\partial x}(S\overline{v'})+\overline{\rho'}\frac{\mathrm{d}}{\mathrm{d}x}(S\overline{v_o})=0 \tag{3.80}$$

$$S\rho_o\left(\frac{\partial \overline{v'}}{\partial t}+\overline{v_o}\frac{\partial \overline{v'}}{\partial x}+\overline{v'}\frac{\mathrm{d}\overline{v_o}}{\mathrm{d}x}\right)+S\overline{v_o}\frac{\mathrm{d}\overline{v_o}}{\mathrm{d}x}\overline{\rho'}+\frac{\partial}{\partial x}\left(S\beta(\overline{v_o})^2\overline{\rho'}\right)+S\frac{\partial \overline{p'}}{\partial x}=0, \tag{3.81}$$

where β is also a function of x. Using the relationship $p'=c_o^2\rho'$, Equation (3.80) can be written in the frequency domain as:

$$-\mathrm{i}k_o\overline{p'}+\bar{M}_o\frac{\partial \overline{p'}}{\partial x}+\rho_o c_o\frac{\partial \overline{v'}}{\partial x}+\rho_o c_o\overline{v'}\frac{\mathrm{d}}{\mathrm{d}x}(\ln S)=0, \tag{3.82}$$

where ln denotes natural logarithm and, as usual, $k_o=\omega/c_o$ and $\bar{M}_o=\overline{v_o}/c_o$. Upon using Equation (3.79), Equation (3.81) may be expressed in the frequency domain as

$$\left(\mathrm{i}k_o+\frac{\mathrm{d}\bar{M}}{\mathrm{d}x}\right)\overline{v'}+\bar{M}_o\frac{\partial \overline{v'}}{\partial x}+\frac{\alpha^2}{\rho_o c_o}\frac{\partial \overline{p'}}{\partial x}=0 \tag{3.83}$$

Equations (3.82) and (3.83) are first put into the state-space form of Equation (3.25):

$$\frac{\partial}{\partial x}\begin{bmatrix}\overline{p'}\\ \overline{v'}\end{bmatrix}=\frac{1}{\alpha^2-\bar{M}_o^2}\begin{bmatrix}-\mathrm{i}k_o\bar{M}_o & \mathrm{i}k_o\rho_o c_o+2\rho_o c_o\bar{M}_o\frac{\mathrm{d}}{\mathrm{d}x}(\ln S)\\ \mathrm{i}k_o\frac{\alpha^2}{\rho_o c_o} & -\mathrm{i}k_o\bar{M}_o-\left(\bar{M}_o^2+\alpha^2\right)\frac{\mathrm{d}}{\mathrm{d}x}(\ln S)\end{bmatrix}\begin{bmatrix}\overline{p'}\\ \overline{v'}\end{bmatrix}, \tag{3.84}$$

where, in view of the continuity equation for the mean flow

$$\frac{\mathrm{d}}{\mathrm{d}x}(\ln\bar{M}_o)=-\frac{\mathrm{d}}{\mathrm{d}x}(\ln S). \tag{3.85}$$

Equation (3.77) can be transformed to the pressure wave components by using Equation (3.29):

$$\frac{\partial}{\partial x}\begin{bmatrix}p^+\\ p^-\end{bmatrix}=\frac{1}{2}\frac{1}{\alpha^2-\bar{M}_o^2}\left[\mathrm{i}k_o\begin{bmatrix}-2\bar{M}_o+\alpha^2+1 & \alpha^2-1\\ -\alpha^2+1 & -2\bar{M}_o-\alpha^2-1\end{bmatrix}\right.$$
$$\left.+\begin{bmatrix}-\bar{M}_o^2+2\bar{M}_o-\alpha^2 & \bar{M}_o^2-2\bar{M}_o+\alpha^2\\ \bar{M}_o^2+2\bar{M}_o+\alpha^2 & -\bar{M}_o^2-2\bar{M}_o-\alpha^2\end{bmatrix}\frac{\mathrm{d}}{\mathrm{d}x}(\ln S)\right]\begin{bmatrix}p^+\\ p^-\end{bmatrix}. \tag{3.86}$$

This equation can be solved numerically by numerical segmentation (see Section 3.3.3). Here, we will present its analytical solutions for mean flow in the subsonic low Mach number range with $\alpha\approx 1$, or for sufficiently gradual cross-sectional area variations. For $\alpha\approx 1$, the matrizant in Equation (3.86) simplifies to:

$$\mathbf{H} = \begin{bmatrix} \frac{\mathrm{i}k_o}{1+\bar{M}_o} & 0 \\ 0 & -\frac{\mathrm{i}k_o}{1+\bar{M}_o} \end{bmatrix} + \frac{1}{2}\begin{bmatrix} -\frac{1-\bar{M}_o}{1+\bar{M}_o} & \frac{1-\bar{M}_o}{1+\bar{M}_o} \\ \frac{1+\bar{M}_o}{1+\bar{M}_o} & -\frac{1+\bar{M}_o}{1+\bar{M}_o} \end{bmatrix} \frac{\mathrm{d}}{\mathrm{d}x}(\ln S). \tag{3.87}$$

The linear differential equation associated with this matrizant, Equation (3.38), is exactly

$$\frac{\mathrm{d}^2 y}{\mathrm{d}x^2} + \left[\frac{\mathrm{i}2k_o}{1-\bar{M}_o^2} + \frac{\mathrm{d}}{\mathrm{d}x}(\ln S) - \frac{\mathrm{d}}{\mathrm{d}x}\left(\ln \frac{\mathrm{d}S}{\mathrm{d}x}\right)\right]\frac{\mathrm{d}y}{\mathrm{d}x} - \left(\frac{1}{2}\frac{\mathrm{d}}{\mathrm{d}x}(\ln S)\right)^2 y = 0. \tag{3.88}$$

Thus, from Equation (3.37), the wave-transfer matrix of the duct can be expressed as:

$$\mathbf{T} = \begin{bmatrix} 1 & 0 \\ 0 & \frac{1}{2}\frac{1-\bar{M}_o}{1+\bar{M}_o}\frac{\mathrm{d}}{\mathrm{d}x}(\ln S) \end{bmatrix}^{-1} \begin{bmatrix} y_2 & y_1 \\ \frac{\mathrm{d}y_2}{\mathrm{d}x} & \frac{\mathrm{d}y_1}{\mathrm{d}x} \end{bmatrix} \times \begin{bmatrix} 1 & 0 \\ 0 & \frac{1}{2}\frac{1-\bar{M}_o}{1+\bar{M}_o}\frac{\mathrm{d}}{\mathrm{d}x}(\ln S) \end{bmatrix} \mathrm{e}^{\int_0^x \left(\frac{\mathrm{i}k_o}{1+\bar{M}_o} - \frac{1}{2}\frac{1-\bar{M}_o}{1+\bar{M}_o}\frac{\mathrm{d}}{\mathrm{d}x}(\ln S)\right)\mathrm{d}x}. \tag{3.89}$$

For subsonic Mach numbers ($\bar{M}_o^2 << 1$), Equation (3.88) becomes independent of the mean-flow Mach number and, therefore, the Wronksian in Equation (3.89) can be determined from the wave-transfer matrix, $\mathbf{T}^o$, say, of the same duct without mean flow. Thus, the Wronksian in Equation (3.89) can be determined from

$$\begin{bmatrix} y_2 & y_1 \\ \frac{\mathrm{d}y_2}{\mathrm{d}x} & \frac{\mathrm{d}y_1}{\mathrm{d}x} \end{bmatrix} = \begin{bmatrix} 1 & 0 \\ 0 & \frac{1}{2}\frac{\mathrm{d}}{\mathrm{d}x}(\ln S) \end{bmatrix} \mathbf{T}^o \begin{bmatrix} 1 & 0 \\ 0 & \frac{1}{2}\frac{\mathrm{d}}{\mathrm{d}x}(\ln S) \end{bmatrix}^{-1} \mathrm{e}^{-\int_0^x \left(\mathrm{i}k_o - \frac{1}{2}\frac{\mathrm{d}}{\mathrm{d}x}(\ln S)\right)\mathrm{d}x}. \tag{3.90}$$

Substituting this back in Equation (3.89) gives the following expression for the wave transfer matrix of the duct with mean flow:

$$\mathbf{T} = \begin{bmatrix} 1 & 0 \\ 0 & \frac{1+\bar{M}_o}{1-\bar{M}_o} \end{bmatrix} \mathbf{T}^o \begin{bmatrix} 1 & 0 \\ 0 & \frac{1-\bar{M}_o}{1+\bar{M}_o} \end{bmatrix} \mathrm{e}^{\int_0^x \left(-\mathrm{i}k_o + \frac{\mathrm{d}}{\mathrm{d}x}(\ln S)\right)\frac{\bar{M}_o \mathrm{d}x}{1+\bar{M}_o}}. \tag{3.91}$$

This result reduces the problem finding the wave transfer matrix of a non-uniform duct with mean flow to that of the same duct without mean flow. The matrizant of the latter problem is given by Equation (3.87) with $\bar{M}_o = 0$; however, for getting exact analytical solutions, it is simpler to proceed with Equations (3.82) and (3.83), which simplify for $\bar{M}_o = 0$ to

$$-\mathrm{i}k_o\overline{p'} + \rho_o c_o \frac{\partial \overline{v'}}{\partial x} + \rho_o c_o \overline{v'} \frac{\mathrm{d}}{\mathrm{d}x}(\ln S) = 0 \tag{3.92}$$

$$-\mathrm{i}k_o \rho_o c_o \overline{v'} + \frac{\partial \overline{p'}}{\partial x} = 0 \tag{3.93}$$

respectively. Eliminating $\overline{v'}$ from these equations, we obtain the wave equation

$$\frac{1}{S}\frac{\partial}{\partial x}\left(S\frac{\partial \overline{p'}}{\partial x}\right) + k_o^2\overline{p'} = 0. \tag{3.94}$$

This is the classical Webster horn equation. There is extensive literature on its solution, motivated not only by the use of horns as radiating or receiving portions of acoustic and electro-acoustic devices, but also because of the vibration problems of non-uniform beams, shafts and strings. Exact analytical solutions are known for large number axial area taper functions [7]. A useful group of exact analytical solutions of the Webster horn equation may be found by applying the transformation $\overline{p'} = \sigma(S)\eta$ in Equation (3.94). If $\mathrm{d}\sigma/\mathrm{d}S = -\sigma/2S$ or, what is the same, $\sigma = 1/\sqrt{S}$, the horn equation reduces to the canonical form

$$\frac{\partial^2\eta}{\partial x^2} + \left(\mp m^2 + k_o^2\right)\eta = 0, \tag{3.95}$$

for area taper functions which obey the equation

$$\frac{\mathrm{d}^2}{\mathrm{d}x^2}\left(\sqrt{S}\right)\mp m^2\sqrt{S} = 0, \tag{3.96}$$

where m denotes a positive constant. If the negative sign is used, the group of taper functions include the hyperbolic, exponential and catenoidal cross-sectional area variations, if $m > 0$; and the conical area variations, if $m = 0$ [8]. If the positive sign is used, a sinusoidal taper function is obtained [9]. For these groups of non-uniform ducts, the wave transfer matrix can be derived by noting that the solution of Equation (3.95) may be expressed as

$$\eta = A\mathrm{e}^{\mathrm{i}\sqrt{k_o^2\mp m^2}x} + B\mathrm{e}^{-\mathrm{i}\sqrt{k_o^2\mp m^2}x}, \tag{3.97}$$

where A and B are integration constants. These constants are determined by requiring that this general solution is compatible with the usual decompositions $\overline{p'} = p^+ + p^-$ and $\rho_o c_o \overline{v'} = p^+ - p^-$, where $\overline{v'}$ is determined from Equation (3.93). Upon carrying out these calculations, the wave-transfer matrix of the duct with no mean flow transpires as:

$$\mathbf{T}^o = \mathbf{\Psi}\begin{bmatrix} \mathrm{e}^{\mathrm{i}\sqrt{k_o^2\mp m^2}L} & 0 \\ 0 & \mathrm{e}^{-\mathrm{i}\sqrt{k_o^2\mp m^2}L} \end{bmatrix}\mathbf{\Psi}^{-1}, \tag{3.98}$$

where

$$\mathbf{\Psi} = \frac{1}{2}\begin{bmatrix} \mathrm{i}\left(k_o + \sqrt{k_o^2\mp m^2}\right) - \frac{1}{2}\frac{\mathrm{d}}{\mathrm{d}x}(\ln S) & \mathrm{i}\left(k_o - \sqrt{k_o^2\mp m^2}\right) - \frac{1}{2}\frac{\mathrm{d}}{\mathrm{d}x}(\ln S) \\ \mathrm{i}\left(k_o - \sqrt{k_o^2\mp m^2}\right) + \frac{1}{2}\frac{\mathrm{d}}{\mathrm{d}x}(\ln S) & \mathrm{i}\left(k_o + \sqrt{k_o^2\mp m^2}\right) + \frac{1}{2}\frac{\mathrm{d}}{\mathrm{d}x}(\ln S) \end{bmatrix} \tag{3.99}$$

The wave-transfer matrices of this family of non-uniform ducts carrying a mean flow may now be determined from Equation (3.91).

3.6.2 High Frequency Approximation

Since $\alpha \approx 1$ for practical sheared mean-flow velocity profiles, the off-diagonal terms of the matrizant in Equation (3.86) are weakly dependent on frequency and can be neglected as small compared to the diagonal terms, if frequency is high enough. Numerical simulations indicate that the threshold frequency for this approximation is weakly sensitive to the mean-flow Mach number in the low subsonic range [10]. The criterion derived by neglecting the mean flow shows that, for the area taper functions obeying Equation (3.89) with $m \geq 0$, the off-diagonal terms of Equation (3.86) may be neglected if

$$k_o^2 >> 2m^2 + \frac{1}{2}\left(\frac{\mathrm{d}}{\mathrm{d}x}(\ln S)\right)^2. \tag{3.100}$$

For example, for a conical duct of length L, small-end diameter d_1 and large-end diameter d_2, this condition can be expressed approximately as $k_o L >> (d_2/d_1 - 1)$. For a slant angle of 5°, this yields the condition $k_o d_1 >> \tan\alpha \approx 0.0875$. Under this approximation, solution of Equation (3.86) may be expressed as

$$p^+(x) = p^+(0)\mathrm{e}^{\frac{\mathrm{i}k_o}{2}\int_0^x \frac{-2\bar{M}_o+\alpha^2+1}{\alpha^2-\bar{M}_o^2}\mathrm{d}\xi}\,\mathrm{e}^{-\frac{1}{2}\int_0^x \frac{\bar{M}_o^2-2\bar{M}_o+\alpha^2}{\alpha^2-\bar{M}_o^2}\frac{\mathrm{d}}{\mathrm{d}\xi}(\ln S)\mathrm{d}\xi} \tag{3.101}$$

$$p^+(x) = p^+(0)\mathrm{e}^{\frac{\mathrm{i}k_o}{2}\int_0^x \frac{-2\bar{M}_o-\alpha^2-1}{\alpha^2-\bar{M}_o^2}\mathrm{d}\xi}\,\mathrm{e}^{-\frac{1}{2}\int_0^x \frac{\bar{M}_o^2+2\bar{M}_o+\alpha^2}{\alpha^2-\bar{M}_o^2}\frac{\mathrm{d}}{\mathrm{d}\xi}(\ln S)\mathrm{d}\xi}. \tag{3.102}$$

For $\alpha = 1$, the inverse Fourier transforms the foregoing solutions are, respectively,

$$p^+(t,x) = p^+\left(t - \frac{1}{c_o}\int_0^x \frac{\mathrm{d}\xi}{1+\bar{M}_o}\right)\mathrm{e}^{-\frac{1}{2}\int_0^x \frac{1-\bar{M}_o}{1+\bar{M}_o}\frac{\mathrm{d}}{\mathrm{d}\xi}(\ln S)\mathrm{d}\xi} \tag{3.103}$$

$$p^-(t,x) = p^-\left(t + \frac{1}{c_o}\int_0^x \frac{\mathrm{d}\xi}{1-\bar{M}_o}\right)\mathrm{e}^{-\frac{1}{2}\int_0^x \frac{1+\bar{M}_o}{1-\bar{M}_o}\frac{\mathrm{d}}{\mathrm{d}x}(\ln S)\mathrm{d}\xi}. \tag{3.104}$$

In this case, therefore, the $p^+(t,x)$ wave travels in the $+x$ direction with local phase velocity of $c_o(1+\bar{M}_o)$, and the $p^-(t,x)$ wave travels in the $-x$ direction with local phase velocity of $c_o(1-\bar{M}_o)$. These waves are attenuated or amplified along the direction of travel. This can be seen clearly for $|\bar{M}_o| << 1$, as then the exponential factors reduce to $\sqrt{S(0)/S(x)}$. Consequently, the $+x$ wave is attenuated (amplified) if the duct is converging (diverging), and the backward wave is attenuated (amplified) if the duct is diverging (converging). A device which utilizes this feature of wave propagation in non-uniform ducts is the acoustic horn, which has been used for transmitting human voice over long distances, or listening to sound sources at long distances, since ancient times. There is an extensive literature on one-dimensional sound propagation in non-uniform ducts, as they are also used widely as nozzles, diffusers and connections between ducts of different sizes.

3.7 Ducts Packed with Porous Material

Engineering theory of sound propagation in a hard-walled duct filled with sound absorbent material is based on the hypothesis that the porous material may be considered as a homogeneous fluid with equivalent bulk acoustic properties. Appendix B describes the basics of this theory and reviews the major practical models developed for the determination of the equivalent density, ρ_e, and speed of sound, c_e, of rigid frame fibrous materials. There may be some degree of mean fluid flow within the porous packing, but the effect of this is included in the equivalent fluid properties. Then, Equations (3.80) and (3.81) simplify to

$$S\frac{\partial \overline{p'}}{\partial t} + \rho_e \frac{\partial}{\partial x}\left(S\overline{v'}\right) = 0 \tag{3.105}$$

$$\rho_e \frac{\partial \overline{v'}}{\partial t} + \frac{\partial \overline{p'}}{\partial x} = 0, \tag{3.106}$$

with acoustic pressure and density being related in the frequency domain as $p' = c_e^2 \rho'$. Therefore, the wave transfer matrices derived in Sections 3.5 and 3.6 can be used directly in this case with $\overline{v_o} = 0$, but with the pressure wave components defined as $p' = p^+ + p^-$, $v' = (p^+ - p^-)/\rho_e c_e$ and with substitutions $c_o \to c_e$, $k_o \to k_e$ and $\rho_o c_o \to \rho_e c_e$.

However, in most applications, the absorbent remains in contact with the fluid in the duct and it is necessary to take into account the characteristic impedance mismatch at the absorbent–fluid interface. In general, at a planar interface of two media of different characteristic impedance, acoustic pressure and particle velocity must be continuous and these conditions are expressed as $p_1^+ + p_1^- = p_2^+ + p_2^-$ and $(p_1^+ - p_1^-)/\rho_1 c_1 = (p_1^+ - p_2^-)/\rho_2 c_2$, where the subscripts "1" and "2" denote the media. Combining these equations, we obtain the following wave transfer equation across the interface:

$$\begin{bmatrix} p_1^+ \\ p_1^- \end{bmatrix} = \frac{1}{2\rho_2 c_2}\begin{bmatrix} \rho_2 c_2 + \rho_1 c_1 & \rho_2 c_2 - \rho_1 c_1 \\ \rho_2 c_2 - \rho_1 c_1 & \rho_2 c_2 + \rho_1 c_1 \end{bmatrix}\begin{bmatrix} p_2^+ \\ p_2^- \end{bmatrix}. \tag{3.107}$$

Thus, if subscripts "1" and "2" denote, respectively, the absorbent packing and the fluid in the duct, the fluid-to-fluid transfer matrix of a packed duct can be obtained by pre-multiplying the absorbent-to-absorbent transfer matrix by $\mathbf{T}_{12}$ and post-multiplying by $\mathbf{T}_{12}^{-1}$, where $\mathbf{T}_{12}$ denotes the wave transfer matrix in Equation (3.107). It is instructive, however, to derive the fluid-to-fluid wave transfer matrix from first principles.

Consider a duct which contains a fluid of characteristic impedance $\rho_o c_o$, except in an axial segment of length L, which is packed with a porous material of characteristic impedance $\rho_e c_e$. Transmission of sound waves within the packing is governed by Equations (3.105) and (3.106). Transmission of sound waves across the packing in the duct may be analyzed by recasting Equations (3.105) and (3.106) in the forms:

$$-\mathrm{i}k_o H\overline{p'} + \frac{\rho_o c_o}{S}\frac{\partial}{\partial x}(S\overline{v'}) = 0 \tag{3.108}$$

$$-\mathrm{i}k_o \rho_o c_o \Sigma \overline{v'} + \frac{\partial \overline{p'}}{\partial x} = 0, \tag{3.109}$$

where $H = \rho_o c_o^2/\rho_e c_e^2$ and $\Sigma = \rho_e/\rho_o$. The foregoing equations may be transformed now as usual to the pressure wave components by using Equation (3.24):

$$\frac{\partial}{\partial x}\begin{bmatrix} p^+ \\ p^- \end{bmatrix} = \frac{\mathrm{i}k_o}{2}\begin{bmatrix} H+\Sigma-\frac{\mathrm{d}}{\mathrm{d}x}(\ln S) & H-\Sigma+\frac{\mathrm{d}}{\mathrm{d}x}(\ln S) \\ -H+\Sigma+\frac{\mathrm{d}}{\mathrm{d}x}(\ln S) & -H-\Sigma-\frac{\mathrm{d}}{\mathrm{d}x}(\ln S) \end{bmatrix}\begin{bmatrix} p^+ \\ p^- \end{bmatrix}. \tag{3.110}$$

For the solution of this equation, we observe that, on elimination of the particle velocity, Equations (3.108) and (3.109) yield the Webster horn equation, Equation (3.94), for the acoustic pressure, with the wavenumber k_o^2 replaced by $k_o^2\Sigma H = k_e^2$. Then, since in this case $\mathbf{T} = \mathbf{T}^o$, the solution of Equation (3.110) in the wave transfer matrix format is given by Equation (3.98), where the axial wavenumbers are now $\mp\sqrt{k_o^2\Sigma H \mp m^2}$ and

$$\mathbf{\Psi} = \frac{1}{2}\begin{bmatrix} \mathrm{i}\left(k_o\Sigma + \sqrt{k_o^2\Sigma H \mp m^2}\right) - \frac{1}{2}\frac{\mathrm{d}}{\mathrm{d}x}\ln S & \mathrm{i}\left(k_o\Sigma - \sqrt{k_o^2\Sigma H \mp m^2}\right) - \frac{1}{2}\frac{\mathrm{d}}{\mathrm{d}x}\ln S \\ \mathrm{i}\left(k_o\Sigma - \sqrt{k_o^2\Sigma H \mp m^2}\right) + \frac{1}{2}\frac{\mathrm{d}}{\mathrm{d}x}\ln S & \mathrm{i}\left(k_o\Sigma + \sqrt{k_o^2\Sigma H \mp m^2}\right) + \frac{1}{2}\frac{\mathrm{d}}{\mathrm{d}x}\ln S \end{bmatrix} \tag{3.111}$$

If the duct is uniform, the wave transfer equation may be expressed explicitly in the form

$$\begin{bmatrix} p^+(x) \\ p^-(x) \end{bmatrix} = \begin{bmatrix} \sqrt{\Sigma}+\sqrt{H} & \sqrt{\Sigma}-\sqrt{H} \\ \sqrt{\Sigma}-\sqrt{H} & \sqrt{\Sigma}+\sqrt{H} \end{bmatrix}\begin{bmatrix} \mathrm{e}^{\mathrm{i}k_o L\sqrt{H\Sigma}} & 0 \\ 0 & \mathrm{e}^{-\mathrm{i}k_o L\sqrt{H\Sigma}} \end{bmatrix} \times \begin{bmatrix} \sqrt{\Sigma}+\sqrt{H} & \sqrt{\Sigma}-\sqrt{H} \\ \sqrt{\Sigma}-\sqrt{H} & \sqrt{\Sigma}+\sqrt{H} \end{bmatrix}^{-1}\begin{bmatrix} p^+(0) \\ p^-(0) \end{bmatrix} \tag{3.112}$$

When the matrix multiplications are carried out, the wave transfer matrix in the foregoing equation may be shown to expand as

$$\mathbf{T} = \begin{bmatrix} \cos(k_e L) + \frac{\mathrm{i}}{2}\left(\frac{\rho_o c_o}{\rho_e c_e} + \frac{\rho_e c_e}{\rho_o c_o}\right)\sin(k_e L) & \frac{\mathrm{i}}{2}\left(\frac{\rho_o c_o}{\rho_e c_e} - \frac{\rho_e c_e}{\rho_o c_o}\right)\sin(k_e L) \\ \frac{\mathrm{i}}{2}\left(\frac{\rho_e c_e}{\rho_o c_o} - \frac{\rho_o c_o}{\rho_e c_e}\right)\sin(k_e L) & \cos(k_e L) - \frac{\mathrm{i}}{2}\left(\frac{\rho_o c_o}{\rho_e c_e} + \frac{\rho_e c_e}{\rho_o c_o}\right)\sin(k_e L) \end{bmatrix} \tag{3.113}$$

where $k_e = \omega/c_e$ and L denotes the length of the duct.

3.8 Acoustic Boundary Conditions on Duct Walls

Acoustic impedance of duct walls is defined, in frequency domain, by the ratio of the local acoustic pressure to the normal component of the acoustic particle velocity (u' in the present one-dimensional theory). Therefore, hard-walled duct elements considered in previous sections can be categorized also as ducts with infinite impedance walls.

Ducts having some kind of non-rigid or permeable walls (or, briefly, soft walls) are widely used for noise control purposes. This section describes, in a prelude to the acoustic modeling of ducts with soft walls, some wall impedance models that can be used with the one-dimensional theory.

3.8.1 Impermeable Walls

If the duct walls are impermeable, then $u_o = 0$ and the mean flow is usually referred to as the grazing mean flow. The finite impedance condition at the duct boundary is applied by introducing an appropriate model which relates the local values of u' and $\bar{p}'$. For this purpose, it is convenient to define the parameter

$$\dot{\mu}' = -\frac{1}{S}\oint_{\ell} \rho_o u' J \mathrm{d}\ell. \tag{3.114}$$

This has physical interpretation as the rate of unsteady fluid mass injected at the duct boundary into per volume of the duct, which explains the use of the over-dot. However, $\dot{\mu}'$ may also be interpreted as acoustic momentum density injected by the duct boundary per unit area of the duct cross section.

Several options exist for the application of the acoustic conditions at a finite impedance boundary with a grazing mean flow.

3.8.1.1 No-Slip Model

The natural option for the duct boundary is the physical surface of the duct walls. Then $\varphi_{SF} = 1$, since the duct walls are assumed to be smooth, and the acoustic boundary condition to be imposed is $Z_W = p'(\ell^o)/u'$, where Z_W denotes the wall impedance proper and the superscript "o" refers to the surface of the duct walls and the dependence of the parameters on ω and x (in the frequency domain) is suppressed for simplicity of notation. This boundary condition can be applied only approximately as $Z_W \approx \overline{p'}/u'$, since $\overline{p'}$, not $p'(\ell^o)$, is a dependent variable of the one-dimensional theory and $p'(\ell^o)$ cannot be extracted from $\overline{p'}$. This approximation is equivalent to the assumption that $p'(\ell^o) \approx \overline{p'}$, which is in line with the restriction of the theory to acoustic fluctuations that do not deviate substantially from their mean values over a duct section. Hence, the no-slip model can be applied by taking

$$\dot{\mu}' = -\frac{A_W}{c_o}\overline{p'}, \tag{3.115}$$

where

$$A_W = \frac{1}{S}\oint \frac{J d\ell}{\zeta_W} \tag{3.116}$$

and $\zeta_W = Z_W/\rho_o c_o$ denotes the normalized wall impedance, and A_W is called the wall admittance parameter. This model is applicable with both uniform and non-uniform ducts. The models considered next are strictly for uniform ducts.

3.8.1.2 Full-Slip Model

In this case, we consider a uniform duct and assume that the mean flow is uniform with velocity $\overline{v_o}$, which implies $\varphi_{SF} = 0$ and $\alpha = 1$, and that it is separated from the physical duct surface by an infinitely thin vortex sheet across which the acoustic pressure and particle displacement is continuous. The duct boundary is taken just at the core side of the vortex sheet. Then the acoustic boundary condition can be expressed as $Z_W = p'(\ell^{\times})/(-\mathrm{i}\omega\eta')$, where the superscript "×" refers to the core side of the vortex sheet and η' denotes the normal displacement of the vortex sheet, which is given by $u' = (-\mathrm{i}\omega + \overline{v_o}\partial/\partial x)\eta'$, since the mean flow is assumed to be uniform in the core. Then, the assumption $\overline{p'} \approx p'(\ell^{\times})$ allows embedding of the acoustic boundary condition $Z_W = p'(\ell^{\times})/(-\mathrm{i}\omega\eta')$ into the linearized equations of the one-dimensional theory. Hence, upon eliminating η' from the foregoing relations, Equation (3.114) gives $\dot{\mu}'$ for the full-slip condition as

$$\dot{\mu}' = -\frac{1}{c_o}\left(1 - \frac{\bar{M}_o}{\mathrm{i}k_o}\frac{\partial}{\partial x}\right) A_W\, \overline{p'}. \tag{3.117}$$

The generalization of this full-slip model in three dimensions is widely used in conjunction with the convected wave equation and is commonly known as Ingard–Myers boundary condition (see Section 6.3.3).

3.8.1.3 Partial-Slip Model

The no-slip model neglects the effects of fluid viscosity and thermal conductivity, which are mainly confined to a thin mean boundary layer for a fully developed turbulent mean flow. On the other hand, the full-slip model assumes that the mean boundary layer is infinitely thin. Therefore, an attractive alternative to these models is the deferral of the duct boundary to the border of the viscous (laminar) sublayer [11]. This sublayer is thin but of finite thickness and viscous shear stress and, hence, refraction, is concentrated largely in this sublayer. Also, in the low Helmholtz number range turbulence scattering is negligible [12]. Then, a uniform duct being assumed, the acoustic boundary condition to be used becomes $Z^+ = p'(\ell^+)/u'$, where the superscript "+" denotes the border of the viscous sublayer. In order to apply this model, again we necessarily assume that $p'(\ell^+) \approx \overline{p'}$. This is, however, a slightly better approximation than $p'(\ell^o) \approx \overline{p'}$ or $\overline{p'} \approx p'(\ell^{\times})$, since the viscous sublayer border is closer to the layer corresponding to $\overline{p'}$.

The viscous sublayer border impedance can thus be defined approximately as $Z^+ = \overline{p'}/u'$, but the determination of Z^+ is not an easy problem. There are several solutions in the literature (see Section 6.3.4), but these are mostly proposed as alternatives to the Ingard-Myers boundary condition when the mean flow is assumed to be uniform. In the present one-dimensional theory, we do not assume uniform mean flow. It is, therefore, necessary to modify the existing models for the border of the viscous sublayer and the model proposed by Starobinski is convenient for this purpose [13]. The main assumptions made in this model are: (i) $\delta_L << \lambda$, where λ denotes the wavelength and δ_L denotes the thickness of the viscous sublayer, (ii) $\delta_L << a$, where a denotes a characteristic dimension of the duct cross section, (iii) shear-stress fluctuations are negligible outside the viscous sublayer, and (iv) $p'(\ell^o) = p'(\ell^+)$, that is, acoustic pressure remains constant across the viscous sublayer. Hence, assuming Z_W is uniform axially and neglecting the effect of thermal diffusion, it can be shown that [13]:

$$u' = \frac{1}{Z_W}\left(1 - \frac{\overline{v_o}}{\mathrm{i}\omega}\left(\frac{w_o}{\overline{v_o}} - \beta_v^+\right)\frac{\partial}{\partial x} + \frac{\beta_v^+ \overline{v_o} w_o}{\omega^2}\frac{\partial^2}{\partial x^2}\right)\overline{p'}, \tag{3.118}$$

where

$$\beta_v^+ = \frac{1}{\overline{v_o}}\int_0^{\delta_L} \frac{\mathrm{d}v_o}{\mathrm{d}\xi} \mathrm{e}^{(-1+\mathrm{i})\frac{\xi}{\delta_v}}\, \mathrm{d}\xi. \tag{3.119}$$

Here, $\delta_v = \sqrt{2\mu_o/\omega\rho_o}$ denotes the viscous acoustic boundary layer thickness [6], μ_o denotes the shear viscosity and ξ denotes the coordinate normal to the surface of the duct walls. In the viscous sublayer ($0 \le \xi \le \delta_L$), the mean-flow velocity varies linearly with ξ as $v_o = \rho_o v_*^2 \xi/\mu_o$, where v_* denotes the friction velocity. The viscous sublayer thickness and the mean velocity at the border of viscous sublayer (w_o of the one-dimensional theory) are calculated from the formulae $\delta_L = 11\mu_o/\rho_o v_*$ and $w_o = 11 v_*$ [14]. The friction velocity is determined from the classical empirical equation $v_o^+ = 2.5 \ln\left(\mathrm{Re}/v_o^+\right) + 5.1$, where $v_o^+ = \hat{v}_o/v_*$, $\mathrm{Re} = \rho_o \hat{v}_o d/\mu_o$ and $\hat{v}_o$ denotes the maximum mean-flow velocity at distance d from the duct walls. Using Equation (3.118) in Equation (3.114), the latter can be expressed as:

$$\dot{\mu}' = -\frac{A_W}{c_o}\left(1 - \frac{\bar{M}_o}{\mathrm{i}k_o}\left(1 - \beta_v^+ - \varphi_{SF}\right)\frac{\partial}{\partial x} + \frac{\beta_v^+ \bar{M}_o^2 (1 - \varphi_{SF})}{k_o^2}\frac{\partial^2}{\partial x^2}\right)\overline{p'}, \tag{3.120}$$

where

$$\beta_v^+ = \frac{1+\mathrm{i}}{2}(1 - \varphi_{\mathrm{SF}})\frac{\delta_v}{\delta_L}\left(1 - \mathrm{e}^{(-1+\mathrm{i})\frac{\delta_L}{\delta_v}}\right). \tag{3.121}$$

Since the duct boundary is deferred to the border of the viscous sublayer, the working cross-sectional area S becomes slightly smaller than the actual duct cross-sectional area, however, the difference is very small and we can use the nominal cross-sectional dimensions in calculations without discernible loss of accuracy.

3.8.1.4 Rough-Wall Model

Duct walls are considered to be smooth if $\rho_o v_* \varepsilon/\mu_o < 5$, where ε denotes the mean roughness of the walls. The viscous sublayer model is no longer applicable for a rough wall, because the sublayer becomes embedded in the roughness structure. A rough wall can be treated as a smooth one having its boundary at the equivalent roughness

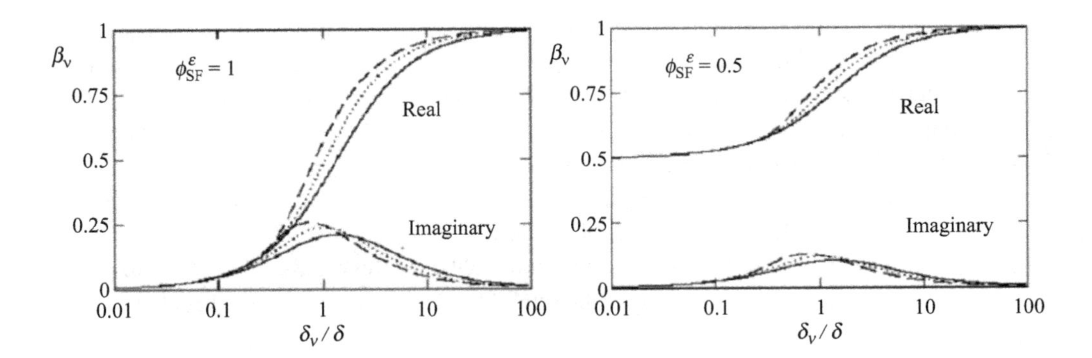

Figure 3.4 Variation of the ratio of a complex weighted average of the mean-flow velocity in the boundary layer to $\overline{v_o}$ as function of δ_v/δ for $\varphi_{\rm SF}^{\varepsilon} = 1$ and $\varphi_{\rm SF}^{\varepsilon} = 0.5$. Mean boundary layer velocity profiles: solid: exponential ($v_o/\overline{v_o} = 1 - \varphi_{\rm SF}^{\varepsilon} {\rm e}^{-\xi}$, $\xi \le 4$); dash: linear ($v_o/\overline{v_o} = 1 - \varphi_{\rm SF}^{\varepsilon}(1-\xi)$, $\xi \le 1$); dot: quadratic ($v_o/\overline{v_o} = 1 - \varphi_{\rm SF}^{\varepsilon}(1-\xi/2)^2$, $\xi \le 2$).

thickness with a slip flow [13, 15], which is largely a heuristic quantity that depends on the roughness elements and the mean flow. The core flow is assumed to be uniform and the mean boundary layer thickness, δ, say, small enough that the core flow can be assumed to fill the duct cross section approximately. This model [15] can be applied with the present one-dimensional theory by taking

$$\dot{\mu}' = -\frac{A_W}{c_o}\left(1 - \frac{\bar{M}_o}{{\rm i}k_o}(1-\beta_v)\frac{\partial}{\partial x}\right)\overline{p'}. \tag{3.122}$$

Here,

$$\beta_v = 1 - \varphi_{SF}^{\varepsilon} + \frac{1}{\overline{v}_o}\int_0^{\delta} \frac{{\rm d}v_o}{{\rm d}\xi}{\rm e}^{(-1+{\rm i})\frac{\xi}{\delta_v}}\,{\rm d}\xi \tag{3.123}$$

where the mean-flow velocity profile in the mean boundary layer may not necessarily be linear. The slip-flow parameter $\varphi_{\rm SF}^{\varepsilon}$ is determined by the slip-flow velocity at the border of the equivalent roughness thickness. It has been necessary to distinguish this by using the superscript "ε," because the core mean flow is assumed to be uniform and, therefore, the linearized one-dimensional equations are to be applied with $\varphi_{\rm SF} = 0$ and $\alpha = 1$.

The variation of β_v as function of δ_v/δ and $\varphi_{\rm SF}^{\varepsilon}$ are shown in Figure 3.4 for three typical velocity profiles. In general, if $\delta_v/\delta << 1, \beta_v \approx 1 - \varphi_{\rm SF}^{\varepsilon}$; else if $\delta_v/\delta >> 1, \beta_v \approx 1$. Note that, if $\delta_v/\delta << 1$ and $\varphi_{SF}^{\varepsilon} = 0$, Equation (3.122) reduces to Equation (3.115).

3.8.1.5 Unified Boundary Condition

It is convenient to express $\dot{\mu}'$ as

$$\dot{\mu}' = -\frac{A_W}{c_o}\left(1 - \frac{\psi\bar{M}_o}{{\rm i}k_o}\frac{\partial}{\partial x}\right)\bar{p}', \tag{3.124}$$

which represents Equation (3.108) for $\psi = 0$, Equation (3.117) for $\psi = 1$, Equation (3.120) for $\psi = 1 - \beta_v^+ - \varphi_{SF}$ and Equation (3.122) for $\psi = 1 - \beta_v$ in a unified form. It should be noted that, this unified expression neglects the third term in the brackets on the right-hand side of Equation (3.120), which is adequately accurate when $\bar{M}_o^2 << 1$. All models reduce to the no-slip model when mean flow is neglected.

If ζ_W is finite only over a part P of perimeter ℓ ($P \supseteq \ell$), then $A_W = P/\zeta_W S$, where $P = P(x)$ in general. So, for A_W to be treated as constant axially, both ζ_W and P should be axially uniform. We refer to this condition briefly as the uniform wall impedance (or uniform liner). When $P \supset \ell$, the segments $\ell - P$ are assumed to be hard wall ($\zeta_W = \infty$). The value of φ_{SF} is strictly relevant only where the duct walls are lined, as $A_W = 0$ over hard walls. It should be noted that, Equation (3.124) assumes a uniform wall impedance, except for the no-slip model. If the wall impedance is non-uniform, the full-slip model should be applied as in Equation (3.117). The partial-slip and rough-wall models are strictly valid for uniform wall impedance.

3.8.2 Permeable Walls

In the case of a duct with permeable walls, it is necessary to take into consideration the coupling of the in-duct acoustic field with the acoustic field in the exterior of the duct walls and the mean flow through the permeable elements (for example, perforations). It is generally adequately accurate to model the unsteady fluid motion in permeable elements as incompressible (constant density) flow under fluctuating pressure differential $p'(\ell_{\text{int}}) - p'(\ell_{\text{ext}})$, where ℓ_{int} and ℓ_{ext} denote, respectively, the inner and the exterior physical boundaries of the duct. But, since $p'(\ell_{\text{int}})$ is not a dependent variable in the one-dimensional theory, we resort again to the approximation $p'(\ell_{\text{int}}) \approx \overline{p'}$, where p' denotes the acoustic pressure in the duct. Then, the equation of motion of the fluid in permeable elements normal to the duct walls can be expressed, in frequency domain, as $\overline{p'} - p'(\ell_{\text{ext}}) = Z_W^o u'$, where Z_W^o denotes the impedance of the wall. Hence, the inner surface impedance of the walls, $Z_W = \overline{p'}/u'$, can be expressed as $Z_W = Z_W^o + Z_{\text{ext}}$, where $Z_{\text{ext}} = p'(\ell_{\text{ext}})/u'$ denotes the exterior surface impedance. This model fits into the one-dimensional theory with

$$\dot{\mu}' = -\frac{\rho_o}{S}\left(\oint_{\ell} \frac{J d\ell}{Z_W}\right)\overline{p'}, \tag{3.125}$$

which is encompassed in Equation (3.124) with $\psi = 0$.

Mean flow through the walls has no effect on the one-dimensional acoustic continuity and momentum equations, Equations (3.17) and (3.19), because the peripheral surface integrals involving ρ' can be neglected in view of the assumption that the fluid motion in permeable elements is of constant density. But, mean through flow does effect Z_W^o (see Appendix C). The rate of mean mass through flow, $\dot{\mu}_o$, can be calculated from

$$\dot{\mu}_o = -\frac{\rho_o}{S}\oint\limits_{\ell} u_o J \mathrm{d}\ell = \frac{1}{S}\frac{\mathrm{d}}{\mathrm{d}x}(S\rho_o\overline{v_o}), \tag{3.126}$$

where the second equality follows from the continuity equation for the mean flow, Equation (3.20).

If both mean through flow and grazing mean flow are present, the overall wall impedance is determined by combining the through flow and grazing flow impedances in series.

3.9 Homogeneous Ducts with Impermeable Finite Impedance Walls

In this section, we continue with the analysis of homogeneous ducts, but this time allow for impermeable finite impedance walls. The formulation proceeds similarly, except that now $u' \neq 0$ (but still $u_o = 0$ since the walls are impermeable). Consequently, the continuity and momentum equations for the mean flow are the same as in Section 3.5, but acoustic continuity and momentum equations are modified to:

$$S\left(\frac{\partial\overline{\rho'}}{\partial t} + \overline{v_o}\frac{\partial\overline{\rho'}}{\partial x}\right) + \rho_o\frac{\partial}{\partial x}(S\overline{v'}) + \overline{\rho'}\frac{\mathrm{d}}{\mathrm{d}x}(S\overline{v_o}) = -\oint\limits_{\ell}\rho_o u' J \mathrm{d}\ell \tag{3.127}$$

$$S\rho_o\left(\frac{\partial\overline{v'}}{\partial t} + \overline{v_o}\frac{\partial\overline{v'}}{\partial x} + \overline{v'}\frac{\mathrm{d}\overline{v_o}}{\mathrm{d}x}\right) + S\overline{v_o}\frac{\mathrm{d}\overline{v_o}}{\mathrm{d}x}\overline{\rho'} + \frac{\partial}{\partial x}\left(S\beta(\overline{v_o})^2\overline{\rho'}\right) + S\frac{\partial\overline{p'}}{\partial x} = \overline{v_o}\oint\limits_{\ell}\rho_o\varphi_{\mathrm{SF}}u' J \mathrm{d}\ell, \tag{3.128}$$

which differ from Equations (3.80) and (3.81) only by their right-hand sides and are still closed by the relationship $p' = c_o^2\rho'$. Hence, Equation (3.127) may be written in frequency domain as

$$-\mathrm{i}k_o\overline{p'} + \bar{M}_o\frac{\partial\overline{p'}}{\partial x} + \rho_o c_o\frac{\partial\overline{v'}}{\partial x} + \rho_o c_o\overline{v'}\frac{\mathrm{d}}{\mathrm{d}x}(\ln S) = c_o\dot{\mu}', \tag{3.129}$$

where $k_o = \omega/c_o$ and $\bar{M}_o = \overline{v_o}/c_o$. Equation (3.128) is simplified by using Equations (3.78) and (3.79) and expressed in frequency domain as

$$\rho_o c_o\left(-\mathrm{i}k_o + \frac{\mathrm{d}\bar{M}_o}{\mathrm{d}x} + \bar{M}_o\frac{\partial}{\partial x}\right)\overline{v'} + \alpha^2\frac{\partial\overline{p'}}{\partial x} = -c_o\bar{M}_o\varphi_{\mathrm{SF}}\dot{\mu}', \tag{3.130}$$

where α is given by Equation (3.66). When Equation (3.124) is substituted, these equations can be expressed as

$$\left(-\mathrm{i}k_o + A_W + \bar{M}_o\left(1 - \frac{A_W\psi}{\mathrm{i}k_o}\right)\frac{\partial}{\partial x}\right)\overline{p'} + \rho_o c_o\left(\frac{\partial}{\partial x} + \frac{\mathrm{d}}{\mathrm{d}x}(\ln S)\right)\overline{v'} = 0 \tag{3.131}$$

$$\rho_o c_o\left(-\mathrm{i}k_o+\frac{\mathrm{d}\bar{M}_o}{\mathrm{d}x}+\bar{M}_o\frac{\partial}{\partial x}\right)\overline{v'}+\left(-\varphi_{SF}\bar{M}_o A_W+\left(\alpha^2+\frac{A_W\varphi_{\mathrm{SF}}\psi\bar{M}_o^2}{\mathrm{i}k_o}\right)\frac{\partial}{\partial x}\right)\overline{p'}=0, \tag{3.132}$$

where Equation (3.85) is still valid for the mean gradient terms and ln denotes natural logarithm. The foregoing equations can be expressed in state-space form of Equation (3.25) as:

$$\frac{\partial}{\partial x}\begin{bmatrix}\overline{p'}\\ \overline{v'}\end{bmatrix}=\frac{\mathrm{i}k_o}{\Delta}\begin{bmatrix}-\bar{M}_o+(1+\varphi_{SF})\bar{M}_o\frac{A_W}{\mathrm{i}k_o} & \rho_o c_o\left(1+\frac{2\bar{M}_o}{\mathrm{i}k_o}\frac{\mathrm{d}}{\mathrm{d}x}(\ln S)\right)\\ \dfrac{\left(1-\frac{A_W}{\mathrm{i}k_o}\right)\alpha^2-\varphi_{\mathrm{SF}}(1-\psi)\bar{M}_o^2\frac{A_W}{\mathrm{i}k_o}}{\rho_o c_o} & \begin{matrix}-\bar{M}_o\left(1-\psi\frac{A_W}{\mathrm{i}k_o}\right)-\frac{1}{\mathrm{i}k_o}\times\\ \left(\alpha^2+\bar{M}_o^2-(1-\varphi_{\mathrm{SF}})\psi\bar{M}_o^2\frac{A_W}{\mathrm{i}k_o}\right)\frac{\mathrm{d}}{\mathrm{d}x}(\ln S)\end{matrix}\end{bmatrix}\begin{bmatrix}\overline{p'}\\ \overline{v'}\end{bmatrix}, \tag{3.133}$$

where

$$\Delta=\alpha^2-\bar{M}_o^2+(1+\varphi_{\mathrm{SF}})\psi\bar{M}_o^2\frac{A_W}{\mathrm{i}k_o}. \tag{3.134}$$

Hence, upon applying the transformation in Equation (3.29), we obtain the matrizant relationship

$$\frac{\partial}{\partial x}\begin{bmatrix}p^+\\ p^-\end{bmatrix}=\frac{1}{\Delta}\begin{bmatrix}A(\bar{M}_o)+C(\bar{M}_o)\frac{\mathrm{d}}{\mathrm{d}x}(\ln S) & B(\bar{M}_o)-C(\bar{M}_o)\frac{\mathrm{d}}{\mathrm{d}x}(\ln S)\\ -B(-\bar{M}_o)-C(-\bar{M}_o)\frac{\mathrm{d}}{\mathrm{d}x}(\ln S) & -A(-\bar{M}_o)+C(-\bar{M}_o)\frac{\mathrm{d}}{\mathrm{d}x}(\ln S)\end{bmatrix}\begin{bmatrix}p^+\\ p^-\end{bmatrix}, \tag{3.135}$$

where the elements of the matrizant are given by

$$A(\bar{M}_o)=\mathrm{i}k_o\left(\frac{\alpha^2+1}{2}-\bar{M}_o\right)+\frac{1}{2}\left(-\alpha^2+\bar{M}_o+\bar{M}_o(\varphi_{\mathrm{SF}}+\psi)-\varphi_{\mathrm{SF}}(1-\psi)\bar{M}_o^2\right)A_W, \tag{3.136}$$

$$B(\bar{M}_o)=\mathrm{i}k_o\left(\frac{\alpha^2-1}{2}\right)+\frac{1}{2}\left(-\alpha^2+\bar{M}_o+\bar{M}_o(\varphi_{\mathrm{SF}}-\psi)-\varphi_{\mathrm{SF}}\bar{M}_o^2(1-\psi)\right)A_W, \tag{3.137}$$

$$C(\bar{M}_o)=\bar{M}_o-\frac{1}{2}\left(\alpha^2+\bar{M}_o^2-(1-\varphi_{\mathrm{SF}})\psi\bar{M}_o^2\frac{A_W}{\mathrm{i}k_o}\right) \tag{3.138}$$

and $A(-\bar{M}_o)$, $B(-\bar{M}_o)$ and $C(-\bar{M}_o)$ are obtained from Equations (3.136)–(3.138) simply by changing the sign of $\bar{M}_o$.

3.9.1 Non-Uniform Duct

For variable-area ducts, Equation (3.135) is used with the no-slip boundary condition at the duct walls. Thus, putting $\psi=0$ and $\varphi_{\mathrm{SF}}=1$, the elements of the matrizant in Equation (3.135) simplify to:

$$\Delta = \alpha^2 - \bar{M}_o^2 \tag{3.139}$$

$$A(\bar{M}_o) = \mathrm{i}k_o\left(\frac{\alpha^2 + 1}{2} - \bar{M}_o\right) - \frac{1}{2}(\alpha^2 - 2\bar{M}_o + \bar{M}_o^2)A_W \tag{3.140}$$

$$B(\bar{M}_o) = \mathrm{i}k_o\left(\frac{\alpha^2 - 1}{2}\right) - \frac{1}{2}\{\alpha^2 - 2\bar{M}_o + \bar{M}_o^2\}A_W \tag{3.141}$$

$$C(\bar{M}_o) = \bar{M}_o - \frac{1}{2}\left(\alpha^2 + \bar{M}_o^2\right), \tag{3.142}$$

where $A_W = A_W(x)$ in general. Since $\alpha^2 \approx 1$ for most practical mean-flow velocity profiles, the matrizant may be simplified to:

$$\mathbf{H} = \begin{bmatrix} \dfrac{\mathrm{i}k_o}{1+\bar{M}_o} & 0 \\ 0 & -\dfrac{\mathrm{i}k_o}{1-\bar{M}_o} \end{bmatrix} + \frac{1}{2}\begin{bmatrix} -\dfrac{1-\bar{M}_o}{1+\bar{M}_o}\left(A_W + \dfrac{\mathrm{d}}{\mathrm{d}x}(\ln S)\right) & \dfrac{1-\bar{M}_o}{1+\bar{M}_o}\left(-A_W + \dfrac{\mathrm{d}}{\mathrm{d}x}(\ln S)\right) \\ \dfrac{1+\bar{M}_o}{1-\bar{M}_o}\left(A_W + \dfrac{\mathrm{d}}{\mathrm{d}x}(\ln S)\right) & -\dfrac{1+\bar{M}_o}{1-\bar{M}_o}\left(-A_W + \dfrac{\mathrm{d}}{\mathrm{d}x}(\ln S)\right) \end{bmatrix} \tag{3.143}$$

Here, both A_W and S are functions of x and, in general, the corresponding wave transfer matrix can be determined by using the numerical segmentation method (Section 3.3.3).

As an alternative approach, we may consider the linear differential equation associated with the foregoing matrizant. It is:

$$\frac{\mathrm{d}^2y}{\mathrm{d}x^2} + \left[\frac{\mathrm{i}2k_o}{1-\bar{M}_o^2} - \left(\frac{1+\bar{M}_o^2}{1-\bar{M}_o^2}\right)A_W - \frac{\mathrm{d}}{\mathrm{d}x}\left(\ln\left(-A_W + \frac{\mathrm{d}}{\mathrm{d}x}(\ln S)\right)\right)\right] \times \frac{\mathrm{d}y}{\mathrm{d}x} - \frac{1}{4}\left(-A_W + \frac{\mathrm{d}}{\mathrm{d}x}(\ln S)\right)^2 y = 0. \tag{3.144}$$

For subsonic low Mach numbers ($\bar{M}_o^2 << 1$) this equation becomes independent of the mean flow and, consequently, its Wronksian can be determined from the wave transfer matrix of the same duct without mean flow, $\mathbf{T}^o$. Then, using the procedure described in Section 3.6.1, the wave transfer matrix of the duct with mean flow may be related to $\mathbf{T}^o$ as

$$\mathbf{T} = \begin{bmatrix} 1 & 0 \\ 0 & \dfrac{1+\bar{M}_o}{1-\bar{M}_o} \end{bmatrix}\mathbf{T}^o\begin{bmatrix} 1 & 0 \\ 0 & \dfrac{1-\bar{M}_o}{1+\bar{M}_o} \end{bmatrix} \mathrm{e}^{\int_0^x\left(-\mathrm{i}k_o + A_W + \frac{\mathrm{d}}{\mathrm{d}x}(\ln S)\right)\frac{\bar{M}_o \mathrm{d}x}{1+\bar{M}_o}}. \tag{3.145}$$

This is similar to Equation (3.91). We may show similarly that Webster horn equation in this case is given by

$$\frac{1}{S}\frac{\partial}{\partial x}\left(\frac{S\partial\overline{p'}}{\partial x}\right) + k_o^2\left(1 - \frac{A_W}{\mathrm{i}k_o}\right)\overline{p'} = 0. \tag{3.146}$$

The transformation $\overline{p'} = \eta/\sqrt{S}$ still reduces this to the canonical form

$$\frac{\partial^2 \eta}{\partial x^2} + k_o^2 \left(1 - \frac{A_W}{\mathrm{i}k_o} \mp \frac{m^2}{k_o^2}\right)\eta = 0, \tag{3.147}$$

for area taper functions which obey Equation (3.96). However, in general, this equation can be solved only numerically. A case that allows an exact solution is a conical duct ($m = 0$) the walls of which are designed such that $\zeta_W = \zeta_W(x)$ and $a(x)\zeta_W(x) =$ constant, where a denotes the radius of duct sections. Then, $A_W =$ constant, since J is constant for a conical duct, and Equation (3.146) admits an exact solution.

3.9.2 Uniform Duct

For a uniform duct, $S =$ constant, $A_W =$ constant, which imply $\zeta_W =$ constant, and the matrizant in Equation (3.135) reduces to

$$\mathbf{H} = \frac{1}{\Delta}\begin{bmatrix} A(\bar{M}_o) & B(\bar{M}_o) \\ -B(-\bar{M}_o) & -A(-\bar{M}_o) \end{bmatrix} \tag{3.148}$$

where $A(\bar{M}_o)$ and $B(\bar{M}_o)$ are given by Equations (3.136) and (3.137). This represents a linear differential system with constant coefficients and, therefore, its solution for a duct with origin at $x = 0$ can always be expressed in the form Equation (3.30) with the wave transfer matrix given by Equation (3.34), that is,

$$\mathbf{T} = \mathbf{\Phi}\begin{bmatrix} \mathrm{e}^{\mathrm{i}k_o K^+ x} & 0 \\ 0 & \mathrm{e}^{\mathrm{i}k_o K^- x} \end{bmatrix}\mathbf{\Phi}^{-1} \tag{3.149}$$

Here, $\mathrm{i}k_o K^{\mp}$ denote the two eigenvalues of the matrizant in Equation (3.148) and the propagation constants K^+ and K^- are given by [11] :

$$K^{\pm} = \frac{\pm 1}{\alpha \pm \bar{M}_o} \frac{1}{(\alpha \mp \bar{M}_o) + \dfrac{A_W \bar{M}_o^2}{\mathrm{i}k_o} \dfrac{\psi(1+\varphi_{SF})}{(\alpha \mp \bar{M}_o)}} \left\{ \mp \bar{M}_o \left(1 + \mathrm{i}\frac{A_W}{k_o}\frac{\varphi_{SF} + \psi + 1}{2}\right) + \sqrt{\alpha^2 - \left(\frac{A_W \bar{M}_o}{2k_o}\right)^2 (\varphi_{SF} - \psi + 1)^2 + \mathrm{i}\frac{A_W}{k_o}\left(\alpha^2 + \varphi_{SF}(1-\psi)\bar{M}_o^2\right)} \right\} \tag{3.150}$$

and the modal matrix $\mathbf{\Phi}$ is given by

$$\mathbf{\Phi} = \begin{bmatrix} 1 & \dfrac{-H_{12}}{H_{11} - \mathrm{i}k_o K^-} \\ \dfrac{-H_{21}}{H_{22} - \mathrm{i}k_o K^+} & 1 \end{bmatrix}, \tag{3.151}$$

where H_{ij}, $i, j = 1, 2$, denote the elements of the matrizant.

As can be seen from Equation (3.150), the propagation constants are determined by the non-dimensional parameters $\psi, \varphi_{\mathrm{SF}}, \bar{M}_o, \alpha$ and A_W/k_o. Of these parameters, A_W/k_o

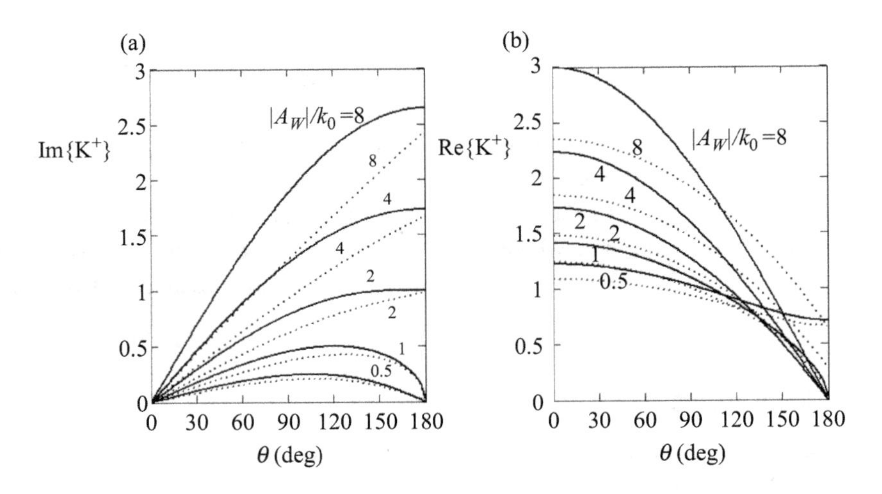

Figure 3.5 Variations of the real and imaginary parts of the propagation constant K^+ with the normalized admittance parameter $|A_W|\mathrm{e}^{\mathrm{i}\theta}/k_o$. Solid: $\bar{M}_o = 0$, dot: $\bar{M}_o = 0.1$. Ingard–Myers boundary condition is assumed.

represents the effect of the wall, $\bar{M}_o$, φ_{SF} and α signify the effect of the mean-flow velocity profile, and ψ depends on the model chosen for the acoustic boundary condition. Variations of the real and imaginary parts of the propagation constant K^+ with the argument, $0 < \theta < \pi$, of the normalized admittance parameter $A_W/k_o = |A_W|\mathrm{e}^{\mathrm{i}\theta}/k_o$ are shown in Figure 3.5 for $\alpha = 1$, $\varphi_{SF} = 0$ and $\psi = 1$, that is, the Ingard-Myers boundary condition. Analytical assessments about such characteristics are difficult to make from Equation (3.150), except for $\bar{M}_o = 0$ (solid curves in Figure 3.5). In this case, Equation (3.150) reduces to $K^\pm = \mp\sqrt{1 + \mathrm{i}A_W/k_o}$, which is tractable to standard explicit analysis. Considering K^+ with the branch cut from -1 to $-\infty$ on the real axis and letting $\mathrm{i}A_W = |A_W|(\cos\theta + \mathrm{i}\sin\theta)$, the partial derivative of $\mathrm{Im}\{K^+\}$ with respect to θ gives a maximum at $\theta = \cos^{-1}(-|A_W|/2k_o)$, $0 < \theta < \pi$, if $|A_W| < 2k_o$; the corresponding maximum of $\mathrm{Im}\{K^+\}$ being $|A_W|/2k_o$. This maximum can be observed in the $|A_W|/k_o = 0.5$ and 1 curves in Figure 3.5a. For $|A_W| \geq 2k_o$, the imaginary part of the propagation constant tends to the maximum $\sqrt{|A_W|/k_o - 1}$ and the real part tends to zero as $\theta \to \pi$; however, they are undefined at $\theta = \pi$. Recalling that $A_W = P/\zeta_W S$ (Section 3.8.1.5), the latter feature means that as $\theta \to \pi$, the wall tends to be purely reactive and the propagation constants tend to be purely imaginary. Note that the real part of the propagation constant K^+ can be greater than unity, its hard-wall value (Figure 3.5b). This means that the phase speed (Section 3.5.2) can be smaller in a duct with finite impedance walls than that in hard-walled ducts by several factors. Some authors refer to this phenomenon as slow sound [16] . The presence of mean flow tends to lower the zero mean-flow characteristics. The effect is regular and rather negligible for relatively small values of the normalized admittance parameter and the Mach number; however, it can be substantial and irregular for the larger values of these parameters and depends on the boundary condition used.

The one-dimensional theory involves no approximation about the distribution of the fluctuating variables over a duct section, apart from the condition that they are limited to the fundamental mode eigenfunctions, which are characterized by having no nodal line. The fundamental mode is not planar except when the duct is hard walled with $v_o =$ constant. The factors which contribute to the non-uniformity of the fundamental mode are refraction in mean shear layers and the finite impedance of the wall. The effect of refraction increases with the mean-flow Mach number and the Helmholtz number. Exact solutions of Pridmore-Brown equation for the fundamental mode of hard-walled ducts with parabolic mean-flow velocity profile [4–5] indicate that the fluctuating pressure profiles deviate only by few percent from $\overline{p'}$ when $k_o a < 1$ and $\bar{M}_o < 0.2$. On the other hand, the hard-wall condition can be relaxed to the condition $k_o a << |\zeta_W|$. For example, in a circular duct with no mean flow, the fluctuating pressure profile of the fundamental mode is $\varphi_{00}(r) = \mathrm{J}_0(\kappa_{00} r/a)$, where r denotes the radial coordinate, J_0 denotes the Bessel function of order zero and $\kappa_{00} \approx \sqrt{\mathrm{i} k_o a/\zeta_W}$, if $k_o a/|\zeta_W| << 1$ (see Section 6.7.1.1). In this case, therefore, the deviations of the fluctuating pressure profile from the planar one are small to the second order. This result also holds approximately for other section shapes and in the presence of a uniform mean flow [17]. However, though the condition $k_o a << |\zeta_W|$ is quite common, walls of smaller impedance are also of interest in many applications. The non-uniformity of the fundamental mode eigenfunction can be more discernible in such ducts, but as the results of applications indicate, the present one-dimensional model can provide useful first estimates for the propagation constants of practical ducts (see Section 3.9.3.1).

The theory does not involve approximation about the shape of the mean-flow velocity profile either. However, the effect of the profile geometry is represented implicitly insofar as it determines the parameters α,$\bar{M}_o$ and φ_{SF}; though the effect essentially culminates on $\bar{M}_o$ and φ_{SF}, since $\alpha \approx 1$ for practical subsonic low Mach number profiles. Therefore, to the accuracy of the one-dimensional theory, any mean-flow velocity profile which matches $\bar{M}_o$ and φ_{SF} of the actual mean flow produces essentially the same effect on the fundamental mode. On this basis, the present one-dimensional duct solution is comparable with solutions of the Pridmore-Brown equation with no-slip and the convective wave equation with full-slip, provided that the same $\bar{M}_o$ is used in calculations. When the partial-slip boundary condition is used with the present theory, since the no-slip condition is implicit in the viscous sublayer model, the results are still comparable with the fundamental mode solutions of the Pridmore-Brown equation with no-slip, provided that the shape of the profile encompasses the viscous sublayer and $\bar{M}_o$ is compatible. However, because of the inevitable mismatch of φ_{SF}, present one-dimensional models with no-slip and partial-slip are not analogous even when they are applied by using the same $\bar{M}_o$, though they both represent approximations to the fundamental mode of the Pridmore-Brown equation with no-slip.

3.9.2.1 Direction of Propagation

Directions of propagation of the principal pressure wave components (see Equation (3.35)) should be determined by using the Briggs criterion, since the propagation constants $K^{\mp}$ are dispersive and inverse Fourier transformation is not easy. The procedure for the application of the Briggs criterion is explained in Section 1.3.2. Briefly, we substitute $\omega + i\varpi$ for ω, with $\varpi \geq 0$, and check if the imaginary part of an axial wavenumber $\kappa(=k_o K^{\mp})$ crosses the real axis as ϖ is varied from 0 to ∞. For the problem at hand, it can be deduced from Equation (3.150) that, as $\varpi \to +\infty$, the sign of $\text{Im}\{k_o K^{\mp}\}$ remains the same as the superscript of $K^{\mp}$. Therefore, the wave modes corresponding to $K^{\mp}$ will decay in $\mp x$ direction, if the sign of the imaginary part of $K^{\mp}$ is the same as the superscript of $K^{\mp}$.

3.9.2.2 Wave Equation

In the case of uniform impedance, Equation (3.133) reduces to

$$\frac{\partial}{\partial x}\begin{bmatrix} p' \\ v' \end{bmatrix} = \frac{ik_o}{\Delta}\begin{bmatrix} -\bar{M}_o + (1+\varphi_{SF})\bar{M}_o \dfrac{A_W}{ik_o} & \rho_o c_o \\ \dfrac{\left(1-\dfrac{A_W}{ik_o}\right)\alpha^2 - \varphi_{SF}(1-\psi)\bar{M}_o^2\dfrac{A_W}{ik_o}}{\rho_o c_o} & -\bar{M}_o\left(1-\psi\dfrac{A_W}{ik_o}\right) \end{bmatrix}\begin{bmatrix} p' \\ v' \end{bmatrix}, \tag{3.152}$$

where A_W = constant. A wave equation for the acoustic pressure can be derived by eliminating the particle velocity from Equation (3.152). It is

$$\left(\alpha^2 - \bar{M}_o^2\left(1-\psi(1+\varphi_{\text{SF}})\frac{A_W}{ik_o}\right)\right)\frac{\partial^2 \overline{p'}}{\partial x^2} + ik_o\bar{M}_o\left(2-(1+\psi+\varphi_{\text{SF}})\frac{A_W}{ik_o}\right)\frac{\partial \overline{p'}}{\partial x} + k_o^2\left(1-\frac{A_W}{ik_o}\right)\overline{p'} = 0. \tag{3.153}$$

If there is no mean flow, this simplifies to

$$\frac{\partial^2 \overline{p'}}{\partial x^2} + k_o^2\left(1-\frac{A_W}{ik_o}\right)\overline{p'} = 0, \tag{3.154}$$

which is same as Molloy's wave equation in the frequency domain [18].

3.9.2.3 Impedance Eduction Formula

Upon invoking the usual $e^{ik_o Kx}$ dependency for the acoustic pressure in Equation (3.153), the resulting dispersion equation can be cast as [11]:

$$\frac{A_W}{ik_o} = \frac{\alpha^2 K^2 + 2K\bar{M}_o - 1}{\bar{M}_o^2(1-\psi(1+\varphi_{\text{SF}}))K^2 + (1+\psi+\varphi_{\text{SF}})K\bar{M}_o - 1}. \tag{3.155}$$

Thus, given the propagation constants, the wall impedance can be educed by using this formula. A relatively simple reliable approach for the measurement of wavenumbers $\kappa^{\mp} = k_o K^{\mp}$ is the Prony method (see Section 12.2.4).

3.9.2.4 Peripherally Non-Uniform Wall Impedance

Although the foregoing analysis assumes A_W = constant, this does not preclude the possibility that the wall impedance may be non-uniformly distributed along the periphery of the duct section. This is evident from the definition of the wall admittance parameter in Equation (3.116), which reduces to $A_W = (1/S)\oint_\ell \mathrm{d}\ell/\zeta_W(\ell)$ for a duct with a uniform cross section. For example, if the wall impedance is distributed in longitudinal uniform bands of breadth $b_1, b_2, \ldots$ of uniform impedance $\zeta_{W1}, \zeta_{W2}, \ldots$, respectively, then $A_W = (1/S)(b_1/\zeta_{W1} + b_2/\zeta_{W2} + \cdots)$, and the duct may be considered as one with constant wall admittance parameter $A_W = P/S\zeta_W$, where $1/\zeta_W = b_1/P\zeta_{W1} + b_2/P\zeta_{W2} + \cdots$ and P denotes the duct perimeter.

3.9.3 Finite Wall Impedance Models

The duct elements described in Sections 3.9.1 and 3.9.2 may be used, given the impedance of the duct walls. This section describes impedance models for some of the more common duct wall structures.

3.9.3.1 Lined Impermeable Walls

Use of impermeable walls that are designed to absorb and dissipate substantial sound power attenuation per unit length is a well-known method of noise control in ductworks. The wall treatment is called liner. A liner can be designed as locally reacting or bulk-reacting. Locally reacting liners are characterized by their surface impedance only and are the subject of this section. Acoustic modeling of bulk-reacting liners is deferred to the next chapter (see Section 4.9), because it is necessary to take into account the effect of coupling between the sound field in the liner with the sound field in the duct.

Locally reacting liners are grouped as absorber and resonator types. Absorber liners (shown in Figure 3.6a) consist of a homogeneous layer(s) of sound-absorbing fibrous material such as rock wool, glass wool or basalt, which is backed by a solid impervious wall and usually protected from the mean flow in the duct by a perforated facesheet. In order to prevent the liner acting as a bulk-reacting one, the axial continuity of the porous packing is broken by rigid baffles.

When there is a facesheet, the wall impedance is given by $Z_W = Z_a/\sigma + Z_{pack}$, where Z_a denotes the acoustic impedance of a perforate hole, σ denotes the open area

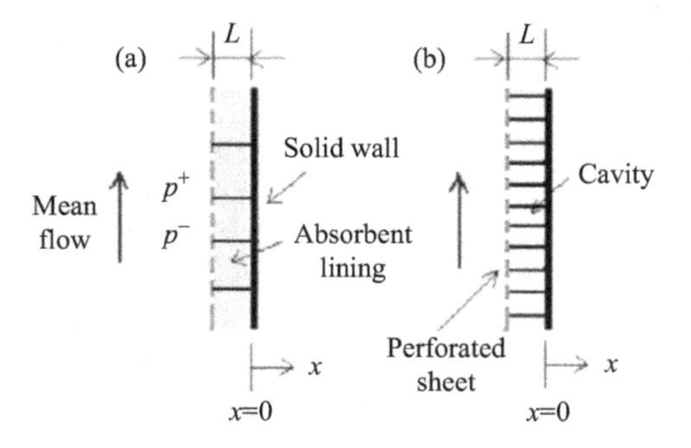

Figure 3.6 Sections of a locally reacting wall: (a) absorber type, (b) resonator type.

porosity of the facesheet (see Section 3.9.3.3) and Z_{pack} denotes the normal surface impedance of the absorber packing. Assuming that the acoustic field in the packing only consists of plane waves propagating along the thickness of the packing, Z_{pack} can be determined from Equation (3.112). Hence, for the packing in Figure 3.6a, $Z_{pack} = p'(-L)/v'(-L)$ and assuming rigid backing at $x = 0$, we find $Z_{pack} = i\rho_e c_e \cot k_e L$, where $k_e = \omega/c_e$, and ρ_e and c_e denote the equivalent fluid density and the equivalent fluid speed of sound of the packing material (see Appendix B).

Several models for the acoustic impedance of apertures, such as circular holes, on a liner facesheet are given in Appendix C. Since the facesheet primarily functions as a screen protecting the porous backing from being drifted with the mean flow, its porosity is selected low enough to achieve this function, but high enough so that it is acoustically transparent to sound waves at the design frequencies and substantial acoustic energy is dissipated in the liner. In ducts with low mean-flow velocity, sometimes a wire mesh is used instead of a perforated facesheet. Then $Z_W = Z_{pack}$, approximately.

In resonator type liners (see Figure 3.6b), the space behind the facesheet is designed as a honeycomb structure, each cavity of which acts, usually as one- or two-degrees-of-freedom resonator, and can be optimized for acoustic performance at discrete frequencies. The impedance of a honeycomb cavity of depth L is given as $Z_{pack} = i\rho_o c_o \cot k_o L$, since the cavity is usually occupied with a fluid, and impedance of the wall is again determined as $Z_W = Z_a/\sigma + Z_{pack}$.

In view of the $e^{ik_o K^{\mp} x}$ dependence of the one-dimensional acoustic field in a duct (see Equation (3.149), the principal pressure wave components decay in the direction of propagation by a factor of $e^{-k_o \mathrm{Im}\{K^{\mp}\}x}$. This is usually expressed as attenuation in dB units by the formula $A^{\mp} = 10 \log \left| e^{-k_o \mathrm{Im}\{K^{\mp}\}x} \right|^2 = 8.686 k_o x |\mathrm{Im}\{K^{\mp}\}|$ and $A^{\mp}/x$ gives the attenuation rate in dB/m of the waves traveling in $\mp x$ directions. It should be noted that, because of the squaring involved in the formula, $A^{\mp}$ actually represents the decay of the squared moduli of the principal pressure wave components, or their time-averaged acoustic power, since these are proportional (see Section 3.4). As can be inferred from Figure 3.5a, liners tend to have optimum attenuation at certain frequencies (see Sections 6.7.4–6.7.5 for further considerations). Exploitation of this feature is important in practical applications where the liner length is limited [19].

Liners used in ducts carrying a grazing mean flow are usually tested in an apparatus which essentially consists of a rectangular duct having solid and impervious walls everywhere, except over a finite length of one side which is treated with the liner to be tested [20–21]. The wavenumber of the least attenuated axial mode in the treated part of the apparatus may be measured by Prony analysis (see Section 12.2.4) of the acoustic pressure signals acquired from sufficiently large number of equidistant microphones mounted centrally on the wall opposite to the treated wall of the apparatus [22]. The liner impedance is usually calculated by using the fundamental mode solution of three-dimensional rectangular duct models (see Chapter 6); however, Equation (3.148) may also be adequately accurate for the duct sizes used in impedance eduction measurements [11].

Compared in Figure 3.7 are the predictions of the one-dimensional theory for the imaginary part of the fundamental propagation constants of a lined duct with the corresponding results calculated by solving the Pridmore-Brown Equation (see Section 6.8), a three-dimensional counterpart of the present one-dimensional theory, with $N = 3/2$ power law profile with no-slip ($\varphi_{\mathrm{SF}} = 1$) and $\bar{M}_o = 0.6\hat{M}_o$), where $\hat{M}_o$ denotes the mid-stream Mach number of the measured mean flow. These characteristics are proportional to the attenuation per unit liner length in dB/m units. The trends of the characteristics are explained by the fact that refraction and convection have opposite effects on sound attenuation per unit length of the liner [12]. This is because refraction causes wavefronts turn into (away from) from the liner for downstream (upstream) propagation and thus tending to increase (decrease) the rate of attenuation with distance, whereas convection causes a decrease (increase) because waves travel faster (slower) and it takes a longer (shorter) distance for the dissipation of an equal amount of acoustic energy. On the other hand, effects of refraction and convection increase with the mean-flow Mach number, but refraction is frequency dependent and becomes discernible at relatively higher frequencies, whereas the effect of convection is present at all frequencies. Therefore, at relatively low frequencies (the effect of convection being dominant) attenuation per unit length can be expected to decrease with the mean-flow Mach number for downstream propagation and increase for upstream propagation. But, as frequency increases, refraction enters with an opposite effect and these variations may be expected to level out approximately as if there is no mean flow.

The differences between the results based on the fundamental mode solutions of the Pridmore-Brown equation and the present one-dimensional models come from not only the approximations involved in representation of refraction and convection, but also the effect of the finite impedance of the liner. The manifestation of the latter can be observed from the results given for $\bar{M}_o = 0$ in Figure 3.7. As already remarked in

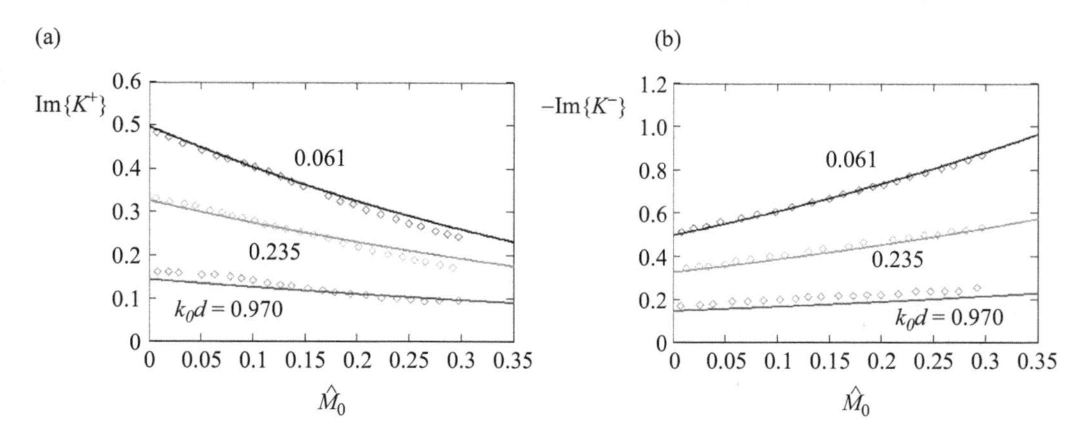

Figure 3.7 Comparison of the imaginary parts of the propagation constants of a 56 × 56 mm rectangular duct ($d = 28$ mm) lined on two opposite faces: ◇◇◇ Pridmore-Brown wave equation with $N = 2/3$ power law and no-slip boundary [12]; solid: present one dimensional model with no-slip boundary condition. (a) Downstream propagation, (b) upstream propagation.

the previous section, Equation (3.150) is a very good approximation if $k_o a/|\zeta_W| << 1$. For the liner in Figure 3.7, $|\zeta_W| = 0.144, 0.237, 0.301$ for $k_o d = 0.061$, 0.235, 0.970, respectively. The foregoing condition is roughly satisfied for $k_o d = 0.061$ and the one-dimensional model accurately represents the solution of the wave equation (Pridmore-Brown equation reduces to the wave equation for $\bar{M}_o = 0$), but for the higher frequencies, the slight discrepancy due to the effect of the liner becomes visible.

3.9.3.2 Permeable Rigid Porous Walls

Ducts made of wire reinforced woven fabric are used in ventilation and air-handling systems and in automotive intake systems. The impedance of the duct wall depends on many details of the wall construction and are usually obtainable from the manufacturers. It is given as function of a parameter called "porous frequency" [23]. This is measured by closing one end of the duct and forcing gas flow from the other end and through the duct walls to the outer environment and measuring the resulting gas volume flow rate. In a consistent system of units, it has the unit of 1/s, and, hence, the term porous frequency, although this has no physical relation to the usual meaning of frequency. The porous frequency of porous hoses used in automotive intake systems are in the range of about 100–600 1/s. The impedance of a wire reinforced woven fabric wall of a typical hose is shown in Figure 3.8 as function of frequency and the porous frequency, f_p. For this wall structure, the porous frequency is given approximately by $f_p = 0.066/(Dh + 0.0077h/L)$ in 1/s units, where L, D and h denote the length, internal diameter and wall thickness of the duct in meters, respectively.

If the environment exterior to the duct wall is a free field, then the outer impedance of the duct wall is given by its radiation impedance. A detailed numerical model of

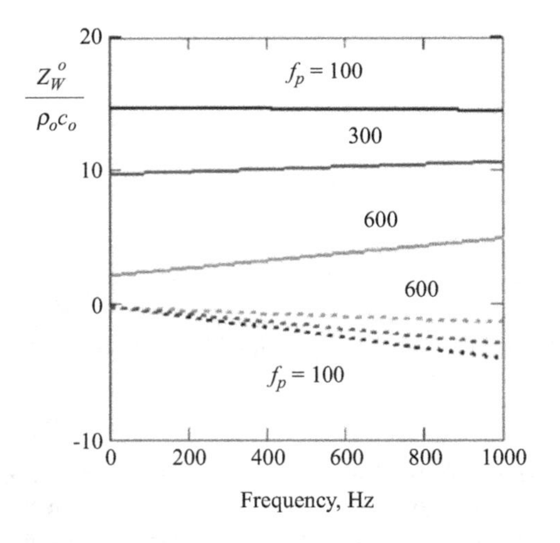

Figure 3.8 Impedance over a wire-reinforced woven fabric wall. Solid: real part, dash: imaginary part.

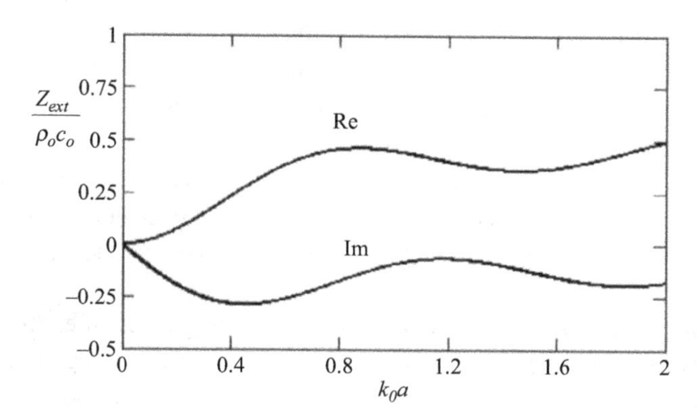

Figure 3.9 The free-field radiation impedance of a duct with permeable walls ($L/2a = 4$).

sound radiation from the outer surface of a circular duct, in which the duct wall is envisaged as swamped by elementary vibrating pistons, is presented by Cummings and Kirby [24]. When the acoustic field in the duct is one-dimensional, however, the theory is not sensitive to circumferential variations and all vibrating pistons lying in the same transverse plane have the same velocity u'. Then, the outer radiation model reduces to that of rings of elementary width embedded in the wall, which vibrate in the breathing mode with radial velocity u'. But, because of the mutual impedance of adjacent vibrating rings, the outer impedance of the duct wall is not locally reacting. Therefore, this model leads to an integro-differential system of equations for the determination of the wave transfer matrix of the duct [25]. However, if the length, L, of the duct is compact enough, that is, $k_o L << 1$, so that the axial distribution of $u' = u'(x)$ can be assumed to be more or less uniform over L, the free-field radiation impedance of the outer surface of the duct wall can be approximated by the formula [25]:

$$\frac{Z_{\text{ext}}}{\rho_o c_o} = \frac{-ik_o}{\pi} \int_0^\infty \frac{H_0^{(1)}\left(a\sqrt{k_o^2 - \lambda^2}\right) \sin(\lambda L) d\lambda}{\lambda\sqrt{k_o^2 - \lambda^2} H_1^{(1)}\left(a\sqrt{k_o^2 - \lambda^2}\right)}, \tag{3.156}$$

where $H_j^{(1)}$ denotes the Hankel function of the first kind of order j, a denotes the external radius of the pipe and $k_o = \omega/c_o$ denotes, as usual, the wavenumber. The variation Z_{ext} with the Helmholtz number $k_o a$ is shown in Figure 3.9 for a duct of length $L = 8a$.

3.9.3.3 Perforated Rigid Walls

Walls of perforated ducts consist of rows of apertures (usually circular holes) separated by solid pipe segments. Aperture rows may be modeled as surfaces having continuously distributed or discretely distributed finite impedance. The continuously distributed model is appropriate if the open area of the perforated surface is large enough, about 10–15% or more of the total surface area. For relatively small open areas, it is more appropriate to model aperture rows discretely (see Section 4.8).

In continuous modeling, the surface of a perforated duct is treated as a permeable wall of continuous impedance $Z_W = Z_W^o + Z_{\text{ext}}$. If the perforate is open to a free-field, Equation (3.156) may still be used for Z_{ext}, but it remains to determine Z_W^o for a given perforate. Here, we will assume that all apertures are identical, as the use of identical aperture is common practice in industry due to the manufacturing considerations.

Let Z_a denote the lumped parameter impedance of a single aperture. An extensive literature exists on empirical, semi-empirical and theoretical lumped impedance models of compact apertures of various shapes (mostly circular) and some models are described in Appendix C. Since the apertures are actually distributed discretely on the duct surface, it is necessary to define an equivalent continuous distribution, $Z_W^o = Z_W^o(x)$. A plausible criterion for this is the condition that the actual discrete distribution and the derived continuous distribution should yield the same $\dot{\mu}'$. To apply this condition, it is more convenient to express Equation (3.114) as

$$\dot{\mu}' = -\frac{\rho_o}{S}\left(\oint_{\ell} \frac{J\mathrm{d}\ell}{Z_W^o}\right)\left(\overline{p'} - p'(\ell_{\text{ext}})\right). \tag{3.157}$$

Then the above stated criterion gives

$$\oint_{\ell} \frac{J\mathrm{d}\ell}{Z_a} = \frac{1}{Z_W^o}\oint_{\ell} J\mathrm{d}\ell, \tag{3.158}$$

where the integral on the right-hand side is equal to the perimeter $P_S = P_S(x)$ of a duct section, whereas the integrand on the left-hand side is non-zero only at apertures. Therefore, Equation (3.158) can be expressed as

$$Z_W^o = \frac{Z_a}{\dfrac{P_{\text{open}}}{P_S}}, \tag{3.159}$$

where $P_{\text{open}} = P_{\text{open}}(x)$ denotes the total open portions of P_S at a section (note that this result still applies for a non-uniform discrete distribution of non-identical apertures). Thus, the perforate impedance can be expressed as $Z_W = Z_a/\sigma + Z_{\text{ext}}$, where the quotient $\sigma = P_{\text{open}}/P_S$ is called perforate porosity. This can be calculated from $\sigma = A_{\text{open}}/A$, where A denotes the total surface area of the perforate and A_{open} denotes the open surface area. If the apertures are not identical, but are compact, it is usually adequately accurate for practical purposes to replace the actual distribution by a distribution of identical apertures of the same A_{open}/A ratio, provided that this ratio is large enough.

Perforated ducts which have their apertures open to an external sound field are sometimes used as noise control elements, however, they are legendary for their use in exhaust silencers and mufflers.

3.9.3.4 Turbulent Boundary Layer Over Rigid Walls

In hard-walled ducts carrying a fully developed mean flow, damping of sound waves occurs due to the straining of vorticity by sound wave interaction in turbulent boundary layers. A measure of the significance of this effect is the normalized viscous acoustic boundary layer thickness, δ_A^+, which is defined by $\delta_A^+ = v_* \sqrt{2\rho_o/\omega\mu_o}$, where v_* denotes the friction velocity. Experiments show that [26], when δ_A^+ is sufficiently small, much less than about 10, say, the viscous sublayer thickness ($\approx 11\mu_o/\rho_o v_*$) of the mean flow is much larger than the acoustic boundary layer thickness δ_v ($= \sqrt{2\mu_o/\omega\rho_o}$), turbulence has negligible effect and the attenuation of sound waves is described adequately accurately by the viscothermal propagation constants (see Section 8.3.1). But, when δ_A^+ is sufficiently large, much larger than about 10, say, the acoustic boundary layer thickness becomes larger than the sublayer thickness and the damping of the acoustic waves is significantly influenced by the action of turbulent stresses.

For circular hard-walled ducts, these effects are fairly accurately predicted by the Howe boundary layer admittance $Y = Y(s, \sigma, \delta_A^+)$ [27], where $s = a\sqrt{\omega\rho_o/\mu_o}$ denotes the shear wavenumber and $\sigma^2 = \mu_o (c_p)_o / \kappa_o$ denotes the Prandtl number. The one-dimensional theory can be applied by modeling the duct as one having a wall impedance of $Z_W = 1/Y$ [28]. Variation of Y with the normalized boundary layer thickness is shown in Figure 3.10 for $\sigma = 0.7$ and typical values of the shear wavenumber.

Howe's boundary layer admittance is based on an empirical model of eddy viscosity with memory effect. A mathematical study of the physical process that links turbulent Reynolds stress with the memory effect is presented by Weng et al. [29].

3.9.3.5 Elastic Walls

Hard but elastic duct walls are characterized by the small structural vibrations they make under the forcing of the in-duct acoustic pressure field. These vibrations are usually highlighted as being a source of noise radiated from the outer surfaces of

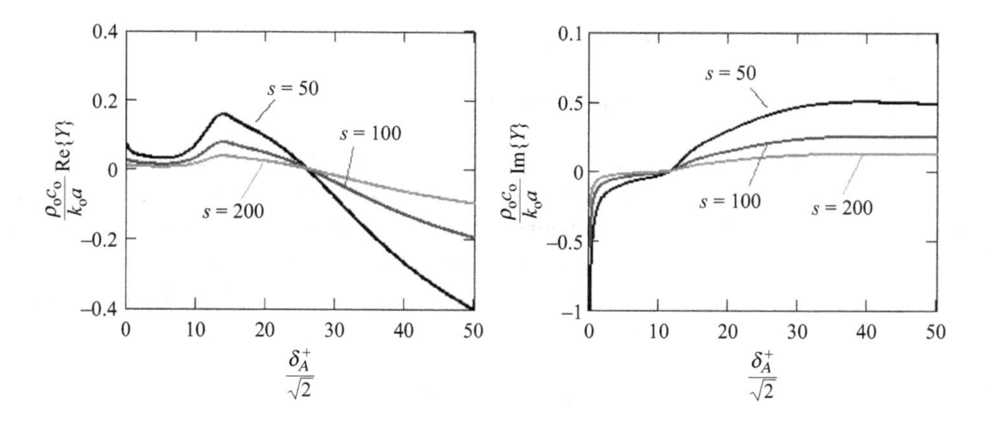

Figure 3.10 Howe's turbulent boundary layer admittance for dry air.

ductworks to the outside environment (see Section 5.8). However, wall vibrations also couple with the in-duct acoustic field and modify the propagation constants predicted for hard-walled ducts, although this may be discernible in ducts with lightweight thin walls or with distensible walls (for example, plastic ducts or mammal arteries). To show this, consider a uniform circular duct as a thin circular shell of uniform thickness. Under the forcing of a one-dimensional internal acoustic field, the mid-plane of a perfectly circular shell undergoes breathing vibrations. Let $w' = w'(t, x)$ denote the radial displacement of the mid-plane of the shell. We assume that the mean flow is uniform and is bounded by an infinitely thin vortex sheet with displacement w' from its unperturbed position. Since the fluid in the core side of the vortex sheet is moving uniformly with velocity $\overline{v_o}$, we have $u' = \partial w'/\partial t + \overline{v_o}\partial w'/\partial x$. Then, from Equation (3.114), $\dot{\mu}' = -\rho_o u' P/S$ and, having thus invoked the Ingard–Myers boundary condition, the vibrating walls may be associated with an equivalent impedance $\zeta_W = (\overline{p'}/w')/(-i\omega\rho_o c_o)$, where the quotient $w'/\overline{p'}$ is a characteristic (receptance) of the shell. It can be determined from the forced vibration analysis of the shell, but it transpires as function of the axial propagation constant of the coupled wave motion of the shell and the fluid in the duct. This implies that elastic walls are not locally reacting. At low frequencies, however, $w'/\overline{p'}$ may be approximated by its static value based on elementary circumferential stress considerations. Consider a thin cylindrical circular shell of mid-plane radius a and thickness h, subject to a uniform static pressure difference, δp_o, say. The circumferential stress is $a\delta p_o/h$, which can be seen by cutting the cylinder longitudinally into two halves and invoking the equilibrium of the internal and external forces. The corresponding circumferential strain is $\delta a/a$, where δa denotes the change in the mid-plane radius of the shell, and from Hooke's law we have $\delta a/a = \delta p_o/hE$, where E denotes the modulus of elasticity of the shell material. Hence, $\delta a = a^2\delta p/hE$ or $w'/\overline{p'} \approx a^2/hE$. To this approximation, elastic walls are locally reacting and their effect on propagation constants can be estimated by using the one-dimensional theory with $\zeta_W = (hE/a^2)/(-i\omega\rho_o c_o)$.

3.10 Inhomogeneous Ducts

3.10.1 Linearized Energy Equation

Since $c_o \neq$ constant for an inhomogeneous duct, the linearized energy equation does not reduce to the relationship $p' = c_o^2\rho'$ (see Section 1.3). It is, however, plausible to assume that the fluctuations of the speed of sound are small to the second order ($c^2 \approx c_o^2$). Then, to close Equations (3.17) and (3.19), we first assume that the transverse gradients are small compared to the axial gradients so that Equation (1.10) can be expressed as

$$\frac{\partial p}{\partial t} + v\frac{\partial p}{\partial x} = c^2\left(\frac{\partial \rho}{\partial t} + v\frac{\partial \rho}{\partial x}\right). \tag{3.160}$$

Using this, Equation (3.5) is recast as

$$\int_{x}^{x+\delta x}\left(\left(\int_{S}\frac{1}{c^2}\left(\frac{\partial p}{\partial t}+v\frac{\partial p}{\partial x}+\rho c^2\frac{\partial v}{\partial x}\right)\mathrm{d}S\right)\right)\mathrm{d}x+\int_{x}^{x+\delta x}\oint_{\ell}\rho uJ\mathrm{d}\ell\mathrm{d}x=0. \tag{3.161}$$

For a perfect gas $\rho c^2=\gamma p$, where γ denotes the ratio of the specific heat coefficients. The latter is a slowly varying function of the temperature for most perfect gases and its fluctuations in an acoustic field can be neglected as being small to the second order. For dry air, for example, the ratio of the specific heat coefficients decreases from 1.40 at 20 °C to 1.32 at 1000 °C. It is, therefore, also adequately accurate to neglect possible variations of γ_o over a duct section. Hence, since $(c^2)'/(c^2)_o << 1$, the differential form of Equation (3.161) may be expressed as

$$S\frac{\partial\bar{p}}{\partial t}+\int_{S}v\frac{\partial p}{\partial x}\mathrm{d}S+\gamma_o\int_{S}p\frac{\partial v}{\partial x}\mathrm{d}S=-c_o^2\oint_{\ell}\rho uJ\mathrm{d}\ell\mathrm{d}x \tag{3.162}$$

where we use the notation $c_o^2=(c^2)_o$. But, since this equation is not tractable to Favre averaging (see Section 3.2.1), we invoke the plane-wave assumption $p\approx p(t,x)$. Then, p can be replaced by $\bar{p}$ and Equation (3.162) gives

$$S\left(\frac{\partial\bar{p}}{\partial t}+\bar{v}\frac{\partial\bar{p}}{\partial x}\right)+\gamma_o\bar{p}\frac{\partial}{\partial x}(S\bar{v})=-c_o^2\oint_{\ell}\rho uJ\mathrm{d}\ell. \tag{3.163}$$

Upon linearization, this yields

$$S\left(\frac{\partial\overline{p'}}{\partial t}+\overline{v_o}\frac{\partial\overline{p'}}{\partial x}+\overline{v'}\frac{\mathrm{d}\overline{p_o}}{\mathrm{d}x}\right)+\gamma_o\overline{p_o}\frac{\partial}{\partial x}(S\overline{v'})+\gamma_o\overline{p'}\frac{\mathrm{d}}{\mathrm{d}x}(S\overline{v_o})=-c_o^2\oint_{\ell}\rho_o u'J\mathrm{d}\ell-c_o^2\oint_{\ell}\rho' u_oJ\mathrm{d}\ell \tag{3.164}$$

for the fluctuations and

$$S\overline{v_o}\frac{\mathrm{d}\overline{p_o}}{\mathrm{d}x}+\gamma_o\overline{p_o}\frac{\mathrm{d}}{\mathrm{d}x}(S\overline{v_o})=-\gamma_o\overline{p_o}\oint_{\ell}u_oJ\mathrm{d}\ell \tag{3.165}$$

for the mean flow. It should be noted that, in view of the a priori assumption $p\approx p(t,x)$ and since $\gamma_o p_o=\rho_o c_o^2$, the first peripheral integral on the right-hand side of Equation (3.164) can be evaluated as $-\gamma_o\oint_{\ell}p_o u'J\mathrm{d}\ell=-\gamma_o\overline{p_o}\oint_{\ell}u'J\mathrm{d}\ell$. If the duct is homogeneous, we can recover the relationship $p'=c_o^2\rho'$ from Equation (3.165) minus c_o^2 times Equation (3.17).

3.10.2 Matrizant of a Duct with Finite Impedance Walls

After substituting $\overline{\rho'}=(\varepsilon'+\bar{p}')/c_o^2$, where ε' denotes the entropy wave component, the frequency domain forms of Equations (3.17), (3.19) and (3.164) can be collected in matrix notation as

$$
\begin{bmatrix}
\bar{M}_o - \dfrac{\psi \bar{M}_o A_W}{\mathrm{i}k_o} & \overline{\rho_o} c_o & \bar{M}_o \\
1 + \beta \bar{M}_o^2 + \dfrac{\psi \varphi_{SF} \bar{M}_o^2 A_W}{\mathrm{i}k_o} & \overline{\rho_o}\,\overline{v_o} & \beta \bar{M}_o^2 \\
\bar{M}_o - \dfrac{\psi \bar{M}_o A_W}{\mathrm{i}k_o} & \overline{\rho_o} c_o & 0
\end{bmatrix}
\frac{\partial}{\partial x}
\begin{bmatrix} \overline{p'} \\ \overline{v'} \\ \varepsilon' \end{bmatrix} +
$$

$$
\begin{bmatrix}
-\mathrm{i}k_o - \bar{M}_o \dfrac{\mathrm{d}}{\mathrm{d}x}(\ln \gamma_o \overline{p_o}) + \alpha_W & -\overline{\rho_o} c_o \dfrac{\mathrm{d}}{\mathrm{d}x}(\ln \overline{v_o}) & -\mathrm{i}k_o - \bar{M}_o \dfrac{\mathrm{d}}{\mathrm{d}x}(\ln \gamma_o \overline{p_o}) \\
\beta \bar{M}_o^2 \left(\dfrac{1}{\beta} \dfrac{\mathrm{d}}{\mathrm{d}x} \ln(\overline{v_o}) + \dfrac{\mathrm{d}}{\mathrm{d}x}\left(\ln \beta \bar{M}_o^2 S \right) - \dfrac{\bar{M}_o \varphi_{SF} A_W}{\beta \bar{M}_o^2} \right) & \overline{\rho_o} c_o \left(-\mathrm{i}k_o + \bar{M}_o \dfrac{\mathrm{d}}{\mathrm{d}x}(\ln \overline{v_o}) \right) & \beta \bar{M}_o^2 \left(\dfrac{1}{\beta} \dfrac{\mathrm{d}}{\mathrm{d}x}(\ln \overline{v_o}) + \dfrac{\mathrm{d}}{\mathrm{d}x} \beta \bar{M}_o^2 S \right) \\
-\mathrm{i}k_o + \gamma_o \bar{M}_o \dfrac{\mathrm{d}}{\mathrm{d}x}(\ln \overline{v_o} S) + A_W & \overline{\rho_o} c_o \left(\dfrac{1}{\gamma_o} \dfrac{\mathrm{d}}{\mathrm{d}x}(\ln \overline{p_o}) + \dfrac{\mathrm{d}}{\mathrm{d}x}(\ln S) \right) & 0
\end{bmatrix}
\begin{bmatrix} \overline{p'} \\ \overline{v'} \\ \varepsilon' \end{bmatrix} = 0
\tag{3.166}
$$

This equation applies for ducts having impermeable or permeable walls. The terms involving fluctuating density on the right-hand-sides of Equations (3.17), (3.19) and (3.164) are discarded in either case, because, in the vicinity of the permeable elements of the wall (typically, compact apertures), the fluid is modeled as incompressible in the lumped impedance models. The effect of the through mean flow, if any, enters the calculations via the impedance models and also through the axial mean-flow velocity and pressure gradients created by the mean through flow. The latter gradients are determined by solving the corresponding mean-flow equations, namely, Equations (3.18), (3.22) and (3.165).

Equation (3.166) can be transformed to the pressure wave components as usual, but the resulting matrizant is too complicated to be of practical use and the transformation in practice is best carried out numerically. For subsonic mean-flow Mach numbers, however, it is adequately accurate to take $\alpha^2 = 1$ ($\beta \bar{M}_o^2 << 1$) and the transformed state-space equations simplify to

$$
\frac{\partial}{\partial x}
\begin{bmatrix} p^+ \\ p^- \\ \varepsilon' \end{bmatrix}
= \left\{ \mathbf{Q}(x) + \frac{1}{2}
\begin{bmatrix} 1 & -1 & 0 \\ -1 & 1 & 0 \\ 0 & 0 & 0 \end{bmatrix}
\frac{\mathrm{d}}{\mathrm{d}x}(\ln \overline{\rho_o} c_o) \right\}
\begin{bmatrix} p^+ \\ p^- \\ \varepsilon' \end{bmatrix},
\tag{3.167}
$$

where

$$
\mathbf{Q} =
\begin{bmatrix}
\dfrac{A(\bar{M}_o) + C(\bar{M}_o)}{\Delta} & \dfrac{B(\bar{M}_o) + D(\bar{M}_o)}{\Delta} & \dfrac{F(\bar{M}_o)}{\Delta} \\
\dfrac{-B(-\bar{M}_o) + D(-\bar{M}_o)}{\Delta} & \dfrac{-A(-\bar{M}_o) + C(-\bar{M}_o)}{\Delta} & \dfrac{F(-\bar{M}_o)}{\Delta} \\
E(\bar{M}_o) & E(-\bar{M}_o) & G(\bar{M}_o)
\end{bmatrix}.
\tag{3.168}
$$

Here,

$$
\Delta = \left(1 - \bar{M}_o^2\right) + \psi(\varphi_{SF} + 1) \frac{\bar{M}_o^2 A_W}{\mathrm{i}k_o}
\tag{3.169}
$$

$$A(\bar{M}_o) = \left(1 - \bar{M}_o + (1 + \varphi_{SF}\bar{M}_o)\frac{\psi\bar{M}_o A_W}{\mathrm{i}2k_o}\right)\mathrm{i}k_o - \frac{1}{2}\left(1 - \bar{M}_o + \varphi_{SF}\frac{\psi\bar{M}_o^2 A_W}{\mathrm{i}k_o}\right)A_W \tag{3.170}$$

$$B(\bar{M}_o) = \frac{1}{2}(\varphi_{SF}\bar{M}_o - 1)\frac{\psi\bar{M}_o A_W}{\mathrm{i}k_o}\mathrm{i}k_o - \frac{1}{2}\left(1 - \bar{M}_o + \varphi_{SF}\frac{\psi\bar{M}_o^2 A_W}{\mathrm{i}k_o}\right)A_W \tag{3.171}$$

$$\begin{aligned} C(\bar{M}_o) = &-\frac{1}{2\gamma_o}\left(1 - \bar{M}_o + \varphi_{SF}\frac{\psi\bar{M}_o^2 A_W}{\mathrm{i}k_o}\right)\frac{\mathrm{d}}{\mathrm{d}x}(\ln\overline{p_o}) \\ &-\frac{\bar{M}_o}{2}\left((1 - \bar{M}_o)(\gamma_o + 1 + \bar{M}_o) + (1 + \bar{M}_o + \gamma_o\varphi_{SF}\bar{M}_o)\frac{\psi\bar{M}_o A_W}{\mathrm{i}k_o}\right)\frac{\mathrm{d}}{\mathrm{d}x}(\ln\overline{v_o}) \\ &-\frac{1}{2}(1 + \bar{M}_o\gamma_o)\left(1 - \bar{M}_o + \varphi_{SF}\frac{\psi\bar{M}_o^2 A_W}{\mathrm{i}k_o}\right)\frac{\mathrm{d}}{\mathrm{d}x}(\ln S) \end{aligned} \tag{3.172}$$

$$\begin{aligned} D(\bar{M}_o) = &\frac{1}{2\gamma_o}\left(1 - \bar{M}_o + \varphi_{SF}\frac{\psi\bar{M}_o^2 A_W}{\mathrm{i}k_o}\right)\frac{\mathrm{d}}{\mathrm{d}x}(\ln\overline{p_o}) \\ &-\frac{\bar{M}_o}{2}\left((1 - \bar{M}_o)(\gamma_o - 1 + \bar{M}_o) + (-1 + \bar{M}_o + \gamma_o\varphi_{SF}\bar{M}_o)\frac{\psi\bar{M}_o A_W}{\mathrm{i}k_o}\right)\frac{\mathrm{d}}{\mathrm{d}x}(\ln\overline{v_o}) \\ &-\frac{1}{2}(\bar{M}_o\gamma_o - 1)\left(1 - \bar{M}_o + \varphi_{SF}\frac{\psi\bar{M}_o^2 A_W}{\mathrm{i}k_o}\right)\frac{\mathrm{d}}{\mathrm{d}x}(\ln S) \end{aligned} \tag{3.173}$$

$$E(\bar{M}_o) = \frac{\mathrm{d}}{\mathrm{d}x}(\ln\gamma_o) + \frac{1 + \gamma_o\bar{M}_o}{\bar{M}_o}\left(\frac{\mathrm{d}}{\mathrm{d}x}(\ln S\overline{v_o}) + \frac{1}{\gamma_o}\frac{\mathrm{d}}{\mathrm{d}x}(\ln\overline{p_o})\right) \tag{3.174}$$

$$F(\bar{M}_o) = -\frac{\bar{M}_o^2}{2}\left(1 - \bar{M}_o + \frac{\psi\bar{M}_o A_W}{\mathrm{i}k_o}\right)\frac{\mathrm{d}}{\mathrm{d}x}(\ln\overline{v_o}) \tag{3.175}$$

$$G(\bar{M}_o) = \frac{\mathrm{i}k_o}{\bar{M}_o} + \frac{\mathrm{d}}{\mathrm{d}x}(\ln\gamma_o\overline{p_o}). \tag{3.176}$$

The solution of Equation (3.167) yields a 3×3 wave transfer matrix in the form of Equation (3.41). If $\varepsilon'(0) = 0$ at the origin, the 2×2 wave transfer matrix for the pressure wave components is obtained as discussed for Equation (3.42). If $\varepsilon'(0) \neq 0$, the pressure wave components can still be decoupled into a 2×2 wave transfer matrix approximately, if $\bar{M}_o^2 << 1$. This follows because the terms $F(\mp\bar{M}_o)$ are then rendered an order of magnitude smaller than the other terms in the first two rows of matrix $\mathbf{Q}$ in Equation (3.168) and eventually lead to terms of $O\left(\bar{M}_o^2\right)$ or smaller in the wave transfer matrix. The entropy wave component is then determined by using the third equation in Equation (3.167).

3.10.3 Hard-Walled Ducts with Mean Temperature Gradient

An application of the foregoing inhomogeneous duct element which is of considerable practical interest is the transmission of sound in hard-walled ducts ($A_W = 0$)

containing a hot gas flow. It arises in a number of applications such as engine exhaust systems, combustor cans, pulse combustor tailpipes and impedance tube tests involving high-temperature flames. Since the mean-flow gradients in these systems are established by complex thermal transport processes, practical solutions have to rely on heuristic data gathered about the mean-flow gradients in the system at hand. In reciprocating internal combustion engine exhaust systems, the mean pressure gradients are usually negligible and the axial temperature gradients are less than 100 °C/m. In combustor applications, the mean temperature gradients can be as high as 200 °C/m.

For hard-walled ducts, Equation (3.166) and (3.167) hold with $\alpha_W = 0$. The mean-flow equations must be solved before the calculation of the system matrizant. Equation (3.18) gives $S\overline{\rho_o}\,\overline{v_o} =$ constant, and for $\alpha^2 = 1$ and $\beta\bar{M}_o^2 << 1$, Equation (3.22) simplifies to

$$\rho_o\overline{v_o}\frac{\mathrm{d}\overline{v_o}}{\mathrm{d}x} + \frac{\mathrm{d}\overline{p_o}}{\mathrm{d}x} = 0. \tag{3.177}$$

If the mean flow is homentropic, Equation (3.165) also holds:

$$S\overline{v_o}\frac{\mathrm{d}\overline{p_o}}{\mathrm{d}x} + \gamma_o\overline{p_o}\frac{\mathrm{d}}{\mathrm{d}x}(S\overline{v_o}) = 0. \tag{3.178}$$

The axial distribution of $\overline{v_o}, \overline{\rho_o}$ and $\overline{p_o}$ (and, hence, $\bar{T}_o$ and c_o) are then determined by solving this set of non-linear differential equations numerically, given the cross-sectional area taper function $S = S(x)$ and the temperature dependence of $\gamma_o = \gamma_o(\bar{T}_o)$.

If the mean flow is non-homentropic, in addition to the cross-sectional area taper function $S = S(x)$, usually $\bar{T}_o = \bar{T}_o(x)$ and the temperature dependence $\gamma_o = \gamma_o(\bar{T}_o)$ are given, and the axial variations of $\overline{v_o}, \overline{\rho_o}, \overline{p_o}$ and c_o are found from Equation (3.178) and the relations $\overline{\rho_o} = \overline{p_o}/R\bar{T}_o$, $\overline{\rho_o}\,\overline{v_o}S =$ constant and $c_o^2 = \gamma_o R\bar{T}_o$, where R denotes the gas constant. For example, if the duct has constant cross-sectional area ($S =$ constant), these equations yield the equation

$$(\overline{v_o}(x))^2 - \left(\frac{R\bar{T}_o(x)}{\overline{v_o}(0)} + \overline{v_o}(0)\right)\overline{v_o}(x) + R\bar{T}_o(x) = 0 \tag{3.179}$$

for the calculation of the mean-flow velocity. Then the mean pressure in the duct is determined from

$$\overline{p_o}(x) = \overline{p_o}(0) + \overline{\rho_o}(0)\overline{v_o}(0)(\overline{v_o}(0) - \overline{v_o}(x)) \tag{3.180}$$

and the mean density then follows from the perfect gas law. In general, however, numerical integration is required for the determination of the axial distribution of the mean-flow velocity and pressure.

Since Equation (3.178) is equivalent to $\mathrm{d}(\ln\overline{p_o})/\mathrm{d}x = \gamma_o\mathrm{d}(\ln\overline{\rho_o})/\mathrm{d}x$, the third equation in Equation (3.167) becomes

$$\frac{\partial\varepsilon'}{\partial x} - \left(\frac{\mathrm{i}k_o}{\bar{M}_o} + \frac{\mathrm{d}}{\mathrm{d}x}\ln(\gamma_o\overline{p_o})\right)\varepsilon' = \overline{p'}\frac{\mathrm{d}}{\mathrm{d}x}(\ln\gamma_o). \tag{3.181}$$

This shows that if $\gamma_o =$ constant, the entropy wave decouples from the pressure wave components and is convected with the mean flow with phase velocity $\overline{v_o}$ in the direction of the mean flow. This kind of wave, which is called hydrodynamic wave, is usually not relevant to acoustic phenomena in ducts.

Numerical solutions of Equation (3.167) by using the segmentation method described in Section 3.3.3 show that the off-diagonal terms of the matrizant in Equation (3.168) become small compared to the diagonal terms if the frequency is high enough. For sufficiently high frequencies, solution of Equation (3.167) for the pressure wave components can be expressed as [10]

$$p^+(x) = p^+(0)\exp\left(\int_0^x \left\{\frac{\mathrm{i}k_o}{1+\bar{M}_o} - \frac{1}{2}\left(\frac{\gamma_o\bar{M}_o}{1+\bar{M}_o} + \bar{M}_o\right)\frac{\mathrm{d}}{\mathrm{d}\xi}(\ln\overline{v_o}) - \frac{1}{2\gamma_o}\frac{1}{1+\bar{M}_o}\frac{\mathrm{d}}{\mathrm{d}\xi}(\ln\overline{p_o}) + \frac{1}{2}\frac{\mathrm{d}}{\mathrm{d}\xi}(\ln\overline{\rho_o}\,c_o) - \frac{1}{2}\frac{1+\gamma_o\bar{M}_o}{1+\bar{M}_o}\frac{\mathrm{d}}{\mathrm{d}\xi}(\ln S)\right\}\mathrm{d}\xi\right) \tag{3.182}$$

$$p^-(x) = p^-(0)\exp\left(\int_0^x \left\{\frac{-\mathrm{i}k_o}{1-\bar{M}_o} + \frac{1}{2}\left(\frac{\gamma_o\bar{M}_o}{1-\bar{M}_o} + \bar{M}_o\right)\frac{\mathrm{d}}{\mathrm{d}\xi}(\ln\overline{v_o}) - \frac{1}{2\gamma_o}\frac{1}{1-\bar{M}_o}\frac{\mathrm{d}}{\mathrm{d}\xi}(\ln\overline{p_o}) + \frac{1}{2}\frac{\mathrm{d}}{\mathrm{d}\xi}(\ln\overline{\rho_o}\,c_o) - \frac{1}{2}\frac{1-\gamma_o\bar{M}_o}{1-\bar{M}_o}\frac{\mathrm{d}}{\mathrm{d}\xi}(\ln S)\right\}\mathrm{d}\xi\right). \tag{3.183}$$

Numerical simulations indicate that the condition for this decoupling is only weakly sensitive to the mean-flow Mach number in the low subsonic range. Neglecting the mean flow on this premise, the condition for the foregoing decoupled solution to be valid is [10]:

$$k_o^2 \gg \chi^2 = \max\left\{\left|\psi\frac{\mathrm{d}}{\mathrm{d}x}(\ln c_o) + \frac{\mathrm{d}\psi}{\mathrm{d}x}\right|,\ \frac{1}{4}\left|2\psi\frac{\mathrm{d}}{\mathrm{d}x}(\ln c_o) + 2\frac{\mathrm{d}\psi}{\mathrm{d}x} + \psi^2\right|\right\}, \tag{3.184}$$

where

$$\psi = \frac{\mathrm{d}}{\mathrm{d}x}\left(\ln\frac{S}{\rho_o c_o}\right) + \frac{1}{\gamma_o}\frac{\mathrm{d}}{\mathrm{d}x}(\ln\bar{p}_o). \tag{3.185}$$

Although this criterion implies a lower limiting frequency of about $k_o > 10\chi$, comparison with numerical simulations shows that Equations (3.182) and (3.183) remain adequately accurate for $k_o > \chi$ in the case of constant area ducts ($S =$ constant) for linear mean temperature distributions. For example, for a 1 m long duct having its cold-end at 600 °C, this approximate analytical solution is almost exact for frequencies larger than 10 Hz, up to hot-end temperatures of 780 °C.

From the inverse Fourier transform of Equations (3.182) and (3.183), it can be seen that, $p^+(t,x)$ and $p^-(t,x)$ represent a pair of waves which propagate in the $+x$ and $-x$ directions with local phase velocities of $c_o(1+\bar{M}_o)$ and $c_o(1-\bar{M}_o)$, respectively.

They are attenuated or amplified in the direction of propagation at rates depending on the exponential factors. In a constant cross-sectional area duct, for example, waves traveling in the direction of decreasing (increasing) characteristic impedance tend to be attenuated (amplified).

The literature on plane-wave propagation in ducts with temperature gradients is rich and exact and approximate analytical solutions are available for uniform ducts with various temperature taper functions [30–34].

3.11 Ducts with Two-Phase Flow

In this section, we consider extension of the one-dimensional theory for ducts carrying a two-phase flow. At low frequencies, the one-dimensional theory may be applied, as a first approximation, by using the equilibrium properties of the composition, but the lag between the responses of the phases to acoustic disturbances becomes significant as frequency increases and it is necessary to consider their motions separately. The basic approach for this is founded on the hypothesis that the motions of the phases are governed by the conservation laws of fluid dynamics and the appropriate force and thermal compatibility conditions at the interfaces of the phases [35]. Thus, assuming for the simplicity of presentation that both phases are homogeneous and the duct is hard-walled, we may write the acoustic continuity and momentum equations in each phase from Equations (3.17) and (3.19) as:

$$S\left(\frac{\partial\overline{\rho_k'}}{\partial t}+\overline{v_{ko}}\frac{\partial\overline{\rho_k'}}{\partial x}+\overline{v_k'}\frac{\mathrm{d}\rho_{ko}}{\mathrm{d}x}\right)+\rho_{ko}\frac{\partial}{\partial x}\left(S\overline{v_k'}\right)+\overline{\rho_k'}\frac{\mathrm{d}}{\mathrm{d}x}\left(S\overline{v_{ko}}\right)=0 \tag{3.186}$$

$$S\rho_{ko}\left(\frac{\partial\overline{v_k'}}{\partial t}+\overline{v_{ko}}\frac{\partial\overline{v_k'}}{\partial x}+\overline{v_k'}\frac{\mathrm{d}\overline{v_{ko}}}{\mathrm{d}x}\right)+S\overline{v_{ko}}\frac{\mathrm{d}\overline{v_{ko}}}{\mathrm{d}x}\overline{\rho_k'}+\frac{\partial}{\partial x}\left(S\beta_k\left(\overline{v_{ko}}\right)^2\overline{\rho_k'}\right)+S_k\frac{\partial\overline{p_k'}}{\partial x}=Sf_{kx}', \tag{3.187}$$

where the subscript $k=1,2$ denote a phase, f_{kx} denotes the body force per unit volume of the duct on phase k and the densities are to be understood as the partial densities of the phases. It should be noted that, we have made two improvisations in writing Equation (3.187) from Equation (3.19). The first is the body force term, which comes from the parent conservation equation, Equation (A.19), and is required here for eventual application of the force condition at interfaces of the two phases. The second is the use of areas S_k with the acoustic pressure gradient. These are the cross-section areas associated with phase k and it is assumed that the phases do not overlap, that is, $S=S_1+S_2$. This modification is required, because the fluid pressure forces are determined by integrals over the surfaces of the volumes occupied by the phases.

For a fluid phase, the foregoing equations are augmented by the acoustic energy equation, Equation (3.164), but we may use here instead the relationship $\overline{p_k'}=(c_{ko})^2\overline{\rho_k'}$, $k\in 1,2$, since the phases are assumed to be homogeneous. There is, however, a large variety of two-phase flows depending on combinations of the two phases and also on the interface structures. Here, in view of its pertinence to the focus of this book, we consider application of the one-dimensional theory for aerosols, as aerosols may sometimes represent a more realistic fluid model for water injected exhaust systems and combustors. Aerosols are dispersive two-phase compositions consisting of a gas mixed randomly with large number of solid particles or droplets of small size, about 1-100 microns, which may be modeled as being rigid relative to the host fluid. The reader is referred to Temkin for a comprehensive treatment of the physics of such suspensions [36].

For aerosols, letting $k = 2$ represent the particulate phase, we may take $\overline{p_2'} = 0$ (note that $\overline{\rho_2'} \neq 0$) and $\overline{v_{ko}} = \overline{v_o}$ and $\beta_k = \beta$, $k = 1, 2$. Also, a plausible model for calculating the force applied by the fluid phase on the particulate is the Stokes drag on a sphere. Assuming that the particulate phase may be represented by identical rigid spheres of notional radius a, application of Stokes drag gives $f_{2x} = -6\pi\mu_1 a(\overline{v_2} - \overline{v_1})n$, where μ_1 denotes the shear viscosity of the fluid and n denotes the number of particles per unit volume [36]. Then, neglecting fluctuations of μ_1 and n:

$$f_{2x}' = -6\pi\mu_{1o} a n_o \left(\overline{v_2'} - \overline{v_1'}\right). \tag{3.188}$$

With these stipulations, Equations (3.186)–(3.187) for $k = 1, 2$ contains four acoustic variables and may be solved numerically for the determination of a wave transfer matrix of the duct. Here, the process is demonstrated for a uniform duct with subsonic low Mach number mean flow with a sufficiently flat mean-flow velocity profile (implying that the terms involving β may be neglected). The partial densities of the phases are given by, $\overline{\rho_1} = \overline{\rho_1^*}(1 - \varphi_2)$ and $\overline{\rho_2} = \overline{\rho_2^*}\varphi_2$, respectively, where $\overline{\rho_1^*}$ and $\overline{\rho_2^*}$ denote the stand-alone densities of the fluid and the particulate and φ_2 denotes the volume occupied by the particles per unit volume of the aerosol. Linearization of these gives $\overline{\rho_1'} = \overline{\rho_1^{*\prime}}(1 - \varphi_{2o}) - \rho_{1o}^*\varphi_2' \approx \overline{\rho_1^{*\prime}} - \underline{\rho_{1o}^*\varphi_2'}$ and $\overline{\rho_2'} = \rho_{2o}^*\varphi_2' + \overline{\rho_2^{*\prime}}\varphi_{2o} = \rho_{2o}^*\varphi_2'$, since $\varphi_{2o} <<< 1$ for dilute aerosols and $\overline{\rho_2^{*\prime}} = 0$, since the compressibility of the particles is neglected. The corresponding mean values are $\rho_{1o} = \rho_{1o}^*(1 - \varphi_{2o}) \approx \underline{\rho_{1o}^*}$ and $\rho_{2o} = \rho_{2o}^*\varphi_{2o}$. Thus, Equations (3.186)–(3.187) may be expressed in terms of $\overline{\rho_1^{*\prime}}$ and φ_2' (instead of $\overline{\rho_1'}$ and $\overline{\rho_2'}$), however, we prefer to use them as they are.

For the fluid phase, Equations (3.186)–(3.187) can be expressed, in frequency domain, as

$$-\mathrm{i}k_{1o}\overline{p_1'} + \bar{M}_{1o}\frac{\partial\overline{p_1'}}{\partial x} + \rho_{1o}c_{1o}\frac{\partial\overline{v_1'}}{\partial x} = 0 \tag{3.189}$$

$$\rho_{1o}c_{1o}\left(-\mathrm{i}k_{1o}\overline{v_1'} + \bar{M}_{1o}\frac{\partial\overline{v_1'}}{\partial x}\right) + \frac{S_1}{S}\frac{\partial\overline{p_1'}}{\partial x} = F_d\left(\overline{v_2'} - \overline{v_1'}\right), \tag{3.190}$$

where c_{1o} denotes the speed of sound in the fluid, $k_{1o} = \omega/c_{1o}$, $\bar{M}_{1o} = \overline{v_o}/c_{1o}$, $F_d = 6\pi\mu_{1o} a n_o$ and we may take $S_1 \approx S$, since $\varphi_{2o} <<< 1$ for dilute aerosols. Similarly, Equations (3.186)–(3.187) yield for the particulate phase:

$$-\mathrm{i}\omega\overline{\rho_2'} + \overline{v_o}\frac{\partial\overline{\rho_2'}}{\partial x} + \rho_{2o}\frac{\partial\overline{v_2'}}{\partial x} = 0 \tag{3.191}$$

$$\rho_{2o}\left(-\mathrm{i}\omega\overline{v_2'}+\overline{v_o}\frac{\partial\overline{v_2'}}{\partial x}\right)=-F_d\left(\overline{v_2'}-\overline{v_1'}\right), \tag{3.192}$$

since $f'_{1x}=-f'_{2x}$. The fluid variables $\overline{p_1'}$ and $\overline{v_1'}$ may be solved from Equations (3.189) and (3.190) after eliminating $\overline{v_2'}$ from Equations (3.190) and (3.192), but this leads to a third order differential equation, which is not tractable to an explicit analysis. For sufficiently low mean-flow speeds, however, the gradient term in Equation (1.192) may be neglected as small compared to the remaining terms. Then the linear relationship

$$\overline{v_2'}\approx\frac{1}{1-\mathrm{i}\omega\tau_2}\overline{v_1'}, \tag{3.193}$$

where $\tau_2=\rho_{20}/F_d$ is called particulate dynamic relaxation time, may be used to simplify the analysis. After substituting this relationship and transforming to the pressure wave components

$$\overline{p_1'}=p^++p^-,\quad \rho_{1o}c_{1o}\overline{v_1'}=p^+-p^-. \tag{3.194}$$

Equations (3.189) and (3.190) can be expressed in the familiar state-space form

$$\frac{\partial}{\partial x}\begin{bmatrix}p^+\\p^-\end{bmatrix}=\mathbf{H}\begin{bmatrix}p^+\\p^-\end{bmatrix}, \tag{3.195}$$

where the matrizant is given by

$$\mathbf{H}=\mathrm{i}k_{1o}\begin{bmatrix}\dfrac{1}{1+\bar{M}_{1o}} & 0\\ 0 & \dfrac{-1}{1-\bar{M}_{1o}}\end{bmatrix}\left[\mathbf{I}+\mathrm{i}\frac{\Psi}{2}\begin{bmatrix}1 & -1\\ -1 & 1\end{bmatrix}\right] \tag{3.196}$$

with

$$\Psi=\left(\frac{\varphi_{2o}}{\omega\tau_2+\mathrm{i}}\right)\frac{\rho^*_{2o}}{\rho^*_{1o}}. \tag{3.197}$$

Thus, the wave transfer matrix of the duct may be expressed as

$$\mathbf{T}=\mathbf{\Phi}\begin{bmatrix}e^{\mathrm{i}k_{1o}K^+x} & 0\\ 0 & e^{\mathrm{i}k_{1o}K^-x}\end{bmatrix}\mathbf{\Phi}^{-1}. \tag{3.198}$$

Here, $\mathrm{i}k_{1o}K^{\mp}$ denote the two eigenvalues of the matrizant in Equation (3.196) and the propagation constants K^+ and K^- are given by

$$K^{\mp}=\frac{\mp1}{1\mp\bar{M}_{1o}}\left(\frac{\sqrt{1-\frac{1}{4}\Psi^2\bar{M}_{1o}^2+\mathrm{i}\Psi}\pm\left(\bar{M}_{1o}+\mathrm{i}\frac{1}{2}\Psi\bar{M}_{1o}\right)}{1\pm\bar{M}_{1o}}\right) \tag{3.199}$$

and the modal matrix $\mathbf{\Phi}$ is given by Equation (3.151) with H_{ij}, $i,j = 1,2$, now denoting the elements of the present matrizant. It may be shown by using the Briggs criterion (see Section 1.5) that, the wave modes corresponding to $K^{\mp}$ decay in $\mp x$ direction, respectively.

In Equation (3.199), the factor in large brackets represents the effect of the particulate phase on the propagation constants. The latter are now complex and the pressure wave components attenuate slightly per unit length of the duct in proportion to $|\text{Im}\{K^{\mp}\}|$. However, as $\omega\tau_2 \to \infty$, then $\Psi \to 0$ and the propagation constants tend to be real and equal to the homogeneous fluid values; that is, the particulate phase has no effect on sound propagation in the aerosol. But this may occur when the frequency is large enough. For spherical particulate, the latter is given by $\tau_2 = 2a^2\rho_{2o}^*/9\pi\mu_{1o}$, since $\varphi_{2o} = n_o(4a^3/3)$. For water droplets in air, for example, the dynamic relaxation time is of the order of microseconds and, therefore, this range is well above the audible frequencies. At the other extreme, as $\omega\tau_2 \to 0$, the propagation constants also tend to be real, but slightly greater in modulus than the corresponding homogeneous fluid values. Between these extremes, as $\omega\tau_2 \to 1$, the modulus of the imaginary part of the propagation constants shows a maximum and the modulus of the real part makes an inflection. The variation of $\text{Im}\{K^{\mp}\}$ with the non-dimensional frequency parameter $\omega\tau_2$ and the mean-flow Mach number is shown in Figure 3.11. It is seen that attenuation decreases (increases) with the mean flow for waves propagating with (against) the mean flow. However, the effect of the particulate phase is not substantial when the radian frequency is sufficiently larger or smaller than $1/\tau_2$ and it should be adequately accurate to use the one-dimensional duct elements based on the equilibrium properties of the aerosol.

In some applications, a substantial mean temperature gradient may be present along the duct. An inhomogeneous duct element with two-phase flow may be derived along the similar lines by using the equations given in Section 3.9, but the corresponding wave transfer matrix can be obtained only numerically. Formation of bubbles causes further complexity, because the compressibility of the particulate phase may no longer

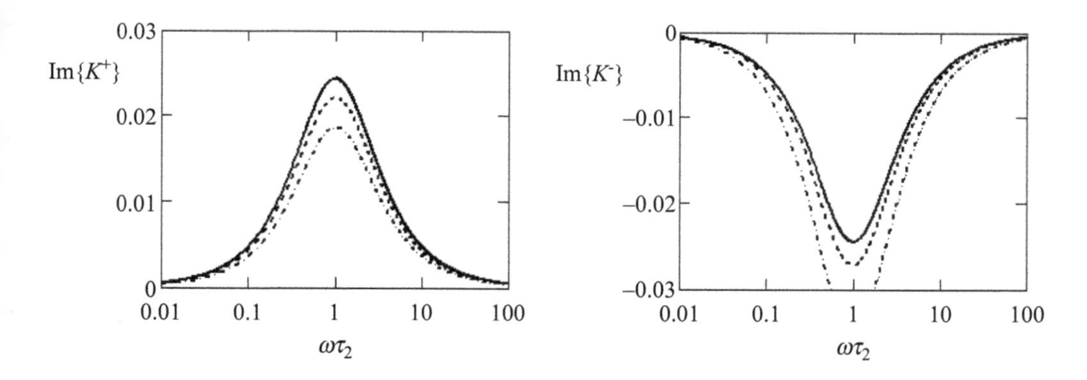

Figure 3.11 Variation of the imaginary part of the propagation constants with $\omega\tau_2$ for an aerosol with $\varphi_{2o} = 0.01$ and $\rho_{2o}^*/\rho_{1o}^* = 10$. Solid: $\bar{M}_{1o} = 0$; dash: $\bar{M}_{1o} = 0.1$; dashdot: $\bar{M}_{1o} = 0.3$.

be neglected [36]. In general, such complications may also be avoided for practical purposes by using, as a first approximation, the equilibrium properties of the mixture.

3.12 Ducts with Time-Variant Mean Temperature

In this section, we consider transmission of sound waves in a uniform duct in which the mean temperature of the gas is uniform along the length of the duct at a given time, but varies periodically with respect to time. The envisaged temperature condition may occur approximately over short lengths of exhaust ducts of reciprocating internal combustion engines. Figure 3.12 shows the measured gas temperature at the exhaust port exit of a steady running engine during an exhaust valve open period [37]. In the blowdown period, the time available for heat transfer is limited and the gas temperature increases substantially. As transition from blowdown flow to displacement occurs, there is more time available for heat transfer and the gas temperature drops until the exhaust valve closes. Thus, the exhaust gas temperature in the runners can vary between high and low values during the open and closed periods of exhaust valves and in a steady running engine these variations may be assumed to be periodic in time of period equal to the cycle period, T, say. Tine-variant temperature conditions also exist in exhaust pipes, particularly under cold-running conditions. However, temporal temperature variations are usually neglected tacitly in practical acoustic calculations on steady running engine exhaust systems on the premise that they are rendered small by heat transfer to the duct walls as the gas moves downstream and eventually have negligible effect on sound propagation.

For the simplicity of the analysis, we consider a hard-walled uniform duct with a perfect gas and assume that the mean pressure, $\overline{p_o}$, is approximately uniform both in space and time, but mean temperature is uniform only spatially and varies periodically with time as $T_o(t) = T_o(t + T)$. In view of heat transfer to duct walls as the gas moves downstream, the assumption that the gas temperature is instantaneously uniform along the duct length is, of course, an approximation. However, as it will transpire

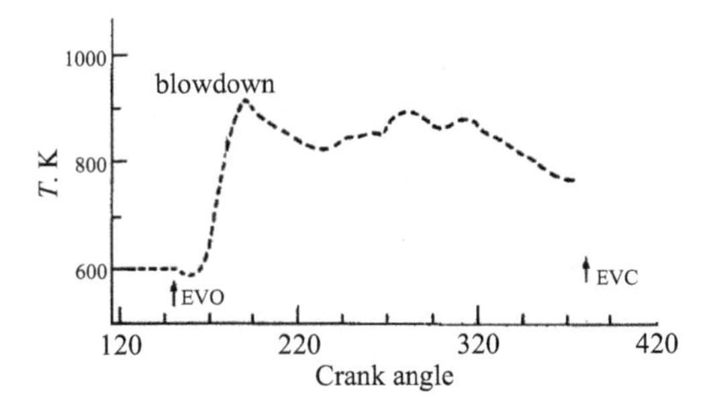

Figure 3.12 Measured gas temperature at the exhaust port exit of a single-cylinder spark-ignition engine at 1000 rpm, EVO: exhaust valve open, EVC: exhaust valve closed (from figure 6.21 in Heywood [37]).

subsequently, the chief phenomenon associated with the effect of time-variant mean temperature is the onset of instability waves and this is not dependent on the duct length. This allows us to work with duct lengths short enough to neglect the effect of axial temperature gradients (Section 3.10.3).

We may also neglect the temporal variations of the ratio of specific heat coefficient, γ_o, as it is a very slowly varying function of the temperature for most perfect gases. Then, for a subsonic low Mach number mean flow ($\alpha \approx 1$), Equations (3.164) and (3.19) reduce to, respectively,

$$\left(\frac{\partial \overline{p'}}{\partial t} + \overline{v_o}\frac{\partial \overline{p'}}{\partial x}\right) + \gamma_o \overline{p_o}\frac{\partial \overline{v'}}{\partial x} = 0 \tag{3.200}$$

$$\overline{\rho_o}\left(\frac{\partial \overline{v'}}{\partial t} + \overline{v_o}\frac{\partial \overline{v'}}{\partial x}\right) + \frac{\partial \overline{p'}}{\partial x} = 0. \tag{3.201}$$

Inspection of the derivation of these equations shows that they are valid also when the mean temperature is time-variant. Hence, eliminating the particle velocity by using the perfect gas relationship $\gamma_o \overline{p_o} = \overline{\rho_o} c_o^2$ gives the wave equation

$$\frac{\partial^2 \overline{p'}}{\partial x^2} - \frac{1}{c_o^2}\left(\frac{\partial \overline{p'}}{\partial t} + \overline{v_o}\frac{\partial \overline{p'}}{\partial x}\right)^2 = 0, \tag{3.202}$$

where $c_o^2 = c_o^2(t) = c_o^2(t+T)$ for periodic mean temperature variations. Note that this model presumes that $\overline{\rho_o} c_o^2$ is constant, implying that the mean gas density is also time-variant, as $\overline{\rho_o} = p_o/RT_o$. But these density fluctuations should not be confused with the acoustic density fluctuations. Indeed, in the reciprocating engine example, the major source of exhaust noise is the same hot exhaust gas blowdown pulse at the ports (see Section 10.2). This process generates acoustic density fluctuations, but these fluctuations propagate with the speed of sound through the exhaust system. What we are concerned here, is the temperature and density of the medium in which acoustic waves propagate after they are generated.

Since Equation (3.202) is not separable for time dependency, it is convenient to introduce the Galilean transformation $y = x - \overline{v_o}t$, $\tau = t$ to reduce it to the form of the classical wave equation

$$\frac{\partial^2 p'}{\partial y^2} - \frac{1}{c_o^2}\frac{\partial^2 p'}{\partial \tau^2} = 0, \tag{3.203}$$

which is separable as $p'(y,\tau) = P(y)\,q(\tau)$, where the separating functions are given by

$$\frac{\mathrm{d}^2 P}{\mathrm{d}y^2} + K^2 P = 0 \tag{3.204}$$

$$\frac{\mathrm{d}^2 q}{\mathrm{d}\tau^2} + K^2 c_o^2(\tau) q = 0, \tag{3.205}$$

where K^2 denotes a real and positive constant. Solution of Equation (3.204) is

$$P(y) = B^{+}\mathrm{e}^{\mathrm{i}Ky} + B^{-}\mathrm{e}^{-\mathrm{i}Ky}, \tag{3.206}$$

where $B^{\pm}$ denote integration constants. Equation (3.205), on the other hand, is the classical Hill equation, solution of which is, by the Floquet theory [38], of the form

$$q(\tau) = \phi(\tau)\mathrm{e}^{\lambda\tau}, \lambda = \frac{\ln\sigma}{T}, \tag{3.207}$$

where σ denotes an eigenvalue of the monodromy matrix, $\mathbf{U}(T)$, of Equation (3.205), which is defined by $\mathbf{q}(\tau + T) = \mathbf{U}(T)\mathbf{q}(\tau)$ with

$$\mathbf{q}(\tau) = \begin{bmatrix} q(\tau) \\ \dfrac{dq}{d\tau} \end{bmatrix}. \tag{3.208}$$

Thus, solution of Equation (3.203) can be expressed as

$$p'(y, \tau) = \left(B^{+}\mathrm{e}^{\mathrm{i}Ky} + B^{-}\mathrm{e}^{-\mathrm{i}Ky}\right)\phi(\tau)\mathrm{e}^{\lambda\tau}. \tag{3.209}$$

An important property of this solution is that the corresponding monodromy matrix, $\mathbf{U}(T)$, is reciprocal. This means that, if an eigenvalue of $\mathbf{U}(T)$ is σ, then $1/\sigma$ is also an eigenvalue. This implies that the characteristic equation of $\mathbf{U}(T)$ is of the form

$$\sigma^2 + a_1\sigma + 1 = 0, \tag{3.210}$$

where a_1 denotes a parameter that is determined by the periodic variations of c_o^2, or the mean temperature variations, since $c_o^2 = \gamma_o R T_o$. Proofs of this reciprocity property are given in books on differential equations with periodic coefficients [38]. Then, if one of the eigenvalues of $\mathbf{U}(T)$ is real and different from unity in absolute value, that is, if $|a_1| > 2$, the wave motion will be unstable (grows indefinitely with time). It should, however, be noted that the blow up of the wave motion with an infinite amplitude is not observed in experiments. The reason for this is that, as amplitudes increase, non-linear effects start taking part and hinder the development of larger amplitudes, resulting in finite amplitude waves [38]. The present linear theory can predict only the onset of instabilities.

If, on the other hand, one of the eigenvalues of $\mathbf{U}(T)$ is complex, that is, if $|a_1| < 2$, then the wave motion will be stable (remains bounded in time). In this case, the two eigenvalues of $\mathbf{U}(T)$ have linearly independent eigenvectors and $\phi(t)$ is a periodic bounded function of period T. The only remaining possibility is that both eigenvalues of $\mathbf{U}(T)$ are real and equal to unity in absolute value, that is, $|a_1| = 2$. In this case, the wave motion will be stable (or unstable) if the corresponding eigenvectors are linearly independent (or dependent). If the eigenvectors are linearly dependent, $\phi(t)$ is a linear polynomial in τ with periodic bounded coefficients of period T and, hence, grows linearly with time. It can be shown that the boundaries separating the regions corresponding to the stable and unstable waves are given by the condition $|a_1| = 2$ [38].

The basic approach for the determination of the stability boundaries is the integration of Equation (3.205) over period T for the determination of its monodromy matrix,

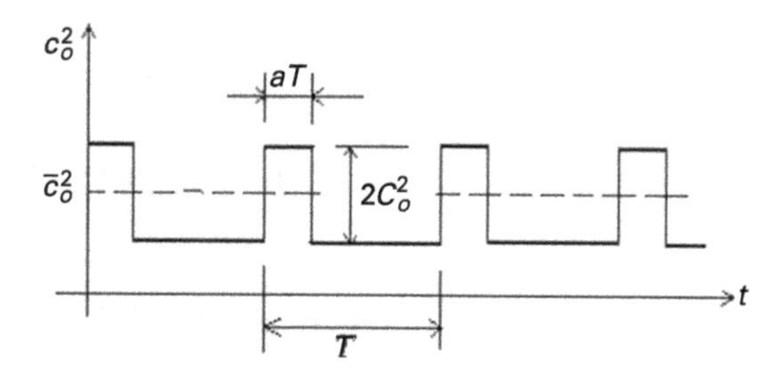

Figure 3.13 Periodic rectangular pulse model of temporal temperature variations.

U(T). Here, we consider, as idealization of actual mean temperature variations like Figure 3.12, the periodic rectangular pulse model shown in Figure 3.13. This assumes that c_o^2 suddenly increases by $2C_o^2$ at the beginning of the period, remains at the maximum value for a fraction a of period T, and then suddenly drops down to its initial value, the mean value of c_o^2 over one period being denoted by $\bar{c}_o^2$. Then, Equation (3.205) can be expressed over one period as:

$$\frac{\mathrm{d}^2 q}{\mathrm{d}t^2} + \omega_1^2 q = 0, \qquad \omega_1^2 = K^2 \bar{c}_o^2 (1 + 2b\beta^2), \qquad 0 < t \leq aT \tag{3.211}$$

$$\frac{\mathrm{d}^2 q}{\mathrm{d}t^2} + \omega_2^2 q = 0, \qquad \omega_2^2 = K^2 \bar{c}_o^2 (1 - 2a\beta^2), \qquad aT < t \leq T \tag{3.212}$$

where $a + b = 1$, $2a\beta^2 < 1$ and $\beta^2 = C_o^2/\bar{c}_o^2$. Solutions of these equations are:

$$q(t) = q(0) \cos \omega_1 t + \frac{\dot{q}(0)}{\omega_1} \sin \omega_1 t, \qquad 0 < t \leq aT \tag{3.213}$$

$$q(t_1) = q(aT) \cos \omega_2 t_1 + \frac{\dot{q}(aT)}{\omega_2} \sin \omega_2 t_1, \quad 0 < t_1 \leq bT \tag{3.214}$$

respectively. Here, $t_1 = t - aT$ and an over-dot denotes differentiation with respect to time. Hence, $q(T)$ and $\dot{q}(T)$ can be calculated from these equation in terms of the initial conditions at $t = 0$ as

$$\begin{bmatrix} q(T) \\ \dot{q}(T) \end{bmatrix} = \begin{bmatrix} U_{11} & U_{12} \\ U_{21} & U_{22} \end{bmatrix} \begin{bmatrix} q(0) \\ \dot{q}(0) \end{bmatrix}, \tag{3.215}$$

where the square matrix is the required monodromy matrix, **U**(T), the elements of which are:

$$U_{11} = \cos \omega_1 aT \cos \omega_2 bT - \frac{\omega_1}{\omega_2} \sin \omega_1 aT \sin \omega_2 bT \tag{3.216}$$

$$U_{12} = \frac{\sin \omega_1 aT \cos \omega_2 bT}{\omega_1} + \frac{\cos \omega_1 aT \sin \omega_2 bT}{\omega_2} \tag{3.217}$$

$$U_{21} = -\omega_2 \cos \omega_1 aT \sin \omega_2 bT - \omega_1 \sin \omega_1 aT \cos \omega_2 bT \tag{3.218}$$

$$U_{22} = -\frac{\omega_2}{\omega_1}\sin\omega_1 aT \sin\omega_2 bT + \cos\omega_1 aT \cos\omega_2 bT. \tag{3.219}$$

It may be shown that the eigenvalues of **U**(T) are given by the roots of Equation (3.210) with

$$a_1 = -2\cos\omega_1 aT\cos\omega_2 bT + \left(\frac{\omega_1}{\omega_2} + \frac{\omega_2}{\omega_1}\right)\sin\omega_1 aT \sin\omega_2 bT. \tag{3.220}$$

Thus, the boundaries separating the regions corresponding to stable and unstable propagating waves may be determined by invoking the condition $|a_1| = 2$.

Representation of stability boundaries on the η versus β^2 plane is called a stability chart. The stability number η is defined as

$$\eta = \frac{K\bar{c}_o}{\Omega/2}, \tag{3.221}$$

where $\Omega = 2\pi/T$. Stability charts for Equation (3.220) are shown in Figure (3.13) for $a = 0.1$ and 0.5. As β^2 tends to zero, the solutions should approach the corresponding frequency domain (stable) solution for constant speed of sound (Section 3.5). This fact – that is, that the $\beta^2 = 0$ axis of a stability chart is stable for all stability numbers – may be used to convert the stability number scale to a frequency scale. Since the eigenvalues $\sigma_{1,2}$ are complex conjugates of modulus unity for stable waves, λ is imaginary, $\lambda = \mp\mathrm{i}\theta/T$, where $\sigma_{1,2} = \mathrm{e}^{\mp\mathrm{i}\theta}$, and, to an observer moving with velocity $\overline{v_o}$, the wave motion will appear as amplitude modulated waves of carrier frequency $\varpi = \mathrm{i}\lambda$. Then, it follows from Equation (3.210) that $a_1 = -2\cos\varpi T$. Upon solving this equation with Equation (3.220) for $\beta^2 = 0$, the required scale transformation transpires as [39]:

$$\eta = \frac{\varpi}{\Omega/2}. \tag{3.222}$$

Thus, the regions of stability occur away from some vicinity of the frequencies $\varpi = n\Omega/2$, $n = 1, 2, \ldots$ Recalling the Doppler effect (Section 3.5.2), the stability number scale may be transformed to the frequency, ω, in the fixed frame by $\omega = (1 + \bar{M}_o)\varpi$, where $\bar{M}_o = \overline{v_o}/\bar{c}_o$.

Prediction of unstable waves is the main feature that distinguishes the present time-variant theory from the usual time-invariant theory. Several conditions increase the likelihood of occurrence of these instabilities in a standard engine exhaust system. The sound waves generated by the exhaust process at a port (see Section 10.2) of a four-stroke engine consist of harmonics of $N/2$ and usually have a decaying spectrum, where N denotes the crankshaft revolutions per second. Thus, for example, in a runner, one has $T = 1/(N/2)$, since the blowdown is the source of both the temperature pulse and the generated sound waves. Accordingly, sound wave motion at frequencies that are in some vicinity of $\omega = 2\pi n(N/2)$, $n = 1, 2, \ldots$, that is, all even harmonics of $N/2$, are likely to be unstable.

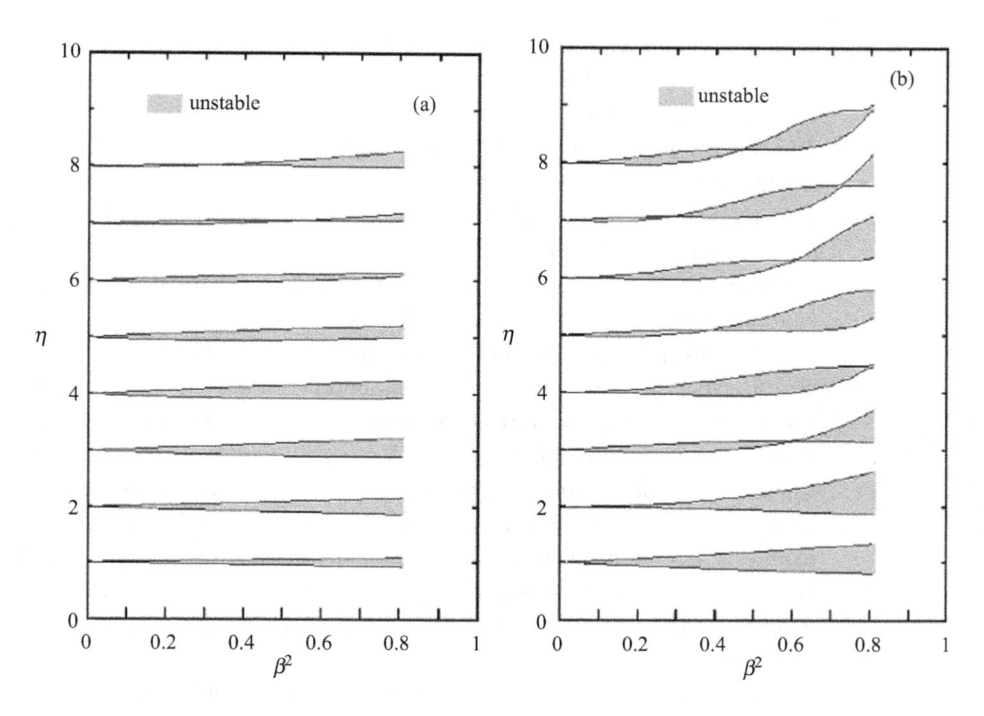

Figure 3.14 Stability charts for a periodic rectangular temperature pulse. (a) $a = 0.1$, (b) $a = 0.5$.

In multi-cylinder engines, as the hot exhaust gas flows downstream of a port it merges with similar flows from the other ports. The actual temperature variations in merged flows are governed by complex heat transfer and fluid dynamic processes. However, for time intervals that are small compared to the thermal time constant associated with heat transfer, the temperature variations in a pipe that carries joining flows may be modeled approximately by juxtaposing, with appropriate phases, the temporal temperature variations of the joining flows. For example, in the exhaust pipe where all port flows merge, assuming that the cylinders are in-line and discharge at equal intervals and that the juxtaposed temperature variations for an exhaust pipe consist of identical elementary pulses that repeat at the same intervals, one has $T \approx 1/(zN/2)$, where z denotes the number of cylinders. For this basic period, it can similarly be concluded that, all even harmonics of $zN/2$ are likely to be unstable. Then, for an engine with an odd number of cylinders, all even harmonics of $N/2$ that are multiples of four are included in the unstable frequencies. For an engine with an even number of cylinders, on the other hand, all harmonics of $N/2$ are included if the number of cylinders is a multiple of four, else, only the even harmonics are included.

The temporal variations of the exhaust gas temperature, and, hence, β^2, will rapidly decrease as the exhaust gas travels towards the tailpipe. Thus, an obvious implication of the stability charts is that wave instabilities are more likely to originate closer to the engine. For a warmed-up engine, β^2 would be relatively small under steady running conditions, but it can be quite large under exacting cold running conditions or very rapid acceleration periods. Also note that widths of instability regions tend to increase with a, the duration of the temperature pulse as a fraction of the period T. For the runners, the periodic rectangular pulse representation with $a \approx 0.25$ may be considered to be a reasonable approximation in a four-stroke cycle under steady running conditions. For an exhaust pipe of a multi-cylinder engine, the value of the parameter a will be substantially larger.

On these premises, it appears that the instabilities are more likely to be excited under cold running or rapid acceleration conditions and close to the engine. When generated, they can travel downstream and excite non-linear resonances in relatively long pipes. The so-called rasping noise (also called abnormal noise [40–41]), in engine exhaust systems may be explained as being such non-linear evolution of this kind of wave instability into finite amplitude waves. Reports suggest that rasping noise (a metallic sound) may be observed in cold running engines and rapidly accelerating engines, and encompass a large number of even harmonics in four-cylinder engines. It usually manifests as non-linear resonances of the pipe between the catalyst and the silencer box and can be eliminated by inserting a small expansion chamber in between. Rasping noise seems to be peculiar to the gasoline engines, which also supports the theory, since diesel engine exhaust temperatures are significantly lower than gasoline engines [40].

The stable waves may be determined by back-substituting the Galilean transformation in Equation (3.209) and noting that Equation (3.222), which is valid for stable waves, implies that $K = \omega/(\bar{c}_o \mp \overline{v_o})$. Then, from $\varpi = \mathrm{i}\lambda$ and the Doppler frequencies, it follows that $\lambda \mp \mathrm{i}K\overline{v_o} = -\mathrm{i}\omega$. Hence, Equation (3.209) may be expressed as the superposition of pressure wave components:

$$p'(x,t) = p^+(0,t)\mathrm{e}^{\mathrm{i}K^+x} + p^-(0,t)\mathrm{e}^{\mathrm{i}K^-x}, \tag{3.223}$$

where

$$p^{\mp}(0,t) = B^{\mp}\mathrm{e}^{-\mathrm{i}\omega t}\phi(t). \tag{3.224}$$

Thus, the one-dimensional acoustic pressure field in the duct is given similarly as in the time-invariant theory (Section 3.4), except for periodic time variation of the pressure wave components at the origin $x = 0$. The particle velocity is given by Equation (3.200) and may be expressed as $\overline{v'}(x,t) = v^+(x,t) + v^-(x,t)$, where

$$v^{\mp}(x,t) = -\frac{1}{\gamma_o \overline{p_o}}\left(\frac{1}{\mathrm{i}K^+}\frac{\partial}{\partial t}p^+(0,t) + \overline{v_o}p^+(0,t)\right)\mathrm{e}^{\mathrm{i}K^+x}. \tag{3.225}$$

Extension of these results to three dimensions and some further aspects of unstable and stable waves are discussed in reference [39].

References

[1] R.A. Frazer, W.J. Duncan and A.R. Collar, *Elementary Matrices and Some Applications to Dynamics and Differential equations*, (Cambridge: Cambridge University Press, 1963).

[2] F.R. Gantmacher, *The Theory of Matrices*, vol. 2, (New York: Chelsea Publishing Company, 1984).

[3] E. Dokumaci, An exact transfer matrix formulation of plane sound wave transmission in inhomogeneous ducts, *J. Sound Vib.* **217** (1998), 869–882.

[4] L.M.B.C. Campos and J.M.G.S. Oliveira, On the acoustic modes in a parabolic shear flow, *J. Sound Vib.* **330** (2011), 1166–1195.

[5] R. Boucheron, H. Bailliet and J.-C. Valiere, Analytical solution of multimodal acoustic propagation in circular ducts with laminar mean flow profile, *J. Sound Vib.* **293** (2006), 504–518.

[6] A.D. Pierce, *Acoustics: An Introduction to Its Physical Principles and Applications,* (New York: McGraw-Hill, 1981).

[7] E. Eisner, Complete solutions of the Webster horn equation, *J. Acoust. Soc. Am.* **46** (1967), 1126–1146.

[8] V. Salmon, A new family of horns, *J. Acoust. Soc. Am.* **17** (1946), 212–218

[9] B.N. Nagarkar and R.D. Finch, Sinusoidal horns, *J. Acoust. Soc. Am.* **50** (1971), 23–31.

[10] E. Dokumaci, An approximate analytical solution for plane sound wave transmission in inhomogeneous ducts, *J. Sound Vib.* **217** (1998), 853–867

[11] E. Dokumaci, A quasi-one-dimensional theory of sound propagation in ducts with mean flow, *J. Sound Vib.* **419** (2018), 1–17.

[12] D.H. Tack and R.F. Lambert, Influence of shear flow on sound attenuation in a lined duct, *J. Acoust. Soc. Am.* **38** (1965), 655–666.

[13] R. Starobinski, Sound propagation in lined duct with essentially non-uniform distribution of velocity and temperature, *Transactions of the Central Institute of Aviation Engine*: *Jet Engine Noise,* CIAM, N 752, Moscow, 1978, pp. 155–181 (an abridged English translation available at: www.researchgate.net/publication/309358647).

[14] H. Schlichting, *Boundary Layer Theory,* (New York: McGraw-Hill, 1979).

[15] Y. Aurégan, R. Starobinski and V. Pagneux, Influence of grazing flow and dissipation effects on the acoustic boundary conditions at a lined wall, *J. Acoust. Soc. Am.* **109** (2001), 59–64.

[16] Y. Aurégan and V. Pagneux, Slow sound in lined flow ducts, *J. Acoust. Soc. Am.* **138** (2015), 605–613.

[17] P.E. Doak and P.G. Vaidya, Attenuation plane wave and higher order mode sound propagation in lined ducts, *J. Sound Vib.* **12** (1970), 201–224.

[18] C.T. Molloy, Propagation of sound in lined ducts, *J. Acoust. Soc. Am.* **16** (1944), 31–37.

[19] R.E. Motsinger and R.E. Craft, Design and performance of duct acoustic treatment, in H.H. Hubbard (ed.), *Aeroacoustics of Flight Vehicles: Theory and Practice, Vol. 2: Noise Control,* NASA RP-1258, (Washington, DC: NASA, 1991), pp. 165–206.

[20] H. Boden, L. Zhou, J. Cordioli, A. Medeiros and A. Spillere, On the effect of flow direction on impedance eduction results, Proceedings of the 22nd AIAA-CEAS Aeroacoustics Conference, No. IAA-2016-2727, Lyon, France, 2016.

[21] W. Watson and M.G. Jones, A comparative study of four impedance eduction methodologies using several test liners, Proceedings of the 19th AIAA/CEAS Aeroacoustics Conference, No. AIAA 2013-2274, Berlin, Germany, 2013.

[22] X. Jing, S. Peng, L. Wang and X. Sun, Investigation of straightforward impedance eduction in the presence of shear flow, *J. Sound Vib.* **335** (2015), 89–104.

[23] C.M. Park, J.G. Ih, Y. Nakayama and S. Kitahara, Single figure rating of porous woven hoses using a non-linear flow resistance model, *J. Sound Vib.* **257** (2002), 404–410.

[24] A. Cummings and R. Kirby, Low frequency sound transmission in ducts with permeable walls, *J. Sound Vib.* 226 (1999), 237–251.

[25] E. Dokumaci, Sound transmission in pipes with porous walls, *J. Sound Vib.* **329** (2010), 5346–5355.

[26] M.C.A. Peters, A. Hirchberg, A.J. Reinjnen and A.P.J. Wijnands, Damping and reflection coefficient measurements for an open pipe at low Mach and low Helmholtz numbers, *J. Fluid Mech.* **256** (1993), 499–534.

[27] M.S. Howe, The damping of sound by wall turbulent shear layers, *J. Acoust. Soc. Am.* **98** (1995), 1725–1730. Also in Howe, *Acoustics of Fluid-Structure Interaction,* (Cambridge: Cambridge University Press, 1998).

[28] E. Dokumaci, On attenuation of plane sound waves in turbulent mean flow, *J. Sound Vib.* **320** (2009), 1131–1136.

[29] C. Weng, S. Boij and A. Hanifi, The attenuation of sound by turbulence in internal flow, *J. Acoust. Soc. Am.* **133** (2013), 3764–3776.

[30] P.O.A.L. Davies, Plane acoustic wave propagation in hot gas flow, *J. Sound. Vib.* **122** (1988), 389–392.

[31] B. Karthik, B.M. Kumar and R.I. Sujith, Exact solutions to one-dimensional acoustic fields with temperature gradient and mean flow, *J. Acoust. Soc. Am.* **108** (2000), 38–43.

[32] A. Cummings, Sound generation and transmission in flow ducts with axial temperature gradients, *J. Sound Vib.* **57** (1978), 261–279.

[33] A. Kapur, A. Cummings and P. Mungur, Sound propagation in a combustion can with axial temperature and density gradients, *J. Sound Vib.* **25** (1972), 129–138.

[34] J. Li and A.S. Morgans, The one-dimensional acoustic field with arbitrary mean axial temperature gradient and mean flow, *J. Sound Vib.* **400** (2017), 248–269.

[35] M. Ishii and T. Hibiki, *Thermo-fluid Dynamics of Two-Phase Flow*, (New York: Springer, 2005).

[36] S. Temkin, *Suspension Acoustics: an Introduction to the Physics of Suspensions*, (Cambridge: Cambridge University Press, 2006).

[37] J.B. Heywood, *Internal Combustion Engine Fundamentals*, (New York: McGraw-Hill, 1988).

[38] V.V. Bolotin *The Dynamic Stability of Elastic Systems*, (San-Francisco: Holden-Day, 1962).

[39] E. Dokumaci, Sound wave motion in pipes having time-variant ambient temperatures, *J. Sound Vib.* **263** (2003), 47–68.

[40] M. Ayadi, S. Ffrikha, P.-Y. Hennion and R. Willats, Characterization of rasping noise in automotive engine exhaust ducts. *J. Sound Vib.* **244** (2001), 79–106.

[41] M. Okada, T. Abe and M. Inaba, Study of the generation mechanism for abnormal exhaust noise, *SAE Transactions* **96** (1987), 943–954.

4 Transmission of One-Dimensional Waves in Coupled Ducts

4.1 Introduction

Some coupled-duct configurations occur recurrently in practical systems. Typical examples are two ducts joined to form a flow expansion or contraction, a junction of a number of ducts or ducts which communicate through apertures on their common walls. Need for these elements arises when we attempt to explode a given duct system into ducts for constructing a block diagram model. In view of the conditions of continuity and compatibility of the acoustic fields, ducts may or may not be exploded as individual units. For example, when ducts are coupled through perforations on their walls, they have to be exploded in packs (Section 4.7 and 4.8). On the other hand, section-wise coupled ducts may, in general, be exploded into the constituent ducts. However, they must be assembled by matching the acoustic fields at their interfaces. This is done conveniently and systematically by formulating the acoustic continuity and compatibility conditions in the format of acoustic block diagram elements (for example, Section 4.3). In this chapter we present one-dimensional acoustic models for several basic configurations of such intermediate acoustic elements and acoustic models for packs of coupled perforated ducts. These are multi-port elements and they can be connected at their ports to the one-dimensional duct elements given in the previous chapter.

4.2 Quasi-Static Theory of Wave Transmission at Compact Junctions

4.2.1 Quasi-Static Conservation Laws

Consider a control volume of volume V of surface σ and assume that: (i) V is acoustically compact and (ii) σ has interfaces with ducts which are modeled by using one-dimensional duct elements such as those described in Chapter 3. Acoustic compactness means that the wavelengths of interest are large compared to characteristic dimensions of the control volume, so that the rates of change of fluctuations of mass, momentum and energy in V may be neglected as small compared to the corresponding fluxes crossing the surface σ. To implement this condition mathematically, we begin with the integral forms of the conservation laws for mass, momentum and energy (see

Appendix A.3), as these equations relate the relevant rate and flux terms. Assuming an inviscid control volume free of sources, they are expressed in time domain as

$$\frac{\partial}{\partial t}\int_V \rho \, \mathrm{d}V + \int_\sigma \rho \, \mathbf{v}\cdot\mathbf{n} \, \mathrm{d}\sigma = 0 \tag{4.1}$$

$$\frac{\partial}{\partial t}\int_V \rho \, \mathbf{v} \, \mathrm{d}V + \int_\sigma \rho\mathbf{v} \;\; \mathbf{v}\cdot\mathbf{n} \, \mathrm{d}\sigma + \int_\sigma p\mathbf{n} \, \mathrm{d}\sigma = 0 \tag{4.2}$$

$$\frac{\partial}{\partial t}\int_V \rho \, e \, \mathrm{d}V + \int_\sigma \rho \, h^o \, \mathbf{v}\cdot\mathbf{n} \, \mathrm{d}\sigma = 0 \tag{4.3}$$

for mass, momentum and energy, respectively. Here, the symbols have the usual meanings: $p, \rho, \mathbf{v}$ denote the fluid pressure, density, particle velocity, respectively, $\mathbf{n}$ denotes the outward unit normal vector of σ, $h^o = h + \mathbf{v}\cdot\mathbf{v}/2$ denotes the specific stagnation enthalpy, h denotes the specific enthalpy and e the specific total energy of the fluid.

Let surface σ be approximated by a number of piecewise contiguous planar surfaces $\sigma_j \subset \sigma$, $j = 1, 2, \ldots$ The first N of these surfaces. $j = 1, 2, \ldots, N$, are the normal sections of the ducts interfacing with the control volume V and represent the ports of the junction. Remaining surfaces of σ are assumed to be impervious but not necessarily hard. The component of the particle velocity normal to each surface is denoted as $\mathbf{v}_j = \mathbf{e}_j v_j$, where $\mathbf{e}_j$ denotes a unit vector (the direction of which is defined by the sign convention used for v_j) normal to surface σ_j, which is parallel to the unit outward normal vector $\mathbf{n}_j$ of σ_j, but not necessarily of the same direction. So, we may write $\mathbf{n}_j = C_j \mathbf{e}_j$, or

$$C_j = \mathbf{e}_j\cdot\mathbf{n}_j, \tag{4.4}$$

where C_j is either -1 or $+1$. Hence, $\mathbf{v}_j\cdot\mathbf{n}_j = C_j v_j$ and, neglecting the contributions of the rate terms as being small compared to the flux terms, Equations (4.1)–(4.3) yield, respectively,

$$\sum_{j=1,2,\ldots} \int_{\sigma_j} \rho_j v_j C_j \mathrm{d}\sigma_j = 0 \tag{4.5}$$

$$\sum_{j=1,2,\ldots} \int_{\sigma_j} \left(p_j + \rho_j v_j^2\right) C_j \mathbf{e}_j \mathrm{d}\sigma_j = 0 \tag{4.6}$$

$$\sum_{j=1,2,\ldots} \int_{\sigma_j} \rho_j v_j h_j^o \, C_j \mathrm{d}\sigma_j = 0, \tag{4.7}$$

where $h_j^o = h_j + v_j^2/2$ and the subscript "j" refers to the properties at the interface σ_j. Integration of Equations (4.5) and (4.6) gives

$$\sum_{j=1,2,\ldots} \overline{\rho_j} \, \overline{\overline{v_j}} S_j C_j = 0 \tag{4.8}$$

$$\sum_{j=1,2,\ldots} \left(\overline{p_j} + \overline{\rho_j}\,\overline{\overline{v_j^2}}\right) S_j C_j \mathbf{e}_j = 0, \tag{4.9}$$

where an overbar denotes ordinary spatial average, a double overbar denotes spatial Favre average, which are defined as in Equations (3.7) and (3.8), and S_j denotes the area of surface σ_j. Linearization of Equation (4.8), as described in Section 3.2.2, yields, for the fluctuations

$$\sum_{j=1}^{N} \left(\bar{\rho}_{jo}\overline{v_j'} + \bar{v}_{jo}\overline{\rho_j'}\right) S_j C_j + \sum_{j=N,N+1,\ldots} \bar{\rho}_{jo}\,\overline{v_j'}\,S_j C_j = 0 \tag{4.10}$$

and

$$\sum_{j=1}^{N} \bar{\rho}_{jo}\,\bar{v}_{jo}\,S_j C_j = 0 \tag{4.11}$$

for the mean flow. Note that, here we use the notation $\overline{q_j} = \left(\overline{q_j}\right)_o + \left(\overline{q_j}\right)' \equiv \bar{q}_{jo} + \overline{q_j'}$, where $q_j \in p_j, \rho_j, v_j, \ldots$, the subscript "$o$" denotes the mean value and a prime denotes acoustic fluctuations, as usual. Equation (4.7) is linearized similarly, giving for the fluctuations and the mean flow

$$\sum_{j=1}^{N} \left(\overline{p_j'} + 2\bar{\rho}_{jo}\bar{v}_{jo}\overline{v_j'} + \left(1+\beta_j\right)\bar{v}_{jo}^2\overline{\rho_j'}\right) S_j C_j \mathbf{e}_j + \sum_{j=N+1,\ldots} \overline{p_j'} S_j C_j \mathbf{e}_j = 0 \tag{4.12}$$

$$\sum_{j=1}^{N} \left(\bar{p}_{jo} + \bar{\rho}_{jo}\left(1+\beta_j\right)\bar{v}_{jo}^2\right) S_j C_j \mathbf{e}_j + \sum_{j=N+1,\ldots} \bar{p}_{jo} S_j C_j \mathbf{e}_j = 0 \tag{4.13}$$

respectively, where

$$1 + \beta_j = \frac{\overline{v_{jo}^2}}{\bar{v}_{jo}^2}. \tag{4.14}$$

Note that, in these equations, the averages are taken over the surfaces σ_j.

In expanding Equation (4.7), we assume that the specific stagnation enthalpy h_j^o is approximately uniformly distributed over the integration surfaces so that it may be replaced by its averages over these surfaces. Then, linearization of Equation (4.7) gives the following equations for the fluctuations and mean parts:

$$\sum_{j=1}^{N} \left(\left(\bar{\rho}_{jo}\,\overline{v_j'} + \bar{v}_{jo}\overline{\rho_j'}\right)(\bar{h}_j^o)_o + \bar{\rho}_{jo}\,\bar{v}_{jo}\,\overline{\left(h_j^o\right)'}\right) S_j C_j + \sum_{j=N+1,\ldots} \bar{\rho}_{jo}\,\overline{v_j'}\,(\bar{h}_j^o)_o S_j C_j = 0 \tag{4.15}$$

$$\sum_{j=1}^{N} \bar{\rho}_{jo}\,\bar{v}_{jo}\,(\bar{h}_j^o)_o S_j C_j = 0, \tag{4.16}$$

respectively. In view of Equation (4.11), the latter equation implies that $(\bar{h}_j^o)_o =$ constant, $j = 1, 2, \ldots, N$. Also noting Equation (4.10), Equation (4.15) may be expressed as

$$\sum_{j=1}^{N} \bar{\rho}_{jo}\,\bar{v}_{jo}\left(\overline{h_j'} + \overline{v_j' v_{jo}}\right) S_j C_j = 0 \tag{4.17}$$

or

$$\sum_{j=1}^{N} \bar{\rho}_{jo}\,\bar{v}_{jo}\left(\overline{T_{jo} s_j'} + \frac{\overline{p_j'}}{\rho_{jo}} + \overline{v_j' v_{jo}}\right) S_j C_j = 0 \tag{4.18}$$

since, from the linearized form of Equation (A.16), the specific enthalpy fluctuations is given by $h' = T_o s' + p'/\rho_o$, where s denotes specific entropy. For the approximately planar distributions envisaged in the one-dimensional theory, Equation (4.18) may be evaluated as

$$\sum_{j=1}^{N} \bar{\rho}_{jo}\,\bar{v}_{jo}\left(\bar{T}_{jo}\,\overline{s_j'} + \frac{\overline{p_j'}}{\bar{\rho}_{jo}} + \bar{v}_{jo}\,\overline{v_j'}\right) S_j C_j = 0. \tag{4.19}$$

In view of Equation (4.11), a sufficient, but not necessary, condition for the satisfaction of this equation is

$$\bar{T}_{jo}\,\overline{s_j'} + \frac{\overline{p_j'}}{\bar{\rho}_{jo}} + \bar{v}_{jo}\,\overline{v_j'} = \text{const}, \qquad j \in \left\{1, 2, \ldots, N \middle| \bar{v}_{jo} S_j \neq 0\right\}. \tag{4.20}$$

4.2.2 Transformation to Pressure Wave Components

The quasi-static junction equations given in the previous section are transformed to the pressure wave components by using Equations (3.23) and (3.24), which are expressed here in the present notation as

$$\overline{\rho_j'} = \frac{\overline{p_j'} + \varepsilon_j'}{c_{jo}^2} \tag{4.21}$$

$$\overline{p_j'} = p_j^+ + p_j^- \tag{4.22}$$

$$\overline{v_j'} = \frac{p_j^+ - p_j^-}{\bar{\rho}_{jo}\,c_{jo}}, \tag{4.23}$$

where c_{jo} denotes the speed of sound at interface σ_j. After substitution of the foregoing three equations, the quasi-static acoustic continuity, momentum and energy equations, Equations (4.10), (4.12) and (4.19) may be expressed, respectively, as

$$\sum_{j=1}^{N}\left(\left[1+\bar{M}_j \quad -1+\bar{M}_j\right]\mathbf{P}_j+\bar{M}_j\varepsilon_j'\right)\frac{S_jC_j}{c_{jo}}+\sum_{j=N,N+1,\ldots}\left[1 \quad -1\right]\frac{S_jC_j}{c_{jo}}\mathbf{P}_j=0 \tag{4.24}$$

$$\sum_{j=1}^{N}\left(\left[\left(1+\bar{M}_j\right)^2+\beta_j\bar{M}_j^2 \quad \left(1-\bar{M}_j\right)^2+\beta_j\bar{M}_j^2\right]\mathbf{P}_j+\left(1+\beta_j\right)\bar{M}_j^2\varepsilon_j'\right)S_jC_j\mathbf{e}_j + \sum_{j=N+1,\ldots}\left[1 \quad 1\right]\mathbf{P}_jS_jC_j\mathbf{e}_j=0 \tag{4.25}$$

$$\sum_{j=1}^{N}\left(\left[1+\bar{M}_j \quad 1-\bar{M}_j\right]\mathbf{P}_j-\frac{\varepsilon_j'}{\gamma_{jo}-1}\right)\bar{v}_{jo}S_jC_j=0, \tag{4.26}$$

where $\bar{M}_j=\bar{v}_{jo}/c_{jo}$ and the port pressure wave components vectors are defined as usual:

$$\mathbf{P}_j=\begin{bmatrix} p_j^+ \\ p_j^- \end{bmatrix}. \tag{4.27}$$

In this notation, Equation (4.20) becomes

$$\frac{1}{\bar{\rho}_{jo}}\left(\left[1+\bar{M}_j \quad 1-\bar{M}_j\right]\mathbf{P}_j-\frac{\varepsilon_j'}{\gamma_{jo}-1}\right)=\text{constant}, \quad j\in\left\{1,2,\ldots,N|\bar{M}_jS_j\neq 0\right\}. \tag{4.28}$$

In subsequent sections, these equations are used in analysis of sound wave transmission at compact junctions of ducts. The entropy wave component at inflow ports of a junction depend on the upstream processes and, therefore, should be specified. For simplicity of the analysis, we will assume that they are zero, as non-zero input values may readily be included in the results by straightforward extension of the algebra. But, since complex flows in junctions usually involve irreversible flow losses, the entropy wave component at an outflow port may be of concern. It can be included in the formulation in some special cases (see Section 4.3), however, there is no general predictive procedure in the framework of the foregoing linearized quasi-static equations. For this reason, they are also neglected, unless stated otherwise. Calculations with the effect of the entropy wave component included show that this is justified at subsonic low Mach number mean flow.

Since we are concerned with subsonic low Mach numbers ($\bar{M}_j^2 << 1$), the terms involving β_j may be neglected in Equation (4.25), for practical mean flow velocity profiles (see Section 3.2.2). Also, we will neglect the variations of the mean temperature and pressure within the control volume. Accordingly, we take $\bar{\rho}_{jo}=\rho_o$ and $c_{jo}=c_o, j=1,2,\ldots$, in the foregoing equations, unless stated otherwise.

4.3 Two Ducts Coupled by Forming Sudden Area Change

In this section we consider two straight ducts of arbitrary cross section, which are coupled by extending one duct partially inside the other, with their axes being parallel

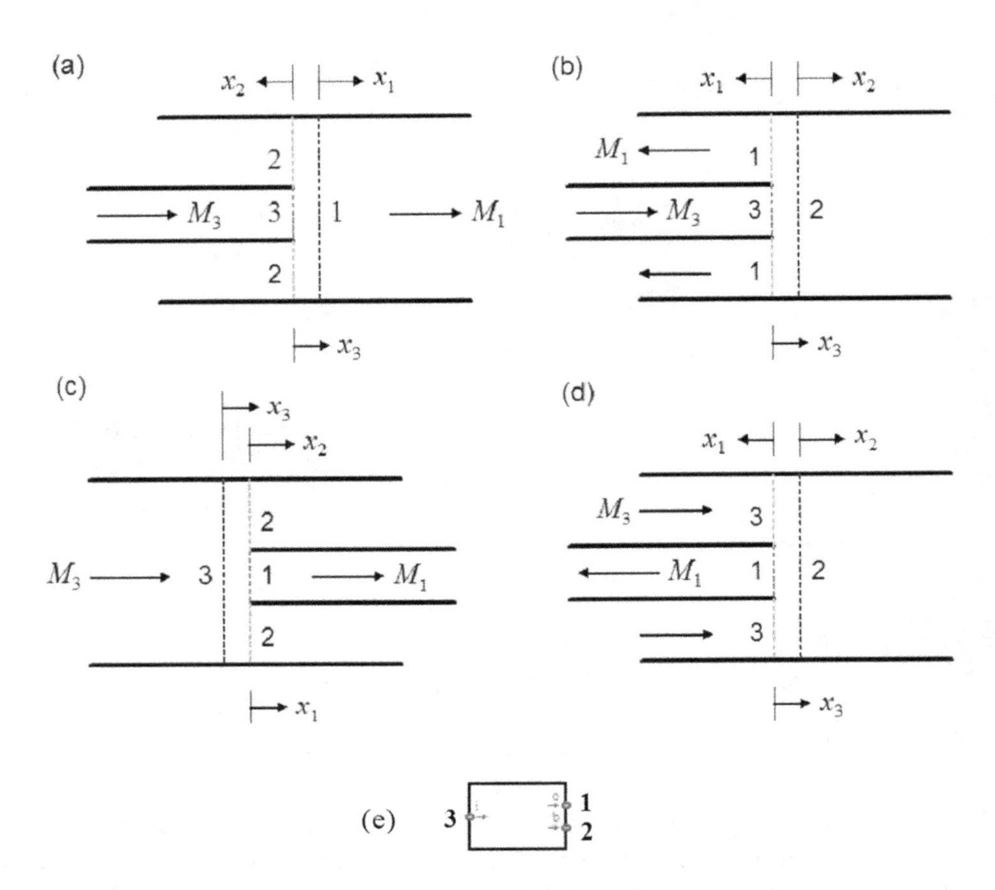

Figure 4.1 Sudden area changes of the open type: (a) Through-flow expansion, (b) flow-reversing expansion, (c) through-flow contraction, (d) flow-reversing contraction and (e) block representation of the four configurations.

but not necessarily collinear. A sudden area change is called expansion or contraction, depending on whether the mean flow enters the junction from the smaller duct or leaves the junction from the smaller duct.

Shown in Figure 4.1 are typical expansion and contraction type area changes. The control volumes, which are assumed to be axially compact and fully enclose the area discontinuity, are indicated by dashed lines. The configuration shown in Figure 4.1a (4.1c) is called through-flow expansion (contraction), because the inlet mean flow in the smaller (larger) duct is assumed to flow out mainly into the larger (smaller) duct. In the configuration of Figure 4.1b (4.1d), on the other hand, the inlet mean flow in the smaller duct (annulus) is assumed to be forced to change direction and flow out mainly through the annulus (smaller duct). For this reason, this configuration is called flow-reversing expansion (contraction). Note that, for the four configurations considered, the ducts are numbered differently, so that the duct containing the inlet and outlet mean flows are always duct 3 and duct 1, respectively, and the duct which may contain negligible secondary mean flow is called duct 2 (also called side-branch duct subsequently). The interfaces of ducts 1, 2 and 3 with the control volume represent port 1 (outlet), port 2 (side-branch) and port 3 (inlet), respectively, of the area change. The default axes of the three ducts are taken as shown in Figure 4.1. The directions of these axes also define the default positive acoustic flow directions at the ports.

Clearly, the sudden area-change configurations shown in Figure 4.1 imply that the side-branch duct terminates with a boundary, such that the inlet mean flow is not diverted into it substantially. Transmission of sound waves can be modeled with or without including the side-branch termination path in the formulation. If the side-branch termination is included in the formulation, the area change is called closed, else it is called open. A closed area change is represented by an acoustic two-port element and can be cascaded directly with other two-port elements. An open area change is represented by an acoustic three-port element and can be assembled with other acoustic elements by using the global matrix method. We present formulation of the wave transmission equation of open area changes first, as the wave transfer matrix of closed area changes can be derived from it.

4.3.1 Open Area Change

Referring to Figure 4.1, let S_1, S_2 and S_3 be the inner cross-sectional areas of ports 1, 2, 3, respectively, and $\bar{M}_1$, $\bar{M}_2$ and $\bar{M}_3$ denote, respectively, the corresponding average mean flow Mach numbers, where $\bar{M}_j = \bar{v}_{jo}/c_o$. The control volume includes also the thickness surface of the smaller duct, S_4, say, and segments of peripheral surfaces of the ducts in the control volume. These surfaces may be assumed to be hard without substantial loss of accuracy. Then, Equation (4.24), reduces to

$$\sum_{j=1}^{3}\left(\begin{bmatrix} 1+\bar{M}_j & -1+\bar{M}_j \end{bmatrix}\mathbf{P}_j + \bar{M}_j\varepsilon_j' \right) S_j C_j = 0, \tag{4.29}$$

where, for the default positive acoustic flow directions shown in Figure 4.1 (see Equation (4.4)): $C_1 = C_2 = 1$ and $C_3 = -1$. Neglecting the terms involving β_j, the quasi-static momentum equation, Equation (4.25), becomes

$$\sum_{j=1}^{3}\left(\left[(1+\bar{M}_j)^2 \quad (1-\bar{M}_j)^2 \right]\mathbf{P}_j + \bar{M}_j^2\varepsilon_j' \right) S_j B_j + \sum_{j=4,5,\ldots} \begin{bmatrix} 1 & 1 \end{bmatrix}\mathbf{P}_j S_j B_j = 0, \tag{4.30}$$

where the coefficients B_j are defined as $B_j = \mathbf{n}_3 \cdot \mathbf{n}_j$, $j = 1, 2, 3$, and for the default positive acoustic flow directions shown in Figure 4.1, have the values given in Table 4.1. Since the acoustic pressure on S_4 and other elementary surfaces in the control volume are not known, Equation (4.30) is not useful, unless we may assume that the second summation on the left-hand side is negligibly small compared to the first one. This may be of adequate accuracy, if the thickness of the inner duct and the thickness of the control volume are sufficiently compact.

Finally, from the quasi-static energy equation, Equation (4.26), we have

Table 4.1 Values of default positive acoustic flow directions shown in Figure 4.1

	Exp (TF) Figure 4.1a	Exp (FR) Figure 4.1b	Con (TF) Figure 4.1c	Con (FR) Figure 4.1d
B_1	-1	$+1$	-1	$+1$
B_2	$+1$	-1	-1	-1
B_3	$+1$	$+1$	$+1$	$+1$

Note: Exp = expansion, Con = contraction, TF = through flow, FR = flow reversal.

$$\sum_{j=1}^{3}\left(\begin{bmatrix}1+\bar{M}_j & 1-\bar{M}_j\end{bmatrix}\mathbf{P}_j-\frac{\varepsilon_j'}{\gamma_{jo}-1}\right)\bar{v}_{jo}\,S_jC_j=0 \tag{4.31}$$

or its sufficient alternative

$$\begin{bmatrix}1+\bar{M}_j & 1-\bar{M}_j\end{bmatrix}\mathbf{P}_j-\frac{\varepsilon_j'}{\gamma_{jo}-1}=\text{constant},\quad j\in\left\{1,2,3|\bar{M}_jS_j\neq 0\right\}. \tag{4.32}$$

In order to derive a wave transmission equation between $\mathbf{P}_1$, $\mathbf{P}_2$ and $\mathbf{P}_3$, we need to have three scalar equations between these pressure wave components. This follows from the fact that a one-dimensional three-port wave transmission equation should reduce to a two-port wave transfer equation when one of the ports is closed.

The following options may be considered for the derivation of a three-port wave transmission equation from Equations (4.29)–(4.32):

(a) Neglect the entropy wave components at the ports (ε_j', $j=1,2,3$) and the hard-surface terms in Equation (4.30). Then, Equations (4.29)–(4.31) reduce, respectively, to

$$\sum_{j=1}^{3}S_jC_j\begin{bmatrix}1+\bar{M}_j & -1+\bar{M}_j\end{bmatrix}\mathbf{P}_j=0 \tag{4.33}$$

$$\sum_{j=1}^{3}S_jB_j\begin{bmatrix}(1+\bar{M}_j)^2 & (1-\bar{M}_j)^2\end{bmatrix}\mathbf{P}_j=0 \tag{4.34}$$

$$\sum_{j=1}^{3}\bar{M}_jS_jC_j\begin{bmatrix}1+\bar{M}_j & 1-\bar{M}_j\end{bmatrix}\mathbf{P}_j=0, \tag{4.35}$$

where $\bar{M}_2=0$. These equations can be expressed in the form of Equation (2.18) with $n=6$ and $k=3$.

(b) Neglect the entropy wave components at the ports (ε_j', $j=1,2,3$), discard Equation (4.30) and use Equation (4.32). The latter yields the two equations

$$\begin{bmatrix}1+\bar{M}_1 & 1-\bar{M}_1\end{bmatrix}\mathbf{P}_1=\begin{bmatrix}1+\bar{M}_2 & 1-\bar{M}_2\end{bmatrix}\mathbf{P}_2=\begin{bmatrix}1+\bar{M}_3 & 1-\bar{M}_3\end{bmatrix}\mathbf{P}_3 \tag{4.36}$$

which are sufficient for closing Equation (4.33). It should be noted that, here Equation (4.32) is invoked on the understanding that $\bar{M}_2$ is not zero, but negligibly small. Again, Equations (4.32) and (4.35) can be put to the form of Equation (2.18) with $n = 6$ and $k = 3$.

(c) Neglect the entropy wave component at the inlet and side-branch ports (ε_j', $j = k, 3$, $k \in 1, 2$) and the hard-surface terms in Equation (4.30). Then, Equation (4.32) similarly yields two equations that are sufficient to close these equations. The resulting set of equations may be expressed as

$$\sum_{j=1}^{3} S_j C_j \left[1 + \bar{M}_j \quad -1 + \bar{M}_j \right] \mathbf{P}_j + S_k C_k \bar{M}_k \varepsilon_k' = 0 \tag{4.37}$$

$$\sum_{j=1}^{3} S_j B_j \left[(1 + \bar{M}_j)^2 \quad (1 - \bar{M}_j)^2 \right] \mathbf{P}_j + S_k C_k B_k \bar{M}_k^2 \varepsilon_k' = 0 \tag{4.38}$$

$$\left[1 + \bar{M}_3 \quad 1 - \bar{M}_3 \right] \mathbf{P}_3 = \left[1 + \bar{M}_{k'} \quad 1 - \bar{M}_{k'} \right] \mathbf{P}_{k'} = \left[1 + \bar{M}_k \quad 1 - \bar{M}_k \right] \mathbf{P}_k - \frac{\varepsilon_k'}{\gamma_{ko} - 1} \tag{4.39}$$

where $k' = \{1, 2| \neq k\}$. After elimination of ε_k', these equations also reduce to the form of Equation (2.18) with $n = 6$ and $k = 3$.

In general, the methods considered above lead to a three-port wave transmission matrix of size 3×6, which can be expressed in the generic form

$$\mathbf{A}_1 \mathbf{P}_1 + \mathbf{A}_2 \mathbf{P}_2 + \mathbf{A}_2 \mathbf{P}_3 = \mathbf{0}. \tag{4.40}$$

Figure 4.1e shows the block representation used in this book for an open area-change. The arrows by the ports indicate, as usual, the default positive acoustic flow directions. These directions may be changed as described in Section 2.3.3. It should be noted that, the wave transmission matrix in Equation (4.40), $[\mathbf{A}_1 \quad \mathbf{A}_2 \quad \mathbf{A}_3]$, has only real elements, which means that a sudden area change modifies the magnitudes of the pressure wave components at the ports, but has no effect on their phases. An example of using the through-flow type open expansion in block diagrams is given in Section 2.4.2 (Figure 2.6).

Experimental justification of these approaches at subsonic low Mach numbers has been provided in the pioneering work of Alfredson and Davies on closed through-flow area changes and confirmed by many later works on low-frequency modeling of silencers and resonators involving sudden area changes [1]. Methods (a) and (b) may be considered for the modeling of a through-flow contraction (Figure 4.1c). In this case, the inlet mean flow converges into the smaller duct and, after entrance, its cross section becomes slightly less than the duct cross section, forming the so-called vena contracta, before it expands to fill the duct cross section. Irreversible flow losses may occur at expansion after vena contracta, but this is not significant. This justifies neglect of the entropy wave component at the outlet port of a contraction. Method (b) seems to be preferred by some authors [1–2], but method (a) also produces similar results.

On the other hand, at a through-flow expansion (Figure 4.1a), the inlet mean flow forms a jet at the outlet of the inlet duct, which expands and fills the duct cross section after a finite distance. In view of the irreversible losses associated with the jet, entropy wave component is generated downstream of the area expansion and convects with the mean flow. If this occurs inside the control volume [3], then ε_1' should be included in the analysis and method (c) is appropriate for this [1–2]. Since four scalar equations are needed now, the quasi-static energy equation is implemented by invoking its sufficient alternative, Equation (4.31b). However, if the entropy wave is assumed to be generated outside the control volume, method (a) or (b) may be used. In fact, method (b) is considered in Alfredson and Davies [1], and found to give as good results as method (c).

4.3.1.1 The Case of Zero Mean Flow

It is instructive to examine how the foregoing methods work when there is no mean flow. Starting with method (a), Equation (4.33) becomes

$$S_1C_1[1 \quad -1]\mathbf{P}_1 + S_2C_2[1 \quad -1]\mathbf{P}_2 + S_3C_3[1 \quad -1]\mathbf{P}_3 = 0. \tag{4.41}$$

Equation (4.35) becomes trivial and Equation (4.34) yields

$$S_1B_1[1 \quad 1]\mathbf{P}_1 + S_2B_2[1 \quad 1]\mathbf{P}_2 + S_3B_3[1 \quad 1]\mathbf{P}_3 = 0. \tag{4.42}$$

But,

$$S_1B_1 + S_2B_2 + S_3B_3 = 0, \tag{4.43}$$

since the thickness of the inner duct is neglected. Hence, under this condition, Equations (4.41) and (4.42) can be solved by assuming, for example,

$$[1 \quad 1]\mathbf{P}_1 = [1 \quad 1]\mathbf{P}_3. \tag{4.44}$$

This solution will satisfy the condition $[1 \quad 1]\mathbf{P}_1 = [1 \quad 1]\mathbf{P}_2$ also.

Since methods (b) and (c) are based on Equation (4.32), they may be applied in the limit of negligible mean flow. Then, in method (b), Equation (4.39) may be complemented with $[1 \quad 1]\mathbf{P}_1 \approx [1 \quad 1]\mathbf{P}_2 \approx [1 \quad 1]\mathbf{P}_3$ as the mean flow tends to zero, which is the same as the solution given by method (a). Method (c) also yields the same solution in the limit of negligible mean flow, because ε_k' vanishes if there is no mean flow.

Thus, in the absence of mean flow, the wave transmission equation of a sudden area change may be expressed as, for example,

$$\begin{bmatrix} S_1 & -S_1 \\ B_1S_1 & B_1S_1 \\ 1 & 1 \end{bmatrix}\mathbf{P}_1 + \begin{bmatrix} S_2 & -S_2 \\ B_2S_2 & B_2S_2 \\ 0 & 0 \end{bmatrix}\mathbf{P}_2 + \begin{bmatrix} S_3 & -S_3 \\ B_3S_3 & B_3S_3 \\ -1 & -1 \end{bmatrix}\mathbf{P}_3 = 0 \tag{4.45}$$

where Equation (4.43) is implied.

4.3.1.2 Core-Flow Model

In general, non-uniform mean flows are common at sudden area-changes. Apart from expanding and contracting core flows with flow sections varying axially over a finite distance, complex vortex flows may be present outside the core flow and in the side-branch duct. Such mean flow non-uniformities become even more pronounced in the flow-reversing configurations (Figure 4.1b,d), due to forcing of the reversal of the mean flow. We neglect the effects of such non-uniform flows, because they are difficult to specify with certainty. However, there arises, as discussed above, the question of whether or not irreversible losses occur within the assumed axially compact control volumes. For simplicity of the analysis, it is advantageous to be able to discard this issue, and one way to do this is to consider an infinitesimally thin control volume that excludes the flow losses. This allows deferral of the effects of the discontinuity flow losses to outside the control volume, where they may be neglected or replaced by simple flows for practical purposes. But now it is necessary to distinguish the mean flow and the duct cross section areas. One possible scheme for a through-flow type area change is to take the mean flow cross section in the control volume equal to the inner cross section area of the smaller duct. This implies that only negligible mean flow crosses the control volume into the side-branch duct. On the other hand, a reasonable scheme for a reversing-flow type area change is to let the mean flow in the control volume have the direction and cross section as the flows in the inlet and outlet ducts.

These mean flow schemes are depicted in Figure 4.2 for the four area-change configurations in Figure 4.1. They can be implemented in a unified manner by splitting the port areas as $S_j = S_j' + S_j''$, $j = 1, 2, 3$, where S_j' and S_j'' denote, respectively, the section areas of core and annular (if any) mean flows, with S_j and the corresponding $\bar{M}_j$ being defined as before. But splitting of flow areas in this way precludes the use of Equation (4.32). Consequently, we proceed with the assumptions of method (a). Thus, Equations (4.33)–(4.35) come out now, respectively, as:

$$\sum_{j=1}^{3} C_j\left(\left[1+M_j' \quad -1+M_j'\right]S_j' + \left[1+M_j'' \quad -1+M_j''\right]S_j''\right)\mathbf{P}_j = 0 \tag{4.46}$$

$$\sum_{j=1}^{3} B_j\left(\left[\left(1+M_j'\right)^2 \quad \left(1-M_j'\right)^2\right]S_j' + \left[\left(1+M_j''\right)^2 \quad \left(1-M_j''\right)^2\right]S_j''\right)\mathbf{P}_j = 0 \tag{4.47}$$

$$\sum_{j=1}^{3} C_j\left(\left[1+M_j' \quad 1-M_j'\right]M_j'S_j' + \left[1+M_j'' \quad 1-M_j''\right]M_j''S_j''\right)\mathbf{P}_j = 0, \tag{4.48}$$

where M_j' and M_j'' have the values defined in Table 4.2. Note that, in this scheme, $\bar{M}_1 S_1 = \bar{M}_3 S_3$ and Equation (4.11) may be written as $\sum_{j=1}^{3} C_j\left(M_j'S_j' + M_j''S_j''\right) = 0$. The entries in Table 4.2 satisfy this equation identically (recall that $C_1 = C_2 = 1$ and $C_3 = -1$). It is also noteworthy that the products $M_j'S_j'$ and $M_j''S_j''$ in Equation (4.48) actually cancel out when this equation is applied to particular area-change configurations with mean flow. Obviously, Equations (4.46)–(4.48) can also be collated in the form of Equation (4.40) with a wave transmission matrix of size 3×6.

The following APL algorithm computes the wave transmission matrix of the open area-change configurations in Figure 4.1 using Equations (4.46)–(4.48).

Table 4.2 Values for M_j' and M_j''

Parameter	Exp(TF) Figure 4.1a	Exp(FR) Figure 4.1b	Con(TF) Figure 4.1c	Con(FR) Figure 4.1d
S_1', M_1'	$S_3, \bar{M}_3$	$S_1, \bar{M}_1$	$S_1, \bar{M}_1$	$S_1, \bar{M}_1$
S_1'', M_1''	S_2,0	0,0	0,0	0,0
S_2', M_2'	$S_2, \approx 0$	$S_3, \bar{M}_3$	$S_2, \approx 0$	$S_1, -\bar{M}_1$
S_2'', M_2''	0,0	$S_1, -\bar{M}_1$	0,0	$S_3, \bar{M}_3$
S_3', M_3'	$S_3, \bar{M}_3$	$S_3, \bar{M}_3$	$S_1, \bar{M}_1$	$S_3, \bar{M}_3$
S_3'', M_3''	0,0	0,0	S_2,0	0,0

Note: Exp = expansion, Con = contraction, TF = through flow, FR = flow reversal.

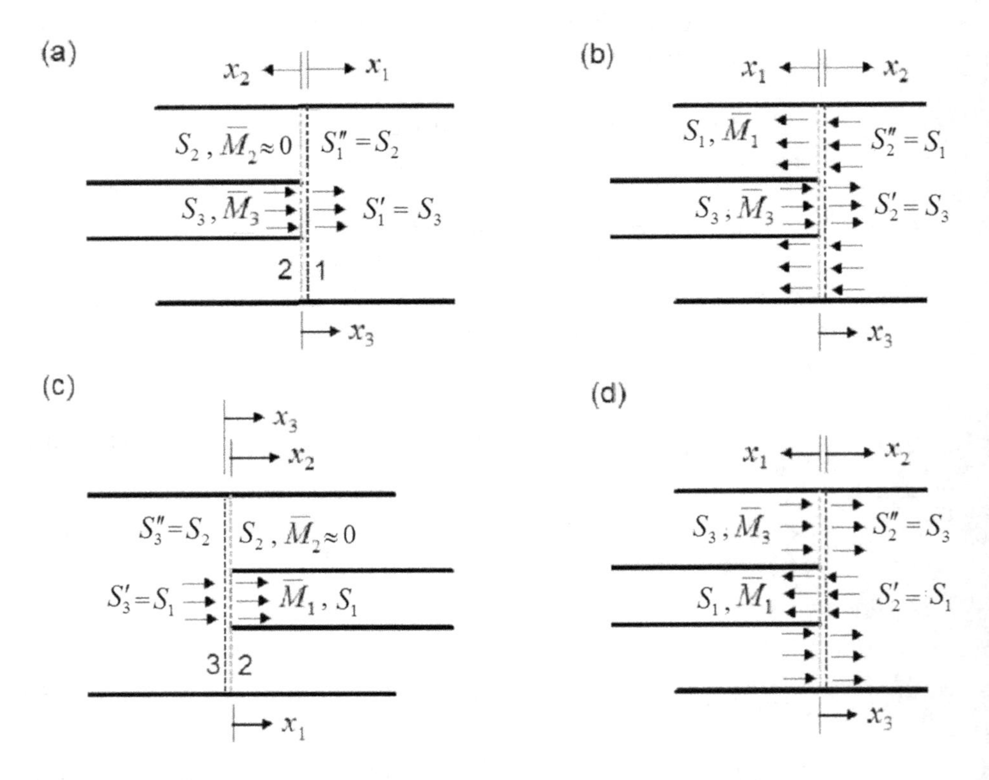

Figure 4.2 Mean flow section areas for the core flow model. (a) Through-flow expansion, (b) flow-reversing expansion, (c) through-flow contraction, (d) flow-reversing contraction.

```
R←OpenAreaChange dat;M;S;B;I;J;K;I1;A;mode;ext;dir
```

⍝ Output: R=wave transmission matrix

⍝ Input: *dat*=vector of data which the program uses to fetch from the input data file the parameters required in calculations.

⍝ Fetch input data from file.

⍝ Determine from input data type of area-change, put in *mode* by following scheme:
⍝ *mode*=1: through-flow expansion (Figure 4.1a)
⍝ *mode*=2: flow-reversing expansion (Figure 4.1c)
⍝ *mode*=3: through-flow contraction (Figure 4.1b)
⍝ *mode*=4: flow-reversing contraction (Figure 4.1d)
⍝ Place incidence of mean flow in Table 4.2 in a 3D array by following scheme: 0 means no mean flow, ∓1 means mean flow: minus sign indicates that mean flow is in opposite direction to the default positive direction at a port:

```
B←4 3 2⍴ 1 0 0 0 1 0  1 0 1 ¯1 1 0  1 0 0 0 1 0  1 0 ¯1 1 1 0
```

⍝ Here, the incidence array has four 2D blocks in the order of *mode*=1,2,3,4:

⍝ $B[1;\,;]=\begin{bmatrix}1 & 0\\ 0 & 0\\ 1 & 0\end{bmatrix},\ B[2;\,;]=\begin{bmatrix}1 & 0\\ 1 & -1\\ 1 & 0\end{bmatrix},\ B[3;\,;]=\begin{bmatrix}1 & 0\\ 0 & 0\\ 1 & 0\end{bmatrix},\ B[4;\,;]=\begin{bmatrix}1 & 0\\ 1 & -1\\ 1 & 0\end{bmatrix}$

⍝ In each block, the first column gives incidence of core flow in the order of ports 1,2,3; the second column gives incidence of annulus flow in the same order.
⍝ Compute S_1, S_2, S_3 from input data and put them in array *S* temporarily:

```
S←S₁,S₂,S₃
```

⍝ Place S_1, S_2, S_3 entries in Table 4.2 in 3D array *S*

```
S←S,0
S←4 3 2⍴ S[3 2 2 4 3 4 1 4 3 1 3 4 1 4 2 4 1 2 1 4 1 3 3 4]
```

⍝ Here, the four 2D blocks of *S* are in the order of *mode*=1,2,3,4:

⍝ $S[1;\,;]=\begin{bmatrix}S_3 & S_2\\ S_2 & 0\\ S_3 & 0\end{bmatrix},\ S[2;\,;]=\begin{bmatrix}S_1 & 0\\ S_3 & S_1\\ S_3 & 0\end{bmatrix},\ S[3;\,;]=\begin{bmatrix}S_1 & 0\\ S_2 & 0\\ S_1 & S_2\end{bmatrix},\ S[4;\,;]=\begin{bmatrix}S_1 & 0\\ S_1 & S_3\\ S_3 & 0\end{bmatrix}$

⍝ In each block, the first column gives core areas in the order of ports 1,2,3; the second column gives the annulus area in the same order.
⍝ Compute $\bar{M}_1, \bar{M}_2(=0), \bar{M}_3$ from input data and put them in array M temporarily:

```
M←M̄₁, M̄₂, M̄₃
```

⍝ Place $\bar{M}_1, \bar{M}_2, \bar{M}_3$ in a 3D array *M* in one-to-one correspondence with *S*

```
M←M,0
M←4 3 2⍴ M[3 2 2 4 3 4 1 4 3 1 3 4 1 4 2 4 1 2 1 4 1 3 3 4]
```

⍝ Place default C_1, C_2, C_3 and B_1, B_2, B_3 (corresponding to sign convention of Figure 4.1) in a 3D array *K* by following scheme:

```
K←4 3 2⍴ 1 ¯1 1 1 ¯1 1 1 1 1 ¯1 ¯1 1 1 ¯1 1 ¯1 ¯1 1 1 1 1 ¯1 ¯1
```

⍝ where *K[mode;;]*= $\begin{bmatrix}C_1 & B_1\\ C_2 & B_2\\ C_3 & B_3\end{bmatrix}_{\text{mode}}$ (see text and Table 4.1)

⍝ Determine from input data the default positive acoustic flow direction change flags for ports 1,2 3, and put them in 3 element vector *dir* in the same order by following scheme:
⍝ *dir[J]*=1 / 0 : changed/unchanged at port *J*=1,2,3
⍝ Update *B* and *K* for +*x* direction changes:

```
:for I :in 1 2 3
     :if dir[I]=1
          B[mode;I;]←-B[mode;I;]
          K[mode;I;1]←-K[mode;I;1]
     :endif
:endfor
```

⍝ Initialize the wave transmission matrix as a 3 by 6 complex matrix:

```
R←2 3 6ρ0
```

⍝ Place continuity equation, Equation (4.34), in the wave transmission matrix:

```
:for I :in 1 2 3
     I1←(¯1+2×I),2×I
     :for J :in 1 2
          A←    B[mode;I;J]×M[mode;I;J]
          A←(1+A),¯1+A
          R[1;1;I1]←R[1;1;I1]+A×K[mode;I;1]×S[mode;I;J]
     :endfor
:endfor
```

⍝ Note that here the inner loop calculates the inner two-term sum in Equation (4.34) and the outer loop does the summation indicated by the summation symbol in Equation (4.34). This iteration is used similarly when placing Equations (4.35) and (4.37).
⍝ Place momentum equation, Equation (4.35), in the wave transmission matrix

```
:for I :in 1 2 3
     I1←(¯1+2×I),2×I
     :for J :in 1 2
          A←1+M[mode;I;J]×M[mode;I;J]+2×B[mode;I;J]
          A ←A,1+M[mode;I;J]×M[mode;I;J]-2×B[mode;I;J]
          R[1;2;I1]←R[1;2;I1]+A×K[mode;I;2]×S[mode;I;J]
     :endfor
:endfor
```

⍝ Place energy equation, Equation (4.37), in the wave transmission matrix

```
:for I :in 1 2 3
     I1←(¯1+2×I),2×I
     :for J :in 1 2
          A←B[mode;I;J]×M[mode;I;J]
          A←(1+A),1-A
```

```
        R [1;3;I1]←R[1;3;I1]+A×K[mode;I;1]×B[mode;I;J]
    :endfor
:endfor
```

⍝ Update the pressure wave components for the current positive acoustic flow directions

```
:for I :in 1 2 3
    I1←(¯1+2×I),2×I
    :if dir[I]=1
        R[;;I1]←R[;;I1]
    :endif
:endfor
```

⍝ Columns of *R* are in the order of ports 1,2,3.

4.3.1.3 Effect of Inner Duct Wall Thickness

If wall thickness of the inner duct is not negligible, Equations (4.46) and (4.47) are modified to

$$\sum_{j=1}^{3} C_j\left(\left[1+M'_j \quad -1+M'_j\right]S'_j+\left[1+M''_j \quad -1+M''_j\right]S''_j\right)\mathbf{P}_j+S_4C_4[1 \quad -1]\mathbf{P}_4=0 \tag{4.49}$$

$$\sum_{j=1}^{3} B_j\left(\left[\left(1+M'_j\right)^2 \quad \left(1-M'_j\right)^2\right]S'_j+\left[\left(1+M''_j\right)^2 \quad \left(1-M''_j\right)^2\right]S''_j\right)\mathbf{P}_j+S_4B_4[1 \quad 1]\mathbf{P}_4=0 \tag{4.50}$$

where S_4 denotes the thickness surface area, $C_4 = C_3$ and $B_4 = B_3$ for expansion type area changes, and $C_4 = C_1$ and $B_4 = B_1$ for contraction type area changes. This neglects the peripheral surfaces of the ducts in the control volume, but this is justified because the control volume is thin. Equations (4.49) and (4.50) may be combined in matrix notation as

$$\mathbf{a}_1\mathbf{P}_1 + \mathbf{a}_2\mathbf{P}_2 + \mathbf{a}_3\mathbf{P}_3 + S_4\mathbf{E}\mathbf{P}_2 = 0, \tag{4.51}$$

where

$$\mathbf{a}_j = \begin{bmatrix} C_j\left(1+\bar{M}'_j\right) & C_j\left(-1+\bar{M}'_j\right) \\ B_j\left(1+\bar{M}'_j\right)^2 & B_j\left(1-\bar{M}'_j\right)^2 \end{bmatrix} S'_j + \begin{bmatrix} C_j\left(1+\bar{M}''_j\right) & C_j\left(-1+\bar{M}''_j\right) \\ B_j\left(1+\bar{M}''_j\right)^2 & B_j\left(1-\bar{M}''_j\right)^2 \end{bmatrix} S''_j \tag{4.52}$$

$$\mathbf{E} = \begin{bmatrix} C_4 & -C_4 \\ B_4 & B_4 \end{bmatrix}. \tag{4.53}$$

Multiplying Equation (4.51) from left by $\mathbf{R}_4\mathbf{E}^{-1}$, where $\mathbf{R}_4 = [r_4 \quad -1]$ and $r_4 = p_4^-/p_4^+$ denotes the reflection coefficient of S_4, we obtain

$$\mathbf{R}_4\mathbf{E}^{-1}(\mathbf{a}_1\mathbf{P}_1 + \mathbf{a}_2\mathbf{P}_2 + \mathbf{a}_3\mathbf{P}_3) = 0 \tag{4.54}$$

This is complemented by Equation (4.48) and, if S_4 is relatively hard, can be closed with Equation (4.46) for the derivation of a three-port wave transmission equation of an open area change.

4.3.2 Closed Area Changes

An open area change is called closed when its wave transmission equation is reduced to a two-port wave transfer equation. This occurs if the side-branch duct (duct 2) in Figure 4.1 terminates (possibly through an arbitrary duct system) by a boundary of known reflection coefficient. In practice, the side-branch duct is usually closed by a rigid end-cap, as shown in Figure 4.3 for the four configurations in Figure 4.1. We can take advantage of this situation to reduce the 3×6 wave transmission matrix of the original open area-change, to a 2×2 wave transfer matrix of a two-port.

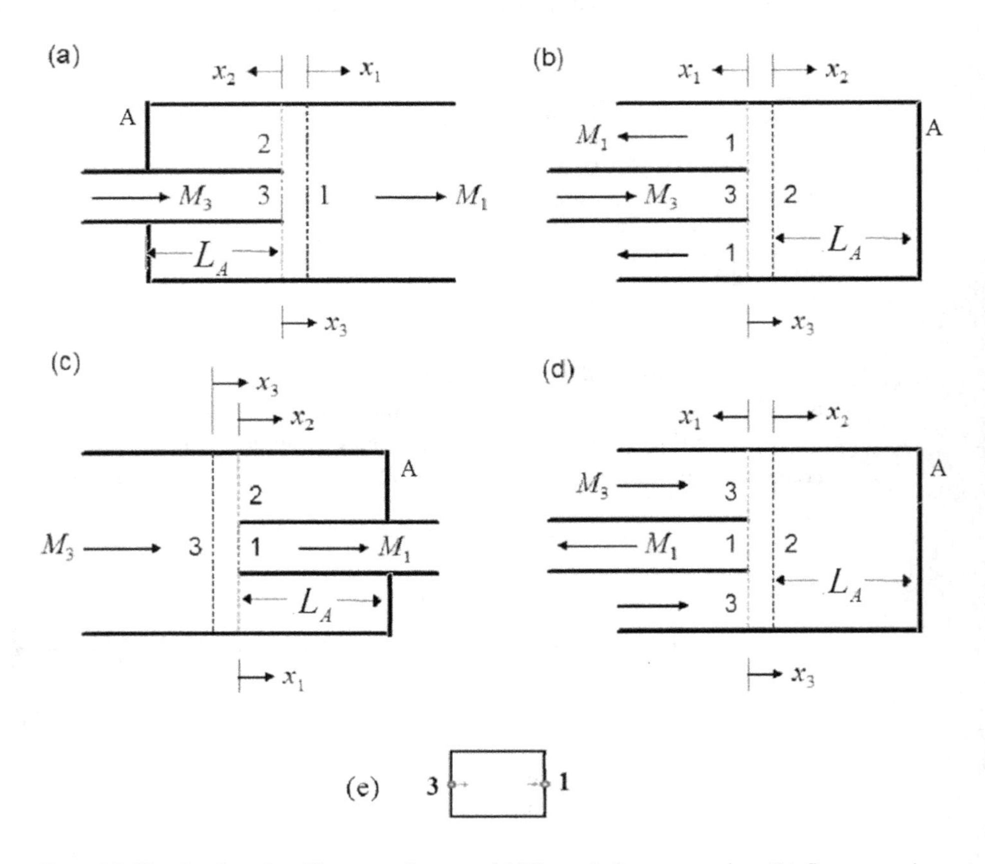

Figure 4.3 Simply closed sudden area-changes: (a) Through-flow expansion, (b) flow-reversing expansion, (c) through-flow contraction, (d) flow-reversing contraction and (e) block representation of the four configurations.

Let the acoustic path between port 2 and the plane of the terminating boundary A of reflection coefficient r_A be defined by the two-port wave transfer matrix $\mathbf{T}_{2A}$. Then, $\mathbf{P}_2 = \mathbf{T}_{2A}\mathbf{P}_A = \mathbf{T}_{2A}\mathbf{R}_A p_A^+$, where $\mathbf{P}_A = \{ p_A^+ \quad p_A^- \}$, $\mathbf{R}_A = \{ 1 \quad r_A \}$ and curly brackets denote a column vector. Using this relationship, Equation (4.40) can be expressed as $\mathbf{A}_1\mathbf{P}_1 + \mathbf{B}_2 p_A^+ + \mathbf{A}_3\mathbf{P}_3 = \mathbf{0}$, where $\mathbf{B}_2 = \mathbf{A}_2\mathbf{T}_{2A}\mathbf{R}_A = \{ b_1 \quad b_2 \quad b_3 \}$ is a 3×1 vector. It is convenient to rearrange this equation so that $b_1, b_2 \neq 0$ and express it in expanded form as:

$$\begin{bmatrix} \mathbf{A}_1^{(1)} \\ \mathbf{A}_1^{(2)} \\ \mathbf{A}_1^{(3)} \end{bmatrix} \mathbf{P}_1 + \begin{bmatrix} b_1 \\ b_2 \\ b_3 \end{bmatrix} p_A^+ + \begin{bmatrix} \mathbf{A}_3^{(1)} \\ \mathbf{A}_3^{(2)} \\ \mathbf{A}_3^{(3)} \end{bmatrix} \mathbf{P}_3 = 0, \tag{4.55}$$

where a superscript "i" in brackets denotes the ith row of the linked matrix (1×2 array). Thus, $p_A^+ = -(\mathbf{A}_1^{(1)}\mathbf{P}_1 + \mathbf{A}_3^{(1)}\mathbf{P}_3)/b_1$. Consequently, the first two rows of Equation (4.55) give $(b_1\mathbf{A}_1^{(2)} - b_2\mathbf{A}_1^{(1)})\mathbf{P}_1 = (b_2\mathbf{A}_3^{(1)} - b_1\mathbf{A}_3^{(2)})\mathbf{P}_3$. Combining this with the third row of Equation (4.55) yields the relationship $\mathbf{P}_3 = \mathbf{T}_{31}\mathbf{P}_1$, where

$$\mathbf{T}_{31} = \begin{bmatrix} b_2\mathbf{A}_3^{(1)} - b_1\mathbf{A}_3^{(2)} \\ b_1\mathbf{A}_3^{(3)} - b_3\mathbf{A}_3^{(1)} \end{bmatrix}^{-1} \begin{bmatrix} b_1\mathbf{A}_1^{(2)} - b_2\mathbf{A}_1^{(1)} \\ b_1\mathbf{A}_1^{(3)} - b_3\mathbf{A}_1^{(1)} \end{bmatrix}. \tag{4.56}$$

This represents the required wave transfer matrix of a closed area change. $\mathbf{P}_2$ is eliminated in the derivation process, but it can be calculated from $\mathbf{P}_2 = \mathbf{T}_{21}\mathbf{P}_1$, where $\mathbf{T}_{21} = -b_1^{-1}\mathbf{T}_{2A}\mathbf{R}_A(\mathbf{A}_1^{(1)} + \mathbf{A}_3^{(1)}\mathbf{T}_{31})$. For an example of using a through-flow type closed area change in block diagrams, the reader is referred to Section 2.4.1 (Figure 2.5).

In general, the wave transfer matrix in Equation (4.56) can be computed only numerically. However, if the mean flow is neglected, it can be expressed explicitly. In this case Equation (4.40) is in the form of (4.45) and Equation (4.56) can be expressed as

$$\mathbf{T}_{31} = \frac{1}{2\zeta_2 B_2 S_3} \begin{bmatrix} S_1(\zeta_2 B_2 - B_1) + S_3(\zeta_2 B_2 - 1) & -S_1(\zeta_2 B_2 + B_1) + S_3(\zeta_2 B_2 - 1) \\ -S_1(\zeta_2 B_2 - B_1) + S_3(\zeta_2 B_2 + 1) & S_1(\zeta_2 B_2 + B_1) + S_3(\zeta_2 B_2 + 1) \end{bmatrix}, \tag{4.57}$$

where ζ_2 denotes the normalized acoustic impedance at port 2. Since $\mathbf{P}_2 = \mathbf{T}_{2A}\mathbf{P}_A$, ζ_2 can always be calculated from r_A. For example, if the side-branch is a uniform duct (as implied in Figure 4.3), then $\mathbf{T}_{2A}$ is given by Equation (3.61) with $\alpha = 1$ and, hence, we get $\zeta_2 = (1 + r_A \mathrm{e}^{-\mathrm{i}2k_o L_A})/(1 - r_A \mathrm{e}^{-\mathrm{i}2k_o L_A})$, where L_A denotes the length of the side-branch. Note that, in view of the sign convention of Figure 4.3, L_A should be input as a negative number. Equation (4.57) applies for the four configurations of Figure 4.3.

4.3.3 End-Correction

One-dimensional theory assumes that planar sound waves incident at an area change are also reflected and transmitted as planar waves. This model is not precise even at low frequencies, because even though the incident wave may be planar, the reflected and transmitted waves may also include many evanescent modes, that is, modes that decay quickly with distance. Evanescent modes are not important sufficiently away from an area change, but may have substantial effect on sound transmission at the area change. In fact, they are necessary for proper application of acoustic continuity and compatibility conditions at an area change (see Section 7.2). For this reason, there has been considerable interest in the possibility of improving the accuracy of one-dimensional area-change models by including the effect of evanescent waves. It has been known for a long time that the effect of the evanescent waves may be taken into account approximately by increasing the length of the smaller duct into the discontinuity by a distance δ, which is called end-correction [4]. This is basically a heuristic parameter, because it can only be determined from a three-dimensional simulation of the problem or by measurements carried out on specific area-change configurations. Therefore, end-correction depends not only on the discontinuity per se but also on the surrounding geometry of the ducts. Results on the effect of mean flow on the end-correction are scarce in the literature; however, from the available data, it appears to be negligible for all practical purposes in the subsonic low Mach number range [5].

Published data on end-correction are limited to area-change configurations formed by uniform hard-walled circular ducts of diameters d and D ($>d$), with side-branch duct closed by a rigid end-cap at or away from the area discontinuity. A classical problem of this type, which is restricted to through-flow configurations, is two ducts of infinite length coupled with zero side-branch length. Consideration of infinitely long ducts ensures that only plane waves are incident at the area change from the inlet duct and that no reflected waves are present in the outlet pipe. If the wavelength is much larger than d, a classical formula for the end-correction is

$$\frac{\delta}{d} = \sum_{m=1}^{\infty} \frac{2\mathrm{J}_1^2\left(\alpha_m \frac{D}{d}\right)}{\left[\alpha_m J_0(\alpha_m)\right]^2 \alpha_m \frac{D}{d}} \tag{4.58}$$

where α_m denotes the mth root of $\mathrm{J}_1(\alpha) = 0$ and J_n denotes the Bessel function of order n. This result, which is due to Karal [4], has been improved by Peat for the effects of subsonic low Mach number mean flow and a rigidly closed side-branch ducts of finite length [5–6]. However, these results are strictly valid for concentric circular ducts. Eccentricity between the axes of the ducts can affect the end-correction significantly [7]. In view of such variations, simplistic empirical formulae are often used for practical purposes. An empirical expression for a through-flow discontinuity having a closed side-branch of zero length is [2]:

$$\delta = 0.3d\left(1 - e^{-\frac{1}{1.5}\left(\frac{D}{d}-1\right)}\right). \tag{4.59}$$

This may also be used to estimate the end-correction if the closed side-branch has finite length. However, the following formula is claimed to be more accurate in this case [8]:

$$\delta = d\left(0.26148 - e^{-1.31906\frac{D}{d}}\right). \tag{4.60}$$

This is based on curve-fitting to a large number of numerical solutions obtained by using the finite element method. An empirical formula is also available for a flow-reversing area discontinuity having a closed side-branch of length L. It is [9]:

$$\delta = \begin{cases} 0 & \text{if} \quad L \leq 0.25d \\ \dfrac{0.8L - 0.4}{d}\delta_D & \text{if} \quad 0.25d < L \leq 1.5d \\ \delta_D & \text{if} \quad L > 1.5d \end{cases} \tag{4.61}$$

where δ_D denotes the end-correction computed by using Equation (4.59). These formulae may be used to estimate the end-correction for non-circular ducts, by using the diameters of circles having equal area.

In general, however, expressions given for the end-correction are subject to several uncontrolled factors and simplifying assumptions and are strictly applicable under the specific conditions they are calculated or measured. In this book, we recommend the use of end-correction not as a decisive predictive method, but rather, for estimating the likely range of fuzzy parameters for sensitivity analyses (see Section 13.3).

4.4 Sudden Area Changes Formed by Multiple Ducts

In this section, we consider some extensions of sudden area changes in Figure 4.1, in that the flow contraction from or expansion to the main duct occur from multiple inner ducts. Typical configurations corresponding to those in Figure 4.1 are shown schematically in Figure 4.4. Here, the area discontinuities formed by the inner ducts are assumed to be enclosed in an axially compact control volume (see Section 4.4.1). The ducts and ports are numbered as in Figure 4.1, except that the additional inner ducts are numbered consecutively as duct 4, duct 5 and so on and the same default positive acoustic flow directions are used at the ports of the inner ducts. The wave transmission equations of these area-change configurations are formulated as described in the previous section, except that there are now $N > 3$ ports and we need to have N scalar equations. For this reason, method (a) may not be used. Method (c) is not practicable in this case, since there are multiple inner ducts. But method (b) may be used and it gives the required equations:

$$\sum_{j=1}^{N} S_j C_j \left[1 + \bar{M}_j \quad -1 + \bar{M}_j\right] \mathbf{P}_j = 0 \tag{4.62}$$

$$\left[1 + \bar{M}_1 \quad 1 - \bar{M}_1\right] \mathbf{P}_1 = \left[1 + \bar{M}_j \quad 1 - \bar{M}_j\right] \mathbf{P}_2, \quad j = 2, 3, \ldots, N \tag{4.63}$$

which are generalizations of Equations (4.33) and (4.36), respectively. If we neglect the mean flow in the side-branch duct (duct 2), we may discard Equation (4.63) for $j = 2$ and replace it with the momentum equation

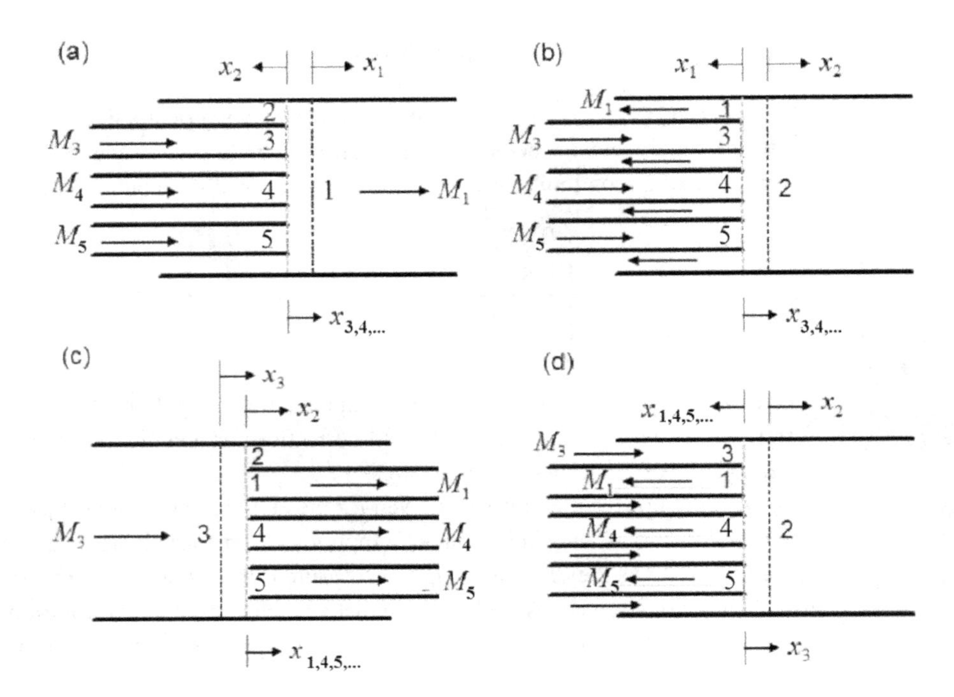

Figure 4.4 Open area changes with multiple ducts: (a) through-flow expansion, (b) flow-reversing expansion, (c) through-flow contraction, (d) flow-reversing contraction.

$$\sum_{j=1}^{N} S_j B_j \left[(1+\bar{M}_j)^2 \quad (1-\bar{M}_j)^2 \right] \mathbf{P}_j = 0, \tag{4.64}$$

which is a generalization of Equation (4.34). In these equations, S_j denotes the inner cross-sectional areas of ports $j = 1, 2, , \ldots, N$, and, as can be seen from Figure 4.4, for expansion type area changes:$C_3 = C_4, C_5 \ldots$, $B_3 = B_4, B_5 \ldots$; and for contraction type area changes: $C_1 = C_4, C_5 \ldots$, $B_1 = B_4, B_5 \ldots$ Either set of the foregoing equations can be collated in matrix notation in the generic form $\mathbf{A}_1\mathbf{P}_1 + \mathbf{A}_2\mathbf{P}_2 + \cdots + \mathbf{A}_N\mathbf{P}_N = 0$, where the wave transmission matrix $[\mathbf{A}_1 \quad \mathbf{A}_2 \quad \ldots \quad \mathbf{A}_N]$ is of size $N \times 2N$. Obviously, this is a generalization of Equation (4.40) and represents a multi-port element with N ports.

4.4.1 Identical Inner Ducts

If the mean flow Mach numbers in the inner ducts are identical, then, for example, Equation (4.63) for $j = 3, 4, \ldots, N$, for a sudden expansion implies that $\mathbf{P}_3 = \mathbf{P}_4 = \cdots = \mathbf{P}_N$. In this case, Equation (4.62) may be expressed as

$$\sum_{j=1}^{2} S_j C_j [1+\bar{M}_j \quad -1+\bar{M}_j] \mathbf{P}_j + S_{3eq} C_3 [1+\bar{M}_3 \quad -1+\bar{M}_3] \mathbf{P}_3 = 0, \tag{4.65}$$

where $S_{3eq} = S_3 + S_4 + \cdots + S_N$. Equation (4.64) may be expressed similarly. This proves that a sudden open area expansion from multiple identical ducts is equivalent to an open area change formed by duct 1 and a single inner duct (duct 3) of cross-sectional area S_{3eq} (Figure 4.1a or 4.1b), provided that this equivalent duct is attributed with the mean flow Mach number in the actual inner ducts. This result applies also for a contraction area change and can be proved similarly. It is useful in applications where N is large, because it reduces the size of the wave transmission matrix to be dealt with from $N \times 2N$ to 3×6 (see Section 4.5.1).

4.4.2 Staggered Inner Duct Extensions

Ends of the inner ducts may be staggered axially with respect to each other by different distances, provided that the control volume enclosing all area discontinuities remains acoustically compact axially. But, if the inner ducts are staggered axially by substantial distances, the resulting area change with multiple ducts can be modeled by using the area-change elements shown in Figure 4.1. This is illustrated in Figure 4.5 for the case of three internal ducts that are axially staggered. As indicated in Figure 4.5a, the overall discontinuity is exploded into two ducts, ducts 6-7 and 8-9,

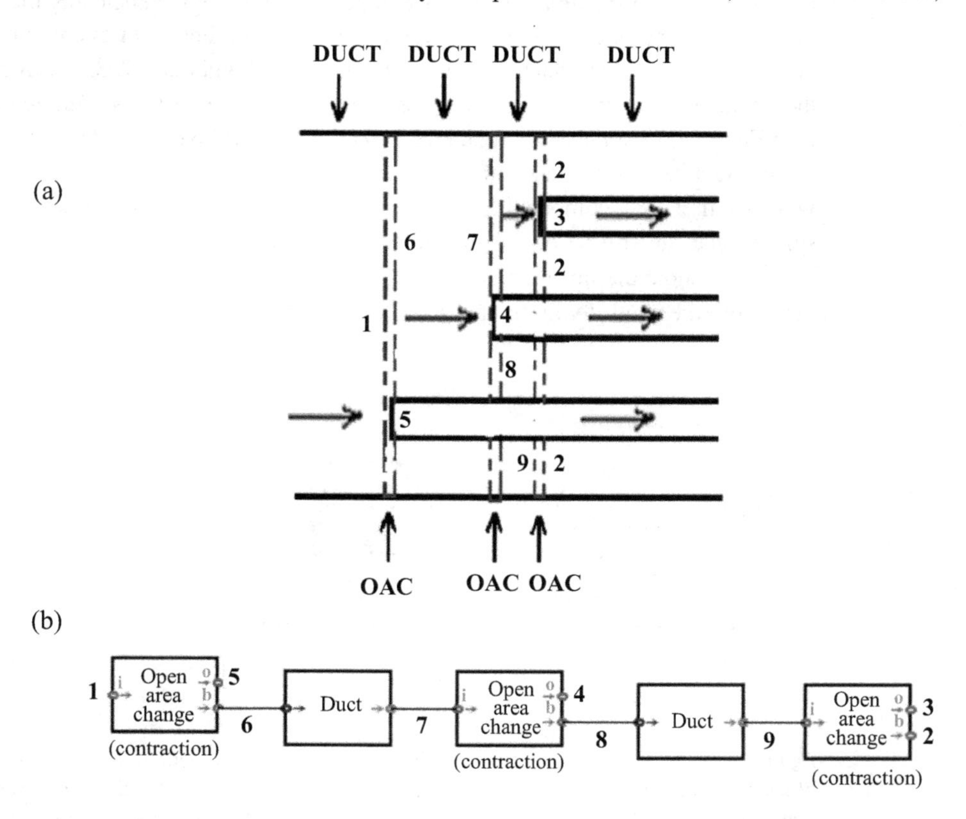

Figure 4.5 Area change with multiple staggered duct extensions. (a) Modeling of the discontinuity as a five-port acoustic element, (b) block diagram of the five-port.

and through-flow open area changes of the contraction type (Figure 4.1c) are introduced for matching them at ports 1, 2, 3, 4 and 5. The block diagram of the resulting five-port is shown in Figure 4.5b and the corresponding global matrix may be derived as discussed in Section 2.4.2. As can be seen, the inner duct which extends most couples with the outer duct in the simple open area-change configuration; the next most extended inner duct couples with the annulus between the outer duct and the most extended inner duct again in the simple open area-change configuration, and so on for the remaining inner ducts. This series of area changes are then coupled through the intermediate side-branch ducts.

4.4.3 Duct Splits

Dissipative silencers are often designed by dividing the cross-sectional area of a rectangular or circular hollow duct into a number of rectangular or circular annular channels, respectively, by using parallel longitudinal splitters, as shown schematically in Figure 4.6. The dissipative action takes place as sound waves propagate along the channels. The faces of splitters are closed at both ends, forming area changes at the inlet and outlet of the unit. An axially compact control volume enclosing the inlet and outlet discontinuities are shown in Figure 4.6 by dashed lines. Ducts are numbered as duct $1, 2, \ldots, N$ where duct 1 is always the main duct and ducts $2, 3, \ldots, N$ represent the channels. The splitter faces are assumed to be plane surfaces (denoted by letter A in Figure 4.6), not necessarily hard, of total surface area S_A and reflection coefficient r_A. In practice, splitter faces may be rounded for streamlining the mean flow. If the wavelength is long compared to the axial extent occupied by this rounded portion, the splitter face can still be modeled as a flat one without substantial loss of accuracy.

Let S_j denote the inner cross-sectional areas of ducts $j = 1, 2, \ldots, N$ at their interfaces with the control volume (ports $1, 2, \ldots, N$, respectively). We assume that effect

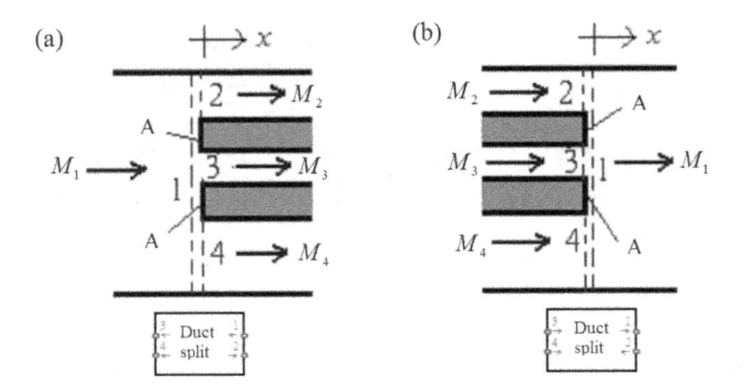

Figure 4.6 Duct split discontinuity consisting of 4 ducts: (a) contraction, (b) expansion and their block diagram representations with positive acoustic flow directions fixed before assembly. The views shown may be conceived as either a rectangular duct divided into three rectangular channels (2,3,4) by two rectangular splitters (in gray), or a circular duct divided into one hollow (3) and one annular (2 or 4) channel by one annular splitter.

of control volume surfaces other than S_j and S_A are negligible. Hence, neglecting the entropy wave components at the ports, Equations (4.24) and (4.25) give, respectively,

$$\sum_{j=1}^{N} C_j S_j \left[1 + \bar{M}_j \quad -1 + \bar{M}_j \right] \mathbf{P}_j + C_A S_A [1 \quad -1] \mathbf{P}_A = 0 \tag{4.66}$$

$$\sum_{j=1}^{N} C_j S_j \left[(1 + \bar{M}_j)^2 \quad (1 - \bar{M}_j)^2 \right] \mathbf{P}_j + C_A S_A [1 \quad 1] \mathbf{P}_A = 0, \tag{4.67}$$

where the mean density and the speed of sound are assumed to be uniform in the control volume and, for the sign convention of Figure 4.6, $C_1 = C_A = 1$ and $C_2 = C_3 = \cdots = -1$ for an expansion discontinuity (Figure 4.6b), and $C_1 = C_A = -1$ and $C_2 = C_3 = \cdots = 1$ for a contraction discontinuity (Figure 4.6a). But $N - 1$ more equations are needed to close the foregoing equations for the determination of a wave transmission matrix for the duct split discontinuity. These equations may be written from Equation (4.28) as

$$\left[1 + \bar{M}_j \; 1 - \bar{M}_j \right] \mathbf{P}_j = \text{constant}, \qquad j = 1, 2, \ldots, N. \tag{4.68}$$

If splitter faces are hard, then Equations (4.66) and (4.68) provide the N scalar equations needed for the determination of the wave transmission equation, since the term involving S_A vanishes in Equation (4.66). Otherwise, Equation (4.68) is complemented with the equation that results from elimination of $\mathbf{P}_A$ between Equations (4.66) and (4.67). This process consists of combining Equations (4.66) and (4.67) in matrix form as

$$C_1 \mathbf{a}_1 \mathbf{P}_1 + C_2 \mathbf{a}_2 \mathbf{P}_2 + \cdots + C_N \mathbf{a}_N \mathbf{P}_N + C_A S_A \mathbf{E} \mathbf{P}_A = 0, \tag{4.69}$$

where

$$\mathbf{a}_j = \begin{bmatrix} 1 + \bar{M}_j & -1 + \bar{M}_j \\ (1 + \bar{M}_j)^2 & (1 - \bar{M}_j)^2 \end{bmatrix} S_j, \quad j = 1, 2, \ldots, N \tag{4.70}$$

$$\mathbf{E} = \begin{bmatrix} 1 & -1 \\ 1 & 1 \end{bmatrix}. \tag{4.71}$$

Upon multiplying from left by $\mathbf{R}_A \mathbf{E}^{-1}$, where $\mathbf{R}_A = [r_A \quad 1]$ and $r_A = p_A^- / p_A^+$ is the reflection coefficient of surface A, Equation (4.69) gives

$$\mathbf{R}_A \mathbf{E}^{-1} (C_1 \mathbf{a}_1 \mathbf{P}_1 + C_2 \mathbf{a}_2 \mathbf{P}_2 + \cdots + C_N \mathbf{a}_N \mathbf{P}_N) = 0. \tag{4.72}$$

This can be collated now with Equation (4.68) to obtain an $N \times 2N$ wave transmission matrix for the duct split. In fact, this wave transmission matrix can also be used when splitter faces are hard, since in this case Equation (4.72) reduces to Equation (4.66).

4.5 Wave Transmission Through a Perforated Rigid Baffle

Perforated baffles are thin plates which partition a duct into two acoustically coupled ducts. A baffle also helps to provide stiffness to the wall structure and must be mounted into the duct tightly enough so that its acoustic performance is not taken over by leaks from randomly distributed circumferential gaps.

4.5.1 Area-Change Model

Based on the results of Section 4.4.1 about open area changes with multiple identical inner ducts, a baffle of uniform thickness having identical circular perforations (Figure 4.7) can be modeled by a sudden closed area contraction from the main duct to a duct of cross-sectional area equal to the total open area of the baffle, followed by a closed sudden area expansion from the latter fictional duct to the main duct. This is illustrated in Figure 4.7 for the through-flow configuration, the side-branch duct being closed rigidly at the discontinuity plane at both the contraction and expansion sides. In this model, we may write Equation (4.56) for the contraction and expansion discontinuities as $\mathbf{P}_U = \mathbf{T}_{U,c}\mathbf{P}_c$ and $\mathbf{P}_D = \mathbf{T}_{D,e}\mathbf{P}_e$, respectively, where the subscripts "U" and "D" denote the just upstream and just downstream planes of the baffle, and the subscripts "c" and "e" denote, respectively, the outlet and inlet ports of sudden contraction and expansion. Since ports "c" and "e" are connected by the fictive duct of area equal the open area of the baffle and length equal to its thickness, L, we have $\mathbf{P}_c = \mathbf{T}_{ce}\mathbf{P}_e$. Hence, eliminating $\mathbf{P}_c$ and $\mathbf{P}_e$ from these equations, we get $\mathbf{P}_U = \mathbf{T}_{U,D}\mathbf{P}_D$, where $\mathbf{T}_{U,D} = \mathbf{T}_{U,c}\mathbf{T}_{ce}\mathbf{T}_{D,e}^{-1}$ is the wave transfer matrix of the baffle.

The effect of local evanescent waves generated at perforations may be taken into account in this model by applying an end-correction to the thickness of the holes, based on single hole area (Section 4.3.3).

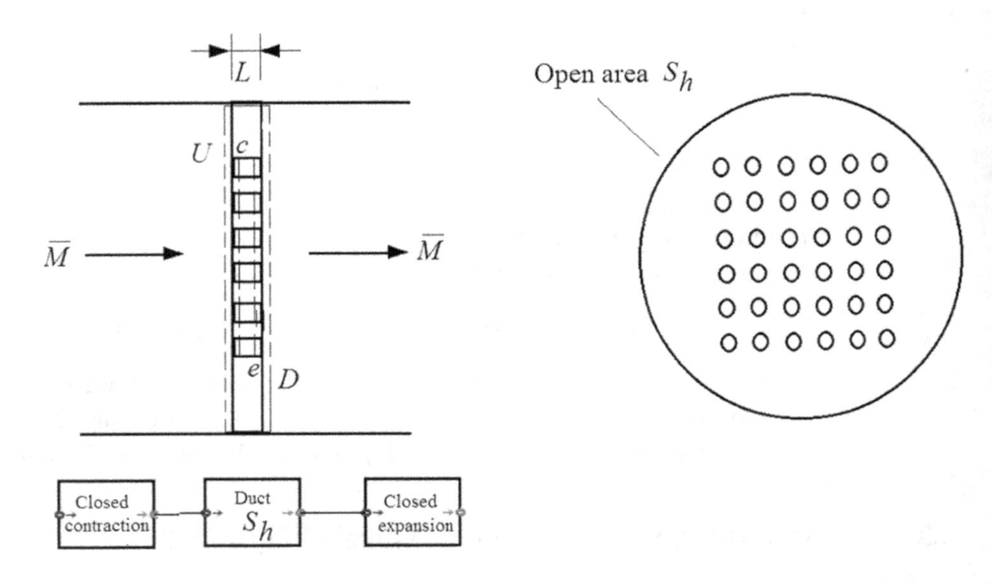

Figure 4.7 A perforated plate and its block diagram for sudden area-change model with multiple ducts.

4.5.2 Lumped Impedance Model

Notwithstanding the evanescent wave effects, the acoustic field in the vicinity of baffle apertures is influenced also by flow-acoustic interactions, diffraction and viscothermal losses. Since these effects are difficult analyze precisely, they are usually deferred to a lumped-parameter model of the fluid motion in compact apertures (see Appendix C). For this reason, an acoustic model of a perforated baffle based on a lumped-parameter model of perforations is a useful alternative to the area-change model described earlier.

The Rayleigh conductivity of an aperture is defined in frequency domain as

$$-\mathrm{i}\omega\rho_{jo}u_j'S_j = K_j\left(\overline{p_U'} - \overline{p_D'}\right), \qquad j = 1, 2, \ldots, N \tag{4.73}$$

where the subscript "j" refers to jth aperture on the baffle, S_j denote the aperture cross-sectional area, K_j denotes its Rayleigh conductivity, u_j' denotes the fluctuating velocity of the fluid in the aperture, ρ_{jo} denotes the constant density of the fluid in the aperture, the subscripts "U" and "D" refer to the duct interfaces just upstream and downstream of the baffle, respectively, and N denotes the number of apertures. Adding Equation (4.73) written for all j, we get

$$-\mathrm{i}\omega\sum_{j=1}^{N}\rho_{jo}u_j'S_j = \left(\overline{p_U'} - \overline{p_D'}\right)K_{eq}, \tag{4.74}$$

where $K_{eq} = K_1 + K_2 + \cdots + K_N$ is the equivalent Rayleigh conductivity of the baffle. But, Equation (4.10) also applies here for the control volume just enclosing the contraction discontinuity upstream of the baffle and the control volume just enclosing the expansion discontinuity downstream of the baffle (Figure 4.7). In the notation of the present section, it can be expressed for these control volumes as, respectively,

$$-\left(\overline{\rho_{Uo}}\,\overline{v_U'} + \overline{\rho_U'}\,\overline{v_{Uo}}\right)S_U + \sum_{j=1}^{N}\rho_{jo}u_j'S_j = 0 \tag{4.75}$$

$$\left(\overline{\rho_{Do}}\,\overline{v_D'} + \overline{\rho_D'}\,\overline{v_{Do}}\right)S_D - \sum_{j=1}^{N}\rho_{jo}u_j'S_j = 0, \tag{4.76}$$

where it is assumed that the baffle surface is hard. Thus, adding the foregoing equations and transforming the result to the pressure wave components, we obtain

$$\left\{(1+\bar{M}_U)p_U^+ + (-1+\bar{M}_U)p_U^-\right\}\frac{S_U}{c_{Uo}} = \left\{(1+\bar{M}_D)p_D^+ + (-1+\bar{M}_D)p_D^-\right\}\frac{S_D}{c_{Do}}, \tag{4.77}$$

where $\bar{M}_U = \overline{v_{Uo}}/c_{Uo}$, $\overline{p_U'} = p_U^+ + p_U^-$, $\overline{v_U'} = \left(p_U^+ - p_U^-\right)/\rho_{Uo}c_{Uo}$, $\overline{p_U'} = c_{Uo}^2\overline{\rho_U'}$ and similarly with the subscript "U" replaced by "D". Next, eliminating the summation term from Equations (4.75) and (4.76) and expressing the resulting equation in terms of the pressure wave components

$$(p_U^+ + p_U^-)K_{eq} = \{K_{eq} - \mathrm{i}k_o S_D(1+\bar{M}_D)\}p_D^+ + \{K_{eq} - \mathrm{i}k_o S_D(-1+\bar{M}_D)\}p_D^-, \tag{4.78}$$

where $k_o = \omega/c_{Do}$. Finally, combining Equations (4.77) and (4.78) in matrix notation, the wave transfer equation of the baffle can be expressed as

$$\begin{bmatrix} p_U^+ \\ p_U^- \end{bmatrix} = \mathbf{T}_{U,D} \begin{bmatrix} p_D^+ \\ p_D^- \end{bmatrix}, \tag{4.79}$$

where

$$\mathbf{T}_{U,D} = \frac{1}{2}\begin{bmatrix} 1-\bar{M}_U & 1 \\ 1+\bar{M}_U & -1 \end{bmatrix} \begin{bmatrix} 1-(1+\bar{M}_D)\dfrac{ik_o S_D}{K_{eq}} & 1+(1-\bar{M}_D)\dfrac{ik_o S_D}{K_{eq}} \\ (1+\bar{M}_D)\dfrac{S_D}{S_U}\dfrac{c_{Uo}}{c_{Do}} & (-1+\bar{M}_D)\dfrac{S_D}{S_U}\dfrac{c_{Uo}}{c_{Do}} \end{bmatrix} \tag{4.80}$$

is the wave transfer matrix of the baffle. This results can be expressed in terms of lumped parameter impedance models of apertures by using Equation (C.23).

4.6 Wave Transmission in Junction Cavities

When several non-parallel ducts are joined together, a cavity is necessarily formed at the junction (Figure 4.8). If such a cavity is specially designed for connection to a number of ducts, it is usually called a collector. In other cases, the geometry of the cavity is determined by the way the junction is constructed. Here, we consider an acoustically compact cavity of arbitrary geometry having a solid impervious surface, except at its interfaces with the connecting ducts. The connecting ducts are numbered

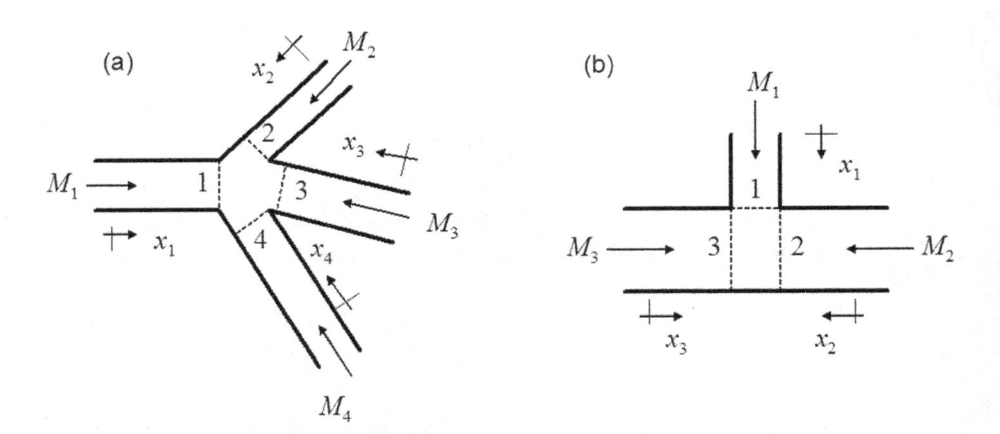

Figure 4.8 Junctions of ducts with different orientations: (a) a four-duct junction, (b) a two-duct junction.

as duct $1, 2, \ldots, N$. The default positive directions of the duct axes are taken into the junction, as shown in Figure 4.8.

4.6.1 Multi-Duct Junction

Consider a control volume enclosed by the solid internal surfaces of the cavity and the duct interfaces (indicated by dashed lines in Figure 4.8). Neglecting the entropy wave component at the ports of the ducts and using the sign convention of Figure 4.8 for the positive directions acoustic flows, Equation (4.24) gives

$$\sum_{j=1}^{N} \frac{S_j}{c_{jo}} \begin{bmatrix} 1+\bar{M}_j & -1+\bar{M}_j \end{bmatrix} \mathbf{P}_j = 0, \tag{4.81}$$

where $\bar{M}_j = \overline{v_{jo}}/c_{jo}, j = 1, 2, \ldots, N,$ S_j denotes the cross-sectional area of duct j at its interface with the control volume and c_{jo} denotes the speed of sound at the same interface. In this case, since Equation (4.25) is not useful in view of the arbitrariness of the solid surfaces of the control volume, we proceed with Equation (4.28), which yields

$$\frac{1}{\bar{\rho}_{jo}} \begin{bmatrix} 1+\bar{M}_j & 1-\bar{M}_j \end{bmatrix} \mathbf{P}_j = \text{constant}, \qquad j = 1, 2, \ldots, N. \tag{4.82}$$

These $N - 1$ equations, together with Equations (4.81) constitute a system of N equations for the $2N$ variables p_j^+, p_j^-, $j = 1, 2, \ldots, N$. These equations can be expressed in matrix notation as

$$\begin{bmatrix} \mathbf{b}_1 & \mathbf{b}_2 & \mathbf{b}_3 & \ldots & \mathbf{b}_N \\ -\mathbf{a}_1 & \mathbf{a}_2 & \mathbf{0} & \ldots & \mathbf{0} \\ -\mathbf{a}_1 & \mathbf{0} & \mathbf{a}_3 & \ldots & \mathbf{0} \\ \vdots & \vdots & \vdots & \ddots & \vdots \\ -\mathbf{a}_1 & \mathbf{0} & \mathbf{0} & \ldots & \mathbf{a}_N \end{bmatrix} \begin{bmatrix} \mathbf{P}_1 \\ \mathbf{P}_2 \\ \mathbf{P}_3 \\ \vdots \\ \mathbf{P}_N \end{bmatrix} = \mathbf{0} \tag{4.83}$$

where

$$\mathbf{a}_j = \frac{1}{\bar{\rho}_{jo}} \begin{bmatrix} 1+\bar{M}_j & 1-\bar{M}_j \end{bmatrix} \tag{4.84}$$

$$\mathbf{b}_j = \frac{S_j}{c_{jo}} \begin{bmatrix} 1+\bar{M}_j & -1+\bar{M}_j \end{bmatrix}. \tag{4.85}$$

Obviously, Equation (4.83) represents an N-port element (Section 2.3.3). According to the sign convention used, the mean flow velocity is always positive if in the $+x_j$ direction. Consequently, Equation (4.82) is independent of the chosen $+x_j$ directions, $j = 1, 2, \ldots, N$ and, if the default $+x_j$ direction (Figure 4.8) of a duct is changed, it suffices to change the sign of the row vector $\mathbf{b}_j$ corresponding to that duct in Equation (4.83).

4.6.2 Two-Duct Junction

Two-duct junctions are usually formed by coupling a straight duct to another duct, the axes of the two ducts not necessarily being perpendicular as implied in Figure 4.8b. Using the port numbering and sign convention of Figure 4.8b, the wave transmission equation of the junction is given by $N = 3$ form of Equation (4.83). Suppose the reflection coefficient at port 3, $r_3 = p_3^-/p_3^+$, is known. Then, the three-duct form of Equation (4.83) reduces to the form

$$\begin{bmatrix} \mathbf{b}_1 + \dfrac{\mathbf{b}_3\mathbf{R}_3}{\mathbf{a}_3\mathbf{R}_3}\mathbf{a}_1 \\ \mathbf{a}_1 \end{bmatrix} \mathbf{P}_1 = \begin{bmatrix} -\mathbf{b}_2 \\ \mathbf{a}_2 \end{bmatrix} \mathbf{P}_2, \tag{4.86}$$

where $\mathbf{R}_3$ denotes the column vector $\mathbf{R}_3 = \{1 \quad r_3\}$. Clearly, the matrices on either side of this equation are 2×2 matrices and this relationship can be recast as a two-port wave transfer equation, for example, $\mathbf{P}_1 = \mathbf{T}_{12}\mathbf{P}_2$. If the reflection coefficient at port 2 is known, the corresponding form of Equation (4.86) can be obtained by interchanging the indices 2 and 3 in this expression. If the reflection coefficient at port 1 is known, then the three-duct form of Equation (4.83) reduces to

$$\begin{bmatrix} \dfrac{\mathbf{a}_2}{\mathbf{a}_1\mathbf{R}_1} + \dfrac{\mathbf{b}_2}{\mathbf{b}_1\mathbf{R}_1} \\ -\dfrac{\mathbf{b}_2}{\mathbf{b}_1\mathbf{R}_1} \end{bmatrix} \mathbf{P}_2 = \begin{bmatrix} \dfrac{-\mathbf{b}_3}{\mathbf{b}_1\mathbf{R}_1} \\ \dfrac{\mathbf{a}_3}{\mathbf{a}_1\mathbf{R}_1} + \dfrac{\mathbf{b}_3}{\mathbf{b}_1\mathbf{R}_1} \end{bmatrix} \mathbf{P}_3, \tag{4.87}$$

where $\mathbf{R}_1 = \{1 \quad r_1\}$ and $r_1 = p_1^-/p_1^+$. Obviously, this relationship also gives a two-port wave transfer equation, for example, $\mathbf{P}_2 = \mathbf{T}_{23}\mathbf{P}_3$.

With appropriate selection of the positive directions of the duct axes, the acoustic two-port element represented by Equation (4.86) can be used for modeling expansion into a chamber from a lateral duct (called side-inlet, Figure 4.9a), or contraction to a lateral duct (called side-outlet, Figure 4.9b) from a chamber. In both cases, the positive direction of duct 3 remains in the default direction, but the positive directions of the

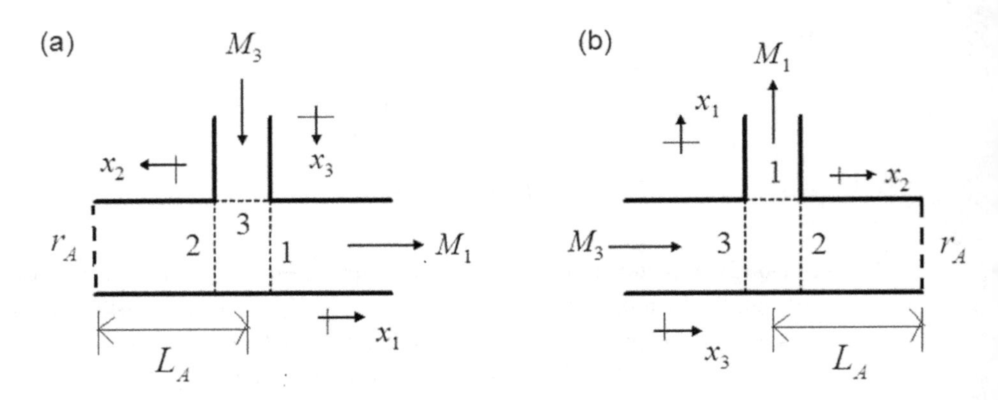

Figure 4.9 Side-inlet expansion (a) and side-outlet contraction (b) elements.

axes connecting to ports 1 and 2 are reversed. Thus, Equation (4.86) applies for both configurations in Figure 4.9 with the subscripts "2" and "3" interchanged and the signs of the vectors $\mathbf{b}_1$ and $\mathbf{b}_2$ changed. As the mean flow tends to zero, the resulting wave transfer matrix $\mathbf{T}_{31}$ can be expressed explicitly as

$$\mathbf{T}_{31} = \frac{1}{2S_3}\begin{bmatrix} S_1 + S_3 + \alpha_2 S_2 & -S_1 + S_3 + \alpha_2 S_2 \\ -S_1 + S_3 - \alpha_2 S_2 & S_1 + S_3 - \alpha_2 S_2 \end{bmatrix}, \tag{4.88}$$

where S_j denotes the cross-sectional area of duct j, $j = 1, 2, 3$, and α_2 $(= \overline{v_2'}/\overline{p_2'})$ denotes the normalized acoustic admittance (inverse of normalized acoustic impedance) at port 2. For example, if port 2 is a terminal of a uniform hard-walled duct which terminates at the other end as indicated in Figure 4.9 by a surface of reflection coefficient r_A, then $\alpha_2 = \left(1 - r_A \mathrm{e}^{-\mathrm{i}2k_o L_A}\right)/\left(1 + r_A \mathrm{e}^{-\mathrm{i}2k_o L_A}\right)$, where, according to the sign convention of Figure 4.9, the side-branch length L_A should be input as a negative number.

The two-port acoustic element represented by Equation (4.87) can be used for modeling of sound transmission across a branch. In this configuration, the main duct cross-sectional area remains the same across the junction and assuming that the flow in the main duct is not restricted downstream of it, the mean flow in the branch duct can be neglected. Hence, putting $S_2 = S_3$, $\bar{M}_2 = \bar{M}_3$, $\bar{M}_1 = 0$, and using the sign convention of Figure 4.9b, the wave transfer matrix of a branch can be expressed explicitly as

$$\mathbf{T}_{32} = \frac{S_1}{2S_2\zeta_1}\begin{bmatrix} -1 + \dfrac{2S_2\zeta_1}{S_1} & -\dfrac{1 - \bar{M}_2}{1 + \bar{M}_2} \\ \dfrac{1 + \bar{M}_2}{1 - \bar{M}_2} & 1 + \dfrac{2S_2\zeta_1}{S_1} \end{bmatrix}, \tag{4.89}$$

where $\zeta_1 = (1 + r_1)/(1 - r_1)$ denotes the normalized acoustic impedance at port 1, the branch interface.

4.7 Continuously Coupled Perforated Ducts

Perforated ducts may be coupled in various ways. Common configurations are shown in Figure 4.10. We will refer to these configurations as perforated n-duct elements, where n denotes the total number of coupled ducts. A perforated n-duct element consists of $n - 1$ perforated ducts enclosed in a duct with impervious walls with arbitrary transverse dispositions. It can be connected to other acoustic elements through either end of the ducts comprising the coupled unit. Thus, a perforated n-duct element has $2n$ ports, but these ports are connected internally in pairs. The ducts may have axially varying cross sections, although they are shown as uniform in the figures for simplicity. The enclosing duct is called the casing or shell. Block diagrams of perforated two-duct and three-duct units are shown in Figure 4.10 for the indicated default positive acoustic flow directions. The numbers by the ports indicate the corresponding duct. The default positive acoustic flow direction at ports may be

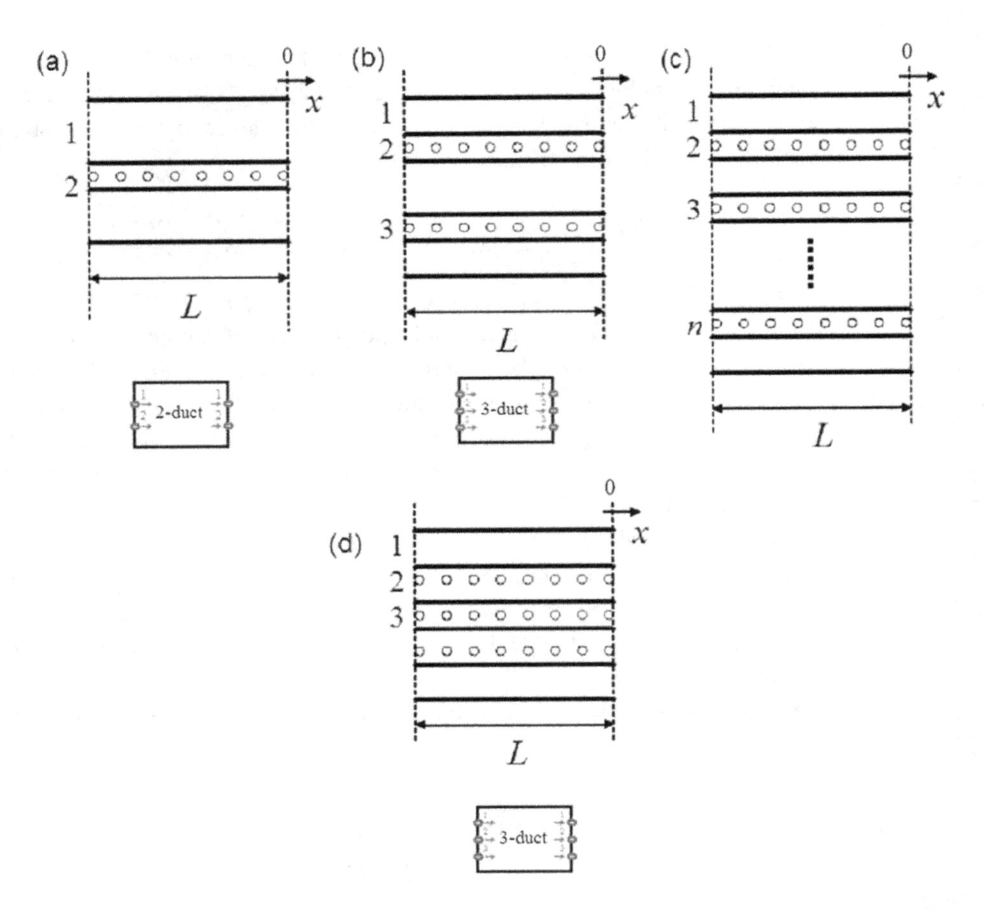

Figure 4.10 Perforated n-duct elements, (a) two-duct, (b) single-coupled three-duct, (c) single-coupled n-duct, (d) double-coupled three-duct.

changed, but they must be changed in pairs, since the ports are internally connected. The simplest n-duct element is the two-duct, which is simply a perforated duct enclosed in a casing (Figure 4.10a).

A perforated n-duct element may consist of single- or double-coupled perforated ducts. A single-coupled perforated duct directly communicates only with the internal acoustic field of the casing (Figure 4.10a,b,c). A double-coupled perforated duct communicates directly with another perforated duct (Figure 4.10d). Multifarious n-duct elements can be formed by appropriate arrangements of single-coupled and double-coupled perforated ducts.

The apertures on a perforated duct can be modeled as continuously or discretely distributed on the duct walls. The continuous model is appropriate if the apertures are sufficiently densely distributed. In this section, we present continuous one-dimensional models of single-coupled and double-coupled perforated duct arrangements. Continuous models may not be realistic if the apertures are scarcely distributed. In the next section, we will describe discrete models which are sensitive to the axial distribution pattern of the apertures.

The acoustic continuity and momentum equations for a duct having finite impedance walls are given by Equations (3.17) and (3.19), respectively. Assuming homogenous ducts (see Section 3.3), these equations are closed by the isentropic relationship $\overline{p'} = c_o^2\overline{\rho'}$ and can be expressed in frequency domain as

$$\left(-ik_o + \frac{d\bar{M}_o}{dx} + \bar{M}_o\frac{dS}{dx}(\ln S)\right)\overline{p'} + \bar{M}_o\frac{\partial\overline{p'}}{\partial x} + \overline{\rho_o}c_o\left(\frac{\partial\overline{v'}}{\partial x} + \overline{v'}\frac{dS}{dx}(\ln S)\right) = -\frac{c_o}{S}\oint_\ell \rho_o u' J d\ell - \underline{\underline{\frac{c_o}{S}\oint_\ell \rho' u_o J d\ell}} \tag{4.90}$$

$$\overline{\rho_o}c_o\left(-ik_o\overline{v'} + \bar{M}_o\frac{\partial\overline{v'}}{\partial x} + \overline{v'}\frac{d\bar{M}_o}{dx}\right) + \underline{\bar{M}_o\frac{d\bar{M}_o}{dx}\overline{p'} + \frac{1}{S}\frac{\partial}{\partial x}\left(S\beta\bar{M}_o^2\overline{p'}\right)} + \frac{\partial\overline{p'}}{\partial x} = \frac{\overline{v_o}}{S}\oint_\ell \rho_o\varphi_{SF}u' J d\ell + \underline{\underline{\frac{\overline{v_o}}{S}\oint_\ell \rho'\varphi_{SF}u_o J d\ell}}, \tag{4.91}$$

respectively. For subsonic low Mach numbers ($\bar{M}_o^2 << 1$), the underlined terms in Equation (4.91) can be neglected without substantial loss of accuracy. Also, the double-underlined terms are neglected on the premises that either $u_o = 0$ on hard sections of the walls or the fluid motion in and about the close vicinity of apertures is of constant density, as assumed in the lumped parameters models of aperture impedance (see Appendix C). With these simplifications, Equations (4.90) and (4.91) reduce to, respectively,

$$\left(-ik_o + \frac{d\bar{M}_o}{dx} + \bar{M}_o\frac{dS}{dx}(\ln S)\right)\overline{p'} + \bar{M}_o\frac{\partial\overline{p'}}{\partial x} + \overline{\rho_o}c_o\left(\frac{\partial\overline{v'}}{\partial x} + \overline{v'}\frac{dS}{dx}(\ln S)\right) = c_o\dot{\mu}' \tag{4.92}$$

$$\overline{\rho_o}c_o\left(-ik_o\overline{v'} + \frac{d\bar{M}_o}{dx}\overline{v'} + \bar{M}_o\frac{\partial\overline{v'}}{\partial x}\right) + \frac{\partial\overline{p'}}{\partial x} = -\varphi_{SF}\overline{v_o}\dot{\mu}', \tag{4.93}$$

where we have also introduced the parameter $\dot{\mu}' = -\oint_\ell \rho_o u' J d\ell / S$ as defined in Equation (3.114). In evaluating $\dot{\mu}'$, we recall from Section 3.8.2 that the equation of the unsteady motion of the fluid in an aperture can be expressed as $\overline{p'_{\text{int}}} - \overline{p'}(\ell_{\text{ext}}) = Z_W^o u'$. Since the exterior environment of a perforated duct is now another duct, this equation is implemented as $\overline{p'_{\text{int}}} - \overline{p'_{\text{ext}}} = Z_W^o u'$, where $\overline{p'_{\text{ext}}}$ denotes the cross section averaged acoustic pressure in the exterior duct. Hence, Equation (3.107) can be expressed as

$$\dot{\mu}' = -\frac{a}{c_o}\left(\overline{p'_{\text{int}}} - \overline{p'_{\text{ext}}}\right), \tag{4.94}$$

where

$$a = \frac{1}{S}\oint_{\ell} \frac{\mathrm{d}\ell}{\zeta_W^o} \tag{4.95}$$

and $\zeta_W^o = Z_W^o/\rho_o c_o$. This integral is difficult to compute, since apertures are actually distributed discretely over the duct surface. However, since the one-dimensional theory is not sensitive to peripheral distributions, we may distribute discrete impedances around the surface of the duct continuously. This process, which is explained in Section 3.9.3.3, allows us to write Equation (4.95) as

$$a = \frac{P\sigma}{S\zeta} \tag{4.96}$$

where P denotes the perimeter of S, $\zeta = Z_a/\rho_o c_o$ denotes the normalized impedance of an aperture and σ denotes the perforate porosity. These parameters may, in general, vary along the duct axis. We will refer to a as the perforate wall admittance parameter.

After substituting Equation (4.94), Equations (4.92) and (4.93) may be recast as

$$\left(-\mathrm{i}k + \frac{\mathrm{d}\bar{M}}{\mathrm{d}x} + \bar{M}\frac{\mathrm{d}}{\mathrm{d}x}(\ln S)\right)\overline{p'} + \bar{M}\frac{\partial\overline{p'}}{\partial x} + \bar{\rho}c\left(\frac{\partial\overline{v'}}{\partial x} + \overline{v'}\frac{\mathrm{d}}{\mathrm{d}x}(\ln S)\right) = -a\left(\overline{p'_{\mathrm{int}}} - \overline{p'_{ext}}\right) \tag{4.97}$$

$$\bar{\rho}c\left(-\mathrm{i}k\overline{v'} + \frac{\mathrm{d}\bar{M}}{\mathrm{d}x}\overline{v'} + \bar{M}\frac{\partial\overline{v'}}{\partial x}\right) + \frac{\partial\overline{p'}}{\partial x} = \varphi\bar{M}a\left(\overline{p'_{\mathrm{int}}} - \overline{p'_{ext}}\right), \tag{4.98}$$

respectively, where we have dropped, for the simplicity of notation, the subscript "o" in k_o, $\overline{\rho_o}$, c_o and $\bar{M}_o$, and also the subscript "SF" in φ_{SF}.

4.7.1 Single-Coupled Perforated Ducts

Referring to Figure 4.10c, ducts are numbered as duct 1, duct 2,..., duct n, where duct 1 is always the annulus between the casing and the perforated ducts. Equations (4.92) and (4.93) apply for each of these ducts. We will use the subscript "j" to refer to the properties of duct j, $j = 1, 2, \ldots, n$. Thus, Equations (4.92) and (4.93) yield for duct $j = 1$:

$$\left\{\left(-\mathrm{i}k_1 + \frac{\mathrm{d}\bar{M}_1}{\mathrm{d}x} + \bar{M}_1\frac{\mathrm{d}}{\mathrm{d}x}(\ln S_1)\right)\overline{p'_1} + \bar{M}_1\frac{\partial\overline{p'_1}}{\partial x}\right\} + \rho_1 c_1\left(\frac{\partial\overline{v'_1}}{\partial x} + \overline{v'_1}\frac{\mathrm{d}}{\mathrm{d}x}(\ln S_1)\right) = \sum_{j=2}^{n}\frac{S_j c_1}{S_1 c_j}a_j\left(\overline{p'_j} - \overline{p'_1}\right) \tag{4.99}$$

$$\bar{\rho}_1 c_1\left\{\left(-\mathrm{i}k_1 + \frac{\mathrm{d}\bar{M}_1}{\mathrm{d}x}\right)\overline{v'_1} + \bar{M}_1\frac{\partial\overline{v'_1}}{\partial x}\right\} + \frac{\partial\overline{p'_1}}{\partial x} = -\bar{M}_1\varphi_1\sum_{j=2}^{n}\frac{S_j c_1}{S_1 c_j}a_j\left(\overline{p'_j} - \overline{p'_1}\right) \tag{4.100}$$

and for ducts $j = 2, 3, \ldots, n$:

$$\left\{\left(-\mathrm{i}k_j + \frac{\mathrm{d}\bar{M}_j}{\mathrm{d}x} + \bar{M}_j \frac{\mathrm{d}}{\mathrm{d}x} \ln S_j\right)\overline{p_j'} + \bar{M}_j \frac{\partial \overline{p_j'}}{\partial x}\right\} + \rho_j c_j \left(\frac{\partial \overline{v_j'}}{\partial x} + \overline{v_j'} \frac{\mathrm{d}}{\mathrm{d}x} \ln S_j\right) = -a_j\left(\overline{p_1'} - \overline{p_j'}\right) \tag{4.101}$$

$$\bar{\rho}_j c_j \left\{\left(-\mathrm{i}k_j + \frac{\mathrm{d}\bar{M}_j}{\mathrm{d}x}\right)\overline{v_j'} + \bar{M}_j \frac{\partial \overline{v_j'}}{\partial x}\right\} + \frac{\partial \overline{p_j'}}{\partial x} = \bar{M}_j \varphi_j a_j \left(\overline{p_1'} - \overline{p_j'}\right) \tag{4.102}$$

where $k_j = \omega / c_j$. In these equations, it is assumed that the apertures on duct j, $j = 2, 3, \ldots, n$, are identical, but different ducts are allowed to have different apertures. It should be noted that S_1 denotes the net (annulus) cross-sectional area of duct 1. Also, the slip-flow parameter φ_1 is assumed to be the same for all perforated ducts.

Equations (4.99)–(4.102) can be collated in state-space form as

$$\frac{\partial}{\partial x}\begin{bmatrix} \mathbf{Q}_1 \\ \mathbf{Q}_2 \\ \vdots \\ \mathbf{Q}_n \end{bmatrix} = -\begin{bmatrix} \mathbf{M}_1\mathbf{B}_{11} & \mathbf{M}_1\mathbf{B}_{12} & \cdots & \mathbf{M}_1\mathbf{B}_{1n} \\ \mathbf{M}_2\mathbf{B}_{21} & \mathbf{M}_2\mathbf{B}_{22} & \cdots & \mathbf{M}_2\mathbf{B}_{2n} \\ \vdots & \vdots & \ddots & \vdots \\ \mathbf{M}_n\mathbf{B}_{n1} & \mathbf{M}_n\mathbf{B}_{n2} & \cdots & \mathbf{M}_n\mathbf{B}_{nn} \end{bmatrix}\begin{bmatrix} \mathbf{Q}_1 \\ \mathbf{Q}_2 \\ \vdots \\ \mathbf{Q}_n \end{bmatrix} \tag{4.103}$$

where

$$\mathbf{Q}_j = \begin{bmatrix} \overline{p_j'} \\ \bar{\rho}_j c_j \overline{v_j'} \end{bmatrix} \tag{4.104}$$

$$\mathbf{M}_j = \frac{1}{1 - \bar{M}_j^2}\begin{bmatrix} 1 & -\bar{M}_j \\ -\bar{M}_j & 1 \end{bmatrix} \quad j = 1, 2, \ldots, n \tag{4.105}$$

$$\mathbf{B}_{jj} = \begin{bmatrix} -\bar{M}_j \varphi_j a_j & -\mathrm{i}k_j + \dfrac{\mathrm{d}\bar{M}_j}{\mathrm{d}x} \\ -\mathrm{i}k_j + a_j + \dfrac{\mathrm{d}\bar{M}_j}{\mathrm{d}x} + \bar{M}_j \dfrac{\mathrm{d}}{\mathrm{d}x} \ln S_j & \dfrac{\mathrm{d}}{\mathrm{d}x} \ln S_j \end{bmatrix}, \quad j = 1, 2, \ldots, n \tag{4.106}$$

$$\mathbf{B}_{1j} = \frac{S_j c_1}{S_1 c_j} a_j \begin{bmatrix} \bar{M}_1 \varphi_1 & 0 \\ -1 & 0 \end{bmatrix}, \qquad \mathbf{B}_{j1} = a_j \begin{bmatrix} \bar{M}_j \varphi_j & 0 \\ -1 & 0 \end{bmatrix}, \quad j = 2, 3, \ldots, n \tag{4.107}$$

$$a_1 = \sum_{j=2}^{n} \frac{S_j c_1 a_j}{S_1 c_j}, \quad a_j = \frac{P_j \sigma_j}{S_j \zeta_j}, \quad j = 2, 3, \ldots, n \tag{4.108}$$

and all remaining $\mathbf{B}_{kl}$ $(k, l = 2, 3, \ldots, n)$ blocks in Equation (4.103) are 2×2 zero matrices. Equation (4.103) is transformed to the pressure wave components in each duct by

$$\mathbf{Q}_j = \begin{bmatrix} 1 & 1 \\ 1 & -1 \end{bmatrix} \mathbf{P}_j, \tag{4.109}$$

where

$$\mathbf{P}_j = \begin{bmatrix} p_j^+ \\ p_j^- \end{bmatrix}, \quad j = 1, 2, \ldots, n. \tag{4.110}$$

Hence, (4.103) can be expressed as

$$\frac{\partial \mathbf{P}}{\partial x} = \mathbf{H}(\omega, x)\mathbf{P}. \tag{4.111}$$

Here, the wave components vector is defined as

$$\mathbf{P} = \begin{bmatrix} \mathbf{P}_1 \\ \mathbf{P}_2 \\ \vdots \end{bmatrix} \tag{4.112}$$

and the matrizant $\mathbf{H} = \mathbf{H}(\omega, x)$ is given by

$$\mathbf{H} = \begin{bmatrix} \mathbf{H}_{11} & \mathbf{H}_{12} & \cdots & \mathbf{H}_{1n} \\ \mathbf{H}_{21} & \mathbf{H}_{22} & \cdots & \mathbf{H}_{2n} \\ \vdots & \vdots & \ddots & \vdots \\ \mathbf{H}_{n1} & \mathbf{H}_{n2} & \cdots & \mathbf{H}_{nn} \end{bmatrix}, \tag{4.113}$$

where

$$\mathbf{H}_{jj} = \frac{1}{2} \begin{bmatrix} A_j & B_j \\ C_j & D_j \end{bmatrix}, \quad j = 1, 2, \ldots, n \tag{4.114}$$

$$A_j = \frac{\mathrm{i}2k_j + (-1 + \varphi_j \bar{M}_j)a_j - (1 + \bar{M}_j)\dfrac{\mathrm{d}}{\mathrm{d}x}\ln S_j - 2\dfrac{\mathrm{d}\bar{M}_j}{\mathrm{d}x}}{1 + \bar{M}_j} \tag{4.114a}$$

$$B_j = \frac{(-1 + \varphi_j \bar{M}_j)a_j + (1 - \bar{M}_j)\dfrac{\mathrm{d}}{\mathrm{d}x}\ln S_j}{1 + \bar{M}_j} \tag{4.114b}$$

$$C_j = \frac{(1 + \varphi_j \bar{M}_j)a_j + (1 + \bar{M}_j)\dfrac{\mathrm{d}}{\mathrm{d}x}\ln S_j}{1 - \bar{M}_j} \tag{4.114c}$$

$$D_j = \frac{-\mathrm{i}2k_j + (1 + \varphi_j \bar{M}_j)a_j + (-1 + \bar{M}_j)\dfrac{\mathrm{d}}{\mathrm{d}x}\ln S_j + 2\dfrac{\mathrm{d}\bar{M}_j}{\mathrm{d}x}}{1 - \bar{M}_j} \tag{4.114d}$$

$$\mathbf{H}_{j1} = \frac{a_j}{2}\begin{bmatrix} \dfrac{1-\varphi_j\bar{M}_j}{1+\bar{M}_j} & \dfrac{1-\varphi_j\bar{M}_j}{1+\bar{M}_j} \\ -\dfrac{1+\varphi_j\bar{M}_j}{1-\bar{M}_j} & -\dfrac{1+\varphi_j\bar{M}_j}{1-\bar{M}_j} \end{bmatrix}, \mathbf{H}_{1j} = \frac{S_j c_1}{S_1 c_j}\mathbf{H}_{j1}, \quad j = 2, 3, \ldots, n \quad (4.115)$$

and all remaining $\mathbf{H}_{kl}$ blocks, $k, l = 2, 3, \ldots, n$, are 2×2 zero matrices.

The literature on single-coupled continuous models of perforated ducts is extensive. The theory appears to have been pioneered by Sullivan and Crocker [10], who considered applications of the two-duct element. Later works were concerned with developing numerical methods for alleviating some ill-conditioning problems inherent to the standard numerical solutions and applying the method for practical muffler configurations [11–19]. Two methods which seem to avoid ill-conditioning problems are the decoupling method (proposed by Peat [13]) and the matrizant method [14]. The latter, which is the method described here, has the advantage of producing the wave transfer matrix directly. Both approaches rely on efficient numerical routines for getting accurate solutions of the underlying algebraic eigenvalue problems.

In general, the wave transfer matrix of the n-duct unit is given by the Peano series defined in Equation (3.32). Equation (4.111) can be solved numerically also by using the segmentation method (see Section 3.3.3). However, single-coupled n-duct units with non-uniform perforated ducts are rare in practice. Furthermore, since forcing the mean flow through perforates involves high flow losses, most perforated n-duct units designed for engine exhaust applications operate with negligible mean flow through the apertures. For a uniform n-duct element with grazing mean flow only, Equation (4.1114) simplifies to [19]:

$$\mathbf{H}_{jj} = \begin{bmatrix} \dfrac{\mathrm{i}k_j + \frac{1}{2}\left(-1+\varphi_j\bar{M}_j\right)a_j}{1+\bar{M}_j} & \dfrac{\frac{1}{2}\left(-1+\varphi_j\bar{M}_j\right)a_j}{1+\bar{M}_j} \\ \dfrac{\frac{1}{2}\left(1+\varphi_j\bar{M}_j\right)a_j}{1-\bar{M}_j} & \dfrac{-\mathrm{i}k_j + \frac{1}{2}\left(1+\varphi_j\bar{M}_j\right)a_j}{1-\bar{M}_j} \end{bmatrix}, \quad j = 1, 2, \ldots, n \tag{4.116}$$

and the matrizant $\mathbf{H}$ becomes independent of the axial coordinate, x. Consequently, the wave transfer matrix of the n-duct element of length L is given by $\mathbf{T} = \mathrm{e}^{\mathbf{H}L}$. We may use the collineatory transformation from Equation (3.34) to express this as $\mathbf{T} = \mathbf{\Phi\Lambda\Phi}^{-1}$, where $\mathbf{\Phi}$ denotes the modal matrix whose columns are the eigenvectors $\boldsymbol{\varphi}_j = \{\varphi_{1j} \quad \varphi_{2j} \quad \cdots\}$ of the matrizant $\mathbf{H}$ and $\mathbf{\Lambda}$ denotes a diagonal square matrix with diagonal elements $\mathrm{e}^{\mathrm{i}\kappa_j x}$, $j = 1, 2, \ldots$, where $\mathrm{i}\kappa_j$ are the eigenvalues of $\mathbf{H}$.

4.7.1.1 Identical Perforated Ducts

A single-coupled perforated n-duct unit is sometimes constructed by inserting a bundle of identical uniform perforated ducts, which are held together by rigid end-plates, into the casing. If the number of ducts is large, a direct application of Equation (4.111) can become problematic because of its size.

A single-coupled n-duct element with identical perforated ducts can be reduced to a two-duct element. To derive this equivalent two-duct element, let $\bar{M}_j = \bar{M}$, $\varphi_j = \varphi$, $k_j = k$, $a_j = a$ and $S_j = S$, where $j = 2, 3, \ldots, n$. Then Equation (4.111) assumes the form

$$\frac{\partial}{\partial x}\begin{bmatrix} \mathbf{P}_1 \\ \mathbf{P}_2 \\ \mathbf{P}_3 \\ \vdots \\ \mathbf{P}_n \end{bmatrix} = \begin{bmatrix} \mathbf{H}_{11} & \mathbf{H}_{12} & \mathbf{H}_{12} & \ldots & \mathbf{H}_{12} \\ \mathbf{H}_{21} & \mathbf{H}_{22} & \mathbf{0} & \ldots & \mathbf{0} \\ \mathbf{H}_{21} & \mathbf{0} & \mathbf{H}_{22} & \ldots & \mathbf{0} \\ \vdots & \vdots & \vdots & \ddots & \vdots \\ \mathbf{H}_{21} & \mathbf{0} & \mathbf{0} & \ldots & \mathbf{H}_{22} \end{bmatrix}\begin{bmatrix} \mathbf{P}_1 \\ \mathbf{P}_2 \\ \mathbf{P}_3 \\ \vdots \\ \mathbf{P}_n \end{bmatrix} \tag{4.117}$$

where $\mathbf{P}_j = \mathbf{P}_2$, $j = 3, \ldots, n$. Therefore, in this case it suffices to solve the much simpler system

$$\frac{\partial}{\partial x}\begin{bmatrix} \mathbf{P}_1 \\ \mathbf{P}_2 \end{bmatrix} = \begin{bmatrix} \mathbf{H}_{11} & (n-1)\mathbf{H}_{12} \\ \mathbf{H}_{21} & \mathbf{H}_{22} \end{bmatrix}\begin{bmatrix} \mathbf{P}_1 \\ \mathbf{P}_2 \end{bmatrix}. \tag{4.118}$$

This result can be generalized for cases where only some of the perforated ducts are identical. For example, assume that all but one of the perforated ducts are identical and let ducts $j = 3, 4, \ldots, n$ be the identical ones. In this case Equation (4.111) reduces to

$$\frac{\partial}{\partial x}\begin{bmatrix} \mathbf{P}_1 \\ \mathbf{P}_2 \\ \mathbf{P}_3 \end{bmatrix} = \begin{bmatrix} \mathbf{H}_{11} & \mathbf{H}_{12} & (n-2)\mathbf{H}_{13} \\ \mathbf{H}_{21} & \mathbf{H}_{22} & \mathbf{0} \\ \mathbf{H}_{31} & \mathbf{0} & \mathbf{H}_{33} \end{bmatrix}\begin{bmatrix} \mathbf{P}_1 \\ \mathbf{P}_2 \\ \mathbf{P}_3 \end{bmatrix} \tag{4.119}$$

and so on if all but 2,3,... perforated ducts are identical.

4.7.2 Double-Coupled Perforated Ducts

A double-coupled three-duct element is shown in Figure 4.10d. This consists of a perforated duct enclosed in another perforated duct, which is contained in a duct (casing) with impervious walls. The generalization of this configuration for $n - 1$ perforated ducts is called double-coupled n-duct element. The ducts are numbered as duct 1, duct 2, ..., duct n, in the order of the outermost to the innermost duct, which may have arbitrary transverse dispositions and need not be uniform. Duct 1 is called the casing or shell of the unit. We use the subscript "j" to refer to the properties of duct j, $j = 1, 2, \ldots, n$. The space between duct j and duct $j + 1$, $j = 1, 2, \ldots, n - 1$, is referred to as annulus j, annulus n being the duct n itself. The annulus parameters $\bar{\rho}_j$ and S_j, $j = 1,2,\ldots,n$, denote the density of the fluid and the cross-sectional area of annulus j, respectively. The duct parameters P_j and σ_j, $j = 1,2,\ldots,n$, denote, respectively, the perimeter of S_j and the porosity of duct j. Note that $\sigma_1 = 0$, since duct 1 is impermeable. Thus, for each annulus of a double-coupled n-duct element, Equations (4.92) and (4.93) yield:

$$\left\{\left(-ik_j+\frac{\mathrm{d}\bar{M}_j}{\mathrm{d}x}+\bar{M}_j\frac{\mathrm{d}}{\mathrm{d}x}\ln S_j\right)\overline{p_j'}+\bar{M}_j\frac{\partial\overline{p_j'}}{\partial x}\right\}+$$
$$+\rho_j c_j\left(\frac{\partial\overline{v_j'}}{\partial x}+\overline{v_j'}\frac{\mathrm{d}}{\mathrm{d}x}\ln S_j\right)=-a_j\left(\overline{p_j'}-\overline{p_{j-1}'}\right)+b_{j+1}\left(\overline{p_{j+1}'}-\overline{p_j'}\right) \tag{4.120}$$

$$\bar{\rho}_j c_j\left\{\left(-ik_j+\frac{\mathrm{d}\bar{M}_j}{\mathrm{d}x}\right)\overline{v_j'}+\bar{M}_j\frac{\partial\overline{v_j'}}{\partial x}\right\}+\frac{\partial\overline{p_j'}}{\partial x}=-\bar{M}_j\varphi_j\left(-a_j\left(\overline{p_j'}-\overline{p_{j-1}'}\right)+b_{j+1}\left(\overline{p_{j+1}'}-\overline{p_j'}\right)\right), \tag{4.121}$$

where $j=1,2,\ldots,n$, $a_{n+1}=0$ by convention, and

$$b_{j+1}=\left(\frac{S_{j+1}\bar{\rho}_j c_j}{S_j\bar{\rho}_{j+1}c_{j+1}}\right)a_{j+1}. \tag{4.122}$$

In these expressions, the parameters and variables with subscript n+1 are superfluous variables with no physical meaning and may be assumed to have arbitrary finite values greater than zero, except for σ_{n+1}, which is zero by definition. For simplicity, the slip-flow parameter,φ_j, is assumed to be the same for each annulus.

In matrix notation, Equations (4.120) and (4.121) can be expressed in the state-space form of Equation (4.103), where the submatrices are now given by

$$\mathbf{B}_{jj}=\begin{bmatrix}-\bar{M}_j\varphi_j(a_j+b_{j+1}) & -ik_j+\dfrac{\mathrm{d}\bar{M}_j}{\mathrm{d}x}\\ -ik_j+a_j+b_{j+1}+\dfrac{\mathrm{d}\bar{M}_j}{\mathrm{d}x}+\bar{M}_j\dfrac{\mathrm{d}}{\mathrm{d}x}\ln S_j & \dfrac{\mathrm{d}}{\mathrm{d}x}\ln S_j\end{bmatrix},\quad j=1,2,\ldots,n \tag{4.123}$$

$$\mathbf{B}_{j(j+1)}=b_{j+1}\begin{bmatrix}\bar{M}_j\varphi_j & 0\\ -1 & 0\end{bmatrix},\quad j=1,2,,\ldots,n-1 \tag{4.124}$$

$$\mathbf{B}_{(j+1)j}=a_j\begin{bmatrix}\bar{M}_j\varphi_j & 0\\ -1 & 0\end{bmatrix},\quad j=2,3,\ldots,n \tag{4.125}$$

and all remaining $\mathbf{B}_{kl}$ blocks are 2×2 matrices with zero elements. Upon transforming similarly to the pressure wave components, we obtain Equation (4.111), where the matrizant is given by Equation (4.113), but with the 2×2 submatrix blocks now given by

$$\mathbf{H}_{j(j+1)}=\frac{b_{j+1}}{2}\begin{bmatrix}\dfrac{1-\varphi_j\bar{M}_j}{1+\bar{M}_j} & \dfrac{1-\varphi_j\bar{M}_j}{1+\bar{M}_j}\\ -\dfrac{1+\varphi_j\bar{M}_j}{1-\bar{M}_j} & -\dfrac{1+\varphi_j\bar{M}_j}{1-\bar{M}_j}\end{bmatrix},\quad j=1,2,\ldots,n-1 \tag{4.126}$$

$$\mathbf{H}_{(j+1)j} = \frac{S_j c_1}{S_1 c_j} \mathbf{H}_{j1}, \quad j = 1, 2, \ldots, n-1. \tag{4.127}$$

The $\mathbf{H}_{jj}$, $j = 1, 2, \ldots, n$, is the same as Equation (4.114), but with a_j replaced by $a_j + b_{j+1}$, and all remaining submatrices of $\mathbf{H}$ are 2×2 zero matrices.

Again, the wave transfer matrix can be determined numerically by evaluating the Peano series or by using the segmentation method. However, when all ducts are uniform and mean through flow through the apertures are negligible, the $\mathbf{H}_{jj}$ blocks of the matrizant reduce Equation (4.116) with a_j replaced by $a_j + b_{j+1}, j = 1, 2, \ldots, n$ and, hence, the matrizant $\mathbf{H}$ becomes independent of the axial coordinate, x. The wave transfer matrix of an n-duct element of length L is then given by $\mathbf{T} = \mathrm{e}^{\mathbf{H}L}$ and can be calculated by applying the collineatory transformation described in Section 3.3.1.1 [20].

4.8 Row-Wise Coupled Perforated Ducts

Continuous models of perforated ducts are apt to be unrealistic if perforations are distributed sparsely over a duct surface. Then we may consider modeling of the distribution of apertures discretely [19, 21–23]. However, in one dimension, the circumferential distribution pattern of apertures cannot be taken into account. Accordingly, the discrete distribution of apertures is modeled as an alternating sequence of discrete aperture rows separated by hard-walled duct segments. Since the apertures are assumed to be compact, aperture rows may be considered to be of elementary width. Figure 4.11 shows generic single- and double-coupled n-duct units with arbitrary aperture row patterns. Here, it suffices to derive a wave transfer matrix which relates the pressure wave components just upstream and just downstream of a transverse section which is assumed to contain an aperture row on every perforated duct (for example, section $x = \xi$ in Figure 4.11). However, it is not necessary for

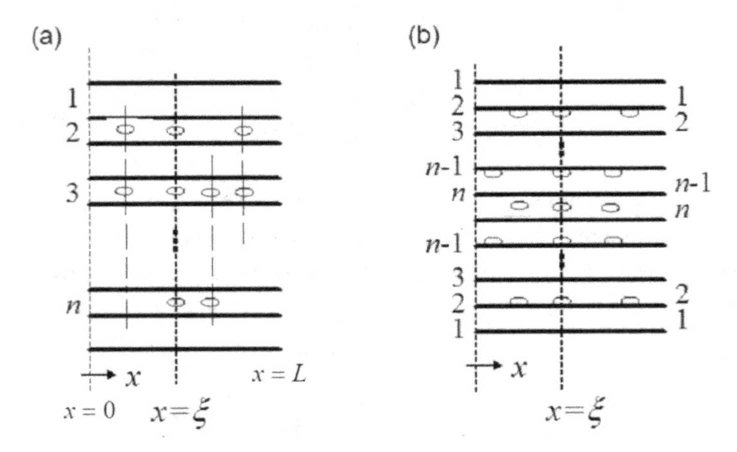

Figure 4.11 Generic aperture row of n-duct elements, (a) single-coupled, (b) double-coupled.

every perforated duct to have an aperture row in the same transverse section. As it will be clear subsequently, the wave transfer matrix derived for such a generic transverse section can be modified for the presence of ducts without an aperture row in that section simply by inputting zero for the aperture area. The overall wave transfer matrix of an n-duct unit is then calculated by cascading the wave transfer matrices across the aperture rows and the one-dimensional wave transfer matrices of the solid duct segments separating them (Section 4.8.4). This approach has the advantage of taking into account the effect of the axial distribution pattern of the aperture rows. A further advantage is that perforated ducts which are coupled by axially staggered aperture distributions can be modeled by a single n-duct element (with the continuous method, each staggered perforate patch has to be modeled by using separate n-duct elements).

4.8.1 Wave Transfer Across a Row of Apertures

Consider a perforated duct having a single aperture row at $x = \xi$. Since the apertures are assumed to be compact, we represent $\dot{\mu}'$ by Dirac function distribution placed at $x = \xi$ as

$$\dot{\mu}'(x) = \frac{\overline{\rho_o}\dot{Q}(x)}{S}\delta(x - \xi) \tag{4.128}$$

where $\delta(x)$ denotes a Dirac function located at $x = 0$, $\overline{\rho_o}\dot{Q}(x)$ denotes the rate of total fluctuating mass flow into the duct from the apertures in the row. After substituting Equation (4.128) in Equation (4.92), we integrate the resulting equation across the interval $\xi_- \leq x \leq \xi_+$ where $\xi_- = \xi - \varepsilon$, $\xi_+ = \xi + \varepsilon$ and ε is ideally zero; however, it may be considered as a compact 'end correction' to account for the diffractive effects in the vicinity of apertures [22].

We define the distribution of the variables in the interval $\xi_- \leq x \leq \xi_+$ by the Heaviside function $H_{1/2}(x)$, which is defined as

$$H_{1/2}(x) = \begin{cases} 1 & \text{if} \quad x > 0 \\ \frac{1}{2} & \text{if} \quad x = 0 \, . \\ 0 & \text{if} \quad x < 0 \end{cases} \tag{4.129}$$

Hence, Equation (4.92) yields

$$[\bar{M}_o]\langle\overline{p'}\rangle + \langle\bar{M}_o\rangle\left[\overline{p'}\right] + \overline{\rho_o}c_o\left[\overline{v'}\right] = \frac{\overline{\rho_o}c_o\dot{Q}(\xi)}{S(\xi)}. \tag{4.130}$$

Here, a pair of square brackets enclosing a variable denote a jump, and angled brackets denote arithmetical average over the discontinuity at $x = \xi$. For example,

$$[p'] = p'_+ - p'_- \tag{4.131}$$

$$\langle p' \rangle = \frac{p'_+ + p'_-}{2}, \tag{4.132}$$

where the subscripts "+" and "−" refer to values at $x = \xi_+$ and $x = \xi_-$, respectively. Jump in $\bar{M}_o$ is considered in order to include the effect of mean flow though the apertures.

Similarly, the momentum equation, Equation (4.83) yields the jump relation

$$\overline{\rho_o} c_o(\langle v' \rangle [\bar{M}_o] + \langle \bar{M}_o \rangle [v']) + [p'] = -\langle \varphi_{SF} \bar{M}_o \rangle \frac{\overline{\rho_o} c_o \dot{Q}(\xi)}{S(\xi)}. \tag{4.133}$$

The jumps and averages of the pressure wave components are defined similarly:

$$\left[\overline{p'}\right] = [p^+] + [p^-], \left\langle \overline{p'} \right\rangle = \langle p^+ \rangle + \langle p^- \rangle \tag{4.134}$$

$$\overline{\rho_o} c_o \left[\overline{v'}\right] = [p^+] - [p^-], \overline{\rho_o} c_o \left\langle \overline{v'} \right\rangle = \langle p^+ \rangle - \langle p^- \rangle. \tag{4.135}$$

With substitution of these, Equations (4.130) and (4.133) can be expressed in matrix notation as

$$\mathbf{P}(\xi_+) = \mathbf{TP}(\xi_-) + \frac{1}{2} \overline{\rho_o} c_o \frac{\dot{Q}(\xi)}{S(\xi)} \mathbf{B}, \tag{4.136}$$

where

$$\mathbf{P}(x) = \begin{bmatrix} p^+(x) \\ p^-(x) \end{bmatrix} \tag{4.137}$$

$$\mathbf{T} = \frac{1}{\Delta} \begin{bmatrix} -\frac{1}{4}[\bar{M}_o]^2 + [\bar{M}_o] + \langle \bar{M}_o \rangle^2 - 1 & 0 \\ 0 & -\frac{1}{4}[\bar{M}_o]^2 - [\bar{M}_o] + \langle \bar{M}_o \rangle^2 - 1 \end{bmatrix} \tag{4.138}$$

$$\mathbf{B} = \frac{1}{\Delta} \begin{bmatrix} (1 - \langle \varphi_{SF} \bar{M}_o \rangle) \left(\frac{1}{2}[\bar{M}_o] + \langle \bar{M}_o \rangle - 1 \right) \\ (1 + \langle \varphi_{SF} \bar{M}_o \rangle) \left(\frac{1}{2}[\bar{M}_o] + \langle \bar{M}_o \rangle + 1 \right) \end{bmatrix} \tag{4.139}$$

$$\Delta = \left(\frac{1}{2}[\bar{M}_o] + \langle \bar{M}_o \rangle \right)^2 - 1. \tag{4.140}$$

To evaluate $\dot{Q}(\xi)$, note that $\dot{Q}(\xi) = -u'(\xi)A(\xi)$ by definition, where $A(\xi)$ denotes the total open cross-sectional area of the aperture row at $x = \xi$. For practical purposes, we may assume that apertures in a row are identical. Then, the equation of the unsteady motion of the fluid in an aperture is given by $\overline{p'_{\text{int}}}(\xi) - \overline{p'_{ext}}(\xi) = Z(\xi)u'(\xi)$, where Z denotes the aperture impedance. Hence, Equation (4.136) may be expressed as

$$\mathbf{P}(\xi_+) = \mathbf{TP}(\xi_-) - \frac{1}{2} \overline{\rho_o} c_o \frac{A(\xi)}{Z(\xi)S(\xi)} \mathbf{BE}(\mathbf{P}(\xi_-) - \mathbf{P}_{\text{ext}}(\xi_-)), \tag{4.141}$$

where $\mathbf{E} = [1 \quad 1]$, $\overline{p'_{\text{ext}}}(\xi) = \mathbf{E}\mathbf{P}_{\text{ext}}(\xi_-)$ and $\overline{p'_{\text{int}}}(\xi) = \overline{p'}(\xi_-)$. This is the required wave transfer relationship across an aperture row of a perforated duct. If the duct carries only a grazing mean flow, then $[\bar{M}_o] = 0$ and $\langle \bar{M}_o \rangle = \bar{M}_o$, and Equations (4.138) and (4.139) simplify, respectively, as [22]:

$$\mathbf{T} = \begin{bmatrix} 1 & 0 \\ 0 & 1 \end{bmatrix} \tag{4.142}$$

$$\mathbf{B} = \begin{bmatrix} \dfrac{1 - \varphi_{\text{SF}} \bar{M}_o}{1 + \bar{M}_o} \\ -\dfrac{1 + \varphi_{\text{SF}} \bar{M}_o}{1 - \bar{M}_o} \end{bmatrix}. \tag{4.143}$$

4.8.2 Single-Coupled *n*-Duct Section

Referring to Figure 4.11a, the ducts are numbered as in Section 4.7. Dropping the subscript "*o*" for simplicity of notation, we write Equation (4.141) for each duct as:

$$\mathbf{P}_1(\xi_+) = \mathbf{T}_1 \mathbf{P}_1(\xi_-) - \frac{1}{2} \sum_{j=2}^{n} \frac{\bar{\rho}_1 c_1}{\bar{\rho}_j c_j} \frac{A_j(\xi)}{\zeta_j(\xi) S_1(\xi)} \mathbf{B}_1 \mathbf{E} \big(\mathbf{P}_1(\xi_-) - \mathbf{P}_j(\xi_-) \big) \tag{4.144}$$

$$\mathbf{P}_j(\xi_+) = \mathbf{T}_j \mathbf{P}_j(\xi_-) - \frac{1}{2} \frac{A_j(\xi)}{\zeta_j(\xi) S_j(\xi)} \mathbf{B}_j \mathbf{E} \big(\mathbf{P}_j(\xi_-) - \mathbf{P}_1(\xi_-) \big), \quad j = 2, 3, \ldots, n \tag{4.145}$$

where the subscript "*j*" refers to pipe *j* as usual and $\zeta_j(\xi) = Z_j(\xi)/\bar{\rho}_j c_j$. The summation in the former equation follows because duct 1 communicates with all perforated pipes and, therefore, $\dot{Q}_1(\xi) = -\sum_{j=2}^{N} \dot{Q}_j(\xi)$. Equations (4.144) and (4.145) can be collected as

$$\begin{bmatrix} \mathbf{P}_1(\xi_+) \\ \mathbf{P}_2(\xi_+) \\ \mathbf{P}_3(\xi_+) \\ \\ \mathbf{P}_n(\xi_+) \end{bmatrix} = \begin{bmatrix} \mathbf{G}_{11} & \mathbf{G}_{12} & \mathbf{G}_{13} & \cdots & \mathbf{G}_{1n} \\ \mathbf{G}_{21} & \mathbf{G}_{22} & 0 & \cdots & 0 \\ \mathbf{G}_{31} & 0 & \mathbf{G}_{33} & \cdots & 0 \\ \vdots & \vdots & \vdots & \ddots & \vdots \\ \mathbf{G}_{n1} & 0 & 0 & \cdots & \mathbf{G}_{nn} \end{bmatrix} \begin{bmatrix} \mathbf{P}_1(\xi_-) \\ \mathbf{P}_2(\xi_-) \\ \mathbf{P}_3(\xi_-) \\ \\ \mathbf{P}_n(\xi_-) \end{bmatrix}. \tag{4.146}$$

Here, the non-zero sub-matrices are given by

$$\mathbf{G}_{11} = \mathbf{T}_1 - \frac{1}{2} \sum_{j=2}^{n} \frac{\bar{\rho}_1 c_1}{\bar{\rho}_j c_j} \frac{A_j(\xi)}{\zeta_j(\xi) S_1(\xi)} \mathbf{B}_1 \mathbf{E} \tag{4.147}$$

$$\mathbf{G}_{jj} = \mathbf{T}_j - \frac{1}{2} \frac{A_j(\xi)}{\zeta_j(\xi) S_j(\xi)} \mathbf{B}_j \mathbf{E}, \quad j = 2, 3, \ldots, n \tag{4.148}$$

$$\mathbf{G}_{j1} = \frac{1}{2}\frac{A_j(\xi)}{\zeta_j(\xi)S_j(\xi)}\mathbf{B}_j\mathbf{E}, \quad j = 2, 3, \ldots, n \tag{4.149}$$

$$\mathbf{G}_{1j} = \frac{1}{2}\frac{\bar{\rho}_1 c_1}{\bar{\rho}_j c_j}\frac{A_j(\xi)}{\zeta_j(\xi)S_1(\xi)}\mathbf{B}_1\mathbf{E}, \quad j = 2, 3, \ldots, n \tag{4.150}$$

The square matrix in Equation (4.146) is the wave transfer matrix across a generic section of single-coupled n-duct unit. Although this is derived by assuming that all perforated ducts contain an aperture row in this section, it is now clear that if a perforated duct(s) $j \in 2, 3, \ldots, n$ does not have an aperture row in this transverse plane, the corresponding wave transfer matrix can be obtained simply by inputting zero for $A_j(\xi)$. This ability to deal with any permutation of the aperture rows in any section enables modeling of communicating perforated ducts which have axially staggered perforate rows by using only a single n-duct element.

4.8.3 Double-Coupled *n*-Duct Section

Referring to Figure 4.11b, the ducts and annuli are numbered as described in Section 4.7.2. In this case, we write Equation (4.141) for each annuli, noting that the rate of fluctuating volume flow into annuli j is $\dot{Q}_j(\xi) - \dot{Q}_{j+1}(\xi)$, $j = 1, 2, \ldots, n$, where $\dot{Q}_1(\xi) = \dot{Q}_{n+1}(\xi) = 0$ by convention, since duct 1 has impervious walls and there are only n ducts. Hence, Equation (4.141) gives

$$\begin{aligned}\mathbf{P}_j(\xi_+) = \mathbf{T}_j\mathbf{P}_j(\xi_-) - \frac{1}{2}\frac{A_j(\xi)}{\zeta_j(\xi)S_j(\xi)}\mathbf{B}_j\mathbf{E}\big(\mathbf{P}_j(\xi_-) - \mathbf{P}_{j-1}(\xi_-)\big) - \\ -\frac{1}{2}\frac{\bar{\rho}_j c_j}{\bar{\rho}_{j+1}c_{j+1}}\frac{A_{j+1}(\xi)}{\zeta_{j+1}(\xi)S_j(\xi)}\mathbf{B}_j\mathbf{E}\big(\mathbf{P}_j(\xi_-) - \mathbf{P}_{j+1}(\xi_-)\big), \quad j = 1, 2, \ldots, n\end{aligned} \tag{4.151}$$

where the above convention is implemented by putting $A_1(\xi) = A_{n+1}(\xi) = 0$. These equations can be expressed in matrix form as

$$\begin{bmatrix}\mathbf{P}_1(\xi_+)\\ \mathbf{P}_2(\xi_+)\\ \mathbf{P}_3(\xi_+)\\ \\ \mathbf{P}_n(\xi_+)\end{bmatrix} = \begin{bmatrix}\mathbf{G}_{11} & \mathbf{G}_{12} & 0 & \cdots & 0\\ \mathbf{G}_{21} & \mathbf{G}_{22} & \mathbf{G}_{23} & \cdots & 0\\ 0 & \mathbf{G}_{32} & \mathbf{G}_{33} & \cdots & 0\\ \vdots & \vdots & \vdots & \ddots & \vdots\\ 0 & 0 & 0 & \cdots & \mathbf{G}_{nn}\end{bmatrix}\begin{bmatrix}\mathbf{P}_1(\xi_-)\\ \mathbf{P}_2(\xi_-)\\ \mathbf{P}_3(\xi_-)\\ \\ \mathbf{P}_n(\xi_-)\end{bmatrix} \tag{4.152}$$

where

$$\mathbf{G}_{jj} = \mathbf{T}_j - \frac{1}{2}\left(\frac{A_j(\xi)}{\zeta_j(\xi)S_j(\xi)} + \frac{\bar{\rho}_j c_j}{\bar{\rho}_{j+1}c_{j+1}}\frac{A_{j+1}(\xi)}{\zeta_{j+1}(\xi)S_j(\xi)}\right)\mathbf{B}_j\mathbf{E}, \quad j = 1, 2, \ldots, n \tag{4.153}$$

$$\mathbf{G}_{j,j-1} = \frac{1}{2} \frac{A_j(\xi)}{\zeta_j(\xi) S_j(\xi)} \mathbf{B}_j \mathbf{E}, \quad j = 1, 2, \ldots, n-1 \tag{4.154}$$

$$\mathbf{G}_{j,j+1} = \frac{1}{2} \frac{\bar{\rho}_j c_j}{\bar{\rho}_{j+1} c_{j+1}} \frac{A_{j+1}(\xi)}{\zeta_{j+1}(\xi) S_j(\xi)} \mathbf{B}_j \mathbf{E}, \quad j = 1, 2, \ldots, n-1 \tag{4.155}$$

The square matrix in Equation (4.152) is the wave transfer matrix of a generic section of the n-duct unit. Again, if a perforated duct does not have an aperture row at position $x = \xi$, we simply input zero for the corresponding total aperture area.

4.8.4 Wave Transfer Matrix of n-Duct Element

Let the transverse sections of an n-duct unit containing at least one aperture row be located at axial positions x_k, $k = 1, 2, \ldots, K$, $x_1 < x_2 < \cdots < x_K$, from origin $x = 0$. We write Equation (4.146), or Equation (4.152), briefly as $\mathbf{P}(x_{k+}) = \mathbf{G}^{(k)} \mathbf{P}(x_{k-})$, $k = 1, 2, \ldots, K$, for each aperture row. The internal ducts have hard walls in the intervals $x_k + \varepsilon \leq x \leq x_{k+1} - \varepsilon$, $k = 1, 2, \ldots, K-1$, and in $0 \leq x \leq x_1 - \varepsilon$ and $x_K + \varepsilon \leq x \leq L$, where L denotes the length of the n-duct unit. The wave transfer equations of these hard-walled segments of the unit can be expressed as $\mathbf{P}\big(x_{(k+1)-}\big) = \mathbf{T}^{(k)} \mathbf{P}(x_{k+})$, where the $\mathbf{T}^{(k)}$ denotes the block-diagonal matrix:

$$\mathbf{T}^{(k)} = \begin{bmatrix} \mathbf{T}_1^{(k)} & 0 & 0 & 0 \\ 0 & \mathbf{T}_2^{(k)} & 0 & 0 \\ 0 & 0 & \ddots & \vdots \\ 0 & 0 & \cdots & \mathbf{T}_n^{(k)} \end{bmatrix} \tag{4.156}$$

and $\mathbf{T}_j^{(k)}$, $k = 1, 2, \ldots, K-1$, denote the individual wave transfer matrices of the segments $x_k + \varepsilon \leq x \leq x_{k+1} - \varepsilon$ of ducts $j = 1, 2, \ldots, n$, which may possibly be packed by sound absorbent material, and $\mathbf{T}_j^{(0)}$ and $\mathbf{T}_j^{(K)}$ similarly denote the wave transfer matrices corresponding to the segments $0 \leq x \leq x_1 - \varepsilon$ and $x_K + \varepsilon \leq x \leq L$, respectively. The wave transfer equation of the n-duct unit is $\mathbf{P}(L) = \mathbf{T}\mathbf{P}(0)$, where the overall wave transfer matrix, $\mathbf{T}$, of the unit is given by

$$\mathbf{T} = \mathbf{T}^{(K)} \mathbf{G}^{(K)} \mathbf{T}^{(K-1)} \mathbf{G}^{(K-1)} \ldots \mathbf{T}^{(2)} \mathbf{G}^{(2)} \mathbf{T}^{(1)} \mathbf{G}^{(1)} \mathbf{T}^{(0)} \tag{4.157}$$

since the aperture sections and solid duct segments are in series. Obviously, this scheme applies for both the single- and double-coupled n-duct elements.

If the aperture rows are identical and separated periodically by identical solid duct segments, then $\mathbf{P}\big(x_{(k+1)-}\big) = \mathbf{S}\mathbf{P}(x_{k-})$, where $\mathbf{S} = \mathbf{T}^{(k)} \mathbf{G}^{(k)}$, and the wave transfer equation of the n-duct unit becomes $\mathbf{P}(L) = \mathbf{S}^K \mathbf{T}(0) \mathbf{P}(0)$. The matrix $\mathbf{S}^K$ can be evaluated by using the decomposition $\mathbf{S} = \mathbf{\Psi}\mathbf{\Gamma}\mathbf{\Psi}^{-1}$, where $\mathbf{\Gamma}$ denotes a $2n \times 2n$ diagonal matrix whose diagonal elements $\gamma_1, \gamma_2, \ldots, \gamma_{2n}$ are the eigenvalues of matrix $\mathbf{S}$ with corresponding right eigenvectors $\psi_1, \psi_2, \ldots, \psi_{2n}$, and $\mathbf{\Psi} = [\psi_1 \quad \psi_2 \quad \cdots \quad \psi_{2n}]$ denotes the modal matrix (see Section 3.3.1.1). Then, for example, $\mathbf{S}^2 = \mathbf{S}\mathbf{S} = \mathbf{\Psi}\mathbf{\Gamma}\mathbf{\Psi}^{-1}\mathbf{\Psi}\mathbf{\Gamma}\mathbf{\Psi}^{-1} = \mathbf{\Psi}\mathbf{\Gamma}^2\mathbf{\Psi}^{-1}$ and in general, $\mathbf{S}^K = \mathbf{\Psi}\mathbf{\Gamma}^K\mathbf{\Psi}^{-1}$. Thus, the sound wave propagation in a periodic n-duct unit are characterized by $\gamma_1^K, \gamma_2^K, \ldots, \gamma_{2n}^K$. Several mathematical and physical aspects of these eigenvalues are discussed by Pachebat and Kergomard for a two-duct unit [23]. In general, however, the eigenvalue problem is not tractable to an analytical study and $\mathbf{S}^K$ has to be computed numerically.

4.9 Dissipative Units and Lined Ducts

Single- and double-coupled perforated n-duct elements described in the previous sections can be used for modeling of various types dissipative silencers. This topic is treated in three-dimensions in Section 7.7.3. Here we will point out briefly how one-dimensional n-duct units may be used in dissipative silencer types. Referring to the continuously coupled models (Section 4.7), filling duct 1 in the single-coupled configuration (Section 4.7.1) by sound absorbing material and channeling the mean flow through the perforated ducts results in a splitter silencer of the type shown in Figure 7.11d. In this case, the matrizant in Equation (4.111) is modified by using the equivalent fluid properties (see Appendix B) of the packing material for ρ_1 and c_1, and putting $\bar{M}_1 = 0$. The alternative design configuration, where the perforated ducts are packed with sound absorbent material and the flow is channeled through duct 1, gives a bar type dissipative silencer (Figure 7.11b). In this case, the equivalent fluid properties of the packing materials are substituted for c_j and ρ_j, and Equation (4.111) is applied by putting $\bar{M}_j = 0$, where j denotes the numbers of the packed perforated ducts. In double-coupled configuration (Section 4.7.2), filling the annuli between concentric perforated ducts alternately with sound absorbing material and letting the flow go through the unfilled annuli, gives a circular splitter silencer. A typical configuration known as a pod silencer is shown in Figure 7.11a. If an annulus $j \in 1, 2, \ldots, n-1$ is packed by a sound absorbing material, it suffices to put in the matrizant $\bar{M}_j = 0$ and use for ρ_j and c_j the corresponding equivalent fluid values of the packing. In fact, the theory of Section 4.7.2 also applies to parallel baffle silencers (Figure 7.11c), if the channels and splitters are numbered consecutively. Thus, the parallel baffle silencer shown in Figure 7.11c can be modeled as a double-coupled five-duct element.

Row-wise coupled models (Section 4.8) are used in the same manner after the aperture rows are combined as described in Section 4.8.4. For any packed duct, for example, duct $j \in 1, 2, \ldots, n$ in a single-coupled configurations or annulus $j \in 2, 3, \ldots, n$ in a double-coupled configuration, we take $\left[\bar{M}_j\right] = 0$, $\left\langle \bar{M}_j \right\rangle = 0$ and use the equivalent fluid values of the packing material for $\bar{\rho}_j$ and c_j in the wave transfer matrices across aperture rows, and model the duct segments separating aperture rows by a packed duct element (Section 3.7).

It should be noted that impedance of perforate apertures may have to be corrected when they are backed with sound absorbent material on either side, since they are usually determined by assuming fluid on both sides (see Appendix B).

Lined ducts may also be modeled by using dissipative n-duct units, if the liner layers are separated by perforated sheets. For example, a duct with a single layer bulk-reacting liner protected by a perforated facesheet can be modeled by a two-duct unit.

In the case of the continuously coupled model, the matrizant of the duct is given by the $n = 2$ case of Equation (4.113), duct 1 being packed with sound absorbent material. If the row-wise coupled model is used, the wave transfer equation across perforate rows is given by the $n = 2$ case of Equation (4.146), or Equation (4.152), again with duct 1 packed. The overall wave transfer matrix is determined as described in Section 4.8.4.

If the liner is reacting locally, then duct 1 becomes superfluous and the matrizant of the continuously coupled model is given by $\mathbf{H}_{22}$ in Equation (4.113) with no through mean flow ($\mathrm{d}\bar{M}_2/\mathrm{d}x = 0$). For a uniform lined duct, $S_2 =$ constant and $\mathbf{H}_{22}$ reduces to the matrizant in Equation (3.141) with $\alpha = 1$, $\psi = 0$ and $A_W = a_2 = P_2\sigma_2/S_2\zeta_2$, with ζ_2/σ_2 representing the liner impedance.

The row-wise coupled model provides an alternative approach for the modeling of a duct with a locally reacting liner. In this case, the wave transfer equation across a row of facesheet holes reduces to $\mathbf{P}_2(\xi_+) = \mathbf{G}_{22}\mathbf{P}_2(\xi_-)$, where $\mathbf{G}_{22}$ is given by Equation (4.148). In most applications, aperture rows may be assumed to be periodic and in calculating the overall wave transfer matrix of a duct of length L, we may use Equation (3.54) for the wave transfer matrix, $\mathbf{T}$, of the solid duct segments separating the aperture rows. Hence, assuming $\alpha = 1$ for simplicity, the periodic element matrix $\mathbf{S}$ (Section 4.8.4) is $\mathbf{S} = \mathbf{TG}_{22}$ an may be expressed explicitly as

$$\mathbf{S} = \begin{bmatrix} \left(1 - \dfrac{1}{2}\dfrac{A_2}{\zeta_2 S_2}\dfrac{1-\varphi_{\mathrm{SF}}\bar{M}_o}{1+\bar{M}_o}\right)\mathrm{e}^{\frac{\mathrm{i}k_o\ell}{1+M_o}} & -\dfrac{1}{2}\dfrac{A_2}{\zeta_2 S_2}\dfrac{1-\varphi_{\mathrm{SF}}\bar{M}_o}{1+\bar{M}_o}\mathrm{e}^{\frac{\mathrm{i}k_o\ell}{1+M_o}} \\ \dfrac{1}{2}\dfrac{A_2}{\zeta_2 S_2}\dfrac{1+\varphi_{\mathrm{SF}}\bar{M}_o}{1-\bar{M}_o}\mathrm{e}^{\frac{-\mathrm{i}k_o\ell}{1-M_o}} & \left(1 + \dfrac{1}{2}\dfrac{A_2}{\zeta_2 S_2}\dfrac{1+\varphi_{\mathrm{SF}}\bar{M}_o}{1-\bar{M}_o}\right)\mathrm{e}^{\frac{-\mathrm{i}k_o\ell}{1-M_o}} \end{bmatrix}, \tag{4.158}$$

where ℓ denotes the length of solid duct segments and ζ_2, normalized impedance of an aperture on the facesheet, is now determined from the impedance of the liner. The two eigenvalues, γ_1 and γ_2, of this matrix can be calculated explicitly. If the mean flow is neglected for simplicity, it can be shown that $\gamma_1\gamma_2 = 1$ and $\gamma_1 + \gamma_2 = \cos k_o\ell - \mathrm{i}A\sin k_o\ell$, where $A = A_2/2\zeta_2 S_2$. These eigenvalues may be categorized as $\gamma_1 = |\gamma|\mathrm{e}^{\mathrm{i}\alpha}$, $\gamma_2 = |\gamma|^{-1}\mathrm{e}^{-\mathrm{i}\alpha}$, where $\alpha > 0$, and for a duct with N aperture rows

$$\mathbf{S}^N = \begin{bmatrix} \psi_{11} & \psi_{21} \\ \psi_{12} & \psi_{22} \end{bmatrix} \begin{bmatrix} |\gamma|^N\mathrm{e}^{\mathrm{i}N\alpha} & 0 \\ 0 & |\gamma|^{-N}\mathrm{e}^{-\mathrm{i}N\alpha} \end{bmatrix} \begin{bmatrix} \psi_{11} & \psi_{21} \\ \psi_{12} & \psi_{22} \end{bmatrix}^{-1}, \tag{4.159}$$

where $\psi_1 = \{\psi_{11} \quad \psi_{12}\}$ and $\psi_2 = \{\psi_{21} \quad \psi_{22}\}$ denote the corresponding eigenvectors. Comparing Equations (4.159) with (3.34), the eigenvalue matrix of the former can be characterized as the wave transfer matrix relating the transmission of principal pressure wave components in the lined duct.

4.10 Wave Transfer Across Adiabatic Pressure-Loss Devices

An important group of devices in intake and exhaust systems of fluid machinery – such as grids and gauzes, bypass and recirculation valves and carburetors – can be modeled in one-dimension as a discontinuity with an adiabatic pressure loss of the form

$$\overline{p_1} - \overline{p_2} = \frac{1}{2} C \overline{\rho_1} (\overline{v_1})^2 \tag{4.160}$$

where the subscripts "1" and "2" denote, respectively, the inlet and outlet sections of the device, the overbar denotes averaging over these sections, and C denotes a constant called pressure loss coefficient, which is usually evaluated empirically in a steady flow rig. Upon linearization by the usual decomposition $q = q_o + q'$, where $q \in \bar{p}, \bar{\rho}, \bar{v}, \ldots$, Equation (4.160) yields, respectively, for the mean flow and fluctuations:

$$\overline{p_1'} - \overline{p_2'} = C\overline{v_{1o}}\left(\overline{\rho_{1o}}\overline{v_1'} + \frac{1}{2}\overline{v_{1o}}\overline{\rho_1'}\right) \tag{4.161}$$

$$\overline{p_{1o}} - \overline{p_{2o}} = \frac{1}{2} C \overline{\rho_{1o}} (\overline{v_{1o}})^2. \tag{4.162}$$

We consider a control volume enclosing the discontinuity and assume that it is quasi-static with mass and energy flux being pertinent only at its inlet and outlet sections. Then, for the conservation of fluctuating and mean mass flux, Equations (4.10) and (4.11) give

$$\left(\bar{\rho}_{1o}\overline{v_1'} + \bar{v}_{1o}\overline{\rho_1'}\right) S_1 = \left(\bar{\rho}_{2o}\overline{v_2'} + \bar{v}_{2o}\overline{\rho_2'}\right) S_2 \tag{4.163}$$

$$\bar{\rho}_{1o}\bar{v}_{1o}S_1 = \bar{\rho}_{2o}\bar{v}_{2o}S_2. \tag{4.164}$$

Using these equations in Equations (4.15) and (4.16), conservation of energy for the fluctuating and mean flows reduce to $\left(\overline{h_1^o}\right)' = \left(\overline{h_2^o}\right)'$ and $\left(\bar{h}_1^o\right)_o = \left(\bar{h}_2^o\right)_o$, respectively. These may be expanded as:

$$\bar{T}_{1o}\overline{s_1'} + \frac{\overline{p_1'}}{\bar{\rho}_{1o}} + \bar{v}_{1o}\overline{v_1'} = \bar{T}_{2o}\overline{s_2'} + \frac{\overline{p_2'}}{\bar{\rho}_{2o}} + \bar{v}_{2o}\overline{v_2'} \tag{4.165}$$

$$\bar{h}_{1o} + \frac{1}{2}\bar{v}_{1o}^2 = \bar{h}_{2o} + \frac{1}{2}\bar{v}_{2o}^2. \tag{4.166}$$

Equations (4.160), (4.162) and (4.164) are transformed to the pressure-wave components and the entropy-wave component by using Equations (4.21)–(4.23):

$$\left(1 - C\bar{M}_1 - \frac{1}{2}C\bar{M}_1^2\right)p_1^+ + \left(1 + C\bar{M}_1 - \frac{1}{2}C\bar{M}_1^2\right)p_1^- - \frac{1}{2}C\bar{M}_1^2\varepsilon_1' = p_2^+ + p_2^- \tag{4.167}$$

$$\left((1+\bar{M}_1)p_1^+ + (-1+\bar{M}_1)p_1^- + \bar{M}_1\varepsilon_1'\right)\frac{S_1}{\bar{c}_{1o}} = \left((1+\bar{M}_2)p_2^+ + (-1+\bar{M}_2)p_2^- + \bar{M}_2\varepsilon_2'\right)\frac{S_2}{\bar{c}_{2o}} \tag{4.168}$$

$$\left((1+\bar{M}_1)p_1^+ + (1-\bar{M}_1)p_1^- + \frac{\varepsilon_1'}{\gamma_{1o}-1}\right)\frac{1}{\bar{\rho}_{1o}} = \left((1+\bar{M}_2)p_2^+ + (1-\bar{M}_2)p_2^- + \frac{\varepsilon_2'}{\gamma_{2o}-1}\right)\frac{1}{\bar{\rho}_{2o}}, \tag{4.169}$$

where $\bar{M}_j = \bar{v}_{jo}/\bar{c}_{jo}$, $j = 1, 2$, denote the mean flow Mach number at the inlet and outlet of the device. In the last equation, we have used the relationship $\varepsilon' = -\rho_o T_o (\gamma_o - 1)s'$, which comes from the linearized form of state equation for specific entropy, Equation (A.15a). We will assume that $\varepsilon_1' = 0$. This is plausible, as flow losses normally occur within the device. Then, ε_2' may be eliminated from Equations (4.167) and (4.168) to obtain

$$(1+\bar{M}_1)\left(\frac{S_1\bar{c}_{2o}}{S_2\bar{c}_{1o}} - \bar{M}_2(\gamma_o-1)\frac{\bar{\rho}_{2o}}{\bar{\rho}_{1o}}\right)p_1^+ - (1-\bar{M}_1)\left(\frac{S_1\bar{c}_{2o}}{S_2\bar{c}_{1o}} + \bar{M}_2(\gamma_o-1)\frac{\bar{\rho}_{2o}}{\bar{\rho}_{1o}}\right)p_1^- = (1+\bar{M}_2)(1-(\gamma_o-1)\bar{M}_2)p_2^+ - (1-\bar{M}_2)(1+(\gamma_o-1)\bar{M}_2)p_2^-, \tag{4.170}$$

where $\gamma_o = \gamma_{1o} \approx \gamma_{2o}$ is assumed for simplicity. Finally, combining this with Equation (4.166)

$$\begin{bmatrix} (1+\bar{M}_1)\left(\frac{S_1\bar{c}_{2o}}{S_2\bar{c}_{1o}} - \bar{M}_2(\gamma_o-1)\frac{\bar{\rho}_{2o}}{\bar{\rho}_{1o}}\right) & -(1-\bar{M}_1)\left(\frac{S_1\bar{c}_{2o}}{S_2\bar{c}_{1o}} + \bar{M}_2(\gamma_o-1)\frac{\bar{\rho}_{2o}}{\bar{\rho}_{1o}}\right) \\ 1 - C\bar{M}_1 - \frac{1}{2}C\bar{M}_1^2 & 1 + C\bar{M}_1 - \frac{1}{2}C\bar{M}_1^2 \end{bmatrix} \begin{bmatrix} p_1^+ \\ p_1^- \end{bmatrix} = \begin{bmatrix} (1+\bar{M}_2)(1-(\gamma_o-1)\bar{M}_2) & (1-\bar{M}_2)(1+(\gamma_o-1)\bar{M}_2) \\ 1 & 1 \end{bmatrix} \begin{bmatrix} p_2^+ \\ p_2^- \end{bmatrix} \tag{4.171}$$

This equation can be expressed as a two-port wave transfer equation of the device with either inlet-to-outlet causality or outlet-to-inlet causality.

The pressure loss coefficient is generally dependent on the downstream mean flow Mach number, $\bar{M}_2$, and some characteristic parameters of the device, such the open area porosity of gauzes and grids and the settings of valves. A relationship between $\bar{M}_1$ and $\bar{M}_2$ can be found from Equations (4.161), (4.163) and (4.165). Assuming a perfect gas, we may substitute in Equation (4.165) $\bar{h}_{jo} = \bar{c}_{jo}^2/(\gamma_o - 1)$, $j = 1, 2$, to obtain the relationship

$$\frac{2}{\gamma_o - 1}(1 - r^2) = r^2\bar{M}_2^2 - \bar{M}_1^2, \tag{4.172}$$

where $r = \bar{c}_{2o}/\bar{c}_{1o}$. On the other hand, Equations (4.161) and (4.163) may be combined in the form

$$\left(1 - r\frac{S_2\bar{M}_2}{S_1\bar{M}_1}\right) = \frac{1}{2}C\bar{M}_1^2. \tag{4.173}$$

Hence, eliminating r from the last two equations gives

$$\bar{M}_2^2 = \frac{S_1^2}{2S_2^2}\frac{(2 - C\bar{M}_1^2)^2\bar{M}_1^2}{2 + (\gamma_o - 1)\left(1 - \frac{S_1^2}{4S_2^2}(2 - C\bar{M}_1^2)^2\bar{M}_1^2\right)}. \tag{4.174}$$

Hence, with $\bar{M}_1$ known, the downstream Mach number can be calculated.

References

[1] R.J. Alfredson and P.O.A.L. Davies, Performance of exhaust silencer components, *J. Sound Vib.* **15** (1971), 175–196.

[2] P.O.A.L. Davies, Practical flow duct acoustics, *J. Sound Vib.* **124** (1988), 91–115.

[3] A. Cummings, Sound transmission at sudden area expansions in circular ducts with superimposed mean flow, *J. Sound Vib.* **38** (1975), 149–155.

[4] F.C. Karal, The analogous acoustic impedance for discontinuities and constrictions of circular cross section, *J. Acoust. Soc. Am.* **25** (1953), 327–333.

[5] K.S. Peat, The acoustical impedance at discontinuities of ducts in the presence of mean flow, *J. Sound Vib.* **127** (1988), 123–132.

[6] K.S. Peat, The acoustical impedance at the junction of an extended inlet or outlet duct, *J. Sound Vib.* **150** (1991), 101–110.

[7] A.D. Sahasrabusdhe, M.L. Munjal and S.A. Ramu, Analysis of inertance due to the higher order mode effects in a sudden area discontinuity, *J. Sound Vib.* **185** (1995), 515–529.

[8] A.J. Torregrosa, A. Broatch, R. Payri and F. Gonzalez, Numerical estimation of end corrections in extended –duct and perforated-duct mufflers, *J. Acoust. Soc. Am.* **121** (1999), 302–308.

[9] A.J. Torregrosa, A. Broatch and R. Payri, The use of transfer matrix for the design of interferential systems in exhaust mufflers, SAE technical paper 2000-01-0728 (2000).

[10] W. Sullivan and M.J. Crocker, Analysis of concentric tube resonators having unpartitioned cavities, *J. Acoust. Soc. Am.* **64** (1978), 207–215.

[11] K. Jayaraman and K. Yam, Decoupling approach to modeling perforated tube mufflers, *J. Acoust. Soc. Am.* **69** (1981), 390–396.

[12] M.L. Munjal, K.N. Rao and A.D. Sahasrabudhe, Aeroacoustic analysis of perforated muffler components, *J. Sound Vib.* **114** (1987), 173–188.

[13] K.S. Peat, A numerical decoupling analysis of perforated pipe silencer elements, *J. Sound Vib.* **123** (1988), 199–212.

[14] E. Dokumaci, Matrizant approach to acoustic analysis of perforated multiple pipe mufflers carrying mean flow, *J. Sound Vib.* **191** (1996), 505–518.

[15] P.O.A.L. Davies, M. Harrison and H.J. Collins, Acoustic modeling of multiple path silencers with experimental validations, *J. Sound Vib.* **200** (1997), 195–225.

[16] S. Selamet, V. Easwaran and A.G. Falkowski, Three-pass mufflers with uniform perforations, *J. Acoust. Soc. Am.* **105** (1999), 1548–1562.

[17] R. Kirby, Simplified techniques for predicting the transmission loss of a circular dissipative silencer, *J. Sound Vib.* **243** (2002), 403–426

[18] T. Kar and M.L. Munjal, Analysis and design of conical concentric tube resonators, *J. Acoust. Soc. Am.* **116** (2004), 74–83.

[19] E. Dokumaci, Effect of sheared grazing mean flow on acoustic transmission in perforated pipe mufflers, *J. Sound Vib.* **283** (2005), 645–663.

[20] E. Dokumaci, Sound transmission in mufflers with multiple perforated co-axial pipes, *J. Sound Vib.* **247** (2001), 379–387.

[21] J.W. Sullivan, A method of modeling perforated tube muffler components: I. Theory, II. Applications, *J. Acoust. Soc. Am.* **66** (1979), 772–788.

[22] E. Dokumaci, A discrete approach for analysis of sound transmission in pipes coupled with compact communicating devices, *J. Sound Vib.* **239** (2001), 679–693.

[23] M. Pachebat and J. Kergomard, Propagation of acoustic waves in two waveguides coupled by perforations. II. Application two periodic lattices of finite length, *Acta Acoustica united with Acoustica* **102** (2016), 611–627.

5 Resonators, Expansion Chambers and Silencers

5.1 Introduction

The previous two chapters dealt with the formulation of one-dimensional basic acoustic elements which may be used for assembling a large variety of practical duct systems. The final acoustic model of the system may require augmentations for the source of noise and acoustic radiation from its open end(s) to the external environment, as discussed in later chapters of the book. At this point, however, we may consider the question of how to evaluate whether a given system is a good transmitter of sound waves or a good attenuator of sound waves, because this information is largely contained in the wave transfer equation of the system proper (with the source and radiation excluded). The present chapter expounds on this basic question of duct-borne noise control by giving an in-depth analysis of resonators, expansion chambers and silencers in the one-dimensional framework of the previous two chapters and discusses the theoretical and practical issues that are relevant to the acoustic design of these devices. Three-dimensional effects are considered in later chapters, where multi-modal elements are introduced.

5.1.1 Mufflers and Silencers

In dictionaries, a silencer is defined as any device which is used to reduce noise, and a muffler is defined as any device for silencing sound. So, these words appear to be synonyms, although the term muffler appears to be used more frequently for automotive exhaust silencers. In this book, we use the word silencer in the sense it includes mufflers.

The first patent on silencers was filed in 1897 by M.O. Reeves and M.T. Reeves of Reeves Pulley Company, Columbus, Indiana, USA, under the title of "Exhaust muffler for engines" (US Patent Office application 582485). It claimed an implementation forcing the gas flow through a plugged perforated duct and concentric annular passages in two chambers which are coupled by an aperture joining the outermost annuli. Experience since then has culminated in a large variety of silencers which vary in length from a few centimeters (for example in a toy helicopter) to over 10 meters (for example in a venting system) and in similar proportion in diameters. In industry, they are usually referred to by names that highlight the application area, although the acoustic principles underlying their design – namely, reflection, absorption and interference of waves – are essentially the same. For example, automotive applications include the intake and exhaust silencers in cars, trucks, motorcycles, locomotives, industrial and agricultural vehicles, airplanes, helicopters, ships, boats and so on. Stationary applications include the intake and exhaust silencers in generator sets and plants, gas turbines, chimneys, cooling towers, compressors and blowers. Venting applications include the silencers used to control the high levels of noise generated in industrial processes involving discharge of a high-pressure gas, usually steam or natural gas and to a lesser extent air, oxygen, nitrogen, carbon, to the atmosphere or a substantially lower pressure. Such processes include steam venting in power generation plants, natural gas compressor station and pipeline blowdowns, process control and relief valves, blowdown tanks and autoclaves, bypass valves on blowers and compressors and steam ejectors and hogging vents. Heating, ventilation and air conditioning (HVAC) silencers and gun silencers are also based on the same principles, but they are usually considered as categories on their own.

Silencers may also be considered in two broad groups as passive and active silencers. An active silencer works on the principle of using an electronically controlled external sound source to attenuate the sound waves created by the primary sound source by destructive interference. Passive silencers are non-active silencers, in the sense that a controlled external source is not involved in their working principles. The basic acoustic wave phenomena that passive silencers rely on are reflection, dissipation and interference. Passive silencers can be divided into two groups as adaptive and non-adaptive silencers. An adaptive silencer is distinguished by having at least one variable parameter, such as a length, a diameter or an area, which can be controlled automatically by the pressure or temperature conditions of the fluid contained, or by a servomechanism, or manually. Acoustic design of an adaptive silencer can, in general, be considered separately from its controller. Acoustic performance of active silencers largely depends on the controller and, consequently, their design is mostly a problem of electronic controllers.

Application of the electrical filter theory in the acoustic design of silencers was already known in the 1920s [1]. The status of this approach in modeling silencers in the 1950s is described in the report by Davis et al. [2], who also presented, for the first time, a technique for the measurement of transmission loss, which is the traditional single parameter acoustic performance metric for silencers (Section 5.2). The 1960s status is contained in the review by Davies [3], who compared the filter method and the then more recently proposed time-domain pulse method. A quasi-static frequency domain approach for including the effect of mean flow on one-dimensional sound transmission at area changes was introduced 1971 by Alfredson and Davies [4]. Today, in view of its simplicity and versatility, the one-dimensional frequency domain theory is widely used in low-frequency acoustic design of duct systems including silencers [5].

5.1.2 The System and Its Environment

In this chapter, we consider duct systems which can be represented by an acoustic two-port model (see Chapter 2). A generic system is shown schematically in Figure 5.1a and

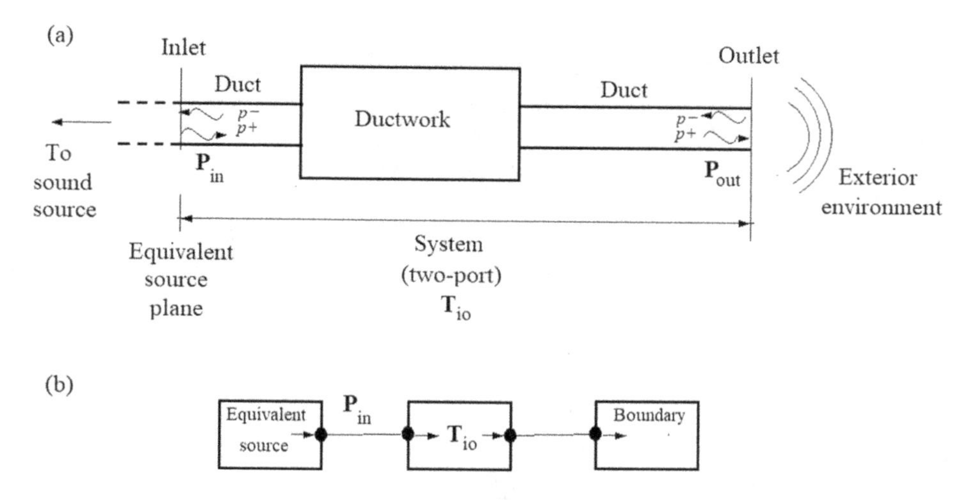

Figure 5.1 (a) A generic flow duct system with single inlet and single outlet ducts. (b) Block diagram of the system.

in block diagram form in Figure 5.1b. It consists of an arbitrary ductwork with an inlet duct and an outlet duct. The acoustic field in the system is sustained by a source located upstream of the inlet duct and is envisaged to be represented by an equivalent source at the inlet section of the inlet duct. The outlet section of the outlet duct represents the boundary (termination) of the system; in this case it is the interface of the system with the exterior environment, where control of the sound radiated from the outlet may be of concern due to environmental noise regulations. The positive direction of acoustic flow is taken in the direction outward from the outlet duct; mean flow may also be in this direction (exhaust system), or it may be in the opposite direction (intake system). It is important to note that we distinguish the inlet and outlet ducts not by the direction of the mean flow, but by the location of the acoustic source driving the system.

In open systems, the exterior environment is usually the atmosphere. For example, consider the exhaust muffler of an automobile (Figure 5.2a). The tailpipe outlet represents the outlet section of the system. We may conceive any section of the exhaust line upstream of this as the inlet section. Of course, if we want to include the muffler in the system, then the inlet section must be taken somewhere upstream of it.

In closed systems, the link with the exterior environment is not as clear-cut as in open systems. As an example, consider the hermetic refrigeration compressor shown in Figure 5.2b. This consists of a fractional power electric motor driving a slider-crank mechanism housed in a cylinder block, the whole unit being hermetically enclosed and resiliently supported in a steel shell (transparent in Figure 5.2b). The refrigerant enters the shell from the evaporator outlet in a gaseous state and is drawn into the intake plenum and, hence, to the cylinder, through the inlet snorkel of the suction muffler, during the suction stroke of the piston. The refrigerant is then compressed and

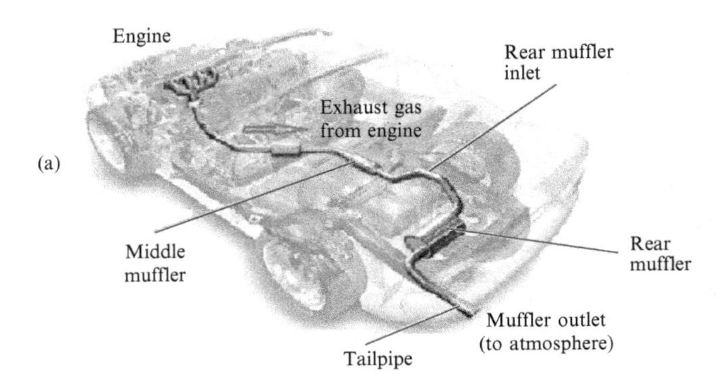

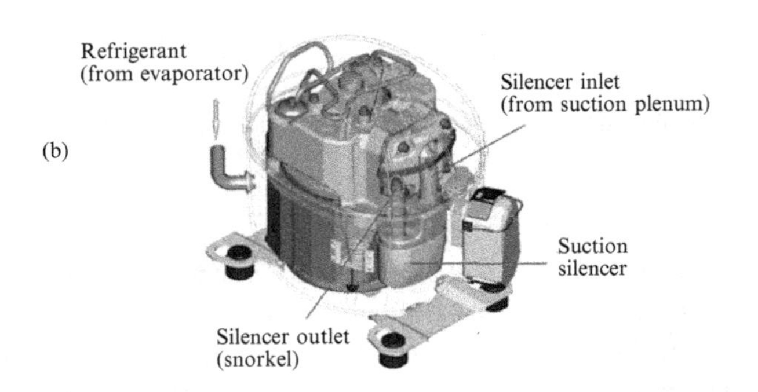

Figure 5.2 Open and closed systems. (a) Exhaust system of a car. (b) Suction muffler of a hermetic refrigeration compressor.

discharged to the condenser in a closed cycle, to return, after condensation, back to the evaporator. The noise of the compressor is rated by the sound radiated by the shell to its exterior environment. Therefore, if we wish to design the suction silencer to control the contribution of the suction noise to the compressor noise radiated to the external environment, we have to include in the system of Figure 5.1 the suction muffler, as well as the cavity inside the shell and the shell structure. The intervention of the shell structure changes the character of the problem from duct acoustics to vibro-acoustics, since the elastodynamics of the compressor shell is coupled to the acoustics of the fluid contained inside and the fluid outside. Fortunately, however, in most similar applications, the intervening structure has much larger characteristic impedance than the fluid inside and outside, so that the effects of the fluid–structure coupling can be neglected. This simplification is adequately accurate for hermetic compressors, since normally a steel shell is used. Then it suffices to consider the system as consisting of the suction muffler and the cavity in the shell and the system outlet section may be taken in the duct coming from the evaporator (see Figure 5.2b). In this case, obviously, the outlet is not directly related to the nuisance for the exterior environment, but the system provides the information about the acoustic pressure loading imposed on the shell due to the suction process. This loading may be used in the vibro-acoustic model of the shell to compute the sound radiated to the exterior by shell vibrations and, hence, evaluate the effectiveness of the suction silencer.

The pressure wave components in the inlet and outlet sections of the system are related by the wave transfer equation $\mathbf{P}_{in} = \mathbf{T}_{io}\mathbf{P}_{out}$, which, assuming one-dimensional sound propagation, may be expressed in frequency domain as

$$\begin{bmatrix} p^+(\omega) \\ p^-(\omega) \end{bmatrix}_{in} = \begin{bmatrix} T_{11}(\omega) & T_{12}(\omega) \\ T_{21}(\omega) & T_{22}(\omega) \end{bmatrix}_{io} \begin{bmatrix} p^+(\omega) \\ p^-(\omega) \end{bmatrix}_{out}. \quad (5.1)$$

Here, the subscripts "in" and "out" refer to the inlet and outlet section of the system, which demarcate the outlet ends of the inlet and outlet ducts, respectively, as described above. In general, the wave transfer matrix in Equation (5.1) can be derived in several ways. Here, we assume that it is derived by using one-dimensional elements described in previous two chapters. But it may also be a contracted three-dimensional model (see Section 2.4.4) or a model determined by measurement (see Section 12.3.4) or by a numerical method such as the finite element method or the boundary element method (see Section 2.5).

5.2 Transmission Loss

Referring to Figure 5.1, the transmission loss of a duct system is defined, in the frequency domain, in dB (decibel) units, by the formula

$$TL = 10\log_{10}\left(\frac{W_{in}^+}{W_{out}^+}\right)_{r_{out}=0}, \quad (5.2)$$

where W^+ denotes the time-averaged sound power of the pressure wave component in the $+x$ direction (into the system at the inlet duct and out of the system at the outlet duct) and TL denotes transmission loss in dB units. In Equation (5.2), the subscript "$r_{out} = 0$" denotes the condition that the outlet section must be anechoic (see Section 2.3). This is a critical requirement, because outlets of real ductwork are never anechoic. In spite of this, transmission loss is a useful parameter. It is relatively easy to calculate and shows the frequencies at which the system acts as a good transmitter or attenuator of acoustic power. This is very useful information for the noise control engineer, because, knowing the frequencies at which noise reduction is necessary, the noise control strategy reduces to matching these frequencies with the frequency regions where transmission loss is sufficiently high (pending later corrections for the error caused by the anechoic outlet assumption).

5.2.1 Single-Frequency Analysis

The time-averaged acoustic power transmitted by one-dimensional sound waves is given by Equation (3.47). Therefore, for a single-frequency wave of radian frequency ω, Equation (5.2) can be expressed as

$$TL(\omega) = 10\log_{10}\left(\frac{|p^+(\omega)|^2_{\text{in}}}{|p^+(\omega)|^2_{\text{out}}}\right)_{r_{\text{out}}=0} + C_{TL}, \tag{5.3}$$

where the correction term C_{TL} is given by

$$C_{TL} = 10\log_{10}\left(\frac{S_{\text{in}}}{S_{\text{out}}}\frac{z_{\text{out}}}{z_{\text{in}}}\frac{(1+\bar{M}_o)^2_{\text{in}}}{(1+\bar{M}_o)^2_{\text{out}}}\right). \tag{5.4}$$

Here, $z = \rho_o c_o$ denotes the characteristic impedance, S denotes the duct cross-sectional area, $\bar{M}_o$ denotes the Mach number of the cross section averaged mean flow velocity. Correction C_{TL} depends only on the area and ambient conditions of the inlet and outlet sections, which are still denoted by the subscripts "in" and "out," and vanishes if these conditions are identical at the two sections.

Upon expanding Equation (5.1), we find $p^+_{\text{in}}(\omega) = T_{11}(\omega)p^+_{\text{out}}$, since the outlet duct is anechoic. Therefore, Equation (5.3) may be expressed as

$$TL(\omega) = 20\log_{10}\left(|T_{11}(\omega)|_{\text{io}}\right)_{r_{\text{out}}(\omega)=0} + C_{TL}. \tag{5.5}$$

Thus, transmission loss can be calculated given the wave transfer matrix.

In the literature, transmission loss is often expressed in terms of the elements of the impedance transfer matrix. For the system in Figure 5.1, impedance transfer matrix is defined in frequency domain as

$$\begin{bmatrix} \overline{p'} \\ \overline{v'} \end{bmatrix}_{\text{in}} = \begin{bmatrix} \tau_{11} & \tau_{12} \\ \tau_{21} & \tau_{22} \end{bmatrix}_{\text{io}} \begin{bmatrix} \overline{p'} \\ \overline{v'} \end{bmatrix}_{\text{out}}, \tag{5.6}$$

where frequency dependency is not shown explicitly for simplicity of notation. Since acoustic pressure and particle velocity are related to the pressure wave components by $\overline{p'} = p^+ + p^-$ and $z\overline{v'} = p^+ - p^-$, the elements of the impedance transfer matrix are related to the elements of the corresponding wave transfer matrix $\mathbf{T}_{\text{io}}$ in Equation (5.1) as:

$$(\tau_{11})_{\text{io}} = \frac{1}{2}(T_{11} + T_{12} + T_{21} + T_{22})_{\text{io}} \tag{5.7}$$

$$(\tau_{12})_{\text{io}} = \frac{1}{2}(T_{11} - T_{12} + T_{21} - T_{22})_{\text{io}} z_{\text{out}} \tag{5.8}$$

$$(\tau_{21})_{\text{io}} = \frac{1}{2}(T_{11} + T_{12} - T_{21} - T_{22})_{\text{io}} \frac{1}{z_{\text{in}}} \tag{5.9}$$

$$(\tau_{22})_{\text{io}} = \frac{1}{2}(T_{11} - T_{12} - T_{21} + T_{22})_{\text{io}} \frac{z_{\text{out}}}{z_{\text{in}}}. \tag{5.10}$$

The inverse relations are:

$$(T_{11})_{\text{io}} = \frac{1}{2}\left(\tau_{11} + \frac{1}{z_{\text{out}}}\tau_{12} + z_{\text{in}}\tau_{21} + \frac{z_{\text{in}}}{z_{\text{out}}}\tau_{22}\right)_{\text{io}} \tag{5.11}$$

$$(T_{12})_{io} = \frac{1}{2}\left(\tau_{11} - \frac{1}{z_{\text{out}}}\tau_{12} + z_{\text{in}}\tau_{21} - \frac{z_{\text{in}}}{z_{\text{out}}}\tau_{22}\right)_{\text{io}} \tag{5.12}$$

$$(T_{21})_{\text{io}} = \frac{1}{2}\left(\tau_{11} + \frac{1}{z_{\text{out}}}\tau_{12} - z_{\text{in}}\tau_{21} - \frac{z_{\text{in}}}{z_{\text{out}}}\tau_{22}\right)_{\text{io}} \tag{5.13}$$

$$(T_{22})_{\text{io}} = \frac{1}{2}\left(\tau_{11} - \frac{1}{z_{\text{out}}}\tau_{12} - z_{\text{in}}\tau_{21} + \frac{z_{\text{in}}}{z_{\text{out}}}\tau_{22}\right)_{\text{io}}. \tag{5.14}$$

Hence, given the impedance transfer matrix, transmission loss can be computed from

$$TL = 20\log_{10}\left(\frac{1}{2}\left|\tau_{11} + \frac{1}{z_{\text{out}}}\tau_{12} + z_{\text{in}}\tau_{21} + \frac{z_{\text{in}}}{z_{\text{out}}}\tau_{22}\right|_{\text{io}}\right)_{r_{\text{out}}=0} + C_{TL}. \tag{5.15}$$

For an anechoic outlet, the quotient $\left(p_{in}^{-}/p_{\text{in}}^{+}\right)_{r_{\text{out}}=0} = T_{21}/T_{11}$ is also determined by the elements of the wave transfer matrix. This gives the reflection coefficient, $[r_{\text{in}}(\omega)]_{r_{\text{out}}=0}$, at the inlet section and is related to the time-averaged acoustic power at the inlet section as

$$\left(\frac{W_{\text{in}}^{-}}{W_{\text{in}}^{+}}\right)_{r_{\text{out}}=0} = \left(\frac{1-\bar{M}_o}{1+\bar{M}_o}\right)_{\text{in}}^{2} |r_{\text{in}}|_{r_{\text{out}}=0}^{2}. \tag{5.15a}$$

This result confirms the obvious fact that reflection of the incident sound waves back to the source can be an effective strategy for the control of duct-borne sound.

5.2.2 Overall Transmission Loss

When the incident sound field is defined over a range of frequencies, a quantity of interest is the fraction of the total incident time-averaged acoustic power that is transmitted to the system outlet. This quantity is determined by evaluating Equation (5.2) by using the total incident time-averaged acoustic power in a given frequency range, $f_1 \leq f \leq f_2$, say, where $f = \omega/2\pi$ denotes the frequency in Hz. The resulting quantity is called overall transmission loss and denoted by $\bar{TL}(f_1,f_2)$.

As we have seen in Section 3.4, in a one-dimensional acoustic field of arbitrary frequency content, the time-averaged acoustic power is given by Equation (3.50). Accordingly, referring to Figure 5.1, the overall transmission loss of a duct system in the frequency range $f_1 \leq f \leq f_2$ is given by

$$\bar{TL}(f_1,f_2) = 10\log_{10}\left(\frac{\int_{f_1}^{f_2}\left|S_{p_{\text{in}}^{+}}(f)\right|^2 \mathrm{d}f}{\int_{f_1}^{f_2}\left|S_{p_{\text{out}}^{+}}(f)\right|^2 \mathrm{d}f}\right)_{r_{\text{out}}=0} + C_{TL}, \tag{5.16}$$

where $S_{p^+}(f)$ denotes the (two-sided or one-sided) power spectral density of the incident acoustic pressure $p^+(t)$. But since the outlet duct is anechoic, $p_{\text{in}}^{+}(f) = [T_{11}(f)]_{\text{io}}p_{\text{out}}^{+}(f)$. Therefore,

$$\overline{TL}(f_1,f_2) = 10\log_{10}\left(\frac{\int_{f_1}^{f_2}\left|S_{p_{\text{in}}^+}(f)\right|^2 \mathrm{d}f}{\int_{f_1}^{f_2}|T_{11}(f)|_{\text{io}}^{-2}\left|S_{p_{\text{in}}^+}(f)\right|^2 \mathrm{d}f}\right) + C_{TL}. \tag{5.17}$$

Thus, the overall transmission loss can be calculated, given the power spectral density of the incident acoustic pressure at the inlet section. In theoretical calculations, it is convenient to assume that the incident power spectral density at the inlet section is constant in the frequency range of interest. Then, Equation (5.17) simplifies to

$$\overline{TL}(f_1,f_2) = 10\log_{10}(f_2 - f_1) - 10\log_{10}\left(\int_{f_1}^{f_2}\frac{\mathrm{d}f}{|T_{11}(f)|_{\text{io}}^2}\right) + C_{TL}, \tag{5.18}$$

where, from Equation (5.5),

$$|T_{11}(f)|_{\text{io}}^2 = 10^{\frac{TL(f)-C_{TL}}{10}}. \tag{5.18a}$$

The integral in Equation (5.18) can be evaluated numerically by computing TL for sufficiently large number of discrete frequencies in the frequency interval $f_1 \le f \le f_2$.

5.3 Duct Resonances

This section discusses the principles underlying acoustic design of resonators and the various basic forms in which they are used in ductworks. Our treatment is based on the recognition of the fact that transmission loss is actually a frequency response function of the system.

5.3.1 Resonance and Anti-Resonance Frequencies

A frequency response function (FRF) of a linear system is defined by the Fourier transform of the system output divided by the Fourier transform of the system input that causes that output (see Section 1.5). The analyst selects the input and output quantities according to the system studied. For example, referring to Figure 5.1, transmission loss of the system is closely related to the FRF:

$$G_{p^+}(\omega) = \frac{p_{\text{out}}^+(\omega)}{p_{\text{in}}^+(\omega)} = \frac{1}{(T_{11})_{\text{io}} + r_{\text{out}}(T_{12})_{\text{io}}}, \tag{5.19}$$

where $r_{\text{out}} = p_{\text{out}}^-/p_{\text{out}}^+$ denotes, as before, the reflection coefficient at the outlet section of the system. Since

$$TL = -20\log_{10}\left|G_{p^+}(\omega)\right|_{r_{\text{out}}(\omega)=0} + C_{TL} \tag{5.20}$$

the modulus of $\left[G_{p^+}(\omega)\right]_{r_{\text{out}}=0}$ carries the same information as TL. But, the FRF view provides an additional advantage: We can now invoke the concepts of resonance and anti-resonance, which are familiar concepts from linear system dynamics [6]. To recollect briefly, the frequencies at which the modulus of an FRF attains maxima (minima) are called the resonance (anti-resonance) frequencies of the system for the corresponding input–output relationship. In undamped systems, the resonance (anti-resonance) frequencies are real and given by the poles (zeroes) of the FRF. In this case, the maxima of the modulus of the FRF tends to infinity (zero) at resonance (anti-resonance) and the phase angle of the FRF changes by 180° as resonance (anti-resonance) frequencies are traversed. The presence of damping modifies these characteristics in that the resonance maxima are rendered finite, resonance frequencies are shifted from their undamped values, and the phase shift becomes gradual and less than 180°, such variations being larger the larger is the damping; for example, the blunter the resonance peaks are, the larger the damping is in the system. The anti-resonance characteristics are similarly influenced by the presence of damping.

Thus, referring to Equation (5.20), we can state that, the transmission loss minima should occur at the resonance frequencies of $\left[G_{p^+}(\omega)\right]_{r_{\text{out}}=0}$. Similarly, the transmission loss maxima occur at anti-resonance frequencies of $\left[G_{p^+}(\omega)\right]_{r_{\text{out}}=0}$. However, the converse statement is not necessarily true; that is, not all transmission loss minima (maxima) necessarily correspond to $\left[G_{p^+}(\omega)\right]_{r_{\text{out}}=0}$ resonances (anti-resonances). It is also necessary that a phase shift occurs at or around these frequencies, but this condition cannot be told from transmission loss, since it does not contain phase information. The phase information must be extracted from the corresponding FRF, in this case $\left[G_{p^+}(\omega)\right]_{r_{\text{out}}=0}$.

5.3.2 Resonators

It transpires from the foregoing discussion that a duct system transmits sound power efficiently (inefficiently) in the vicinity of the resonance (anti-resonance) frequencies of the corresponding FRF. This result is general and applies also for subsystems of the system. In fact, with appropriate redefinitions of the source and boundary, Figure 5.1 may also be considered as a generic representation of any subsystem with two ports. Thus, an internal component of a system transmits the sound waves incident at its inlet efficiently at the resonance frequencies of its FRF and, hence, can divert sound waves from their main path. Components which are designed for creating acoustic paths which divert sound waves from their main path of propagation are called resonators.

Resonators may be designed in a variety of configurations and the common ones are considered separately in the following subsections. Since these are generic configurations, we will use a suitable local FRF in discussing their design principles. An FRF that is convenient for the subsequent discussion is

$$G_{p'}(\omega) = \frac{\overline{p'_{\text{out}}}}{\overline{p'_{\text{in}}}} = \frac{1}{(\tau_{11})_{\text{io}} + \dfrac{(\tau_{12})_{\text{io}}}{Z_{\text{out}}}}, \tag{5.21}$$

where $Z_{\text{out}} = \overline{p'_{\text{out}}}/\overline{v'_{\text{out}}}$ denotes the acoustic impedance at the outlet and the second equality follows from Equation (5.6). With the use of Equations (5.7) and (5.8), this FRF can be written in terms of the elements of the wave transfer matrix as:

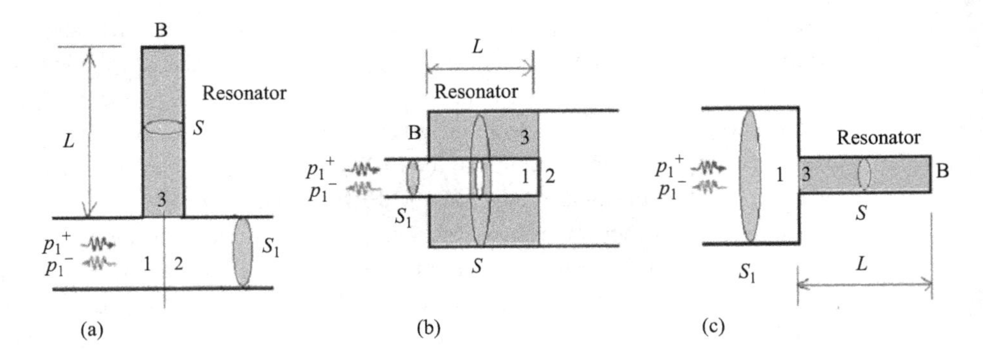

Figure 5.3 Single-duct resonators. (a) Branch resonator. (b) Closed in-line side-branch resonator. (c) Closed in-line resonator. Gray denotes the resonator duct (3B).

$$G_{p'}(\omega) = \frac{2}{(T_{11}+T_{12}+T_{21}+T_{22})_{\text{io}} + \left(\dfrac{1-r_{\text{out}}}{1+r_{\text{out}}}\right)(T_{11}-T_{12}+T_{21}-T_{22})_{\text{io}}}, \tag{5.22}$$

where $r_{\text{out}} = p^-_{\text{out}}/p^+_{\text{out}}$ denotes the reflection coefficient at the outlet section. The acoustic design problem may then be stated as the problem of tuning the resonance frequencies of the FRF to some given target frequencies. In general, this problem can be solved only numerically. However, some simple models are tractable to analysis and are considered in subsequent subsections in view of their practical importance.

5.3.3 Single-Duct Resonator

The simplest resonator is a single duct and is convenient for explaining the basic procedure involved in resonator design. Typical configurations are shown in Figure 5.3. According to the notation of this figure, $\mathbf{P}_{\text{in}} = \mathbf{P}_3$ and $\mathbf{P}_{\text{out}} = \mathbf{P}_{\text{B}}$. Then, the system FRF in Equation (5.22) represents the quotient $G_{p'}(\omega) = \overline{p'_{\text{B}}}/\overline{p'_3}$ and can be determined from the elements of the wave transfer matrix $\mathbf{T}_{3\text{B}}$ of the appropriate duct model. In order to keep the problem amenable for an explicit analysis, we assume that the duct is uniform and has solid and impervious walls.[1] Then the wave transfer matrix is given by Equation (3.65) with $\alpha = 1$, and Equation (5.22) simplifies to

$$G_{p'}(\omega) = \frac{2}{\mathrm{e}^{\mathrm{i}k_oK^+L} + \mathrm{e}^{\mathrm{i}k_oK^-L} + \left(\dfrac{1-r_{\text{B}}}{1+r_{\text{B}}}\right)\left(\mathrm{e}^{\mathrm{i}k_oK^+L} - \mathrm{e}^{\mathrm{i}k_oK^-L}\right)}, \tag{5.23}$$

since $r_{\text{out}} = r_{\text{B}}$. With substitution for the propagation constants from Equation (3.66), this becomes

[1] Note that, for this duct model, the duct resonances are not detectable by the FRF in Equation (5.19), because the incident wave is transmitted to the outlet as is.

$$G_{p'}(\omega) = \frac{1}{e^{\frac{-ik_o\bar{M}_oL}{1-\bar{M}_o^2}}\left[\cos\left(\frac{k_oL}{1-\bar{M}_o^2}\right) - i\left(\frac{1-r_B}{1+r_B}\right)\sin\left(\frac{k_oL}{1-\bar{M}_o^2}\right)\right]}. \tag{5.24}$$

If the outlet of this duct is anechoic, then $r_B = r_3 = 0$ and, hence, $|G_{p'}(\omega)| = 1$, implying that, as can be expected, the duct does not operate as a resonator. If the duct outlet is rigidly closed, $r_B = 1$ and Equation (5.24) reduces to $G_{p'}(\omega) = 1/\cos(k_oL)$, where we assume that $\bar{M}_o = 0$ since the duct is closed. Hence, a closed duct acts as a resonator with resonance frequencies given by $\cos(k_oL) = 0$, or

$$f_n = n\frac{c_o}{4L}, \qquad n = 1, 3, 5, \ldots \tag{5.25}$$

The phase of $G_{p'}(\omega)$ changes by 180° each time these frequencies are traversed on the frequency axis. The minima of $|G_{p'}(\omega)|$ occur at frequencies $\cos(k_oL) = \mp 1$, but these do not correspond to anti-resonances.

If the duct outlet is a pressure release boundary, then $r_B = -1$ and $G_{p'}(\omega) = 0$. Resonance is not observable with this FRF, because the output variable corresponds to the pressure release boundary condition. It can be observed by switching the FRF to $G_{v'p'}(\omega) = \overline{v'_B}/\overline{p'_3} = G_{p'}/Z_B$. Then, from Equation (5.24),

$$G_{v'p'}(\omega) = \frac{1}{z_B e^{\frac{-ik_o\bar{M}_oL}{1-\bar{M}_o^2}}\left[\left(\frac{1+r_B}{1-r_B}\right)\cos\left(\frac{k_oL}{1-\bar{M}_o^2}\right) - i\sin\left(\frac{k_oL}{1-\bar{M}_o^2}\right)\right]}, \tag{5.26}$$

where z_B denotes the characteristic impedance at B. For $r_B = -1$, this reduces to $|G_{v'p'}(\omega)| = 1/z_B \sin\left(k_oL/\left(1-\bar{M}_o^2\right)\right)$, where we allow for the possibility that $\bar{M}_o \neq 0$, since the duct is open. Hence, an open duct acts as a resonator with resonance frequencies

$$f_n = n\frac{c_o\left(1-\bar{M}_o^2\right)}{2L}, \qquad n = 1, 2, 3, \ldots \tag{5.27}$$

The phase of $G_{v'p'}(\omega)$ changes by 180° each time these frequencies are traversed on the frequency axis. The minima of $|G_{v'p'}(\omega)|$ occur at frequencies $\sin(k_oL) = \mp 1$, but these do not correspond to anti-resonances.

Comparing the foregoing results with Table 3.1, we observe that the resonance frequencies of the resonator with an ideally closed end are equal to the natural frequencies of an open–closed duct, and the resonance frequencies of the resonator with a pressure release (open) end are equal to the natural frequencies of an open–open duct. In fact, it is a general result of the theory of vibrations that the resonance frequencies of an undamped linear system occur at the natural frequencies of the system. However, there is a subtle point which deserves a comment.

Reconsider the single duct resonator with the pressure release end B, $r_B = -1$. We have shown above that the resonance frequencies of this resonator are given by Equation (5.27), if the FRF is taken as $G_{v'p'}(\omega) = \overline{v'_B}/\overline{p'_3}$. Now, let us calculate the resonance frequencies of the same resonator for the FRF $G_{v'}(\omega) = v'_B/v'_3$. This is

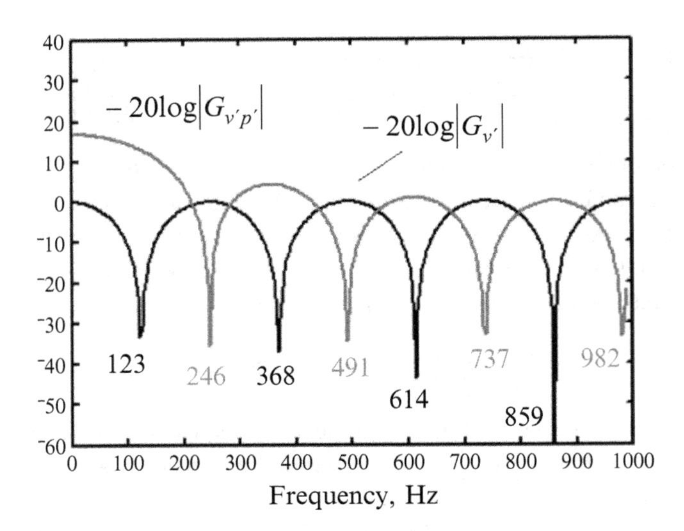

Figure 5.4 Resonance frequencies of a duct of length 0.7 m with pressure release boundary under ideal acoustic pressure forcing and ideal acoustic particle velocity forcing.

evaluated similarly from Equation (5.8) and, for $r_B = -1$, it simplifies to $|G_{v'}(\omega)| = 1/\left|\cos\left(k_o L/\left(1-\bar{M}_o^2\right)\right)\right|$, which yields the resonance frequencies $f_n = nc_o\left(1-\bar{M}_o^2\right)/4L$, $n = 1, 3, 5, \ldots$. These are obviously different than the resonance frequencies in Equation (5.27). This paradoxical situation can be explained by considering the implied input (forcing) variable for the FRFs considered, namely, $\overline{p_3'}$ for $G_{v'p'}(\omega)$ and $\overline{v_3'}$ for $G_{v'}(\omega)$. The paradox unfolds when we realize that $\overline{v_3'}$ can be imposed only by a rigid plane (or piston) executing small axial vibrations about section 3. In other words, $G_{v'}(\omega)$ could be used only if the resonator was closed at its moving input boundary. Indeed, the resonance frequencies deduced from $G_{v'}(\omega)$ are the same as the natural frequencies of a closed–open duct (see Table 3.1, mean flow being neglected in view of the closed end). A similar situation holds when the resonator is rigidly closed at B, $r_B = -1$. The resonance frequencies given in Equation (5.25) are deduced from $G_{p'}(\omega) = \overline{p_B'}/\overline{p_3'}$ for $r_B = -1$. If we use $G_{p'v'}(\omega) = \overline{p_B'}/\overline{v_3'}$, the resonance frequencies transpire as $f_n = nc_o/2L$, $n = 1, 2, 3, \ldots$, which are, as can be expected now, the resonance frequencies of an closed–closed duct (see Table 3.1). Thus, $G_{p'v'}(\omega)$ must be used only when the input boundary 3 of the resonator is a closed vibrating boundary.

The foregoing discussion shows that the resonance frequencies of a resonator depend on the way it is forced to operate. Figure 5.4 demonstrates this graphically for a duct having a pressure release boundary. In most applications, however, resonators are forced by acoustic pressure, as they are attached at passive sections of duct systems.

5.3.4 Resonators with Open Outlet

Resonators that are envisaged as noise control devices are seldom designed with an end open to the exterior environment, because this may defeat its purpose by radiating efficiently to the exterior environment at its resonance frequencies. However, when we want to transmit sound over long distances, efficient radiation to the exterior environment from an open duct end becomes the main design objective and devices commonly known as megaphones have been used for this purpose since ancient times. Classic megaphones employ non-uniform ducts. The elements described in Section 3.6 can be used to calculate the resonance frequencies of megaphones for various cross-sectional area taper functions.

Referring to Figure 5.1, we can now comment on the effect of reflections at the system outlet, which is neglected in transmission loss considerations. The outlet duct can be conceived as a single-duct resonator. It does not act as such when we assume that it is anechoic for transmission loss calculation. But, real outlet ducts are seldom anechoic and do act as resonators, transmitting sound efficiently to the exterior environment at their resonance frequencies.

5.3.5 Helmholtz Resonator

The disadvantage of using a uniform duct as a resonator is that it provides only one parameter, its length, for tuning its resonance frequency. For example, assuming acoustic pressure forcing, the length of a closed duct resonator has to be equal to one-quarter of the targeted wavelength, which may imply an impractical duct length if the frequency is low enough (for example, 1 m at 100 Hz and $c_o = 400$ m/s). The resonance frequencies can also be altered by using non-uniform ducts, but multi-duct resonators are usually preferred, because they are more versatile and can be tuned in a wider range of frequencies.

A popular configuration, which is known as the Helmholtz resonator, consists of a duct coupled to a cavity, in which the length of the duct and essentially the volume of the cavity act as the main parameters for tuning at low frequencies. Figures 5.5a,b,c show the use of a Helmholtz resonator concept in (a) branch, (b) in-line and (c) folded-duct configurations, and Figure 5.5d shows a block diagram which is applicable for the three configurations. It should be noted that, in these configurations, single duct resonators are integrated to the Helmholtz resonator concept. For example, in the branch configuration in Figure 5.5a, the extension of the branch duct into the chamber (an annular duct) is a resonator on its own.

According to the notation of Figure 5.5, $\mathbf{P}_{\text{in}} = \mathbf{P}_3$ and $\mathbf{P}_{\text{out}} = \mathbf{P}_{\text{B}}$. Thus, the wave transfer equation of the three resonators is $\mathbf{P}_3 = \mathbf{T}_{3\text{B}}\mathbf{P}_{\text{B}}$ and, as it is clear from the block diagram, the wave transfer matrix is given by $\mathbf{T}_{3\text{B}} = \mathbf{T}_{34}\mathbf{T}_{45}\mathbf{T}_{5\text{B}}$. Here, $\mathbf{T}_{34}$ is the wave transfer matrix of the duct of length L and cross-sectional area S (duct 3-4), which is called the neck of the resonator, $\mathbf{T}_{45}$ is the wave transfer matrix of the area expansion and $\mathbf{T}_{5\text{B}}$ is the wave transfer matrix of the duct of length L_c and cross-sectional area S_c, which may be called cavity or chamber, since it represents the traditional Helmholtz resonator cavity as the side-branch length L_A tends to zero. For impermeable finite impedance boundaries A and B, the wave transfer matrix $\mathbf{T}_{3\text{B}}$ can be calculated numerically by using the duct and area-change elements described in previous chapters. $\mathbf{T}_{45}$ represents a closed area change and is given by Equation

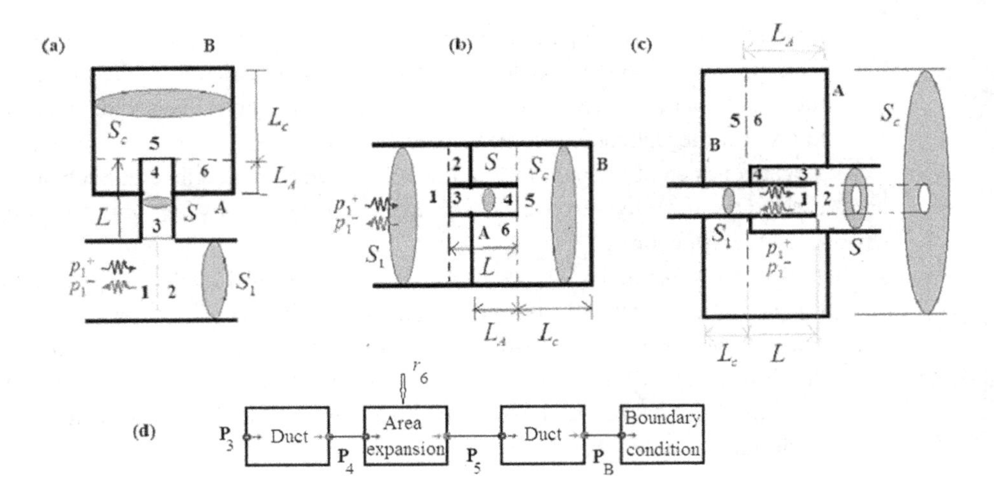

Figure 5.5 Helmholtz resonators. (a) Branch resonator. (b) In-line resonator. (c) Folded-duct resonator. (d) Block diagram of the three resonators.

(4.40); and an appropriate duct element from Chapter 3 can be used for $\mathbf{T}_{34}$ and $\mathbf{T}_{5\mathrm{B}}$. The problem simplifies considerably if we assume that the ducts are uniform and carry negligible mean flow and that the resonator outlet is rigidly closed, that is, $r_{\mathrm{B}} = r_{\mathrm{out}} = 1$. Then, from Equation (4.41):

$$\mathbf{T}_{45} = \frac{1}{2\zeta_6 S}\begin{bmatrix} S_c(\zeta_6+1)+S(\zeta_6-1) & -S_c(\zeta_6-1)+S(\zeta_6-1) \\ -S_c(\zeta_6+1)+S(\zeta_6+1) & S_c(\zeta_6-1)+S(\zeta_6+1) \end{bmatrix} \tag{5.28}$$

and from Equation (3.65):

$$\mathbf{T}_{34} = \begin{bmatrix} \mathrm{e}^{-\mathrm{i}k_oL} & 0 \\ 0 & \mathrm{e}^{\mathrm{i}k_oL} \end{bmatrix} \tag{5.29}$$

$$\mathbf{T}_{5B} = \begin{bmatrix} \mathrm{e}^{-\mathrm{i}k_oL_c} & 0 \\ 0 & \mathrm{e}^{\mathrm{i}k_oL_c} \end{bmatrix}, \tag{5.30}$$

where $L > 0$ and $L_c > L_A \geq 0$. Hence,

$$\mathbf{T}_{3\mathrm{B}} = \frac{1}{2\zeta_6}\begin{bmatrix} (m-1+(m+1)\zeta_6)\mathrm{e}^{\mathrm{i}k_o(-L-L_c)} & (m-1+(-m+1)\zeta_6)\mathrm{e}^{\mathrm{i}k_o(-L+L_c)} \\ (-m+1+(-m+1)\zeta_6)\mathrm{e}^{\mathrm{i}k_o(L-L_c)} & (-m+1+(m+1)\zeta_6)\mathrm{e}^{\mathrm{i}k_o(L+L_c)} \end{bmatrix}, \tag{5.31}$$

where $m = S_c/S$.

A convenient FRF to demonstrate the calculation of the resonance frequencies is $G_{p'}(\omega) = p'_{\mathrm{B}}/p'_3$; alternative FRFs can be analyzed similarly. Using Equation (5.22), where the elements of the wave transfer matrix are now given by Equation (5.31), and assuming that end B is rigidly closed, this FRF can be expressed as:

$$G_{p'}(\omega) = \frac{1}{\cos(k_oL)\cos(k_oL_c) - m\sin(k_oL)\sin(k_oL_c) - \mathrm{i}\frac{1}{\zeta_6}(m-1)\cos(k_oL)\sin(k_oL_c)}, \tag{5.32}$$

where $\zeta_6 = \left(1 + r_A \mathrm{e}^{\mathrm{i}2k_oL_A}\right)/\left(1 - r_A \mathrm{e}^{\mathrm{i}2k_oL_A}\right)$. If the side-branch end-cap A is also rigidly closed, $r_A = 1$, $\zeta_6 = \mathrm{i}\cot k_oL_A$ and Equation (5.32) simplifies, after some trigonometric manipulations, to

$$G_{p'}(\omega) = \frac{1}{\cos(k_oL)\cos(k_o(L_c - L_A)) - m\sin(k_oL_c)\sin(k_o(L + L_A))}. \tag{5.33}$$

Consequently, the resonance frequencies, f_n, are given by the nth root of the equation

$$\cos\left(\frac{2\pi f}{c_o}L\right)\cos\left(\frac{2\pi f}{c_o}(L_c - L_A)\right) = m\sin\left(\frac{2\pi f}{c_o}L_c\right)\sin\left(\frac{2\pi f}{c_o}(L + L_A)\right). \tag{5.34}$$

According to this model, the resonance frequencies can be controlled by tuning the duct length L and the non-dimensional parameters $m, L_c/L$ and L_A/L. It should be noted that if end-correction is applied to the neck (duct of length L), the correction to the chamber side should also be included in L_A.

If the frequencies of interest are low enough so that $k_oL_c << 1$ and $k_o(L + L_A) << 1$, then retaining only the first terms in the Taylor series expansions of the two sides of Equation (5.34) yields, for the smallest resonance frequency, the approximation

$$f_1 \approx \frac{c_o}{2\pi}\sqrt{\frac{S}{(L + L_A)S_cL_c}}. \tag{5.35}$$

A similar formula is ubiquitous in books on acoustics in the form $f_1 = (c_o/2\pi)\sqrt{S/LV_c}$ where V_c denotes the total volume of the resonator cavity. This is based on the low frequency approximation discussed in Section 3.5.4. In this approximation, the fluid in the chamber is modeled as an ideal spring of stiffness coefficient $k = \rho_o c_o^2/V_c$, connected to a rigid mass $m = \rho_o L/S$, which represents the fluid in the connecting duct (the "neck"), and the resonance frequency follows from the well-known single-degree-of-freedom system formula $f_1 = (1/2\pi)\sqrt{k/m}$ [6]. Equation (5.35) reduces to this form if $L_A = 0$.

5.3.6 Transmission Loss of Resonators

Transmission loss of the branch and folded-duct configurations (Figures 5.5a,c) may be defined as $TL = 10\log_{10}\left(W_1^+/W_2^+\right)_{r_2=0}$. For the branch type (Figure 5.5a), this can be computed from the wave transfer matrix of the branch given in Equation (4.83). With conversion of the notation of Figure 4.6b to that of Figure 5.5a, the transmission loss of the resonator can be expressed as

$$TL = 10\log_{10}\left(\left|1 - \frac{S}{2S_1\zeta_3}\right|^2\right)_{r_2=0}, \tag{5.36}$$

where $\zeta_3 = (1 + r_3)/(1 - r_3)$. To derive an expression for the reflection coefficient $r_3 = p_3^-/p_3^+$, we note that $G_{p'} = \overline{p'_B}/\overline{p'_3} = 2(p_B^+/p_3^+)/(1 + r_3)$, since $r_B = 1$. But, from the wave transfer equation corresponding to the wave transfer matrix in Equation (5.31), it follows that

$$\frac{p_3^+}{p_B^+} = \mathrm{e}^{-\mathrm{i}k_oL}\left(\left(1 + \frac{m-1}{\zeta_6}\right)\cos(k_oL_c)\right) - \mathrm{i}m\sin(k_oL_c). \tag{5.37}$$

Hence, r_3 can be computed now from $r_3 = -1 + 2(\overline{p'_B}/\overline{p'_3})/G_{p'}$, where $G_{p'}$ is given by Equation (5.33). This formula for r_3 is also valid for the branch duct resonator in Figure 5.3a and, consequently, the *TL* of a uniform branch duct resonator can be computed similarly, except that the system FRF is now $G_{p'}(\omega) = 1/\cos(k_oL)$, which could also be obtained from Equation(5.33) by putting $L_A = L_c = 0$. Equation (5.36) is general and can be used with any branch resonator of normalized inlet impedance ζ_3. According to this formula, the mean flow in the main duct has no effect on the *TL* of the resonator, although the wave transfer matrix T_{12} of a branch resonator is dependent on the mean flow (see Equation (4.89)).

Shown in Figure 5.6 are the *TL* and overall transmission loss of a Helmholtz resonator with $m = 10$, $L_A/L = 0$, $L_c/L = 0.5$ and $S/S_1 = 0.5$. The *TL* peaks occur at the anti-resonance frequencies of the system FRF in Equation (5.33). It is observed

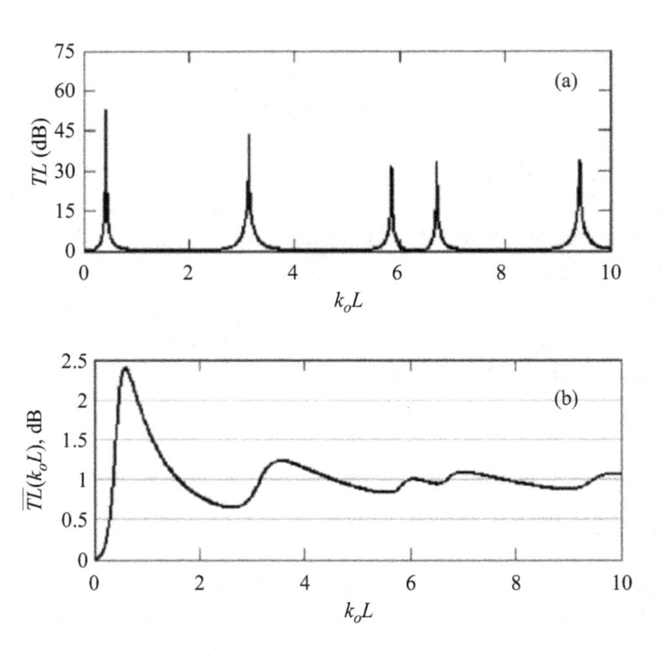

Figure 5.6 (a) *TL* and (b) overall *TL* of a Helmholtz resonator. $m = 10$, $L_A/L = 0$, $L_c/L = 0.5$ and $S/S_1 = 0.5$.

that the TL levels in the vicinity of these frequencies are much larger than the overall transmission loss of the resonator. In general, the overall transmission loss characteristics display an overshoot at the first anti-resonance frequency and then converge in an oscillatory manner to a lower value. For $L_A/L \approx 0$, the overshoot increases with m, S/S_1 and L_c/L, and the overall transmission loss approaches $2S/S_1$ as the frequency is increased. For this reason, single Helmholtz resonators can be effective in the vicinity of the smallest anti-resonance frequencies rather than over a finite frequency range. However, a wider frequency band can be covered by using an array of contiguously tuned Helmholtz resonators [7].

For the calculation of the transmission loss of a folded-duct resonator (Figure 5.2c), the counterpart of Equation (5.36) can be obtained from Equation (4.45). Then, using the notation of Figure 5.3c, the transmission loss of the folded-duct resonator can be expressed as

$$TL = 10\log_{10}\left(\left|1 + \frac{1}{2}\frac{S}{S_1}\left(1 + \frac{1}{\zeta_3}\right)\right|^2\right)_{r_2=0}, \tag{5.38}$$

where ζ_3 is still calculated as described above. This result neglects the effect of the mean flow.

The transmission loss of the in-line closed side-branch resonator shown in Figure 5.5b may be defined, for example, as $TL = 10\log_{10}\left(W_1^+/W_B^+\right)_{r_R=0}$, but this would be meaningful only for the effect of the auxiliary side branch as a resonator. However, this configuration is useful for showing the sound absorption function of a resonator. This can be seen from the wave transfer equation $\mathbf{P}_1 = \mathbf{T}_{1B}\mathbf{P}_B$, where the wave transfer matrix $\mathbf{T}_{1B}$ can be calculated by using the assembly methods described in Chapter 2. Since the reflection coefficient r_B is known, this wave transfer equation fixes the reflection coefficient at the resonator inlet, r_1. This means that one-dimensional sound waves incident on the resonator see it as a plane surface of finite normalized impedance $\zeta_1 = (1 + r_1)/(1 - r_1)$. Then, a quantity which is descriptive of the role of the resonator is $\alpha(\omega) = \left(W_1^+ - W_1^-\right)/W_1^+$. This is called the absorption coefficient and gives the power absorbed by the resonator as fraction of the incident acoustic power. Substituting from Equation (3.50) for the time-averaged power terms, we get $\alpha(\omega) = 1 - |r_1|^2$ or, in terms of the inlet impedance, $\alpha(\omega) = 4\zeta_{1R}/\left[(1 + \zeta_{1R})^2 + \zeta_{1I}^2\right]$, where the subscripts "$R$" and "$I$" denote the real and imaginary parts of a complex quantity. Therefore, if $\zeta_{1R} > 0$, the resonator acts as an acoustic absorber. In fact, this absorptive role is inherent to all real resonators, since some dissipation is always present due to viscothermal effects, but it can also be achieved in a controlled manner by treating the resonator cavity with sound absorbent material. This reduces the high noise levels that may build up in the resonator cavity at its resonance frequency and, hence, the noise transmitted from the walls of the cavity to the exterior environment as shell noise.

5.3.7 Interferential Resonator

Consider the block diagram in Figure 5.7a. This represents a duct system, the inlet and outlet ducts of which bifurcate into two parallel duct paths. The simplest implementation

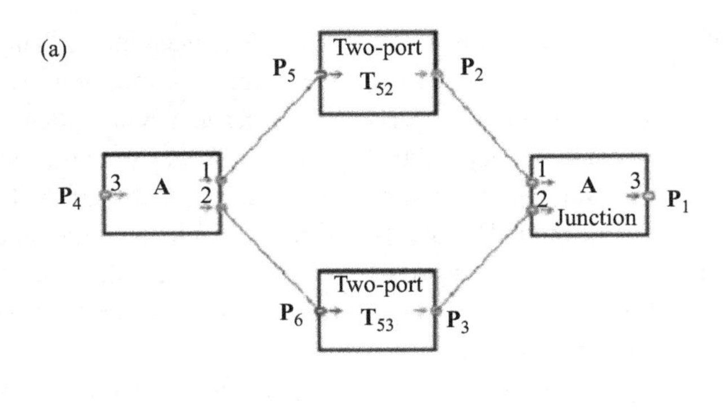

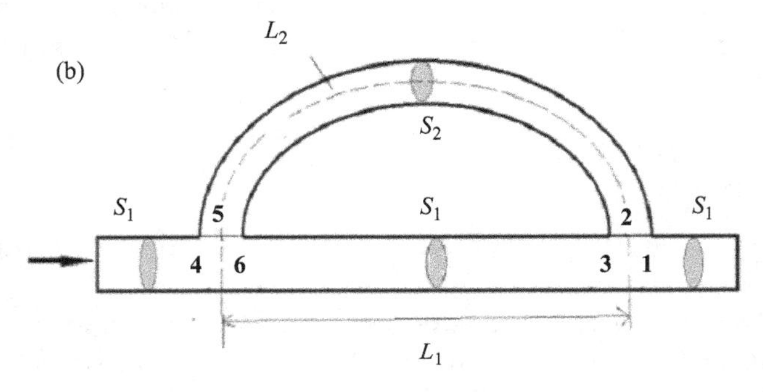

Figure 5.7 Two components connected in parallel. (a) General block diagram of parallel interferential flow ducts, (b) Herschel–Quincke tube configuration.

of this system is a bypass duct which joins two sections of a main duct that are a finite distance apart, as shown in Figure 5.7b. The bypass duct is commonly known as the Herschel–Quincke (HQ) tube. The destructive interferential capacity of such a bypass duct for one-dimensional waves propagating in the main duct is almost intuitive. Waves which split at the upstream junction into the HQ tube join the main duct again at the downstream junction. So, if the length of the HQ tube is chosen so that a frequency component of the wave in the HQ tube arrives at the downstream junction 180° out of phase with that in the main duct, the two waves cancel each other upon superposition and that frequency component is said to be annihilated by interference. However, this action is characteristic of expansion chambers (see Section 5.4) and the use of the HQ tube as a resonator is not immediately evident. A convenient FRF for studying the performance of the HQ tube is $G_{p^+}(\omega) = p_4^+(\omega)/p_1^+(\omega)$. So, we derive first the wave transfer matrix $\mathbf{T}_{41}$.

5.3.7.1 Wave Transfer Matrix of Parallel Two-Ports

Referring to the block diagram model in Figure 5.7a, suppose that the wave transfer matrices $\mathbf{T}_{52}$ and $\mathbf{T}_{63}$ of the two acoustic paths 5-2 and 6-3 are known. Then we have $\mathbf{P}_5 = \mathbf{T}_{52}\mathbf{P}_2$ and $\mathbf{P}_6 = \mathbf{T}_{63}\mathbf{P}_3$. The wave transfer relations for the junctions 1-2-3 and 4-5-6 are, from Equation (4.83),

$$\begin{bmatrix} \mathbf{b}_4 & -\mathbf{b}_5 & -\mathbf{b}_6 \\ \mathbf{a}_4 & \mathbf{a}_5 & \mathbf{0} \\ \mathbf{a}_4 & \mathbf{0} & \mathbf{a}_6 \end{bmatrix} \begin{bmatrix} \mathbf{P}_4 \\ \mathbf{P}_5 \\ \mathbf{P}_6 \end{bmatrix} = \mathbf{0} \tag{5.39}$$

$$\begin{bmatrix} -\mathbf{b}_1 & \mathbf{b}_2 & \mathbf{b}_3 \\ \mathbf{a}_1 & \mathbf{a}_2 & \mathbf{0} \\ \mathbf{a}_1 & \mathbf{0} & \mathbf{a}_3 \end{bmatrix} \begin{bmatrix} \mathbf{P}_1 \\ \mathbf{P}_2 \\ \mathbf{P}_3 \end{bmatrix} = \mathbf{0} \tag{5.40}$$

respectively, where the sign convention of Figure 5.7a is applied and the row vectors $\mathbf{a}_j$ and $\mathbf{b}_j$, $j = 1,2,\ldots,6$ are defined by Equations (4.84) and (4.85). Using the wave transfer matrices of the two paths, Equation (5.39) may be recast as

$$\begin{bmatrix} \mathbf{b}_4 & -\mathbf{b}_5\mathbf{T}_{52} & -\mathbf{b}_6\mathbf{T}_{63} \\ \mathbf{a}_4 & \mathbf{a}_2\mathbf{T}_{52} & \mathbf{0} \\ \mathbf{a}_4 & \mathbf{0} & \mathbf{a}_6\mathbf{T}_{63} \end{bmatrix} \begin{bmatrix} \mathbf{P}_4 \\ \mathbf{P}_2 \\ \mathbf{P}_3 \end{bmatrix} = \mathbf{0}. \tag{5.41}$$

This equation is expanded as

$$\mathbf{B}_4\mathbf{P}_4 + \mathbf{B}_5\mathbf{P}_2 + \mathbf{B}_6\mathbf{P}_3 = \mathbf{0} \tag{5.42}$$

$$\mathbf{c}_4\mathbf{P}_4 + \mathbf{c}_5\mathbf{P}_2 + \mathbf{c}_6\mathbf{P}_3 = \mathbf{0}, \tag{5.43}$$

where $\mathbf{B}_4 = \{\mathbf{b}_4 \ \ \mathbf{a}_4\}$, $\mathbf{B}_5 = \{-\mathbf{b}_5\mathbf{T}_{52} \ \ \mathbf{a}_2\mathbf{T}_{52}\}$ and $\mathbf{B}_6 = \{-\mathbf{b}_6\mathbf{T}_{63} \ \ 0\}$ are 2×2 matrices (curly brackets denote compound column vectors), and $\mathbf{c}_4 = \mathbf{a}_4$, $\mathbf{c}_5 = \mathbf{0}$ and $\mathbf{c}_6 = \mathbf{a}_6\mathbf{T}_{63}$. Equation (5.40) is similarly expanded as

$$\mathbf{B}_1\mathbf{P}_1 + \mathbf{B}_2\mathbf{P}_2 + \mathbf{B}_3\mathbf{P}_3 = \mathbf{0} \tag{5.44}$$

$$\mathbf{c}_1\mathbf{P}_1 + \mathbf{c}_2\mathbf{P}_2 + \mathbf{c}_3\mathbf{P}_3 = \mathbf{0}, \tag{5.45}$$

where $\mathbf{B}_1 = \{-\mathbf{b}_1 \ \ \mathbf{a}_1\}$, $\mathbf{B}_2 = \{\mathbf{b}_2 \ \ \mathbf{a}_2\}$, $\mathbf{B}_3 = \{\mathbf{b}_3 \ \ 0\}$ and $\mathbf{c}_1 = \mathbf{a}_1$, $\mathbf{c}_2 = \mathbf{0}$, $\mathbf{c}_3 = \mathbf{a}_3$. Multiplying Equation (5.44) from the left by $\mathbf{B}_1^{-1}$ and substituting the resulting $\mathbf{P}_1$ in Equation (5.45), gives us:

$$\left(\mathbf{c}_2 - \mathbf{c}_1\mathbf{B}_1^{-1}\mathbf{B}_2\right)\mathbf{P}_2 + \left(\mathbf{c}_3 - \mathbf{c}_1\mathbf{B}_1^{-1}\mathbf{B}_3\right)\mathbf{P}_3 = \mathbf{0}. \tag{5.46}$$

Similarly, from Equations (5.42) and (5.43):

$$\left(\mathbf{c}_5 - \mathbf{c}_4\mathbf{B}_4^{-1}\mathbf{B}_5\right)\mathbf{P}_2 + \left(\mathbf{c}_6 - \mathbf{c}_4\mathbf{B}_4^{-1}\mathbf{B}_6\right)\mathbf{P}_3 = \mathbf{0}. \tag{5.47}$$

Equations (5.46) and (5.47) are combined in a single matrix equation as

$$\mathbf{A}_2\mathbf{P}_2 + \mathbf{A}_3\mathbf{P}_3 = \mathbf{0}, \tag{5.48}$$

where

$$\mathbf{A}_2 = \begin{bmatrix} \mathbf{c}_2 - \mathbf{c}_1\mathbf{B}_1^{-1}\mathbf{B}_2 \\ \mathbf{c}_5 - \mathbf{c}_4\mathbf{B}_4^{-1}\mathbf{B}_5 \end{bmatrix} \tag{5.49}$$

$$\mathbf{A}_3 = \begin{bmatrix} \mathbf{c}_3 - \mathbf{c}_1\mathbf{B}_1^{-1}\mathbf{B}_3 \\ \mathbf{c}_6 - \mathbf{c}_4\mathbf{B}_4^{-1}\mathbf{B}_6 \end{bmatrix}. \tag{5.50}$$

Hence, from Equations (5.44) and (5.48):

$$\mathbf{P}_1 + \left(\mathbf{B}_1^{-1}\mathbf{B}_2 - \mathbf{B}_1^{-1}\mathbf{B}_3\mathbf{A}_3^{-1}\mathbf{A}_2\right)\mathbf{P}_2 = \mathbf{0} \tag{5.51}$$

and finally from Equations (5.41), (5.48) and (5.51), $\mathbf{P}_4 = \mathbf{T}_{41}\mathbf{P}_1$, where

$$\mathbf{T}_{41} = \mathbf{B}_4^{-1}\left(\mathbf{B}_5 - \mathbf{B}_6\mathbf{A}_3^{-1}\mathbf{A}_2\right)\left(\mathbf{B}_2 - \mathbf{B}_3\mathbf{A}_3^{-1}\mathbf{A}_2\right)^{-1}\mathbf{B}_1. \tag{5.52}$$

5.3.7.2 Wave Transfer Matrix of the HQ Tube

Referring now to the Herschel–Quincke configuration in Figure 5.7b, Equation (5.52) can be calculated numerically for arbitrary parallel acoustic paths. If the paths are uniform hard-walled ducts with no mean flow, the result can be expressed by reasonably simple explicit formulae. Here, we consider the case of single bypass duct of length L_2 and cross-sectional area S_2, mounted to control a one-dimensional sound field in a straight duct of length L_1 and cross-sectional area S_1. So, it is reasonable to assume that $L_2 > L_1$ and $S_2 < S_1$, although this is not strictly necessary for the validity of the subsequent analysis. Then, the matrices in Equation (5.52) are given by

$$\mathbf{B}_1 = \mathbf{B}_4 = \begin{bmatrix} -S_1 & S_1 \\ 1 & 1 \end{bmatrix}, \mathbf{B}_2 = \begin{bmatrix} S_2 & -S_2 \\ 1 & 1 \end{bmatrix}, \mathbf{B}_3 = \begin{bmatrix} S_1 & -S_1 \\ 0 & 0 \end{bmatrix} \tag{5.53}$$

$$\mathbf{B}_5 = \begin{bmatrix} -S_2\mathrm{e}^{-\mathrm{i}k_oL_2} & S_2\mathrm{e}^{\mathrm{i}k_oL_2} \\ \mathrm{e}^{-\mathrm{i}k_oL_2} & \mathrm{e}^{\mathrm{i}k_oL_2} \end{bmatrix} \tag{5.54}$$

$$\mathbf{B}_6 = \begin{bmatrix} -S_1\mathrm{e}^{-\mathrm{i}k_oL_1} & S_1\mathrm{e}^{\mathrm{i}k_oL_1} \\ 0 & 0 \end{bmatrix}, \tag{5.55}$$

where $L_1 > 0$. Also,

$$\mathbf{c}_1 = \mathbf{c}_3 = \mathbf{c}_4 = [1 \quad 1], \mathbf{c}_2 = \mathbf{c}_5 = \mathbf{0}, \mathbf{c}_6 = \left[\mathrm{e}^{-\mathrm{i}k_oL_1} \quad \mathrm{e}^{\mathrm{i}k_oL_1}\right] \tag{5.56}$$

$$\mathbf{A}_2 = -\begin{bmatrix} 1 & 1 \\ \mathrm{e}^{-\mathrm{i}k_oL_2} & \mathrm{e}^{\mathrm{i}k_oL_2} \end{bmatrix}, \mathbf{A}_3 = \begin{bmatrix} 1 & 1 \\ \mathrm{e}^{-\mathrm{i}k_oL_2} & \mathrm{e}^{\mathrm{i}k_oL_2} \end{bmatrix}. \tag{5.57}$$

Hence, Equation (5.52) can be expressed explicitly as

$$\mathbf{T}_{41} = \frac{1}{m\sin(k_oL_1) + \sin(k_oL_2)}\begin{bmatrix} A + \mathrm{i}(B+C) & \mathrm{i}(B+D) \\ -\mathrm{i}(B+D) & A - \mathrm{i}(B+C) \end{bmatrix}, \tag{5.58}$$

where $m = S_2/S_1$ and

$$A = m\cos(k_oL_2)\sin(k_oL_1) + \sin(k_oL_2)\cos k_oL_1 \tag{5.59}$$

$$B = m(1 - \cos(k_oL_1)\cos(k_oL_2)) \tag{5.60}$$

$$C = \left(1 + \frac{m^2}{2}\right)\sin(k_oL_1)\sin(k_oL_2) \tag{5.61}$$

$$D = \frac{m^2}{2} \sin(k_o L_1) \sin(k_o L_2). \tag{5.62}$$

5.3.7.3 Herschel–Quincke Tube Resonator

Let us take $G_{p^+}(\omega) = p_1^+(\omega)/p_4^+(\omega)$ as the FRF of the system in Figure 5.7b. From Equations (5.19) and (5.58):

$$G_{p^+}(\omega) = \frac{\sin(k_o L_2) + \frac{S_2}{S_1} \sin(k_o L_1)}{X + \mathrm{i}Y}, \tag{5.63}$$

where

$$X = \frac{S_2}{S_1} \cos(k_o L_2) \sin(k_o L_1) + \sin(k_o L_2) \cos k_o L_1 \tag{5.64}$$

$$Y = (1 + r_1)\left\{\frac{S_2}{S_1}(1 - \cos(k_o L_1)\cos(k_o L_2)) + \left(\frac{1}{1 + r_1} + \frac{S_2^2}{2S_1^2}\right) \sin(k_o L_1) \sin(k_o L_2)\right\}. \tag{5.65}$$

The anti-resonances occur at frequencies f given by the roots of the equation

$$\sin\left(\frac{2\pi f L_2}{c_o}\right) = -\frac{S_2}{S_1} \sin\left(\frac{2\pi f L_1}{c_o}\right). \tag{5.66}$$

Roots of this equation can be expressed in closed form only in the special cases of $S_2 = S_1$ or $L_2 = L_1$ or $L_2 = 2L_1$ (see Table 5.1). Although $S_2 = S_1$ or $L_2 = L_1$ are not compatible with conditions $L_2 > L_1$ and $S_2 < S_1$, they are helpful in providing useful

Table 5.1 Anti-resonance frequencies of a bypass duct

Case	Anti-resonance frequency (Hz)
$S_2 = S_1$	$f_n^{(1)} = \frac{n c_o}{2(L_2 - L_1)}, n = 1, 3, 5, \ldots$
	$f_n^{(2)} = \frac{n c_o}{2(L_2 - L_1)}, n = 2, 4, 6, \ldots$
$L_2 = L_1$	No anti-resonance
	$G_{p^+}(\omega) = \left[1 + \frac{1}{4}\left(1 + \frac{S_2}{S_1} - \frac{1}{1 + \frac{S_2}{S_1}}\right)^2 \sin^2\left(\frac{\omega L^2}{c_o}\right)\right]^{-1}$
$L_2 = 2L_1$	$f_n^{(1)} = \frac{n c_o}{2L_1}, n = 1, 3, 5, \ldots$
	$f_n^{(2)} = \frac{c_o}{2\pi L_1}\left(n\pi \mp \arccos\left(-\frac{S_2}{2S_1}\right)\right), n = 1, 3, 5, \ldots$
	e.g., if $S = S_1$: $f_n^{(2)} = \frac{c_o}{2L_1}\left(n \mp \frac{1}{3}\right)$

Note: Superscripts in brackets denote branches of roots of Equation (5.66).

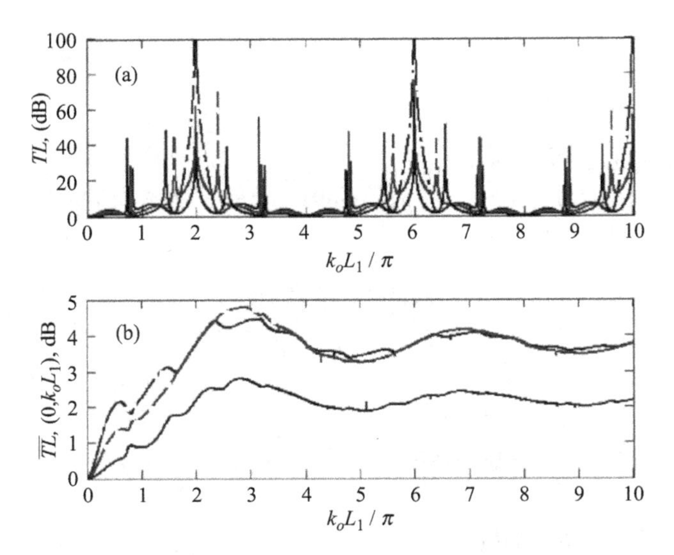

Fig 5.8 (a) *TL* and (b) overall *TL* of an interferential resonator with single bypass duct in Figure 5.7b: $L_2/L_1 = 1.5$ and $S_2/S_1 = 0.5$ (solid), 1 (dash) and 1.5 (dash-dot).

estimates for anti-resonance frequencies when $S_2 \approx S_1$ or $L_2 \approx L_1$. This is demonstrated in Figure 5.8a, which gives the transmission loss characteristics ($r_1 = 0$, Equation (5.20)) of the resonator for $L_2/L_1 = 1.5$ and $S_2/S_1 = 0.5$, 1 and 1.5. The anti-resonance frequencies given in Table 5.1 are strictly valid for $S_2/S_1 = 1$, but it is seen that these are not substantially altered when the area ratio is varied even by $\mp 50\%$. It is also observed that, as $L_2 \rightarrow L_1$, the anti-resonances tend to be annihilated, which is confirmed by the second case entry in Table 5.1. In the limit of $L_2 = L_1$, the FRF reduces to the form given in Table 5.1, which does not have resonance or anti-resonance. We will see in Section 5.4 that this form of FRF is very similar to that of a simple expansion chamber with an anechoic outlet.

Special consideration is also be required when L_2 is a higher integer-multiple of L_1. For example, in the case given in Table 5.1 for $L_2 = 2L_1$, Equation (5.66) determines only the second branch of the anti-resonance frequencies. The first branch comes from the condition that the denominator of the FRF becomes infinitely large. In general, for a given area ratio, as the length ratio L_2/L_1 is increased, the *TL* characteristics tend to be shifted toward the lower end of the frequency axis.

The overall transmission-loss characteristics of the resonator are shown in Figure 5.8b. In general, as for Helmholtz resonators, the overall transmission-loss levels are relatively small compared to the values in the vicinity of the peaks. In this case, however, the resonator provides possibilities for tuning over larger bandwidths than Helmholtz resonators.

There is a rich literature on the theory and applications of this type of interferential resonator. More recent studies include configurations with arrays of bypass ducts and mean flow effects [8–9].

5.3.8 Straight-Through Resonator

A straight-through resonator and its block diagram model is shown in Figure 5.9. This consists of a perforated duct having one or more rows of apertures enclosed in another duct having closed ends. The equations of the elements from which this configuration can be assembled are given in Section 2.4.2. We repeat the elements that are pertinent to the present discussion using the node numbers in the block diagram in Figure 5.9:

Side-branch ducts:

$$\mathbf{P}_3 = \mathbf{T}_{3A}\mathbf{P}_A, \mathbf{P}_4 = \mathbf{T}_{4B}\mathbf{P}_B. \tag{5.67}$$

Side-branch closed boundaries:

$$\mathbf{r}_A\mathbf{P}_A = \mathbf{0}, \mathbf{r}_B\mathbf{P}_B = \mathbf{0}. \tag{5.68}$$

The 2-duct (from Section 4.7 or 4.8.4):

$$\begin{bmatrix} \mathbf{P}_3 \\ \mathbf{P}_1 \end{bmatrix} = \begin{bmatrix} \mathbf{S}_{11} & \mathbf{S}_{12} \\ \mathbf{S}_{21} & \mathbf{S}_{22} \end{bmatrix} \begin{bmatrix} \mathbf{P}_4 \\ \mathbf{P}_2 \end{bmatrix}. \tag{5.69}$$

Again, it is convenient to use $G_{p^+}(\omega) = p_2^+(\omega)/p_1^+(\omega)$ as the system FRF. Accordingly, we proceed with the determination of the wave transfer matrix $\mathbf{T}_{12}$. First, expand the 2-duct equation as $\mathbf{P}_3 = \mathbf{S}_{11}\mathbf{P}_4 + \mathbf{S}_{12}\mathbf{P}_2$ and $\mathbf{P}_1 = \mathbf{S}_{21}\mathbf{P}_4 + \mathbf{S}_{22}\mathbf{P}_2$. Using the foregoing side-branch relations, these equations can be expressed, respectively, as

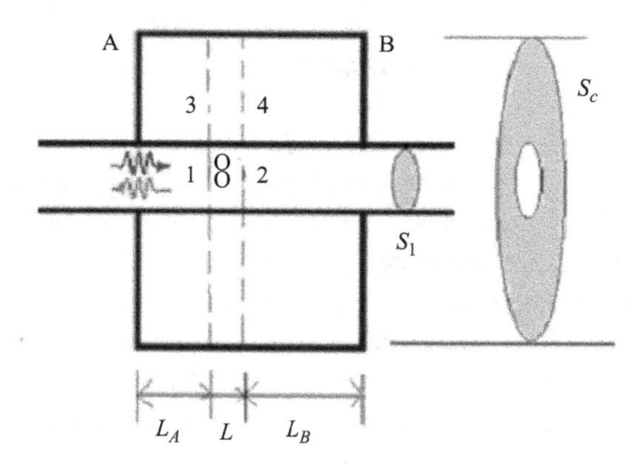

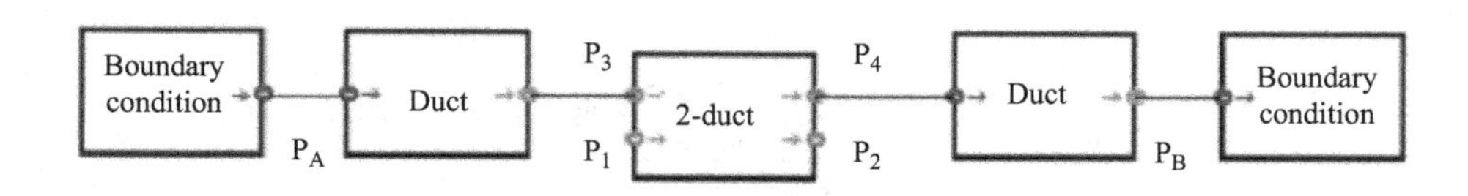

Figure 5.9 Straight-through resonator and its block diagram.

$$\mathbf{P}_A = \mathbf{T}_{3A}^{-1}\mathbf{S}_{11}\mathbf{T}_{4B}\mathbf{P}_B + \mathbf{T}_{3A}^{-1}\mathbf{S}_{12}\mathbf{P}_2 \tag{5.70}$$

$$\mathbf{P}_1 = \mathbf{S}_{21}\mathbf{T}_{4B}\mathbf{P}_B + \mathbf{S}_{22}\mathbf{P}_2. \tag{5.71}$$

Let $\mathbf{r}_B = \{1 \quad r_B\}$ and $\mathbf{R}_A = [r_A \quad -1]$, where r_A and r_B denote the reflection coefficients of the end-cap A and B, respectively. Pre-multiplication of Equation (5.70) by $\mathbf{R}_A$ gives

$$0 = \mathbf{R}_A\mathbf{T}_{3A}^{-1}\mathbf{S}_{11}\mathbf{T}_{4B}\mathbf{r}_B p_B^+ + \mathbf{R}_A\mathbf{T}_{3A}^{-1}\mathbf{S}_{12}\mathbf{P}_2. \tag{5.72}$$

Solving p_B^+ from this equation and substituting the result in Equation (5.71) yields $\mathbf{P}_1 = \mathbf{T}_{12}\mathbf{P}_2$, where

$$\mathbf{T}_{12} = \mathbf{S}_{22} - \frac{\mathbf{S}_{21}\mathbf{T}_{4B}\mathbf{r}_B\mathbf{R}_A\mathbf{T}_{3A}^{-1}\mathbf{S}_{12}}{\mathbf{R}_A\mathbf{T}_{3A}^{-1}\mathbf{S}_{11}\mathbf{T}_{4B}\mathbf{r}_B}. \tag{5.73}$$

This wave transfer matrix can in general only be calculated numerically. If the ducts are assumed to be uniform and carry no mean flow, then, for the case of single row of apertures and rigid end-caps ($r_A = r_B = 1$), Equation (5.73) can be expressed explicitly as

$$\mathbf{T}_{12} = \begin{bmatrix} 1+\alpha & \alpha \\ -\alpha & 1-\alpha \end{bmatrix}, \tag{5.74}$$

where

$$\alpha = \frac{(\tan(k_o L_A) + \tan(k_o L_B))\dfrac{S_a}{\mathrm{i}2\zeta S_1}}{(\tan(k_o L_A) + \tan(k_o L_B)) - \dfrac{S_a}{\mathrm{i}\zeta S_c}}. \tag{5.75}$$

Here, $L_A > 0$ and $L_B > 0$ denote the side-branch lengths, S_a denotes the total area of the apertures in the row (all aperture being assumed to be identical), S_c and S_1 denote the cross-sectional areas of the inner duct and the annulus (see Figure 5.8), and ζ denotes the normalized aperture impedance (see Appendix C). In deriving Equation (5.74), the perforated duct is assumed to have only one aperture row and the wave transfer matrix of the two-duct is obtained from the two-duct form of Equation (4.146). Thus, in Figure 5.8, $L \approx 2\varepsilon$ may be understood as a two-sided end-correction. However, the foregoing result can still be used with adequate accuracy if the apertures are not distributed in a single row, provided that they remain in a compact segment of length L, $k_o L << 1$.

From Equation (5.74), FRF of the system, $G_{p^+}(\omega) = p_2^+(\omega)/p_1^+(\omega)$, is found as

$$G_{p^+}(\omega) = \frac{\dfrac{1}{\mathrm{i}\zeta}\dfrac{S_a}{S_c} - (\tan(k_o L_A) + \tan(k_o L_B))}{\dfrac{1}{\mathrm{i}\zeta}\dfrac{S_a}{S_c} - (\tan(k_o L_A) + \tan(k_o L_B)) - \dfrac{1+r_2}{2\zeta}\dfrac{S_a}{S_1}(\tan(k_o L_A) + tan(k_o L_B))}, \tag{5.76}$$

where r_2 denotes the outlet reflection coefficient. Inspection of this equation indicates that this FRF has no resonances. In general, ζ is complex, but, its resistive (real) part is small and can be neglected in the determination of the anti-resonance frequencies with adequate accuracy. For example, a commonly used expression is $\zeta = -ik_o(t+\delta)$, where t denotes the wall thickness of the pipe and δ denotes an end-correction (see Appendix C for more detailed models). For this aperture impedance, the anti-resonance frequencies are given by the roots of the equation:

$$\tan\left(\frac{2\pi f}{c_o}L_A\right) + \tan\left(\frac{2\pi f}{c_o}L_B\right) = \frac{c_o}{2\pi f(t+\delta)}\frac{S_a}{S_c}. \tag{5.77}$$

If the side-branch lengths are compact, then $\tan(k_oL_A) \approx k_oL_A$ and $\tan(k_oL_B) \approx k_oL_B$, and the following approximate expression can be derived from Equation (5.77) for the smallest anti-resonance frequency, namely,

$$f_1 = \frac{c_o}{2\pi}\sqrt{\frac{S_a}{S_c(L_A+L_B)(t+\delta)}}, \tag{5.78}$$

which is similar to the expression for the resonance frequency of the traditional Helmholtz resonator (compare with Equation (5.35)). Thus, if the side-branch lengths are compact, the FRF in Equation (5.76) applies for any cavity of volume $S_c(L_A+L_B)$ and the actual side-branch lengths are not relevant.

5.4 Expansion Chambers

A pure expansion chamber may be defined as a duct arrangement that includes no resonator elements (that is, the system FRF has no anti-resonances), but still has frequency tunable transmission loss characteristics. A duct arrangement which conforms to this definition is the bypass duct in Figure 5.7 with $L_2 = L_1$. The transmission loss of this system is $TL = -20\log_{10}|G_{p^+}(\omega)|$, where $G_{p^+}(\omega)$ is given by the $L_2 = L_1$ case of Table 5.1. This FRF has no anti-resonance (and resonance for that matter) and the corresponding transmission loss characteristics are frequency dependent, as shown in Figure 5.10a, for several area ratios $m = 1 + S_2/S_1$. The maxima and minima of the transmission loss characteristics occur at frequencies satisfying $\sin(k_oL) = 1$ and $\sin(k_oL) = 0$, and are given by $TL_{\max} = 10\log_{10}\left[1+(m-1/m)^2/4\right]$ and $TL_{\min} = 0$, respectively. The overall transmission-loss characteristics (Figure 5.10b) are oscillatory about $TL_{\max}/2$ for Helmholtz numbers $k_oL > \pi$, which delimits the range of the first lobe of the transmission loss characteristics, and closes on this value as the Helmholtz number is increased.

In practice, however, an expansion chamber is usually understood to be a relatively large volume enclosed within impermeable walls, with at least one inlet and one outlet duct connections. They are often manufactured as a duct with closed end-caps and regular section geometry such as circle, rectangle, oval, ellipse or racetrack. Depending on the application, the inlet and outlet ducts can be connected to the

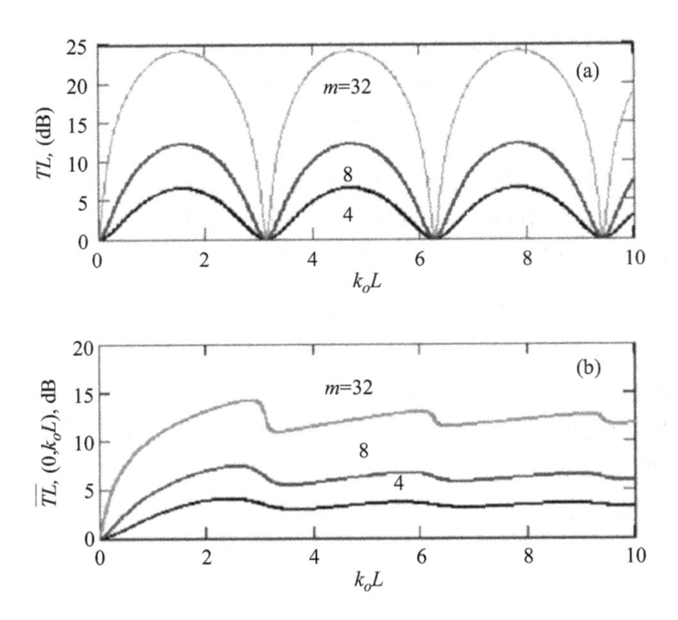

Figure 5.10 (a) *TL* and (b) overall *TL* of the bypass duct in Figure 5.7b with $L = L_2 = L_1$ and $m = 1 + S_2/S_1$. The same characteristics also hold for a pure through-flow expansion chamber shown in Figure 5.11a with $m = S_c/S_1$.

chamber on its lateral surfaces or on end-cap surfaces. Some typical expansion chamber configurations are shown schematically in Figure 5.11. They differ essentially by the arrangement of the inlet and outlet ducts relative to the chamber. For prismatic chambers, the inlet and outlet ducts are usually referred to as side-inlet or side-outlet, if they are connected to the chamber from its lateral surface (Figure 5.11d, e), or axial inlet or outlet if connected from the end-caps (Figure 5.11a,b,c). The chamber in Figure 5.11a is usually referred to as the through-flow chamber, because the direction of the mean flow is not subjected to a reversal. The configuration in Figure 5.11c is known as the flow-reversing expansion chamber. In the chambers in Figure 5.11b,d,e, the mean flow also changes direction in some way. Obviously, further similar configurations can be added to Figure 5.11 by permutation of inlet and outlet duct connections.

5.4.1 Through-Flow Expansion Chambers

The expansion chambers shown in Figure 5.11 can be modeled by cascading the duct and area-change elements described in Chapters 3 and 4. This cascade is shown in Figure 5.11f. The wave transfer matrix of the cascade is $\mathbf{T}_{12} = \mathbf{T}_{13}\mathbf{T}_{34}\mathbf{T}_{42}$, where $\mathbf{T}_{13}$ and $\mathbf{T}_{42}$ denote the wave transfer matrices of area expansion and contraction elements, respectively, and $\mathbf{T}_{34}$ denotes that of a duct element. The area expansion elements are of the through-flow type shown in Figure 5.11a; flow-reversing type in Figure 5.11b,c, e; and junction type in Figure 5.11d; whilst the area contraction elements are of the

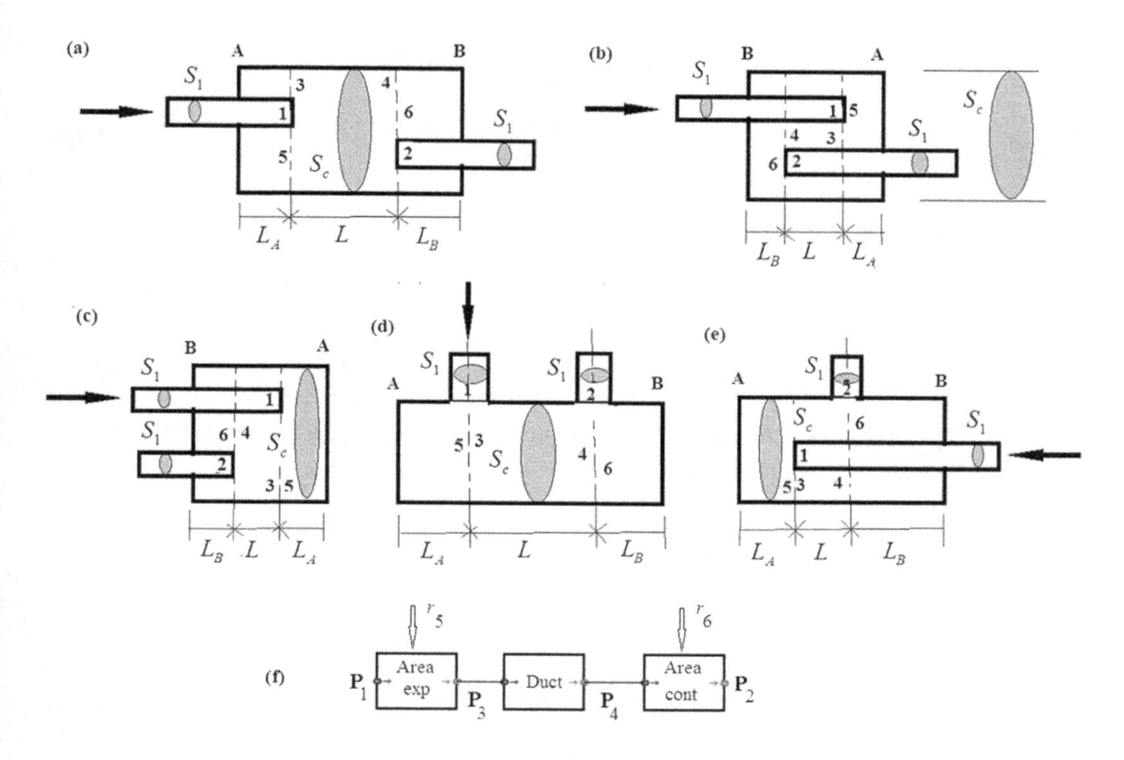

Figure 5.11 Simple expansion chambers. (a) Through-flow chamber. (b) Through-flow chamber with overlapping inlet and outlet extensions. (c) Flow-reversing chamber. (d) Side-inlet and side-outlet chamber. (e) Axial inlet and side-outlet chamber. (f) Block diagram of these chambers.

through-flow type in Figure 5.11a,c; flow-reversing type in Figure 5.11b and junction type in Figure 5.11d,e.

The chamber wave transfer matrix $\mathbf{T}_{12}$ can be determined only numerically if these elements are non-uniform or inhomogeneous. For the discussion of the basic acoustic features of the chambers, however, it is convenient to consider chambers made of uniform hard-walled ducts and neglect the effect of the mean flow to begin with. Then the elements of $\mathbf{T}_{12}$ can be expressed by reasonably simple explicit formulae, which are useful insights into the basic physics of chambers.

It suffices to demonstrate the derivation of $\mathbf{T}_{12}$ for the through-flow expansion chamber (Figure 5.11a), as the other configurations in Figure 5.11 can be treated similarly. For simplicity of the analysis, we assume that the inlet and outlet ducts are of the same cross-sectional area.

The synthesis of the through-flow chamber from basic acoustic elements is discussed in Sections 2.4.1 and 2.4.2. The elements that are pertinent to the present discussion are summarized below, using the node numbers shown in Figure 5.11a for their identification.

Duct (from Section 3.5.1):

$$\mathbf{T}_{34} = \begin{bmatrix} \mathrm{e}^{-\mathrm{i}k_o L} & 0 \\ 0 & \mathrm{e}^{\mathrm{i}k_o L} \end{bmatrix}. \tag{5.79}$$

Area expansion (from Section 4.3.2 or Section 4.6.2):

$$\mathbf{T}_{13} = \frac{1}{2\zeta_5 S_1}\begin{bmatrix} S_c(\zeta_5+1)+S_1(\zeta_5-1) & -S_c(\zeta_5-1)+S_1(\zeta_5-1) \\ -S_c(\zeta_5+1)+S_1(\zeta_5+1) & S_c(\zeta_5-1)+S_1(\zeta_5+1) \end{bmatrix}. \tag{5.80}$$

Area contraction (from Section 4.3.2 or Section 4.6.2):

$$\mathbf{T}_{42} = \frac{-1}{2\zeta_6 S_c}\begin{bmatrix} S_1(-\zeta_6+1)+S_c(-\zeta_6-1) & -S_1(-\zeta_6-1)+S_c(-\zeta_6-1) \\ -S_1(-\zeta_6+1)+S_c(-\zeta_6+1) & S_1(-\zeta_6-1)+S_c(-\zeta_6+1) \end{bmatrix}. \tag{5.81}$$

Hence, upon calculation of the matrix product $\mathbf{T}_{13}\mathbf{T}_{34}\mathbf{T}_{42}$, the wave transfer matrix of the chamber is found as:

$$\mathbf{T}_{12} = \begin{bmatrix} T_{11} & T_{12} \\ T_{21} & T_{22} \end{bmatrix}, \tag{5.82}$$

where

$$\begin{aligned} T_{11} = {} & \cos(k_oL) - \frac{\mathrm{i}}{2}\left(m+\frac{1}{m}\right)\sin(k_oL) - \frac{\mathrm{i}}{2\zeta_5\zeta_6}\left(\sqrt{m}-\sqrt{\frac{1}{m}}\right)^2\sin(k_oL) \\ & +\frac{1}{2}\left(\frac{1}{\zeta_5}+\frac{1}{\zeta_6}\right)\left((m-1)\cos(k_oL)+\mathrm{i}\left(\frac{1}{m}-1\right)\sin(k_0L)\right) \end{aligned} \tag{5.83}$$

$$\begin{aligned} T_{12} = {} & -\frac{\mathrm{i}}{2}\left(m-\frac{1}{m}\right)\sin(k_oL) - \frac{\mathrm{i}}{2\zeta_5\zeta_6}\left(\sqrt{m}-\sqrt{\frac{1}{m}}\right)^2\sin(k_oL) \\ & +\frac{1}{2}\left(\frac{1}{\zeta_5}+\frac{1}{\zeta_6}\right)(m-1)\cos(k_oL) - \frac{\mathrm{i}}{2}\left(\frac{1}{\zeta_5}-\frac{1}{\zeta_6}\right)\left(\frac{1}{m}-1\right)\sin(k_oL) \end{aligned} \tag{5.84}$$

$$\begin{aligned} T_{21} = {} & \frac{\mathrm{i}}{2}\left(m-\frac{1}{m}\right)\sin(k_oL) + \frac{\mathrm{i}}{2\zeta_5\zeta_6}\left(\sqrt{m}-\sqrt{\frac{1}{m}}\right)^2\sin(k_oL) \\ & -\frac{1}{2}\left(\frac{1}{\zeta_5}+\frac{1}{\zeta_6}\right)(m-1)\cos(k_oL) + \frac{\mathrm{i}}{2}\left(\frac{1}{\zeta_5}-\frac{1}{\zeta_6}\right)\left(\frac{1}{m}-1\right)\sin(k_oL) \end{aligned} \tag{5.85}$$

$$\begin{aligned} T_{22} = {} & \cos(k_oL) + \frac{\mathrm{i}}{2}\left(m+\frac{1}{m}\right)\sin(k_oL) + \frac{\mathrm{i}}{2\zeta_5\zeta_6}\left(\sqrt{m}-\sqrt{\frac{1}{m}}\right)^2\sin(k_oL) \\ & +\frac{1}{2}\left(\frac{1}{\zeta_5}+\frac{1}{\zeta_6}\right)\left(-(m-1)\cos(k_oL)+\mathrm{i}\left(\frac{1}{m}-1\right)\sin(k_oL)\right). \end{aligned} \tag{5.86}$$

Here, $L > 0$, $m = S_c/S_1$, and ζ_5 and ζ_6 denote the normalized acoustic impedance at side-branch interface planes 5 and 6, which are given by

$$\zeta_5 = \frac{1 + r_A e^{-i2k_oL_A}}{1 - r_A e^{-i2k_oL_A}} \tag{5.87}$$

$$\zeta_6 = \frac{1 + r_B e^{-i2k_oL_B}}{1 - r_B e^{-i2k_oL_B}}, \tag{5.88}$$

where r_A and r_B denote, respectively, the reflection coefficients of end-caps A and B. Note that, in view of the sign convention of Figure 4.1, the side-branch lengths L_A and L_B must be input as negative quantities.

5.4.2 Transmission Loss of Expansion Chambers

The wave transfer matrices $\mathbf{T}_{12}$ of the other chambers in Figure 5.11 can be derived similarly by using the relevant area-change elements. In fact, it turns out that the wave transfer matrices of the chambers in Figure 5.11, and any other configuration which conforms to the block diagram in Figure 5.11f, can be expressed in a unified form. Here, we give the unified formula for the transmission loss of this family of chambers with rigid end-caps, that is, $r_A = r_B = 1$, which implies that $\zeta_5 = 1/\mathrm{i}\tan(k_oL_A)$ and $\zeta_6 = 1/\mathrm{i}\tan(k_oL_B)$:

$$\begin{aligned} TL = 20\log_{10}\Bigg| &\cos(k_oL) - \frac{\mathrm{i}}{2}\left(m + \frac{1}{m}\right)\sin(k_oL) - \frac{\mathrm{i}(B_1 + m)}{2A_2B_2\zeta_5\zeta_6}\left(A_1 + \frac{1}{m}\right)\sin(k_oL) \\ &+ \frac{1}{2A_2\zeta_5}\left(-(1 + A_1m)\cos(k_oL) + \mathrm{i}\left(A_1 + \frac{1}{m}\right)\sin(k_0L)\right) \\ &+ \frac{1}{2B_2\zeta_6}\left(-(B_1 + m)\cos(k_oL) + \mathrm{i}\left(1 + \frac{B_1}{m}\right)\sin(k_0L)\right)\Bigg|, \end{aligned} \tag{5.89}$$

where $m = S_2/S_1$ and S_2 denotes the cross-sectional area of the duct connecting the two area-change elements of the chamber (see Table 5.2). The parameters A_1, A_2 and B_1, B_2 come from Equation (4.45), or Equation (4.83) and have the values given in Table 5.2 for the chambers in Figure 5.11. It should be noted that the expression in modulus brackets in Equation (5.89) is the inverse of the system FRF, which is $G_{p^+}(\omega) = p_2^+(\omega)/p_1^+(\omega)$.

Table 5.2 Characteristic parameters of chambers in Figure 5.11 (used in Equation (5.89))

Chamber	A_1	A_2	B_1	B_2	S_2
(a)	-1	$+1$	-1	-1	S_c
(b)	$+1$	-1	$+1$	-1	$S_c - 2S_1$
(c)	$+1$	-1	-1	-1	$S_c - S_1$
(d)	$1 - S_1/S_2$	-1	0	-1	S_c
(e)	$+1$	-1	0	-1	$S_c - S_1$

5.4.3 Pure Expansion Chambers

Whether an expansion chamber is pure or not can be tested by removing the resonators that are integral to it, if any. For example, for the through-flow expansion chamber (Figure 5.11a), the side-branch resonators can be removed by taking $L_A = L_B = 0$. Then, Equation (5.89) simplifies to

$$TL = 10\log_{10}\left(1 + \frac{1}{4}\left(m - \frac{1}{m}\right)^2 \sin^2(k_o L)\right), \tag{5.90}$$

where $m = S_c/S_1$. This proves that this is a pure expansion chamber. The foregoing formula, which is ubiquitous in books on acoustics, is exactly of the same form as the transmission loss of the bypass duct with $L_2 = L_1$, except for the definition of the area ratio m (see Table 5.1). Thus, the transmission loss characteristics of the chamber are given by Figure 5.10 and $TL_{\min} = 0$ and $TL_{\max} = 10\log_{10}\left(1 + (m - 1/m)^2/4\right)$ occur at frequencies $k_o L = n\pi$ and $k_o L = (2n+1)\pi/2$, respectively, where $n = 0, 1, 2, \ldots$. Since $k_o = 2\pi f/c_o$, these critical frequencies can be expressed as $f_n = nc_o/2L$ and $\frown{f} = (2n+1)c_o/4L$, respectively, where an over-arc is used to distinguish the $TL_{\max}$ frequencies.

The overall transmission loss in the frequency band $0 \le f \le f_1$ is equal to $TL_{\max}/2$ and, therefore, depends only on the area ratio m. The overall transmission-loss characteristics in Figure 5.10b coalesce as shown in Figure 5.12. It is seen that as each transmission loss lob is traversed, overall transmission loss fluctuates about $TL_{\max}/2$ and that the fluctuations are smoothed out as frequency is increased.

The side branches in the overlapping-duct configuration of Figure 5.11b can be removed only if the overlap is undone, in which case we revert to the through-flow chamber.

The side branches of the side-inlet and side-outlet chamber (Figure 5.11d) cannot be removed completely because the inlet and outlet ducts sections must remain on the lateral surface of the chamber. Nevertheless, we may still assume $L_A = L_B = 0$ if the side-branch resonance frequencies are higher than the frequencies of interest. The transmission loss of the chamber can then be determined by using Equation (5.90).

The flow-reversing chamber (Figure 5.11c) is interesting, because one of the area changes must be of the flow-reversing type and the side branch associated with this

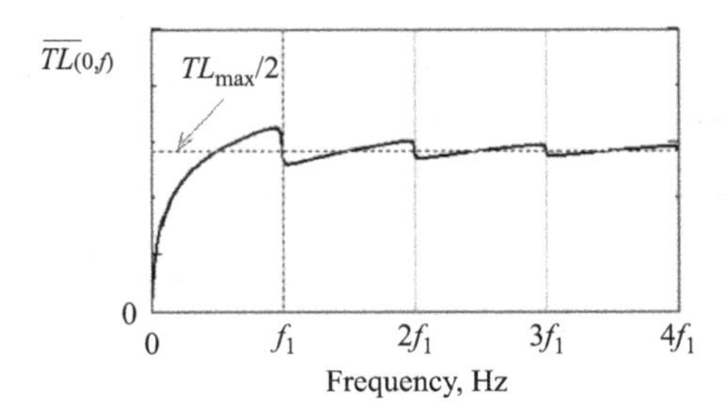

Figure 5.12 Overall transmission loss of a pure expansion chamber ($f_1 = c_o/2L$).

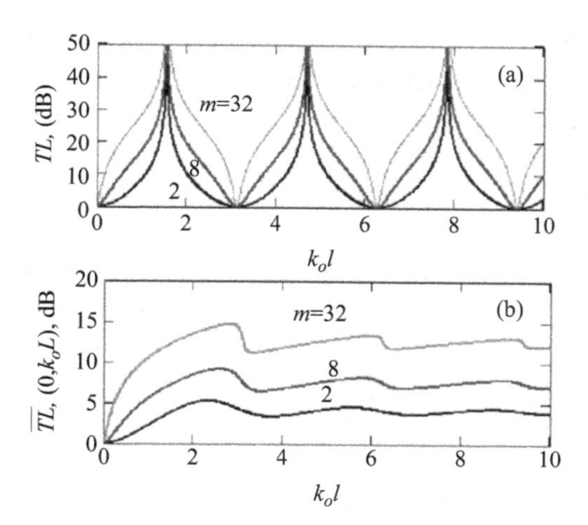

Figure 5.13 (a) TL and (b) overall transmission toss of a flow-reversing chamber in Figure 5.10c with $L_B = L = 0$, $m = S_2/S_1$.

area-change cannot be removed. For example, let $L_B = 0$ in the configuration of Figure 5.11c. This removes the outlet side branch. If the inlet duct has no extension, then $L = 0$ and $L_A = \ell$, where ℓ denotes the chamber length, and Equation (5.89) simplifies to

$$TL = 10\log_{10}\left(1 + \frac{1}{4}\left(1 + \frac{S_2}{S_1}\right)^2 \tan^2(k_o\ell)\right). \tag{5.91}$$

Obviously, the corresponding FRF, $G_{p^+}(\omega) = p_2^+(\omega)/p_1^+(\omega)$, has no resonance, but it has anti-resonances at $\tan(k_o\ell) = \infty$, which occur at frequencies $f_n = nc_o/4\ell$, $n = 1, 3, 5, \ldots$ Therefore, in this specific configuration, the flow-reversing chamber actually acts as a resonator. The transmission loss and overall transmission loss characteristics of the chamber are shown in Figure 5.13. The $TL_{\min} = 0$ frequencies are given by $f_n = nc_o/2\ell$, $n = 0, 1, 2, \ldots$, but these do not correspond to FRF resonances. The transmission-loss characteristics are now cusped to infinity at the anti-resonance frequencies, but the overall transmission-loss characteristics are similar to that of a pure through-flow expansion chamber. In fact, Figure 5.12 still applies, but $TL_{\max}$ must be interpreted as the maximum TL of a pure through-flow expansion having the same area ratio.

5.4.4 The Strongest Pure Expansion Chamber

Ideally, the transmission loss of a pure expansion chamber should cover the frequency range of interest in the first lob of transmission loss characteristics. This is possible if $c_o/2L$ is large enough, where L denotes the chamber length. Thus, it appears that the frequency range can be made as large as desired by reducing the chamber length. But the chamber length can be reduced as long as only one-dimensional waves propagate in the chamber. As we shall see in the next chapter, one-dimensional propagation is only one of the possible modes of sound propagation in ducts. If frequency is high enough, sound waves can also propagate in ducts in higher order modes for which one-dimensional approximation is not valid.

Let f^* denote the smallest frequency at which the first higher-order mode begins to propagate. The one-dimensional theory and, hence, Equation (5.89), remain valid while $f < f^*$. The value of f^* depends mainly on the size and shape of the duct cross section. For example, for a circular duct of diameter D carrying no mean flow, it can be expressed as $f^* = (2 \times 1.84118L/\pi D)c_o/2L$ or $f^* = (L/0.853D)c_o/2L$ for the first non-axisymmetric mode (see Table 6.1), and as $f^* = (L/0.41D)c_o/2L$ for the first axisymmetric mode. Consequently, if $L > 0.853D$ (or $L > 0.41D$, if only axisymmetric waves are excited), the higher-order modes will not intervene in the frequency range of the first lobe of the transmission loss characteristics of the chamber based on one-dimensional approximation. For this reason, we may consider a circular chamber of length $L = 0.853D$ (or $L = 0.41D$, if only axisymmetric waves are excited) to be the strongest one-dimensional pure chamber because its first lobe covers the largest frequency range without intervention of the higher-order modes.

The effect of the chamber length on the frequency range of the first lobe of transmission loss characteristics of a circular axisymmetric chamber (inlet and outlet ducts centrally in line) is shown in Figure 5.14. These characteristics are computed by using the three-dimensional elements described in the next two chapters of this book.

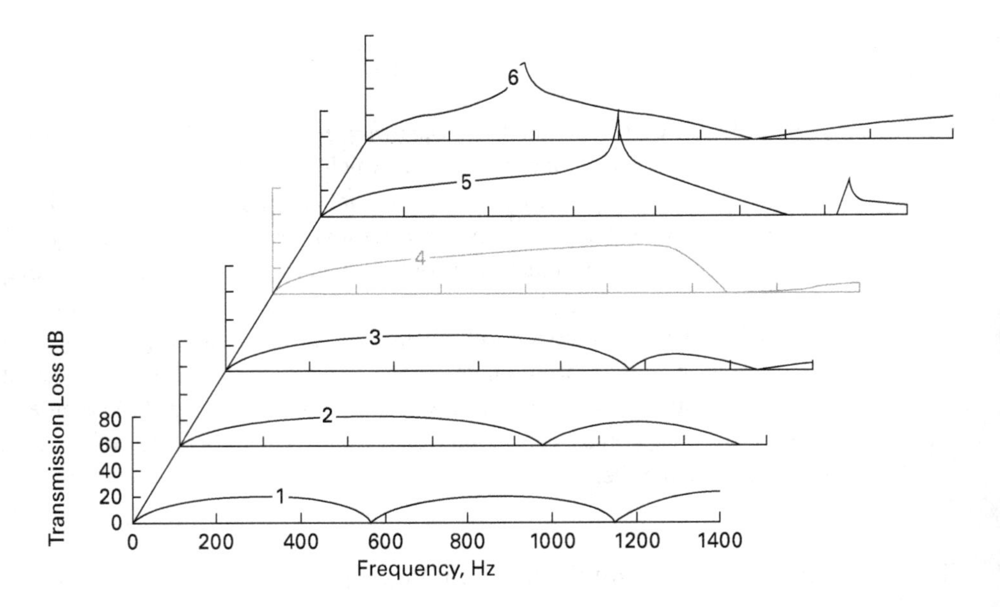

Figure 5.14 Transmission loss characteristic of an axisymmetric circular expansion chamber with inlet and outlet ducts of diameter 57 mm and $c_o = 343.7$ m/s. Curve 1: $c_o/2L = 572.9$ Hz, $L/D = 1.162$; curve 2: $c_o/2L = 829.3$ Hz, $L/D = 0.632$; curve 3: $c_o/2L = 954.8$ Hz, $L/D = 0.540$; curve 4: $c_o/2L = 1074$ Hz, $L/D = 0.453$; curve 5: $c_o/2L = 1228$ Hz, $L/D = 0.370$; curve 6: $c_o/2L = 1719$ Hz, $L/D = 0.224$.

It is seen that as the length of the chamber is reduced, the first transmission-loss lobe is predicted fairly accurately by the one-dimensional theory, up to about the strongest chamber length ($0.41D$, since the chamber is axisymmetric). However, the periodic multi-lobe character of the transmission loss characteristics of the one-dimensional chamber displayed in Figure 5.10a is destroyed by three-dimensional effects for about $L/D < 1$, and the symmetry of the first lobe becomes distorted gradually.

5.4.5 Tuning Inlet and Outlet Duct Extensions

Figure 5.15a shows the computed transmission loss characteristics of a pure through-flow expansion chamber. (See the figure caption for the salient chamber parameters.)

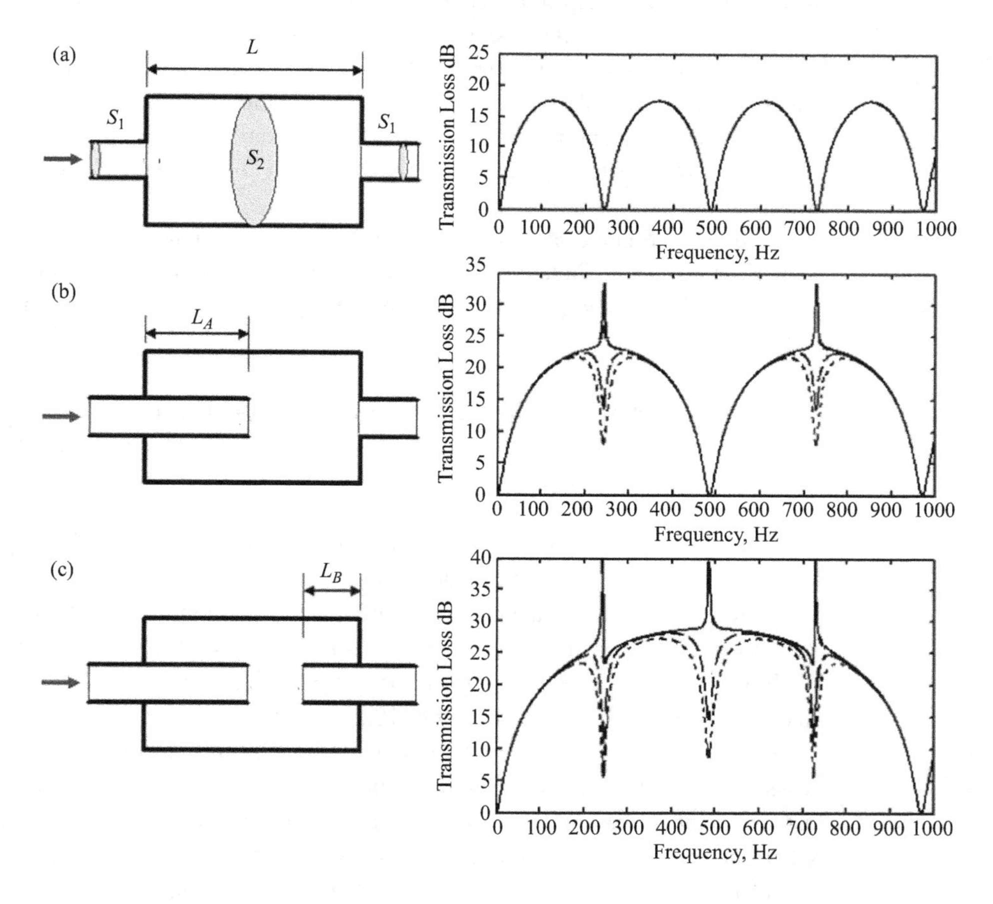

Figure 5.15 Transmission loss characteristics of expansion chambers, inlet and outlet cross-sectional area (csa) = 31.67 cm^2, chamber csa = 471.35 cm^2, L = 71 cm, L_A = 35.5 cm, L_B = 17.75 cm, c_o = 344.9 m/s, ρ_o = 1.1935 kg/m^3; solid $\bar{M}_o = 0$, long dash $\bar{M}_o = 0.1$, short dash $\bar{M}_o = 0.2$, where $\bar{M}_o$ is Mach number of the average mean flow velocity in the inlet and outlet ducts. (a) Pure chamber, (b) inlet side-branch resonance tuned to the first critical frequency of the pure chamber and (c) additionally, outlet side-branch resonance tuned to the second critical frequencies of the pure chamber.

This figure also includes the transmission loss characteristics of the same chamber with mean flow, which is neglected in Equation (5.89), but these are barely discernible from the zero mean flow characteristic, implying that the effect is negligible.

If we wish to improve the transmission loss of this expansion chamber in the vicinity of its first non-zero minimum at frequency $f_1 = c_o/2L$, we may consider adding a resonator tuned to this frequency. There are various ways of doing this, but the simplest is to extend the inlet or the outlet duct into the chamber by a suitable length. From Equation (5.27), we find that the resonance frequencies of this resonator should be $nc_o/4\ell$, $n = 1, 3, 5, \ldots$, where ℓ denotes the length of the resonator (possibly including an end-correction). So, we can match the critical chamber frequency $f_1 = c_o/2L$ by extending either the inlet or the outlet duct into the chamber by a length of $\ell = L/2$. Implementation of this resonator on the inlet side is shown in Figure 5.15b. It is seen that the effect of tuning is manifested by sharp transmission loss peaks, and that this resonator also tunes to the resonance frequencies of the chamber at odd multiples of $f_1 = c_o/2L$. However, the transmission loss characteristics computed by taking into account the effect of a uniform mean flow show that the effect of tuning is strongly sensitive to mean flow. This behavior is also observable in tests.

Of course, we could also see the effect of using an extended inlet or outlet side branch directly from Equation (5.89). For example, if an inlet side branch is used, $\zeta_5 = 1/\mathrm{i}\tan(k_oL_A)$ and $\zeta_6 = \infty$, since the end-caps A and B are rigid. Then Equation (5.89) reduces to

$$TL = 20\log_{10}\left|\cos(k_oL) + \frac{1}{2}\left(1 - \frac{1}{m}\right)\tan(k_oL_A)\sin(k_oL) + \frac{i}{2}\left(-\left(m + \frac{1}{m}\right)\sin(k_oL) + (m-1)\tan(k_oL_A)\cos(k_oL)\right)\right|, \tag{5.92}$$

which at once indicates the chamber anti-resonances at $\cos(k_oL_A) = 0$, or $f_n = nc_o/4L_A$, $n = 1, 3, 5, \ldots$, confirming the above observations.

Figure 5.15c shows the application of the same idea to improve the performance of chamber in the vicinity of its second minimum transmission loss frequency $f_2 = 2c_o/2L$. In this case, the needed length of the outlet side branch is $\ell = L/4$, which also tunes to transmission loss minima at 6, 10,... multiples of $f_1 = c_o/2L$.

Tuning of the inlet and/or outlet duct extensions also improves the overall transmission loss of the chamber. Compared in Figure 5.16 are the overall transmission-loss of chambers having only an inlet duct extension tuned and both the inlet and outlet duct extensions tuned as described above. Also shown for comparison is the overall transmission loss characteristic of the pure expansion chamber. The dotted horizontal lines correspond to the overall transmission loss in the frequency range of the first lobe for each case. Since tuning one of the duct extensions merges two lobes of the pure expansion chamber, according to the rule of decibel addition, an improvement of overall transmission loss in the first lobe frequency range of about 3 dB can be expected when the first (non-zero) transmission loss minimum is tuned and another 3 dB when the second one is tuned. This is confirmed by Figure 5.16.

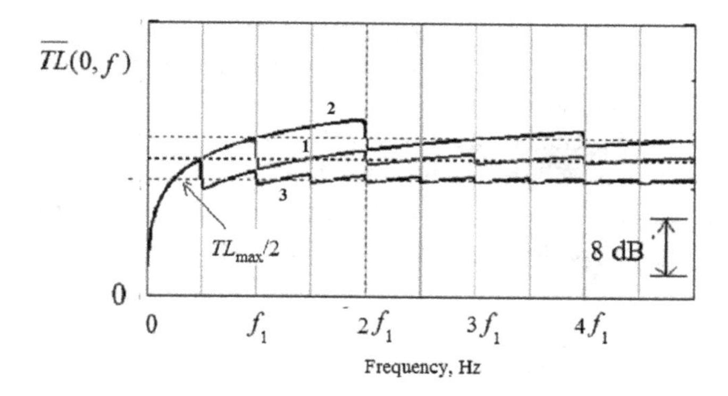

Figure 5.16 Overall transmission loss of expansion chambers with tuned inlet and outlet ducts. Curve 1: Only inlet duct extension tuned to $f_1 = c_o/2L$. Curve 2: Inlet and outlet duct extensions tuned to $f_1 = c_o/2L$ and $f_1 = c_o/L$, respectively. Curve 3: Pure expansion chamber (as in Figure 5.12).

The inlet and outlet duct extensions (L_A and L_B) of the other chamber configurations in Figure 5.11 can be tuned similarly by starting from the transmission loss characteristics of the respective chamber with no physical inlet and outlet extensions. For example, Figure 5.16 also applies for the overall transmission-loss characteristics of a tuned flow-reversing chamber of the same area ratio.

5.4.6 Effect of Inlet and Outlet Duct Configurations

The expansion chambers shown in Figure 5.11 differ essentially by the arrangement of the inlet and outlet ducts relative to the chamber. As a matter of fact, insofar as it can be predicted to the accuracy of Equation (5.89), these chambers have the same transmission-loss characteristics as the through-flow expansion chamber, if they have the same values for the equivalent physical parameters. However, the use of a specific configuration may be dictated by the available space and mounting conditions of particular applications. It should, therefore, be kept in mind that the acoustic performance of different configurations can be different, even though they may appear to be equivalent on the basis of transmission loss predicted by Equation (5.89). Some factors which may have to be considered are:

- Effects of mean flow on transmission loss can be different for through-flow and flow-reversing type of area changes.
- The accuracy of the assumed mean flow model in acoustic elements can be different for different chambers.
- Higher-order mode effects in different chamber configurations will also be different (recall that, the end-corrections required for one-dimensional area-change elements are dependent on the environment of the discontinuity).

On the other hand, a one-dimensional model works only for the specific propagation path it is planned for. When the axes of the inlet and outlet ducts are not aligned, it is usually possible to conceive two one-dimensional wave paths for a given chamber configuration. For example, the flow-reversing chamber in Figure 5.11c can also be conceived as a side-inlet and side-outlet chamber (Figure 5.11d). This model may be more appropriate if the lateral distance between the axes of the inlet and outlet ducts is substantially larger than the length of the chamber. This configuration can be more difficult to analyze because an axially uniform circular or oval chamber is a non-uniform one for waves traveling in the transverse direction.

5.4.7 Division of a Pure Expansion Chamber

Vibration and strength considerations in design of expansion chambers may necessitate stiffening of chamber walls and rigid supports for duct extensions. Such stiffening and support are usually provided by using baffle plates.

A baffle plate that fully closes the chamber section partitions it into two sub-chambers. Effects of partitioning on transmission-loss characteristics of the unbaffled chamber depend on how the sub-chambers are connected. Commonly, adjacent partitions are connected by using a duct mounted on a solid baffle (Figure 5.17) or, if the baffle is perforated, through its apertures. A thin perforated baffle, or a very short connecting duct, is transparent to sound waves at relatively low frequencies and does not modify the undivided transmission-loss characteristics of the chamber substantially. This can be seen from the wave transfer matrix of a pure through-flow expansion chamber, which is, from Equation (5.82),

$$T_{\text{io}}(k_oL, m) = \begin{bmatrix} \cos(k_oL) - \frac{\text{i}}{2}\left(m + \frac{1}{m}\right)\sin(k_oL) & -\frac{\text{i}}{2}\left(m - \frac{1}{m}\right)\sin(k_oL) \\ \frac{\text{i}}{2}\left(m - \frac{1}{m}\right)\sin(k_oL) & \cos(k_oL) + \frac{\text{i}}{2}\left(m + \frac{1}{m}\right)\sin(k_oL) \end{bmatrix} \tag{5.93}$$

A feature of this matrix is that

$$T_{\text{io}}\left((k_oL)_1, m\right)T_{\text{io}}\left((k_oL)_2, m\right) = T_{\text{io}}\left((k_oL)_1 + (k_oL)_2, m\right) \tag{5.93a}$$

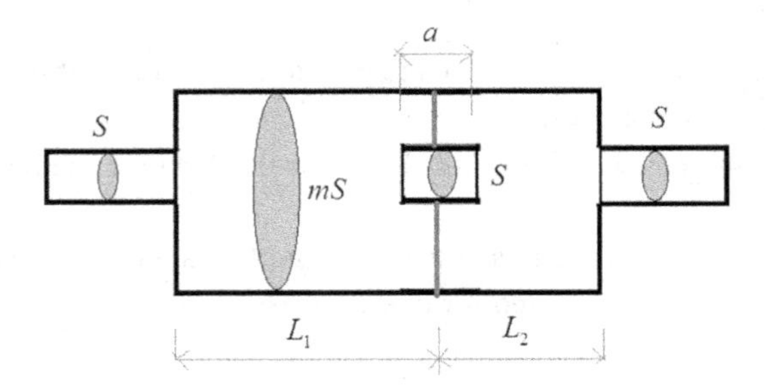

Figure 5.17 Division of a pure expansion chamber by a baffle into two chambers.

Therefore, if N uniform chambers of the same area ratio are connected by infinitesimally thin diaphragms ($a \to 0$ in Figure 5.17), the resulting compound chamber is equivalent to a single chamber of $c_o L = (c_o L)_1 + (c_o L)_2 + \cdots + (c_o L)_N$ or, if the speed of sound, c_o, is the same for all chambers, of length $L = L_1 + L_2 + \cdots + L_N$. Thus, partitioning a pure expansion chamber by thin baffles into N chambers of equal area ratio has no effect on the transmission loss characteristics of the undivided chamber.

Coupling by a duct of finite length (a in Figure5.17) impresses the characteristic frequencies ($f_1 = c_o/2L_i$) of the sub-chambers created by the division, as well as the resonance frequencies of the connecting ducts and the anti-resonances due to the side branches they create. Since the acoustic fields in the sub-chambers and connecting ducts are all coupled with each other, these impressions occur close to but not exactly at the corresponding stand-alone frequencies of the components.

The lengths of sub-chambers can be selected in many ways. Common strategies are to use equal-length sub-chambers or to tune the maximum transmission loss frequency of one sub-chamber to the minimum transmission loss frequency of the adjacent sub-chamber. In the latter case, the required lengths of the sub-chambers can be calculated as follows: Let ℓ_i, $i = 1, 2, \ldots, N$, denote the length of the ith sub-chamber. The minimum and the maximum transmission loss frequencies of the ith sub-chamber are $c_i/2\ell_i$ and $c_i/4\ell_i$, respectively, where c_i denotes the speed of sound in sub-chamber i. Since the strategy requires that $c_i/2\ell_i = c_{i+1}/4\ell_{i+1}$, $i = 1, 2, \ldots, N-1$, the length of the ith sub-chamber can be expressed as

$$\ell_i = \frac{L}{\sum_{j=1}^{N} \left(\frac{1}{2^{j-i}} \frac{c_j}{c_i} \right)}, \qquad i = 1, 2, \ldots, N \tag{5.94}$$

where $L = \ell_1 + \ell_2 + \cdots + \ell_N$. For example, for three sub-chambers having the same speed of sound, $\ell_1 = 4L/7$, $\ell_2 = 2L/7$ and $\ell_3 = L/7$. In general, the shortest sub-chamber is given by $\ell_N = Lc_N / \sum_{j=1}^{j=N} 2^{N-j} c_j$, or $\ell_N = L/\left(2^N - 1\right)$ if the speed of sound is constant. This should not be less that the strongest chamber length; for example, if the main chamber is circular of diameter D, we must have $\ell_N > 0.853D$ (Section 5.4.4). This limit also determines the maximum number of sub-chambers, if they are chosen to be of equal length.

Once the lengths of the sub-chambers are determined, the lengths of the connecting ducts can be selected by tuning the side branches they create as described in Section 5.4.5. For example, Figure 5.18 shows an application involving the division of the through-flow chamber in Figure 5.15a into two and three sub-chambers.

In the case of division into two sub-chambers (Figure 5.18a), if the sub-chamber lengths are selected according to Equation (5.94), the minimum transmission-loss frequencies of the sub-chamber are given by integer multiples of 364 Hz and 729 Hz. Since the inlet and outlet duct extensions are tuned to odd multiples of these frequencies, we observe in Figure 5.18a the anti-resonances close to these frequencies. In this case, the resonance frequency of the connecting duct happens to be 729 Hz and is also annihilated by side-branch tuning. However, this effect is strongly wiped out by the presence of mean flow.

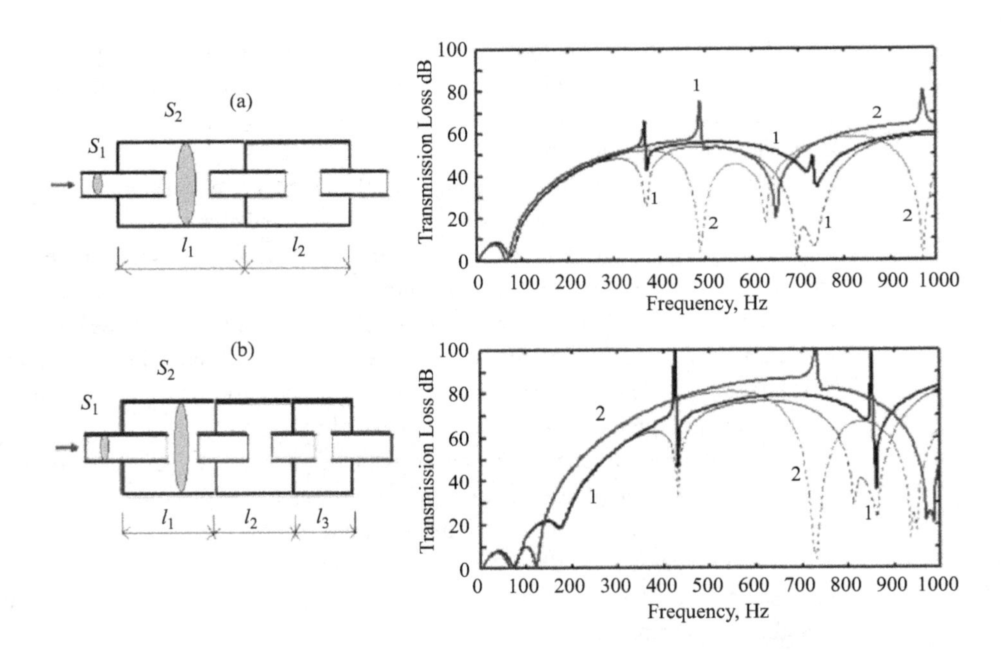

Figure 5.18 Transmission loss characteristics of the chamber in Figure 5.15a after it is partitioned into (a) two sub-chambers, (b) three sub-chambers. Curve 1: Sub-chamber lengths are given by Equation (5.94). Curve 2: Sub-chamber lengths are equal. Solid: $\bar{M}_o = 0$. Dotted: $\bar{M}_o = 0.2$. Sub-chamber inlet side branches $= \ell_i/2$, outlet side branches $= \ell_i/4$, $i = 1, 2, 3$. (See caption of Figure 5.15 for the dimensions of the main chamber.)

If the sub-chambers have equal length, the minimum TL of the chambers now will occur at 486 Hz and 972 Hz, which are manifested in Figure 5.18a as anti-resonances due to side-branch tuning. However, the connecting duct resonance at 648 Hz now stands out. Again, the presence of mean flow drastically modifies the characteristics in the vicinity of the tuning frequencies.

Ideally, baffle plates should be tightly fitted to the chamber walls and ducts. However, random leaks are unavoidable with common manufacturing methods. For this reason, the open area through which the sub-chambers on two sides of a baffle communicate should be much larger than the expected leak, so that the contribution of the acoustic effects of random leaks is rendered indiscernible in the deterministic acoustic model of the chamber.

5.4.8 Low-Frequency Response of Chambers

The low-frequency transmission-loss minimum of the divided chamber observed below 100 Hz in Figure 5.18a is not tractable directly to similar arguments. This is because this minimum is associated with the coupled response of the sub-chambers and the ducts at low frequencies. Recall from Section 3.5.4 that, at low frequencies, a uniform duct can be modeled by ideal springs and ideal masses. For example, at low frequencies, the two sub-chamber volumes in the divided chamber in Figure 5.18a may be modeled as ideal springs and the connecting duct as an ideal mass. From Section 3.5.4, the analogous stiffness coefficients of the sub-chambers are given by $k_1 = \rho_o c_o^2/\ell_1 S_2$ and $k_2 = \rho_o c_o^2/\ell_2 S_2$, and the analogous mass of the connecting duct by $m = \rho_o(\ell_1/4 + \ell_2/2)/S_1$. Since the springs are in parallel, their total stiffness coefficient is $K = k_1 + k_2$. Now, it is ubiquitous in books on dynamics that the resonance frequency of this kind of a one-degree-of-freedom spring + mass system is given by $f_o = (1/2\pi)\sqrt{K/m}$ (effect of damping being neglected). Therefore, the low-frequency resonance of the divided chamber in terms of the chamber parameters is:

$$f_o = \frac{c_o}{\pi}\sqrt{\frac{S_1}{\left(1 - \frac{\ell_2^2}{L^2}\right) S_2 L \ell_2}}. \tag{5.95}$$

If numerical values are substituted, we find $f_o = 74.6$ Hz for division according to Equation (5.94), and $f_o = 65.5$ Hz for division into sub-chambers of equal length. These values are quite close to the corresponding values observed in Figure 5.18a.

The transmission-loss characteristics for division into three sub-chambers (Figure 5.18b) display similar features. A notable feature is the occurrence of two low-frequency resonances. In this case, the three sub-chambers may be modeled by ideal springs and the two connecting ducts by ideal masses. This is a two degrees-of-freedom mass–spring system and, hence, we have two low-frequency resonances. These can be calculated by using the following formula, which is proved in books on mechanical vibrations:

$$f_o = \frac{1}{2\pi}\sqrt{\frac{1}{2}\left(\frac{(k_1+k_2)m_2+(k_2+k_3)m_1}{m_1 m_2}\right) \mp \frac{1}{2}\sqrt{\left(\frac{(k_1+k_2)m_2+(k_2+k_3)m_1}{m_1 m_2}\right)^2 - 4\left(\frac{(k_1+k_2)(k_2+k_3)-k_2^2}{m_1 m_2}\right)}} \tag{5.96}$$

where

$$k_1 = \frac{\rho_o c_o^2}{\ell_1 S_2}, \; k_2 = \frac{\rho_o c_o^2}{\ell_2 S_2}, \; k_3 = \frac{\rho_o c_o^2}{\ell_3 S_2} \tag{5.97}$$

$$m_1 = \left(\frac{\ell_1}{4} + \frac{\ell_2}{2}\right)\rho_o S_1, \; m_2 = \left(\frac{\ell_2}{4} + \frac{\ell_3}{2}\right)\rho_o S_1. \tag{5.98}$$

Upon substituting the numerical values of the parameters, we find $f_o = 73.3$ Hz and 177.6 Hz for division according to Equation (5.94), and $f_o = 69.4$ Hz and 120.2 Hz for division into sub-chambers of equal length. Again, these results are quite close to the corresponding values observed in Figure 5.18b.

5.4.9 Chambers with a Perforated Duct Bridge

The presence of mean flow can have adverse effects on transmission loss of a chamber that are not accounted for in linear acoustic elements. Apart from the mean flow

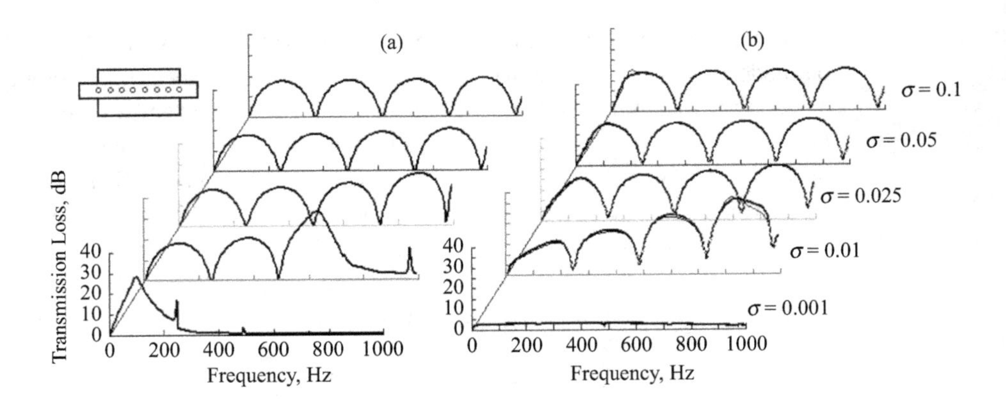

Figure 5.19 Effect of the perforate bridge on *TL* of the chamber of Figure 5.15a. (a) $\bar{M}_1 = \bar{M}_2 = 0$; (b)$\bar{M}_1 = 0,\ \ \bar{M}_2 = 0.1$. Acoustic model is based on the theory of Section 4.7 and the Cummings model for perforate impedance (Appendix C). Results with mean flow without slip ($\varphi = 1$, solid lines) and with slip ($\varphi = 0.5$, dashed lines) are superimposed, but are hardly distinguishable.

velocity field in the chamber being far more irregular than that assumed in the theory, the breaking down of instable vortex sheet at the flow expansion into the chamber is often of more concern, because the accompanying vortex generation can be synchronized to the acoustic field in the chamber, setting up strong self-sustained pressure oscillations which derive their energy from the mean flow. The contribution of these oscillations to the sound transmitted by the chamber can be measured [10–11], but it is difficult to include them in predictive calculations, since the underlying phenomena are non-linear (see Section 5.6.4).

A practical method for avoiding such a self-excited expansion chamber is to bridge the inlet and outlet ducts by a perforated duct. Figure 5.19 shows the transmission-loss characteristics of the expansion chamber in Figure 5.15a, when the inlet and outlet ducts are bridged by a perforated duct of wall-thickness 1.5 mm, hole diameter of 3 mm and various values of open area porosity, σ. The acoustic modeling of this configuration of an expansion chamber, which is usually referred to as the straight-through muffler, is described in Section 2.4.2. It is seen that, as the open area porosity increases, the transmission-loss characteristics tend to become similar to that of a simple through-flow expansion chamber.

This result indicates that a perforated duct becomes transparent to sound waves if the porosity is high enough. On this premise, perforated ducts are sometimes modeled by cutting off the perforated segments. If the side branches are effective in the frequency range of interest, this approach should be applied with caution. The effect of the side branches is illustrated in Figure 5.20, which is the counterpart of Figure 5.19 for the expansion chamber of Figure 5.15c. Increasing porosity further does not result in noteworthy changes in these characteristics. When the porosity is very low, the chamber acts as a resonator (Figure 5.20a), but its efficacy is counteracted by the presence of mean flow (Figure 5.20b).

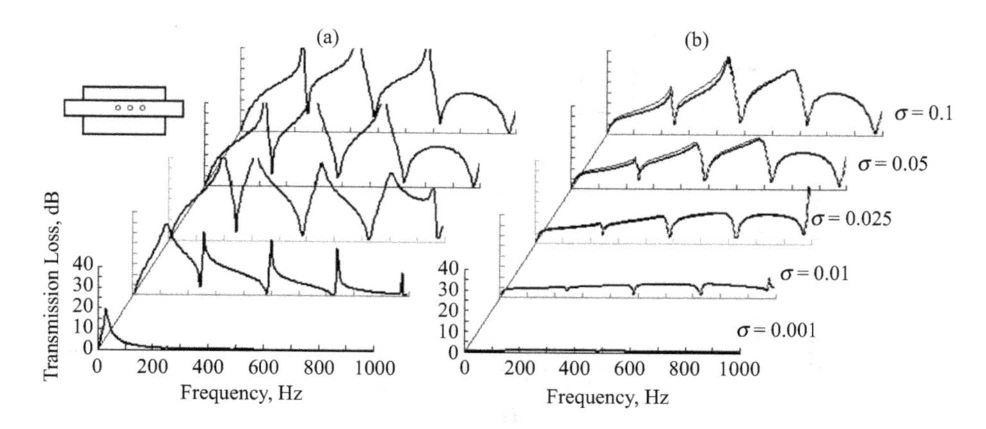

Figure 5.20 Effect of the perforate bridge on *TL* of the chamber of Figure 5.15c. (a) $\bar{M}_1 = \bar{M}_2 = 0$, (b) $\bar{M}_1 = 0$, $\bar{M}_2 = 0.1$: Solid lines: Mean flow without slip ($\varphi = 1$). Dotted lines: Mean flow with slip parameter of $\varphi = 0.5$. The acoustic model is based on the theory of Section 4.7 and the Cummings model for perforate impedance (Appendix C).

5.4.10 Packed Chambers

Walls of expansion chambers containing hot gases have to be treated with thermal insulation material in order to keep the temperature of the outer wall surface at safe values. The insulation material is protected from being drifted by the mean flow by using a perforated facesheet. Thermal insulation material such as glass wool is also good absorber of sound wave energy and can improve the transmission loss of the chamber. A further advantage of insulated walls is the reduction of the structure-borne sound radiation from the walls of the chamber. In general, however, for thicknesses required for thermal insulation of the walls, the hard-walled chamber transmission-loss lobes are not changed substantially at relatively low frequencies. But substantial changes can occur if the chamber volume is filled with fibrous material such as glass wool, rock wool or basalt. In fact, this is the principle of a versatile silencer configuration known as a straight-through muffler. This is essentially an expansion chamber having a perforate bridge, except that the chamber volume is packed with a fibrous material. In general, fibrous packing changes the acoustic character of the chamber, in that, the transmission loss resonances and anti-resonances of the unpacked chamber are annihilated to become an almost monotonous increasing function of frequency, as shown in Figure 5.21. Although the low-frequency performance may be reduced to some extent, the presence of packing is advantageous for making the system more robust in a wider frequency range and improving the high-frequency performance, which usually also improves the psychoacoustic quality of the transmitted sound.

5.5 Reciprocal Two-Ports

The explicit formulae given in this chapter suggest that the elements of the wave transfer matrices may be related by some general relations. This is most suggestive,

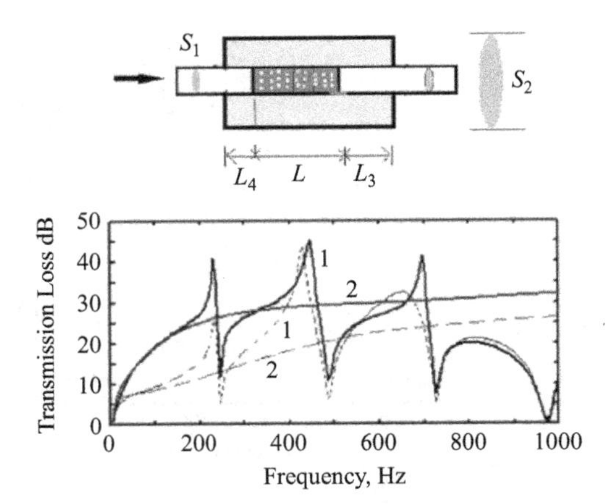

Figure 5.21 Curve 1: *TL* characteristics of the chamber in Figure 5.15c with the inlet and outlet duct bridged by a perforate of porosity 0.3, hole diameter of 3 mm and wall thickness of 1.5 mm, and the Cummings model for perforate impedance (see Appendix C). Curve 2: As for curve 1, the chamber is filled with glass wool E, at density 120 kg/m^3, with bulk acoustic properties as per the Kirby model (Appendix B). Solid: $\bar{M}_o = 0$. Dashed: $\bar{M}_o = 0.2$

for example, in Equation (5.58). To see the underlying principle, consider an acoustic two-port element with zero mean flow, but otherwise arbitrary. The pressure wave component vectors at the inlet and outlet ports of the element are $\mathbf{P}_i = \{p_i^+ \quad p_i^-\}$ and $\mathbf{P}_o = \{p_o^+ \quad p_o^-\}$, where the subscripts "*i*" and "*o*" refer to inlet and outlet ports, respectively. The pressure wave components are related by the wave transfer equation:

$$\begin{bmatrix} p_i^+ \\ p_i^- \end{bmatrix} = \begin{bmatrix} T_{11} & T_{12} \\ T_{21} & T_{22} \end{bmatrix} \begin{bmatrix} p_o^+ \\ p_o^- \end{bmatrix}, \tag{5.99}$$

where outlet-to-inlet causality is assumed. At a given frequency, the time-averaged acoustic power at the inlet port is given by Equation (3.203) and can be expressed as

$$\begin{aligned} \left(\frac{\rho_o c_o}{2S}\right)_i W_i = {} & \left(|T_{11}|^2 - |T_{21}|^2\right) |p_o^+|^2 + \left(|T_{12}|^2 - |T_{22}|^2\right) |p_o^2|^2 \\ & + \left(\breve{T}_{11} T_{12} - \breve{T}_{21} T_{22}\right) \breve{p}_o^+ p_o^- + \left(\breve{T}_{12} T_{11} - \breve{T}_{22} T_{21}\right) \breve{p}_o^- p_o^+ \end{aligned} \tag{5.100}$$

The time-averaged acoustic power at the outlet port is expressed, after reversing the causality of Equation (5.99), similarly as:

$$\begin{aligned} \left(\frac{\rho_o c_o}{2S}\right)_o W_o \Delta = {} & \left(|T_{22}|^2 - |T_{21}|^2\right) |p_i^+|^2 + \left(|T_{12}|^2 - |T_{11}|^2\right) |p_i^2|^2 \\ & + \left(\breve{T}_{22} \, T_{12} - \breve{T}_{21} \, T_{11}\right) \breve{p}_i^+ p_i^- + \left(\breve{T}_{12} \, T_{22} - \breve{T}_{11} \, T_{21}\right) \breve{p}_i^- p_i^+, \end{aligned} \tag{5.101}$$

where Δ denotes the determinant of the wave transfer matrix in Equation (5.99).

A two-port element is said to be reciprocal if interchanging the subscripts "i" and "o" in Equation (5.100) yields Equation (5.101), and vice versa. This interchange of subscripts can be interpreted physically as the reversal of the "source" and "receiver," which are envisaged to be located at the inlet and outlet ports, respectively.

Thus, if an acoustic two-port is reciprocal, the time-averaged power at the receiver should not change if the source and receiver positions are reversed. The conditions required for such reciprocity (when there is no mean flow) can be derived by comparing Equations (5.100) and (5.101), after the subscripts "i" and "o" are interchanged in one of these equations. These conditions are: (1) $\Delta = 1$, (2) $T_{11} = \breve{T}_{22}$ and (3) $T_{12} = \breve{T}_{21}$. For example, consider the wave transfer matrix in Equation (5.58); conditions 2 and 3 are obvious. Condition 1 is not so obvious at once; however, it can also be shown to be true after some algebra. This proves that this system is reciprocal, which is, of course, just a mathematical confirmation of what is intuitive from the geometry of the system in Figure 5.7b.

However, the situation is somewhat subtle for the wave transfer matrix in Equation (5.82). Inspection of the elements of this wave transfer matrix shows that, conditions 2 and 3 can be satisfied only if the side-branch interface impedances are either infinitely large or purely imaginary. Then condition 1 is also satisfied, but again, some algebra is required to verify this. A similar situation holds for the wave transfer matrix in Equation (5.74), whose determinant is unity, but is reciprocal only if the normalized hole impedance is purely imaginary.

When mean flow is present, the direction of the mean flow velocity relative to the "source" and "receiver" should remain the same when they are reversed. This means that, the mean flow velocity directions at the corresponding ports of the element should also be reversed. But this will change the mean flow directions in all internal components of the two-port and, hence, the terms of its wave transfer matrix. Apart from this, the above stated conditions for reciprocity remain valid. The manifestation of this flow reversal concept can be seen most easily in a uniform duct element carrying a uniform mean flow. In this case, conditions 2 and 3 can be expressed as $T_{11}(\bar{M}_o) = \breve{T}_{22}(-\bar{M}_o)$ and $T_{12}(\bar{M}_o) = \breve{T}_{21}(-\bar{M}_o)$, where $\bar{M}_o$ is the Mach number of the mean flow velocity in the duct. In general, however, the flow reversal principle cannot be depicted as conveniently as in this example, because the mean flow velocities are modified, as result of flow reversal, in different ways in different elements.

The reciprocity principle is both of theoretical and practical importance in duct acoustics and is extensively documented in the literature for ducts with or without mean flow [12–13].

5.6 Some Practical Issues

5.6.1 Irregular Geometry

The geometry of resonators and expansion chambers is often dictated by space requirements. When the space is limited, an irregular geometry may be the only

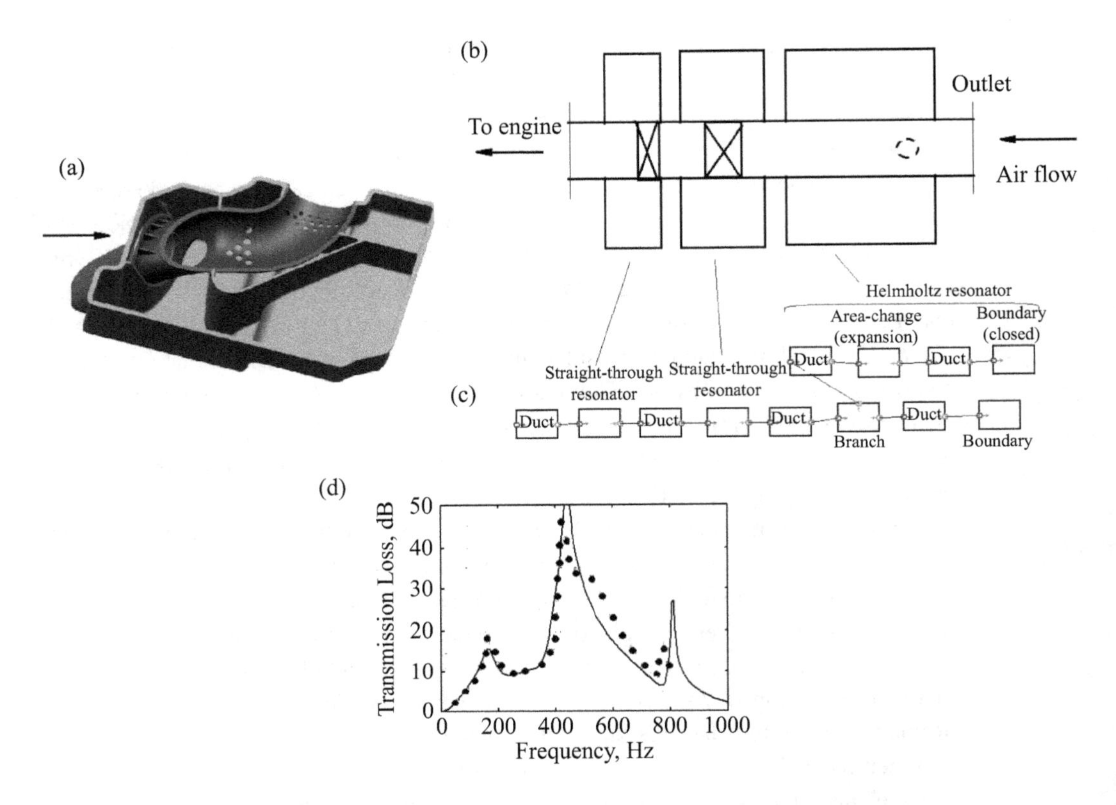

Figure 5.22 (a) Half-section view of an air duct prototype for engine intake of an automobile (courtesy of Mark IV, France). Duct diameter = 65 mm, overall dimensions are about 250 × 200 × 100 mm. (b, c) Simple model consisting of one Helmholtz resonator and two straight-through resonators in series (cavity volumes are about 4, 1.5 and 0.5 liters. (d) Transmission loss of the system at air flow rate of 0.072 kg/s at 40 °C; dotted line: measured (digitized from measured characteristics); solid line: computed by using one-dimensional acoustic elements.

option for the best use of the volume available. A typical situation occurs, for example, in the air intake system of passenger cars, which must fit into the space available under the hood. Figure 5.22a shows an air duct envisaged for noise control in the intake system of an automobile. The geometry is quite irregular and can be modeled with all details only by using a finite element model. Nevertheless, this air duct has uniform diameter and the three irregular cavities it communicates with can be represented approximately as in Figure 5.22b: one of the cavities is modeled as a Helmholtz resonator and the other two as straight-through resonators, all having assumed regular geometry with volumes equal to the actual volumes the respective irregular cavities. A block diagram of this model is shown in Figure 5.22c and the transmission loss of the air duct computed by using one-dimensional acoustic elements in this model is compared with the measured results in Figure 5.22d. It is seen that, in spite of the rather arbitrary representation of the cavities, such a simple one-dimensional model can yield adequately accurate results in the low frequency range, which is of primary interest in this application.

Another application where irregular geometry is typical is the suction silencer of a hermetic reciprocating compressor of the indirect suction type shown in Figure 5.2b. Since the space available in the shell cavity is very limited, these silencers tend to be of irregular geometry. Furthermore, the shell cavity, which is itself very irregular due to the presence of the compressor body, couples with the suction muffler acoustically and has to be included in the acoustic model of the suction muffler (see Section 5.1.2).

In general, sound waves are not sensitive to geometric details of chambers and resonators which are small compared to the wavelength. This is often useful to keep in mind when modeling irregular chambers by one-dimensional elements, for it allows replacing the actual geometry by a regular one having the same volume and nominal cross-sectional area.

5.6.2 Variable Mean Flow Conditions

Automobile engines operate over a wide range of speeds and the mean temperature and the gas mass flow rate in the muffler varies with the engine speed. Consequently, it is usually of interest to know how the transmission loss of an intake or exhaust silencer varies with the engine speed. For example, consider the exhaust silencer in Figure 5.23a, which is commonly known as the three-pass muffler. It consists of a packed central chamber and two flow-reversing chambers, which are coupled by perforated baffles and a traversing solid walled duct, which is called resonance duct. A block diagram of the muffler is shown in Figure 5.23b. The central chamber can be modeled by using a continuous (or row-wise discrete) 3-duct element with singly coupled perforated ducts and packed annulus. In the particular application considered here, perforates are of 1 mm wall thickness, 3.5 mm hole diameter and 20% porosity. Baffles are of 1 mm thickness, perforated with 4 mm holes to a porosity of 1.5%. The middle chamber is packed with rockwool to the density of 120 kg/m^3. The temperature in the muffler may be assumed to be approximately uniform, but varies with the engine speed as in Figure 5.24 in this particular application. Also shown in this figure is the variation of the mean exhaust gas mass flow rate with the engine speed. These data may be used to determine the speed of sound and the mean flow Mach number in the silencer in the speed range of the engine.

The transmission loss (*TL*) characteristics of the muffler at a given engine speed is calculated as usual. The computed *TL* spectra for all engine speeds is displayed conveniently on the N vs. f plane, where N is the engine rotational speed and f is the frequency, color intensity being used to show the *TL* levels in dB. Figure 5.25 shows this display for the *TL* of the three-pass muffler in Figure 5.23a for the mean flow data of Figure 5.24. A feature this type of display is that it enables us to distinguish the levels controlled by the rotational speed of steady running fluid machinery, since these levels occur at frequencies which are integer multiples of the cycle frequency. Hence, if they are dominant, we should observe high color intensity along lines passing through the origin with slopes $60Y/n$ rpm/Hz, $n = 1, 2, \ldots$, where Y denotes the number of rotations per work cycle. Figure 5.25 confirms that transmission loss is not controlled by the engine rotational speed. This is to be expected, because transmission loss is not

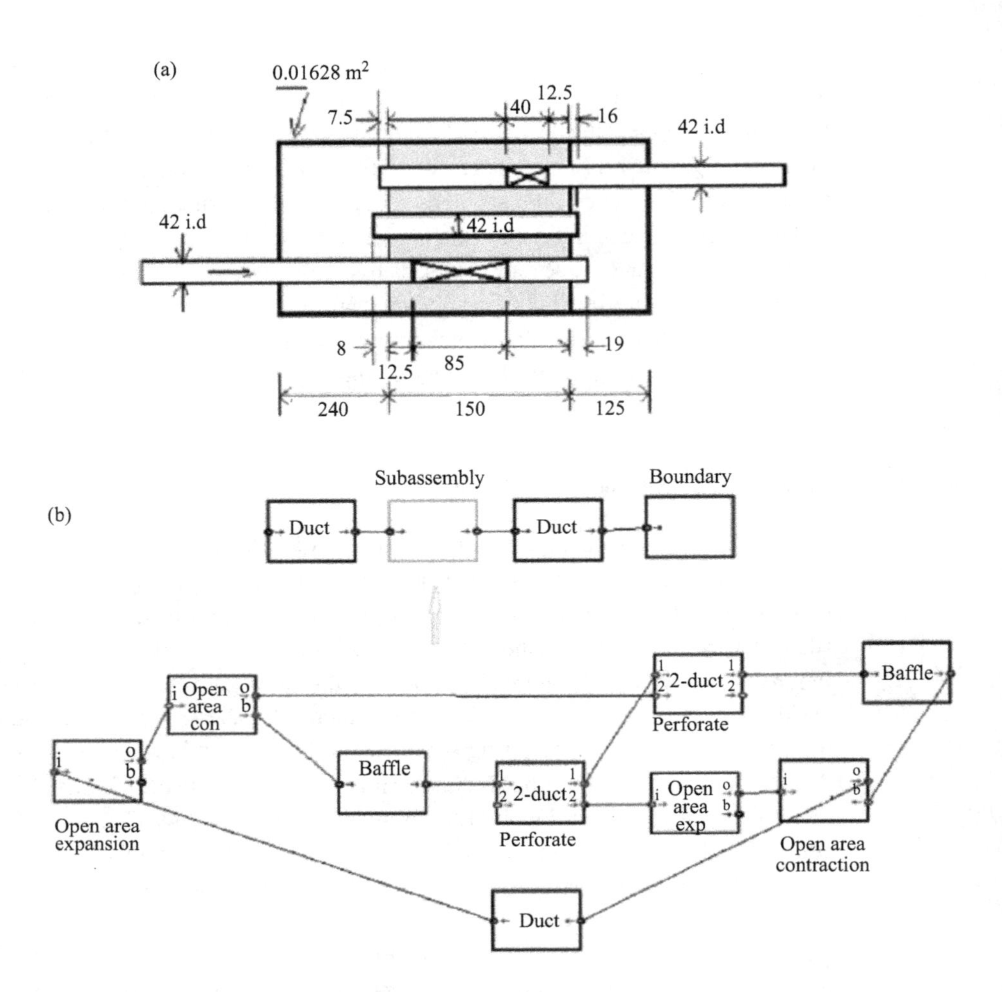

Figure 5.23 A contemporary automobile exhaust muffler and its block diagram model.

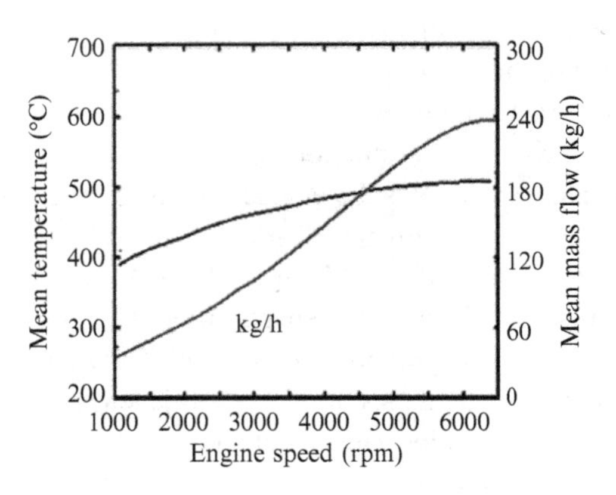

Figure 5.24 Variation of mean temperature in the muffler (Figure 5.23a) and the exhaust gas mass flow rate as function of engine speed.

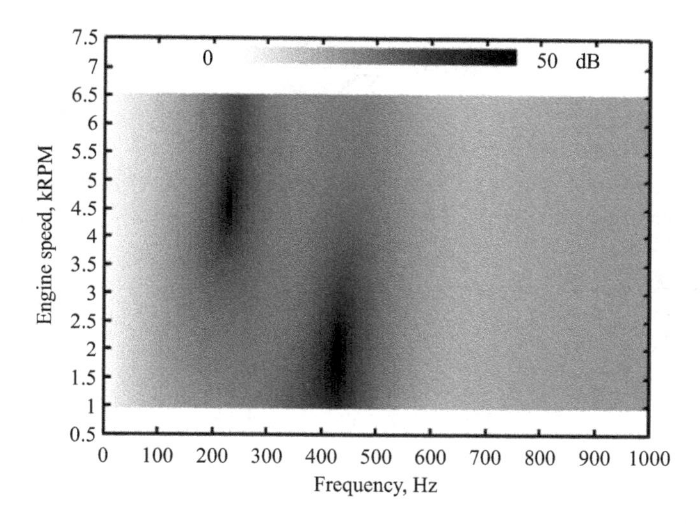

Figure 5.25 Transmission loss spectra of the muffler in Figure 5.23a, for the mean flow data in Figure 5.24.

dependent on the sound source and, hence, the engine rotational speed.[2] Consequently, high-intensity regions in Figure 5.25 are observed as approximately constant frequency lines which correspond to the transmission loss maxima. The slight inclination of these high-intensity regions is due to the effect of the mean flow.

In multi-cylinder engines with appropriately designed manifolds, the acoustic design of the muffler usually reduces to obtaining adequate attenuation at frequencies corresponding to a few multiples of the rotational speed. For example, for the muffler under consideration, the engine is a four-stroke four-cylinder in-line one and the spectrum of the exhaust noise at a steady rotational speed is dominated by discrete harmonic frequencies at integer multiples of $2N$, the remaining harmonics being practically annihilated by interference in the exhaust manifold (see Section 10.3). Then, it suffices to examine the transmission loss of the muffler at engine speeds corresponding to these major harmonics. These characteristics can be obtained by taking cuts of Figure 5.25 along inclined lines with slopes $120/n$ rpm/Hz, $n = 2, 4, 6, \ldots$, which are denoted as H2, H4, H6,. . ., respectively. Figure 5.26 shows the transmission loss of the muffler as function of the engine speed for the first four major harmonics.

5.6.3 Multiple Outlet Ducts

Transmission loss of systems having multiple outlets may be calculated by modifying the definition given in Equation (5.2) as:

[2] This assumes that the source is one-dimensional (see Section 10.5).

$$TL = 10\log_{10}\left(\frac{W_{\text{in}}^{+}}{\sum\left(W_{\text{out}}^{+}\right)_{r_{\text{out}}=0}}\right), \tag{5.102}$$

where the summation is over all outlets.

Figure 5.27 shows a sketch of an exhaust muffler of a helicopter powered by a horizontally opposed six-cylinder four-stroke reciprocating engine [14]. The volume

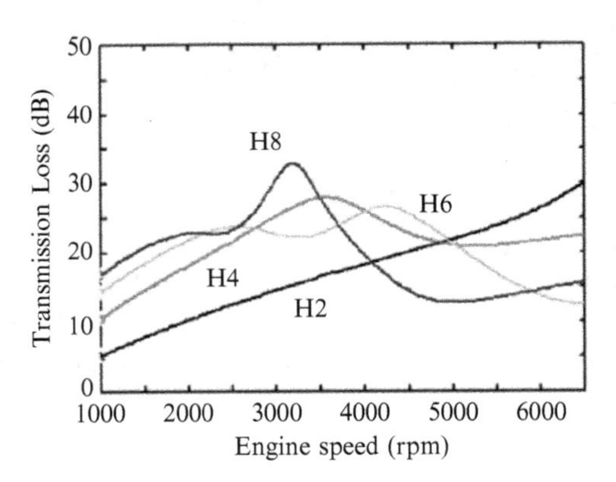

Figure 5.26 Variation of the transmission loss of the muffler in Figure 5.23a with the engine speed for the first four major harmonics.

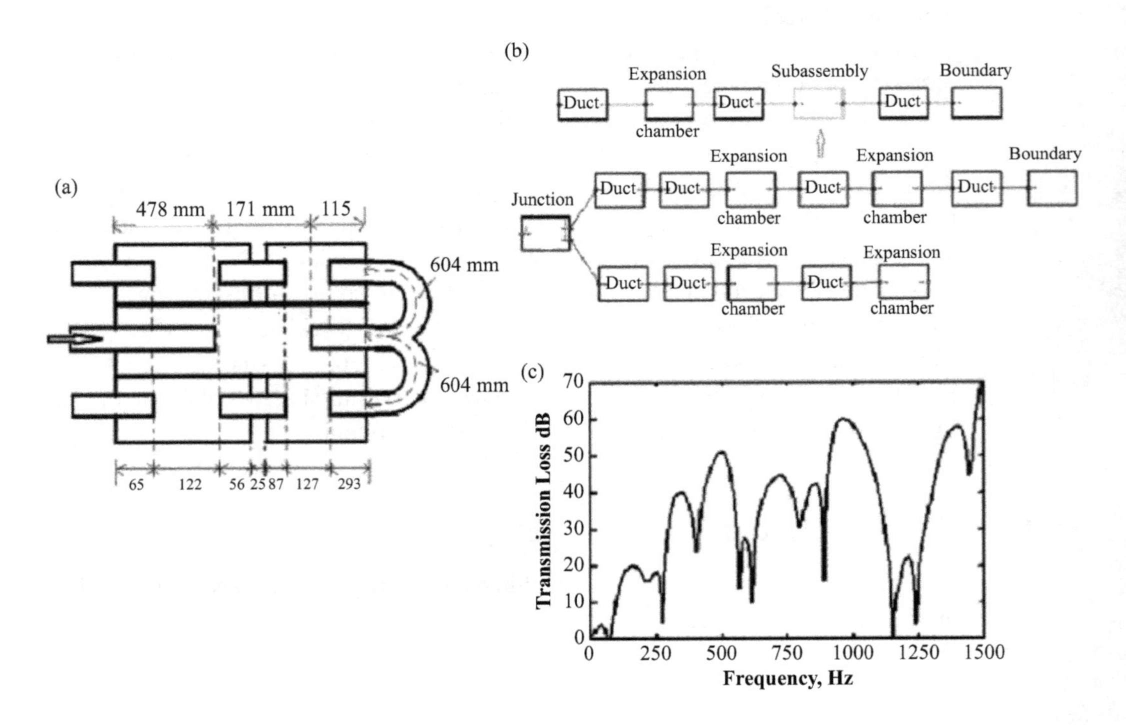

Figure 5.27 Dual tailpipe exhaust muffler of a helicopter [14]. (a) Sketch of the exhaust muffler. (b) Block diagram model of the muffler. (c) Predicted transmission loss spectrum.

of the muffler is about 28 liters. The exhaust gas flow from the two-cylinder banks of the engine joins and flows into a single duct before it enters the inlet chamber, but is then divided into two identical paths, each containing two chambers and one tailpipe. All chambers have the same cross-sectional area and all ducts are of 50.8 mm diameter. A block diagram of the muffler is shown in Figure 5.27b. The transmission loss characteristics shown in Figure 5.27c are computed for steady operation at which the mean gas flow velocity at the tailpipes is about 60 m/s, corresponding to a Mach number of 0.1. The mean temperature is assumed to be uniform in the silencer volume.

In some multi-cylinder engines, the exhaust manifold does not connect to a single exhaust duct before entering the silencer (as in the foregoing example), but divides into two exhaust ducts (typically, from the two banks of a V engine), which then enter the silencer separately. Transmission loss of such silencers with two inlets can be calculated by applying the principle of superposition. To see the procedure, consider a silencer with multiple inlet ducts and a single outlet duct. Transmission loss of the silencer is given by

$$TL = 10\log_{10}\left(\frac{1}{W_{\text{out}}^+}\sum_{j=1}^{N} W_{\text{in}j}^+\right)_{r_{\text{out}}=0}, \tag{5.102a}$$

where the summation is over all inlets. To determine this, it is convenient to calculate first the partial transmission losses

$$TL_k = 10\log_{10}\left(\frac{W_{\text{in}k}^+}{W_{\text{out},k}^+}\right)_{r_{\text{out}}=0}, \quad k = 1, 2, \ldots, N \tag{5.102b}$$

where $W_{\text{out},k}^+$ denotes the time-averaged acoustic power at the outlet due to only the incident sound power at inlet k, $W_{\text{in}k}^+$. Using Equation (3.203), this may be expressed as

$$TL_k = 10\log_{10}\left(\frac{\dfrac{2S_{\text{in}k}}{(\rho_o c_o)_{\text{in}k}}\left(1+(\bar{M}_o)_{\text{in}k}\right)^2\left|p_{\text{in}k}^+\right|^2}{\dfrac{2S_{\text{out}}}{(\rho_o c_o)_{\text{out}}}\left(1+(\bar{M}_o)_{\text{out}}\right)^2\left|p_{\text{out},k}^+\right|^2}\right)_{r_{\text{out}}=0}, \tag{5.102c}$$

where $p_{\text{in}k}^+ \neq 0$, but $p_{\text{in}j}^+ = 0, j \in 1, 2, \ldots, k-1, k+1, \ldots, N$; that is, all inlets except inlet k are assumed to be anechoic (the positive direction of the inlet duct axes being assumed to be outwards from the silencer). Thus, partial transmission losses can be computed as usual for single-inlet-single-outlet silencers. But by the principle of superposition, $p_{\text{out}}^+ = p_{\text{out},1}^+ + p_{\text{out},2}^+ + \cdots + p_{\text{out},N}^+$. Hence, substituting from Equation (3.50) for the incident time-averaged powers and using Equation (5.102c), the transmission loss of the silencer may be expressed as

$$TL = 10\log_{10}\left(\frac{\sum_{j=1}^{N} 10^{\frac{TL_j}{10}}\left|\dfrac{\sigma_1}{\sigma_j}\dfrac{p_{\text{in}j}^+}{p_{\text{in}1}^+}\right|^2}{\left|\sum_{j=1}^{N}\dfrac{\sigma_1}{\sigma_j}\dfrac{p_{\text{in}j}^+}{p_{\text{in}1}^+}\right|^2}\right)_{r_{\text{out}}=0}, \tag{5.102d}$$

where the transfer functions $\sigma_j = p^+_{\text{in}j}/p^+_{\text{out},j}$ are computed as in partial transmission loss calculations. This shows that the transmission loss of silencers with multiple inlets can be computed, provided that the relative magnitudes and phases of incident pressure wave components at the inlets are given. These parameters, which are in general frequency dependent, are usually not readily available at the design stage, as they are determined by the acoustic source driving the system. In the absence of any information, Equation (5.102d) is usually implemented as a first approximation, by assuming that the incident pressure wave components at the inlets are identical, that is, $p^+_{\text{in}j} = p^+_{\text{in}1}$.

A popular variation of two main exhaust ducts is known as a dual exhaust system. This is obtained by connecting the two exhaust ducts to two separate silencers in H or X configurations. In the H type, the two main exhaust ducts are joined by a bypass duct before they enter the silencers. In the X type, the two exhaust ducts are first joined and then bifurcated to the silencers. Transmission loss of a dual exhaust system may be computed by modeling it as a silencer with two inlets (ducts upstream of H or X joint) and two outlets (tailpipes of the two silencers). In this case, generalizing for M inlet ducts, Equation (5.102a) applies with $W^+_{\text{out}} = W^+_{\text{out1}} + W^+_{\text{out2}} + \cdots + W^+_{\text{out}M}$, where the number subscripts denote the outlets, and the partial transmission losses are defined as

$$TL_k = 10\log_{10}\left(\frac{\dfrac{2S_{\text{in},k}}{(\rho_o c_o)_{\text{in},k}}\left(1+(\bar{M}_o)_{\text{in},j}\right)^2 \left|p^+_{\text{in},k}\right|^2}{\sum_{j=1}^{M}\left(\dfrac{2S_{\text{out}j}}{(\rho_o c_o)_{\text{out}j}}\left(1+(\bar{M}_o)_{\text{out}j}\right)^2 \left|p^+_{\text{out}j,k}\right|^2\right)}\right)_{r_{\text{out}j}=0}. \tag{5.102e}$$

This is computed as for a silencer with single inlet and multiple outlets (Figure 5.27), with all inlets except inlet k being assumed to be anechoic. In general, the transmission loss is not tractable to a simple expression, but, when the outlets are identical, it can be shown to be (assuming N outlets),

$$TL = \frac{\sum_{j=1}^{M}\left|\dfrac{\sigma_{1,\text{out}1}}{\sigma_{1,\text{out}j}}\right|^2 \left(\sum_{k=1}^{N} 10^{\frac{TL_k}{10}}\left|\dfrac{\sigma_{1,\text{out}1}}{\sigma_{k,\text{out}1}}\right|^2 \left|\dfrac{p^+_{\text{in}k}}{p^+_{\text{in}1}}\right|^2\right)}{\sum_{j=1}^{M}\left|\dfrac{\sigma_{1,\text{out}1}}{\sigma_{1,\text{out}j}}\right| \left|\sum_{k=1}^{N}\dfrac{\sigma_{1,\text{out}j}\, p^+_{\text{in}k}}{\sigma_{k,\text{out}j}\, p^+_{\text{in}1}}\right|^2}, \tag{5.102f}$$

where the transfer functions $\sigma_{k,\text{out}\,j} = p^+_{\text{out}j,k}/p^+_{\text{in}k}$ are again computed along with the partial transmission loss calculations.

It can be seen that using multiple inlets not only complicates transmission loss calculations substantially, but also introduces some uncertainty to the results – in view of the information required about the relative magnitudes and phases of the incident pressure wave components. In general, single-inlet-single-outlet or single-inlet-multiple-outlet silencers can be as effective as silencers with multiple inlets. For this reason, unless the equivalent source characteristics at the silencer inlets are known with accuracy or there are ample opportunities for prototype testing, consideration of a multiple inlet silencer may not be a good design strategy.

5.6.4 Flow Excited Resonators and Chambers

An important effect, which is necessarily neglected in our analysis of resonators and chambers due to its non-linear nature, is the coupling of acoustic wave motion with periodic flow disturbance resulting from shear layer instabilities, as a result of which energy may be extracted from the flow to sustain large amplitude acoustic oscillations.

In general, whenever a flow leaves a sharp downstream edge, it separates and forms a thin shear layer (also called a vortex sheet). This causes very unstable transverse fluctuations in streamwise flow velocity and quickly develops waves, which grow rapidly, until the sheet rolls up to form a train of vortices (flow separation may also occur at upstream facing corners, but the flow may re-attach before an ordered set of vortices forms). Periodically generated vortices are not themselves strong radiators of sound, but they can excite resonators and chambers strongly into self-excited oscillations.

In a branch resonator (Figure 5.3a) or a Helmholtz resonator (Figure 5.5a and also Figure 5.9), for example, the phenomenon may occur when the mean flow in the main duct separates at the upstream edge of the resonator neck, generating a train of vortices. Flow visualizations show that each of the discrete vortices induces an inflow to the resonator ahead of itself and an outflow from the resonator behind itself [15]. The net inflow or outflow becomes maximum when the width of the resonator neck is about an odd multiple of the half of the spacing between discrete vortices [16]. Consequently, if the vortices generated at the upstream edge are assumed to convect on a straight line with velocity U_c, the resonator will be subject to strongest periodic flow excitation at the neck at frequencies $f \approx (n - 1/2)U_c/L$, where $n = 1, 2, \ldots$ and L denotes the width of the neck and the convection velocity of vortices shed at the upstream edge is usually estimated by the empirical relationship $U_c \approx 0.4U_o$ [17], where U_o denotes the free-stream velocity of the mean flow. If this frequency corresponds to one of its resonance frequencies, a resonator may become a significant noise source and may manifest as a whistle, which can be subjectively disturbing in practical applications.

The literature on self-excited resonators is extensive and encompasses experimental, analytical and numerical simulation studies, which also contain further references on the subject [15–19]. A review of the previous work on suppressing self-excited oscillations in resonators is given by Knotts and Selamet [20], who also present detailed experimental results on the effect of sharp, beveled, ramped and radiused edges. It appears that, for low Mach numbers, the latter three types of edges can be effective in reducing amplitudes of self-excited oscillations in the sharp-edge case, provided that they are appropriately dimensioned relative to the size and configuration of the side branch.

Expansion chambers, too, possess the potential to become a generator rather than attenuator of sound waves by a mechanism of self-excitation [10]. Davies gives a description of the vortex structure associated with the flow separation at a circular sudden area expansion [16]. Instability of the generated vortex sheet degenerates into a series of convecting large-scale vortices, with a convection velocity at about half of the mean flow velocity in the inlet duct, and smaller vortex structures. These smaller vortex structures combine to form the turbulent mixing region, which eventually attaches with the duct walls, but in the case of a standard muffler, its length would be too small for a fully developed pipe flow to be established. When the incident sound field from the inlet duct impinges on the vortex structures, for certain frequencies, the sound field changes the dominant frequency of the generation of vortices and the vortices are shed at the acoustic excitation frequency. Davies argues that although these locked vortices are not strong generators of sound, the traveling irrotational field induces extra pressure disturbances [16]. Finding values of its amplitude and attenuation coefficient from experiment, he shows good agreement with the measured wall pressures along the outlet duct and suggests that the transmission loss of an expansion chamber is determined by the sum of the contributions of the linear acoustic field and the irrotational field of traveling vortices. The effect of the latter rapidly dies away as the distance from the sudden area expansion junction exceeds about $\ell = 5d$, where d denotes the diameter of the inlet duct. Hence, in expansion chambers, the main parameter controlling sound generation by self-excited flow-acoustic coupling seems to be, apart from the mean flow Mach number, the relative gap length x/d, where x denotes the distance between the sudden area expansion and the sudden area contraction planes. Davies demonstrates this experimentally on a series of expansion chambers[16]. Based on the test results, it is suggested that the intensity of the flow noise generated is proportional to $(x/d)^{-4.25}$ in the range $4 \leq x/d \leq 12$ and to $M^{4.5}$, where $M < 0.3$ denotes the Mach number in the inlet duct [10], and that the traveling vortices have potential to excite the tailpipe resonances, perhaps with a mechanism similar to that in a flute [21], if they reach the outlet duct junction. It is also argued that increasing the level of excitation moves the distance ℓ towards the separation point without changing the maximum strength of the irrotational field of traveling vortices [16]. Thus, the flow-acoustic coupling tends to saturate above a certain level of excitation. Davies' results also verify the bridging of inlet and outlet ducts by a perforated duct (see Section 5.4.9) as a practical method for avoiding self-excited expansion chambers [16].

5.7 Flow Rate and Back-Pressure Calculation

In this section, we consider the problem of calculation of mean mass flow rates and pressure drops in silencers. These calculations are required prior to frequency domain acoustic performance calculations in order to provide the mean flow data required by acoustic elements. They are also required for another important reason, namely, to make sure that a silencer to be mounted on the intake or exhaust ductwork of a fluid machine does not introduce additional back pressure that causes the total to exceed a permissible value required for efficient operation of the machine.

For subsonic low Mach number flows, the lumped parameter hydraulic network approach is commonly used for the estimation of flow rates [22]. In this approach, mean flow is assumed to be of constant density and ductwork is considered to be made up of some basic two-port elements in the same way as we would do if we were to construct the one-dimensional acoustic model. But, these elements are now associated with a lumped parameter resistance model of the form $\Delta p_o = f(\dot{Q}_o)$, where Δp_o denotes the mean pressure drop across the element and $\dot{Q}_o$ denotes the volume flow rate, on the understanding that $\dot{Q}_o$ and Δp_o have the same sign. The function $f(\dot{Q}_o)$ can be determined by measurement or taken from the results published in the literature. The latter are usually presented for specific ductwork components as experimental correlations for the flow loss coefficient $K = (\Delta p_s)/\left(\rho_o v_o^2/2\right)$, where Δp_s denotes the stagnation pressure drop, ρ_o denotes the mean density and v_o denotes the mean flow velocity on which K is based. Accordingly, the static pressure drop can be expressed as $\Delta p_o = \left(K + (\Delta v_o^2)/v_o^2\right)\rho_o v_o^2/2$, where $(\Delta v_o^2)\rho_o/2$ represents the dynamic pressure drop. Hence, noting that $\dot{Q}_o = v_o A$ is conserved, where A denotes the cross-sectional area associated with v_o, the lumped parameter resistance model is characterized by $f(\dot{Q}_o) = \left(K + (\Delta A^2)/A^2\right)\rho_o \dot{Q}_o^2/2A^2$. Since the flow loss coefficient is normally not dependent on the volume flow rate, this implies a lumped parameter resistance model of the form $\Delta p_o = R\dot{Q}_o^2$, where $R > 0$ denotes a constant parameter. This is usually written as $\Delta p_o = R\dot{Q}_o|\dot{Q}_o|$ in order to impart sign causality to the model.

Thus, each element may be characterized mathematically by a lumped resistance parameter, $R_{1,2}$, as

$$p_{o1} - p_{o2} = R_{1,2}\dot{Q}_o|\dot{Q}_o|, \tag{5.103}$$

where the subscripts "1" and "2" denote the ports through which $\dot{Q}_o$ is maintained and it is implied by the sign convention that $p_{o1} > p_{o2}$ when $\dot{Q}_o > 0$. In this section, we use this model to describe essentials of flow rate and pressure drop calculations using lumped resistances. Some well-known correlations for K are given in Table 5.3 to aid the discussion. An extensive list of correlations is given in Fried and Idel'chik [23].

First consider an expansion chamber which is mounted on an exhaust line with inlet mean mass flow rate $\dot{m}_{o1} = \rho_o \dot{Q}_{o1}$ and the outlet pipe (tailpipe) open to the atmosphere (Figure 5.28a). The chamber is divided into elements as is usual for acoustic analysis, as depicted by the port numbers in Figure 5.28a. Ducts 1-2 and 5-6 are uniform and their loss coefficients are given in the first row of Table 5.3. Similarly, the loss coefficients of the sudden expansion 2-3 and sudden contraction 4-5 are determined from the last two rows of Table 5.3. For these elements, Equation (5.103) gives $p_1 - p_2 = R_{1,2}\dot{Q}_1|\dot{Q}_1|$, $p_2 - p_3 = R_{2,3}\dot{Q}_1|\dot{Q}_1|$, $p_4 - p_5 = R_{4,5}\dot{Q}_1|\dot{Q}_1|$, $p_5 - p_6 = R_{1,6}\dot{Q}_1|\dot{Q}_1|$ and $p_6 - p_7 = R_{6,7}\dot{Q}_1|\dot{Q}_1|$, where (and subsequently) the subscript "o" signifying mean flow is dropped for the simplicity of notation. Here, $R_{6,7}$ represents the flow resistance parameter of the open end. The corresponding loss coefficient may be approximated by that of a sudden expansion with a very large outlet cross-sectional area. The loss coefficient correlation at sudden expansion is based on the assumption of re-attachment of the separated flow. Since practical chamber lengths are not usually long enough for this, the region 3-4 is ambiguous and the losses associated with the complex flows in it are difficult to specify with certainty. For this reason, we neglect $R_{3,4}$ and assume that $p_3 = p_4$. Then, adding the foregoing resistance equations gives

Table 5.3 Flow loss coefficients of basic ductwork elements

Element	Loss coefficient (K)	Definitions
Uniform hard-walled duct turbulent mean flow [22]	$K = \dfrac{fL}{D_H}$ $f = \left(-1.8\log\left(\dfrac{6.9}{\mathrm{Re}} + \left(\dfrac{r}{3.7D_H}\right)^{1.11}\right)\right)^{-2}$	A = duct csa $D_H = \sqrt{4A/\pi}$ $\mathrm{Re} = \rho_o \dot{Q}_o D_H / A\mu_o$ L = duct length r = wall roughness height
Sharp-edged thin orifice	$K = \dfrac{1}{C_d^2}$	C_d = discharge coefficient (typically = 0.7)
Sudden expansion [22]	$K = \left(1 - \dfrac{S_1}{S_2}\right)^2$	S_1, S_2
Sudden contraction [22]	$K = \begin{cases} 0.42\left(1 - \dfrac{S_1}{S_2}\right) & \text{for } \dfrac{S_1}{S_2} \le \sqrt{0.76} \\ \left(1 - \dfrac{S_1}{S_2}\right)^2 & \text{for } 1 > \dfrac{S_1}{S_2} > \sqrt{0.76} \end{cases}$	Vena contracta, S_2, S_1

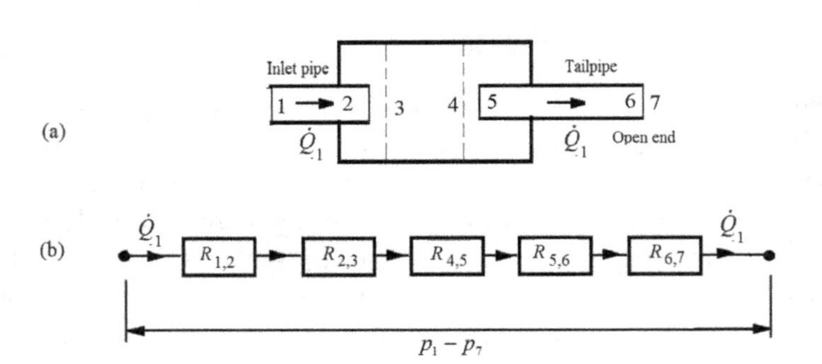

Figure 5.28 An expansion chamber with tailpipe (a) and its lumped resistance circuit model (b).

$p_1 - p_7 = R_{1,7}\dot{Q}_1^2$ for the total pressure drop between ports 1 and 6, where $R_{1,7} = R_{1,2} + R_{2,3} + R_{4,5} + R_{5,6} + R_{6,7}$ denotes the equivalent flow resistance parameter of path 1-6.

This result can be deduced directly if an electrical analogy is established for Equation (5.103) by taking the static pressure difference as analogous to voltage difference and the $\dot{Q}_o|\dot{Q}_o|$ as analogous to current. Then, the ductwork elements and their connectivity may be represented by resistance blocks in series as shown in the circuit diagram in Figure 5.28b. The latter yields the above equivalent resistance parameter immediately, as the resistance blocks are in series.

Actually, since the inlet duct and the tailpipe are normally essential components of an exhaust line, the pressure drop of the chamber proper may be stated to be $p_2 - p_5 = (R_{2,3} + R_{4,5})\dot{Q}_1^2$.

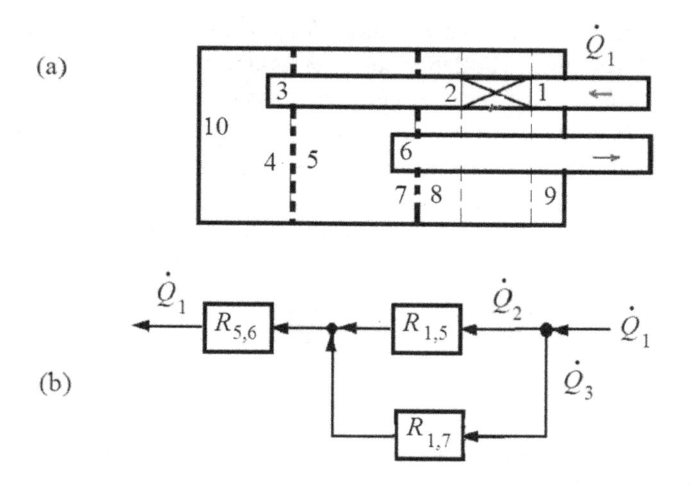

Figure 5.29 Mean flow circuit diagram of commercial automotive muffler.

Back pressure criterion of fluid machinery is important in the acoustic design of silencers, because better acoustic performance and smaller pressure loss targets pose conflicting design requirements. This trade-off can be shown relatively easily for a pure expansion chamber, the transmission loss of which is given by Equation (5.90), or Figure 5.10. The latter shows that the larger the area ratio, m, the larger is the transmission loss. But, as noted above, the pressure drop of the chamber proper is dominated by $R_{2,3}$ and $R_{4,5}$. As can be deduced from Table 5.3, the flow loss coefficients of the corresponding sudden area expansion and the sudden area contraction are proportional to $(1 - 1/m)$ or $(1 - 1/m)^2$ and, obviously, the requirement of small pressure drop is in conflict with the large m requirement for better transmission loss.

Next, consider the calculation of the pressure drop of the silencer shown in Figure 5.29a. This consists of three chambers coupled with perforated baffles and a perforated two-duct unit. Again, we have divided it into elements as we would do for acoustic analysis. Referring to the port numbers in Figure 5.29a, the elements consist of a perforated two-duct unit with ports 1-2-7-8, uniform duct 2-3, flow-reversing open expansion with ports 3-10-4, open contraction with ports 5-6-7 and two perforated baffles 4-5 and 7-8. Since the inlet mass flow rate $\dot{m}_1 = \rho_o \dot{Q}_1$ divides into two paths at the perforated duct, we have

$$\dot{Q}_1 = \dot{Q}_2 + \dot{Q}_3, \tag{5.104}$$

where $\dot{Q}_2$ denotes the grazing flow rate in the perforated duct and $\dot{Q}_3$ denotes the through (bias) flow rate through the perforate holes. Grazing flow resistance, $R_{1,2}$, is usually approximated by using a rough wall correlation (first row, Table 5.3). Correlations for the through-flow resistance, $R_{1,8}$, say, should be obtained empirically or semi-empirically. An estimate may be obtained by assuming that the total through-flow rate is uniformly distributed to the perforations. Then $R_{1,8}$ is given by the flow resistance of single perforation and can be calculated by using a thin orifice correlation (second row, Table 5.3). Applying Equation (5.103), we get $p_1 - p_2 = R_{1,2}\dot{Q}_2|\dot{Q}_2|$ and $p_1 - p_8 = R_{1,8}\dot{Q}_3|\dot{Q}_3|$. The latter equation assumes that $p_1 \approx p_2$ and that the mean pressure in annulus of the rightmost chamber is approximately uniform and equal to p_8. (See Payri et al. for including the effect of pressure gradients [24].) The flow resistance of the perforated baffles, $R_{4,5}$ and $R_{8,7}$, may be estimated similarly, if an experimental correlation is not available.

Following the grazing flow path in the perforated duct, resistance of duct 2-3, $R_{2,3}$, resistance of the sudden expansion, $R_{3,4}$, and resistance of baffle 4-5, $R_{4,5}$, are in series with $R_{1,2}$. Hence, we represent them by the equivalent resistance parameter $R_{1,5} = R_{1,2} + R_{2,3} + R_{3,4} + R_{4,5}$.. Then,

$$p_1 - p_5 = R_{1,5}\dot{Q}_2|\dot{Q}_2|. \tag{5.105}$$

Similarly, since in the through-flow path $\dot{Q}_3$, the resistance of baffle 7-8, $R_{8,7}$ is in series with $R_{1,8}$, we have

$$p_1 - p_7 = R_{1,7}\dot{Q}_3|\dot{Q}_3|, \tag{5.106}$$

where $R_{1,7} = R_{1,6} + R_{8,7}$. Again, we assume that the mean pressure in the middle chamber is approximately uniform. Then, $p_7 \approx p_5$ and from Equations (5.105) and (5.106), it follows that

$$R_{1,5}\dot{Q}_2|\dot{Q}_2| = R_{1,7}\dot{Q}_3|\dot{Q}_3|. \tag{5.107}$$

Since both flow rates are positive, the solution of Equations (5.104) and (5.107) gives

$$\dot{Q}_2 = \frac{\dot{Q}_1\sqrt{R_{1,7}}}{\sqrt{R_{1,7}} + \sqrt{R_{1,5}}} \tag{5.108}$$

$$\dot{Q}_3 = \frac{\dot{Q}_1\sqrt{R_{1,5}}}{\sqrt{R_{1,7}} + \sqrt{R_{1,5}}}. \tag{5.109}$$

Thus, the pressure drop of the silencer is

$$p_1 - p_6 = \left(R_{5,6} + \left(\frac{\sqrt{R_{1,5}R_{1,7}}}{\sqrt{R_{1,7}} + \sqrt{R_{1.5}}}\right)^2\right)\dot{Q}_1^2, \tag{5.110}$$

where, clearly, the squared brackets represent an equivalent resistance parameter, R_{eq}, say, referred to $\dot{Q}_1$ of the two parallel branches in Figure 5.29b. So

$$\frac{1}{\sqrt{R_{\text{eq}}}} = \frac{1}{\sqrt{R_{1,5}}} + \frac{1}{\sqrt{R_{1,7}}}. \tag{5.111}$$

The foregoing considerations are summarized by the lumped resistance circuit shown in Figure 5.29b. Such circuitry is characterized mathematically by the Kirchhoff laws for electrical circuits or, what is the same, by the node and loop equations [6]. In this case, the latter are represented by, respectively, Equations (5.104) and (5.107).

In general, silencer configurations are analyzed more conveniently by drawing the lumped resistance circuitry first and then invoking the circuitry laws. For example,

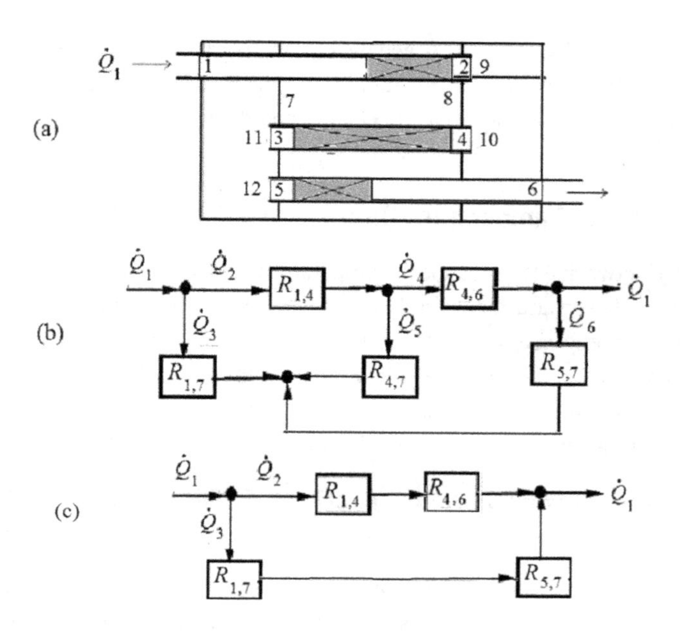

Figure 5.30 Lumped resistance model of a three-pass muffler.

consider the muffler shown in Figure 5.30a. This is known as a three-pass muffler and, as indicated in the figure, its acoustic model can be constructed as an assembly of a perforated four-duct unit with ports 1-2-3-4-5-6-7-8, two sudden area expansions, 2-9 and 3-11, and two sudden area contractions 4-10 and 12-5. The lumped resistance circuit of the muffler is shown in Figure 5.30b. It is constructed by assuming that the mean pressures in the chambers are uniform and all frictional losses in the ducts are neglected for simplicity. Then $R_{1,4} = R_{2,9} + R_{10,4}$ and $R_{3,6} = R_{3,11} + R_{12,5}$. Hence, we can write the following node and loop equations:

$$\dot{Q}_1 = \dot{Q}_2 + \dot{Q}_3 \tag{5.112}$$

$$\dot{Q}_2 = \dot{Q}_4 + \dot{Q}_5 \tag{5.113}$$

$$\dot{Q}_4 = \dot{Q}_6 + \dot{Q}_1 \tag{5.114}$$

$$R_{1,4}\dot{Q}_2|\dot{Q}_2| + R_{4,7}\dot{Q}_5|\dot{Q}_5| - R_{1,7}\dot{Q}_3|\dot{Q}_3| = 0 \tag{5.115}$$

$$R_{4,6}\dot{Q}_4|\dot{Q}_4| + R_{5,7}\dot{Q}_6|\dot{Q}_6| - R_{4,7}\dot{Q}_5|\dot{Q}_5| = 0. \tag{5.116}$$

Possible solutions for the flow rates can be found in terms of $\dot{Q}_1$ by solving these equations numerically. The pressure drop can then be calculated from

$$p_1 - p_6 = R_{1,4}\dot{Q}_2|\dot{Q}_2| + R_{4,6}\dot{Q}_4|\dot{Q}_4|. \tag{5.117}$$

For algebraic solutions, it is necessary to make trial assumptions about the directions of perforate through flows. A combination which is tractable to an explicit solution is $\dot{Q}_1, \dot{Q}_2, \dot{Q}_3, \dot{Q}_4 > 0$, $\dot{Q}_5 = 0$ and $\dot{Q}_6 < 0$. In this case, the node equations yield

$\dot{Q}_4 = \dot{Q}_2, \dot{Q}_3 = \dot{Q}_1 - \dot{Q}_2, |\dot{Q}_6| = \dot{Q}_3$ and the circuit diagram degenerates to that shown in Figure 5.30c. This consists of two parallel paths the results of the previous example (Figure 5.29) may be invoked to calculate $\dot{Q}_2$, $\dot{Q}_3$ and the pressure drop $p_1 - p_6$.

5.7.1 Calculation of Mean Temperature Drop

In hot exhaust ductworks, the mean flow temperature may vary considerably along the system, however, these variations mainly occur in relatively long connection ducts and can be estimated by engineering heat transfer considerations. We proceed by equating the rate of heat loss of the gas in an elementary segment of the duct of length δx to the heat transferred to the walls of the duct and also to the heat transferred from the walls of the duct to the exterior medium, as depicted mathematically by the following equation:

$$-\dot{m}_o c_{p_o} \delta T_g = h_{g,w}\left(T_g - T_w\right)P\delta x = h_{w,e}(T_w - T_e)P\delta x + \varepsilon_e \sigma\left(T_w^4 - T_e^4\right)P\delta x, \tag{5.118}$$

where T denotes the absolute temperature, the subscripts "g," "w" and "e" denote the gas, the duct walls and the exterior respectively, c_{po} denotes the specific heat coefficient at constant pressure, P denotes the perimeter of the duct cross section, $h_{g,w}$ and $h_{w,e}$ denote the convective heat transfer coefficients from gas to the walls and from walls to the exterior, ε_e denotes the emissivity of the exterior surface of the duct and $\sigma = 5.67 \times 10^{-8}$ W/m^2 K^4 is the Stefan–Boltzmann constant. The radiation heat transfer from walls to the gas and the conduction heat transfer in the walls are neglected in Equation (5.118) for the simplicity of presentation. The heat transfer coefficients are taken from experimental correlations published in the literature. For example, the Chilton–Colburn analogy may be used for $h_{g,w}$[25], giving

$$\frac{h_{g,w}d}{\kappa_o} = \frac{1}{2}\psi \mathrm{Re} \times \mathrm{Pr}^{\frac{1}{3}}, \tag{5.119}$$

and the Churchill–Chu correlation for $h_{w,e}$ [25]:

$$\sqrt{\frac{h_{w,e}d}{\kappa_o}} = 0.6 + 0.387\mathrm{Gr}^{\frac{1}{6}} \times \mathrm{Pr}^{\frac{1}{6}}\left(1 + \left(\frac{0.56}{\mathrm{Pr}}\right)^{\frac{9}{16}}\right)^{-\frac{8}{27}}. \tag{5.120}$$

Here, $\mathrm{Re} = \rho_o d v_o/\mu_o$ is the Reynolds number, $\mathrm{Pr} = \mu_o c_{po}/\kappa_o$ is the Prandtl number, $\mathrm{Gr} = g\beta_o(T_w - T_e)d^3/(\mu_o/\rho_o)^2$ is the Grashof number, d denotes the hydraulic diameter of the duct, v_o denotes the mean flow velocity, κ_o denotes the thermal conductivity of the gas, β_o denotes the coefficient of thermal expansion, g denotes acceleration due to gravity and ψ denotes the Prandtl friction factor which is solved from $1/\sqrt{\psi} = 4\log\left(2\mathrm{Re}\sqrt{\psi}\right) - 1.6$.

Hence, from the first and second equalities in Equation (5.118) we obtain, respectively,

$$\frac{\mathrm{d}T_g}{\mathrm{d}x} = f_1\left(T_g, T_w\right) = -\frac{h_{g,w}P}{\dot{m}_o c_{p_o}}\left(T_g - T_w\right) \tag{5.121}$$

$$T_w = f_2(T_g, T_w) = \frac{h_{g,w}T_g + h_{w,e}T_e + \varepsilon_e\sigma\left(T_e^4 - T_w^4\right)}{h_{w,e} + h_{g,w}}. \tag{5.122}$$

These equations can be solved by using a numerical method, such as the Newton–Raphson method, to calculate the axial distribution of the gas temperature $T_g(x)$ in a duct, given the initial temperature $T_g(0)$. For example, solutions predict that, in a duct of diameter 50 mm, the temperature of dry air entering the duct at 900 °C drops to 730 °C in 1 m, if the mass flow rate is 500 kg/hr and $T_e = 300$ K. This result is not affected substantially by radiative heat transfer and wall conduction.

5.8 Shell Noise

If the casing of a silencer is flexible enough, the internal acoustic pressure field can excite it into small vibrations, which manifest as sound radiation to the exterior environment. This is usually called shell noise (also break-out noise) and can be a source of annoyance if it affects noise sensitive areas, in addition to the potential of the original mechanical vibrations causing structural damage by fatigue failure. It is usually more perceptible at low frequencies and with casings made of thin sheet metal. Shell noise of vehicle exhaust silencers may be transmitted to the passenger compartment and from heating, ventilation and air conditioning ducts to offices and other spaces. Substantial shell noise may also occur from downstream ducts of control valves, where high-pressure drops occur under sonic or near sonic conditions, even though ducts walls are thicker than sheet metal ducts.

The prediction of shell noise is complicated because it depends on how the casing is supported and excited into vibration. It may be caused by local in-duct evanescent waves generated at discontinuities, the mean flow momentum changes at bends and discontinuities or due to turbulence, or by inappropriate support conditions. However, the normal source mechanism is the forced vibrations of the casing under internal acoustic pressure loading and the resulting shell noise can be predicted to some extent. Casing vibrations may also be affected by the exterior acoustic pressure loading, but this effect is generally discernible in underwater applications and can be neglected in terrestrial conditions. A general numerical solution of the problem may be obtained by coupling a finite element model of the internal acoustic field with a finite element model of the elastodynamics of the casing, which may then be used in a boundary element model of the exterior acoustic field to predict the radiated sound power. Such computation of intensive models and testing may be necessary for final conclusions. However, understanding of the physics of shell noise is important in preliminary design studies and extensive work has been carried out on simpler models, usually uniform ducts, to understand the roles of the physical parameters involved.

At low frequencies, the internal acoustic field in a duct may be modeled by using the continuity and momentum equations, Equations (3.17) and (3.19), of the one-dimensional theory described in Chapter 3. We will review these equations, because when the duct walls are elastic, they are not, in general, applicable as described there for ducts with finite or infinite impedance walls. For a uniform homogeneous duct with a subsonic mean flow, Equations (3.17) and (3.19) reduce to, respectively,

$$\frac{\partial \overline{\rho'}}{\partial t} + \overline{v_o}\frac{\partial \overline{\rho'}}{\partial x} + \rho_o \frac{\partial \overline{v'}}{\partial x} = -\frac{\rho_o}{S}\oint_{\ell} u' \mathrm{d}\ell \tag{5.123}$$

$$\rho_o\left(\frac{\partial \overline{v'}}{\partial t} + \overline{v_o}\frac{\partial \overline{v'}}{\partial x}\right) + \frac{\partial \overline{p'}}{\partial x} = \frac{\overline{v_o}\rho_o\varphi_{SF}}{S}\oint_{\ell} u' \mathrm{d}\ell, \tag{5.124}$$

where we have put $\rho_o = \overline{\rho_o}$ for the simplicity of notation. Substituting $\overline{\rho'} = \overline{p'}/c_o^2$, these equations may be expressed in frequency domain as

$$-\mathrm{i}k_o\overline{p'} + \bar{M}_o\frac{\partial \overline{p'}}{\partial x} + \rho_o\frac{\partial \overline{v'}}{\partial x} = -\frac{\rho_o c_o}{S}\oint_{\ell} u' \mathrm{d}\ell \tag{5.125}$$

$$\rho_o c_o\left(-\mathrm{i}k_o\overline{v'} + \bar{M}_o\frac{\partial \overline{v'}}{\partial x}\right) + \frac{\partial \overline{p'}}{\partial x} = \frac{\bar{M}_o\rho_o c_o\varphi_{\mathrm{SF}}}{S}\oint_{\ell} u' \mathrm{d}\ell. \tag{5.126}$$

Now, we have to introduce a mathematical model of the duct structure and a thin cylindrical shell theory, or the thin plate theory for rectangular ducts, is commonly used for this purpose. Small elastic displacements of thin shells are defined by the displacements of points on its midplane from their unperturbed positions. Thus, u' in the foregoing equations is determined by the first time-derivative of the radial component of the midplane displacement, w, say, of a plate or shell modeling the duct walls under the forcing of $\overline{p'}$. Then, for example, if we adopt the Ingard–Myers boundary condition at the duct walls, we may write, in the frequency domain, $u' = -\mathrm{i}\omega w + \overline{v_o}\partial w/\partial x$, or, alternatively, if we assume the no-slip condition, then simply $u' = -\mathrm{i}\omega w$ (see Section 3.8.1). Hence, Equations (5.125) and (5.126) may be recast as

$$-\mathrm{i}k_o\overline{p'} + \bar{M}_o\frac{\partial \overline{p'}}{\partial x} + \rho_o c_o\frac{\partial \overline{v'}}{\partial x} = -\frac{\rho_o c_o^2}{S}\oint_{\ell}\left(-\mathrm{i}k_o w + \psi\bar{M}_o\frac{\partial w}{\partial x}\right)\mathrm{d}\ell \tag{5.127}$$

$$\rho_o c_o\left(-\mathrm{i}k_o\overline{v'} + \bar{M}_o\frac{\partial \overline{v'}}{\partial x}\right) + \frac{\partial \overline{p'}}{\partial x} = \frac{\bar{M}_o\rho_o c_o^2\varphi_{\mathrm{SF}}}{S}\oint_{\ell}\left(-\mathrm{i}k_o w + \psi\bar{M}_o\frac{\partial w}{\partial x}\right)\mathrm{d}\ell, \tag{5.128}$$

where $\psi = 1$ for the Ingard–Myers condition and $\psi = 0$ for the no-slip condition. In order to close these equations, we have to determine the receptance, $w/\overline{p'}$, of the duct walls from the elastodynamics of the shell model. This part of the analysis depends on the cross-sectional shape of the duct and has to be carried out separately for different section geometry. Usually rectangular and circular ducts are considered, as they are relatively more tractable to elastodynamic analysis.

It is instructive, however, to begin with a very simple elastostatic model of a perfectly circular shell. As discussed in Section 3.9.3, this model, which is appropriate at low frequencies, gives the radial receptance of the shell as $w/\overline{p'} = a^2/hE$, where a denotes the midplane radius, h denotes the wall thickness and E denotes the modulus of elasticity of the duct material. To this approximation, Equations (5.127) and (5.128) reduce to the uniform duct case of Equations (3.129) and (3.130), respectively, with $\zeta_W = (hE/a^2)/(-\mathrm{i}\omega\rho_o c_o)$ and $\alpha = 1$. Thus, from Section 3.9.2, the acoustic pressure field in the duct consists of a superposition of the pressure wave components $p^{\mp} = A^{\mp}\mathrm{e}^{\mathrm{i}k_o K^{\mp} x}$, where $k_o = \omega/c_o$ an the axial propagation constants $K^{\mp}$ are given by Equation (3.150). It should be noted that these propagation constants now apply for acoustic waves both in the fluid and in the shell.

The shell noise may be calculated by using the results available in the literature for sound radiation from vibrating uniform cylindrical shells or by an equivalent acoustic line source with the same axial volume velocity distribution [26]. For this part of the analysis, we will borrow the results given in the literature. Using the vibrating uniform cylindrical shell model, Cummings et al. calculated the time-averaged far-field sound power radiated per unit surface area of a circular thin shell of length L, W_{rad}, say, due to an internal plane pressure wave component, p^+, traveling in the positive direction of the duct axis [27]. For a perfectly circular shell, this excites only the breathing mode of the shell (all points of the midplane vibrating radially in phase) and W_{rad} for this case may be expressed as

$$W_{\mathrm{rad}} = \frac{a}{2L}\rho_E c_E |\omega w|^2 F(k_E L). \tag{5.129}$$

Here, $k_E = \omega/c_E$, ρ_E and c_E denote, respectively, the density and speed of sound of the exterior medium and

$$F(k_E L) = -1 + \cos(2k_E L) + 2k_E L \int_0^{2k_E L} \frac{\sin y}{y}\mathrm{d}y. \tag{5.130}$$

The time-averaged acoustic power transmitted in the duct is given by $W^+ = 2S(1+\bar{M}_o)^2|p^+|^2/\rho_o c_o$, where S denotes cross-sectional area of the duct (Section 3.4). Therefore, the ratio of the radiated acoustic power to the in-duct specific (per unit area) acoustic power, that is, $\tau = (W^+/S)/W_{\mathrm{rad}}$, may be expressed as

$$\tau = \frac{h^2}{a^2}\frac{L}{a}\frac{\rho_S c_S}{\rho_E c_E}\frac{\rho_S c_S}{\rho_o c_o}\frac{4(1+\bar{M}_o)^2(1-\nu)^2}{k_S^2 a^2}\frac{1}{F(k_E L)}, \tag{5.131}$$

where $c_S^2 = E/\rho_S(1-\nu_S^2)$ is the square of the longitudinal speed of sound in the shell material, E is the modulus of elasticity, ρ_S the density and ν_S Poisson's ratio of the shell material. This quotient is called the transmission ratio or, if expressed in decibel (dB) units by the formula $10\log_{10}\tau$, the transmission loss, of the duct wall. Equation (5.131) predicts increase in transmission ratio proportional to the thickness to radius, length to radius and the characteristic impedance ratios. It also predicts that the transmission ratio is inversely proportional to the third power of the frequency, since outside the vicinity of $k_E L = 0$, the function $F(k_E L)$ is approximately proportional to $k_E L$. This implies that the transmission loss of the duct wall increases by about 9 dB per halving of the frequency and, consequently, can attain very high values at low frequencies. But, the measured transmission loss of cylindrical shells is known to be substantially lower than that predicted by Equation (5.131) at low frequencies. Such a discrepancy varies with individual ducts, because it may be caused by non-acoustic forces or by virtue of the duct cross sections not being perfectly circular.

We turn now to the determination of the dynamic radial receptance of the shell. There are many cylindrical shell theories that can be used for this purpose. For example, Donnell–Mushtari thin circular shell equations are in the frequency domain [28]:

$$\left(\frac{\partial^2}{\partial x^2}+\frac{1-\nu_S}{2a^2}\frac{\partial^2}{\partial \theta^2}+\frac{\omega^2}{c_S^2}\right)u+\frac{1+\nu_S}{2a}\frac{\partial^2 v}{\partial x\partial \theta}+\frac{\nu_S}{a}\frac{\partial w}{\partial x}=0 \tag{5.132}$$

$$\frac{1+\nu_S}{2a}\frac{\partial^2 u}{\partial x\partial \theta}+\left(\frac{1-\nu_S}{2}\frac{\partial^2}{\partial x^2}+\frac{1}{a^2}\frac{\partial^2}{\partial \theta^2}+\frac{\omega^2}{c_S^2}\right)v+\frac{1}{a^2}\frac{\partial w}{\partial \theta}=0 \tag{5.133}$$

$$\frac{\nu_S}{a}\frac{\partial u}{\partial x}+\frac{1}{a^2}\frac{\partial v}{\partial \theta}+\left(\frac{1}{a^2}+\frac{h^2}{12a^4}\left(a^2\frac{\partial^2}{\partial x^2}+\frac{\partial^2}{\partial \theta^2}\right)^2-\frac{\omega^2}{c_S^2}\right)w=\frac{\overline{p'}}{c_S^2\rho_S h}. \tag{5.134}$$

Here, u, v, w denote, respectively, the axial, tangential and radial displacements of the shell midplane, θ denotes the angular coordinate and the remaining symbols have the meanings defined previously. Since the internal acoustic field is one-dimensional, we may assume that the shell executes axisymmetric vibrations. Then, all terms involving the angular coordinate can be discarded and the Donnell–Mushtari equations reduce to:

$$\left(\frac{\partial^2}{\partial x^2}+\frac{\omega^2}{c_S^2}\right)u+\frac{\nu_S}{a}\frac{\partial w}{\partial x}=0 \tag{5.135}$$

$$\frac{\nu_S}{a}\frac{\partial u}{\partial x}+\left(\frac{1}{a^2}+\frac{h^2}{12a^4}\left(a^2\frac{\partial^2}{\partial x^2}\right)^2-\frac{\omega^2}{c_S^2}\right)w=\frac{\overline{p'}}{c_S^2\rho_S h}. \tag{5.136}$$

Radial receptance of the shell is determined from these equations by assuming wave-like solutions for u, v, w and $\overline{p'}$, with $e^{ik_o Kx}$ dependence on the axial coordinate, where K denotes the propagation constant (to be determined). Hence, the foregoing equations yield the relationship

$$\frac{w}{\overline{p'}}=\frac{a^2}{hE}\left(1-\frac{(k_o aK)^2}{1-\nu_S^2}\left(\frac{h^2}{12a^2}(k_o aK)^2+\frac{k_S^2a^2}{(k_o aK)^2}\right)\right)^{-1}. \tag{5.137}$$

It is seen that the radial receptance of the shell is dependent on the propagation constant. This highlights the fact that elastic duct walls are not locally reacting. Therefore, the propagation constants given by Equation (3.152) are not valid now. To determine the propagation constant, we substitute the above defined wave-like solution in Equations (5.127) and (5.128), and obtain, respectively,

$$-(1-\bar{M}_oK)+\rho_o c_o K\frac{\overline{v'}}{\overline{p'}}=\frac{\rho_o c_o^2 P}{S}(1-\psi\bar{M}_oK)\frac{w'}{\overline{p'}} \tag{5.138}$$

$$-\rho_o c_o(1-\bar{M}_oK)\frac{\overline{v'}}{\overline{p'}}+K=-\frac{\bar{M}_o\rho_o c_o^2\varphi_{SF}P}{S}(1-\psi\bar{M}_oK)\frac{w'}{\overline{p'}}. \tag{5.139}$$

Hence, a dispersion equation for K may be derived by eliminating the circumferential receptance $\overline{v'}/\overline{p'}$, from these equations. For example, assuming the Ingard–Myers full-slip boundary condition ($\psi=1$ and $\varphi_{SF}=0$), the dispersion equation is found as:

$$\frac{K^2}{(1-\bar{M}_oK)^2}=1+2\frac{\rho_o c_o^2}{E}\frac{a}{h}\left(1-\frac{(k_o aK)^2}{1-\nu_S^2}\left(\frac{h^2}{12a^2}(k_o aK)^2+\frac{k_S^2a^2}{(k_o aK)^2}\right)\right)^{-1}. \tag{5.140}$$

If $h^2 << a^2$ and $k_S^2a^2 << 1$, which are valid in many cases, this equation may be simplified to $K^2/(1-\bar{M}_oK)^2=1+2(\rho_o c_o^2/E)a/h$. If, furthermore, $2(\rho_o c_o^2/E)a/h << 1$, we recover the uniform hard-walled duct propagation constants $K=\mp 1/1\mp\bar{M}_o$. Accordingly, this condition may be used to assess the effect of internal fluid loading on the propagation constant. For example, when the fluid is air at standard conditions, it implies that $h/2a \gg 700\times 10^{-9}$ for steel ducts and $h/2a \gg 2000\times 10^{-9}$ for aluminum ducts. If the fluid is water, the corresponding values are $h/2a \gg 1.1\times 10^{-2}$ and $h/2a \gg 3.2\times 10^{-2}$, respectively. These figures show that the effect of internal fluid loading on the propagation constants can be neglected in industrial applications. This considerably simplifies the problem of shell noise prediction, because the usual propagation constants determined by ignoring the fluid loading (see Chapters 3 and 6) may be used in the calculation of shell vibrations. For example, we may put $K=1/1+\bar{M}_o$ in Equation (5.137) and use the result in Equation (5.129) to obtain an improved estimate of Equation (5.131).

Shell noise of rectangular ducts can be analyzed similarly, but it is more involved because of the necessity of splitting the peripheral integrals over four surfaces with appropriate edge conditions. However, an analytical formulation is possible and charts are available for quick estimation of the radiated noise [29].

In general, circular ducts have far the better theoretical wall transmission ratio compared to other duct shapes, but their performance may be degraded by the presence of shape distortions due to welding seams or other causes [27, 30–31]. This effect and various other important aspects of shell noise and its modeling are discussed in the review article by Cummings [32]. It seems that it is possible to obtain some success in predicting the sound radiated from uniform ducts. However, more pragmatic methods based on resonance tests and modal analysis and vibration control techniques, such as adjusting the mass and stiffness characteristics to avoid critical resonances and introducing damping for controlling vibration amplitudes, are routine in the industry for decisive conclusions on the shell noise of silencer casings. The internals of a silencer also require attention, since they usually couple with the casing and affect its vibrations. End plates and baffle plates can provide stiffness to the casing, but they may also contribute to the shell noise and modify the acoustic performance of a silencer predicted on the assumption that they are perfectly rigid.

References

[1] G.W. Stewart and R.B. Lindsay, *Acoustics*, (D. Van Nostrand Co., Inc., 1930).

[2] D.D. Davis, G.M. Stokes, D. Moore and J.L. Stevens, Theoretical and experimental investigation of mufflers with comments on engine-exhaust muffler design, *NACA Report* (1954), 1192.

[3] P.A.O.L. Davies, The design of silencers for internal combustion engines, *J. Sound Vib.* **1** (1964), 185–201.

[4] R.J. Alfredson and P.A.O.L. Davies, Performance of exhaust silencer components, *J. Sound Vib.* **15** (1971), 175–196.

[5] P.A.O.L. Davies, Practical flow duct acoustics *J. Sound Vib.* **241** (1988), 91–115.

[6] J.L. Shearer, A.T. Murphy and H.H. Richardson, *Introduction to System Dynamics*, (Reading: Addison-Wesley, 1971).

[7] S.-H. Seo and Y.-H. Kim, Silencer design by using array resonators for low frequency band noise reduction, *J. Acoust. Soc. of Am.* **118** (2005), 2332–2338.

[8] A.J. Torregrosa, A. Broatch and R. Payri, A study of the influence of mean flow on the acoustic performance of Herschel-Quincke tubes, *J. Acoust. Soc. Am.* **107** (2000), 1874–1879.

[9] R.F. Hallez, R.A. Burdisso, Analytical modeling of Herschel-Quincke concept applied to inlet turbofan engines, NASA/CR-2002-211429 (2002).

[10] P.O.A.L. Davies and K.R. Holland, The observed aeroacoustic behaviour of some flow-excited expansion chambers, *J. Sound Vib.* **239** (2001), 695–708.

[11] A.J. Torregrossa, A. Broatch, H. Climent and I. Andres, A note on Srouhal number dependence of the relative importance of internal and external sources in IC engine exhaust systems, *J. Sound Vib.* **282** (2005), 1255–1263.

[12] P.O.A.L. Davies, Transmission matrix representation of exhaust system acoustic characteristics, *J. Sound Vib.* **151** (1991), 333–338.

[13] W. Eversman, A reverse flow theorem and acoustic reciprocity in compressible potential flows in ducts, *J. Sound Vib.* **246** (2001), 71–95.

[14] T.L. Parrott, An improved method for design of expansion chamber mufflers with application to an operational helicopter, NASA TN D-7309 (1973).

[15] P.A. Nelson, N.A. Halliwell and P.E. Doak, Fluid dynamics of a flow excited resonance: part I: experiment, *J. Sound Vib.* **78** (1981), 15–38.

[16] P.O.A.L. Davies, Flow-acoustic coupling in ducts, *J. Sound Vib.* **77** (1981), 191–209.

[16] J.C. Bruggerman, A. Hirschberg, M.E.H. van Dongen, A.P.J. Wijnands and J. Gorter, Self-sustained aero-acoustic pulsations in gas transport systems: experimental study of the influence of closed side branches, *J. Sound Vib.* **150** (1991), 371–393.

[17] P.A. Nelson, N.A. Halliwell and P.E. Doak, Fluid dynamics of a flow excited resonance: part II: flow acoustic interaction, *J. Sound Vib.* **91** (1983), 375–402.

[18] P.M. Radavich and A. Selamet, A computational approach for flow-acoustic coupling in closed side branches, *J. Acoust. Soc. Am.* **109** (2001), 1343–1353.

[19] S.M.N. Dequand, Duct aeroacoustics: from technological application to the flute, (Doctoral Thesis, Eindhoven: Technische Universiteit Eindhoven, 2001) (doi: 10.6100/IR550318).

[20] B.D. Knotts and A. Selamet, Suppression of flow-acoustic coupling in sidebranch ducts by interface modification, *J. Sound Vib.* **265** (2003), 1025–1045.

[21] M.S. Howe, Edge, cavity and aperture tones at very low Mach numbers, *J. Fluid Mech.* **330** (1997), 61–84.

[22] R.M. White, *Fluid Mechanics,* (Boston: McGraw-Hill, 2008).

[23] E. Fried and I.E. Idel'chik, *Flow Resistance: A Design Guide for Engineers*, (Philadelphia: Taylor and Francis Ltd, 1989).
[24] F. Payri, A.J. Torregrosa, A. Broatch and J-P. Brunel, Pressure loss characterization of perforated ducts, International Congress and Exposition, Detroit, Michigan, (SAE Technical Paper Series 980282) (1998).
[25] D.K. Dixit, *Heat and Mass Transfer*, (New Delhi: McGraw-Hill, 2016).
[26] A. Cummings, Low frequency acoustic radiation from duct walls, *J. Sound Vib.* **71**(2) (1980), 201–226
[27] A. Cummings, I.-J. Chang and R.J. Astley, Sound transmission at low frequencies through the walls of distorted circular ducts, *J. Sound Vib.* **97** (1984), 261–286.
[28] A.W. Leissa, Vibration of shells, NASA SP-288, (1973).
[29] A. Cummings, Design charts for low frequency acoustic transmission through the walls of rectangular ducts, *J. Sound Vib.* **78** (1981), 269–289.
[30] G.F. Kuhn and C.L. Morfey, Transmission of low frequency internal sound through pipe walls, *J. Sound Vib.* **47** (1976), 147–161
[31] S.N. Yousri and F.J. Fahy, Distorted cylindrical shell response to internal acoustic excitation below the cut-off frequency, *J. Sound Vib.* **52** (1977), 441–452.
[32] A. Cummings, Sound transmission through duct walls, *J. Sound Vib.* **239** (2001), 731–765.

6 Multi-Modal Sound Propagation in Ducts

6.1 Introduction

The need for three-dimensional models of sound propagation in ducts arises when one-dimensional models are not justified. Typically, this is the case if frequencies of interest are not low enough or characteristic dimensions of duct sections are not small compared to the wavelength. A three-dimensional model may also be dictated if a major feature of a duct cannot be accommodated with adequate accuracy in a one-dimensional acoustic model, such as the curvature effect of a curved duct or the swirl component of the mean flow.

As can be expected, acoustic modeling of ductworks in three dimensions is considerably more difficult than in one dimension. For this reason, three-dimensional acoustic models are usually considered when there is a definite indication that this is really warranted due to size, geometry or frequency considerations.

The basic problem of three-dimensional duct acoustics is again the transmission of acoustic wave motion in a straight uniform hard-walled duct with an inviscid homogeneous mean flow (Section 6.5). This model captures most of the essential features of three-dimensional wave propagation in ducts and is still simple enough to be handled analytically. More complicated cases, such as curved ducts, non-uniform ducts, inhomogeneous ducts, and lined ducts, are considered after a detailed study of this basic model. Coupling of acoustic wave motion with turbulent fluctuations is neglected [1], except for their role in thin acoustic boundary layers near the duct walls (Section 6.3) and the study of the effects of the fluid viscosity and thermal conductivity is deferred to Chapter 8.

The basic state variables which are used to define three-dimensional acoustic wave fields in ducts are still the acoustic fluctuations of pressure, p', density, ρ', and particle velocity, $\mathbf{v}'$. The acoustic continuity and momentum equations which govern these variables in a source-free region of a duct containing an inviscid fluid are given in Section 1.1 as Equations (1.5) and (1.8), respectively, and the corresponding mean flow equations are Equations (1.4) and (1.7). We repeat these equations here for convenience: the acoustic continuity and momentum equations are

$$\frac{\partial \rho'}{\partial t} + \mathbf{v}_o \cdot \nabla \rho' + \rho_o \nabla \cdot \mathbf{v}' + \mathbf{v}' \cdot \nabla \rho_o + \rho' \nabla \cdot \mathbf{v}_o = 0 \qquad (6.1)$$

$$\rho_o\left(\frac{\partial \mathbf{v}'}{\partial t} + \mathbf{v}_o\cdot\nabla\mathbf{v}' + \mathbf{v}'\cdot\nabla\mathbf{v}_o\right) + \rho'\,\mathbf{v}_o\cdot\nabla\mathbf{v}_o + \nabla p' = 0. \tag{6.2}$$

The corresponding mean flow equations are

$$\nabla\cdot(\rho_o\mathbf{v}_o) = 0 \tag{6.3}$$

$$\rho_o\mathbf{v}_o\cdot\nabla\mathbf{v}_o + \nabla p_o = 0, \tag{6.4}$$

where, as usual, the subscript "o" denotes a mean part. Equations (6.1) and (6.2) are closed by Equation (1.14) if the mean flow is homentropic, and by Equation (1.15) if it is homogeneous. Under certain conditions, these equations can be reduced to a wave equation in one acoustic variable. Most of the three-dimensional duct models described in this chapter are based on solutions of such wave equations in the frequency domain. The corresponding modal wave transfer matrices are formulated, when possible, in the format of Equation (2.15) by arranging the modal solutions in matrix notation. The procedure is described in Section 6.4.2 for the convected wave equation and applied similarly when a similar modal solution is available. Solutions of Equations (6.1) and (6.2) may also be obtained by using a numerical method and transformed to modal wave transfer matrix format of Equation (2.15) as described in Section 2.5 for the finite element and boundary element methods. However, some ducts, for example, ducts with axially varying cross-sectional area, are not tractable to this procedure due to the coupling of the axial and transverse modes. The derivation of a wave transfer matrix in such situations is discussed in Section 6.9.

6.2 Uniform Ducts with Axial Mean Flow

Consider a uniform duct and let the duct geometry be defined with respect to a fixed rectangular Cartesian coordinate system x_1, x_2, x_3 where the x_1 is the duct axis and the duct sections lying in planes parallel to the (x_2, x_3) plane (normal to the duct axis). The mean flow is homogeneous with velocity distribution $\mathbf{v}_o = \mathbf{e}_1 v_{o1}(x_2, x_3)$, where $\mathbf{e}_j$, $j = 1, 2, 3$, denote fixed unit vectors in the positive direction of the x_j axis. Since $\mathbf{v}_o\cdot\nabla\mathbf{v}_o = 0$, Equation (6.4) gives p_o =constant. Then Equation (1.12) implies $\rho_o =$ constant and Equation (6.3) is satisfied identically, since the mean flow is also solenoidal, that is, $\nabla\cdot\mathbf{v}_o = 0$. Thus, as shown in Section 1.1.2, the relationship $p' = c_o^2\rho'$ between the acoustic pressure and density fluctuations is valid, where c_o denotes the speed of sound. Under these conditions, Equations (6.1) and (6.2) simplify to

$$\frac{\partial p'}{\partial t} + \mathbf{v}_o\cdot\nabla p' + \rho_o c_o^2\nabla\cdot\mathbf{v}' = 0 \tag{6.5}$$

$$\rho_o\left(\frac{\partial \mathbf{v}'}{\partial t} + \mathbf{v}_o\cdot\nabla\mathbf{v}' + \mathbf{v}'\cdot\nabla\mathbf{v}_o\right) + \nabla p' = 0, \tag{6.6}$$

respectively. From these equations, a wave equation for the acoustic pressure can be derived as follows: Take the divergence of Equation (6.6):

$$\rho_o \frac{\partial}{\partial t} \nabla \cdot \mathbf{v}' + \rho_o \nabla \cdot (\mathbf{v}_o \cdot \nabla \mathbf{v}') + \rho_o \nabla \cdot (\mathbf{v}' \cdot \nabla \mathbf{v}_o) = -\nabla^2 p'. \tag{6.7}$$

But

$$\nabla \cdot (\mathbf{v}_o \cdot \nabla \mathbf{v}') = \nabla \cdot (\mathbf{v}' \cdot \nabla \mathbf{v}_o) + \nabla \cdot \mathbf{v}' \nabla \cdot \mathbf{v}_o + (\nabla \nabla \cdot \mathbf{v}') \cdot \mathbf{v}_o \tag{6.8}$$

holds as vector identity and

$$(\nabla \nabla \cdot \mathbf{v}') \cdot \mathbf{v}_o = v_{o1} \frac{\partial}{\partial x_1} (\nabla \cdot \mathbf{v}') = \mathbf{v}_o \cdot \nabla \nabla \cdot \mathbf{v}', \tag{6.9}$$

since $\mathbf{v}_o = \mathbf{e}_1 v_{o1}(x_2, x_3)$. Hence, noting that the mean flow velocity is solenoidal, Equation (6.7) can be written as

$$\rho_o \left(\frac{\partial}{\partial t} + \mathbf{v}_o \cdot \nabla \right) \nabla \cdot \mathbf{v}' + 2\rho_o \nabla \cdot (\mathbf{v}' \cdot \nabla \mathbf{v}_o) = -\nabla^2 p'. \tag{6.10}$$

Next, operating on both sides of Equation (6.5) by $\partial / \partial t + \mathbf{v}_o \cdot \nabla$

$$\left(\frac{\partial}{\partial t} + \mathbf{v}_o \cdot \nabla \right)^2 p' + \rho_o c_o^2 \left(\frac{\partial}{\partial t} + \mathbf{v}_o \cdot \nabla \right) \nabla \cdot \mathbf{v}' = 0. \tag{6.11}$$

Substituting this in Equation (6.10), we get

$$\left(\nabla^2 - \frac{1}{c_o^2} \left(\frac{\partial}{\partial t} + \mathbf{v}_o \cdot \nabla \right)^2 \right) p' = -2\rho_o \nabla \cdot (\mathbf{v}' \cdot \nabla \mathbf{v}_o), \tag{6.12}$$

where $\nabla^2 = \nabla \cdot \nabla$ denotes the Laplacian in three dimensions. Now define the following vector decompositions:

$$\mathbf{v}' = \mathbf{e}_1 v_1' + \mathbf{v}_\perp', \nabla = \mathbf{e}_1 \frac{\partial}{\partial x_1} + \nabla_\perp, \tag{6.13}$$

where the subscript "$\perp$" denotes the transverse components of a three-dimensional vector and $\nabla_\perp$ is called the gradient operator on the duct cross section. Since the vector $\mathbf{v}_o = \mathbf{e} v_o(x_2, x_3)$ is normal to $\nabla_\perp$, we get

$$\nabla \cdot (\mathbf{v}' \cdot \nabla \mathbf{v}_o) = \frac{\partial \mathbf{v}_\perp'}{\partial x_1} \cdot \nabla_\perp v_{o1}. \tag{6.14}$$

Using this in Equation (6.12):

$$\left(\nabla^2 - \frac{1}{c_o^2} \left(\frac{\partial}{\partial t} + \mathbf{v}_o \cdot \nabla \right)^2 \right) p' = -2\rho_o \frac{\partial \mathbf{v}_\perp'}{\partial x_1} \cdot \nabla_\perp v_{o1}. \tag{6.15}$$

Now consider the transverse component of the acoustic momentum equation:

$$\rho_o \left(\frac{\partial}{\partial t} + \mathbf{v}_o \cdot \nabla \right) \mathbf{v}_\perp' + \nabla_\perp p' = 0. \tag{6.16}$$

Taking derivative of this with respect to x_1

$$\rho_o\left(\frac{\partial}{\partial t}+\mathbf{v}_o\cdot\nabla\right)\frac{\partial \mathbf{v}'_\perp}{\partial x_1}+\nabla_\perp\frac{\partial p'}{\partial x_1}=0 \tag{6.17}$$

The required wave equation for the acoustic pressure is obtained now by eliminating $\partial \mathbf{v}'_\perp/\partial x_1$ from Equations (6.15) and (6.17):

$$\left(\frac{\partial}{\partial t}+v_{o1}\frac{\partial}{\partial x_1}\right)\left(\nabla^2 p'-\frac{1}{c_o^2}\left(\frac{\partial}{\partial t}+v_{o1}\frac{\partial}{\partial x_1}\right)^2 p'\right)-2(\nabla_\perp v_{o1})\cdot\nabla_\perp\frac{\partial p'}{\partial x_1}=0. \tag{6.18}$$

This can be written as

$$\left(\frac{1}{c_o}\frac{\partial}{\partial t}+M_o\frac{\partial}{\partial x_1}\right)(\nabla_\perp^2+\mathcal{J})p'-2(\nabla_\perp M_o)\cdot\nabla_\perp\frac{\partial p'}{\partial x_1}=0, \tag{6.19}$$

where $M_o = v_{o1}/c_o$ denotes the Mach number of the local mean flow velocity and the differential operator $\mathcal{J}$ is given by

$$\mathcal{J}=-\frac{1}{c_o^2}\frac{\partial^2}{\partial t^2}+\left(1-M_o^2\right)\frac{\partial^2}{\partial x_1^2}-2M_o\frac{1}{c_o}\frac{\partial^2}{\partial t\partial x_1}. \tag{6.20}$$

We will study Equation (6.19) in the frequency domain. The transformation involves simply replacing the operator $\partial/\partial t$ by $-\mathrm{i}\omega$, where ω denotes the radian frequency and i denotes the unit imaginary number (Section 1.2). Also, since the mean flow is assumed to be axially uniform, we take advantage of this by transforming the result to the wavenumber domain (Section 1.3) by the ansatz:

$$p'(x_1,\bullet)=\varphi(\bullet)\mathrm{e}^{\mathrm{i}k_oKx_1}, \tag{6.21}$$

where $k_o = \omega/c_o$, $\varphi(\bullet)$ denotes a function of the transverse coordinates, for example, $\varphi(\bullet) = \varphi(x_2, x_3)$ for rectangular ducts (see Figure 6.3) defined in an orthogonal rectangular coordinate system (x_1, x_2, x_3), or $\varphi(\bullet) = \varphi(r, \theta)$ for circular ducts (see Figure 6.4) in a cylindrical coordinate system (x_1, r, θ); K denotes an undetermined parameter, which is called propagation constant, and k_oK is called the axial wavenumber. Upon inserting Equation (6.21), Equation (6.19) yields the following equation for $\varphi(\bullet)$:

$$(1-KM_o)(\nabla_\perp^2+\alpha^2)\varphi(\bullet)+2K(\nabla_\perp M_o)\cdot\nabla_\perp\varphi(\bullet)=0, \tag{6.22}$$

where

$$\alpha^2=k_o^2\left((1-KM_o)^2-K^2\right). \tag{6.23}$$

Solution of Equation (6.22), which is commonly referred to as the Pridmore-Brown equation [2], involves some difficulty, not only because of the coefficients that depend on the transverse coordinates, but also because the actual shape of the mean flow velocity profile is usually not known a priori with certainty and substantial effort is required for its experimental determination. For this reason, in predictive calculations the mean flow velocity profile is usually modeled by using heuristic distributions.

A fully developed turbulent mean flow in a smooth-walled duct may be characterized by three contiguous regions [3–4], called the viscous sublayer, the logarithmic zone and the core flow, with distinct velocity distributions. For simplicity, such profiles are often approximated by the $1/n$ power law (see Section 3.2.2). A laminar mean flow, on the other hand, is modeled by using a parabolic profile. The Pridmore-Brown equation includes both the convective and refractive effects of the mean flow. This is best seen from Equation (6.15), where the material derivative on the left-hand side accounts for the convection of acoustic waves with the mean flow and the term on the right-hand side, which corresponds to the second term of Equation (6.22), represents refraction of acoustic waves in shear layers.

The most widely used ducts in industry are of rectangular or circular sections and are modeled conveniently in Cartesian coordinates and cylindrical polar coordinates, respectively. The Pridmore-Brown equation can be transformed to these coordinates by noting that the gradient and Laplacian operators on the duct cross section in Cartesian coordinates x_1, x_2, x_3 are

$$\nabla_\perp = \mathbf{e}_2 \frac{\partial}{\partial x_2} + \mathbf{e}_3 \frac{\partial}{\partial x_3}, \nabla_\perp^2 = \frac{\partial^2}{\partial x_2^2} + \frac{\partial^2}{\partial x_3^2} \tag{6.24}$$

and in cylindrical polar coordinates x_1, r, θ, they are

$$\nabla_\perp = \mathbf{e}_r \frac{\partial}{\partial r} + \mathbf{e}_\theta \frac{1}{r}\frac{\partial}{\partial \theta}, \nabla_\perp^2 = \frac{1}{r}\frac{\partial}{\partial r} + \frac{\partial^2}{\partial r^2} + \frac{1}{r^2}\frac{\partial^2}{\partial \theta^2}, \tag{6.25}$$

where the unit vectors are given by

$$\mathbf{e}_r = \mathbf{e}_2 \cos\theta + \mathbf{e}_3 \sin\theta, \mathbf{e}_\theta = -\mathbf{e}_2 \sin\theta + \mathbf{e}_2 \cos\theta. \tag{6.26}$$

Upon substitution of these operators, Equation (6.22) can be expressed in Cartesian and cylindrical polar coordinates, respectively,

$$(1 - KM_o)\left(\frac{\partial^2 \varphi}{\partial x_2^2} + \frac{\partial^2 \varphi}{\partial x_3^2} + \alpha^2 \varphi\right) + 2K\left(\frac{\partial M_o}{\partial x_2}\frac{\partial \varphi}{\partial x_2} + \frac{\partial M_o}{\partial x_3}\frac{\partial \varphi}{\partial x_3}\right) = 0 \tag{6.27}$$

$$(1 - KM_o)\left(\frac{1}{r}\frac{\partial \varphi}{\partial r} + \frac{\partial^2 \varphi}{\partial r^2} + \frac{1}{r^2}\frac{\partial^2 \varphi}{\partial \theta^2} + \alpha^2 \varphi\right) + 2K\left(\frac{\partial M_o}{\partial r}\frac{\partial \varphi}{\partial r} + \frac{1}{r^2}\frac{\partial M_o}{\partial \theta}\frac{\partial \varphi}{\partial \theta}\right) = 0. \tag{6.28}$$

For rectangular ducts, if the mean flow is two-dimensional, that is, $M_o = M_o(x_j)$ where $j = 2$ or 3, solutions of Equation (6.27) may be searched in the form

$$\varphi(x_2, x_3) = X_2(x_2)X_3(x_3), \tag{6.29}$$

where

$$(1 - KM_o)\left(X_1 \frac{\partial^2 X_2}{\partial x_2^2} + X_2 \frac{\partial^2 X_3}{\partial x_3^2} + \alpha^2 X_2 X_3\right) + 2K\frac{\partial M_o}{\partial x_j}\frac{\partial X_j}{\partial x_j} = 0, \quad j = 2 \text{ or } 3. \tag{6.30}$$

For circular ducts (hollow or annular), on the other hand, if the duct and the mean flow are axisymmetric, that is, $M_o = M_o(r)$, solutions of Equation (6.28) may be searched in the form

$$\varphi(r,\theta) = R(r)\mathrm{e}^{\mathrm{i}m\theta}, m = 0, \mp 1, \mp 2, \ldots, \tag{6.31}$$

since φ must be periodic of period 2π. Then the radial function $R(r)$is given by

$$(1 - KM_o)\left(\frac{1}{r}\frac{\partial R}{\partial r} + \frac{\partial^2 R}{\partial r^2} + \left(\alpha^2 - \frac{m^2}{r^2}\right)R\right) + 2K\left(\frac{\partial M_o}{\partial r}\frac{\partial R}{\partial r}\right) = 0. \tag{6.32}$$

A substantial portion of this chapter is devoted to discussion of solutions of these equations under appropriate boundary conditions (Section 6.3) and properties of the solutions.

6.3 Boundary Condition on Impermeable Walls

Several options exist for the application of the acoustic conditions at boundaries of ducts with impermeable walls. These are conceptually similar to the conditions described in Section 3.8.1 for the one-dimensional theory and are extended here to three dimensions.

6.3.1 No-Slip Model

No-slip is an essential condition for Newtonian fluid flow over smooth surfaces [3]. It implies that the velocity of both the mean flow and the acoustic wave motion vanish at the duct walls. The acoustic condition over smooth locally reacting physical duct wall surfaces, Γ, of acoustic impedance $Z_W = Z_W(\omega)$ may be expressed, in frequency domain, as

$$\mathbf{n}\cdot\mathbf{v}' = \frac{p'}{Z_W} \text{on } \Gamma, \tag{6.33}$$

where $\mathbf{n}$ denotes the unit outward normal vector of Γ. To write this condition in terms of acoustic pressure, we make use of the component of the acoustic momentum equation, Equation (6.6) in the normal direction $\mathbf{n}$, which is

$$\rho_o\left(-\mathrm{i}\omega + v_{o1}\frac{\partial}{\partial x_1}\right)\mathbf{n}\cdot\mathbf{v}' + \mathbf{n}\cdot\nabla_\perp p' = 0 \text{ on } \Gamma, \tag{6.34}$$

Here, in view of the no-slip condition, $v_{o1} = 0$ on Γ, but we will keep it in the analysis for the sake of generality. Then, eliminating the particle velocity from Equations (6.33) and (6.34) gives the boundary condition as

$$\mathbf{n}\cdot\nabla_\perp p' = \mathrm{i}k_o\left(1 - \frac{M_o}{\mathrm{i}k_o}\frac{\partial}{\partial x_1}\right)\frac{p'}{\zeta} \text{ on } \Gamma, \tag{6.35}$$

where $\zeta = Z_W/\rho_o c_o$ is called the normalized wall impedance and $\rho_o c_o$ is the characteristic impedance of the fluid. If the wall impedance is axially uniform, Equation (6.35) can be expressed as

$$\mathbf{n}\cdot\nabla_{\perp}\varphi(\bullet) = \frac{ik_o}{\zeta}(1 - M_o K)\varphi(\bullet) \text{ on } \Gamma, \tag{6.36}$$

where the $e^{ik_o K x_1}$ dependence on the axial coordinate is introduced from Equation (6.21).

6.3.2 Partial-Slip Model

The no-slip model involves a discrepancy for the mean flow in that the no-slip condition is a consequence of the viscosity of the fluid, but viscosity is neglected in derivation of the Pridmore-Brown equation. For a fully developed turbulent mean flow, this discrepancy may be removed by deferral of the duct boundary to the border of the viscous (laminar) sublayer. Using the model described in Section 3.8.1.3, the normal component of the particle velocity at the border of the viscous sublayer is given by Equation (3.118), which is written in the present notation as

$$\mathbf{n}\cdot\mathbf{v}' = \frac{1}{Z_W}\left(1 - \frac{\overline{v_{o1}}}{i\omega}\left(\frac{v_{o1}^+}{\overline{v_{o1}}} - \beta_v^+\right)\frac{\partial}{\partial x} + \frac{\beta_v^+ \overline{v_{o1}} v_{o1}^+}{\omega^2}\frac{\partial^2}{\partial x^2}\right)\overline{p'} \text{ on } \Gamma^+, \tag{6.37}$$

where v_{o1}^+ denotes the mean flow velocity at the border of the viscous sublayer, Γ^+ denotes the surface of the viscous sublayer border and β_v^+ is given by Equation (3.121). Using Equation (6.34) and invoking $e^{ik_o K x_1}$ dependence on the axial coordinate, the foregoing condition may be expressed as

$$\mathbf{n}\cdot\nabla_{\perp}\varphi(\bullet) = \frac{ik_o}{\zeta}(1 - KM_o)\left(1 - (1 - \varphi_{\text{SF}} - \beta_v^+)K\bar{M}_o - \beta_v^+(1 - \varphi_{\text{SF}})K^2(\bar{M}_o)^2\right)\varphi(\bullet) \text{ on } \Gamma^+, \tag{6.38}$$

where $\bar{M}_o = \overline{v_{o1}}/c_o$, φ_{SF} denotes the slip flow parameter defined by Equation (3.21), which is, in the present notation, $\varphi_{\text{SF}} = 1 - v_{o1}^+/\overline{v_{o1}}$. Note that the condition $\Gamma^+ \to \Gamma$ as $Z_W \to \infty$ is required for consistency of this model with the hard-wall case.

6.3.3 Ingard–Myers Model

This boundary condition has its origin in the fact that at the interface of two fluids moving with distinct uniform velocities, the normal components of the particle velocities on either side of the interface are not continuous across the boundary, because they are given by the material derivative with respect to time of the normal displacement of the boundary. Based on this fact, it was first proposed by Ingard for uniform ducts [5] and later generalized by Myers for non-uniform ducts [6] that at the walls of ducts with mean flow, the acoustic boundary condition should be the continuity of displacement and acoustic pressure across the fluid–wall interface. This condition, however, raises some questions. Firstly, since the mean flow velocity is assumed to be uniform, refraction in shear layers in the core flow is tacitly assumed to be negligible. The value to use for the velocity of the assumed uniform mean flow remains largely a heuristic matter, although the cross-section average of the actual mean flow velocity, $\overline{v_{o1}}$, is often suggested as a plausible value. Secondly, the mean boundary layer is approximated by an infinitely thin vortex sheet, but this is not a realistic model of true boundary layers, which are of finite thickness and dominated by viscothermal and turbulence absorption. However, in spite of these drawbacks, the Ingard–Myers theory is used extensively.

Let η denote displacement of the vortex sheet at the fluid–wall interface from its unperturbed position. Since the medium on the mean flow side of the vortex sheet is presumed to be moving with uniform velocity $\overline{\mathbf{v}_o} = \mathbf{e}_1\overline{v_{o1}}$, the normal velocity of the vortex sheet on this side is given (relative to a fixed observer) by the material derivative $(\partial/\partial t + \overline{\mathbf{v}_o}\cdot\nabla)\eta$. But this should be compatible with the particle velocity of the acoustic field on the flow side. This condition is expressed, in frequency domain, as

$$\mathbf{n}\cdot\mathbf{v}' = \left(-\mathrm{i}\omega + \overline{v_{o1}}\frac{\partial}{\partial x_1}\right)\eta \text{ on } \Gamma, \tag{6.39}$$

where Γ still denotes the surface of the duct walls, since the vortex sheet is infinitesimally thin. Assuming that the transverse displacements of the vortex sheet is small, the normal component of the particle velocity on the wall side of the vortex sheet is given by $\partial\eta/\partial t$, since that side of the vortex sheet is stationary. Then, by definition, the impedance of the duct wall is $Z_W = p'/(-\mathrm{i}\omega\eta)$, because the acoustic pressure is continuous across the vortex sheet. Substituting this relationship in Equation (6.39) and using Equation (6.34), which now holds with v_{o1} replaced by $\bar{v}_o$, to eliminate the particle velocity:

$$\mathbf{n}\cdot\nabla_\perp p' = \mathrm{i}k_o\left(1 - \frac{\bar{M}_o}{\mathrm{i}k_o}\frac{\partial}{\partial x_1}\right)^2\frac{p'}{\zeta} \text{ on } \Gamma. \tag{6.40}$$

In view of the $\mathrm{e}^{\mathrm{i}k_oKx_1}$ dependence of the acoustic field on the axial coordinate, this can be expressed as

$$\mathbf{n}\cdot\nabla_\perp\varphi(\bullet) = \frac{\mathrm{i}k_o}{\zeta}(1 - \bar{M}_oK)^2\varphi(\bullet) \text{ on } \Gamma, \tag{6.41}$$

if the wall impedance is axially uniform.

A crucial feature of the Ingard–Myers theory is the assumption that the particle displacement is continuous across the vortex sheet. If we assume instead that the particle velocity is continuous across the vortex sheet, then Equation (6.40) becomes redundant and the boundary condition comes out in the form of Equation (6.36) with M_o replaced by $\bar{M}_o(\neq 0)$.

6.3.4 Modified Ingard–Myers Models

It has been proposed by several authors that the Ingard–Myers theory can be improved by assuming that the mean boundary layer to be thin, but of finite thickness, and deferring the acoustic boundary condition to the border of the mean boundary layer, the core flow outside boundary layer still being assumed to be uniform. Several asymptotic analytical solutions, with or without viscothermal effects taken into account, are available in the literature for the acoustic impedance at the border of the mean boundary layer [7–12]. Two of these models are discussed in Section 6.8.2.

An advantage of this approach over the Ingard–Myers theory is the inclusion of refractive effect to some extent, since shear gradients are greatest in the mean boundary layer. A more subtle advantage is related to the liability of the infinitely thin vortex sheet envisaged in the Ingard–Myers theory to Kelvin–Helmholtz type instability [13], if the duct walls have finite impedance. Thin but finite thickness boundary layer border impedance models seem to alleviate this problem [11].

The details of the various thin asymptotic boundary layer impedance models are given in the original articles. Here, we mention the extension of the partial slip model described in in Section 6.3.2. The mean boundary layer extends down to a core flow of uniform velocity $\overline{v_{o1}}$ (not necessarily the same as $\overline{v_{o1}}$ used in the application of the Ingard–Myers boundary condition). The normal component of the particle velocity at the border of the mean boundary layer is still given by Equation (3.118), but the integration in Equation (3.119) now extends through the thickness, δ, of the mean boundary layer. Hence, using Equation (6.34) for the elimination of the particle velocity:

$$\mathbf{n}\cdot\nabla_{\perp}p' = \frac{\mathrm{i}k_o}{\zeta}\left(1 - \frac{\bar{M}_o}{\mathrm{i}k_o}\frac{\partial}{\partial x_1}\right)^2\left(1 + \frac{\beta_v\bar{M}_o}{\mathrm{i}k_o}\frac{\partial}{\partial x_1}\right)p' \text{ on } \Gamma_\delta, \tag{6.42}$$

where Γ_δ denotes the surface of the mean boundary layer border and

$$\beta_v = \frac{1}{\bar{v}_o}\int_0^\delta \frac{\mathrm{d}v_{o1}}{\mathrm{d}\xi}\mathrm{e}^{(-1+\mathrm{i})\frac{\xi}{\delta_v}}\mathrm{d}\xi. \tag{6.43}$$

On invoking the $\mathrm{e}^{\mathrm{i}k_oKx_1}$ dependence of the acoustic field on the axial coordinate, this can be expressed as

$$\mathbf{n}\cdot\nabla_{\perp}\varphi(\bullet) = \frac{\mathrm{i}k_o}{\zeta}(1 - \bar{M}_oK)^2(1 + \beta_v\bar{M}_oK)\varphi(\bullet) \text{ on } \Gamma_\delta, \tag{6.44}$$

where $\Gamma_\delta \rightarrow \Gamma$ as $Z_W \rightarrow \infty$ for consistency with the hard-wall case.

The factor β_v depends on the mean flow velocity profile in the mean boundary layer and is plotted in Figure 3.4 ($\varphi_{\mathrm{SF}}^{\varepsilon} = 1$ case) as function of δ_v/δ. It is seen that if $\delta_v/\delta << 1$, then $\beta_v \approx 0$ and Equation (6.44) reduces to the form of Ingard–Myers boundary condition. However, if $\delta_v/\delta >> 1$, then $\beta_v \approx 1$ and Equation (6.44) reduces, for subsonic low Mach numbers ($\bar{M}_o^2 << 1$), to the no-slip form, that is, Equation (6.36) with M_o replaced by $\bar{M}_o$.

This model can be extended to the rough wall case as described in Section 3.8.1.4.

6.4 Wave Transmission in a Uniform Duct with Uniform Mean Flow

6.4.1 General Solution of the Convected Wave Equation

If the mean flow is uniform, Equation (6.19) reduces to

$$\left(\frac{1}{c_o}\frac{\partial}{\partial t}+\bar{M}_o\frac{\partial}{\partial x_1}\right)\left(\nabla^2-\left(\frac{1}{c_o}\frac{\partial}{\partial t}+\bar{M}_o\frac{\partial}{\partial x_1}\right)^2\right)p'=0, \tag{6.45}$$

where the overbar on $\bar{M}_o$ is used to remind us that mean flow is assumed to be uniform in the sense defined in the boundary conditions discussed in Sections 6.3.3 and 6.3.4. Solution of Equation (6.45) is given by superposition of the solutions of

$$\left(\frac{1}{c_o}\frac{\partial}{\partial t}+\bar{M}_o\frac{\partial}{\partial x_1}\right)p'=0 \tag{6.46}$$

and

$$\left(\nabla^2-\left(\frac{1}{c_o}\frac{\partial}{\partial t}+\bar{M}_o\frac{\partial}{\partial x_1}\right)^2\right)p'=0. \tag{6.47a}$$

The former gives a pressure wave that travels with phase velocity of $\overline{v_{o1}}=\bar{M}_o c_o$ and, hence, can exist only if the mean flow exists. This kind of wave is called a hydrodynamic wave. It may be important in aerodynamic design of fluid machinery, but may be ignored in acoustic analyses, since it is convected with the mean flow and plays no role in acoustic phenomena (for example, reflection). So, it suffices to proceed with the solution of Equation (6.47), which is known as the convected wave equation and is usually expressed as

$$\left(\nabla_\perp^2-\frac{1}{c_o^2}\frac{\partial^2}{\partial t^2}+\left(1-\bar{M}_o^2\right)\frac{\partial^2}{\partial x_1^2}-2\bar{M}_o\frac{1}{c_o}\frac{\partial^2}{\partial t\partial x_1}\right)p'=0. \tag{6.47}$$

Inspection of Equation (6.22) shows that, in frequency and wavenumber domain Equation (6.47) is simply

$$(\nabla_\perp^2+\alpha^2)\varphi(\bullet)=0, \tag{6.48}$$

where α^2 is given by Equation (6.23). This is known as the Helmholtz equation in two dimensions. The problem of solving this equation for given boundary conditions at the duct walls is a classical eigenvalue problem. In general, for a given duct shape and acoustic boundary conditions on the duct boundaries, there exists an infinite number of values of the parameter α^2, which are denoted by $\alpha_1^2, \alpha_2^2, \ldots$, which satisfy Equation (6.48). These are called transverse duct eigenvalues. To each eigenvalue, there corresponds a function φ, denoted by $\varphi_1, \varphi_2, \ldots$ in the same order as the eigenvalues, which are called transverse duct eigenfunctions. The eigenfunctions are linearly independent, but not unique, because, if φ is an eigenfunction, then any real multiple of this is also a valid solution of Equation (6.48) for the corresponding eigenvalue. For every eigenvalue α_μ^2, $\mu = 1, 2, \ldots$, Equation (6.23) gives two propagation constants K_μ^+ and K_μ^- which can be expressed as

$$K_\mu^\mp = \frac{\mp a_\mu - \bar{M}_o}{1 - \bar{M}_o^2}, \tag{6.49}$$

where

$$a_\mu = \sqrt{1 - \frac{\alpha_\mu^2}{k_o^2}\left(1 - \bar{M}_o^2\right)}. \tag{6.50}$$

Hence, in the frequency domain, to each eigenvalue α_μ^2, $\mu = 1, 2, \ldots$, there correspond two solutions of the convected wave equation, $p' = C_\mu^+ \varphi_\mu(\bullet) e^{ik_o K_\mu^+ x_1}$ and $p' = C_\mu^- \varphi_\mu(\bullet) e^{ik_o K_\mu^- x_1}$, where $C_\mu^\mp$ denotes integration constants. It is convenient to express these solutions as

$$p'_\mu(x_1, \bullet) = p_\mu^+(x_1)\varphi_\mu(\bullet) \tag{6.51}$$

$$p'_\mu(x_1, \bullet) = p_\mu^-(x_1)\varphi_\mu(\bullet), \tag{6.52}$$

where

$$p_\mu^+(x_1) = p_\mu^+(0) e^{ik_o K_\mu^+ x_1} \tag{6.53}$$

$$p_\mu^-(x_1) = p_\mu^-(0) e^{ik_o K_\mu^- x_1} \tag{6.54}$$

and, obviously, $C_\mu^+ = p_\mu^+(0)$, $C_\mu^- = p_\mu^-(0)$. Such solutions with specific propagation constants are called axial propagation modes and $p_\mu^\mp(x_1)$ are called acoustic pressure wave components of mode μ. The latter are also denoted as $p^{\mp(\mu)}$ when it is necessary to leave space for a subscript.

The general solution of the convected wave equation for the acoustic pressure field in the duct is obtained by superimposition of all possible axial mode solutions:

$$p'(x_1, \bullet) = \sum_{\mu=1}^{\infty} \varphi_\mu(\bullet)\left(p_\mu^+(x_1) + p_\mu^-(x_1)\right). \tag{6.55}$$

The corresponding solution for the particle velocity can be obtained from the acoustic momentum equation, Equation (6.6). The axial component of the latter is

$$\rho_o\left(-i\omega + \overline{v_{o1}}\frac{\partial}{\partial x_1}\right)v'_1 + \frac{\partial p'}{\partial x_1} = 0, \tag{6.56}$$

where v'_1 is the axial component of the particle velocity. Since the acoustic pressure of the axial modes have $e^{ik_o K x_1}$ dependence on the axial coordinate ($K \in K_\mu^\mp$, $\mu = 1, 2, \ldots$), it can be deduced from Equation (6.56) that v'_1 and $p'$are related in each mode as

$$v'_1 = \frac{K p'}{\rho_o c_o (1 - K\bar{M}_o)}, \tag{6.57}$$

where $K \in K_\mu^\mp, \mu = 1, 2, \ldots$. Accordingly, the general solution of the convected wave equation for the axial component of the particle velocity can be expressed as

$$v_1'(x_1, \bullet) = \frac{1}{\rho_o c_o} \sum_{\mu=1}^{\infty} \varphi_\mu(\bullet) \left(A_\mu^+ p_\mu^+(x_1) + A_\mu^- p_\mu^-(x_1) \right), \tag{6.58}$$

where

$$A_\mu^\mp = \frac{\mp a_\mu - \bar{M}_o}{1 \pm \bar{M}_o a_\mu} \tag{6.59}$$

are called axial modal admittance coefficients. It can be similarly deduced from Equation (6.16) that the transverse component of the particle velocity, $\mathbf{v}_\perp'$, is related to the acoustic pressure in each mode as

$$\mathbf{v}_\perp' = \frac{1}{\mathrm{i}\omega\rho_o(1 - K\bar{M}_o)} \nabla_\perp p', \tag{6.60}$$

with $K \in K_\mu^\mp, \mu = 1, 2, \ldots$. So, the general solution of the convected wave equation for this component is

$$\mathbf{v}_\perp'(x_1, \bullet) = \frac{1}{\mathrm{i}\omega\rho_o} \sum_{\mu=1}^{\infty} \left(\frac{p_\mu^+(x_1)}{1 - \bar{M}_o K_\mu^+} + \frac{p_\mu^-(x_1)}{1 - \bar{M}_o K_\mu^-} \right) \nabla_\perp \varphi_\mu(\bullet). \tag{6.61}$$

6.4.2 Modal Wave Transfer Matrix

The foregoing solution of the convected wave equation can be expressed concisely by using matrix notation. For this purpose, it is convenient to define the following arrays: Transverse eigenfunctions matrix (a one-row array):

$$\mathbf{\Phi}(\bullet) = [\varphi_1(\bullet) \quad \varphi_2(\bullet) \quad \ldots]. \tag{6.62}$$

Axial propagation constants matrices:

$$\mathbf{K}^\mp = \begin{bmatrix} K_1^\mp & 0 & \cdots \\ 0 & K_2^\mp & \cdots \\ \vdots & \vdots & \ddots \end{bmatrix}. \tag{6.63}$$

Modal admittance matrices, from Equation (6.59):

$$\mathbf{A}^\mp = \begin{bmatrix} A_1^\mp & 0 & \cdots \\ 0 & A_2^\mp & \cdots \\ \vdots & \vdots & \ddots \end{bmatrix}. \tag{6.64}$$

Modal pressure wave components vectors in frequency domain:

$$\mathbf{P}^{\mp}(x_1) = \begin{bmatrix} p_1^{\mp}(x_1) \\ p_2^{\mp}(x_1) \\ \vdots \end{bmatrix} = \mathrm{e}^{ik_o\mathbf{K}^{\mp}x_1}\mathbf{P}^{\mp}(0). \tag{6.65}$$

Then, from Equation (6.55), the acoustic pressure field in the duct can be written as

$$p'(x_1, \bullet) = \mathbf{\Phi}(\bullet)\ \left(\mathbf{P}^{+}(x_1) + \mathbf{P}^{-}(x_1)\right) \tag{6.66}$$

and, similarly, the axial component of the particle velocity becomes, from Equation (6.58):

$$v_1'(x_1, \bullet) = \frac{1}{\rho_o c_o}\Phi(\bullet)\ \left(\mathbf{A}^{+}\mathbf{P}^{+}(x_1) + \mathbf{A}^{-}\mathbf{P}^{-}(x_1)\right). \tag{6.67}$$

The modal pressure wave components vectors $\mathbf{P}^{\mp}$at any section of the duct are related as

$$\mathbf{P}^{-}(x_1) = \mathbf{R}(x_1)\mathbf{P}^{+}(x_1), \tag{6.68}$$

where $\mathbf{R}$ is called reflection coefficient matrix. It follows from Equation (6.65) that

$$\mathbf{R}(x_1) = \mathrm{e}^{-ik_o\mathbf{K}^{+}x_1}\mathbf{R}(0)\mathrm{e}^{ik_o\mathbf{K}^{-}x_1}. \tag{6.69}$$

The modal wave transfer matrix, $\mathbf{T}$, of a duct element of length x_1 may be defined now as

$$\begin{bmatrix} \mathbf{P}^{+}(x_1) \\ \mathbf{P}^{-}(x_1) \end{bmatrix} = \mathbf{T} \begin{bmatrix} \mathbf{P}^{+}(0) \\ \mathbf{P}^{-}(0) \end{bmatrix}, \tag{6.70}$$

since the continuity of the modal pressure wave components is sufficient to ensure the continuity of the acoustic pressure and particle velocity in the duct. Then, from Equation (6.65), it follows that

$$\mathbf{T} = \begin{bmatrix} \mathrm{e}^{ik_o\mathbf{K}^{+}x_1} & \mathbf{0} \\ \mathbf{0} & \mathrm{e}^{ik_o\mathbf{K}^{-}x_1} \end{bmatrix}. \tag{6.71}$$

This is the three-dimensional generalization of the one-dimensional duct element described in Section 3.5.1 for the case of uniform mean flow ($\alpha = 1$). In order to implement this duct element, it is necessary to know the solution of Equation (6.48) for specific section geometry and wall structure. Hard-walled ducts and ducts with locally reacting finite impedance walls are considered separately in the following sections.

6.5 Hard-Walled Ducts with Uniform Mean Flow

Multi-modal sound transmission in a uniform hard-walled duct with uniform mean flow is a basic problem of duct acoustics. It is an adequately accurate model in many practical applications with or without mean flow and, being tractable to relatively straightforward analysis, provides insight into the physics of multi-modal sound propagation in ducts. In this section we will discuss the general features of the transverse modes, derive expressions for the time-averaged acoustic power transmitted by the modes, and describe analytical and numerical calculation of the eigenvalues and eigenfunctions, which are required for the formulation of the modal wave transfer matrices for ducts and duct systems.

6.5.1 Eigenvalues and the Orthogonality of Eigenfunctions

If the duct is hard-walled, $Z_W \to \infty$, both the Ingard–Myers boundary condition (Section 6.3.3) and the modified Ingard–Myers boundary condition (Section 6.3.4), which are relevant here because the mean flow is assumed to be uniform, reduce to

$$\mathbf{n}\cdot\nabla_\perp \ \varphi = 0 \ \text{on} \ \Gamma. \tag{6.72}$$

In this case, it can be proved that the eigenvalues of the two-dimensional Helmholtz equation, Equation (6.48), under this boundary condition are real and positive [14]. They are usually given in ascending order as $\alpha_1^2 \le \alpha_2^2 \le \cdots \le \alpha_\mu^2 \le \cdots$ and the corresponding eigenfunctions are denoted as φ_μ, $\mu = 1, 2, \ldots$ Depending on the mode number, μ, eigenfunctions vanish along a number of lines or curves on a duct section, which are called nodal lines. According to Courant's nodal line theorem [14], the nodal lines of the μth eigenfunction divide the duct cross section into no more than $\mu - 1$ subregions. The smallest eigenvalue is always $\alpha_1^2 = 0$ and corresponding eigenfunction is $\varphi_1(\bullet)$ =constant. Eigenfunctions are usually normalized so as to have a mean-squared cross-section area average of unity; that is,

$$\int_S \varphi_\mu^2 \mathrm{d}S = S, \quad \mu = 1, 2, \ldots \tag{6.73}$$

where S denotes the duct cross-sectional area.

Another general property of eigenfunctions can be deduced by noting that, if φ_μ and φ_η are two eigenfunctions which satisfy Equation (6.48) for a given duct, then $\phi_\eta \nabla_\perp^2 \phi_\mu = -\alpha_\mu^2 \phi_\mu \phi_\eta$ and $\varphi_\mu \nabla_\perp^2 \varphi_\eta = -\alpha_\eta^2 \varphi_\eta \varphi_\mu$. Integrating the difference of these equations over the duct cross-sectional area gives

$$\int_S \left(\varphi_\eta \nabla_\perp^2 \varphi_\mu - \varphi_\mu \nabla_\perp^2 \varphi_\eta\right) \mathrm{d}S = -\left(\alpha_\mu^2 - \alpha_\eta^2\right) \int_S \varphi_\mu \varphi_\eta \mathrm{d}S. \tag{6.74}$$

After substitution of the identity $\nabla_\perp\cdot\left(\varphi_\eta \nabla_\perp \varphi_\mu - \varphi_\mu \nabla_\perp \varphi_\eta\right) = \varphi_\eta \nabla_\perp^2 \varphi_\mu - \varphi_\mu \nabla_\perp^2 \varphi_\eta$, Stoke's theorem is used to transform the area integral on the left-hand side of Equation (6.74) to a line integral:

$$\oint_\ell \left(\varphi_\eta \nabla_\perp \varphi_\mu - \varphi_\mu \nabla_\perp \varphi_\eta\right)\cdot\mathbf{n}\mathrm{d}\ell = -\left(\alpha_\mu^2 - \alpha_\eta^2\right) \int_S \varphi_\mu \varphi_\eta \mathrm{d}S, \tag{6.75}$$

where ℓ denotes the boundary curve of S. But in view of Equation (6.72), this line integral vanishes. Therefore, if $\alpha_\eta \neq \alpha_\mu$, then

$$\int_S \varphi_\mu \varphi_\eta \mathrm{d}S = 0. \tag{6.76}$$

This result is usually stated as the orthogonality property of the eigenfunctions. It is possible that two or more independent eigenfunctions correspond to the same eigenvalue. However, the eigenfunctions can always be selected, for example, by Schmidt orthogonalization [4], so that Equation (6.76) is satisfied by any two independent eigenfunctions.

An important corollary to Equation (6.76) is that all higher order eigenfunctions ($\mu > 1$) vanish when integrated over the duct cross section, which follows from the fact that φ_1 is a constant.

6.5.2 Propagating and Evanescent Modes

The propagation constants $K_\mu^\mp$ given by Equation (6.49) are depicted graphically in Figure 6.1 as function of the wavenumber $k_o = \omega/c_o$ for an arbitrary mode $\mu \in 1, 2, \ldots$ They are real for $k_o > k_\mu = \alpha_\mu \sqrt{1 - \bar{M}_o^2}$. This condition is called the cut-on condition for mode μ and $f_\mu = c_o k_\mu / 2\pi$ is called cut-on frequency of mode μ. If this condition is satisfied for mode μ, it is called cut-on or propagating mode. Application of Briggs' criterion (Section 1.3.2) to the dispersion equation, Equation (6.23), shows that, the $\mp$signs in $K_\mu^\mp$ then correspond to acoustic modes propagating in the $\mp x_1$ directions, respectively. The signs of $K_\mu^\mp$ are also the same as the superscripts "$\pm$," except for $k_\mu < k_o < \alpha_\mu$. In this interval, both acoustic propagation constants have the same sign as the sign of $-\bar{M}_o$, however, Briggs' criterion shows that they still correspond to acoustic modes propagating in opposite directions.

The smallest transverse eigenvalue, α_1^2, is real and equal to zero. Consequently, the $\mu = 1$ mode, which is called fundamental or plane-wave mode, is always cut-on. For the plane-wave mode, $\alpha_1 = 0$, ϕ_1 =constant and $K_1^\mp = \mp 1/(1 \mp \bar{M}_o)$. In Figure 6.1,

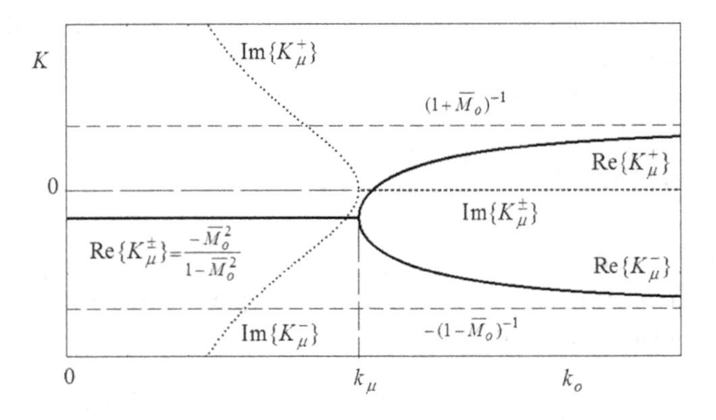

Figure 6.1 Display of the propagation constants on the K vs. k_o plane ($\mu = 1, 2, \ldots$).

the dashed horizontal asymptotes represent the propagation constants of the plane-wave mode. As $k_o \to \infty$, cut-on propagation constants approach the acoustic plane-wave propagation constants asymptotically.

Propagation constants are complex in the range $k_o < k_\mu$ and correspond to decaying (evanescent) acoustic modes. At frequencies below the cut-on frequency f_μ, the propagation constants for $\mu > 1$ have a non-zero imaginary part and, therefore, the corresponding modes decay axially with distance in the direction of propagation. At the cut-on frequency, the two propagation constants become identical and equal to $-\bar{M}_o/\left(1-\bar{M}_o^2\right)$ for all modes.

Thus, the modal pressure wave components p_μ^+ and p_μ^- can be sharply classified as propagating and evanescent (decaying) modes depending on whether the frequency is greater or less than the cut-on frequency of the mode in question [15]. The quotient k_μ/k_o is called the cut-off ratio. The greater the cut-off ratio from unity is, the more rapid is the decay of the mode with axial distance. In fact, the rate of decay of a cut-off mode is determined by the exponential factors in Equations (6.53) and (6.54). Hence, attenuation of a cut-off mode per unit duct length in the $\mp x_1$ directions can be estimated from $8.686k_o|\mathrm{Im}\{K^\mp\}|$ in dB/m units. A well cut-off mode with $k_\mu >> k_o$ attenuates at the rate of 8.686 dB/m and may usually be neglected in practical calculations. However, when the mode is near to cut-off, the rate of attenuation is relatively small and it may have to be included in calculations.

The axial propagation constants K_μ^+ and K_μ^- can be determined graphically [16]. The construction is based on the observation that Equation (6.23) can be recast as

$$\frac{\left(K+\dfrac{\bar{M}_o}{1-\bar{M}_o^2}\right)^2}{\left(\dfrac{1}{1-\bar{M}_o^2}\right)^2}+\frac{\left(\dfrac{\alpha}{k_o}\right)^2}{\dfrac{1}{1-\bar{M}_o^2}}=1. \tag{6.77}$$

This is an equation of the ellipse in the α/k_o vs. K plane, with its center at $(-\bar{M}_o/(1-\bar{M}_o^2), 0$ and having a semi-major axis of $1/(1-\bar{M}_o^2)$ and semi-minor axis of $1/\sqrt{1-\bar{M}_o^2}$, as shown in Figure 6.2. The coordinates of the extremities of the major axis are $(-1/1-\bar{M}_o, 0)$ and $(1/1+\bar{M}_o, 0)$, which correspond to the propagation constants K_1^+ and K_1^-, respectively, of the plane-wave mode. For the higher-order modes, the propagation constants are given by the intersections of this basic ellipse with the horizontal line drawn from the corresponding α_μ/k_o point on this axis. Obviously, no intersection indicates no cut-on. Since the squared semi-minor axis of the ellipse is $1/1-\bar{M}_o^2$, if a mode is cut-on for $\bar{M}_o = M$, it will also cut-on for any $\bar{M}_o > M$. For $\bar{M}_o = 0$, the basic ellipse becomes a circle of unit radius.

6.5.3 Modal Propagation Angles

Propagating modes may be envisaged as plane waves propagating at certain angles to the duct axis. To see this, consider a plane wave with a real wavenumber vector

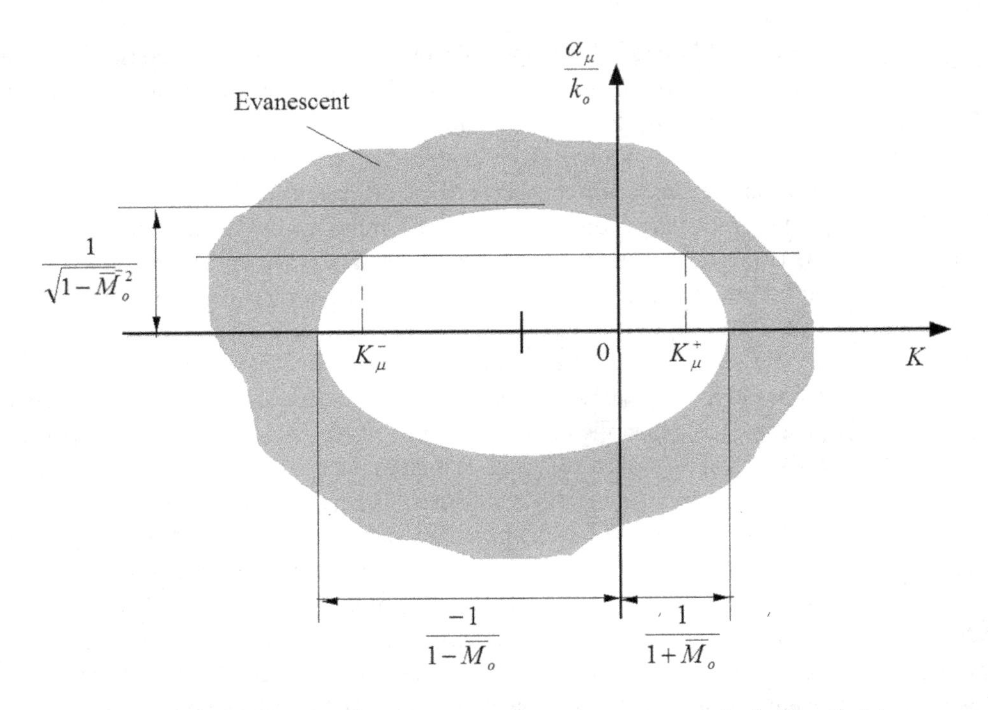

Figure 6.2 Graphical determination of the axial propagation constants of a hard-walled duct.

$\boldsymbol{\kappa} = \mathbf{e}_1\kappa_1 + \mathbf{e}_2\kappa_2 + \mathbf{e}_3\kappa_3$, where $\mathbf{e}_j$ denotes unit vectors in the positive direction of axes x_j, $j = 1, 2, 3$, of a rectangular frame (x_1, x_2, x_3), x_1 being the duct axis, and κ_j denotes the component of the wave number vector in the $\mathbf{e}_j$ direction. In terms of its direction cosines, $\boldsymbol{\kappa}$ may be expressed as $\boldsymbol{\kappa} = |\boldsymbol{\kappa}|(\mathbf{e}_1 \cos\theta_1 + \mathbf{e}_2 \cos\theta_2 + \mathbf{e}_3 \cos\theta_3)$, where θ_j denote the angle between the wavenumber vector and the axis x_j, that is, $\cos\theta_j = \kappa_j/|\boldsymbol{\kappa}|$. In the frequency domain, the spatial dependence of this plane wave is of the form $e^{i\boldsymbol{\kappa}\cdot\mathbf{x}}$, where $\mathbf{x} = \mathbf{e}_1x_1 + \mathbf{e}_2x_2 + \mathbf{e}_3x_3$ denotes the position vector of a point in the coordinate system considered. Thus, $e^{i\boldsymbol{\kappa}\cdot\mathbf{x}} = e^{i\kappa_1x_1/|\boldsymbol{\kappa}|}e^{i\kappa_2x_2/|\boldsymbol{\kappa}|}e^{i\kappa_3x_3/|\boldsymbol{\kappa}|}$ and we observe that the axial factor $e^{i\kappa_1x_1/|\boldsymbol{\kappa}|}$ is in the form of Equation (6.53) or (6.54). The idea now is to identify κ_1 and $|\boldsymbol{\kappa}|$, so that the latter are expressed in the form of the axial plane-wave factor $e^{i\kappa_1x_1/|\boldsymbol{\kappa}|}$. For this purpose, we recast the dispersion equation, Equation (6.23), in the form $\alpha^2 + k_o^2K^2 = k_o^2(1 - K\bar{M}_o)^2$. Since K is the axial propagation constant, it follows that $\kappa_1 = k_oK$ and it is plausible to assume that $\alpha^2 = \kappa_2^2 + \kappa_3^2$, so that $|\boldsymbol{\kappa}|^2 = k_o^2(1 - K\bar{M}_o)^2$. Hence, $\cos\theta_1 = K/(1 - K\bar{M}_o)$ and substituting for K from Equation (6.49), we find $\cos\theta_1 = A_\mu^\mp$ for mode $\mu \in 1, 2, \ldots$, where $A_\mu^\mp$ denotes the modal admittance coefficients given in Equation (6.59). Angles θ_2 and θ_3 may be determined similarly, but the foregoing decomposition of α^2 depends on the shape of the duct section. For example, it is given by Equation (6.87) in the case of rectangular sections, but it may not be as clear cut as this for other section shapes.

Since $A_1^\mp = \mp 1$ for the fundamental mode, we see that its angle of propagation is $\theta_{1,1} = 0$ for the forward wave and $\theta_{1,1} = \pi$ for the backward wave, which of course implies that the fundamental mode propagates with plane wavefronts perpendicular to the duct axis and, therefore, with no reflections at the duct walls. Higher-order modes ($\mu > 1$) also propagate as plane waves, but with wavefront normals at an angle of $\theta_{1,\mu} \neq 0, \pi$ to the duct axis, which means that they propagate by making multiple reflections at the duct walls.

Clearly, the propagation angle concept, which appears to have been pioneered by Rice [17], provides insight into the role of the duct walls in multi-modal propagation, which is not evident from the modal solutions. It leads to ray models of sound transmission in ducts, which have more significant usage at high frequencies when the duct walls have finite impedance. Counting the number of wave reflections at the walls provides an alternative approach for calculation of sound attenuation of the duct walls, as each bounce on the walls adds to the dissipation of acoustic energy transmitted in the duct. Following ray paths may also provide insight into the physics of radiation of higher-order wave modes from open duct terminations and the directivity patterns [18].

6.5.4 Transverse Modes of Common Duct Sections

For section geometries that can be described in a separable coordinate system, the transverse eigenvalues and eigenfunctions can be determined analytically; however, only the rectangular and circular sections are tractable to practical calculations. Elliptical and parabolic ducts have also received the attention of some authors [19–20], but the analyses are rather cumbersome. The following sections describe the eigen-characteristics of rectangular and circular hollow and annular ducts, which are the most common duct section geometries used in practice. The eigen characteristics of other common regular section geometry such as oval, elliptic and racetrack, are more easily determined by using a numerical method. Section 6.5.4.1 gives a detailed exposition of the numerical solution of the transverse duct eigenvalue problem by using the finite element method.

6.5.4.1 Rectangular Ducts

The geometry of a rectangular duct is described in rectangular coordinates (x_1, x_2, x_3). Let the transverse coordinates x_2 and x_3 represent, respectively, the depth and width coordinates of a rectangular section occupying the region $0 \leq x_2 \leq b$, $0 \leq x_3 \leq d$ (Figure 6.3). The transverse eigenfunctions are of the form of Equation (6.29), that is, $\varphi(x_2, x_3) = X_2(x_2)X_3(x_3)$, and as can be seen from Equation (6.30), in this case Equation (6.48) can be expressed as

$$\frac{1}{X_2}\frac{d^2X_2}{dx_2^2} = -\left(\frac{1}{X_3}\frac{d^2X_3}{dx_3^2} + \alpha^2\right) = -\lambda^2, \tag{6.78}$$

where λ^2 denotes a positive constant. Hence,

$$X_2 = A_2 \cos \lambda x_2 + B_2 \sin \lambda x_2 \tag{6.79}$$

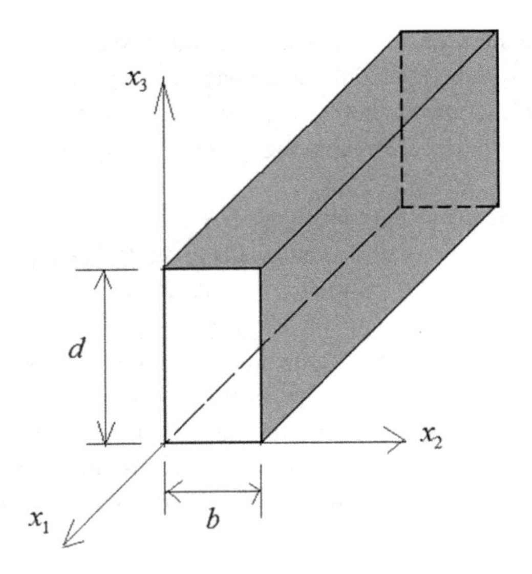

Figure 6.3 Rectangular uniform duct.

$$X_3 = A_3 \cos \beta x_3 + B_3 \sin \beta x_3 \tag{6.80}$$

where

$$\beta^2 = \alpha^2 - \lambda^2. \tag{6.81}$$

From Equations (6.29) and (6.72), the hard-wall boundary conditions on the sides of a rectangular section can be expressed as

$$\frac{dX_2}{dx_2} = 0 \quad \text{at} \quad x_2 = 0, b \tag{6.82}$$

$$\frac{dX_3}{dx_3} = 0 \quad \text{at} \quad x_3 = 0, d. \tag{6.83}$$

Substituting Equations (6.79) and (6.80), we find

$$B_2 = 0, \quad \lambda = \frac{m\pi}{b}, \quad m = 0, 1, \ldots \tag{6.84}$$

$$B_3 = 0, \quad \beta = \frac{n\pi}{d}, \quad n = 0, 1, \ldots \tag{6.85}$$

Thus, the transverse eigenfunctions are given by (the integration constants being suppressed):

$$\varphi_{mn}(x_2, x_3) = \cos\left(\frac{m\pi x_2}{b}\right) \cos\left(\frac{n\pi x_3}{d}\right) \qquad m, n = 0, 1, 2, \ldots \tag{6.86}$$

The corresponding transverse eigenvalues follow from Equation (6.81) as

$$\alpha_{mn}^2 = \left(\frac{m^2}{b^2} + \frac{n^2}{d^2}\right)\pi^2. \tag{6.87}$$

Thus, the eigenfunctions are characterized by m nodal lines spanning the cross section along its width and n nodal lines spanning along its depth.

In this solution, the transverse eigenvalues occur naturally in double index notation (m,n). They can be converted to the single index notation ($\mu = 1, 2, \ldots$), by ordering them in ascending order. For example, for the plane-wave mode, $\alpha_1^2 = \alpha_{00}^2 = 0$.

6.5.4.2 Hollow Circular Ducts

The geometry of circular ducts is described in cylindrical polar coordinates (x_1, r, θ) centered at the duct center, where r denotes the radial coordinate and θ denotes the angular coordinate measured anti-clockwise from the x_2 axis (Figure 6.4). The transverse eigenfunctions are of the form of Equation (6.31), that is, $\varphi(r, \theta) = R(r)\mathrm{e}^{\mathrm{i}m\theta}$, $m = 0, \mp 1, \mp 2, \ldots$ and, as can be seen from Equation (6.32), in this case Equation (6.48) gives for the radial function $R(r)$

$$r^2 \frac{\mathrm{d}^2 R}{\mathrm{d}r^2} + r \frac{\mathrm{d}R}{\mathrm{d}r} + \left(\alpha^2 r^2 - m^2\right) R = 0. \tag{6.88}$$

Solution of this equation can be expressed as

$$R(r) = A_m \mathrm{J}_m(\alpha r) + B_m \mathrm{Y}_m(\alpha r), \qquad m = 0, 1, 2, \ldots \tag{6.89}$$

where A_m and B_m denote the integration constants, and J_m and Y_m denote the Bessel functions of the first and second kind of order m. Strictly speaking, this solution also applies for $m = -1, -2, \ldots$; however, since $\mathrm{J}_{-m} = (-1)^m \mathrm{J}_m$, it suffices to consider only the positive values of m.

From Equations (6.31) and (6.72), the hard-wall boundary conditions on the duct wall reduce to

$$\left.\frac{\mathrm{d}R}{\mathrm{d}r}\right|_{r=a} = 0, \tag{6.90}$$

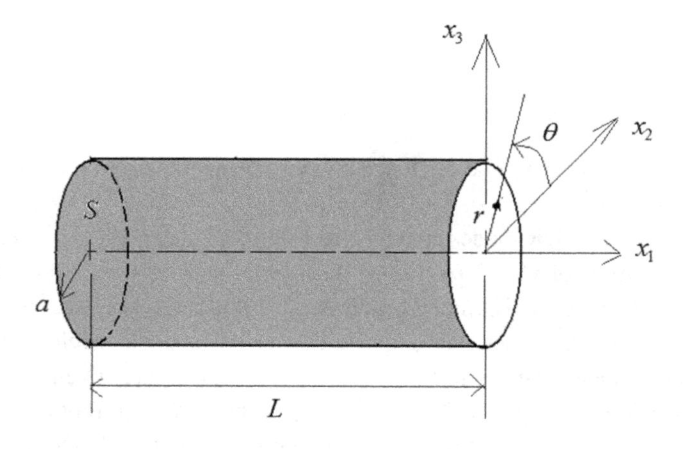

Figure 6.4 Coordinate system for circular ducts.

where a denotes the duct radius. But, $R(r)$ must be finite for $0 \leq r \leq a$. Therefore, $B_m = 0$, since $Y_m(\alpha r) \to \infty$ as $r \to 0$. Then, Equation (6.90) gives

$$\left.\frac{\mathrm{d}J_m(\alpha r)}{\mathrm{d}r}\right|_{r=a} = 0, \tag{6.91}$$

This condition is usually expressed as $J'_m(\alpha a) = 0$, where the prime denotes differentiation with respect to the argument of the function (not to be confused with the prime notation used for the acoustic variables). The roots of Equation (6.91) for a given m are denoted by $j'_{mn} = \alpha_{mn}a$, $n = 0, 1, 2, \ldots$, in ascending order. (Some authors prefer the $n = 1, 2, \ldots$ scheme.) The smallest eigenvalue occurs for $m = 0$ and $n = 0$ and is $\alpha_{00} = 0$, since $j'_{00} = 0$. Zero roots also occur for $m \neq 0$, but these should be discarded since they correspond to trivial eigenfunctions.

The transverse eigenfunctions corresponding to eigenvalue α^2_{mn} are, except for a constant factor,

$$\varphi_{mn}(r,\theta) = J_m(\alpha_{mn}r)e^{im\theta} \qquad m, n = 0, 1, 2, \ldots \tag{6.92}$$

or in real form

$$\varphi_{mn}(r,\theta) = J_m(\alpha_{mn}r)\begin{cases} \cos m\theta, \\ \sin m\theta, \end{cases} \qquad m, n = 0, 1, 2, \ldots \tag{6.93}$$

which implies that the eigenvalues of the non-axisymmetric $m \neq 0$ modes are in fact double roots of Equation (6.91) with independent eigenfunctions.

The integral over the duct cross-sectional area of all possible combinations of products of any two eigenfunctions vanish (Section 6.5.1), except for the squares of the individual eigenfunctions, which is

$$\int_S \varphi^2_{mn}(r,\theta)\mathrm{d}S = S\lambda_{mn}, \tag{6.94}$$

where $S = \pi a^2$, $\mathrm{d}S = r\mathrm{d}r\mathrm{d}\theta$ and

$$\lambda_{mn} = \varepsilon_m\left(1 - \left(\frac{m}{j'_{mn}}\right)^2\right)J^2_m(j'_{mn}) \tag{6.95}$$

and $\varepsilon_m = 1$ if $m = 0$, else $\varepsilon_m = 0.5$. Hence, the eigenfunctions can be normalized by dividing them by $\sqrt{\lambda_{mn}}$.

Roots j'_{mn} for the first few modes are given in Table 6.1. In general, $j'_{m0} > m$ and lies between positive roots of $x^4 - 3(m+2)^2x^2 + 2m(m+1)(m+3)(m+4) = 0$.

The $m = 0$ modes are axisymmetric and have n nodal circles, but no diametral nodal lines. The $m > 0$ modes have, in addition to the n nodal circles, m equally spaced diametral nodal lines. The radii of the nodal circles are given by $r_{mn}/a = j_{mn}/j'_{mn}$, $m = 0, 1, 2, \ldots,$ $n = 1, 2, \ldots$, where j_{mn} denotes the nth root of $J_m(z) = 0$. These roots are also given in Table 6.1 (the numbers in brackets). It should be noted that the nodal circles for a value of n consist of n nodal circles corresponding j_{mn} for

Table 6.1 Zeroes of $dJ_m(z)/dz = 0$ and $J_m(Z) = 0$

n	$j'_{0n}(j_{0n})$	$j'_{1n}(j_{1n})$	$j'_{2n}(j_{2n})$	$j'_{3n}(j_{3n})$	$j'_{4n}(j_{4n})$
0	0	1.84118	3.05424	4.20119	5.31755
1	3.83170	5.33144	6.70613	8.01524	9.28240
	(2.40483)	(3.83171)	(5.13562)	(6.38016)	(7.58834)
2	7.01558	8.53632	9.96947	11.34592	12.68191
	(5.52008)	(7.1559)	(8.41724)	(9.76102)	(11.06471)
3	10.17346	11.70600	13.17037	14.58585	15.96411
	(8.65372)	(10.17347)	(11.61984)	(13.01520)	(14.37254)
4	13.32367	14.86359	16.34752	17.78875	19.19603
	(11.79153)	(13.32369)	(14.79595)	(16.22347)	(17.61597)

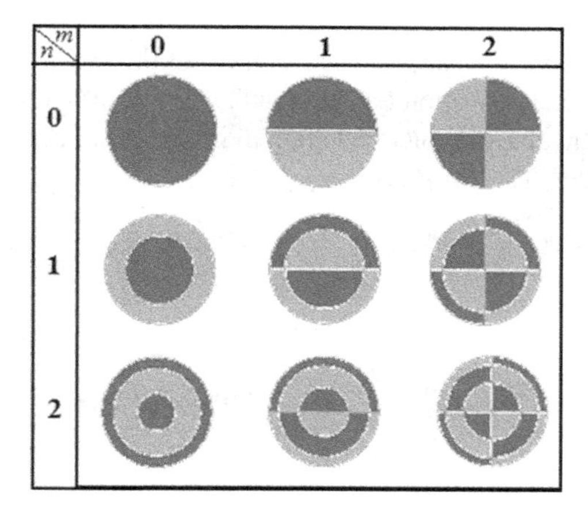

Figure 6.5 Nodal lines of the transverse eigenfunctions of circular ducts. (Only the $\cos(m\theta)$ modes are shown.) Change in the tone of gray indicates change of sign. (See Table 6.1 for the corresponding eigenvalues.)

$n, n-1, \ldots, 1$. For example, for the axisymmetric mode $m = 0$, $n = 2$, the radii of the nodal circles are $r_{01}/a = 2.40483/3.83170 = 0.62761a$ and $r_{02}/a = 5.52008/7.01558 = 0.78683$.

So except for the axisymmetric $m = 0$ modes, for which the non-trivial eigenfunctions are $\varphi_{0n}(r, \theta) = J_0(\alpha_{0n} r)$, all remaining eigenvalues of a circular duct are characterized by having multiplicity of two; that is, there are two linearly independent eigenfunctions for each of them, which are distinguished by the $\cos(m\theta)$ and $\sin(m\theta)$ factors in Equation (6.93). Consequently, if the transverse eigenvalues are to be placed in ascending order for single index representation, each double indexed non-axisymmetric eigenvalue will have to be counted twice.

Figure 6.5 shows the nodal line configurations for some modes graphically, with the radii of the nodal circles drawn to scale. Only the $\cos(m\theta)$ modes are shown, as the sin $(m\theta)$ modes can be obtained from them by rotating the nodal diameters by an angle of $\pi/2m$ radians. Eigenfunctions have the same sign in a region bounded by nodal lines or arcs, but the sign changes as each nodal circle or arc is crossed in passing to the adjacent region.

6.5.4.3 Annular Circular Ducts

The solution for a hollow circular section can be extended to a circular annulus $b \le r \le a$ (Figure 6.6) by noting that Equation (6.91) must be satisfied both at $r = a$ and $r = b$. Application of these conditions gives two linear homogeneous equations for the integration constants A_m and B_m, and the condition for determinant of the coefficients matrix should vanish for non-trivial solutions, giving the following equation for the transverse eigenvalues:

$$\alpha^2 \left(\mathrm{J}'_m(\alpha a)\mathrm{Y}'_m(\alpha b) - \mathrm{J}'_m(\alpha b)\mathrm{Y}'_m(\alpha a) \right) = 0, \tag{6.96}$$

where a prime is used to denote differentiation with respect to the argument.

Some roots, $\alpha_{mn}a$, of this eigen equation are given in Table 6.2 for radius ratios of b/a =0.1, 0.3 and 0.5. Eigenvalues at intermediate radius ratios can be determined approximately by interpolation. The corresponding modes are given by

$$\varphi_{mn}(r,\theta) = \psi_m(\alpha_{mn}r)\mathrm{e}^{\mathrm{i}m\theta}, \qquad m,n = 0,1,2,\ \ldots \tag{6.97}$$

where

$$\psi_m(\alpha_{mn}r) = \mathrm{J}_m(\alpha_{mn}r) - \frac{\mathrm{J}'_m(\alpha_{mn}a)}{\mathrm{Y}'_m(\alpha_{mn}a)} Y_m(\alpha_{mn}r). \tag{6.98}$$

Except for the squares of the individual eigenfunctions, the integral over the duct cross-sectional area of all possible combinations of the product of any two of these eigenfunctions also vanish and Equation (6.95) holds in this case with

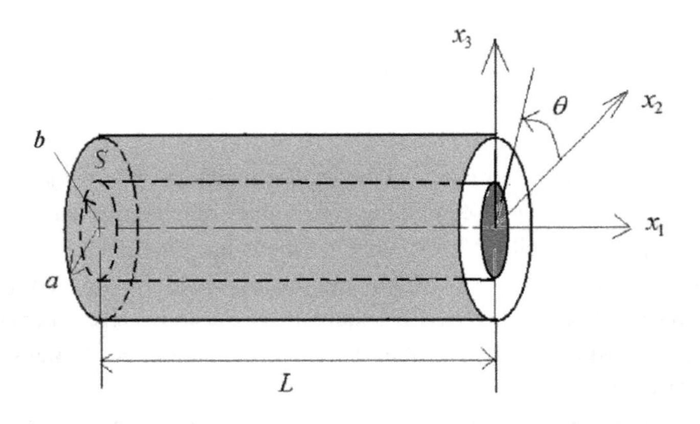

Figure 6.6 Circular annular duct of radii a, b.

Table 6.2 Transverse eigenvalues of circular annular ducts

n	$\alpha_{on}a$	$\alpha_{1n}a$	$\alpha_{2n}a$	$\alpha_{3n}a$
		$b/a = 0.1$		
0	0	1.80347	3.05294	4.20115
1	3.94094	5.13714	6.68669	8.01416
2	7.33057	11.35879	9.88752	11.33836
3	10.74838	14.63436	12.96963	14.55603
		$b/a = 0.3$		
0	0	1.58206	2.96850	4.18011
1	4.70578	5.13739	6.27367	7.72126
2	9.10423	9.30827	9.91800	10.92015
3	13.55317	13.68364	14.07466	14.72716
		$b/a = 0.5$		
0	0	1.35467	2.68120	3.95775
1	6.39316	6.56494	7.06258	7.84011
	(4.90541)	(5.05565)	(5.49637)	(6.19964)
2	12.62470	12.70642	12.94941	13.34760
	(7.95752)	(8.01660)	(8.19339)	(8.48684)
	(11.09475)	(11.17121)	(11.39890)	(11.77313)
3	18.88893	18.94266	19.10316	19.36843
	(11.06603)	(11.10126)	(11.20680)	(11.38221)
	(14.20515)	(14.25026)	(14.38522)	(14.60901)
	(17.34516)	(17.39658)	(17.55027)	(17.80449)

$$\lambda_{mn} = \frac{\varepsilon_m a^2}{(a^2 - b^2)}\left(\left(1 - \frac{m^2}{\alpha_{mn}^2 a^2}\right)\psi_m^2(\alpha_{mn}a) + \psi'^2_m(\alpha_{mn}a) - \frac{b^2}{a^2}\left\{\left(1 - \frac{m^2}{\alpha_{mn}^2 b^2}\right)\psi_m^2(\alpha_{mn}b) + \psi'^2_m(\alpha_{mn}b)\right\}\right) \tag{6.99}$$

As in the hollow circle case, the $m = 0$ modes are axisymmetrical and have n nodal circles, but no diametral nodal lines $(n = 1, 2, \ldots)$. The $m > 0$ modes have, in addition to the n nodal circles, m equally spaced diametral nodal lines. The radii of the nodal circles can be determined from the roots of $R_{mn}(r) = 0,\ \ n = 1, 2, \ldots$ Some of these roots, $\alpha_{mn} r_{mn}$, are given in Table 6.2 in brackets.

Circular annular ducts of relatively large b/a ratio are sometimes approximated by a rectangular duct of infinite width and of depth equal to the gap thickness $b - a$. In this approximation, Equation (6.48) reduces to $(\partial^2/\partial x_3^2 + \alpha^2)\varphi(x_3) = 0$ and, therefore, the transverse eigenfunctions are given by $\varphi_n(x_3) = \cos(n\pi x_3/d)$, $n = 0, 1, 2, \ldots$.

6.5.4.4 Spinning Modes

The acoustic field in a circular annular duct is given by the modal expansion depicted generally in Equation (6.55) in single index notation. With the use of double index notation (m, n) of the previous sections for eigenvalues and eigenfunctions of circular ducts, Equation (6.55) can be expressed explicitly as

$$p' = \sum_{m=0}^{\infty}\sum_{n=0}^{\infty}\left(p_{mn}^{+}(0)\mathrm{e}^{\mathrm{i}k_oK_{mn}^{+}x_1} + p_{mn}^{-}(0)\mathrm{e}^{\mathrm{i}k_oK_{mn}^{-}x_1}\right)\psi_m(\alpha_{mn}r)\begin{cases}\cos m\theta\\ \sin m\theta\end{cases}, \tag{6.100}$$

where $\psi_m(\alpha_{mn}r)$ is given by Equation (6.98) or equal to $\mathrm{J}_m(\alpha_{mn}r)$ if the duct is hollow (see Equation (6.93)). The propagation constants are written from Equation (6.49) also in double index notation as:

$$K_{mn}^{\mp} = \frac{a_{mn}^{\mp} - \bar{M}_o}{1 - \bar{M}_o^2}, \tag{6.101}$$

with

$$a_{mn} = \sqrt{1 - \frac{\alpha_{mn}^2}{k_o^2}\left(1 - \bar{M}_o^2\right)} \tag{6.102}$$

and, therefore, the cut-on condition for mode (m,n) becomes $k_o > \alpha_{mn}\sqrt{1-\bar{M}_o^2}$.

Equation (6.100) shows that, for $m = 0$, the propagating modes (which correspond to real $K_{mn}^{\mp}$) have constant phase over the duct cross sections $x_1 =$ constant. This is illustrated in Figure 6.7a, which shows the acoustic pressure pattern of a propagating (0,n)-mode over a cylindrical surface $r =$ constant. But, for a propagating $m > 0$ mode, the constant phase points on such cylindrical surface $r =$ constant are given by $k_oK_{mn}^{\mp}x_1 \pm m\theta =$ constant, which is the equation of a helix. Here, the sign is selected so that the helix is right-handed of pitch $2\pi m/k_oK_{mn}^{\mp}$ for a mode propagating in the $+x_1$ direction.

The helical constant phase pattern is illustrated in Figure 6.7b. Thus, whilst the axisymmetrical acoustic pressure modes propagate in planar constant phase patterns, the non-axisymmetrical modes propagate in helical constant phase patterns. For this reason, the non-axisymmetrical modes are often referred to as rotating or spinning modes.

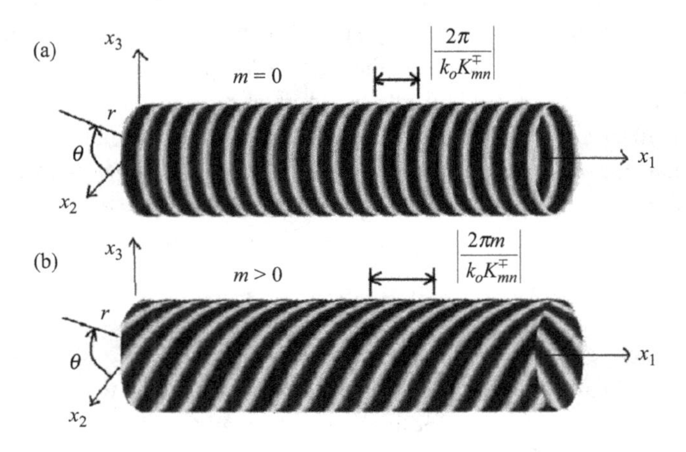

Figure 6.7 Loci of constant phase points over a cylindrical surface for propagating modes.

6.5.5 Numerical Determination of Transverse Duct Modes

For ducts having other than rectangular or circular section geometry, it is usually more convenient to use a numerical method for the determination of the transverse eigenvalues and eigenvectors. In general, any numerical method for the solution of partial differential equations may be used for this purpose. However, the finite element method is the most popular contemporary method. In this section we describe its application for the solution of Equation (6.48), $(\nabla_\perp^2 + \alpha^2)\varphi = 0$, by using triangular elements.

Let the duct cross section S of perimeter Γ be divided into N contiguous triangles of area S^Δ, $\Delta = 1, 2, \ldots, N$. As a result of this, the boundary Γ is divided into M contiguous lines Γ^l, $l = 1, 2, \ldots, M$. Since Equation (6.48) is of the same form as Equation (2.58), except that the Laplacian is now two-dimensional, its weak form may be written directly in analogy with Equation (2.60) as

$$\sum_{\Delta=1}^{N} \int_{S^\Delta} (-\nabla_\perp w \cdot \nabla_\perp \varphi + \alpha^2 w\varphi)\mathrm{d}S + \sum_{l=1}^{M} \int_{\Gamma^l} w\mathbf{n}\nabla_\perp \varphi \mathrm{d}\Gamma = 0, \tag{6.103}$$

where $\mathbf{n}$ denotes the outward unit normal vector of Γ. In view of Equation (6.60), this equation may be written as

$$\sum_{\Delta=1}^{N} \int_{S^\Delta} (-\nabla_\perp w \cdot \nabla_\perp \varphi + \alpha^2 w\varphi)\mathrm{d}S + i\omega\rho_o(1 - K\bar{M}_o) \sum_{l=1}^{M} \int_{\Gamma^l} w v_n' \,\mathrm{d}\Gamma = 0, \tag{6.104}$$

where $v_n' = \mathbf{n} \cdot \mathbf{v}'$ denotes the component of the particle velocity normal to Γ.

It is assumed that the distribution of $\varphi = \varphi(x_2, x_3)$ over a triangular element may be approximated as $\varphi(x_2, x_3) \approx \varphi^\Delta(x_2, x_3) = c_1 + c_2 x_2 + c_3 x_3$, where c_j, $j = 1, 2, 3$, are undetermined constants, and x_2 and x_3 denote the rectangular coordinates spanning a duct cross section (see Figure 6.3). This is actually the simplest complete polynomial that can be used for the approximation of Equation (6.104). Let $\varphi_j^\Delta, j = 1, 2, 3$, denote the values of φ at the three vertices of a triangular element. It is convenient to express φ^Δ in matrix notation as $\varphi^\Delta = \mathbf{gc}$, where $\mathbf{g} = [1 \;\; x_2 \;\; x_3]$ and $\mathbf{c} = [c_1 \;\; c_2 \;\; c_3]^\mathrm{T}$, the superscript "T" denoting the matrix transpose. Let $\boldsymbol{\varphi}^\Delta = [\varphi_1^\Delta \;\; \varphi_2^\Delta \;\; \varphi_3^\Delta]^\mathrm{T}$. Upon interpolating φ_j^Δ, this vector can be expressed as $\boldsymbol{\varphi}^\Delta = \mathbf{Gc}$, where

$$\mathbf{G} = \begin{bmatrix} 1 & (x_2)_1 & (x_3)_1 \\ 1 & (x_2)_2 & (x_3)_2 \\ 1 & (x_2)_3 & (x_3)_3 \end{bmatrix}. \tag{6.105}$$

Here, $(x_i)_j$, $i = 2, 3$ and $j = 1, 2, 3$ denotes value of the x_i coordinate at vertex j. Hence, the assumed distribution of φ over an element can be expressed as

$$\varphi^\Delta = \mathbf{g}\mathbf{G}^{-1}\boldsymbol{\varphi}^\Delta. \tag{6.106}$$

This is of the form of $\varphi^\Delta = \sum_{i=1}^{3} \psi_i^\Delta(x_2, x_3)\phi_i^\Delta$, where $\psi_j^\Delta((x_2)_i, (x_3)_i) = 1$, if $i = j$, else 0 (compare with Equation (2.61)). Similarly, we take $v_n' \approx v_n'^l = \mathbf{g}\mathbf{G}^{-1}\mathbf{v}^l$ with

$\mathbf{v}^l = \begin{bmatrix} v_1^l & v_2^l & v_3^l \end{bmatrix}^T$ where v_j^l denotes value of v_n' at node $j \in 1, 2$ on Σ^l. Substituting these approximations in Equation (2.60) and selecting the n independent functions for w as ψ_i^Δ, $i = 1, 2, \ldots, n^\Delta$, gives

$$\sum_{\Delta=1}^{N} \left[\mathbf{G}^{-1}\right]^{\mathrm{T}} \left(\int_{S^\Delta} \left[\frac{\partial \mathbf{g}^{\mathrm{T}}}{\partial x_2} \frac{\partial \mathbf{g}}{\partial x_2} + \frac{\partial \mathbf{g}^{\mathrm{T}}}{\partial x_3} \frac{\partial \mathbf{g}}{\partial x_3} - \alpha^2 \mathbf{g}^{\mathrm{T}} \mathbf{g} \right] \mathrm{d}x_2 \mathrm{d}x_3 \right) \mathbf{G}^{-1} \boldsymbol{\varphi}^\Delta = \mathrm{i}\omega\rho_o (1 - K\bar{M}_o) \sum_{l=1}^{M} \left[\mathbf{G}^{-1}\right]^T \left(\int_{\Gamma^l} \mathbf{g}^T \mathbf{g} \mathbf{G}^{-1} \mathbf{v}^l \mathrm{d}\Gamma \right). \tag{6.107}$$

The integrals over Δ can be evaluated exactly to obtain

$$\sum_{\Delta=1}^{N} (\mathbf{K}^\Delta - \alpha^2 \mathbf{M}^\Delta) \varphi^\Delta = \sum_{l=1}^{M} \mathbf{F}^l \mathbf{v}^l, \tag{6.108}$$

where

$$\mathbf{K}^\Delta = \left[\mathbf{G}^{-1}\right]^{\mathrm{T}} \begin{bmatrix} 0 & 0 & 0 \\ 0 & S^\Delta & 0 \\ 0 & 0 & S^\Delta \end{bmatrix} \mathbf{G}^{-1} \tag{6.109}$$

$$\mathbf{M}^\Delta = \frac{S^\Delta}{4} \left[\mathbf{G}^{-1}\right]^{\mathrm{T}} \begin{bmatrix} 4 & 4\bar{x}_2 & 4\bar{x}_3 \\ 4\bar{x}_2 & \overline{x_2^2} + 3\bar{x}_2^2 & \overline{x_2 x_3} + 3\bar{x}_2\bar{x}_3 \\ 4\bar{x}_3 & \overline{x_2 x_3} + 3\bar{x}_2\bar{x}_3 & \overline{x_3^2} + 3\bar{x}_3^2 \end{bmatrix} \mathbf{G}^{-1}. \tag{6.110}$$

Here, an over-bar denotes averaging over the three vertices of Δ, for example, $\bar{x}_2 = \left((x_2)_1 + (x_2)_2 + (x_2)_3\right)/3$. The matrix $\mathbf{F}^l$ need not be computed, as we are considering a hard-walled duct and v_n' vanishes at all nodes on Γ. Hence, defining the global vector

$$\boldsymbol{\varphi} = \begin{bmatrix} \varphi_1 \\ \varphi_2 \\ \vdots \\ \varphi_n \end{bmatrix}, \tag{6.111}$$

the summations in Equation (6.108) may be assembled as a single matrix equation

$$\left(\mathbf{K} - \alpha^2 \mathbf{M}\right) \boldsymbol{\varphi} = \mathbf{0}, \tag{6.112}$$

where $\mathbf{K}$ and $\mathbf{M}$ are symmetric $n \times n$ matrices formed from $\mathbf{K}^\Delta$ and $\mathbf{M}^\Delta$, $\Delta = 1, 2, \ldots, N$ and n denotes the number of nodes of the triangular mesh over a duct section. Equation (6.112) is known as an algebraic eigenvalue equation. It is a set of n homogeneous equations in n unknowns and can have non-trivial solutions for the vector $\boldsymbol{\varphi}$ only if the determinant of the coefficient matrix vanishes, that is, if

$|\mathbf{K} - \alpha^2\mathbf{M}| = 0$. This condition gives an equation of degree n in α^2, the roots of which represent approximations for the first (smallest) n transverse eigenvalues of the duct section. The corresponding vectors $\boldsymbol{\varphi}$, which are called eigenvectors, are then determined from Equation (6.118) and represent approximations to the corresponding eigenfunctions of the duct section. A useful property of the approximate eigenvalues given by Equation (6.111) is that, when in ascending order, they are always greater than the corresponding eigenvalues which Equation (6.48) will yield if solved exactly. This is because the triangular element described above is a complete and conforming one [21].

Although the foregoing finite element formulation is quite simple, its implementation will require further consideration of the fast numerical solution of large algebraic eigenvalue problems and the automatic triangularization of the duct cross section. The solution of the algebraic eigenvalue problem by expanding the determinant of Equation (6.116) is only of theoretical interest. In practice, the eigenvalues and eigenvectors are computed by using numerical algorithms which make full use of the fact that the matrices $\mathbf{K}$ and $\mathbf{M}$ are symmetrical. On the other hand, the accuracy of the results depends on the number and locations of the nodes used. In general, the nodes should be distributed over the duct section so that all triangular elements have approximately equal area. A rule of thumb for the minimum size of the elements is $\alpha d << 1$, where d is the maximum size of elements in meters. The accuracy of the finite element method can also be improved by using higher-order polynomials and rectangular elements. Such issues are discussed in further detail in the extensive literature on the finite element method (for example, the FreeFem++ site, www3.freefem.org/, which also contains open source resources for the solution of the Helmholtz equation in two dimensions, including automatic triangularization codes).

Shown in Figure 6.8 is the triangularization of an oval section having a circular cutout region and the nodal lines of the first nine modes of order higher than the plane-wave mode computed by using the previously described finite element formulation with this mesh. The numbers underneath each mode shape are the corresponding values of α in 1/m units. An oval section is constructed by joining tangentially two pairs of symmetric circular arcs. The numerical solution gives values of the normalized eigenfunctions at each node of the triangular mesh.

6.5.6 Time-Averaged Acoustic Power

The time-averaged acoustic power transmitted normal to a duct section is given by Equation (1.49). For a homogeneous straight duct with axial mean flow, we have $\mathbf{v}_o = \mathbf{e}_1 v_{o1}, \rho' = p'/c_o^2$ and with the normal vector $\mathbf{n}$ of a duct section in the direction of the unit vector $\mathbf{e}_1$, Equation (1.49) yields

$$W = 2\mathrm{Re}\int_S \left(\left(1+\ \bar{M}_o^2\right)\left(\breve{p}'\right)^{\mathrm{T}} v_1' + \rho_o c_o \bar{M}_o \left(\breve{v}_1'\right)^{\mathrm{T}} v_1' + \frac{\bar{M}_o}{\rho_o c_o}\left(\breve{p}'\right)^{\mathrm{T}} p'\right) \mathrm{d}S, \quad (6.113)$$

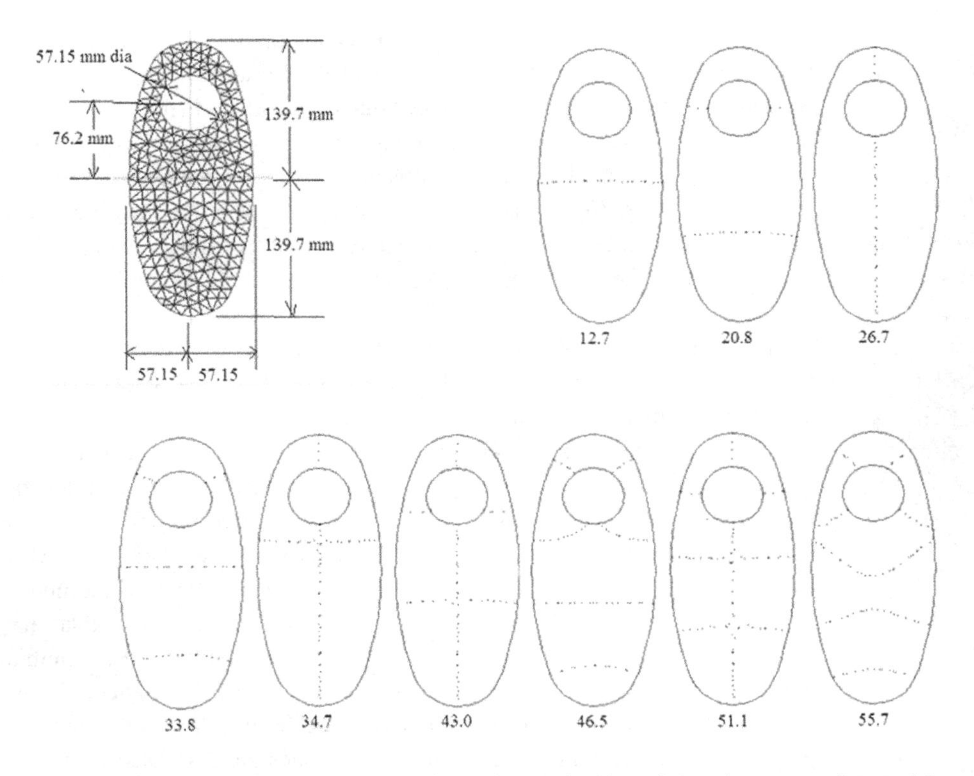

Figure 6.8 First nine non-zero transverse eigenvalues (α in 1/m units) and corresponding nodal lines of a duct section having an oval section and a circular cut-out.

where $\bar{M}_o = \bar{v}_{o1}/c_o$. With substitution of the matrix representation of the multi-modal solutions of the acoustic variables (Section 6.4.2), this may be expressed as

$$W = \frac{2}{\rho_o c_o}\mathrm{Re}\left\{\left(1+\bar{M}_o^2\right)\left(\breve{\mathbf{P}}^+ + \breve{\mathbf{P}}^-\right)^{\mathrm{T}}\left(\int_S \breve{\mathbf{\Phi}}^{\mathrm{T}}\mathbf{\Phi}\mathrm{d}S)(\mathbf{A}^+\mathbf{P}^+ + \mathbf{A}^-\mathbf{P}^-\right)\right.$$
$$+\bar{M}_o\left(\breve{\mathbf{A}}^+\breve{\mathbf{P}}^+ + \breve{\mathbf{A}}^-\breve{\mathbf{P}}^-\right)^{\mathrm{T}}\left(\int_S \breve{\mathbf{\Phi}}^{\mathrm{T}}\mathbf{\Phi}\mathrm{d}S\right)(\mathbf{A}^+\mathbf{P}^+ + \mathbf{A}^-\mathbf{P}^-)$$
$$\left.+\bar{M}_o\left(\breve{\mathbf{P}}^+ + \breve{\mathbf{P}}^-\right)^{\mathrm{T}}\left(\int_S \breve{\mathbf{\Phi}}^{\mathrm{T}}\mathbf{\Phi}\mathrm{d}S\right)(\mathbf{P}^+ + \mathbf{P}^-)\right\}. \tag{6.114}$$

For hard-walled ducts, since the eigenfunctions are real and orthogonal (Section 6.5.1), the integral of matrix $\breve{\mathbf{\Phi}}^{\mathrm{T}}\mathbf{\Phi}$ is diagonal, with diagonal elements $\int_S \varphi_\mu^2(\bullet)\mathrm{d}S$, $\mu = 1, 2, \ldots$, which are equal to S, the duct cross-sectional area, if the eigenfunctions are normalized as in Equation (6.76). Therefore, Equation (6.114) has the canonical form $W = W_1 + W_2 + \cdots + W_\mu + \cdots$, where each term represents the mode contributions:

$$W_\mu = \frac{2S}{c_o\rho_o}\left\{ \left(\left(1+\bar{M}_o^2\right)\mathrm{Re}\left\{A_\mu^+\right\} + \bar{M}_o + \bar{M}_o\left|A_\mu^+\right|^2\right)\left|p_\mu^+(x_1)\right|^2 \right.$$
$$+ \left(\left(1+\bar{M}_o^2\right)\mathrm{Re}\left\{A_\mu^-\right\} + \bar{M}_o + \bar{M}_o\left|A_\mu^-\right|^2\right)\left|p_\mu^-(x_1)\right|^2$$
$$+ \left(\left(1+\bar{M}_o^2\right)\mathrm{Re}\left\{A_\mu^+ + A_\mu^-\right\} + 2\bar{M}_o + 2\bar{M}_o\mathrm{Re}\left\{A_\mu^+\breve{A}_\mu^-\right\}\right)\mathrm{Re}\left\{p_\mu^+(x_1)\breve{p}_\mu^-(x_1)\right\}$$
$$\left. - \left(\left(1+\bar{M}_o^2\right)\mathrm{Im}\left\{A_\mu^+ - A_\mu^-\right\} + 2\bar{M}_o\mathrm{Im}\left\{A_\mu^+\breve{A}_\mu^-\right\}\right)\mathrm{Im}\left\{p_\mu^+(x_1)\breve{p}_\mu^-(x_1)\right\} \right\}, \tag{6.115}$$

where $S = \int_S \varphi_\mu^2(\bullet)\mathrm{d}S$, $\mu = 1, 2, \ldots$, and A_μ are given by Equation (6.59). If mode μ is cut-on, then a_μ and, hence, A_μ, is real and it can be verified that Equation (6.114) simplifies to

$$W_\mu = \frac{2S}{c_o\rho_o}\left\{ \left(\left(1+\bar{M}_o^2\right)A_\mu^+ + \bar{M}_o + \bar{M}_o A_\mu^+ A_\mu^+\right)\left|p_\mu^+(x_1)\right|^2 + \left(\left(1+\bar{M}_o^2\right)A_\mu^- + \bar{M}_o + \bar{M}_o A_\mu^- A_\mu^-\right)\left|p_\mu^-(x_1)\right|^2 \right\}. \tag{6.116}$$

Thus, the time-averaged acoustic power of a cut-on mode is given by the algebraic sum of the powers transmitted by the positive and negative wave systems. If mode μ is evanescent, then a_μ is imaginary and the modal admittances are given by

$$\mathrm{Re}\left\{A_\mu^+\right\} = \mathrm{Re}\left\{A_\mu^-\right\} = -\frac{\bar{M}_o\left(1+\left|a_\mu\right|^2\right)}{1+\bar{M}_o^2\left|a_\mu\right|^2} \tag{6.117a}$$

$$\mathrm{Im}\left\{A_\mu^+\right\} = -\mathrm{Im}\left\{A_\mu^-\right\} = \frac{\left(1-\bar{M}_o^2\right)\left|a_\mu\right|}{1+\bar{M}_o^2\left|a_\mu\right|^2} \tag{6.117b}$$

and Equation (6.114) reduces to

$$W_\mu = \frac{2S}{c_o\rho_o}\left\{ \left(\left(1+\bar{M}_o^2\right)\mathrm{Re}\left\{A_\mu^+ + A_\mu^-\right\} + 2\bar{M}_o + 2\bar{M}_o\mathrm{Re}\left\{A_\mu^+\breve{A}_\mu^-\right\}\right)\mathrm{Re}\left\{p_\mu^+(x_1)\breve{p}_\mu^-(x_1)\right\} \right.$$
$$\left. - \left(\left(1+\bar{M}_o^2\right)\mathrm{Im}\left\{A_\mu^+ - A_\mu^-\right\} + 2\bar{M}_o\mathrm{Im}\left\{A_\mu^+\breve{A}_\mu^-\right\}\right)\mathrm{Im}\left\{p_\mu^+(x_1)\breve{p}_\mu^-(x_1)\right\} \right\}. \tag{6.118}$$

This means that an evanescent mode does not contribute to the acoustic power transmitted in the duct unless it is maintained as a reflective field, that is, both the positive and negative wave systems are present in the acoustic field. Since evanescent waves decay rapidly after they are generated, it is usually hypothesized that they are too weak to reflect when they reach the next discontinuity and, hence, do not discernibly contribute to the transmitted time-averaged acoustic power.

6.6 Hard-Walled Uniform Ducts Packed with Porous Material

As discussed in Section 3.7, the acoustic field in a duct packed with sound absorbent material can be determined from the results given for ideal fluids in frequency domain, by putting $\mathbf{v}_o = 0$ and replacing the speed of sound c_o and the mean density ρ_o by the corresponding equivalent bulk acoustic properties of the porous medium, which are denoted by c_e and ρ_e, respectively. Thus, the multi-modal acoustic pressure field in the duct is still given by Equation (6.55), but with the wavenumber $k_o = \omega/c_o$ replaced by $k_e = \omega/c_e$. Since Equation (6.23) now reduces to $k_e^2 K^2 = k_e^2 - \alpha^2$, the modal pressure wave components are given by

$$p_\mu^{\mp}(x_1) = p_\mu^{+}(0)\mathrm{e}^{\mp\mathrm{i}\sqrt{k_e^2-\alpha_\mu^2}x_1}, \quad \mu = 1, 2, \ldots \tag{6.119}$$

But, the classification of the wave modes as propagating and evanescent modes is not clear-cut now, since c_e (and ρ_e) are complex numbers dependent on frequency. The nature of this wave system can be deduced from the identity

$$\sqrt{k_e^2 - \alpha_\mu^2} \equiv \frac{1}{\sqrt{2}}\left(\sqrt{\left|k_e^2 - \alpha_\mu^2\right| + \left(k_e^2\right)_\mathrm{I}} + \mathrm{i}\left[\mathrm{sgn}\left(k_e^2\right)_\mathrm{I}\right]\sqrt{\left|k_e^2 - \alpha_\mu^2\right| - \left(k_e^2\right)_\mathrm{I}}\right), \tag{6.120}$$

where subscript "I" ("R") denotes the imaginary (real) part of a complex quantity. Since the square roots on the right of this equation are positive, we conclude that the pressure wave components $p_\mu^{\mp}$ still correspond to waves traveling in the $\mp x_1$ directions, respectively. On the other hand, from the equivalent fluid models of porous media, we know that $\mathrm{sgn}\left(k_e^2\right)_\mathrm{I} > 0$. Therefore, $p_\mu^{\mp}$ represent waves which decay in $\mp x_1$ directions, respectively, at the rate $\mathrm{e}^{-\sqrt{\left|k_e^2-\alpha_\mu^2\right|-\left(k_e^2\right)_\mathrm{I}}}$ per unit distance. Since

$$\left|k_e^2 - \alpha_\mu^2\right| = \sqrt{\left(\left(k_e^2\right)_\mathrm{R} - \alpha_\mu^2\right)^2 + \left(k_e^2\right)_\mathrm{I}^2} \tag{6.121}$$

there is no decay if $\left(k_e^2\right)_\mathrm{R} = \alpha_\mu^2$. The frequency at which this condition occurs can be defined as the cut-on frequency of mode μ. As the frequency is increased beyond this cut-on frequency, the rate of decay of the corresponding mode increases. So, at a given frequency, the mode which is propagating with the smallest rate of decay is the one whose cut-on frequency is closest to that given frequency.

The wave transfer matrix of a uniform duct packed with a sound absorbing material is still given by Equation (6.69), but with k_o replaced by k_e and the propagation constants defined by

$$K_\mu^{\mp} = \mp\sqrt{1 - \alpha_\mu^2/k_e^2}, \quad \mu = 1, 2, \ldots \tag{6.122}$$

Attention should be paid to the fact that this wave transfer matrix describes the acoustic wave transfer within a porous medium. If the porous medium is in contact with the fluid in the duct, the modal pressure wave components must be transferred from one medium to the other by requiring that the acoustic pressure and particle velocity are continuous at the interface. This condition can be expressed mathematically as

$$\begin{bmatrix} \mathbf{E} & \mathbf{E} \\ \dfrac{\mathbf{A}_e^+}{\rho_e c_e} & \dfrac{\mathbf{A}_e^-}{\rho_e c_e} \end{bmatrix} \begin{bmatrix} \mathbf{P}^+ \\ \mathbf{P}^- \end{bmatrix}_e = \begin{bmatrix} \mathbf{E} & \mathbf{E} \\ \dfrac{\mathbf{A}_o^+}{\rho_o c_o} & \dfrac{\mathbf{A}_o^-}{\rho_o c_o} \end{bmatrix} \begin{bmatrix} \mathbf{P}^+ \\ \mathbf{P}^- \end{bmatrix}_o, \tag{6.123}$$

where $\mathbf{E}$ denotes a unit matrix of a conforming size, $\mathbf{A}^{\mp}$ denote the diagonal modal admittance matrices given by Equation (6.64) and the subscripts "o" and "e" refer to the fluid and porous medium. Obviously, Equation (6.123) is a generalization of Equation (3.107).

6.7 Lined Uniform Ducts with Uniform Mean Flow

In this section, we continue with the analysis of Section 6.4, but now assume that duct walls are lined with locally reacting liners (Section 3.9.3.1). In most studies, the liner structure is modeled as a continuous distribution of locally reacting impedance over the duct walls and we follow this approach in this section. The liner structure may also be modeled as discrete impedance distribution, but analysis of this case is deferred to Section 7.7.4.

Axial propagation constants, K, are still related to the transverse eigenvalues as in Equation (6.49). But, since the eigenvalues are in general complex now, it is convenient to recast Equation (6.49) into the following form, in which the real and imaginary parts of the propagation constants appear explicitly:

$$K(\alpha) = \frac{\mp\sqrt{A} - \bar{M}_o}{1 - \bar{M}_o^2} + \mathrm{i}\frac{\mp\sqrt{B}}{1 - \bar{M}_o^2}, \tag{6.124}$$

where

$$A = \frac{1}{2}\left(X + \sqrt{X^2 + 4Y^2}\right) \tag{6.125}$$

$$B = \frac{1}{2}\left(-X + \sqrt{X^2 + 4Y^2}\right) \tag{6.126}$$

$$X = 1 - \frac{|\alpha|^2}{k_o^2}\left(1 - \bar{M}_o^2\right)\cos 2\theta \tag{6.127}$$

$$Y = \frac{1}{2}\frac{|\alpha|^2}{k_o^2}\left(1 - \bar{M}_o^2\right)\sin 2\theta \tag{6.128}$$

and $\alpha = |\alpha|\mathrm{e}^{\mathrm{i}\theta}$. Equation (6.124) may be proved by equating the square root in Equation (6.49) to the complex number $\sqrt{A} + \mathrm{i}\sqrt{B}$, where A and B are real and positive. In the case of a hard-walled duct ($\theta = 0$), $A = X = \alpha_{\mu}^2$, $B = 0$ and Equation (6.140) reduces to the standard form of Equation (6.49).

The transverse duct eigenvalues, α^2, and the eigenfunctions $\varphi(\bullet)$, too, are still determined by the Helmholtz equation in two dimensions, Equation (6.48), but now the Ingard–Myers boundary condition (Section 6.3.3) or its modified form (Section 6.3.4) applies. This difference introduces non-trivial difficulties in the determination and classification of the transverse duct eigenvalues and eigenfunctions.

6.7.1 Dispersion Equation for Uniformly Lined Circular Ducts

The general solution of Equation (6.48) for a circular duct (hollow or annular) is in the form of Equation (6.89). Therefore, substituting this in the Ingard–Myers boundary condition and noting from Equation (6.25) that $\mathbf{n}\cdot\nabla_{\perp} = \partial/\partial r$, Equation (6.41) gives

$$\alpha\left(\mathrm{J}'_m(\alpha r) + \frac{B_m}{A_m}\mathrm{Y}'_m(\alpha r)\right) = \frac{\mathrm{i}k_o}{\zeta}(1 - \bar{M}_o K(\alpha))^2\left(\mathrm{J}_m(\alpha r) + \frac{B_m}{A_m}\mathrm{Y}_m(\alpha r)\right), m = 0, 1, 2, \ldots \tag{6.129}$$

where $r = a$ and $B_m = 0$ for a hollow duct of radius a, $r \in a, b$ for an annular duct of outer radius a and inner radius b, and the propagation constant K is related to the eigenvalues as in Equation (6.124). For a hollow duct, Equation (6.125) simplifies to

$$\kappa \mathrm{J}'_m(\kappa) = \frac{\mathrm{i}k_o a}{\zeta}(1 - \bar{M}_o K(\kappa))^2 \mathrm{J}_m(\kappa), \quad m = 0, 1, 2, \ldots \tag{6.130}$$

where $\kappa = \alpha a$ (this κ not be confused with the axial wavenumber $\kappa = k_o K$). The transverse eigenvalues of the lined duct, α_{mn}^2, are given, by $\alpha_{mn} = \kappa_{mn}/a$, where κ_{mn} is the nth root of Equation (6.130).

For an annular duct lined on both inner and outer surfaces, it may be shown that

$$\frac{B_m}{A_m} = \frac{-\kappa \mathrm{J}'_m(\kappa) + \frac{\mathrm{i}k_o a}{\zeta}(1 - \bar{M}_o K(\kappa))^2 \mathrm{J}_m(\kappa)}{\kappa \mathrm{Y}'_m(\alpha a) - \frac{\mathrm{i}k_o a}{\zeta}(1 - \bar{M}_o K(\kappa))^2 \mathrm{Y}_m(\kappa)}. \tag{6.131}$$

Then, the dispersion equation is given by

$$\kappa\left(\mathrm{J}'_m(\varepsilon\kappa) + \frac{B_m}{A_m}\mathrm{Y}'_m(\varepsilon\kappa)\right) = \frac{\mathrm{i}k_o a}{\zeta}(1 - \bar{M}_o K(\kappa))^2\left(\mathrm{J}_m(\varepsilon\kappa) + \frac{B_m}{A_m}\mathrm{Y}_m(\varepsilon\kappa)\right), \ m = 0, 1, 2, \ldots \tag{6.132}$$

where $\varepsilon = b/a < 1$ and $\kappa = \alpha a$.

6.7.1.1 Hard-Liner Solution for Hollow Ducts

If the liner is relatively hard such that $|\zeta| >> k_o a$, it may be assumed that small changes in $\mathrm{i}k_o a/\zeta$ about zero also result in small changes of the eigenvalues about the corresponding hard-walled duct values, j'_{mn}, $m, n = 0, 1, 2, \ldots$, which satisfy $\mathrm{J}'(j'_{mn}) = 0$ (Section 6.5.4.2). Then, assuming $h = \kappa - j'_{mn}$ is small, we may set out to determine the lined duct eigenvalues κ by expanding the dispersion equations in Taylor series about j'_{mn}. Let Equation (6.130) be written as $\kappa \mathrm{J}'_m(\kappa) = (\mathrm{i}k_o a/\zeta) f(\kappa) \mathrm{J}_m(\kappa)$, where $f(\kappa) = (1 - \bar{M}_o K(\kappa))^2$. The following Taylor expansions may be shown:

$$\mathrm{J}_m(\kappa) = \mathrm{J}_m(j'_{mn}) + \frac{1}{2}\mathrm{J}''_m(j'_{mn})h^2 - \cdots \tag{6.133}$$

$$\mathrm{J}'_m(\kappa) = -\mathrm{J}''_m(j'_{mn})h + \frac{1}{2}\mathrm{J}''_m(j'_{mn})h^2 - \cdots \tag{6.134}$$

$$\begin{aligned} f(\kappa) = f(j'_{mn}) \mp \frac{2\bar{M}_o}{k_o^2 a^2 a_{mn}(j'_{mn})}\kappa_0 (1 - \bar{M}_o K_{mn}(j'_{mn}))h + \\ \frac{\mp \bar{M}_o}{k_o^2 a^2 a_{mn}(j'_{mn})}\left[\left(1 + \frac{j'^2_{mn}(1 - \bar{M}_o^2)}{k_o^2 a^2 a_{mn}^2(j'_{mn})}\right)(1 - \bar{M}_o K_{mn}(j'_{mn})) + \frac{\bar{M}_o j'^2_{mn}}{k_o^2 a^2 a_{mn}(j'_{mn})}\right]h^2 - \cdots, \end{aligned} \tag{6.135}$$

where K_{mn} and a_{mn} are given, respectively, by Equations (6.49) and (6.50) in the double index notation of Section 6.5.3, and the $\mp$ signs correspond to the signs in Equation (6.124), respectively. Substituting the foregoing expansions, the dispersion equation may be expressed as

$$\begin{aligned} &\frac{\mathrm{i}k_o a}{\zeta} f(j'_{mn})\mathrm{J}_m(j'_{mn}) + \left[\mp \frac{\mathrm{i}k_o a}{\zeta}\frac{2\bar{M}_o \mathrm{J}_m(j'_{mn})}{k_o^2 a^2 a_{mn}(j'_{mn})}(1 - \bar{M}_o K(j'_{mn})) + \mathrm{J}''{}_m(j'_{mn})\right] j'_{mn} h \\ &+ \left[-\frac{\mathrm{i}k_o a}{\zeta}\left\{ f(j'_{mn})\frac{1}{2}\mathrm{J}''_m(j'_{mn}) \mp \frac{\bar{M}_o}{k_o^2 a^2 a_{mn}(j'_{mn})}\left(\left(1 + \frac{(1 - \bar{M}_o^2)j'^2_{mn}}{k_o^2 a^2 a_{mn}^2(j'_{mn})}\right)(1 - \bar{M}_o K(j'_{mn}))\right.\right.\right. \\ &\left.\left.\left. + \frac{\bar{M}_o j'^2_{mn}}{k_o^2 a^2 a_{mn}(j'_{mn})}\right)\mathrm{J}_m(j'_{mn})\right\} - \frac{1}{2}\mathrm{J}''_m(j'_{mn})j'_{mn} + \mathrm{J}''_m(j'_{mn})\right] h^2 + \cdots = 0. \end{aligned} \tag{6.136}$$

Thus, unless $j'_{mn} = 0$, we may neglect the h^2 and higher order terms to get

$$h = \frac{\frac{\mathrm{i}k_o a}{\zeta} f(j'_{mn}) j'_{mn}}{j'^2_{mn} - m^2 \pm \frac{\mathrm{i}k_o a}{\zeta}\frac{2\bar{M}_o j'^2_{mn}}{k_o^2 a^2 a_{mn}(j'_{mn})}(1 - \bar{M}_o K(j'_{mn}))}, \quad m = 1, 2, \ldots, \quad n = 0, 1, 2, \ldots \tag{6.137}$$

where use is made of the relationship $\mathrm{J}''_m(z) = -(1 - m^2/z^2)\mathrm{J}_m(z)$. For $j'_{mn} = 0$, which corresponds to $m = n = 0$, we neglect $h^3$and higher-order terms. In this case, we have the fundamental mode relations $K(j'_{00}) = \mp 1/(1 \mp \bar{M}_o)$, $f(j'_{00}) = 1/(1 \mp \bar{M}_o)^2$ and $a_{00} = 1$. Upon using these relations, Equation (6.136) gives

$$\kappa_{00}^2 = \frac{\mathrm{i}k_o a}{\zeta}\left[(1\mp\bar{M}_o)^2 + \frac{\mathrm{i}k_o a}{2\zeta}\left(1 \pm \frac{2\bar{M}_o(1\mp\bar{M}_o)}{k_o^2 a^2}\right)\right]^{-1}. \tag{6.138}$$

For no mean flow, this reduces to $\kappa_{00} \approx \sqrt{\mathrm{i}k_o a/\zeta}$ and from Equation (6.137) it can be inferred that this is the greatest of all axisymmetric ($m = 0$) modes.

6.7.1.2 Iterative Graphical Solution

The evolution of the modes of a lined duct as frequency and liner characteristics are varied may be studied on a polar plane. The procedure is based on recasting Equation (6.130) in the form

$$H(k_o a, |\kappa|, \theta) = g(k_o a), \tag{6.139}$$

where $\kappa = |\kappa|\mathrm{e}^{1\theta}$ and

$$H(k_o a, |\kappa|, \theta) = \frac{|\kappa|\mathrm{e}^{\mathrm{i}\theta}}{(1 - K(k_o a, |\kappa|\mathrm{e}^{\mathrm{i}\theta})\bar{M}_o)^2}\frac{\mathrm{J}_m'(|\kappa|\mathrm{e}^{\mathrm{i}\theta})}{\mathrm{J}_m(|\kappa|\mathrm{e}^{\mathrm{i}\theta})} \tag{6.140a}$$

$$g(k_o a) = \frac{\mathrm{i}k_o a}{\zeta(k_o a)}. \tag{6.140b}$$

The function H also depends on the mean flow Mach number, but this is suppressed for simplicity of notation. It is convenient to introduce the polar forms

$$H(k_o a, |\kappa|, \theta) = |H(k_o a, |\kappa|, \theta)|\mathrm{e}^{\mathrm{i}\varphi(k_o a, |\kappa|, \theta)} \tag{6.141}$$

$$g(k_o a) = |g(k_o a)|\mathrm{e}^{\mathrm{i}\gamma(k_o a))}. \tag{6.142}$$

The function H is independent of the liner and as θ varies in the principal interval $0 \le \theta < 2\pi$, its magnitude, $|H|$, will depict a closed curve in the complex plane for given $k_o a$, $|\kappa|$ and $\bar{M}_o$. For ease of reference, let this curve be denoted by Λ. It should be noted that, in general, $\Lambda = \Lambda(k_o a, |\kappa|, \bar{M}_o)$, but, if $\bar{M}_o = 0$, then $\Lambda = \Lambda(|\kappa|)$.

For a given liner, the function g is represented by a single point $g(k_o a)$ on the polar plane for a given value of the frequency parameter $k_o a$. This point should fall on Λ for the same value of $k_o a$, provided that the value of $|\kappa|$ satisfies the real equation

$$|H(k_o a, |\kappa|, \theta)| = |g(k_o a)|. \tag{6.143}$$

The correct value of $|\kappa|$ can be found by keeping $k_o a$ fixed and iterating Λ by varying $|\kappa|$ until it passes through the point $g(k_o a)$. When this occurs, the following equality of the polar angles must be true:

$$\varphi(k_o a, |\kappa|, \theta) = \gamma(k_o a). \tag{6.144}$$

For an application of this graphical procedure, consider a lined duct with $m = 1$, $\bar{M}_o = 0.5$, $k_o a = 10$ and $\zeta = 2 + \mathrm{i}$. Shown in Figure 6.9 is the polar plot $|H|$ vs φ for this case with the + sign in Equation (6.124), and $|\kappa|$ varied in steps of 0.1 starting at 2. The intersection point of the dotted radius and the dotted circle represents the point

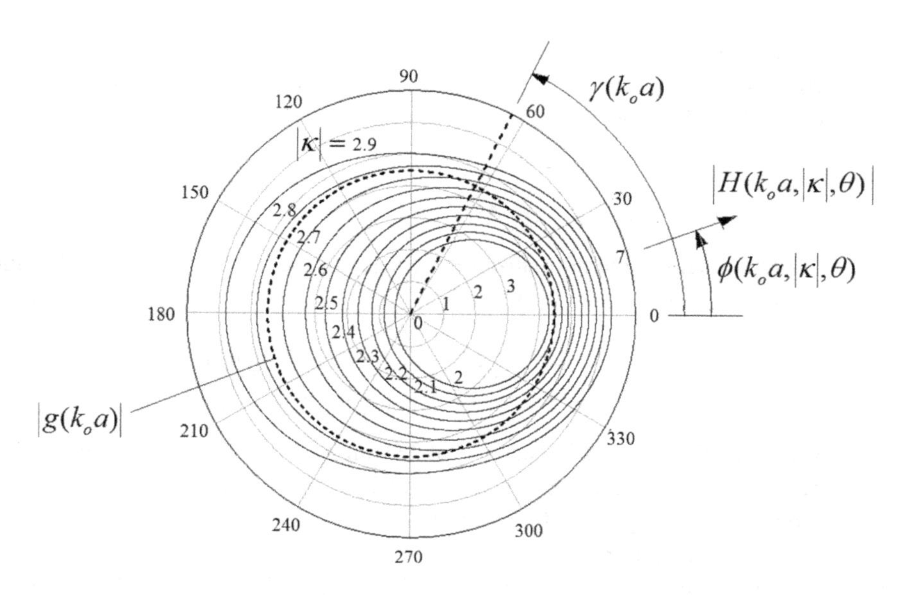

Figure 6.9 Graphical representation of Equation (6.156) on the complex plane, $m = 1, \overline{M_o} = 0.5$, $k_o a = 10$, $\xi = 2 + \mathrm{i}$ for the + sign in Equation (6.124).

$g(10) = 2 + \mathrm{i}4$. Then, the $\Lambda = \Lambda(k_o a, |\kappa|, \bar{M}_o)$ curve which passes through this point corresponds to the solution of Equation (6.130) for the problem parameters under consideration. From Figure 6.9, the solution for $|\kappa|$ is observed to lie in the range $2.6 < |\kappa| < 2.7$. This interval can be refined to the accuracy desired by increasing the resolution used in scanning $|\kappa|$. The argument of κ at the limits of the $|\kappa|$ interval are determined most easily by solving Equation (6.144) numerically. For example, simple Newton iteration gives $\theta = 2.514$ rad at $|\kappa| = 2.6$ and $\theta = 2.5034$ rad at $|\kappa| = 2.7$ (exact values are $\theta = 2.5085$ rad, $|\kappa| = 2.6501$). The corresponding value of the propagation constant is equal to $K^+ = 0.6565 + \mathrm{i}0.0338$ for both limits of the bracket to the accuracy of the number of digits given. It should be noted that if κ is a root of Equation (6.130), then $-\kappa$ is also a root, since Equation (6.130) is an even function of κ and both signs yield the same propagation constant.

Polar plots such as Figure 6.9, which can be generated relatively easily by using general purpose mathematical computation software, are useful for the visual display of the topology of the roots of Equation (6.130) and can provide estimates of the propagation constants to the desired accuracy. In this case, the performance of different liners can be compared on the same polar plot, since the Λ curves are not dependent on the liner.

6.7.2 Dispersion Equations for Uniformly Lined Rectangular Ducts

A rectangular duct may be lined on a single side or a number of sides and Equations (6.79) and (6.80) are still valid for the eigenfunctions for all possible lining combinations. Here we demonstrate the derivation of the dispersion equations for the Ingard–Myers boundary condition. If the side $x_3 = 0$ is lined and the remaining walls are hard, then Equation (6.41) links the integration constants in Equation (6.80) as

$$\frac{B_3}{A_3} = -\frac{\mathrm{i}k_o(1 - K\bar{M}_o)^2}{\zeta_1 \beta}, \tag{6.145}$$

where ζ_1 denotes the normalized impedance of the liner on side $x_3 = 0$. Similarly, if the side $x_3 = d$ is lined, then

$$\frac{B_3}{A_3} = \frac{\mathrm{i}k_o(1 - K\bar{M}_o)^2 + \zeta_2 \beta \tan \beta d}{\zeta_2 \beta - \mathrm{i}k_o(1 - K\bar{M}_o)^2 \tan \beta d}, \tag{6.146}$$

where ζ_2 denotes the impedance of the lining on side $x_3 = d$. The corresponding relations for the hard-wall case are obtained by letting the liner impedance go to infinity, that is, $B_3/A_3 = 0$ for side $x_3 = 0$ and $B_3/A_3 = \tan \beta d$ for side $x_3 = d$.

As it can be inferred from these results, if the side $x_2 = 0$ or $x_2 = b$ is lined, then the quotient of the integration constants in Equation (6.79), B_2/A_2, will be given, respectively, by the right-hand side of Equations (6.145) and (6.146) with β replaced by λ, and d replaced by b. These results enable the determination of the transverse eigenvalues of a rectangular duct having any combination of lined sides as summarized in Table 6.3. Typically, two opposite sides or all sides are lined uniformly. For example, if only the sides $x_3 = 0, d$ are lined, Equations (6.145) and (6.146) yield the dispersion equation

$$\left[1 + \frac{k_o^2(1 - K\bar{M}_o)^4}{\beta^2 \zeta_1 \zeta_2}\right] \frac{\beta \tan \beta d}{(1 - K\bar{M}_o)^2} = -\frac{\mathrm{i}k_o(\zeta_1 + \zeta_2)}{\zeta_1 \zeta_2}, \tag{6.147}$$

where $\beta^2 = \alpha^2 - (m\pi/b)^2$, $m = 0, 1, 2, \ldots$, since, in view of the hard-wall condition on sides $x_3 = 0, b$, Equation (6.79) still applies. If the duct is lined on only one wall, the corresponding form of Equation (6.147) is obtained by letting $\zeta_1 \to \infty$ or $\zeta_2 \to \infty$. If only the sides $x_2 = 0, b$ are lined, Equation (6.147) applies with β replaced by λ, d replaced by b, ζ_1, ζ_2 replaced by ζ_3, ζ_4 and $\lambda^2 = \alpha^2 - (n\pi/d)^2$, $n = 0, 1, 2, \ldots$ Finally, if all sides are lined, the transverse eigenvalues are given by $\alpha^2 = \beta^2 + \lambda^2$, where β is given by Equation (6.147) and λ by the above described counterpart of Equation (6.147) for lined sides $x_2 = 0, b$.

The graphical approach described in Section 6.7.1.2 may be used similarly for lined rectangular ducts. In this case, for example, Equation (6.147) is expressed as

$$|H(k_o d|k|, \theta)\mathrm{e}^{\mathrm{i}\varphi(k_o d|k|, \theta)} = |g(k_o d)|\mathrm{e}^{\mathrm{i}\gamma(k_o d))}, \tag{6.148}$$

where $\kappa = \beta d = |\kappa|\mathrm{e}^{1\theta}$ and the expressions on the left- and right-hand sides of Equation (6.163) are denoted by $H(k_o d, |\kappa|, \theta)$ and $g(k_o d)$, respectively. As θ varies in the principal interval $0 \le \theta < 2\pi$, $|H|$ depicts a closed curve, Λ, in the complex plane for given $k_o d$, $|\kappa|$ and $\bar{M}_o$; however, in this case, Λ is also dependent on the wall impedances ζ_1 and ζ_2. Then, since the function g is represented by a single point

Table 6.3 Dispersion equations of rectangular ducts lined on various walls

Liner	Dispersion equation	Parameters
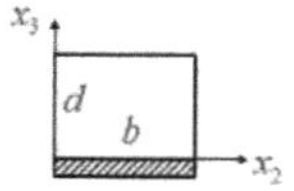 Wall $x_3=0$ lined	$\tan\beta d = -\frac{ik_o}{\zeta_{x_3=0}}(1-K\bar{M}_o)^2\frac{1}{\beta}$	$\beta^2=\alpha^2-\frac{m^2\pi^2}{b^2}$, $m=0,1,2..$
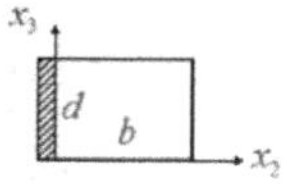 Wall $x_2=0$ lined	$\tan\lambda b = -\frac{ik_o}{\zeta_{x_2=0}}(1-K\bar{M}_o)^2\frac{1}{\lambda}$	$\lambda^2=\alpha^2-\frac{n^2\pi^2}{d^2}$, $n=0,1,2,\ldots$
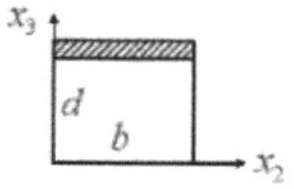 Wall $x_3=d$ lined	$\tan\beta d = -\frac{ik_o}{\zeta_{x_3=d}}(1-K\bar{M}_o)^2\frac{1}{\beta}$	$\beta^2=\alpha^2-\frac{m^2\pi^2}{b^2}$, $m=0,1,2,\ldots$
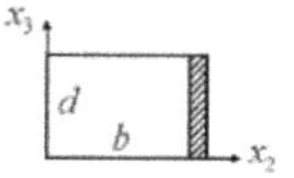 Wall $x_2=b$ lined	$\tan\lambda b = -\frac{ik_o}{\zeta_{x_2=b}}(1-K\bar{M}_o)^2\frac{1}{\lambda}$	$\lambda^2=\alpha^2-\frac{n^2\pi^2}{d^2}$, $n=0,1,2,\ldots$
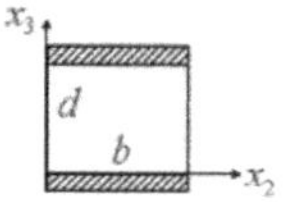 Walls $x_3=0$ and $x_3=d$ lined	$\tan\beta d = \frac{-ik_o(\zeta_{x_3=d}+\zeta_{x_3=0})(1-K\bar{M}_o)^2\beta}{\beta^2\zeta_{x_3=d}\zeta_{x_3=0}+k_o^2(1-K\bar{M}_o)^4}$	$\beta^2=\alpha^2-\frac{m^2\pi^2}{b^2}$, $m=0,1,2,\ldots$
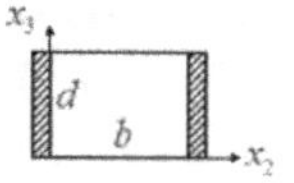 Walls $x_2=0$ and $x_2=b$ lined	$\tan\lambda b = \frac{-ik_o(\zeta_{x_2=b}+\zeta_{x_2=0})(1-K\bar{M}_o)^2\lambda}{\lambda^2\zeta_{x_2=b}\zeta_{x_2=0}+k_o^2(1-K\bar{M}_o)^4}$	$\lambda^2=\alpha^2-\frac{n^2\pi^2}{d^2}$, $n=0,1,2,\ldots$
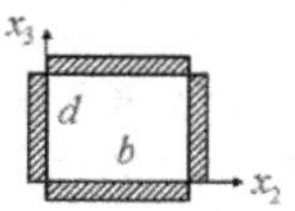 All walls lined	$\tan\beta d =$ as $x_3=0, d$ lined $\tan\lambda b =$ as $x_2=0, b$ lined	$\alpha^2=\beta^2+\lambda^2$

$g(k_o d)$ in the complex plain, the curve Λ which passes through this point corresponds to a root of Equation (6.148).

Efficacy of the lining is usually discussed in the literature in relation to two-dimensional ducts lined on one wall or both walls, since the results for all possible combinations of lined walls do not coalesce in a useful manner. For a two-dimensional duct lined on both walls, the transverse eigenvalues are given by the roots of Equation (6.147) with $\beta^2=\alpha^2$. In this case, if $\zeta_1=\zeta_2=\zeta$, we may consider the even and odd modes in Equation (6.80) separately and decompose the dispersion equation into the two dispersion equations as $\alpha\tan\alpha d = -(ik_o/\zeta)(1-\bar{M}_oK)^2$ and $\alpha\cot\alpha d = (ik_o/\zeta)(1-\bar{M}_oK)^2$, respectively, which have been often used in previous studies due to their relative simplicity. Hard-liner solutions of these dispersion equations can be derived by using the Taylor series expansion approach described in Section 6.7.1.1.

Shown in Figure 6.10 is the polar plot for a two-dimensional duct with $\bar{M}_o=0.4$, $k_o d/2=9.3915$ and $\zeta_1^{-1}=0.72+i0.42$, $\zeta_2^{-1}=0.2813-i0.121$, for the $+$ sign in Equation (6.141). It is seen that the solution for $|\kappa|$ is in the interval of $2.7<|\kappa|<2.8$. Simple Newton iteration gives $\theta=2.59201$ rad at $|\kappa|=2.7$, and $\theta=2.58738$ rad at

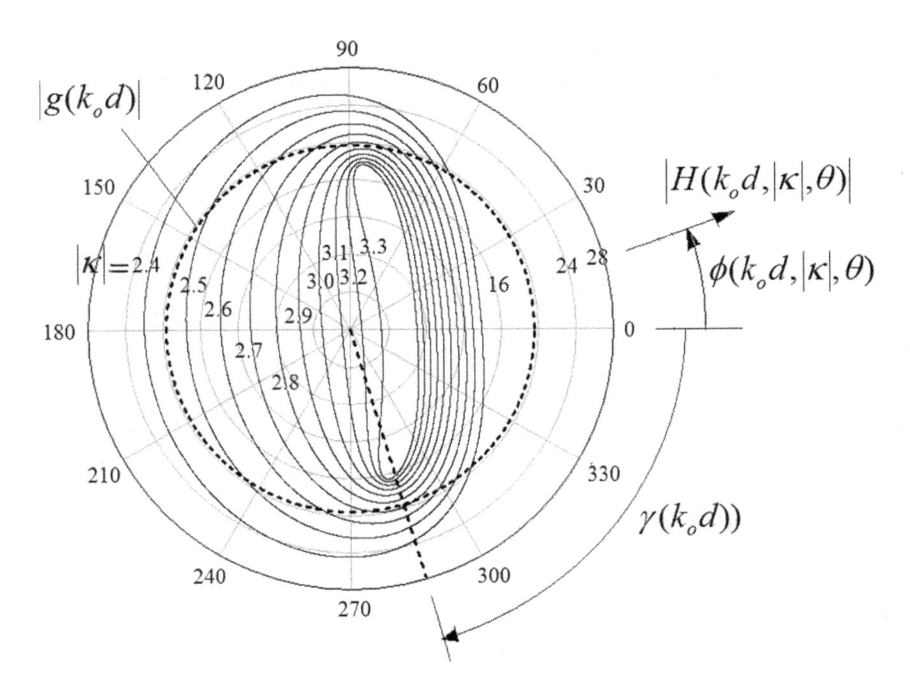

Figure 6.10 Graphical representation of Equations (6.126) on the complex plane, $M_o = 0.4$, $k_o d/2 = 9.3915$, $\zeta_1^{-1} = 0.72 + \mathrm{i}0.42$, $\zeta_2^{-1} = 0.2813 - \mathrm{i}0.121$ for the + sign in Equation (6.124).

$|\kappa| = 2.8$. The corresponding values of the propagation constants are $K^+ = 0.7096 + \mathrm{i}0.00924$ and $K^+ = 0.7094 + \mathrm{i}0.00999$, the exact values, found by Newton iteration are $|\kappa| = 2.72834$, $\theta = 2.59062\,\mathrm{rad}$ and $k_o dK^+ = 13.32742 + \mathrm{i}0.177748$. These results are in agreement with the corresponding results of Brooks for this duct [22].

6.7.3 Discussion of Transverse Modes

In contrast with the transverse eigenvalues of ducts with rigid impervious walls, the transverse eigenvalues of lined ducts are complex, except for the rather insignificant case of $\bar{M}_o = 0$ and $\mathrm{Re}\{\zeta\} = 0$, and are dispersive (dependent on frequency). Behavior of the modes is largely governed by the imaginary part of the propagation constant, which is, from Equation (6.124), $\mathrm{Im}\{K\} = \sqrt{B}/\left(1 - \bar{M}_o^2\right)$. Recalling that the modal pressure wave components have $\mathrm{e}^{\mathrm{i}k_o K x_1}$ dependence on the axial coordinate, their amplitudes decay by a factor of $\mathrm{e}^{k_o \mathrm{Im}\{K\} L}$ over a liner of length L. In fact, we may calculate the transmission loss of a single mode, μ, say, by using Equation (5.2), where W_{in}^+ and $\left(W_{\mathrm{out}}^+\right)_{r_{\mathrm{out}}=0}$ are now given by Equation (6.116). Then, since the duct is homogeneous, $TL_\mu = 20\log_{10}\left|p_\mu^+(0)/p_\mu^+(L)\right|$ and, upon substituting Equation (6.53) we find $TL_\mu = 8.686(k_o L)\text{Im}\{K_\mu^+\}$. This is usually referred to in the literature as the attenuation, or power attenuation, of a liner.

If the wall is not dissipative, that is, if $\text{Re}\{\zeta(\omega)\} = 0$, some modes may have $\sqrt{B} = 0$, which implies no decay. In fact, X and Y in Equations (6.127) and (6.128) vary on an ellipse centered at point (1,0) on the X, Y plane, having a major axis of $|\alpha|^2\left(1 - \bar{M}_o^2\right)/k_o^2$ and a minor axis equal to the half of the major axis. In the vicinity of the points where this ellipse cuts the major axis, $Y \approx 0$ and $K(\alpha)$ has very small imaginary part. This occurs when $|\sin 2\theta| \approx 0$, or $\theta \approx 0, \pi/2, \pi, 3\pi/2$, If $\theta \approx 0, \pi$, then $\text{Im}\{\alpha\} \approx 0$; and, if $\theta \approx \pi/2, 3\pi/2$, then $\text{Re}\{\alpha\} \approx 0$. But, in both cases, $K(\alpha)$ has a small imaginary part. If the imaginary part of $K(\alpha)$ is zero, the direction of propagation of a mode can be deduced by adding a small resistive term to the liner impedance and observing the direction of decay as the added term tends to zero.

The propagation constants corresponding to the modes which decay in the $+x_1$ direction are denoted by K^+, and the propagation constants which decay in the opposite direction are denoted by K^-. This convention is similar to that used for the propagation constants of hard-walled ducts. But the acoustic modes of lined ducts are ordered in increasing order of attenuation, that is, in increasing order of $|\text{Im}\{K\}|$. With all modes being of decaying type, the concept of propagating and evanescent modes is not as clear cut as for hard-walled ducts.

Since the decay of acoustic modes implies absorption of acoustic power by the duct walls, it is plausible to expect that the modes corresponding to $\text{Im}\{K\} > 0$ propagate in the $+x_1$ direction and those corresponding to $\text{Im}\{K\} < 0$ in the $-x_1$ direction. But, the direction of propagation of some lined duct modes may not be consistent with the direction of decay indicated by the sign of $\text{Im}\{K\}$[23]. Such modes are considered to be an artifact of the infinitely thin vortex sheet assumption of the Ingard–Myers boundary condition. If the problem parameters are favorable, some of the acoustic energy absorbed by the liner is returned back to the acoustic field and the vortex sheet becomes unstable. This is usually convective instability, which means that acoustic pressure amplitudes grow with axial distance but remain bounded in time. The impact of convective instability can be predicted by using the Briggs criterion (Section 1.3.2).

The eigenfunctions corresponding to the transverse eigenvalues have the same function forms as the corresponding hard-walled duct eigenfunctions, but since the eigenvalues are now complex, dispersive and dependent on the liner characteristics, they cannot be categorized in a generally useful manner. The positions of nodal lines are as in ducts with hard walls, but the nodal circles are in general different for the real and imaginary parts of the eigenfunctions. Figure 6.11 shows the normalized eigenfunction corresponding to the mode of the duct considered in Figure 6.9. Except for the imaginary part, this is akin to the transverse mode shapes of ducts with hard walls. An infinite number of the transverse eigenfunctions of lined ducts are of this type.

The modal wave transfer matrix formulation of Section 6.4 is applicable also for lined ducts, except for the ordering of the modes in ascending order of the imaginary part of the propagation constants in the $\mp x_1$ directions of propagation. However, systematic determination of the least attenuated modes to be considered in the wave transfer matrix can be problematic. The most obvious method for calculating the roots of dispersion

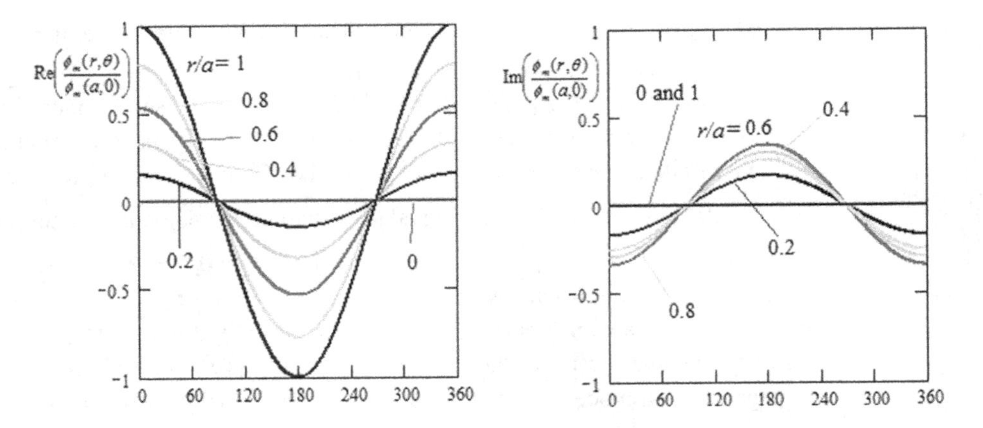

Figure 6.11 Normalized eigenfunction corresponding to the mode of the duct in Figure 6.9.

equations of lined ducts is the Newton–Raphson method. The success of this method, or similar iterative methods, depends largely on the availability of accurate starting values. Development of algorithms for systematic extraction of eigenvalues has been the subject of extensive research in this field [24], but simple general numerical methods for engineering predictions are not available. In general, it is inevitable to become involved in a detailed study of the topology of the dispersion equation and efficient eigenvalue tracking algorithms for preventing jumping of iterations to a different mode. Computer codes in Fortran language for general analysis of lined hollow and annular circular ducts and also for the optimization of practical liners are given in Syed [25].

6.7.3.1 Surface Modes

The modes which have relatively large imaginary part decay quickly away from the walls. For example, for $m = 2$ and the + sign in Equation (6.124), one of the transverse eigenvalues of the duct considered in Figure 6.9 is $k = 64.2226 + \mathrm{i}70.7412$. The corresponding eigenfunction is given in Figure 6.12. It is seen that, this eigenfunction is essentially localized to the duct wall and decays rapidly by the radial distance away from the wall. For this reason, such modes are usually referred to as surface modes.

Since surface modes appear to occur for eigenvalues with relatively large imaginary part, their impact in circular hollow ducts is studied by invoking the asymptotic approximation of the Bessel functions as $\mathrm{J}_m(z) = \sqrt{2/\pi z}\cos(z - m\pi/2 - \pi/4)$, which is valid for $z \to \infty$ and $|\arg(z)| < \pi$. When z is large by virtue of its imaginary part, this formula yields $\mathrm{J}'_m(z)/\mathrm{J}_m(z) = \imath$, where $\imath = \mathrm{i}$ if $\mathrm{Im}(z) < 0$ and $\imath = -\mathrm{i}$ if $\mathrm{Im}(z) > 0$, where i denotes the unit imaginary number. Then Equation (6.142) simplifies to $\imath\kappa\zeta = k_o a\left(1 - K\bar{M}_o\right)^2$, which may be recast as

$$\frac{\bar{M}_o^2\lambda^2}{\zeta^2} + \left(1 - \bar{M}_o^2\right)\lambda - 2\sqrt{\lambda} + 1 = 0, \tag{6.149}$$

here $\lambda = \imath\kappa\zeta/k_o a$, $\kappa = \alpha a$ and eigenvalues are given by $\kappa^2 = \left(k_o a\lambda/\zeta\right)^2$. Under some special conditions, these roots can be expressed in simple analytical forms. Some of

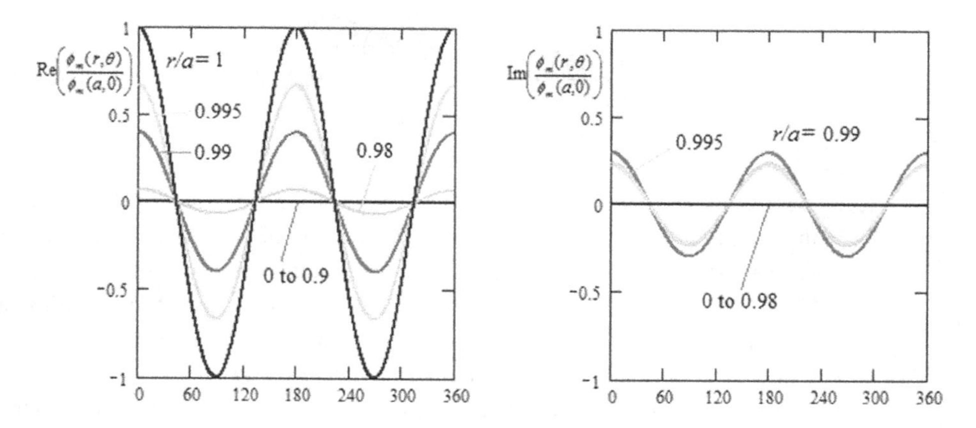

Figure 6.12 Normalized eigenfunction for $m = 2$, $\bar{M}_o = 0.5$, $k_o a = 10$, $\zeta = 2 + \mathrm{i}$ for the + sign in Equation (6.124). The corresponding eigenvalue is $k = 64.2226 + \mathrm{i}70.74125$.

these are listed and discussed in detail by Rienstra [26].[1] It should be noted that the large frequency limit $k_o a \to \infty$ tacitly assumes that the circumferential order, m, is not of the same order as or greater order than $k_o a$. The reason for this is that the asymptotics of Bessel functions of orders as large as its argument are different than those of a fixed order. Surface modes are also studied on this basis by Brambley and Peake [27].

6.7.3.2 Orthogonality of Modes

For a lined duct, the orthogonality property of the eigenfunctions is not valid in general. In this case, since the eigenfunctions are complex, we begin by writing Equation (6.48) for two eigenfunctions φ_μ and φ_η as $\breve{\varphi}_\eta\ \nabla_\perp^2 \varphi_\mu = -\alpha_\mu^2 \varphi_\mu \breve{\varphi}_\eta$, where an inverted over-arc denotes the complex conjugate. Next, exchanging the roles of μ and η and conjugating, $\varphi_\mu \nabla_\perp^2 \breve{\varphi}_\eta = -\breve{\alpha}_\eta^2 \breve{\varphi}_\eta\ \varphi_\mu$. Then, integrating the difference of these equations over the duct cross-sectional area and invoking the Stokes theorem as in Section 6.4.1 gives the line integral

$$\oint_\ell \left(\breve{\varphi}_\eta\ \nabla_\perp \varphi_\mu - \varphi_\mu \nabla_\perp \breve{\varphi}_\eta \right) \cdot \mathbf{n} \mathrm{d}\ell = -\left(\alpha_\mu^2 - \breve{\alpha}_\eta^2 \right) \int_S \varphi_\mu \breve{\varphi}_\eta\ \mathrm{d}S \tag{6.150}$$

and, upon substituting Equation (6.41), we obtain

[1] In the subsonic low Mach number mean flow case, the roots are:
$\lambda = (\mathrm{i}\zeta/2\bar{M}_o)\left(1 \mp \sqrt{1 - 4\bar{M}_{o1}/\mathrm{i}\zeta}\right)$ and $\lambda = (-\mathrm{i}\zeta/2\bar{M}_{o1})\left(1 \mp \sqrt{1 + 4\bar{M}_{o1}/\mathrm{i}\zeta}\right)$.

$$\mathrm{i}k_o \oint_{\ell} \left(\frac{\left(1-\bar{M}_o K_\mu\right)^2}{\zeta(\alpha_\mu)} - \frac{\left(1-\bar{M}_o \breve{K}_\eta\right)^2}{\breve{\zeta}\left(\breve{\alpha}_\mu\right)} \right) \varphi_\mu \breve{\varphi}_\eta \, \mathrm{d}\ell = -\left(\alpha_\mu^2 - \breve{\alpha}_\eta^2\right) \int_S \varphi_\mu \breve{\varphi}_\eta \, \mathrm{d}S. \tag{6.151}$$

Obviously, several conditions are necessary for the vanishing of the left-hand side of this equation. Firstly, we must have $\bar{M}_o = 0$. Secondly, the liner impedance should not be dependent on the duct modes. Thirdly, liner impedance should be purely real. Therefore, transverse eigenfunctions of lined ducts are in general non-orthogonal. This point has important bearing on the calculation of the time-averaged acoustic power transmitted in the duct. The latter is still given by Equation (6.114), but since the transverse eigenfunctions are not orthogonal, the area integral of the square matrix $\breve{\mathbf{\Phi}}^{\mathrm{T}}\mathbf{\Phi}$ is not diagonal and, therefore, the total time-averaged acoustic power may not be expressed as the sum of the contributions of single modes; it must be calculated directly from Equation (6.114) by including all significant modes.

It should be noted that if we omit conjugation in the above standard procedure, we get the orthogonality-like condition (called bi-orthogonality) $\int_S \varphi_\mu \varphi_\eta \mathrm{d}S = 0$, $\mu \neq \eta$, if $\bar{M}_o = 0$ and the liner impedance is independent of the modes. This is not orthogonality in the sense of scalar products of functions in the Hilbert space, but it has been used by some authors as being valid asymptotically for higher-order modes [28].

6.7.4 Liner Optimization

Substantial noise reduction may be achieved with the use of sound absorbent locally reacting liners on walls of inlet and outlet ducts of fluid machinery. But the transmission loss is proportional to the treated length and, since the duct lengths are available for the lining are usually limited, selection of the liner impedance for maximum attenuation becomes an important acoustic design issue. A widely used method for the determination of an optimal liner impedance is the Cremer–Tester method [23, 29–30]. This method is based on the finding of these authors that transmission loss of a mode attains a maximum when the corresponding transverse eigenvalue of the duct is a branch point (double-root) of the dispersion equation. To illustrate application of the method, consider a rectangular duct lined uniformly only on the $x_3 = 0$ side. For this case Equation (6.147) reduces to

$$\frac{\alpha \tan(\alpha d)}{\left(1 - K\bar{M}_o\right)^2} = -\frac{\mathrm{i}k_o}{\zeta_1}, \tag{6.152}$$

where $K = K(\alpha)$ is given by Equation (6.140). Assuming further that the mode in question is well cut-on, that is, $k_o >> \alpha$, we have $1 - \bar{M}_o K \approx 1/(1 \mp \bar{M}_o)$ and the foregoing equation simplifies to

$$\left(1 \mp \bar{M}_o\right)^2 \alpha \tan(\alpha d) = -\frac{\mathrm{i}k_o}{\zeta_1}. \tag{6.153}$$

Then, the branch-point condition becomes $\partial(\alpha \tan(\alpha d))/\partial\alpha = 0$, or $2\alpha d + \sin(2\alpha d) = 0$. The latter equation has an infinity of solutions which correspond to increasing order of sequential pairs of merging transverse modes. The solution corresponding to the mode with the largest transmission loss (also called the least attenuated mode) is $\alpha d = 2.1062 - \mathrm{i}1.12536$. When this is substituted in Equation (6.153), the Cremer–Tester optimum liner impedance is found as

$$\zeta_1 = \frac{k_o d}{(1+\bar{M}_o)^2}(0.29563 + \mathrm{i}0.23688), \tag{6.154}$$

which is the solution given by Tester and Cremer [23, 29, 30]. We may compare this with the predictions of the one-dimensional theory in Figure 3.5. In terms of the parameters used in this figure, the solution given in Equation (6.154) for $\bar{M}_o = 0$ corresponds to $|A_W|/k_o = 2.64/k_o^2 d^2$ and $\theta = 141.3^\circ$, since $P = b$ (only the $x_3 = 0$ side is lined) and $S = bd$. This is quite close to the optimum points in Figure 3.5a for about $k_o d \leq 1$.

A similar result can be derived for circular ducts [30]; however, in view of the well cut-on mode assumption, such solutions are not accurate at relatively low frequencies. In order to encompass low frequencies, the branch-point condition should be applied to the exact dispersion equation, which would be Equation (6.152) in the example considered above. Such applications, which have been considered recently [31, 32], show that the predicted optimum liner impedance may come out with negative real part at low frequencies, implying a need for active liners.

Optimization of the wall impedance may also be posed as one of finding the stationary values of $\sqrt{B}$ in Equation (6.124), which is, explicitly,

$$2B = -1 + \frac{|\alpha|^2}{k_o^2}\left(1-\bar{M}_o^2\right)\cos 2\theta + \sqrt{1 - 2\frac{|\alpha|^2}{k_o^2}\left(1-\bar{M}_o^2\right)\cos 2\theta + \left(\frac{|\alpha|^2}{k_o^2}\left(1-\bar{M}_o^2\right)\right)^2}. \tag{6.155}$$

Since the arguments are constrained by the dispersion equation of the duct, the maxima of this function should be determined by using a constrained optimization technique, such as the method of Lagrange multipliers [33]. The latter entails solution of the equations

$$\frac{\partial B}{\partial|\alpha|} + \lambda\frac{\partial D}{\partial|\alpha|} = \frac{\partial B}{\partial\theta} + \lambda\frac{\partial D}{\partial\theta} = 0, \tag{6.156}$$

where $D(|\alpha|,\theta) = 0$ denotes the dispersion equation of the duct and λ denotes an undetermined parameter. But because of the complexity of the derivatives, this approach is not tractable to analysis. It is much simpler, however, to iterate to a solution by checking if the eigenvalues computed for a given liner are optimal or not. If a liner is optimal, then either $\partial B/\partial\theta = 0$, or $\partial B/\partial|\alpha| = 0$, and, as may be found by straightforward calculus, either of these equations yield the condition for maxima of B as

$$\cos 2\theta = \frac{1}{2}\frac{|\alpha|^2}{k_o^2}\left(1-\bar{M}_o^2\right). \tag{6.157}$$

For example, in the case of Equation (6.153), this equation gives $\theta \approx \pi/4$ and $3\pi/4$ (recall that, in deriving Equation (6.153), a mode is assumed to be well cut-on). For $\theta \approx 3\pi/4$, Equation (6.154) becomes

$$\zeta_1 = \frac{k_o d}{(1 \mp \bar{M}_o)^2} \frac{-\mathrm{i}}{\frac{-1+\mathrm{i}}{\sqrt{2}} |\alpha d| \tan\left(\frac{-1+\mathrm{i}}{\sqrt{2}} |\alpha d|\right)}. \tag{6.158}$$

The Cremer–Tester optimum for the least attenuated mode in Equation (6.154) corresponds to $|\alpha d| = 2.388$ and for this value, the second quotient in the foregoing equation becomes $0.273 + \mathrm{i}0.281$, which is reasonably close to the corresponding factor in the Cremer–Tester optimum. The option $\theta \approx \pi/4$ may be discarded because it gives a liner with a negative resistance. Note that, according to Equation (6.157), a liner which is optimal for a particular mode is also optimal for all other modes having eigenvalues of equal modulus, if any such modes exist.

Such results are useful in that they provide insight into the optimization of single modes at given frequencies. In multi-modal propagation, however, the optimization problem is more complicated because, in general, different modes require different optimal liners. We may still use transmission loss as the cost function for optimization of a liner over multiple cut-on modes, however, this requires knowledge of the distribution of the cut-on modes at a given frequency. This usually presents a dilemma because it depends on the spectrum content of the sound source, which is rarely available precisely at the design stage in most instances. For this reason, one usually proceeds by making an educated guess and the hypothesis of equal power per cut-on mode is commonly used for this purpose. Such issues in calculation of the transmission loss in general are discussed in Section 11.5.

Liners achieving optimal overall transmission loss over the cut-on modes can be determined by using a numerical search method (Section 13.2). In practical applications, it is more convenient to work with the actual liner parameters than the real and imaginary parts of its impedance, because the optimal solutions found for the latter may not always be translated to a physical liner. For example, in resonator type liners, the salient parameters are the porosity of the facesheet, volume of the backing cavity, thickness of the facesheet and the diameter of perforations. Using these as optimization variables will lead directly to a physically realizable liner structure. Application of such considerations are described by Law and Dowling [34], who also propose the use of a cost function which relates to the environmental impact of the noise radiated from a lined duct.

6.7.5 Multi-Modal Attenuation Characteristics

In this section, we present some numerical results on multi-modal transmission loss characteristics of practical locally reacting liners. Our discussion borrows from the results of Syed [25] and is limited to circular hollow ducts; however, liner performance will be more or less similar qualitatively in rectangular and annular ducts. We

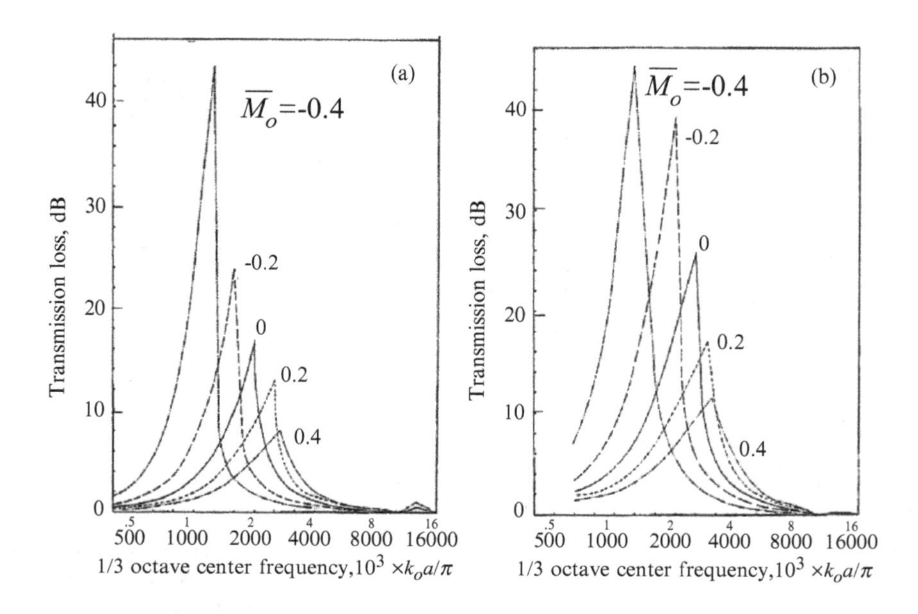

Figure 6.13 Variation of the 1/3 octave modal transmission loss spectra of a lined hollow circular duct with mean flow Mach number. (a) $m = n = 0$ mode, (b) $m = 1, n = 0$ mode. Liner parameters are: $R^* = 1.5$, $m^* = 0.01$, $\ell/2a = 0.05$, $L/2a = 1$ (Syed [25], cc by-nc-nd 4.0, labels adapted to present notation).

consider single layer resonator type liners (Figure 3.6b), the impedance of which is of the form $Z_W = Z_a/\sigma + i\rho_o c_o \cot k_o \ell$ (Section 3.9.3.1), where Z_a denotes the impedance of apertures on the facesheet, σ denotes the facesheet porosity and ℓ denotes the depth of the backing cavity. For compatibility with Syed [25], this is written as $\zeta_W = R^* + i(-2k_o am^* + \cot k_o \ell)$, where m^* is determined by the impedance of facesheet perforations (see Appendix C) and is assumed constant ($= 0.01$), since the facesheet was not varied. The liner resistance, R^*, depends on the facesheet and the mean flow Mach number. It may also depend on frequency, but this is neglected.

Typical convective effects of the mean flow on the transmission loss of the $m = n = 0$ and $m = 1, n = 0$ modes of a circular duct of radius a treated with this type of liner of length L are shown in Figure 6.13 for $\bar{M}_o = 0$ and $\bar{M}_o = \mp 0.2, \mp 0.4$, where plus and minus signs denote, respectively, exhaust and intake flows. Transmission loss of single modes is calculated as function of the Helmholtz number $k_o a/\pi$, but plotted as dB level averages in 1/3 octave bands. This representation is usually adequate in high frequency applications and aids assessment of the performance of a liner against environmental noise criteria, which are usually given in sound levels in octave bands. For the same reason, when multiple modes are incident at the duct inlet, the transmission loss based on the total time-averaged acoustic power input will be of interest. (See Section 11.5 for further notes on calculation of total transmission loss.) Such total transmission loss spectra are shown in Figure 6.14 for the liner in Figure 6.13, again in 1/3 octave bands. These characteristics are based on the

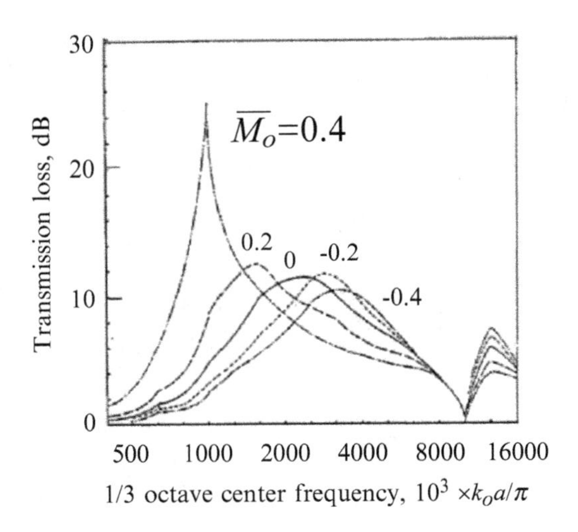

Figure 6.14 Variation of the 1/3 octave total transmission loss spectra of a lined hollow circular duct with mean flow Mach number. Liner parameters are: $R^* = 1.5$, $m^* = 0.01$, $\ell/2a = 0.05$, $L/2a = 1$ (Syed [25], cc by-nc-nd 4.0, labels adapted to present notation).

assumption that all incident modes at the duct inlet carry equal time-averaged power. Then, as it may be shown by writing inverse of Equation (5.2) for the incident modes at the inlet and adding them, the total transmission loss is given by $TL = 10\log_{10}N - 10\log_{10}\sum_{\mu=1}^{\mu=N} 10^{-TL_\mu/10}$ dB, where N denotes the number of incident modes and TL_μ denotes the transmission loss of mode $\mu \in 1, 2, \ldots$ In the calculation of Figure 6.14 presented by Syed [25], the incident modes at the duct inlet at a given frequency were determined by using the cut-on criterion of the hard-walled duct modes. Thus, the lined duct modes are classified in double index notation as (m, n) on this basis.

Of course, the typical feature of the characteristics in Figure 6.13a, b is the transmission loss peaks. These peaks are largely determined by the resonance frequency of the liner ($\approx c_o/4\ell$, the backing depth being assumed as an open–closed duct) and the mean flow Mach number. It is seen that as the algebraic value of the mean flow Mach number increases, the peaks decrease and move towards increasingly higher frequencies, whilst the transmission loss at lower frequencies decreases at lower frequencies and increase at higher frequencies. It becomes very small at the anti-resonance frequency of the liner ($\approx c_o/2\ell$), occurs in about $k_o a/\pi = 10$ band and the mean flow does not have discernible effect on its level. These features also apply for the total transmission loss characteristics, except for the rounding of the peaks due to the averaging effect of the equal power per mode assumption.

Similar calculations carried out for the effects of the liner parameters show that, in general, transmission loss of single modes increases both with m and n (based on hard-wall classification); however, at certain frequencies an increase in n may result in a decrease in modal transmission loss and when this occurs, an (m, n) mode may have higher transmission loss than the $(m, n+1)$mode. The resistive part of the liner, R^*, has little effect of the frequency of the peaks. But, as it increases, transmission loss peak levels decrease, whilst the levels at lower and higher frequencies increase. This is important from the point of view of design, because it may sometimes be desirable to have the transmission loss sufficiently high over an as large frequency range as possible, rather than a sharp peak in a narrow frequency range. However, the latter case will be desirable if the noise source spectrum contains dominant discrete frequency components.

6.7.6 Non-Uniformly Lined Ducts

Attenuation of lined ducts may also be optimized by using liners the impedance of which vary piecewise uniformly along the duct axis or around the duct periphery. The analysis of the latter situation may also be necessary with uniform liners, because of the necessity of using longitudinal hard-walled splices due to manufacturing considerations on large diameter ducts. The effect of arbitrary circumferential and/or axial non-uniformity of the liner impedance is included in the one-dimensional model described in Section 3.9. In this section, we consider modelıng of uniform lined ducts with circumferentially varying impedance in three dimensions. Modeling of axially segmented liners is deferred to Section 7.3, as this requires consideration of mode scattering at interfaces of uniform segments with different liner impedances. It should be noted that we have already treated the case of rectangular ducts having different liners on different faces (Section 6.7.2) and this covers most likely practical applications of rectangular ducts with peripherally non-uniform liners.

In one dimension, the effect of a peripherally non-uniform liner on attenuation culminates on the effective impedance of the treatment (Section 3.9.2.4). This may be expected to be also approximately valid for the axisymmetric modes. In general, however, the characteristics of wave propagation in uniformly lined ducts manifest with the added complexity of the presence of splices, which are known to have potential to significantly influence the acoustic performance of lined ducts.

For a circular hollow duct, impedance of a circumferentially non-uniform liner may be expressed as $\zeta = \zeta(\theta)$, where θ denotes the angular coordinate (Figure 6.4). Assuming uniform mean flow, the Ingard–Myers boundary condition is appropriate and the problem is still the solution of Equation (6.19) under the boundary condition in Equation (6.40). Since the liner is assumed to be axially uniform, we may search wave-like solutions of Equation (6.19) in the frequency domain by the ansatz $p'(x_1, r, \theta) = \varphi(r, \theta)\mathrm{e}^{\mathrm{i}k_o K x_1}$. Accordingly, Equations (6.41) and (6.48) still hold, but we cannot proceed with their solution as in previous sections because the separation of $\varphi(r, \theta)$ is not possible now. It is, therefore, necessary to use a numerical method. A complete review of the previous methods used is given in the article by Brambley et al. [35], who also present an integral equation approach, which includes the effect of uniform and sheared mean flow, and discuss the effects of the splices on surface modes. The finite element method [36, 37] and functional approximations based on expansion of $\varphi(r, \theta)$ into an infinite series in hard-walled duct eigenfunctions [35, 38–42], appear to be the methods more commonly used by previous authors. The latter approaches have some numerical variations which are beyond the scope of the present discussion. However, the triangular finite element formulation described in Section 6.5.5 may be extended readily for the determination of the axial propagation constants and eigenfunctions of circumferentially non-uniform liners (and this would also be applicable for a duct with any cross-section geometry). The analysis deviates from Section 6.5.5 in that the boundary integral in Equation (6.104) is to be evaluated now by using the Ingard–Myers boundary condition, Equation (6.41), which may be expressed for this purpose, using Equation (6.16), in the form

$$v_n' = \frac{1}{\rho_o c_o \zeta}(1 - \bar{M}_o K)\varphi \text{ on } \Gamma. \tag{6.159}$$

Then it is only necessary to modify the boundary integral in Equation (6.107) and, as may be verified by inspection of the steps following this equation, Equation (6.108) now becomes

$$\sum_{\Delta=1}^{N} (\mathbf{K}^{\Delta} - \alpha^2 \mathbf{M}^{\Delta})\boldsymbol{\varphi}^{\Delta} = -ik_o(1 - \bar{M}_o K)^2 \sum_{l=1}^{M} \mathbf{F}^l \boldsymbol{\varphi}^l, \tag{6.160}$$

where

$$\mathbf{F}^l = \left[\mathrm{G}^{-1}\right]^T \left(\int_{\Gamma^l} \frac{\mathbf{g}^T \mathbf{g}}{\zeta} \mathrm{d}\Gamma \right) \mathrm{G}^{-1}. \tag{6.161}$$

Hence, fitting Equation (6.160) to the global vector $\boldsymbol{\varphi}$ defined by Equation (6.110), we obtain the single matrix equation

$$(\mathbf{K} - \alpha^2 \mathbf{M})\boldsymbol{\varphi} = -ik_o(1 - \bar{M}_o K)^2 \mathbf{F} \boldsymbol{\varphi}_{\Sigma}, \tag{6.162}$$

where the sub-vector $\boldsymbol{\varphi}_{\Sigma}$ includes all terms of the global vector corresponding to the nodes on Γ. The working form of the foregoing equation is obtained by partitioning it as

$$\begin{bmatrix} \mathbf{K}_{11} - \alpha^2 \mathbf{M}_{11} + ik_o(1 - \bar{M}_o K)^2 \mathbf{F}_{11} & \mathbf{K}_{12} - \alpha^2 \mathbf{M}_{12} \\ \mathbf{K}_{21} - \alpha^2 \mathbf{M}_{21} + ik_o(1 - \bar{M}_o K)^2 \mathbf{F}_{21} & \mathbf{K}_{22} - \alpha^2 \mathbf{M}_{22} \end{bmatrix} \begin{bmatrix} \boldsymbol{\varphi}_{\Sigma} \\ \boldsymbol{\varphi}_S \end{bmatrix} = 0, \tag{6.163}$$

where the sub-vector $\boldsymbol{\varphi}_S$ includes all terms of $\boldsymbol{\varphi}$ not on Γ. Since α^2 is related to K by Equation (6.23), this constitutes a quadratic eigenvalue problem for the propagation constant, K, Once a propagation constant is found, the corresponding eigenvalue is determined from Equation (6.163) by back-substitution. If the mean flow is neglected, Equation (6.162) reduces to a simple eigenvalue problem for α^2. Note that the circumferential variation of the wall impedance is embedded in the submatrices $\mathbf{F}_{11}$ and $\mathbf{F}_{21}$ through Equation (6.161). In most applications, ζ would be piecewise uniform circumferentially and the integral in Equation (6.161) can be calculated exactly on lined circumferential segments of the boundary.

6.8 Uniform Ducts with Sheared Mean Flow

The convective wave equation with the infinitely thin boundary layer hypothesis of the Ingard–Myers boundary condition is convenient for studying the convective effect of the mean flow on propagation of sound waves in ducts. However, this model neglects the effect of refraction in mean flow shear layers. This effect is included in the one-dimensional theory described in Chapter 3. In this chapter we consider solution of the Pridmore-Brown equation, Equation (6.22), which is actually the simplest extension of the convective wave equation for the refractive effects in three dimensions. Since the velocity profile is still assumed to be uniform axially, solutions for the acoustic pressure field may be searched in the frequency domain by the ansatz $p'(x_1, \bullet) = \varphi(\bullet)\mathrm{e}^{\mathrm{i}k_o K x_1}$, but the transverse eigenfunctions $\varphi(\bullet)$ must satisfy now the Pridmore-Brown equation and the boundary condition at the duct walls. For the latter, either the no-slip condition (Section 6.3.1) or the partial-slip condition (Section 6.3.2) may be used. Equation (6.60) still holds for the transverse component of the particle velocity and the axial component of the particle velocity follows from Equation (6.6) as

$$v_1' = \frac{1}{\rho_o c_o(1 - KM_o)}\left(K - \frac{1}{k_o^2(1 - KM_o)}(\nabla_\perp M_o)\cdot\nabla_\perp\right)p'. \qquad (6.164)$$

In rectangular ducts (Figure 6.3), the mean flow velocity profile is usually assumed to be one-dimensional with $M_o = M_o(x_j)$, where $j = 2$ or 3. In circular ducts (Figure 6.4), it is common to assume an axisymmetric profile, $M_o = M_o(r)$. The specific form of the profile to be used usually presents a dilemma, since its actual form is seldom known precisely. In the laminar range, the mean flow velocity profile is usually modeled by using a parabolic profile and in the turbulent range by assuming a $1/n$ power law (see Figure 3.2); however, more complex profiles based on experimental correlations are also used frequently.

The transverse eigenfunctions of rectangular ducts may be expressed as $\varphi(x_2, x_3) = X_2(x_2)X_3(x_3)$, where the separating functions X_2 and X_3 satisfy Equation (6.30). In most studies, however, the acoustic field is assumed to be two-dimensional, so that Equation (6.30) simplifies to

$$(1 - KM_o)\left(\frac{\partial^2 X_j}{\partial x_j^2} + \alpha^2 X_j\right) + 2K\frac{\partial M_o}{\partial x_j}\frac{\partial X_j}{\partial x_j} = 0, \quad j = 2 \text{ or } 3 \qquad (6.165)$$

where $\alpha^2 = k_o^2\left((1 - KM_o)^2 - K^2\right)$. In circular ducts with an axisymmetric mean flow, the transverse eigenfunctions may be searched in the form $\varphi(r, \theta) = R(r)\mathrm{e}^{\mathrm{i}m\theta}$, $m = 0, \mp1, \mp2, \ldots$, where the radial function $R(r)$ is given by Equation (6.32), which may be expressed now as

$$(1 - KM_o)\left(\frac{1}{r}\frac{\partial R}{\partial r} + \frac{\partial^2 R}{\partial r^2} + \left(k_o^2\left((1 - KM_o)^2 - K^2\right) - \frac{m^2}{r^2}\right)R\right) + 2K\left(\frac{\partial M_o}{\partial r}\frac{\partial R}{\partial r}\right) = 0. \qquad (6.166)$$

An important feature of this equation is that for every $r = r^*$ such that $a \le r^* \le b$, where a and b denote the outer and inner radii of a circular duct (Figures 6.4, 6.6), there exists a propagation constant $K = 1/M_o(r^*)$ for which its solution is simply $R =$ constant. Similarly, in a rectangular duct (Figure 6.3), solution of Equation (6.165) becomes $X_j =$constant for every K satisfying $K = 1/M_o(x_j{}^*)$, where $0 \le x_3{}^* \le d$ if $j = 3$, and $0 \le x_2{}^* \le b$ if $j = 2$. Obviously, this propagation constant corresponds to a hydrodynamic mode, as it exists in such an interval, which is usually referred to as the critical layer, only if there is a mean flow. It is of a somewhat different nature than the hydrodynamic mode associated with the convected wave equation (Section 6.4.1), in that it has a continuous spectrum. In contrast with the convected wave equation, the boundary value problems presented by the Pridmore-Brown equation are not of the regular Sturm–Liouville type [14], which implies that orthogonality and completeness of its eigen characteristics are not guaranteed, but Swinbanks showed that this applies only for the eigenvalue spectrum associated with the critical layer [43]. Since hydrodynamic modes are not significant in acoustic phenomena at subsonic low Mach number mean flows, the critical layer spectrum may be filtered out in the formulation of the wave transfer matrix of a duct element.

6.8.1 Solution of the Pridmore-Brown Equation

On the understanding that $KM_o \ne 1$, that is, that the critical layer is outside the duct, Equations (6.165) and (6.166) may be written as

$$\frac{\partial^2 X_j}{\partial x_j^2} + \frac{2K}{1 - KM_o}\frac{\mathrm{d}M_o}{\mathrm{d}x_j}\frac{\partial X_j}{\partial x_j} + k_o^2\left((1 - KM_o)^2 - K^2\right)X_j = 0, \quad j = 2 \text{ or } 3 \quad (6.167)$$

$$\frac{\mathrm{d}^2 R}{\mathrm{d}r^2} + \left(\frac{1}{r} + \frac{2K}{1 - KM_o}\frac{\mathrm{d}M_o}{\mathrm{d}r}\right)\frac{\mathrm{d}R}{\mathrm{d}r} + \left(k_o^2\left((1 - M_oK)^2 - K^2\right) - \frac{m^2}{r^2}\right)R = 0 \quad (6.168)$$

respectively. Solutions of these equations have to satisfy the acoustic boundary condition on the duct walls. It is usual to assume the no-slip condition (Section 6.3.1) and we will follow this trend for simplicity. Then, the acoustic boundary conditions on the walls of rectangular and circular ducts may be expressed as, respectively,

$$\frac{\partial X_j}{\partial x_j} = \frac{\mathrm{i}k_o}{\zeta}\left(1 - KM_o\left(x_j\right)\right)X_j \text{ at } x_3 = 0, d \text{ or } x_2 = 0, b \quad (6.169)$$

$$\frac{\partial R}{\partial r} = \frac{\mathrm{i}k_o}{\zeta}(1 - KM_o(r))R \text{ at } r = a. \quad (6.170)$$

Analytical solutions of these boundary value problems are known only for laminar (parabolic) mean flows (see References [4–5] in Chapter 3). However, these solutions are difficult to compute and, in general, it is necessary to use a numerical method. Several methods, such as the Runge–Kutta integration, collocation and the finite element method, have been proposed by workers on the subject. Reviews of the works prior to the 1990s are presented by Eversman [24] and Nayfeh et al. [44]. In view of applications in aircraft jet engine nacelle acoustics, there have been considerable advances in recent years in the numerical solution of the Pridmore-Brown equation and reviews of more recent works may be found in thesis works on the subject, for example, Brooks [22], Oppeneer [45] and Sánchez [46]. The finite element method is applied by a weighted residual approximation of the bilinear functional of Equation (6.167) or (6.168), which is obtained by integration of their product with a test function over the duct cross section, on a mesh of triangular elements, in much the same way as described in Section 6.7.6. However, from the point of view of computer programming requirements, perhaps the simplest numerical method to use is the Runge–Kutta method [47]. In this method, considering a hollow circular duct of radius a, for example, Equation (6.168) is split into two first order equations

$$\frac{\mathrm{d}R}{\mathrm{d}r} = Q \tag{6.171}$$

$$\frac{\mathrm{d}Q}{\mathrm{d}r} = -\left(\frac{1}{r} + \frac{2K}{1 - KM_o}\frac{\mathrm{d}M_o}{\mathrm{d}r}\right)Q - \left(k_o^2\left((1 - M_oK)^2 - K^2\right) - \frac{m^2}{r^2}\right)R. \tag{6.172}$$

The solution region $0 \le r \le a$ is divided into N steps and integration is started for given k_o, m and $M_o(r)$, from $r = a$ by assuming values for K and $R(a)$, and usually proceeded by using fourth order difference formulae. A plausible value for K for starting the iteration is the propagation constant of the same duct with uniform mean flow of Mach number $\bar{M}_o$ or $M_o(0)$. Then integration continues up to just short of the duct center line, because of the singularity at $r = 0$. If the initial guess is correct, we must have $Q(0) = 0$ for axisymmetric modes ($m = 0$), or $R(0) = 0$ for non-axisymmetric modes ($m \ge 1$). If this condition is not satisfied, integration is repeated by using different starting values for K (it is not necessary to change $R(a)$, because K does not depend on it) until satisfactory agreement occurs. During integration, it is necessary to keep track of the sign changes in $R(r)$ for the identification of the radial mode index n of the converged solution. This process becomes more difficult when K is complex, as it is for lined ducts or the evanescent modes in hard-walled ducts, because the previously mentioned checks must be made for both the real and imaginary parts of the relevant quantities.

In general, for a given frequency, there will be two values of K corresponding to mode propagation in positive and negative directions of the duct axis. The directions of propagation can be identified in most cases by checking the sign of the imaginary part of the propagation constants, but, as has been noted in the previous section, it is possible to have situations where this approach does not work and that require further investigation. An application of the Runge–Kutta method for hard-walled circular ducts with a turbulent mean flow velocity profile consisting of a viscous sublayer, a buffer layer and a logarithmic region is described by Agarwal and Bull [4]. Their results show that the cut-on frequencies of the higher-order modes are not significantly different than those of the same duct with uniform mean flow, if the mean flow Mach number is taken as $\bar{M}_o$. This implies that the characteristics in Figure 6.1 are only slightly disturbed when the mean flow has a turbulent flow profile.

In general, whilst the effect of convection on sound attenuation is relevant for all frequencies, the effect of refraction in shear layers in the mean flow enters only at relatively high frequencies and acts in opposition to the effect of convection (see Section 3.9.3.1). Thus, insofar as the prediction of sound attenuation of a lined duct is concerned, the assumption of uniform mean flow can be adequately accurate at relatively low frequencies and the effect of the mean flow velocity profile needs to be taken into account when the frequencies of interest are relatively high. A high-frequency approximation for solutions of Equation (6.167) or (6.168) may be derived by assuming $X_j = A(x_j)e^{i\phi(x_j)}$or $R = A(r)e^{i\phi(r)}$, respectively, where A and ϕ denote real functions which are considered to be slowly varying functions of x_j or r. The rational for this approximation, which is commonly known as the WKB (Wentzel–Kramers–Brillouin) method [48], may be explained, for example for circular ducts, with reference to Equation (6.172), and the same argument applies also for rectangular ducts. As the frequency increases to the high limit, the R term in Equation (6.173) dominates over the Q term and Equation (6.168) tends to the form $d^2R/dr^2 + k_o^2 r = 0$, the solution of which is harmonic with a constant amplitude and phase. It is plausible to expect this situation to be perturbed to some extent when the frequency is high enough and, hence, amplitude and phase are allowed to be slowly varying functions of r. Before we apply this approximation, it is convenient to recast Equation (6.168) in the form

$$\frac{d^2R}{dr^2} + \frac{d}{dr}\left(\ln \frac{r}{(1-KM_o)^2}\right)\frac{dR}{dr} + \left(k_o^2\left((1-M_oK)^2 - K^2\right) - \frac{m^2}{r^2}\right)R = 0. \quad (6.173)$$

Substituting the ansatz $R = A(r)e^{i\phi(r)}$ and equating the real and imaginary parts of the resulting expression to zero, this equation yields

$$A = C\frac{1-KM_o}{\sqrt[4]{k_o^2\left((1-M_oK)^2 - K^2\right)\, r^2 - m^2}} \quad (6.174)$$

$$\phi = \mp\int^r \sqrt{k_o^2\left((1-M_oK)^2 - K^2\right) - \frac{m^2}{r^2}}dr, \quad (6.175)$$

where C denotes an integration constant, $m = 0, 1, 2, \ldots$ and, in getting the former equation, derivatives of $A(r)$ are neglected on the premise that it is a slowly varying function. This result may now be used in Equation (6.170) to derive a dispersion equation for the propagation constants. The efficacy of the WKB approximation is demonstrated by Vilenski and Rienstra [49], who apply the method to a more difficult problem of an annular circular duct with a mean flow having non-uniform axial and swirl velocity components.

6.8.2 Effect of the Mean Boundary Layer Thickness

Boundary layer thickness is the most common single parameter metric used for the characterization of a parallel sheared mean flow velocity profile in a duct. This is defined as the distance from the duct walls at which the mean flow velocity attains 99% of the maximum velocity. Since this accounts for almost all variation of the mean flow velocity, the mean flow velocity outside the boundary layer region, which is briefly called the core region, is generally assumed to be uniform for practical purposes. Accordingly, the effect of a sheared mean flow on sound propagation in ducts is studied by many authors by solving the Pridmore-Brown equation in the boundary layer and combining the solution with the solution of the convected wave equation for the core region by impedance matching at the boundary layer interface [24, 25]. An advantage of this approach is that it provides a characterization of the effects of a sheared mean flow as function of its boundary layer thickness. In fact, Nayfeh et al. solving Equation (6.167) by using the Runge–Kutta method for lined ducts with various mean flow velocity profiles showed that the attenuation with different mean flow velocity profiles are essentially the same, if they have the same displacement thickness and a shape factor that is appropriate for the nature of the mean flow [50]. Here, we do not intend to delve into a discussion on definitions of displacement thickness and shape factor, as the calculation of these parameters is described in detail in books on duct flows [51]. In general, for the same boundary layer thickness, the greater the velocity gradient at the wall, the larger the displacement thickness and the smaller the shape factor.

If the mean boundary layer is thin, asymptotic analytical solutions of the Pridmore-Brown equation may be obtained by using a small parameter expansion technique. We have considered one such model in Section 6.3.4 [9], and, as can be seen from Equation (6.44), it is tantamount to the Ingard–Myers boundary condition

$$\mathbf{n}\cdot\nabla_{\perp}\varphi = \frac{\mathrm{i}k_o}{\zeta_\delta}(1 - M_c K)^2\varphi, \text{on } \Gamma_\delta, \tag{6.176}$$

where $\zeta_\delta = \zeta/(1 + \beta_v \bar{M}_o K)$ denotes the impedance at the edge of the boundary layer, Γ_δ, and M_c denotes the Mach number of the mean core flow. In this section, we look at the asymptotic models proposed by Eversman and Beckemeyer [7] and Brambley [11], as this will be instructive from the point of view of illustrating solution of the Pridmore-Brown equation in thin mean boundary layers.

Considering a circular duct of outer radius a and a mean boundary layer of thickness δ, Eversman and Beckemeyer, who seem to be pioneers of this approach, recast Equation (6.168) in the form [7]:

$$\frac{\mathrm{d}}{\mathrm{d}r}\left(\frac{r}{(1 - KM_o)^2}\frac{\mathrm{d}R}{\mathrm{d}r}\right) + \frac{1}{r}\left(k_o^2 r^2 - \frac{k_o^2 r^2 K^2 + m^2}{(1 - KM_o)^2}\right)R = 0 \tag{6.177}$$

and introduce the transformation $\xi = (a - r)/\delta$. They expand this by treating δ/a as a small parameter as

$$\frac{\mathrm{d}}{\mathrm{d}\xi}\left(\frac{1}{(1 - KM_o)^2}\frac{\mathrm{d}R}{\mathrm{d}\xi}\right) = \frac{\delta}{a}\frac{\mathrm{d}}{\mathrm{d}\xi}\left(\frac{\xi}{(1 - KM_o)^2}\frac{\mathrm{d}R}{\mathrm{d}\xi}\right) - \frac{\delta^2}{a^2}\left(k_o^2 a^2 - \frac{k_o^2 a^2 K^2 + m^2}{(1 - KM_o)^2}\right)R + O\left(\frac{\delta^3}{a^3}\right), \tag{6.178}$$

where $R = R(\xi)$ and $M_o = M_o(\xi)$ with $0 \leq \xi \leq 1$, and $\xi = 1$ and $\xi = 0$ correspond to the edge of the boundary layer and the duct wall, respectively, and $M_o(0) = 0$, since the no-slip condition holds at the duct walls. A solution of the foregoing equation is sought in the form

$$R(\xi) = R_0 + \frac{\delta}{a} R_1 + \frac{\delta^2}{a^2} R_2 + O\left(\frac{\delta^3}{a^3}\right). \tag{6.179}$$

Substituting this expansion in Equation (6.178) and equating the coefficients of powers of δ/a to zero gives zeroth, first and second order equations

$$\frac{\mathrm{d}R_0}{\mathrm{d}\xi} = C_0(1 - KM_o)^2 \tag{6.180}$$

$$\frac{\mathrm{d}R_1}{\mathrm{d}\xi} = \xi \frac{\mathrm{d}R_0}{\mathrm{d}\xi} + C_1(1 - KM_o)^2 \tag{6.181}$$

$$\frac{\mathrm{d}R_2}{\mathrm{d}\xi} = \xi \frac{\mathrm{d}R_1}{\mathrm{d}\xi} - (1 - KM_o)^2 \int \left(k_o^2 a^2 - \frac{k_o^2 a^2 K^2 + m^2}{(1 - KM_o)^2} \right) R_0 \mathrm{d}\xi + C_2(1 - KM_o)^2 \tag{6.182}$$

respectively, where $C_j,\ j = 0, 1, 2$ denote integration constants. The boundary condition at the duct walls, Equation (6.170), is written now as

$$\left.\frac{\mathrm{d}R}{\mathrm{d}\xi}\right|_{\xi=0} = -\frac{\mathrm{i}k_o\delta}{\zeta}(1 - KM_o(0))^2 R(0), \tag{6.183}$$

is expanded similarly to obtain $\mathrm{d}R_0/\mathrm{d}\xi|_{\xi=0} = 0$, $\mathrm{d}R_1/\mathrm{d}\xi|_{\xi=0} = -\mathrm{i}k_o a R_0(0)/\zeta$ and $\mathrm{d}R_2/\mathrm{d}\xi|_{\xi=0} = -\mathrm{i}k_o a R_1(0)/\zeta$ to zeroth, first and second order, respectively. Eversman and Beckemeyer determine the constant C_j by requiring that $R_j,\ j = 0, 1, 2$, satisfy these conditions respectively [7]. Thus, Equation (6.179) is found as

$$R(\xi) = R_0 - \left(\frac{\mathrm{i}k_o a R_0}{\zeta} \int_0^\xi (1 - KM_o)^2 \mathrm{d}\xi \right) \frac{\delta}{a} + O(\delta^2/a^2), \tag{6.184}$$

where $R_o =$ constant and $R_1(0) = 0$ is set as definition in deriving this result. On the core side (with uniform mean flow), solution of the convective wave equation is required to satisfy the acoustic boundary condition at the edge of the mean boundary layer ($\xi = 1$):

$$\left.\frac{\mathrm{d}R}{\mathrm{d}\xi}\right|_{\xi=1} = -\frac{\mathrm{i}k_o\delta}{\zeta_\delta}(1 - KM_c)^2 R(1). \tag{6.185}$$

Upon substitution of Equation (6.184), it may be shown that

$$\zeta_\delta = \zeta \frac{1 - \left(\frac{ik_o a}{\zeta} \int_0^1 (1 - KM_o)^2 \mathrm{d}\xi \right) \frac{\delta}{a}}{1 + \left(1 - \mathrm{i}k_o a\zeta - \frac{\zeta}{\mathrm{i}k_o a} (k_o^2 a^2 K^2 + m^2) \int_0^1 \frac{\mathrm{d}\xi}{(1 - KM_o)^2} \right) \frac{\delta}{a}}. \tag{6.186}$$

From the results presented in Eversman and Beckemeyer [7], it transpires that the use of ζ_δ as the boundary impedance in solutions of the convected wave equation for the core side produces results which are almost the same as the Runge–Kutta solutions of Equation (6.168) for the corresponding full mean flow velocity profiles, if $\delta/a \leq 0.05$ and are of satisfactory accuracy up to about $\delta/a = 0.1$. The graphical solution technique described in Section 6.7.1.2 may be used similarly when ζ_δ is used. For example, for the duct considered in Figure 6.9, the polar trajectories and the propagation constants are modified only slightly for up to about $\delta/a = 0.05$; the propagation constants for $\delta/a = 0.01$ and 0.05 being $K^+ = 0.6546 + \mathrm{i}0.0317$ and $K^+ = 0.6512 + \mathrm{i}0.0237$, respectively.

An alternative approach, which is proposed by Brambley [11], makes use of the finite general solution of the convective wave equation for the core side of the boundary layer, which is $R(r) = \mathrm{J}_m(\alpha r)$, $m = 0, 1, 2, \ldots$, except for a constant factor (Section 6.5.4.2). Substituting the transformation $r = a - \delta\xi$ and expanding into the Taylor series around $\xi = 1$, this solution may be expressed as

$$R(\xi) = \mathrm{J}_m(\alpha a_\delta) - \delta\alpha\xi \mathrm{J}'_m(\alpha a_\delta) + \frac{1}{2}\delta^2\alpha^2\xi^2 \mathrm{J}''_m(\alpha a_\delta) + \cdots, \tag{6.187}$$

where $a_\delta = a - \delta$ and a prime denotes differentiation with respect to a function argument. In this approach, the coefficients $C_j,\ j = 0, 1, 2$ are determined by matching equations (6.180)–(6.182) with the corresponding order terms of Equation (6.187) at $\xi = 1$. Hence, it may be shown that

$$R(\xi) = \mathrm{J}_m(\alpha_c a_\delta) - \left(\alpha_c a \mathrm{J}'_m(\alpha_c a_\delta) \left[1 + \int_1^\xi \left(\frac{1 - KM_o}{1 - KM_c} \right)^2 \mathrm{d}\xi \right] \right) \frac{\delta}{a} + O(\delta^2/a^2). \tag{6.188}$$

Using this in Equation (6.183) and putting $\mathrm{J}''_m(z) = -(1 - m^2/z^2)\mathrm{J}_m(z)$, we obtain

$$\alpha_c a \frac{\mathrm{J}'_m(\alpha_c a_\delta)}{\mathrm{J}_m(\alpha_c a_\delta)} = \frac{\mathrm{i}k_o a}{\zeta} (1 - KM_c)^2 \left(1 - \alpha_c a \frac{\mathrm{J}'_m(\alpha_c a_\delta)}{\mathrm{J}_m(\alpha_c a_\delta)} I_0 \frac{\delta}{a} \right) + \left(m^2 \left(\frac{a^2}{a_\delta^2} - 1 \right) + (k_o^2 a^2 K^2 + m^2) I_1 \right) \frac{\delta}{a}, \tag{6.189}$$

where

$$I_0 = \int_0^1 \left[1 - \left(\frac{1 - KM_o}{1 - KM_c} \right)^2 \right] \mathrm{d}\xi \tag{6.190}$$

$$I_1 = \int_0^1 \left[1 - \left(\frac{1 - KM_c}{1 - KM_o}\right)^2\right] \mathrm{d}\xi. \tag{6.191}$$

Equation (6.189) is a dispersion equation of a circular hollow duct with a uniform mean core flow and thin boundary layer of thickness δ and can be used as such for the determination of propagations constants. The integrals I_0 and I_1 are determined from the profile of the mean boundary layer. For example, for a linear profile, $M_o = M_c - M_c(1 - \xi)$ and these integrals are given by $I_0 = KM_c(-1 + 2KM_c/3)/(1 - KM_c)^2$ and $KI_1 = KM_c$. These formulae can also be used to evaluate the integrals in Equation (6.186) for a linear boundary layer profile.

Equation (6.189) is given, under slightly more general conditions, by Brambley [11], who went on to deduce from it a new boundary condition at the duct walls and demonstrate its efficacy relative to some other models. In fact, if we recall that for a circular duct $\varphi = \mathrm{J}_m(\alpha_c a_\delta)$ and $\mathbf{n}\cdot\nabla_\perp\varphi = \alpha_c \mathrm{J}'_m(\alpha_c a_\delta)$ at the core boundary, $\xi = 1$, where φ denotes a solution of the two dimensional Helmholtz equation, Equation (6.48), we may recast Equation (6.189) in the general form of Equation (6.176) with

$$\zeta_\delta = \zeta\left(1 + \frac{\mathrm{i}k_o a}{\zeta}(1 - KM_c)^2 I_0 \frac{\delta}{a}\right)\left(1 + \frac{m^2\left(\dfrac{a^2}{a_\delta^2} - 1\right) + (k_o^2 a^2 K^2 + m^2)I_1}{\dfrac{\mathrm{i}k_o a}{\zeta}(1 - KM_c)^2}\frac{\delta}{a}\right)^{-1}. \tag{6.192}$$

Therefore, Brambley's boundary condition is also tantamount, in the frequency–wavenumber domain, to a boundary layer impedance model and reduces to the Ingard–Mayer boundary condition as $\delta/a \to 0$. Although Equations (6.186) and (6.192) are derived for circular ducts, they may be extended to rectangular ducts by understanding a as the length of a side normal to a lined side and replacing m^2, depending on the side in question, by β^2 or λ^2 (Section 6.5.4.1). The dispersion equation for a given combination of lined sides may then be derived as described in Section 6.7.2.

The concept of mean boundary layer impedance is quite convenient, because the solutions of the convective wave equations described in previous sections may still be used. However, at the present, the data available on relative merits of different boundary layer impedance models are not conclusive, although the results of a comparative study carried out by Gabard [52] by considering the Ingard-Myers model and the thin boundary layer impedance model proposed in Rienstra and Darau [10] and Brambley [11] seem to be in favor of the latter model.

6.9 Ducts with Axially Non-Uniform Cross-Sectional Area

Sound propagation in ducts with axially non-uniform cross sections is covered by the one-dimensional theory described in Chapter 3. In three dimensions, this problem is

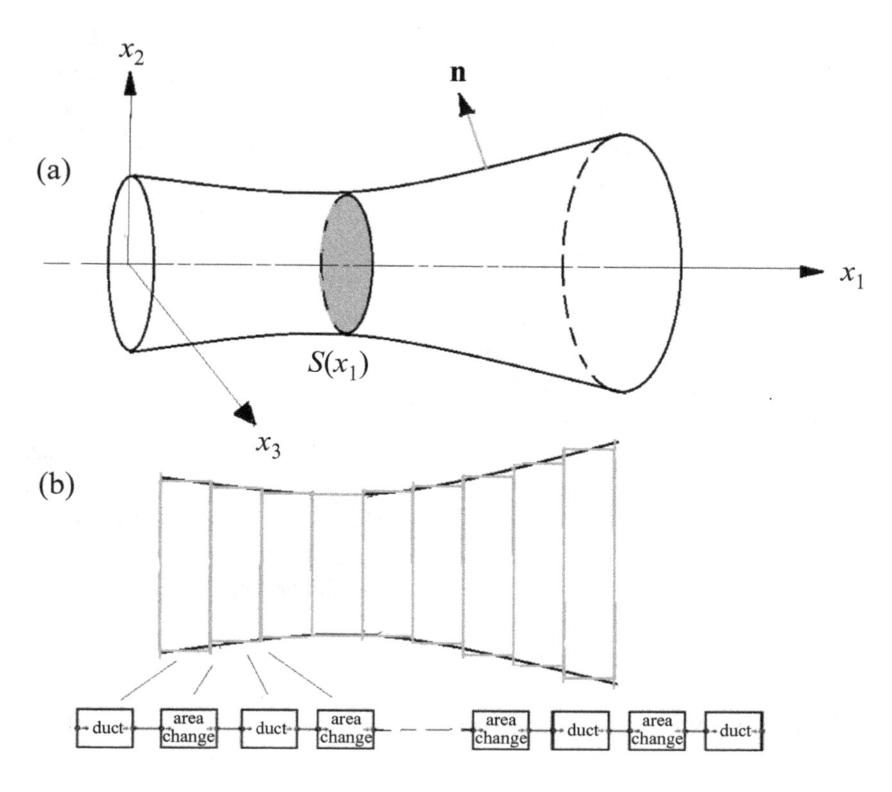

Figure 6.15 Non-uniform straight duct and its stepped model.

governed by the continuity and momentum equations, Equations (6.1) and (6.2), and the energy equation, Equation (1.11). The degree of difficulty of solving these equations varies with the form ascribed to the mean flow. If the mean flow is negligible, then Equations (1.12) and (6.4) imply a homogeneous medium and Equations (6.1) and (6.2) reduce, in the frequency domain, to $-\mathrm{i}\omega\rho' + \rho_o\nabla\cdot\mathbf{v}' = 0$ and $-\mathrm{i}\omega\rho_o\mathbf{v}' + \nabla p' = 0$, respectively. Thus, since Equation (1.15) holds also, the problem reduces to the solution of the classical wave equation $\left(\nabla^2 + k_o^2\right)p' = 0$, where $k_o = \omega/c_o$. The boundary condition to be satisfied on the duct walls is still the no-slip condition, Equation (6.33), that is, $\mathbf{n}\cdot\mathbf{v}' = p'/Z_W$, but the wave equation is no longer tractable to an exact analysis for non-uniform ducts. Numerical solutions may be obtained by using the finite element method or the boundary element method (Section 2.5); however, an approach which has been used by many researchers due to its relative simplicity and appeal to intuition is the stepped duct method. In this method, a non-uniform duct (Figure 6.15a) is represented by a sequence of uniform duct segments with step changes in the cross-sectional area in between (Figure 6.15b). This approximation is classical in plane-wave analysis of non-uniform waveguides, although it is not considered in detail in Chapter 3, because it is superseded by the numerical matrizant method (Section 3.3.3). In three dimensions, there is no counterpart of the matrizant theory that backs up the convergence of the stepped duct approximation to the corresponding continuous model. However, all previous applications indicate that it can produce results which are accurate enough for practical purposes, provided that sufficiently large numbers of segments are used and both the segments and the area discontinuities between them are modeled by using a sufficient number of higher order modes [53].[2]

A significant advantage of this approach is that the relatively simpler uniform duct models, such as those described in previous sections, may be used for the wave transfer matrices of the segments. Consecutive uniform duct segments are assembled by mode matching at the area discontinuity in between. Application of this procedure may be systemized by formulating sudden area-change elements in three dimensions (Section 7.2). Then, a stepped duct model of a non-uniform duct is represented by a block diagram consisting of a cascade of uniform duct and area-change elements (Figure 6.15b) and can be assembled by the techniques described in Chapter 2. This constitutes a significant advantage of the stepped duct method when it is necessary to include non-uniform ducts in the block diagram of a duct system. Note that if one-dimensional duct and area-change elements (Section 4.3) are used for the blocks, the same block diagram becomes a one-dimensional stepped duct model of the non-uniform duct.

Convective effect of the mean flow in non-uniform ducts is usually studied by assuming a potential (irrotational) mean flow. In general, this requires that the mean flow equations are solved separately by using a numerical method and provided as data for acoustic analysis. For the application of the acoustic boundary conditions at the duct walls, it is usual to envisage an infinitely thin boundary layer model, but the Ingard–Myers boundary condition, Equation (6.39), may not be used, because it presumes an axially uniform duct. The correct boundary condition, derived by Myers [6], is

$$\mathbf{n}\cdot\mathbf{v}' = (-\mathrm{i}\omega + \mathbf{v}_o\cdot\nabla - \mathbf{n}\cdot(\mathbf{n}\cdot\nabla)\mathbf{v}_o)\eta, \text{ on } \Gamma, \tag{6.193}$$

which is applicable for arbitrary flows over smoothly curved walls. Here, as in Section 3.6.2, the displacement η is related to the impedance of duct walls, Z_W, as $Z_W = p'/(-\mathrm{i}\omega\eta)$ and Γ denotes the surface of the duct walls. When the duct is uniform, Equation (6.193) reduces to Equation (6.39). Also, we have to pay attention to the effect of duct geometry on the mean flow, since the prevalence of sonic or subsonic high Mach number flows can be influential on the scope of acoustic analysis. In many applications, however, the area taper angles are kept small, typically about 5–7°, in order to prevent flow separation and minimize flow losses. This brings forth the possibility of using a perturbation method [55], in which the axial cross-sectional area gradient is considered as a small parameter, for the solution of the acoustic equations. A discussion of previous applications is given in Nayfeh et al. [44]. More recently, applications of the multiple scales method to turbofan ducts are described by Rienstra for irrotational mean flows [56, 57], by Cooper and Peake

[2] Ducts that are actually constructed from stepped ducts are sometimes used as silencing components [54].

and Lloyd and Peake for non-irrotational flows [58, 59], and by Brambley and Peake for curved non-uniform ducts [60].

The stepped duct approximation may still be used in the same way as described for ducts with no mean flow, but it is necessary to assume a locally axial mean flow. The radial component of the mean flow velocity is required for a consistent description of the mean flow in a non-uniform duct, nevertheless, this assumption may be acceptable for practical calculations in subsonic low Mach number ductworks containing non-uniform duct components. The effect of the mean flow should be included also in the transition elements modeling the sound wave transmission at sudden area changes between the uniform segments (Section 7.2).

We conclude this section with an informal description of a small parameter solution, which is instructive for the understanding of the coupled multi-modal character of the acoustic field in a non-uniform duct. The analysis is simplified considerably by assuming an approximately constant density mean flow ($\rho_o \approx$ constant). This implies, through Equation (6.3), an approximately solenoidal mean flow ($\nabla\cdot\mathbf{v}_o \approx 0$) and reduces Equation (6.1) to Equation (6.5), approximately. It also implies, through Equation (1.12), that $p_o \approx$ constant, which in turn implies, through Equation (6.4), $\mathbf{v}_o\cdot\nabla\mathbf{v}_o \approx 0$, reducing Equation (6.2) to Equation (6.6). But, $\nabla\times\mathbf{v}_o = 0$ for an irrotational mean flow. This allows introduction of velocity potentials as $\mathbf{v}_o = \nabla\vartheta_o$ and $\mathbf{v}' = \nabla\vartheta'$ for the mean flow velocity and acoustic particle velocity, respectively. Thus, Equations (6.5) and (6.6) may be written as

$$\frac{\partial\rho'}{\partial t} + \mathbf{v}_o\cdot\nabla\rho' + \rho_o\nabla^2\vartheta' = 0 \tag{6.194}$$

$$\rho_o\left(\nabla\frac{\partial\vartheta'}{\partial t} + \mathbf{v}_o\cdot\nabla\nabla\vartheta' + \nabla\vartheta'\cdot\nabla\mathbf{v}_o\right) + \nabla p' = 0. \tag{6.195}$$

Using Equation (1.14), the former equation may be expressed in frequency domain as

$$(-ik_o + \mathbf{M}_o\cdot\nabla)p' + \rho_o c_o\nabla^2\vartheta' = 0, \tag{6.196}$$

where $k_o = \omega/c_o$ and $\mathbf{M}_o = \mathbf{v}_o/c_o$ denotes the Mach number of the local mean flow velocity. Making use of the vector identity $\nabla\vartheta'\cdot\nabla\mathbf{v}_o = \nabla\nabla\vartheta'\cdot\mathbf{v}_o - \mathbf{v}_o\cdot\nabla\nabla\vartheta'$, Equation (6.195) may be simplified to

$$p' = -\rho_o c_o(-ik_o + \mathbf{M}_o\cdot\nabla)\vartheta'. \tag{6.197}$$

Hence, Equation (6.196) yields

$$\left(\nabla^2 - (-ik_o + \mathbf{M}_o\cdot\nabla)^2\right)\vartheta' = 0 \tag{6.198}$$

This is a convective wave equation for the acoustic velocity potential. When the duct is uniform, it reduces to Equation (6.47a). The approximations leading to this wave equation may be quantified for ducts with slow cross-sectional area variations. Such a duct is characterized by a small parameter, for example, $\varepsilon = \left|\mathrm{d}\sqrt{S}/\mathrm{d}x_1\right| << 1$, where $S = S(x_1)$ denotes the area of a duct section. Then, the approximations made in deriving Equation (6.198), which are exact for uniform ducts (Section 6.2), remain valid to $O(\varepsilon)$ when the uniformity of the duct is perturbed slightly.

When the mean flow is in the subsonic low Mach number range, $|\mathbf{M}_o|^2 << 1$, Equation (6.198) simplifies to

$$\nabla^2 \vartheta' + \mathrm{i}2\frac{k_o}{c_o}\nabla\vartheta_o \cdot \nabla\vartheta' + k_o^2\vartheta' = 0. \tag{6.199}$$

It may be shown that the solution of this equation is

$$\vartheta' = \chi(x_1, \bullet)\mathrm{e}^{-\frac{\mathrm{i}k_o}{c_o}\vartheta_o(x_1,\bullet)}, \tag{6.200}$$

where χ denotes a solution of the Helmholtz equation

$$(\nabla^2 + k_o^2)\chi = 0. \tag{6.201}$$

We look for wave-like solutions of this equation by the ansatz

$$\chi = \varphi(x_1, \bullet)\mathrm{e}^{\mathrm{i}k_o \int^{x_1} K(x_1)\mathrm{d}x_1}, \tag{6.202}$$

where "•" denotes the transverse coordinates. Evaluation of the second partial derivative of χ with respect to x_1 gives

$$\frac{\partial^2 \chi}{\partial x_1^2} = \left(\left(\frac{\partial^2}{\partial x_1^2}(\ln \varphi) + \mathrm{i}k_o\frac{\mathrm{d}K}{\mathrm{d}x_1}\right) + \left(\frac{\partial}{\partial x_1}(\ln \varphi) + \mathrm{i}k_o K\right)^2\right)\chi. \tag{6.203}$$

Since $\left|\mathrm{d}\sqrt{S}/\mathrm{d}x_1\right|$ is of $O(\varepsilon)$ and $|\partial/\partial x_1| = \varepsilon\left|\partial/\partial\sqrt{S}\right|$, the products of χ, which is small to first order, with the derivatives on the right-hand side of Equation (6.203) can be neglected as quantities small to the second order. Then, Equation (6.203) may be simplified to $\partial^2\chi/\partial x_1^2 = -k_o^2K^2\chi$ and Equation (6.201) gives

$$(\nabla_\perp^2 + \alpha^2)\varphi(x_1, \bullet) = 0, \tag{6.204}$$

where $\alpha^2 = k_o^2(1 - K^2)$ and $\nabla_\perp^2$ denotes the Laplacian on the duct cross section. The eigenvalues, α^2, are determined from the condition that the solutions of Equation (6.204) for given duct section geometry satisfy the Myers boundary condition, Equation (6.193). This problem may also be simplified because, when the cross-sectional area gradients are small, the third term in brackets in Equation (6.193) becomes negligible compared to the other terms. Therefore, the eigen-characteristics of Equation (6.204) are locally (that is, for a particular section of the duct) the same as those of a uniform duct (Section 6.5) of the same section and all previous results about the eigenvalues, α_μ, $\mu = 1, 2, \ldots$, and the corresponding eigenfunctions of specific section geometry are applicable. But, since the cross-section dimensions vary along the duct axis, eigenvalues and the corresponding eigenfunctions also vary along the duct axis, that is, mathematically, $\alpha_\mu = \alpha_\mu(x_1)$ and $\varphi_\mu = \varphi_\mu(x_1, \bullet)$. For example, the eigenvalues and eigenfunction of a non-uniform hollow circular duct of radius $a = a(x_1)$ are given in double index notation by $\alpha_{mn}(x_1) = j'_{mn}/a(x_1)$ and

$\varphi_{mn}(x_1, r, \theta) = \mathrm{J}_m(j'_{mn} r/a(x_1))\mathrm{e}^{\mathrm{i}m\theta}$, respectively, where j'_{mn} has the values given in Table 6.2. Accordingly, the propagation constants for mode μ are

$$K_\mu^\mp(x_1) = \mp\sqrt{1 - \frac{\left(\alpha_\mu^\mp(x_1)\right)^2}{k_o^2}}, \quad \mu = 1, 2, \ldots \tag{6.205}$$

An interesting implication of this result for hard-walled ducts is that if the equation $\alpha_\mu^\mp(x_1) - k_o = 0$ has a root x_1^* in $0 \le x_1 \le L$ for a given mode, where L denotes the duct length, this mode becomes evanescent in $x > x_1^*$, if it is propagating (cut-on) in $x < x_1^*$ and, conversely, it becomes propagating in $x > x_1^*$, if it is evanescent in $x < x_1^*$. Root x_1^* is usually called the turning point.

Thus, combining Equations (6.197), (6.200) and (6.202), there corresponds two solutions to each eigenvalue α_μ^2

$$p_\mu^\mp(x_1, \bullet) = \mathrm{i}\omega\rho_o\left(1 - \frac{1}{\mathrm{i}k_o}\mathbf{M}_o\cdot\nabla\right)\varphi_\mu(x_1, \bullet)\mathrm{e}^{\mathrm{i}k_o\int^{x_1} K_\mu^\mp(x_1)\mathrm{d}x_1}\mathrm{e}^{-\frac{\mathrm{i}k_o}{c_o}\vartheta_o}, \quad \mu = 1, 2, \ldots \tag{6.206}$$

and the acoustic pressure field in the duct is obtained by summing the contributions of all modes

$$p'(x_1, \bullet) = \sum_{\mu=1}^{\infty}\left(a_\mu^+ p_\mu^+(x_1, \bullet) + a_\mu^- p_\mu^-(x_1, \bullet)\right), \tag{6.207}$$

where $a_\mu^\mp$ denote constant coefficients. Note that the modal pressure wave components are not separable now from the transverse eigenfunctions. The eigenfunctions $\varphi(x_1, \bullet)$ are velocity potential eigenfunctions; that is, they have the unit and phase of the velocity potential. They are converted to acoustic pressure eigenfunctions by multiplying by $\mathrm{i}\omega\rho_o$.

6.10 Circularly Curved Ducts

In many instances, the use of curved ducts is unavoidable in practical flow duct systems. Typically, ducts are bent in-plane in circular arcs, since this geometry is less susceptible to mean flow losses and more convenient for manufacturing. They usually provide a transition of flow between uniform rectangular or circular ducts of identical sections, but of different orientations. Circularly curved duct geometry may be described in a Cartesian frame (x_1, x_2, x_3) centered at the center of curvature of the duct center line (a circular arc), by cylindrical polar coordinates ϕ, R, x_3. It is also convenient to define the local Cartesian frame y_1, y_2, y_3, the (y_1, y_3) plane containing the duct sections normal to the center line, the y_2 axis being in the ϕ direction. These coordinate systems are shown in Figure 6.16 for a duct with rectangular sections, for which they are most convenient. Ducts with different cross sections may also be

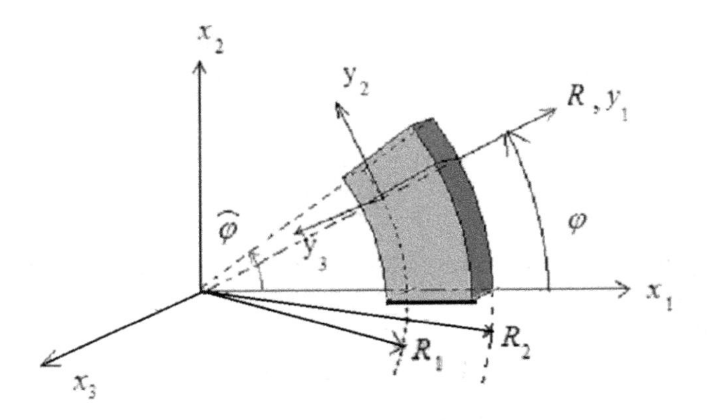

Figure 6.16 Circularly curved rectangular duct (back face on $x_3 = 0$ plane).

described similarly with R_1 and R_2 interpreted as the innermost and outermost radii of the bend and the centers of the coordinate systems dispositioned as appropriate for the section geometry.[3]

The acoustic field in a curved duct containing an inviscid fluid is governed by the continuity and momentum equations, Equations (6.1) and (6.2), the energy equation, Equation (1.11), and the corresponding mean flow equations. In most previous studies, however, they are simplified by assuming a constant density mean flow ($\rho_o =$ constant). Then Equation (6.3) is satisfied identically and Equation (6.4) implies $p_o =$ constant. Consequently, Equations (6.1) and (6.2) reduce, respectively, to Equations (6.5) and (6.6). The monograph by Rostafinski presents a comprehensive review of the work carried out on solution of these equations for curved ducts without mean flow before 1991 [61]. A later work by Félix and Pagneux proposes a solution for circular ducts in toroidal coordinates by expansion of the acoustic field in the curved duct in terms of the eigenfunctions of the straight uniform duct of the same cross section [62]. It transpires from previous studies that the sound propagation in circularly curved ducts is characterized by a dimensionless angular wavenumber, which is generally determined by using a numerical method, an analytical solution being possible only for rectangular ducts (see Section 6.10.1).

The number of publications on the effect of mean flow is relatively few. Fuller and Bies proposed conformal mapping of a straight uniform two-dimensional rectangular duct with a uniform mean flow to a circularly curved duct for including the convective effect of the mean flow [63]. Since the solution is assumed to approximate a solution

[3] Sections of circular ducts with center line radius $R_o$are usually defined in analytic studies by local polar coordinates r, θ, which are related to the y_1, y_3 coordinates of Figure 6.16 as $y_3 = r\cos\theta$ and $y_1 = r\sin\theta$. The ϕ, r, θ system is known as toroidal coordinates, which are related to the Cartesian coordinates as $x_1 = R_o\cos\phi + r\cos\theta\cos\phi$, $x_2 = R_o\sin\phi + r\cos\theta\sin\phi$ and $x_3 = r\sin\theta$. But the Helmholtz equation is not separable in toroidal coordinates.

of the Laplace equation on the duct cross section, this approach is limited to low frequencies. Capelli used the finite element method for determination of both the potential flow field and the acoustic flow field in a right-angled bend [64]. More recently, Brambley and Peake studied, with application to a military aircraft engine intake, a strongly curved slightly non-uniform circular duct with a potential mean flow and presented a numerical solution by using a spectral collocation method [60].

Steady flows in curved ducts, a comprehensive discussion of which is available in the review article by Berger et al. [65], are rather complex. The flow structure is characterized by secondary flows which are vortical in character and depends on several factors, including the entrance condition and the development status of the flow, the curvature of the duct, density stratification due to centrifugal force field, shape of the duct cross section and the Dean number, $\mathrm{De} = \mathrm{Re}\sqrt{a/\mathrm{Re}}$, where Re is an equivalent Reynolds number, which is based on a, a characteristic size of the duct section, and the maximum steady flow velocity in a straight duct that is identical to the curved duct, except for the curvature. Nevertheless, the velocity of the fluid normal to the duct sections, which determines the rate of fluid mass flow through the duct, has profiles which are akin to the straight duct profiles, except that the velocity peak may be shifted outward due to the effect of centrifugal forces and the boundary layer structures modified by curvature effects [66]. This suggests modeling of a turbulent mean flow in a curved uniform duct, as a first approximation, by an irrotational core flow which is separated from the duct walls by an infinitely thin boundary layer. However, the existing literature is inconclusive about an engineering theory on the convective and refractive effects of the mean flow on sound propagation in curved ducts.

In general, the mean flow field in an irrotational mean core flow with transverse and azimuthal (in the ϕ direction) velocity components, $\mathbf{v}_o = \mathbf{e}_R v_{oR} + \mathbf{e}_3 v_{o3} + \mathbf{e}_\phi v_{o\phi}$, say, may be expressed as $\mathbf{v}_o = \nabla\vartheta_o$, where ϑ_o denotes the mean flow velocity potential.[4] Here $\nabla = \mathbf{e}_R\partial/\partial R + \mathbf{e}_3\partial/\partial x_3 + \mathbf{e}_\phi\partial/R\partial\phi$ is the gradient operator and $\mathbf{e}_3, \mathbf{e}_R, \mathbf{e}_\phi$ denote the unit vectors in the positive directions of the x_3, R, ϕ axes, respectively. As in the previous section, we introduce also the acoustic velocity potential by $\mathbf{v}' = \nabla\vartheta'$ and transform Equations (6.5) and (6.6) to Equations (6.194) and (6.195), respectively, to derive the wave equation governing the acoustic velocity potential, Equation (6.199), for a subsonic low Mach number mean flow. General solution of this wave equation is of the form $\vartheta' = \chi\mathrm{e}^{-\mathrm{i}k_o\vartheta_o/c_o}$, where $k_o = \omega/c_o$ and χ is a solution of the Helmholtz equation $(\nabla^2 + k_o^2)\chi = 0$, which transforms in the coordinate system x_3, R, ϕ (Figure 6.16) to the form

$$\frac{\partial^2\chi}{\partial R^2} + \frac{1}{R}\frac{\partial\chi}{\partial R} + \frac{1}{R^2}\frac{\partial^2\chi}{\partial\phi^2} + \frac{\partial^2\chi}{\partial x_3^2} + k_o^2\chi = 0. \tag{6.208}$$

[4] For constant density mean flow, the continuity equation, Equation (6.3), gives $\nabla\cdot\mathbf{v}_o = 0$. Therefore, the mean flow velocity potential satisfies the Laplace equation $\nabla^2\vartheta_o = 0$. For a fully developed mean flow, $v_{o\phi} = \partial\vartheta_o/R\partial\phi = F(R, x_3)$; that is, $\vartheta_o = Rv_{o\phi}\phi$.

Wave-like solutions of this equation may be sought by the ansatz $\chi = \varphi(R, x_3)\mathrm{e}^{\mathrm{i}k_\phi \phi}$, where k_ϕ is called the angular wavenumber, and the transverse field function φ is given by

$$\frac{\partial^2 \varphi}{\partial R^2} + \frac{1}{R}\frac{\partial \varphi}{\partial R} + \frac{\partial^2 \varphi}{\partial x_3^2} + \left(k_o^2 - \frac{k_\phi^2}{R^2}\right)\varphi = 0. \tag{6.209}$$

Thus, the acoustic velocity potential acquires the form

$$\vartheta' = \varphi(R, x_3)\mathrm{e}^{-\mathrm{i}\frac{k_o}{c_o}\vartheta_o}\mathrm{e}^{\mathrm{i}k_\phi \phi}. \tag{6.210}$$

The corresponding acoustic pressure field follows from Equation (6.197) as

$$p' = \rho_o(\mathrm{i}\omega - \nabla\vartheta_o \cdot \nabla)\vartheta' \tag{6.211}$$

and the components of the particle velocity in x_3, R, ϕ directions, $\mathbf{v}' = \mathbf{e}_R v_R' + \mathbf{e}_3 v_3' + \mathbf{e}_\phi v_\phi'$, are found from $\mathbf{v}' = \nabla\vartheta'$.

The angular wavenumber k_ϕ and the transverse function $\varphi(R, x_3)$ are determined from the condition that the solution of Equation (2.209) satisfies the acoustic boundary condition, Equation (6.193), on the duct walls. For a uniform duct with in-plane curvature, the third term on the right of Equation (6.193) vanishes, reducing it to the Ingard–Myers form $\mathbf{n}\cdot\nabla_\perp\vartheta' = (-\mathrm{i}\omega + \nabla\vartheta_o\cdot\nabla)\eta$ with $\eta = p'/(-\mathrm{i}\omega Z_W)$, where $\nabla_\perp = \mathbf{e}_R\partial/\partial R + \mathbf{e}_3\partial/\partial x_3$ denotes the gradient operator on a section of the duct, $\mathbf{n}$ denotes the outward unit normal vector of the duct boundary and Z_W is the wall impedance. Upon using Equations (6.210)–(6.211), the Myers boundary condition can be expressed as

$$\mathbf{n}\cdot\nabla_\perp\vartheta' = \frac{\mathrm{i}k_o}{\zeta}\left(1 - \frac{1}{\mathrm{i}\omega}\left(\nabla_\perp\vartheta_o\cdot\nabla_\perp + \frac{1}{R^2}\frac{\partial\vartheta_o}{\partial\phi}\frac{\partial}{\partial\phi}\right)\right)^2\vartheta', \tag{6.212}$$

where $\zeta = Z_W/\rho_o c_o$ is the normalized wall impedance, which for simplicity is assumed here to be uniformly distributed. We may assume that solutions in the form of Equation (6.210) satisfying Equations (6.209) and (6.212) exist for given ϑ_o, for an infinite number discrete values of k_ϕ, however, such solutions may be obtained only by using a numerical method.

The neglect of the mean flow is often justified in practice for simplicity or because of lack of knowledge about the actual mean flow velocity field. For example, mean flows in short segments of curved ducts, which are ubiquitous in ductworks, are difficult to determine with certainty, as they are almost always developing flows with rotational structures. When mean flow is neglected, Equations (6.210) and (6.212) simplify to

$$\vartheta' = \varphi(R, x_3)\mathrm{e}^{\mathrm{i}k_\phi \phi} \tag{6.213}$$

$$\mathbf{n}\cdot\nabla_\perp\varphi = \frac{\mathrm{i}k_o}{\zeta}\varphi \tag{6.214}$$

respectively, where the normalized wall impedance need not be uniform. The eigenvalue problem constituted by Equations (6.209) and (6.214) may be solved numerically for arbitrary section geometry by using the finite element method (Section 6.10.2).

However, for insight on the characteristics of transverse modes in curved ducts, it is convenient to consider first the case of rectangular ducts, because this geometry is tractable to an exact analytical solution.

6.10.1 Rectangular Ducts

Consider a circularly curved rectangular duct of section $0 \le y_1 \le d,\ 0 \le y_2 \le b$ (Figure 6.16). The surface of the duct is defined by the surfaces $R = R_1,\ R_2(> R_1)$ and $x_3 = 0, b$, where $d = R_2 - R_1$. Equation (6.209) is separable in this geometry as $\varphi(R, x_3) = F(R)X(x_3)$. As may be shown by substitution, the separating functions are given by solutions of

$$R^2 \frac{\partial^2 F}{\partial R^2} + R \frac{\partial F}{\partial R} + \left(R^2 k_R^2 - k_\varphi^2\right) F = 0 \tag{6.215}$$

$$\frac{\partial^2 X}{\partial x_3^2} + (k_o^2 - k_R^2) X = 0, \tag{6.216}$$

where k_R is called the radial wavenumber. The former equation is a Bessel equation and its general solution is $F = A\mathrm{J}_{k_\phi}(k_R R) + B\mathrm{Y}_{k_\phi}(k_R R)$, where A and B are integration constants. The general solution of Equation (6.216) is $X = \mathrm{C}_1 \cos(k_3 x_3) + C_2 \sin(k_3 x_3)$, where $k_3^2 = k_o^2 - k_R^2$ and C_1, C_2 are integration constants. Thus, general solution of Equation (6.209) is then of the form

$$\varphi = \left(A\mathrm{J}_{k_\phi}(k_R R) + B\mathrm{Y}_{k_\phi}(k_R R)\right)\left(\mathrm{C}_1 \cos(k_3 x_3) + C_2 \sin(k_3 x_3)\right). \tag{6.217}$$

The condition that Equation (6.214) must be satisfied on surfaces $R = R_1,\ R_2$ and $x_3 = 0, b$ gives four equations:

$$\left.\frac{\mathrm{d}F}{\mathrm{d}R}\right|_{R=R_1} = Ak_R\mathrm{J}'_{k_\phi}(k_R R_1) + Bk_R\mathrm{Y}'_{k_\phi}(k_R R_1) = \frac{\mathrm{i}k_o}{\zeta_1}\left(A\mathrm{J}_{k_\phi}(k_R R_1) + B\mathrm{Y}_{k_\phi}(k_R R_1)\right) \tag{6.218}$$

$$\left.\frac{\mathrm{d}F}{\mathrm{d}R}\right|_{R=R_2} = Ak_R\mathrm{J}'_{k_\phi}(k_R R_2) + Bk_R\mathrm{Y}'_{k_\phi}(k_R R_2) = -\frac{\mathrm{i}k_o}{\zeta_2}\left(A\mathrm{J}_{k_\phi}(k_R R_2) + B\mathrm{Y}_{k_\phi}(k_R R_2)\right) \tag{6.219}$$

$$\left.\frac{\mathrm{d}X}{\mathrm{d}x_3}\right|_{x_3=0} = C_2 k_3 = -\frac{\mathrm{i}k_o}{\zeta_3}\mathrm{C}_1 \tag{6.220}$$

$$\left.\frac{\mathrm{d}X}{\mathrm{d}x_3}\right|_{x_3=b} = -\mathrm{C}_1 k_3 \sin(k_3 b) + C_2 k_3 \cos(k_3 b) = \frac{\mathrm{i}k_o}{\zeta_4}\left(\mathrm{C}_1 \cos(k_3 b) + C_2 \sin(k_3 b)\right), \tag{6.221}$$

where ζ_1, ζ_2 denote the normalized impedance of liners on sides $R = R_1, R_2$, and ζ_3, ζ_4 denote the impedance of sides $x_3 = 0, b$, respectively. The first two and the last two of the foregoing equations yield, respectively, the following dispersion equations:

$$\frac{A}{B}=\frac{-k_R Y'_{k_\phi}(k_R R_1)+\dfrac{ik_o}{\zeta_1}Y_{k_\phi}(k_R R_1)}{k_R J'_{k_\phi}(k_R R_1)-\dfrac{ik_o}{\zeta_1}J_{k_\phi}(k_R R_1)}=-\frac{k_R Y'_{k_\phi}(k_R R_2)+\dfrac{ik_o}{\zeta_2}Y_{k_\phi}(k_R R_2)}{k_R J'_{k_\phi}(k_R R_2)+\dfrac{ik_o}{\zeta_2}J_{k_\phi}(k_R R_2)} \tag{6.222}$$

$$\frac{C_2}{C_1}=-\frac{ik_o}{\zeta_3 k_3}=\frac{k_3\sin(k_3 b)+\dfrac{ik_o}{\zeta_4}\cos(k_3 b)}{k_3\cos(k_3 b)-\dfrac{ik_o}{\zeta_4}\sin(k_3 b)}. \tag{6.223}$$

The wave numbers k_3,k_ϕ and k_R can be computed by applying these conditions for any combination of lined and rigid sides. In general, Equation (6.223) gives one dispersion equation for the wave numbers k_R and k_ϕ, and Equation (6.222) gives a dispersion equation for k_3. The three wavenumbers can be determined from these two equations, since $k_3^2=k_o^2-k_R^2$.

In the case of hard-walled ducts, Equation (6.223) gives

$$k_3=\frac{m\pi}{b},\qquad m=0,1,2,\ldots \tag{6.224}$$

and Equation (6.222) reduces to

$$\frac{J'_{k_\phi}(k_R R_1)}{Y'_{k_\phi}(k_R R_1)}=-\frac{J'_{k_\phi}(k_R R_2)}{Y'_{k_\phi}(k_R R_2)}. \tag{6.225}$$

Using the identity $Y_\nu(z)\sin\pi\nu=J_\nu(z)\cos\pi\nu-J_{-\nu}(z)$, this is usually written in the form

$$J'_{k_\phi}(k_R R_1)J'_{-k_\phi}(k_R R_2)-J'_{k_\phi}(k_R R_2)J'_{-k_\phi}(k_R R_1)=0. \tag{6.226}$$

This equation may have real or imaginary roots. The real roots will occur when $k_R>0$. From the symmetry of the equation, it is clear that, if k_ϕ is a root, then $-k_\phi$ is also a root. So it suffices to determine only the positive real or imaginary roots. On the other hand, $k_\phi=0,\mp1,\mp2,\ldots$ are also roots, since $J_{-\nu}(z)=(-1)^\nu J_\nu(z)$, if ν is an integer. The non-zero integer roots correspond to an acoustic field which is periodic in $\phi$of period 2π. These roots are, therefore, relevant for ring-like ducts and may be discarded when a circular bend is of concern. The zero root can also be discarded, because it implies that the acoustic field in a duct with a quiescent medium is not dependent on the angular coordinate, which is not physically viable. On the other hand, Equation (6.225) is satisfied identically also when $k_R=0$, in which case the dispersion equation seems to be valid for any value of k_ϕ. The correct interpretation of this situation comes from the limiting form of Equation (6.226) for $k_R\to0$, when the Bessel function is accurately represented by the first term in its Taylor series expansion and yields for the dispersion equation$(R_1/R_2)^{2k_\phi}=1$, which has the solution $k_\phi=0$ for an arbitrary radius ratio. Thus, for a given m, if k_o is increased from zero, the positive non-integer roots of Equation (6.226) will occur for $k_o>k_3=m\pi/b$, because, for $k_o<m\pi/b$, k_R becomes imaginary and the dispersion equation has only imaginary roots.

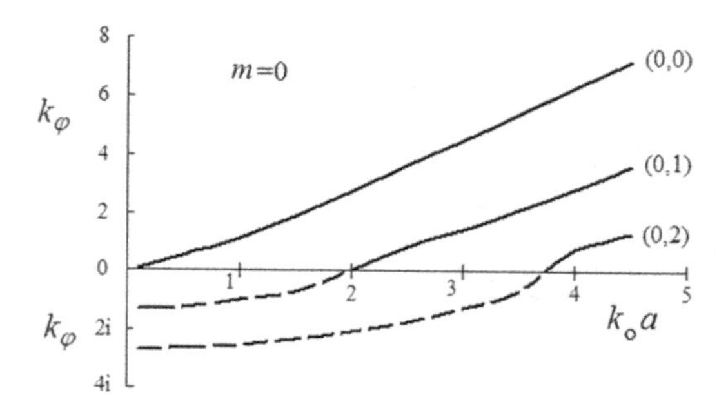

Figure 6.17 The angular wavenumbers of the (0,n) modes of a square section duct of $R_1 = 0.0125$ m, $R_2 = 0.1275$ m and $b = 0.115$ m.

The non-integer roots of Equation (6.226) are denoted by $(k_\phi)_{m,n}$, where the index m refers to the roots corresponding to $k_R^2 = k_o^2 - (m\pi/b)^2$, $m = 0, 1, 2, \ldots$. For a fixed m, these roots (eigenvalues) are classified in ascending order as $(k_\phi)^2_{m,0} < (k_\phi)^2_{m,1} < \cdots < (k_\phi)^2_{m,n} < \cdots$. These eigenvalues are associated in one-to-one correspondence with the transverse duct modes (eigenfunctions) $\varphi_{mn}(x_3, R) = F_{mn}(R)X_m(x_3)$, $m, n = 0, 1, 2, \ldots$, which represent propagating or non-propagating (evanescent) acoustic wave modes according to whether the corresponding $(k_\phi)_{m,n}$ is real or imaginary. Whether a mode is propagating or evanescent depends on the frequency. The cut-on frequency of a mode is defined as the smallest frequency at which it occurs as a propagating mode. Thus, referring to the discussion of the previous paragraph, it can be stated that the cut-on condition for a width-wise mode m is $k_o > m\pi/b$. However, this condition is precise only for $(m, 0)$ modes, which correspond to the smallest positive non-integer root of the dispersion equation for a fixed m. The cut-on frequency for the (m, n) mode for $n > 0$ is always greater than that of $(m, 0)$ mode and increases with n, but it can be determined only by solving the roots of the dispersion equation numerically.

For example, consider a square section duct of $R_1 = 0.0125$ m, $R_2 = 0.1275$ m and $b = 0.115$ m. The angular wavenumbers for the (0,n) modes of this duct are shown in Figure 6.17 for $n = 0, 1, 2$ as functions of the Helmholtz number $k_o a$, where a denotes the radius of a circle having the same area as the duct section, that is, for this duct, $a = 0.064882$ m. The (0,0) mode is always cut-on. The (0,1) and (0,2) modes are cut-on for $k_o a > 2.033$ and $k_o a > 3.724$, respectively. Below these frequencies they are imaginary, that is, evanescent.

Similar cut-on characteristics apply also for the $m > 0$ modes. For example, Figure 6.18 shows the angular wavenumbers of the same duct for the first three $(1, n)$ modes. In this case, the (1,0) mode cuts-on at the predicted value of $k_o a = \pi a/b = 1.772$.

As the radius of the duct center line increases to infinity, the angular wavenumber characteristics coalesce with the frequency axis, the cut-on frequencies being dispositioned in the order of the transverse modes of the straight rectangular ducts (Section 6.5.4.1). In fact, this result is general: Sound transmission in any curved duct having a

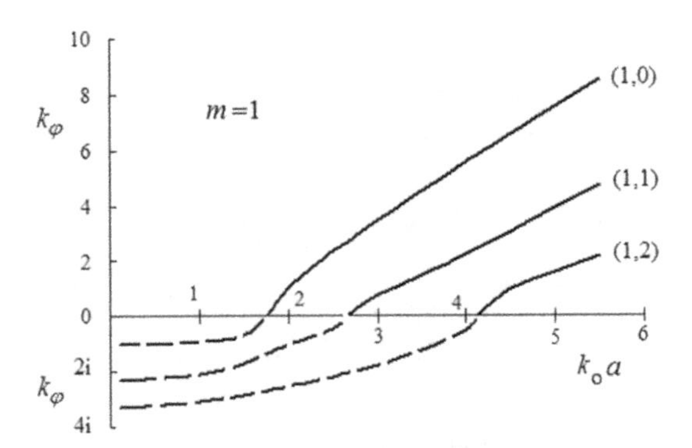

Figure 6.18 The angular wavenumbers of the (1,*n*) modes of a square section duct of $R_1 = 0.0125$ m, $R_2 = 0.1275$ m and $b = 0.115$ m.

radius of curvature large compared to the characteristic dimensions of the duct section may be modeled with adequate accuracy as a straight duct of length equal to the median curve length, and having the same cross section and mean flow.

The acoustic velocity potential in the duct consists of a sum of contributions of all modes. Since the contribution of each mode is in the form of Equation (6.213) and the angular wavenumbers occur in $\mp$ pairs, the acoustic potential can be expressed, in the frequency domain, as

$$\vartheta' = \sum_{m,n=0}^{\infty} A_{mn}^{+}\varphi_{mn}^{+}(R,x_3)e^{i(k_\phi)_{m,n}\phi} + \sum_{m,n=0}^{\infty} A_{mn}^{-}\varphi_{mn}^{-}(R,x_3)e^{i(-k_\phi)_{m,n}\phi}, \tag{6.227}$$

where

$$\varphi_{mn}^{\mp}(R,x_3) = \left(J_{(\mp k_\phi)_{m,n}}(k_{R,m}R) + \frac{J'_{(\mp k_\phi)_{m,n}}(k_{R,m}R_1)}{Y'_{(\mp k_\phi)_{m,n}}(k_{R,m}R_1)} Y_{(\mp k_\phi)_{m,n}}(k_{R,m}R) \right) \cos\left(\frac{m\pi x_3}{b}\right), \tag{6.228}$$

where $k_{R,m}^2 = k_o^2 - (m\pi/b)^2$. The acoustic pressure in the duct follows from Equation (6.211) as $p' = i\omega\rho_o\vartheta'$ and, clearly may be decomposed as $p' = p^+ + p^-$ into sum of positive and negative wave systems, as in straight uniform ducts.

For the formulation of the modal wave transfer matrix of a duct bend, it is convenient to order the angular wavenumbers in single index notation in descending order as $(k_\phi)_1^2 > (k_\phi)_2^2 > \cdots > (k_\phi)_\mu^2 > \cdots$ at a given frequency (see for example Figure 6.17), while keeping track of the mode numbers m and n for identification of the corresponding eigenfunctions. So, at a given frequency, let the ordered angular wavenumbers and corresponding transverse eigenfunctions be collected in matrix notation as

$$\mathbf{\Phi}^{\mp} = \begin{bmatrix} \varphi_1^{\mp}(R,x_3) & \varphi_2^{\mp}(R,x_3) & \ldots \end{bmatrix} \tag{6.229}$$

$$\mathbf{K} = \begin{bmatrix} (k_\phi)_1 & 0 & \cdots \\ 0 & (k_\phi)_2 & \cdots \\ \vdots & \vdots & \ddots \end{bmatrix}. \quad (6.230)$$

In this notation, the acoustic pressure field in the duct is expressed as

$$p' = \mathbf{\Phi}^+(R, x_3)\ \mathbf{P}^+(\phi) + \mathbf{\Phi}^-(R, x_3)\mathbf{P}^-(\phi). \quad (6.231)$$

Here, $\mathbf{P}^\mp$ denote the modal pressure wave components vectors defined by

$$\mathbf{P}^\mp(\phi) = \begin{bmatrix} p_1^\mp(\phi) \\ p_2^\mp(\phi) \\ \vdots \end{bmatrix} = \mathrm{e}^{\mp\mathrm{i}\mathbf{K}\phi}\mathbf{P}^\mp(0). \quad (6.232)$$

Then,

$$\begin{bmatrix} \mathbf{P}^+(\phi) \\ \mathbf{P}^-(\phi) \end{bmatrix} = \mathbf{T} \begin{bmatrix} \mathbf{P}^+(0) \\ \mathbf{P}^-(0) \end{bmatrix}, \quad (6.233)$$

where the modal wave transfer matrix of a bend subtending angle $\hat{\phi}$ is

$$\mathbf{T} = \begin{bmatrix} \mathrm{e}^{\mathrm{i}\mathbf{K}\hat{\phi}} & \mathbf{0} \\ \mathbf{0} & \mathrm{e}^{-\mathrm{i}\mathbf{K}\hat{\phi}} \end{bmatrix}, \quad (6.234)$$

Thus, the modal expansion of the acoustic field in a curved duct is formally similar to a straight uniform duct, except for the dispersive nature of the angular wavenumber. However, the modal wave transfer matrix, Equation (6.234), may not be combined directly to a uniform duct, because of mode mismatch at the connection plane (see Section 7.3, see also Section 6.10.3).

These results apply also when the duct has one or more lined surfaces, except that the wavenumbers are now given by appropriate forms of Equations (6.222) and (6.223). In general, Equation (6.222) may be expressed in the form

$$\begin{aligned} & k_R^2\Big(\mathrm{J}'_{-k_\phi}(k_R R_1)\mathrm{J}'_{k_\phi}(k_R R_2) - \mathrm{J}'_{k_\phi}(k_R R_1)\mathrm{J}'_{-k_\phi}(k_R R_2)\Big) \\ & \quad + \frac{\mathrm{i}k_o}{\zeta_1}k_R\Big(-\mathrm{J}'_{k_\phi}(k_R R_2)\mathrm{J}_{-k_\phi}(k_R R_1) + \mathrm{J}_{k_\phi}(k_R R_1)\mathrm{J}'_{-k_\phi}(k_R R_2)\Big) \\ & \quad + \frac{\mathrm{i}k_o}{\zeta_2}k_R\Big(\mathrm{J}_{k_\phi}(k_R R_2)\mathrm{J}'_{-k_\phi}(k_R R_1) - \mathrm{J}'_{k_\phi}(k_R R_1)\mathrm{J}_{-k_\phi}(k_R R_2)\Big) \\ & \quad + \frac{k_o^2}{\zeta_1\zeta_2}\big(-\mathrm{J}_{k_\phi}(k_R R_1)\mathrm{J}_{-k_\phi}(k_R R_2) + \mathrm{J}_{-k_\phi}(k_R R_1)\mathrm{J}_{k_\phi}(k_R R_2)\big) = 0, \end{aligned} \quad (6.235)$$

which shows that the angular wavenumbers still occur in plus and minus pairs. However, they are in general complex numbers and it is more convenient to classify them in the order least attenuated to the most attenuated modes, as in straight lined ducts. Calculation of the roots of the foregoing dispersion equation (for k_3 satisfying

Equation (6.223) also requires special considerations similar to those in the lined straight duct case (Section 6.7), with the added complexity of dealing with Bessel functions of complex orders.

The time-averaged acoustic power transmitted in the duct is, from Equation (1.49), $W = 2\mathrm{Re}\int_S \left(\breve{p}'\right)^{\mathrm{T}} v'_\phi \mathrm{d}S$, since mean flow is neglected. Hence, using the above described matrix description of the acoustic field

$$W = \frac{2}{\rho_o c_o}\mathrm{Re}\left\{\left[\left(\breve{\mathbf{P}}^+\right)^T \ \left(\breve{\mathbf{P}}^-\right)^T\right]\left(\int_S \begin{bmatrix} \left(\breve{\mathbf{\Phi}}^+\right)^T \mathbf{\Phi}^+ & -\left(\breve{\mathbf{\Phi}}^+\right)^T \mathbf{\Phi}^- \\ \left(\breve{\mathbf{\Phi}}^-\right)^T \mathbf{\Phi}^+ & -\left(\breve{\mathbf{\Phi}}^-\right)^T \mathbf{\Phi} \end{bmatrix} \frac{\mathrm{d}S}{k_o R}\right)\begin{bmatrix} \mathbf{K} & 0 \\ 0 & \mathbf{K} \end{bmatrix}\begin{bmatrix} \mathbf{P}^+ \\ \mathbf{P}^- \end{bmatrix}\right\} \tag{6.236}$$

For the evaluation of the integrals, we first inquire if the transverse eigenfunctions satisfy an orthogonality condition. Using the procedure described in Section 6.7.3.2 for Equation (6.209), it may be shown that any two transverse modes of indices μ and η satisfying Equation (6.209) also satisfy

$$\breve{\varphi}_\eta \nabla_\perp^2 \varphi_\mu - \varphi_\mu \nabla_\perp^2 \breve{\varphi}_\eta + \frac{1}{R}\left(\breve{\varphi}_\eta \frac{\partial \varphi_\mu}{\partial R} - \varphi_\mu \frac{\partial \breve{\varphi}_\eta}{\partial R}\right) = \frac{1}{R^2}\left(\left(k_\phi^2\right)_\mu - \breve{k}_{\phi\eta}^{\,2}\right)\varphi_\mu \breve{\varphi}_\eta . \tag{6.237}$$

Upon integration over a duct section, the first two terms on the left-hand side reduce, by Stokes' theorem, to the line integral $\oint_\ell \left(\breve{\varphi}_\eta \nabla_\perp \varphi_\mu - \varphi_\mu \nabla_\perp \breve{\varphi}_\eta\right)\cdot \mathbf{n}\mathrm{d}\ell$, which can be evaluated by using Equation (6.213) to see that it vanishes only when the wall impedance is purely real and is independent of modes. These conditions are generally accepted to be valid for hard-walled ducts, and Equation (6.237) then yields the condition $0 = \left(\left(k_\phi^2\right)_\mu - \left(k_\phi^2\right)_\eta\right)\int_S \left(\varphi_\mu \varphi_\eta / R\right)\mathrm{d}S$, which implies the orthogonality property $\int_S \left(\varphi_\mu \varphi_\eta / R\right)\mathrm{d}S = 0$ unless $\mu = \eta$. In this case, therefore, the integral of the matrix in Equation (6.236) is diagonal and the time-averaged acoustic power may be expressed as $W = W_1 + W_2 + \cdots + W_\mu + \cdots$, where

$$W_\mu = \frac{2}{\rho_o c_o}\frac{\left(k_\phi\right)_\mu}{k_o}\left(\left(\int_S \frac{\left(\varphi_\mu^+\right)^2}{R}\mathrm{d}S\right)\left|p_\mu^+(\phi)\right|^2 - \left(\int_S \frac{\left(\varphi_\mu^-\right)^2}{R}\mathrm{d}S\right)\left|p_\mu^-(\phi)\right|^2\right), \tag{6.238}$$

the time-averaged power of mode μ. The acoustic field in a curved hard-walled duct is also canonical, which means in that W_μ consists of the algebraic sum of the contributions of wave systems propagating in opposite azimuthal directions, $W_\mu = W_\mu^+ + W_\mu^-$. This is similar to the acoustic field in a straight uniform duct; however, the plane-wave mode does not exist in curved ducts, although the (0,0) mode still propagates at all frequencies (Figure 6.17), and integrals of the eigenfunctions have to be calculated numerically.

If the duct has one or more lined surfaces, the orthogonality condition does not apply and the acoustic power transmitted in the duct is computed by using Equation

(6.236) by taking into account all modes propagating at the frequency of interest. The transmission loss in a lined curved duct may be calculated by using Equation (5.2), where the acoustic power at the inlet and outlet of the duct may be computed from the form of Equation (6.238) for an anechoic outlet, that is,

$$W^{+} = \frac{2}{\rho_o c_o} \mathrm{Re}\left\{ \left(\breve{\mathbf{P}}^{+}\right)^{T} \left(\int_{S} \left(\breve{\boldsymbol{\Phi}}^{+}\right)^{T} \boldsymbol{\Phi}^{+} \frac{\mathrm{d}S}{k_o R} \right) \mathbf{K}\mathbf{P}^{+} \right\}. \tag{6.239}$$

The attenuation characteristics of lined curved ducts are in general qualitatively similar to those in straight lined ducts. Specific results are discussed in the monograph by Rostafinski and the source articles therein [61].

6.10.2 Numerical Determination of Angular Wavenumbers

The results given in the previous section for rectangular ducts also apply for non-rectangular ducts, except for the angular wavenumbers and the corresponding transverse eigenfunctions, which now have to be computed numerically, with the finite element method being the most common choice. The weak form of Equation (6.209) may be derived by recalling that the weak form of its parent equation, Equation (6.208), is in fact given by Equation (2.59) with p' replaced by χ, and, in view of the ansatz $\chi = \varphi \mathrm{e}^{\mathrm{i}k_\phi \phi}$, the gradient operator may be written as $\nabla = \mathbf{e}_R \partial/\partial R + \mathbf{e}_3 \partial/\partial x_3 + \mathbf{e}_\phi \mathrm{i}k_\phi/R$. Substitution of this in Equation (2.59) and integrating, the weak form of Equation (6.209) may be shown to be

$$\int_{S} \left(\frac{\partial w}{\partial R}\frac{\partial \varphi}{\partial R} + \frac{\partial w}{\partial x_3}\frac{\partial \varphi}{\partial x_3} + \left(k_o^2 - \frac{k_\phi^2}{R^2}\right) w\varphi \right) R\mathrm{d}S = \int_{\Gamma} w\mathbf{n}\cdot\nabla_{\perp}\varphi R \mathrm{d}\Gamma, \tag{6.240}$$

where S is the duct cross-sectional area, Γ denotes boundary of S and $\mathbf{n}$ denotes the outward unit normal vector of Γ. As in Section 6.5.5, using a triangular finite element mesh of N contiguous triangles of area S^Δ, $\Delta = 1, 2, \ldots, N$, with M contiguous lines Γ^l, $l = 1, 2, \ldots, M$, on the boundary Γ, Equation (6.240) may be written as

$$\sum_{\Delta=1}^{N} \int_{S^\Delta} \left(\frac{\partial w}{\partial R}\frac{\partial \varphi}{\partial R} + \frac{\partial w}{\partial x_3}\frac{\partial \varphi}{\partial x_3} + \left(k_o^2 - \frac{k_\phi^2}{R^2}\right) w\varphi \right) R\mathrm{d}S = -\mathrm{i}k_o \sum_{l=1}^{M} \int_{\Gamma^l} \frac{w\varphi}{\zeta} \mathrm{d}\Gamma. \tag{6.241}$$

In this case, the distribution of $\varphi = \varphi(R, x_3)$ over a triangular element Δ is approximated by the test function $\varphi^\Delta = c_1 + c_2 R + c_3 x_3$, where c_j, $j = 1, 2, 3$, are some undetermined parameters. The formulae given in Section 6.5.5 for interpolation over the vertices of a triangular element also apply for a curved duct, except that x_2 is now replaced by R. Thus, substituting these approximations in the foregoing equation and selecting the n independent functions for w as ψ_i^Δ, $i = 1, 2, \ldots, n^\Delta$, we obtain:

$$\sum_{\Delta=1}^{N} \left[\mathbf{G}^{-1}\right]^{\mathrm{T}} \left(\int_{S^\Delta} \left[\frac{\partial \mathbf{g}^{\mathrm{T}}}{\partial R}\frac{\partial \mathbf{g}}{\partial R} + \frac{\partial \mathbf{g}^{\mathrm{T}}}{\partial x_3}\frac{\partial \mathbf{g}}{\partial x_3} - \left(k_o^2 - \frac{k_\phi^2}{R^2}\right)\mathbf{g}^{\mathrm{T}}\mathbf{g} \right] R\ \mathrm{d}R\ \mathrm{d}x_3 \right) \mathbf{G}^{-1}\boldsymbol{\varphi}^\Delta = -\mathrm{i}k_o \sum_{l=1}^{M} \mathbf{F}^l \boldsymbol{\varphi}^l, \tag{6.242}$$

where $\mathbf{g}$ and $\mathbf{G}$ are as defined in Section 6.5.5, but with x_2 replaced by R, and

$$\mathbf{F}^l = \left[\mathbf{G}^{-1}\right]^T \left(\int_{\Gamma^l} \frac{\mathbf{g}^T\mathbf{g}}{\zeta} \mathrm{d}\Gamma \right) \mathbf{G}^{-1}. \tag{6.243}$$

Equation (6.242) may be expressed as

$$\sum_{\Delta=1}^{N} \left(\mathbf{K}^\Delta + k_\phi^2 \mathbf{M}^\Delta\right) \boldsymbol{\varphi}^\Delta = -\mathrm{i}k_o \sum_{l=1}^{M} \mathbf{F}^l \boldsymbol{\varphi}^l, \tag{6.244}$$

where

$$\mathbf{K}^\Delta = \left[\mathbf{G}^{-1}\right]^\mathrm{T} \begin{bmatrix} 0 & 0 & 0 \\ 0 & S^\Delta \bar{R} & 0 \\ 0 & 0 & S^\Delta \bar{R} \end{bmatrix} \mathbf{G}^{-1} - k_o^2 \left[\mathbf{G}^{-1}\right]^\mathrm{T} \left(\int_\Delta \mathbf{g}^\mathrm{T}\mathbf{g} R \mathrm{d}R \mathrm{d}x_3 \right) \mathbf{G}^{-1} \tag{6.245}$$

$$\mathbf{M}^\Delta = \left[\mathbf{G}^{-1}\right]^\mathrm{T} \left(\int_\Delta \frac{\mathbf{g}^\mathrm{T}\mathbf{g}}{R} \mathrm{d}R \mathrm{d}x_3 \right) \mathbf{G}^{-1} \tag{6.246}$$

and $\bar{R}$ denotes the average value of the R coordinates of the three vertices of a triangular element. The rest of the formulation proceeds as in Section 6.5.5 for the assembly of the triangular elements and leads to the algebraic eigenvalue problem $\left(\mathbf{K} + k_\phi^2 \mathbf{M}\right)\boldsymbol{\varphi} = -\mathrm{i}k_o \mathbf{F} \boldsymbol{\varphi}_\Sigma$ for the angular wavenumbers and the associated eigenfunctions, where $\mathbf{M}$ and $\mathbf{K}$ denote the assembled $\mathbf{M}^\Delta$ and $\mathbf{K}^\Delta$ matrices, respectively, and the sub-vector $\boldsymbol{\varphi}_\Sigma$includes all terms of the global vector $\boldsymbol{\varphi}$ corresponding to the nodes on Γ. If the duct is lined, this equation may be solved as described in Section 6.7.6. In the case of hard-walled ducts, the eigenvalues problem reduces to $\left(\mathbf{K} + k_\phi^2 \mathbf{M}\right)\boldsymbol{\varphi} = 0$ and may be solved as discussed in Section 6.5.5.

6.10.3 Fundamental Mode Approximation

The effect of duct curvature on sound wave propagation cannot be modeled in a one-dimensional framework. However, it is desirable to have a curved duct model which can be used in low-frequency applications with the one-dimensional duct models described in Chapter 3. Such an element may be defined by the wave transfer matrix of the (0,0) mode of a circularly curved duct, which may be written from Equation (6.234) as

$$\mathbf{T} = \begin{bmatrix} \mathrm{e}^{\mathrm{i}(k_\phi)_{0,9}\phi} & 0 \\ 0 & \mathrm{e}^{-\mathrm{i}(k_\phi)_{0,9}\phi} \end{bmatrix}. \tag{6.247}$$

But this is valid for ducts with no mean flow. When there is mean flow, it is plausible to expect its convective effect on the fundamental mode be reflected on the angular wavenumber in the same manner as in the straight uniform ducts. On this basis,

Equation (6.247) may be extended ad hoc for the presence of mean flow by introducing the Doppler factors similarly:

$$\mathbf{T} = \begin{bmatrix} \mathrm{e}^{\frac{\mathrm{i}}{1+\bar{M}_o}\frac{(k_\phi)_{0,0}}{k_\phi}s} & 0 \\ 0 & \mathrm{e}^{\frac{-\mathrm{i}}{1-\bar{M}_o}\frac{(k_\phi)_{0,0}}{R_o}s} \end{bmatrix}, \tag{6.248}$$

where $s = R_o\phi$ denotes the length of the duct center line, R_o denotes the radius of the center line and $\bar{M}_o$ denoted the Mach number of the cross-section averaged mean flow velocity. The fundamental angular wavenumber may be determined numerically for arbitrary duct sections. In practical calculations, it may be estimated by using the Rostafinski formula [61]:

$$(k_\phi)_{0,0} = \sqrt{2}\left(1 + \left(\frac{4}{k_o^2R_1^2} + \frac{R_2^2}{R_1^2} + 1\right)\left(\frac{R_2^2}{R_1^2} - 1\right)^{-1} \ln\frac{R_2}{R_1}\right)^{-\frac{1}{2}}. \tag{6.249}$$

Although this is based on rectangular ducts, it may be used also for curved ducts of arbitrary cross section, since it is dominated by the Helmholtz number, k_oR_1, at low frequencies. The corresponding eigenfunctions (assuming they are not affected by the mean flow) are:

$$\varphi_{0,0}^{\mp}(R) = \mathrm{J}_{(\mp k_\phi)_{0,0}}(k_oR) + \frac{\mathrm{J}'_{(\mp k_\phi)_{0,0}}(k_oR_1)}{\mathrm{Y}'_{(\mp k_\phi)_{0,0}}(k_oR_1)}\mathrm{Y}_{(\mp k_\phi)_{0,0}}(k_oR), \tag{6.250}$$

which are obviously not planar. Therefore, the acoustic pressure and the azimuthal component of the particle velocity decompose into fundamental pressure wave components $p^{\mp} = p^{\mp}(s)$, where the (0,0) mode index is omitted for simplicity, as

$$p' = \varphi_{0,0}^{+}p^{+} + \varphi_{0,0}^{-}p^{-} \tag{6.251}$$

$$\rho_oc_ov'_\phi = \frac{(k_\phi)_{0,0}}{k_oR}\left(\varphi_{0,0}^{+}p^{+} - \varphi_{0,0}^{-}p^{-}\right). \tag{6.252}$$

Since these are not compatible with the corresponding decomposition for straight ducts, Equations (4.22) and (4.23), it is necessary to match the pressure wave components at the connection plane when a curved duct is to be connected to a straight one-dimensional duct. We may do this by applying the quasi-static conservation equations (Section 4.2) to a control volume enclosing the connection plane. Let the control surfaces on the curved and straight sides of the discontinuity be denoted by subscripts "1" and "2," respectively. Applying Equation (4.10) gives

$$\rho_o\overline{v_1'} + \bar{v}_{2o}\frac{\overline{p_1'}}{c_o^2} = \rho_o\overline{v_2'} + \bar{v}_{2o}\frac{\overline{p_2'}}{c_o^2}. \tag{6.253}$$

With substitution of Equations (6.251) and (6.252) and their counterparts for the uniform duct, this may be expressed as

$$\left((k_\phi)_{0,0}\overline{\left(\frac{\varphi_{0,0}^+}{k_oR}\right)}+\bar{M}_o\overline{\varphi_{0,0}^+}\right)p_1^+ + \left(-(k_\phi)_{0,0}\overline{\left(\frac{\varphi_{0,0}^-}{k_oR}\right)}+\bar{M}_o\overline{\varphi_{0,0}^-}\right)p_1^-$$

$$= (1+\bar{M}_o)p_2^+ + (-1+\bar{M}_o)p_2^-, \tag{6.254}$$

where $\bar{M}_o = \bar{v}_{2o}/c_o$. Similarly, using Equation (4.20) for the energy balance and neglecting the entropy fluctuations

$$\frac{\overline{p_1'}}{\rho_o} + \bar{v}_{2o}\overline{v_1'} = \frac{\overline{p_2'}}{\rho_o} + \bar{v}_{2o}\overline{v_2'} \tag{6.255}$$

or

$$\left(\overline{\varphi_{0,0}^+} + \bar{M}_o(k_\phi)_{0,0}\overline{\left(\frac{\varphi_{0,0}^+}{k_oR}\right)}\right)p_1^+ + \left(\overline{\varphi_{0,0}^-} - \bar{M}_o(k_\phi)_{0,0}\overline{\left(\frac{\varphi_{0,0}^-}{k_oR}\right)}\right)p_1^-$$

$$= (1+\bar{M}_o)p_2^+ + (1+\bar{M}_o)p_2^-. \tag{6.256}$$

Equations (6.254) and (6.256) can now be solved to find $p_1^{\mp}$ in terms of $p_2^{\mp}$ or vice versa. The averages of the transverse function have to be computed by numerical integration, since no closed formula is available for integrals of Bessel functions of arbitrary real order.

6.11 Uniform Ducts with Mean Swirl

The swirl component of mean flow velocity is largely of interest in turbomachinery ducts and to some extent in combustion chambers and engine intake systems. The literature on the subject is extensive, treating various aspects of wave phenomena in circular ducts. Typically, the mean flow is modeled as combination of uniform axial flow (for example, of the same average axial velocity as the actual mean flow) and a rotation of constant angular velocity (for example, of the same angular momentum as the actual mean swirl) [67]. The governing wave equation for circular ducts encompasses coupled acoustic and hydrodynamic wave modes [68]. When the main interest is on duct-borne sound propagation, the nearly acoustic modes are of major importance. In this section we describe a solution which gives these modes directly whilst retaining the mean flow effects with adequate accuracy. An approximate separation of the nearly acoustic modes is described also by Nijboer [69]

Consider a uniform straight duct of circular cross section carrying an inviscid fluid flow. The duct geometry is defined in cylindrical polar coordinates x_1, r, θ, where x_1 is the duct axis and r, θ are the radial and angular coordinates, respectively (see Figure 6.4). The duct is assumed to contain a homentropic axisymmetric mean flow of velocity distribution

$$\mathbf{v}_o = \mathbf{e}_1 v_{o1} + \mathbf{e}_\theta \varpi_o r. \tag{6.257}$$

Here, $\mathbf{e}_j$, $j = 1, r, \theta$, denotes a unit vector in the positive x_1, r, θ directions, respectively, v_{o1} denotes the axial component of the mean flow velocity, which is assumed to be uniform (v_{o1} = constant) and $v_{o\theta} = \varpi_o r$ denotes the circumferential component of the mean flow velocity due to solid body rotation of constant angular velocity ϖ_o. The latter component is assumed to be subsonic of low Mach number, so that $(\Omega_o r)^2 << 1$, $r \leq a$, where $\Omega_o = \varpi_o / c_o$ and a denotes the duct radius (or the outer radius if the duct is annular).

The mean flow is governed by Equations (6.3) and (6.4). Equation (6.257) implies that $\nabla \cdot \mathbf{v}_o = 0$ and $\mathbf{v}_o \cdot \nabla \mathbf{v}_o = -\mathbf{e}_r \varpi_o^2 r$, since the gradient operator in cylindrical coordinates is $\nabla = \mathbf{e}_1 \partial / \partial x_1 + \mathbf{e}_r \partial / \partial r + \mathbf{e}_\theta \partial / r \partial \theta$. Therefore, Equation (6.4) gives

$$\frac{\mathrm{d}p_o}{\mathrm{d}r} = \rho_o \varpi_o^2 r \tag{6.258}$$

or, $p_o = p_o(r)$, and Equation (6.3) implies $\rho_o = \rho_o(r)$. For homentropic mean flow (Section 1.1), on the other hand, $\nabla p_o = c_o^2 \nabla \rho_o$, which reduces in this case to $\mathrm{d}p_o / \mathrm{d}r = c_o^2 \mathrm{d}\rho_o / \mathrm{d}r$. Using these relationships, it may be shown that for a perfect gas

$$\frac{\rho_o(r)}{\rho_o(a)} = \left(1 - \frac{1}{2}(\gamma_o - 1) \left(1 - \frac{r^2}{a^2} \right) (\Omega_o a)^2 \right)^{1/(\gamma_o - 1)}, \tag{6.259}$$

where γ_o denotes the ratio of the specific heat coefficients and a denotes the duct radius (outer radius if the duct is annular). Therefore, under the restriction $(\Omega_o r)^2 << 1$, it is adequately accurate to assume that ρ_o = constant. A convenient value for this constant is the average mean density $(1/2a^2) \int \rho_o(r) r \mathrm{d}r$. A corollary to this is

$$\frac{c_o^2(r)}{c_o^2(a)} = 1 + 0.5\gamma_o \left(1 - \frac{r^2}{a^2} \right) (\Omega_o a)^2. \tag{6.260}$$

Therefore, the restriction $(\Omega_o r)^2 << 1$ also permits us to assume that the speed of sound c_o is constant and, hence, the relationship $p' = c_o^2 \rho'$.

The acoustic field in the duct is given by Equations (6.1) and (6.2). With the above mean flow approximations, these equations simplify, respectively, to

$$\left(\frac{\partial}{\partial t} + \mathbf{v}_o \cdot \nabla \right) p' + \rho_o c_o^2 \nabla \cdot \mathbf{v}' = 0 \tag{6.261}$$

$$\rho_o \left(\frac{\partial \mathbf{v}'}{\partial t} + \mathbf{v}_o \cdot \nabla \mathbf{v}' + \mathbf{v}' \cdot \nabla \mathbf{v}_o \right) + \nabla p' = 0, \tag{6.262}$$

where the term $\rho' \mathbf{v}_o \cdot \nabla \mathbf{v}_o$ of the acoustic momentum equation, Equation (6.2), is neglected on the grounds that $|\mathbf{v}_o / c_o|^2 << 1$. The components of Equation (6.262) in x_1, r, θ directions are

$$\rho_o \left(\frac{\partial v_1'}{\partial t} + \mathbf{v}_o \cdot \nabla v_1' \right) + \frac{\partial p'}{\partial x_1} = 0 \tag{6.263}$$

$$\rho_o\left(\frac{\partial v_r'}{\partial t}+\mathbf{v}_o\cdot\nabla v_r'-2\varpi_o v_\theta'\right)+\frac{\partial p'}{\partial r}=0 \tag{6.264}$$

$$\rho_o\left(\frac{\partial v_\theta'}{\partial t}+\mathbf{v}_o\cdot\nabla v_\theta'+2\varpi_o v_r'\right)+\frac{1}{r}\frac{\partial p'}{\partial \theta}=0 \tag{6.265}$$

respectively. In frequency domain, a wave-like solution of these equations may be sought as

$$p'=R(r)\mathrm{e}^{\mathrm{i}m\theta}\mathrm{e}^{\mathrm{i}k_oKx_1} \tag{6.266}$$

$$v_j'=Q_j(r)\mathrm{e}^{\mathrm{i}m\theta}\mathrm{e}^{\mathrm{i}k_oKx_1},\qquad j=1,r,\theta \tag{6.267}$$

where $k_o=\omega/c_o$, K denotes the axial propagation constants (to be determined) and $m=0,\mp1,\mp2,\ldots$, since the acoustic field in a circular duct must be circumferentially periodic of period 2π. Under the foregoing ansatz, the operator $-\mathrm{i}\omega+\mathbf{v}_o\cdot\nabla$ can be replaced by $-\mathrm{i}\omega G$, where G is given by

$$G=1-KM_o-\frac{m\Omega_o}{k_o}, \tag{6.268}$$

where $M_o=v_{o1}/c_o$ and G, the Doppler shift factor, is not dependent on r to present approximation. Hence, Equation (6.263)–(6.265) can be expressed as, respectively,

$$v_1'=\frac{1}{\rho_oc_o}\frac{K}{G}p' \tag{6.269}$$

$$v_r'=\frac{1}{\rho_o}\frac{1}{4\varpi_o^2-\omega^2G^2}\left(\mathrm{i}\omega G\frac{1}{R}\frac{\mathrm{d}R}{\mathrm{d}r}-\frac{\mathrm{i}2m\varpi_o}{r}\right)p' \tag{6.270}$$

$$v_\theta'=\frac{1}{\rho_o}\frac{1}{4\varpi_o^2-\omega^2G^2}\left(2\varpi_o\frac{1}{R}\frac{\mathrm{d}R}{\mathrm{d}r}-\frac{\omega Gm}{r}\right)p'. \tag{6.271}$$

The radial function $R=R(r)$ is determined now by substituting these results in Equation (6.261):

$$r^2\frac{\mathrm{d}^2R}{\mathrm{d}r^2}+r\frac{\mathrm{d}R}{\mathrm{d}r}+\left(\alpha^2r^2-m^2\right)R=0, \tag{6.272}$$

where

$$\frac{\alpha^2}{k_o^2}=\left(G^2-K^2\right)\left(G^2-4\frac{\Omega_o^2}{k_o^2}\right). \tag{6.273}$$

Then, the functions $Q_j=Q_j(r)$, $j=1,r,\theta$, are determined from Equations (6.269)–(6.271) in terms of the radial function.

A fourth-order algebraic dispersion equation in K may be derived by eliminating G from Equations (6.268) and (6.273). The nature of the roots of this equation as $\omega\rightarrow\infty$ can be seen by noting that in this limit Equation (6.273) reduces to $\left(G^2-K^2\right)G^2=0$.

Therefore, to first order in $1/k_o$ (that is, for $1/k_o << 1$), the four roots of the dispersion equation are given by

$$K^{\mp}(\Omega_o) \approx \frac{\mp 1 \pm \left(\frac{m\Omega_o}{k_o}\right)}{1 \mp M_o} \tag{6.274}$$

$$K^{(1,2)}(\Omega_o) \approx \frac{1 - \left(\frac{m\Omega_o}{k_o}\right)}{M_o}, \tag{6.275}$$

which are the roots of $G^2 - K^2 = 0$and $G^2 = 0$, respectively. This result also shows that, eventually, as $k_o \to \infty$, the four roots of the dispersion equation approach the asymptotes $K^{\mp}(0) = \mp 1/(1 \mp M_o)$ and $K^{(1,2)}(0) = 1/M_o$. We recall that, $\mp 1/(1 \mp M_o)$ are the fundamental acoustic modes of the same duct with only uniform axial mean flow and $1/M_o$ is the hydrodynamic propagation constant of the same duct (Section 6.5). Hence, it can be inferred that, if the frequency is sufficiently large, so that $k_o^2 >> 4\Omega_o^2/G^2$, the acoustic propagation constants can be determined approximately from the dispersion equation based on Equation (6.272) with

$$\alpha^2 = k_o^2(G^2 - K^2) \tag{6.276}$$

Then, upon eliminating G between this equation and Equation (6.268), the acoustic propagation constant may be expressed as

$$K^{\mp}(\Omega_o) = \left(1 - \frac{m\Omega_o}{k_o}\right)\left(\frac{\mp\sigma - M_o}{1 - M_o^2}\right), \tag{6.277}$$

where

$$\sigma = \sqrt{1 - \frac{\alpha^2}{k_o^2}\frac{1 - M_o^2}{\left(1 - \frac{m\Omega_o}{k_o}\right)^2}} \tag{6.278}$$

and the eigenvalues α^2 are determined by the usual procedure, that is, by requiring that the general solution of Equation (6.73) for the radial function, $R = C_1 \mathrm{J}_m(\alpha r) + C_2 \mathrm{Y}_m(\alpha r)$, where J_m and Y_m are first and second kind Bessel functions of order m, and C_1, C_2 are integration constants, satisfy the Ingard–Myers boundary condition (Section 6.3.3) $v_r' = (-\mathrm{i}\omega + \mathbf{v}\cdot\nabla)p'/(-\mathrm{i}\omega Z)$. Using Equation (6.270) and recalling the restriction $k_o^2 >> 4\Omega_o^2/G^2$ the latter may be expressed as

$$\frac{\mathrm{d}R}{\mathrm{d}r} = \left(\frac{\mathrm{i}k_o G^2}{\zeta} + \frac{2m\Omega_o}{Gk_o r}\right) R \text{ on } \Gamma, \tag{6.279}$$

where $\zeta = Z/\rho_o c_o$ is assumed to be uniform. For a hollow circular duct of radius a, $C_2 = 0$ and Equation (6.279) yields the dispersion equation

$$\alpha a \frac{\mathrm{J}_m'(\alpha a)}{\mathrm{J}_m(\alpha a)} = \frac{\mathrm{i}k_o a}{\zeta} G^2 + \frac{2m\Omega_o}{Gk_o} \tag{6.280}$$

and the transverse eigenfunctions corresponding to the nth root of this equation for given m, α_{mn}, are (except for an arbitrary constant factor):

$$\varphi_{mn}(r,\theta) = |\mathrm{J}_m(\alpha_{mn} r)|\mathrm{e}^{\mathrm{i}(m\theta+\lambda_{mn})}), \tag{6.281}$$

where λ_{mn} denotes the argument of the expression in the modulus bracket. This result shows that, the presence of mean swirl has no effect on the axisymmetric ($m = 0$) modes and the spinning modes of orders $m << Gk_o/2\Omega_o$ of a uniform circular duct with uniform mean flow (Section 6.5.1).

The dispersion equation for an annular duct of inner radius b and outer radius a may be derived similarly by requiring that Equation (6.279) must be satisfied both at $r = a$ and $r = b$.

The time-averaged acoustic power in the duct is given by Equation (1.49). Since the mean flow velocity is given by Equation (6.257), it expands now as

$$\begin{aligned} W = {} & 2\mathrm{Re}\int_S \left(\left(1+M_o^2\right)\left(\breve{p}'\right)^{\mathrm{T}} v_1' + \rho_o c_o \left(\breve{v}_1'\right)^{\mathrm{T}} v_1' + \frac{M_o}{\rho_o c_o}\left(\breve{p}'\right)^{\mathrm{T}} p' \right) \mathrm{d}S + \\ & + 2\mathrm{Re}\int_S \left(\rho_o c_o \left(\breve{v}_1'\right)^{\mathrm{T}} v_\theta' + M_o \Omega_o r \left(\breve{p}'\right)^{\mathrm{T}} v_\theta' \right) \mathrm{d}S, \end{aligned} \tag{6.282}$$

where the second integral represents the effects of the mean swirl. The acoustic particle velocity components are given by Equations (6.269)–(6.271).

6.12 Ducts with Mean Temperature Gradient

The acoustic field in a duct with mean temperatures gradients is still governed by Equations (6.1) and (6.2). However, since hot gases can be modeled with adequate accuracy by assuming the perfect gas laws, it is more convenient to use a modified form of the acoustic continuity equation, Equation (6.1). Going back to the continuity equation for the unsteady motion of an inviscid fluid, Equation (1.1), substitute Equation (1.9), the corresponding energy equation. This yields

$$\frac{\partial p}{\partial t} + \mathbf{v}\cdot\nabla p + \gamma p \nabla\cdot\mathbf{v} = 0, \tag{6.283}$$

since $\rho c^2 = \gamma p$ for a perfect gas, where γdenotes the ratio of the specific heat coefficients. In linearizing this equation, it is adequately accurate to assume $\gamma = \gamma_o =$ constant, since γ is a very slowly varying function of temperature for perfect gases. Hence, linearizing Equation (6.283), the required modified acoustic continuity equation is obtained as

$$\frac{\partial p'}{\partial t} + \mathbf{v}_o\cdot\nabla p' + \mathbf{v}'\cdot\nabla p_o + \gamma_o p_o \nabla\cdot\mathbf{v}' + \gamma_o p' \nabla\cdot\mathbf{v}_o = 0. \tag{6.284}$$

The corresponding mean flow equation is $\mathbf{v}_o \cdot \nabla p_o + \gamma_o p_o \nabla \cdot \mathbf{v}_o = 0$ where $\gamma_o p_o = \rho_o c_o^2$. In general, Equations (6.284) and (6.2) can be solved only numerically. In few cases, however, they may be reduced to a single wave equation.

6.12.1 Ducts without Mean Flow

In applications such as pulse combustors and rocket motors where mean temperature gradients are of concern, the mean flow Mach number is small and can be neglected in practical calculations. Then $\mathbf{v}_o = 0$ and Equation (6.4) implies $p_o =$ constant and, hence, Equations (6.284) and (6.2) simplify drastically to

$$\frac{1}{\gamma_o p_o} \frac{\partial p'}{\partial t} + \nabla \cdot \mathbf{v}' = 0 \tag{6.285}$$

$$\rho_o \frac{\partial \mathbf{v}'}{\partial t} + \nabla p' = 0 \tag{6.286}$$

respectively. Eliminating the particle velocity between these equations yields the wave equation

$$-\frac{1}{c_o^2} \frac{\partial^2 p'}{\partial t^2} + \rho_o \nabla \cdot \left(\frac{1}{\rho_o} \nabla p' \right) = 0. \tag{6.287}$$

This is known as the Bergmann wave equation [70]. In the frequency domain, it can be expressed as

$$(\nabla^2 + k_o^2) p' - \frac{1}{\rho_o} \nabla \rho_o \cdot \nabla p' = 0, \tag{6.288}$$

where $k_o = \omega / c_o$. A form of the Bergmann wave equation which is useful for the present analysis is

$$\left(\nabla^2 + k_o^2 + \frac{1}{2} \frac{\nabla^2 \rho_o}{\rho_o} - \frac{3}{4} \frac{\nabla \rho_o \cdot \nabla \rho_o}{\rho_o^2} \right) \left(\frac{p'}{\sqrt{\rho_o}} \right) = 0. \tag{6.289}$$

Consider a straight uniform duct in which mean temperature varies very little over a duct section so that the mean absolute temperature distribution in the duct is predominantly axial, that is, $T_o = T_o(x_1)$, where x_1 denotes the duct axis ($0 \leq x_1 \leq L$). In terms of a given mean temperature taper function, Equation (6.289) can be expressed as

$$\left(\nabla^2 + k_o^2 + \frac{1}{4} \left(\frac{1}{T_o} \frac{\mathrm{d}T_o}{\mathrm{d}x_1} \right)^2 - \frac{1}{2} \frac{1}{T_o} \frac{\mathrm{d}^2 T_o}{\mathrm{d}x_1^2} \right) \left(\frac{p'}{\sqrt{\rho_o}} \right) = 0, \tag{6.290}$$

where $\rho_o = p_o / RT_o$, $c_o^2 = \gamma_o RT_o$ and R denotes the gas constant. Solutions of this equations may be searched in the form

$$p' = \varphi(\bullet) u(x_1) \sqrt{\rho_o(x_1)}, \tag{6.291}$$

where $\varphi(\bullet)$ denotes a function of the transverse coordinates, for example, $\varphi(\bullet) = \varphi(x_2, x_3)$ for rectangular ducts (see Figure 6.3) and $\varphi(\bullet) = \varphi(r, \theta)$ for circular ducts (see Figure 6.4). Substituting Equation (6.291) in Equation (6.290) and separating the variables, we find for $\varphi(\bullet)$

$$(\nabla_\perp^2 + \alpha^2)\varphi(\bullet) = 0, \tag{6.292}$$

where $\nabla_\perp^2$ denotes the Laplacian on the duct cross section, and for u

$$\left(\frac{d^2}{dx_1^2} + \kappa(x_1)\right)u = 0 \tag{6.293}$$

where

$$\kappa(x_1) = k_o^2 - \alpha^2 + \frac{1}{4}\left(\frac{1}{T_o}\frac{dT_o}{dx_1}\right)^2 - \frac{1}{2}\frac{1}{T_o}\frac{d^2T_o}{dx_1^2}. \tag{6.294}$$

Equation (6.292) is the same as Equation (6.48). This shows that for given duct section shapes and boundary conditions at the duct walls, the eigenvalues α^2 and the corresponding eigenfunctions $\varphi(\bullet)$ are independent of the temperature gradient. Consequently, we can use solutions of the two-dimensional Helmholtz equation described previously, for example, in Section 6.5.3 for hard-walled rectangular and circular hollow and annular ducts. For hard-walled ducts, the eigenvalues α^2 are real and can be ordered in ascending order as $0 = \alpha_1^2 < \alpha_2^2 \leq \cdots$, the corresponding eigenfunctions, which are denoted as $\varphi_1(\bullet)$, $\varphi_2(\bullet)$,... in the order of the eigenvalues, satisfy the orthogonality condition (Section 6.5.1). This section is limited, for the simplicity of the presentation, to hard-walled ducts, but soft-walled ducts can be treated similarly, except for the differences in the properties of the transverse eigenvalues and eigenfunctions.

Since α is defined over a discrete spectrum of infinite extent, there are, in fact, infinite number of solutions of these equations corresponding to $\alpha^2 = \alpha_1^2, \alpha_2^2, \ldots$, which are referred to as the axial modes of the duct, as usual.

Exact analytical solutions of Equation (6.293) for $\alpha > 0$ can be found by fitting some specific temperature distributions to the possible solution templates given in handbooks on ordinary differential equations. However, the cut-on conditions of the higher-order modes are not usually immediately evident from these solutions. We consider mean temperature distribution functions which are tractable to relatively simple exact analytical solutions and, as such, we consider the following linear distribution:

$$T_o(x_1) = T_o(0)\left(1 + \varepsilon\frac{x_1}{L}\right), \tag{6.295}$$

where L denotes the length of the duct and $\varepsilon = -1 + T_o(L)/T_o(0)$ is called the temperature change parameter. A relatively simple analytical solution of Equation (6.293) can be found for this taper function if the mean temperature gradient is low so that $\varepsilon^2 << 1$. This condition is satisfied with less than 10% error for $\varepsilon < 0.3$, which covers the frequency ranges relevant to oscillatory higher-order modes and temperature gradients of interest in many applications. Then, neglecting $O(\varepsilon^2)$ terms, Equation (6.293) becomes [71]:

$$\frac{\mathrm{d}^2 u}{\mathrm{d}x_1^2} + \left(k_o^2(0) - \alpha^2 - \varepsilon k_o^2(0)\frac{x_1}{L}\right)u = 0. \tag{6.296}$$

For $\varepsilon \neq 0$,[5] the transformation

$$\sigma(x_1) = \left(\frac{\varepsilon k_o^2(0)}{L}\right)^{-\frac{2}{3}} \left(-k_o(0)^2 + \alpha^2 + \varepsilon k_o^2(0)\frac{x_1}{L}\right) \tag{6.297}$$

reduces Equation (6.296) to the Airy equation

$$\frac{\mathrm{d}^2 u}{\mathrm{d}\sigma^2} - \sigma u = 0, \tag{6.298}$$

the general solution of which is

$$u(\sigma) = C_1 \mathrm{Ai}(\sigma) + C_2 \mathrm{Bi}(\sigma), \tag{6.299}$$

where $\mathrm{Ai}(\sigma), \mathrm{Bi}(\sigma)$ are the Airy functions of the first and second kinds [72], respectively, and C_1, C_2 denote the integration constants. In contrast with the corresponding solution in the absence of a mean temperature gradient (see Footnote 5), the terms on the right of Equation (6.299) are not decomposable as independent waves traveling in the $+x_1$ and $-x_1$ directions. A physical explanation of this situation follows from the fact that sound waves are reflected at impedance discontinuities. In the presence of an axial mean temperature gradient, characteristic impedance varies continuously axially and, hence, an incident sound wave is reflected continuously as it propagates along the duct.

Since σ is a function of α and α is defined over a discrete spectrum of infinite extent, the general solution of Equation (6.290) can be expressed as a sum of infinite number of modes as

$$p'(x_1, \bullet) = \sqrt{\rho_o(x_1)} \sum_{q=1}^{\infty} \left(C_{1q}\mathrm{Ai}(\sigma_q(x_1)) + C_{2q}\mathrm{Bi}(\sigma_q(x_1))\right)\varphi_q(\bullet), \tag{6.300}$$

where σ_q denote the values of σ for $\alpha = \alpha_q$, $q = 1, 2, \ldots$ We will generally omit the use of the subscript "q" for the mode numbers in the subsequent analysis, and assume that, that α and the quantities dependent on it are defined over a discrete spectrum of infinite extent, is understood from the context of the analysis.

[5] If $\varepsilon = 0$, then k_o =constant and the solution of Equation (6.296) degenerates to

$$u(x_1) = C_1 \cos\left(\sqrt{k_o^2 - \alpha^2 x_1}\right) + C_2 \sin\left(\sqrt{k_o^2 - \alpha^2 x_1}\right) \text{ if } \alpha^2/k_o^2 < 1;$$

$$u(x_1) = C_1 \cosh\left(\sqrt{\alpha^2 - k_o^2 x_1}\right) + C_2 \sinh\left(\sqrt{\alpha^2 - k_o^2 x_1}\right) \text{ if } \alpha^2/k_o^2 > 1$$

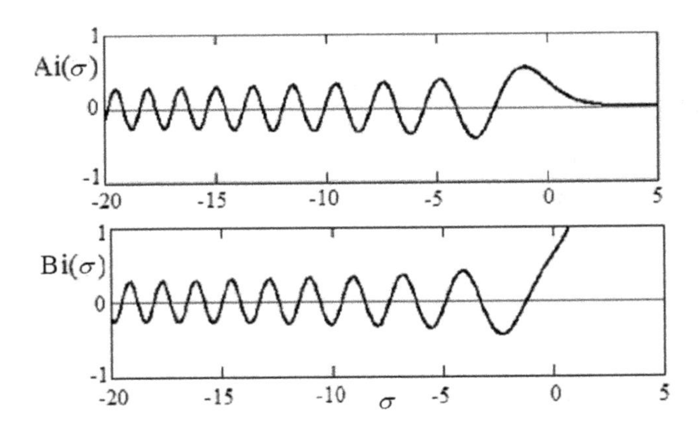

Figure 6.19 Airy functions of the first and second kind.

The Airy functions are shown in Figure 6.19 as function of σ. It is seen that they are not oscillatory if $\sigma > 0$. In fact, $\mathrm{Ai}(\sigma)$ decreases with σ and $\mathrm{Bi}(\sigma)$ increases with σ exponentially for large σ. If $\sigma < 0$, the Airy functions are oscillatory, but not purely sinusoidal; their amplitudes increase gradually with σ. Note that σ varies linearly with x_1 and its sign is determined by $\kappa(x_1)$, the expression in the second bracket in Equation (6.297). Consequently, the amplitudes of the oscillating (cut-on) modes in Equation (6.300) are increasing with x if $\varepsilon > 0$ and decreasing with x if $\varepsilon < 0$. This shows that, the feature of the plane waves being assisted (resisted) by the presence of a positive (negative) mean temperature gradient applies for all cut-on modes (Section 3.9.3).

If mode q is non-oscillatory at the origin, then $\sigma_q(0) > 0$ $(\alpha_q^2/k_o^2(0) > 1)$. It will remain non-oscillatory throughout L, if $\varepsilon > 0$. But if $\varepsilon < 0$, the condition $\sigma_q > 0$ turns over at the duct sections $x/L \geq \left(\alpha_q^2/k_o^2(0) - 1\right)/|\varepsilon|$ to $\sigma_q < 0$; that is, the mode becomes oscillatory (cut-on) in this part of the duct. This phenomenon will not occur, i.e., the mode remains non-oscillatory if $\alpha_q^2/k_o^2(0) \geq 1 + |\varepsilon|$.

The case of $\sigma_q(0) < 0$ can be analyzed similarly. If mode q is oscillatory (cut-on) at the origin, $\sigma_q(0) < 0$ $(\alpha_q^2/k_o^2(0) < 1)$, it will remain so throughout L if $\varepsilon < 0$. But, if $\varepsilon > 0$, the condition $\sigma_q < 0$ turns over at the section $x/L > \left(1 - \alpha_q^2/k_o^2(0)\right)/\varepsilon$ to $\sigma_q > 0$ the mode becomes non-oscillatory in this part of the duct. This phenomenon will not occur, i.e., the mode remains oscillatory (cut-on) if $\alpha_q^2/k_o^2(0) \leq 1 - \varepsilon$.

It can be observed from these results that, the fundamental mode ($q = 1$) is always oscillatory, since $\alpha_1 = 0$ and, if $\varepsilon = 0$, we get the cut-on condition for the higher order modes in a duct with constant mean temperature (Section 6.5.2).

The simplicity of the low gradient solution and the knowledge of exact cut-on condition for the higher-order modes, makes it suitable for calculation of the wave modes in ducts with non-linear axial mean temperature distribution by piecewise linear segmentation of the duct. This approach can yield adequately accurate results if a sufficiently large number of segments are used and the condition $\varepsilon^2 << 1$ is satisfied in every segment, the oscillation criterion given above applying for each segment. It is implemented by multiplication of the impedance transfer matrices of the linear segments in cascade, which are derived similarly as for ducts with no mean temperature gradient, with the exception that the axial component of the particle velocity is now obtained from the linearized momentum equation $-i\omega\rho_o\mathbf{v}'+\nabla p' = 0$ to first order in ε. Theoretically, the impedance transfer matrix encompasses an infinite number of modes, but, in practice, it is used in a truncated form, because, depending on the frequency range of interest, there will be only a finite number of cut-on modes.

6.12.2 Effect of Mean Flow

For subsonic low Mach number mean flows, a plausible approximation is to treat mean flow velocity and mean pressure gradients in the duct as quantities small to the first order. Then, Equations (6.284) and (6.2) can be simplified, respectively, to the forms

$$\rho_o\left(\frac{\partial \mathbf{v}'}{\partial t} + \mathbf{v}_o\cdot\nabla\mathbf{v}'\right) + \nabla p' = 0 \tag{6.301}$$

$$\frac{\partial p'}{\partial t} + \mathbf{v}_o\cdot\nabla p' + \gamma_o p_o \nabla\cdot\mathbf{v}' = 0 \tag{6.302}$$

and, upon eliminating the particle velocity, these equations yield the following convected wave equation for the acoustic pressure

$$-\frac{1}{c_o^2}\left(\frac{\partial}{\partial t} + \mathbf{v}_o\cdot\nabla\right)^2 p' + \rho_o\nabla\cdot\left(\frac{1}{\rho_o}\nabla p'\right) = 0. \tag{6.303}$$

Assuming axial mean flow $\mathbf{v}_o = \mathbf{e}_1 v_o$, this can be expressed in frequency domain as

$$\left(\nabla^2 - M_o^2\frac{\partial^2}{\partial x_1^2} + k_o^2 + \mathrm{i}2k_o M_o\frac{\partial}{\partial x_1} + \frac{1}{2}\frac{\nabla^2\rho_o}{\rho_o} - \frac{3}{4}\left(\frac{\nabla\rho_o}{\rho_o}\right)^2\right)\left(\frac{p'}{\sqrt{\rho_o}}\right) = 0, \tag{6.304}$$

where $M_o = v_o/c_o$. For a straight uniform duct with axial mean absolute temperature distribution $T_o = T_o(x_1)$, Equation (6.304) may be written as

$$\left(\nabla_\perp^2 + (1 - M_o^2)\frac{\partial^2}{\partial x_1^2} + k_o^2 + \mathrm{i}2k_o M_o\frac{\partial}{\partial x_1} + \frac{1}{4}\left(\frac{1}{T_o}\frac{\mathrm{d}T_o}{\mathrm{d}x_1}\right)^2 - \frac{1}{2}\frac{1}{T_o}\frac{\mathrm{d}^2T_o}{\mathrm{d}x_1^2}\right)\left(\frac{p'}{\sqrt{\rho_o}}\right) = 0. \tag{6.305}$$

Upon substituting Equation (6.291) and invoking the condition $M_o^2 << 1$, we still obtain Equation (6.292) for the transverse duct eigenvalues, but the function $u = u(x_1)$ is now given by

$$\left(\frac{\mathrm{d}^2}{\mathrm{d}x_1^2} + \mathrm{i}2k_o M_o\frac{\mathrm{d}}{\mathrm{d}x_1} + \kappa(x_1)\right)u = 0. \tag{6.306}$$

This equation can be solved numerically when the axial variations of the mean flow velocity are provided as data (Section 3.10.3).

References

[1] C. Weng, S. Boij and A. Hanifi, The attenuation of sound by turbulence in internal flows, *J. Acoust. Soc. Am.* **133** (2013), 3764–3776.

[2] D.C. Pridmore-Brown, Sound propagation in a fluid flowing through an attenuating duct, *J. Fluid Mech.* **4** (1958) 393–406.

[3] H. Schlichting, *Boundary Layer Theory*, (New York: McGraw-Hill, 1979).

[4] N.K. Agarwal and M.K. Bull, Acoustic wave propagation in a pipe with fully developed turbulent flow, *J. Sound Vib.* **132** (1989), 275–298.

[5] U. Ingard, Influence of fluid motion past a plane boundary on sound reflection, absorption, and transmission, *J. Acoust. Soc. Am.* **31** (1959), 1035–1036.

[6] M.K. Myers, On the acoustic boundary condition in the presence of flow, *J. Sound Vib.* **71** (1980), 429–434.

[7] W. Eversman and R.J. Beckemeyer, Transmission of sound in ducts with shear layers: Convergence to the uniform flow case, *J. Acoust. Soc. Am.* **52** (1972), 216–220.

[8] A.H. Nayfeh, Effect of acoustic boundary layer on the wave propagation in ducts, *J. Acoust. Soc. Am.* **54** (1973), 1737–1742.

[9] Y. Auregan, R. Starobinski and V. Pagneux, Influence of grazing flow and dissipation effects on the acoustic boundary conditions at a lined wall, *J. Acoust. Soc. Am.* **109** (2001), 59–64.

[10] S. Rienstra and M. Darau, Boundary layer thickness effects of the hydrodynamic instability along an impedance wall, *J. Fluid Mech.* **671** (2011), 559–573.

[11] E.D. Brambley, Well-posed boundary condition for acoustic liners in straight ducts with flow, *AIAA J.* **49** (2011), 1272–1282.

[12] D. Khamis and E.J. Brambley, Acoustic boundary conditions at an impedance lining in inviscid shear flow, *J. Fluid Mech.* **796** (2016) 386–416.

[13] P.Drazin, *Introduction to Hydraudynamic Instability*, (Cambridge: Cambridge University Press, 2002).

[14] R. Courant and D. Hilbert, *Methods of Mathematical Physics,* Vol.1, (New York: Interscience, 1953).

[15] C.L. Morfey, Sound transmission and generation in ducts with flow, *J. Sound Vib.* 14 (1971), 37–55.

[16] F. Farassat and M.K. Myers, A graphical approach to wave propagation in a rigid duct, *J. Sound Vib.* **200** (1997), 729–735.

[17] E.J. Rice, Modal propagation angles in ducts with soft walls and their connection with suppressor performance, NASA-TM-79081, (1979).

[18] C.J. Chapman, Sound radiation from a cylindrical duct: part I. Ray structure of the duct modes and the external field, *J. Fluid Mech.* **281** (1994), 293–311.

[19] M.V. Lowson and S. Baskaran, Propagation of sound in elliptic ducts, *J. Sound Vib.* **38** (1975), 185–194.

[20] M. Willatzen and L.C.L.Y. Voon, Theory of acoustic eigenmodes in parabolic cylindrical enclosures, *J. Sound Vib.* **286** (2005), 251–254.

[21] J.N. Reddy, *An Introduction to the Finite Element Method*, (New York: McGraw-Hill Inc., 1993).

[22] C.J. Brooks, Prediction and control of sound propagation in turbofan engine bypass ducts, PhD thesis, Institute of Sound and Vibration Research, University of Southampton, UK, (2007).

[23] B.J. Tester, The propagation and attenuation of sound in lined ducts containing uniform or "plug" flow, *J. Sound Vib.* **28** (1973), 151–203.

[24] W. Eversman, Theoretical models for duct acoustic propagation and radiation, in H.H. Hubbard (ed.), *Aeroacoustics of Flight Vehicles: Theory and Practice, Vol. 2: Noise Control,* NASA RP-1258, (Washington, DC: *NASA*, 1991), pp. 101–163.

[25] A.A. Syed, On the prediction of sound attenuation in acoustically lined circular ducts, PhD thesis, Loughborough University, UK, (1980), available online at https://bit.ly/30rp5Kl.

[26] S.W. Rienstra, A classification of duct modes based on surface waves, *Wave Motion* **37** (2003), 119–135.

[27] E.J. Brambley and N. Peake, Classification of aeroacoustically relevant surface modes in cylindrical lined ducts, *Wave Motion* **43** (2006) 301–310.

[28] E. Redon, A.-S. Bonnet-Ben Dhia, J. F. Mercier and S.P. Sari, Non-reflecting boundary conditions for acoustic propagation in ducts with acoustic treatment and mean flow, *Int. J. Num. Meth. Engng.* **86** (2011), 1360–1378.

[29] L. Cremer, Theory regarding the attenuation of sound transmitted by air in a rectangular duct with an absorbing wall, and the maximum attenuation constant produced during this process, *Acustica* 3 (1953), 249–263.

[30] B. J. Tester, The optimization of modal sound attenuation in ducts, in absence of mean flow, *J.Sound Vib.* **27** (1973), 477–513.

[31] R. Kabral, L. Du, M. Åbom, Optimum sound attenuation in flow ducts based on the exact Cremer impedance, *Acta Acustica united with Acustica,* **102** (2016), 851–860.

[32] Z. Zhang, H. Bodén and M. Åbom, The Cremer impedance: An investigation of the low frequency behavior, *J. Sound Vib.* **549** (2019) 114844.

[33] R.P. Gillespie, *Partial Diffrentiation*, (Edinburgh: Oliver and Boyd, 1954).

[34] T.R. Law and A.P. Dowling, Optimization of traditional and blown liners for a silent aircraft, 12th AIAA/CEAS Aeroacoustics Conference, Cambridge, Massachusettes, (2006), (AIAA paper 2006-2525).

[35] E.J. Brambley, A.M.J. Davis and N. Peake, Eigenmodes of lined flow ducts with rigid splices, *J. Fluid Mech.* **690** (2012), 388–425.

[36] B. Regan and J. Eaton, Modeling the influence of acoustic liner nonuniformities on duct modes, *J. Sound Vib.* **219** (1999), 859–879.

[37] M. C. M. Wright and A. McAlpine, Calculation of modes in azimuthally non-uniform lined ducts with uniform flow, *J. Sound Vib.* **302** (2007), 403–407.

[38] W. R. Watson, Circumferentially segmented duct liners optimized for axisymmetric and standing-wave sources, NASA Rep. No. 2075 (1982).

[39] L. M. B. C. Campos and J. M. G. S. Oliveira, On the acoustic modes in a cylindrical duct with an arbitrary wall impedance distribution, *J. Acoust. Soc. Am.* **116** (2004), 3336–3346.

[40] W. P. Bi, V. Pagneux, D. Lafarge, and Y. Aurégan, Modelling of sound propagation in a non-uniform lined duct using a multi-modal propagation method, *J. Sound Vib.* **289** (2006) 1091–1111.

[41] W.-P. Bi, V. Pagneux, D. Lafarge and Y. Auregan, An improved multimodal method for sound propagation in nonuniform lined ducts, *J. Acoust. Soc. Am.* 122 (2007) 280–290.

[42] E. J. Brambley, M. Darau and S. W. Rienstra, The critical layer in linear-shear boundary layers over acoustic linings, *J. Fluid Mech.* **710** (2012), 545–568.

[43] M.A. Swinbanks, The sound field generated by a source distribution in a long duct carrying shear flow, *J. Sound Vib.* **40** (1975), 51–76.

[44] A.H. Nayfeh, J.E. Kaiser and D.P. Telionis, Acoustics of aircraft engine-duct systems, *AIAA Journal* **13** (1975), 130–153.
[45] M. Oppeneer, Sound propagation in lined ducts with parallel flow, PhD thesis, Technische Universiteit Eindhoven, Eindhoven, the Netherlands, (2014).
[46] J.R. Sánchez, Étude théorique et numérique des modes propres acoustiques dans un conduit avec écoulement et parois absorbantes. Modélisation et simulation, Doctorat de l'Université de Toulouse, Institut Supérieur de l'Aeronautique et de l'Espace, France, (2016).
[47] F. B. Hilderbrand, *Introduction to Numerical Analysis*, (New York: McGraw-Hill, 1956).
[48] C.M. Bender and S.A. Orszag, *Advanced Numerical Methods for Scientists and Engineers,* (New York: Springer-Verlag, 1999).
[49] G.G. Vilenski and S.W. Rienstra, Numerical acoustic modes in a ducted shear flow, 11th AIAA/CEAS Aeroacoustics Conference, Monterey, CA, USA, (2005).
[50] A.H. Nayfeh, J.E. Kaiser and B.S. Shaker, Effect of mean-velocity profile shape on sound transmission through two dimensional ducts, *J. Sound Vib.* **34** (1974), 413–423.
[51] R.P. Benedict, *Fundamentals of Pipe Flow*, (New York: John Wiley and Sons, 1980).
[52] G. Gabard, A comparison of boundary conditions for flow acoustics, *J. Sound Vib.* **332** (2013), 714–724.
[53] V. Pagneux, N. Amir and J. Kergomard, A study of wave propagation in varying cross-section waveguides by modal decomposition. Part I: Theory and validation, *J. Acoust. Soc. Am.* **100** (1996), 2034–2048.
[54] F.P. Metchel, Modal analysis in lined wedge-shaped ducts, *J. Sound Vib.* **216** (1998), 673–696.
[55] A.H. Nayfeh, *Introduction to Perturbation Techniques*, (John Wiley & Sons, 1981).
[56] S.W. Rienstra, Sound transmission in slowly varying circular and annular lined ducts with flow, *J. Fluid Mech.* **380** (1999), 279–296.
[57] S.W. Rienstra, Sound propagation in slowly varying lined flow ducts of arbitrary cross-section. *J. Fluid Mech.* **495** (2003), 157–173.
[58] A.J. Cooper, N. Peake, Propagation of unsteady disturbances in a slowly varying duct with mean swirling flow, *J. Fluid Mech.* **445** (2001), 207–234.
[59] A.E.D. Lloyd and N. Peake, The propagation of acoustic waves in a slowly varying duct with radially sheared axial mean flow, *J. Sound Vib.* **332** (2013), 3937–3946.
[60] E.J. Brambley and N. Peake, Sound transmission in strongly curved slowly varying cylindrical ducts with flow, *J. Fluid Mech.* **596** (2008), 387–412.
[61] W. Rostafinski, *Monograph on Propagation of Sound Waves in Curved Ducts*, NASA Reference Publication 1248, (Washington, DC: NASA, 1991).
[62] S. Felix and V. Pagneux, Sound attenuation in lined bends, *J. Acoust. Soc. Am.* **116** (2004), 1921–1931.
[63] G.D. Furnell and D.A. Bies, Characteristics of modal wave propagation within longitudinally curved acoustic waveguides, *J. Sound Vib.* **130** (1989), 405–423.
[64] A. Capelli, The influence of flow on the acoustic characteristics of duct bend for highr order modes: A numerical study, *J. Sound Vib.* **82** (1982), 131–149.
[65] S.A. Berger, L. Talbot and L.-S. Yao, Flow in curved pipes, *Ann. Rev. Fluid Mech.* **15** (1983), 461–512.
[66] S.V. Patankar, V.S. Pratap and D.B. Spalding, Prediction of turbulent flow in curved pipes, *J. Fluid Mech.* **67** (1975), 583–595.
[67] J.L. Kerrobrock, Small disturbances in turbomachine annuli with swirl, *AIAA J.* **15** (1977), 794–803.

[68] C.K.W. Tam and L. Auriault, The wave modes in ducted swirling with mean vortical swirling, *J. Fluid Mech.* 371 (1998), 1–20.

[69] R.J. Nijboer, Eigenvalues and eigenfunctions of ducted swirling flows, National Aerospace Laboratory NLR, NLR-TP-2001-141, (2001), (also AIAA-2001-2178).

[70] P.G. Bergmann, The wave equation in a medium with a variable index of refraction, *J. Acoust. Soc. Am.* 17 (1946), 329–333.

[71] E. Dokumaci, A simple approach for evaluation of cut-on condition of higher order modes in inhomogeneous uniform waveguides, *J. Sound Vib.* **440** (2019), 231–238.

[72] M. Abromovitch, I.A. Stegun, *Handbook of Mathematical Functions*, (Washington: National Bureau of Standards, Applied Mathematics Series 55, 1972).

7 Transmission of Wave Modes in Coupled Ducts

7.1 Introduction

This chapter presents modal acoustic models of some basic coupled-duct configurations, such as area changes, junctions and perforates, which occur recurrently in practical systems. These are similar to the coupled-duct arrangements considered in Chapter 4, but they are now modeled by taking into account the higher-order duct modes. Only the fundamental mode or few higher-order modes may be propagating in ducts themselves at the frequency of interest, but evanescent modes (or modes decaying at increasingly higher rates) are generated at a discontinuity formed by the coupling and a sufficient number of these need also be taken into account in order to model the acoustic continuity and compatibility conditions at the interface adequately accurately.

The generic duct system concept depicted in Figure 5.1 remains valid in modeling chambers, resonators and silencers using modal duct elements, except that the acoustic fields in the inlet and outlet ducts may now include multiple modes and the time-averaged acoustic powers of all propagating modes are taken into account in transmission loss calculations (see Section 11.6). In many practical systems, however, effects of multi-modal propagation may be important in the internals of the system, although only the fundamental mode propagates at frequencies of interest in the inlet and outlet ducts (see Figure 5.1). The size of the modal wave transfer matrices of such systems can be reduced significantly by contraction (Section 2.4.4), while retaining the effects of higher-order modes in the internals. In this chapter, we make use of contracted modal models to demonstrate, in sequel to Chapter 5, some typical effects of the higher-order modes on transmission loss characteristics of typical chambers and resonators.

In order to develop simple and versatile modal acoustic models of coupled ducts, it is necessary to make some simplifying assumptions. The main assumption made in this chapter is that, in the vicinity of the discontinuity formed by coupled ducts, the sound propagation in the coupled ducts is governed by the convected wave equation (see Section 6.4). Thus, the coupled ducts can be of arbitrary section shape with uniform mean flow, but effects of refraction and flow-acoustic interactions in the vicinity of the discontinuity are neglected.

7.2 Weak Form of the Convected Wave Equation

Modal models of two straight ducts may be connected in series directly if they have identical transverse modes. Mode discrepancies occur, for example, when transverse geometry or wall impedance or mean temperature is different in the ducts. Ducts with non-identical transverse modes can be coupled, but it is necessary to match the acoustic wave modes in each duct over their interface. This process, which is called mode-matching, may be applied by using the weak (variational) form of the convected wave equation. This is derived in this section from first principles.

Using the notation of Equation (6.12), the convected wave equation is

$$\left(\nabla^2 - \frac{1}{c_o^2}\left(\frac{\partial}{\partial t} + \mathbf{v}_o\cdot\nabla\right)^2\right)p' = 0. \tag{7.1}$$

For a uniform straight duct $\mathbf{v}_o = \mathbf{e}_1 v_{o1}$, where v_{o1} is constant and $\mathbf{e}_1$ denotes the unit vector in the positive direction of the duct axis. Let w be a sufficiently differentiable arbitrary function and consider the following weighted integral of Equation (7.1) over an arbitrary volume V of the duct in an arbitrary time interval (t_1, t_2):

$$\int_V \int_{t_1}^{t_2} w\left(\nabla^2 - \frac{1}{c_o^2}\frac{\mathrm{D}^2}{\mathrm{D}t^2}\right)p'\,\mathrm{d}t\,\mathrm{d}V = 0, \tag{7.2}$$

where $\mathrm{D}/\mathrm{D}t = \partial/\partial t + \mathbf{v}_o\cdot\nabla$ denotes the material derivative. Integrating this by parts with respect to time gives

$$\int_V \left\{\int_{t_1}^{t_2}\left[c_o^2 w\nabla^2 p' + \frac{\partial w}{\partial t}\frac{\mathrm{D}p'}{\mathrm{D}t} - w\mathbf{v}_o\cdot\nabla\frac{\mathrm{D}p'}{\mathrm{D}t}\right]\mathrm{d}t - \left|w\frac{\partial p'}{\partial t}\right|_{t_1}^{t_2}\right\}\mathrm{d}V = 0. \tag{7.3}$$

Applying the differentiation rule for the products of functions to $\mathbf{v}_o\cdot\nabla w\mathrm{D}p'/\mathrm{D}t$ and then to $\nabla\cdot w\nabla p'$, this may be expanded as

$$\begin{aligned}\int_V \Bigg\{\int_{t_1}^{t_2}\Bigg[&c_o^2\nabla\cdot w\nabla p' - c^2\nabla w\cdot\nabla p' + \frac{\partial w}{\partial t}\frac{\mathrm{D}p'}{\mathrm{D}t} - \mathbf{v}_o\cdot\nabla\left(w\frac{\mathrm{D}p'}{\mathrm{D}t}\right)\\ &+ (\mathbf{v}_o\cdot\nabla w)\frac{\mathrm{D}p'}{\mathrm{D}t}\Bigg]\mathrm{d}t - \left|w\frac{\partial p'}{\partial t}\right|_{t_1}^{t_2}\Bigg\}\mathrm{d}V.\end{aligned} \tag{7.4}$$

Partial integration of the first term in square brackets by application of the divergence theorem over the volume V then yields

$$\begin{aligned}&\int_V \left\{\int_{t_1}^{t_2}\left[-c_o^2\nabla w\cdot\nabla p' + \frac{\mathrm{D}w}{\mathrm{D}t}\frac{\mathrm{D}p'}{\mathrm{D}t}\right]\mathrm{d}t - \left|w\frac{\partial p'}{\partial t}\right|_{t_1}^{t_2}\right\}\mathrm{d}V\\ &\qquad + \int_\Sigma \left\{\int_{t_1}^{t_2} w\left[c_o^2\mathbf{n}\cdot\nabla p' - \mathbf{v}_o\cdot\mathbf{n}\frac{\mathrm{D}p'}{\mathrm{D}t}\right]\mathrm{d}t\right\}\mathrm{d}S = 0,\end{aligned} \tag{7.5}$$

where Σ denotes the boundary surface of V.

We take for the test function w small variations $\delta p'$ of p', which is assumed to be fixed at times t_1 and t_2. Then, substituting $w = \delta p'$ and noting that the variation operator δ commutes with the differential and integral operators, Equation (7.5) may be cast in the form of the following variational statement

$$\int_{t_1}^{t_2} \left\{ \frac{1}{2}\delta \int_V \left(-c_o^2 (\nabla p')^2 + \left(\frac{\mathrm{D}p'}{\mathrm{D}t} \right)^2 \right) \mathrm{d}V + \int_\Sigma \delta p' \left(c_o^2 \mathbf{n} \cdot \nabla p' - \mathbf{v}_o \cdot \mathbf{n} \frac{\mathrm{D}p'}{\mathrm{D}t} \right) \mathrm{d}S \right\} \mathrm{d}t = 0 \tag{7.6}$$

This is the required weak form of the convected wave equation in the time domain. We will implement this equation by using exact solutions of the convected wave equation. Then, variation of the domain integral vanishes and Equation (7.6) reduces to the surface integral, which may be expressed in the frequency domain as

$$\int_\Sigma \delta \breve{p}' \left(c_o^2 \mathbf{n} \cdot \nabla p' - \mathbf{v}_o \cdot \mathbf{n} (-\mathrm{i}\omega + \mathbf{v}_o \cdot \nabla) p' \right) \mathrm{d}S = 0, \tag{7.7}$$

where an inverted over-arc denotes the complex conjugate. This form is more convenient, because the solutions of the convected wave equation which we have derived in Chapter 6 can be substituted directly.

7.3 Ducts with Identical Sections

A typical coupled-duct configuration which is often encountered in practice is a uniform hard-walled straight duct with an inserted lined duct of the same cross-section (Figure 7.1a). For simplicity, we assume that the speed of sound, c_o, and the

(a)

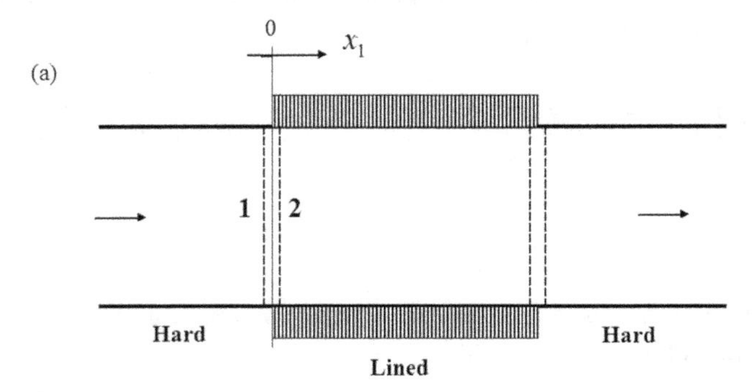

(b)

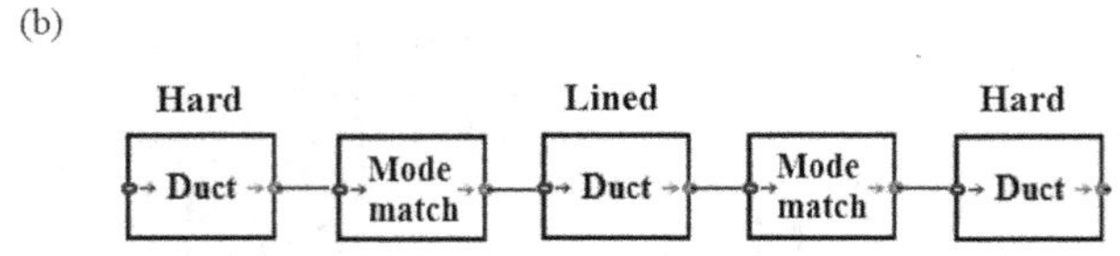

Figure 7.1 A lined duct inserted in a hard-walled duct.

mean flow velocity $\mathbf{v}_o = \mathbf{e}_1 v_{o1}$ are uniform throughout. The modal wave transfer matrices of both the lined and hard-walled sections are given by Equation (6.71); however, these cannot be assembled as they are since the transverse modes of the hard-walled and lined ducts do not match per mode. (For example, recall that, the fundamental mode of a hard-walled duct is planar, whereas it is non-planar in the lined duct case.) To illustrate how these modes may be matched, it suffices to consider the inlet discontinuity between ducts 1 and 2 in Figure 7.1, as the outlet discontinuity can be handled in exactly the same way.

Let ducts 1 and 2 be defined in local coordinate frames $(x_1, \bullet)$, with the axes of the ducts taken as shown in Figure 7.1. Here, $(\bullet)$ denotes the transverse coordinates (not shown in Figure 7.1), for example, $\bullet \equiv (x_2, x_3)$ if ducts are rectangular (Figure 6.3), or $\bullet \equiv (r, \theta)$ if ducts are circular (Figure 6.4). In matrix notation (see Section 6.4.2), exact solutions of the convected wave equation for the acoustic pressure at points $(0, \bullet)$ in ducts $j = 1, 2$ are given, in frequency domain, by

$$p_j'(0, \bullet) = \mathbf{\Phi}_j(\bullet)\left(\mathbf{P}_j^+(0) + \mathbf{P}_j^-(0)\right), \tag{7.8}$$

where $\mathbf{P}_j^{\mp}(x_1)$ denote the modal pressure wave components vectors for the $\mp x_1$ wave systems. Differentiation of Equation (7.8) with respect to the axial coordinate gives

$$\frac{\partial p_j'}{\partial x_1}(0, \bullet) = \mathrm{i}k_o \mathbf{\Phi}_j(\bullet)\left(\mathbf{K}_j^+ \mathbf{P}_j^+(0) + \mathbf{K}_j^- \mathbf{P}_j^-(0)\right), \tag{7.9}$$

where $k_o = \omega / c_o$ and $\mathbf{K}_j^{\mp}$ denote the corresponding axial propagations constants matrices, which are defined by Equation (6.63).

Consider an infinitesimally thin control volume which just encloses the hard-lined wall discontinuity, as indicated by broken lines $x_1 = 0^{\mp}$ in Figure 7.1. Let V denote the control volume, Σ the surface of V and $\mathbf{n}$ the unit outward normal vector of Σ, and apply Equation (7.8) by splitting the surface integral over the surfaces of Σ:

$$\begin{aligned} &-\int_S \left(\delta \breve{p}_1'\right)^{\mathrm{T}}\left(\left(1 - M^2\right)\frac{\partial p_1'}{\partial x_1} + \mathrm{i}k_o M p_1'\right)\mathrm{d}S \\ &\qquad + \int_S \left(\delta \breve{p}_2'\right)^{\mathrm{T}}\left(\left(1 - M^2\right)\frac{\partial p_2'}{\partial x_1} + \mathrm{i}k_o M p_2'\right)\mathrm{d}S = 0, \qquad x_1 = 0 \end{aligned} \tag{7.10}$$

where S denotes the duct cross-sectional area, $M = v_{o1}/c_o$ and the effect of the lateral duct surface included in Σ is neglected on the premise that the control volume is infinitesimally thin (note that this contribution vanishes exactly on the side of the hard-walled duct). The superscript "T" denotes array transposition and is introduced for the conformity of matrix operations after eventual substitution of Equations (7.8) and (7.9). Since the variations $\delta p'$ are arbitrary, it is essential that continuity of acoustic pressure is imposed at the interfaces of the ducts:

$$p_1'(0, \bullet) = p_2'(0, \bullet). \tag{7.11}$$

Hence, upon substituting Equations (7.8) and (7.9) and using the foregoing condition, Equation (7.10) may be expressed as

$$-\mathbf{H}_{11}\left(\left(1-M^2\right)\left[\mathbf{K}_1^+\mathbf{P}_1^+ + \mathbf{K}_1^-\mathbf{P}_1^-\right] + M\left[\mathbf{P}_1^+ + \mathbf{P}_1^-\right]\right) + \mathbf{H}_{12}\left(\left(1-M^2\right)\left[\mathbf{K}_2^+\mathbf{P}_2^+ + \mathbf{K}_2^-\mathbf{P}_2^-\right] + M\left[\mathbf{P}_2^+ + \mathbf{P}_2^-\right]\right) = 0, \tag{7.12}$$

where we put $\mathbf{P}_j^{\mp} \equiv \mathbf{P}_j^{\mp}(0)$ for brevity and

$$\mathbf{H}_{11} = \int_S \breve{\boldsymbol{\Phi}}_1^{\mathrm{T}}(\bullet)\boldsymbol{\Phi}_1(\bullet)\mathrm{d}S \tag{7.13}$$

$$\mathbf{H}_{11} = \int_S \breve{\boldsymbol{\Phi}}_1^{\mathrm{T}}(\bullet)\boldsymbol{\Phi}_2(\bullet)\mathrm{d}S \tag{7.14}$$

On rearranging, Equation (7.12) may be recast as

$$-\mathbf{H}_{11}\mathbf{a}_1\left(\mathbf{P}_1^+ - \mathbf{P}_1^-\right) + \mathbf{H}_{12}\mathbf{a}_2\left(\mathbf{P}_2^+ - \mathbf{P}_2^-\right) = 0, \tag{7.15}$$

where $\mathbf{a}_1$ denotes a diagonal matrix, the diagonal elements of which are the mode cut-off ratios defined by Equation (6.50) of duct 1 in the order of the eigenfunctions in $\boldsymbol{\Phi}_1(\bullet)$, and $\mathbf{a}_2$ is defined similarly for duct 2. On the other hand, substituting Equation (7.8), pre-multiplying by $\breve{\boldsymbol{\Phi}}_1^{\mathrm{T}}(\bullet)$and integrating over S, Equation (7.11) gives

$$-\mathbf{H}_{11}\left(\mathbf{P}_1^+ + \mathbf{P}_1^-\right) + \mathbf{H}_{12}\left(\mathbf{P}_2^+ + \mathbf{P}_2^-\right) = 0. \tag{7.16}$$

Finally, combining Equations (7.15) and (7.16)

$$\begin{bmatrix} \mathbf{P}_1^+ \\ \mathbf{P}_1^- \end{bmatrix} = \frac{1}{2}\begin{bmatrix} \mathbf{a}_1^{-1}\mathbf{H}_{11}^{-1}\mathbf{H}_{12}\mathbf{a}_2 + \mathbf{H}_{11}^{-1}\mathbf{H}_{12} & -\mathbf{a}_1^{-1}\mathbf{H}_{11}^{-1}\mathbf{H}_{12}\mathbf{a}_2 + \mathbf{H}_{11}^{-1}\mathbf{H}_{12} \\ -\mathbf{a}_1^{-1}\mathbf{H}_{11}^{-1}\mathbf{H}_{12}\mathbf{a}_2 + \mathbf{H}_{11}^{-1}\mathbf{H}_{12} & \mathbf{a}_1^{-1}\mathbf{H}_{11}^{-1}\mathbf{H}_{12}\mathbf{a}_2 + \mathbf{H}_{11}^{-1}\mathbf{H}_{12} \end{bmatrix}\begin{bmatrix} \mathbf{P}_2^+ \\ \mathbf{P}_2^- \end{bmatrix}. \tag{7.17}$$

This is the required mode matching equation at the inlet discontinuity between the hard and lined ducts in Figure 7.1. The good thing about it is that it is formally a modal two-port element and can be implemented as such in the block diagrams as illustrated in Figure 7.1b. Furthermore, as an inspection of the foregoing analysis will show, duct 1 need not necessarily be a hard-walled duct, although we have imagined it to be so until now to fix the ideas. Equation (7.17) applies also if duct 1 is another lined duct, too. Such interfaces of lined ducts occur in ducts treated with axially segmented liners for optimal noise control purposes. If duct 1 is hard-walled, then $\mathbf{H}_{11}$ is diagonal and real (Section 6.5.1). If both ducts have identical wall impedance, the modal wave transfer matrix in Equation (7.17) reduces, as expected, to a unit matrix.

Mode-matching in straight uniform ducts with a parallel sheared flow of arbitrary velocity profile is discussed by Gabard and Astley [1], who use the weak form of the Pridmore-Brown equation with solutions determined numerically by the finite element method. Rienstra argues that the singularity involved at a hard-lined wall edge due to transition to the Ingard–Myers infinitely thin vortex sheet model may give rise to instability waves and presents an analysis of the problem by using the Wiener–Hopf method [2].

The foregoing modal two-port mode-matching model may be extended readily to the case where the ambient conditions in ducts 1 and 2 are uniform, but different. It

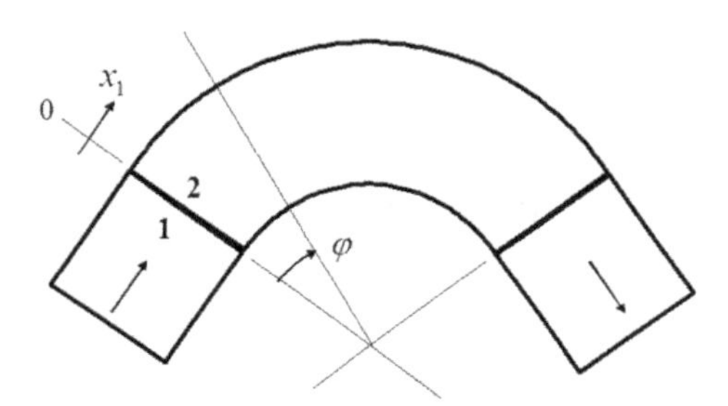

Figure 7.2 Curved duct with straight duct extensions.

can be shown that, in this case Equation (7.17) applies with the cut-off ratio matrices $\mathbf{a}_1$ and $\mathbf{a}_2$ replaced by $k_1\mathbf{a}_1$ and $k_2\mathbf{a}_2$, respectively, where $k_1 = \omega/c_1$, $k_2 = \omega/c_2$, and c_1 and c_2 denote the speed of sound in ducts 1 and 2. The mean flow Mach numbers will also be different in the two ducts. An application of this case is a uniform duct with an axial temperature gradient (Section 6.12). This problem may be modeled numerically by replacing the actual ambient by a large number of axially stratified segments with uniform properties. The modal wave transfer matrices of the segments are then connected by mode matching, as in Figure (7.1b). Note that, in this case, the modal wave transfer matrix in Equation (7.17) will not reduce to a unit matrix when the ducts have identical wall impedance.

Another coupled-duct configuration, which is common in practical ductworks, occurs when two uniform straight ducts with identical sections, but different axes orientations, are connected by a curved uniform duct (Figure 7.2). For a constant density potential mean flow, the acoustic velocity potential in the curved duct is governed by the convected wave equation and the weak form in Equation (7.6) is applicable; however, a general exact solution with a mean flow is not available (Section 6.10). The essentials of mode-matching at a straight-curved duct interface can be explained, however, by neglecting the mean flow.

Referring to Figure 7.2, consider a uniform duct curved circularly in-plane (duct 2) with a straight duct extension (duct 1). We assume, for simplicity, that ambient conditions are the same in both ducts. Let duct 1 be defined in a local coordinate frame $(x_1, \bullet)$, and duct 2 in a local frame $(\phi, *)$, with axes of the ducts taken as shown in Figure 7.2. The transverse coordinates $(\bullet)$ and $(*)$ are different (see Section 6.10), but they span the same cross-sectional area at $x_1 = 0$ and $\phi = 0$. Consider an infinitesimally thin control volume V of surface Σ just enclosing the interface between ducts 1 and 2. On the $x_1 = 0^-$ face of Σ, the acoustic pressure and its axial derivative are given by Equations (7.8) and (7.9) with $j = 1$. But, on the $\phi = 0^+$ face, the acoustic pressure is given by Equation (6.232), which we write now as

$$p_2'(0, *) = \mathbf{\Phi}_2^+(*)\ \mathbf{P}_2^+(0) + \mathbf{\Phi}_2^-(*)\mathbf{P}_2^-(0). \tag{7.18}$$

The azimuthal derivative is then, from Equation (6.233):

$$\frac{\partial p_2'}{\partial \phi}(0, *) = \mathrm{i}\big(\mathbf{\Phi}_2^+(*)\ \mathbf{K}\mathbf{P}_2^+(0) - \mathbf{\Phi}_2^-(*)\mathbf{K}\mathbf{P}_2^-(0)\big). \tag{7.19}$$

Again, we apply Equation (7.7) by splitting the surface integral to $x_1 = 0^-$ and $\phi = 0^+$ faces:

$$-\int_S \delta\breve{p}_1'(0,\bullet)\left(c_o^2 \frac{\partial p_1'}{\partial x_1}(0,\bullet)\right)\mathrm{d}S + \int_S \delta\breve{p}_2'(0,*)\left(c_o^2 \frac{1}{R}\frac{\partial p'}{\partial \phi}(0,*)\right)\mathrm{d}S = 0. \quad (7.20)$$

This yields

$$-k_o\mathbf{H}_{11}\left(\mathbf{K}_1^+\mathbf{P}_1^+(0) + \mathbf{K}_1^-\mathbf{P}_1^-(0)\right) + \mathbf{G}_{12}^+\mathbf{K}\mathbf{P}_2^+(0) - \mathbf{G}_{12}^-\mathbf{K}\mathbf{P}_2^-(0) = 0, \quad (7.21)$$

since the essential condition $p_1'(0,\bullet) = p_2'(0,*)$ must hold. Upon projecting on the transverse modes of duct 1, the latter condition gives

$$\mathbf{H}_{11}\left(\mathbf{P}_1^+(0) + \mathbf{P}_1^-(0)\right) = \mathbf{H}_{12}^+\mathbf{P}_2^+(0) + \mathbf{H}_{12}^-\mathbf{P}_2^-(0), \quad (7.22)$$

where

$$\mathbf{G}_{12}^{\mp} = \int_S \frac{1}{R}\mathbf{\Phi}_1^{\mathrm{T}}(\bullet)\mathbf{\Phi}_2^{\mp}(*)\mathrm{d}S \quad (7.23)$$

$$\mathbf{H}_{12}^{\mp} = \int_S \mathbf{\Phi}_1^{\mathrm{T}}(\bullet)\mathbf{\Phi}_2^{\mp}(*)\mathrm{d}S. \quad (7.24)$$

Hence, the mode-matching modal two-port element is obtained by combining Equations (7.21) and (7.22):

$$\begin{bmatrix} k_o\mathbf{H}_{11}\mathbf{K}_1^+ & k_o\mathbf{H}_{11}\mathbf{K}_1^- \\ \mathbf{H}_{11} & \mathbf{H}_{11} \end{bmatrix}\begin{bmatrix} \mathbf{P}_1^+ \\ \mathbf{P}_1^- \end{bmatrix} = \begin{bmatrix} \mathbf{G}_{12}^+\mathbf{K} & -\mathbf{G}_{12}^-\mathbf{K} \\ \mathbf{H}_{12}^+ & \mathbf{H}_{12}^- \end{bmatrix}\begin{bmatrix} \mathbf{P}_2^+ \\ \mathbf{P}_2^- \end{bmatrix}. \quad (7.25)$$

The integrals in Equations (7.23) and (7.24) can be calculated only numerically. A modal two-port for mode matching at a curved-straight duct transition may be derived similarly. In general, most curved uniform ductwork may be modeled approximately by subdividing it into a series of straight and circularly curved segments separated by modal mode-matching two-ports [3].

7.4 Sudden Area Changes

In this section, we consider mode-matching when two straight uniform ducts of different section size and possibly of different section shape are coupled with their axes being parallel but not necessarily collinear. One-dimensional modeling of several forms of this configuration, which are quite common in mufflers and silencers, are discussed in Section 4.3. The mode-matching approach is used extensively in the literature in multi-modal analysis of wave reflection at specific area discontinuities [4–12], and the calculation of the transmission loss of specific chambers and resonators involving sudden area changes [13–24]. Here, we present modal acoustic models for sudden area-change configurations, which can be used as transitional mode-matching elements to connect the modal models of straight uniform ducts in block diagrams.

7.4.1 Open Sudden Expansion

Shown in Figure 7.3 is an expansion type open sudden area change, with possible closure (end-caps) options that will turn it to a through-flow or flow-reversing type closed area change indicated. The inlet, annulus and outlet ducts are denoted by using the subscripts 1, 2 and 3, respectively. These ducts are defined in local coordinate frames $(x_1^{(j)}, \bullet^{(j)})$ where the superscripts $j = 1, 2, 3$ in brackets refer to the duct numbers, $x_1^{(j)}$ denote the duct axes, the positive directions of which are taken as shown in Figure 7.3, and "$\bullet^{(j)}$" denote, as usual in this book, the transverse coordinates of the ducts, which may be of different shapes and sizes. Consider a control volume V, of surface Σ with outward unit normal vector $\mathbf{n}$, which encloses the area discontinuity, and let the transverse surfaces of the control volume approach each other, while enclosing the discontinuity, until they degenerate to the interface plane, $x_1 = 0$ (Figure 7.3).

From Section 6.4.2, the exact solutions of the convected wave equation for the acoustic pressure and its axial derivative at this interface are, in frequency domain:

$$p_j'(0, \bullet^{(j)}) = \mathbf{\Phi}_j(\bullet^{(j)}) \left[\mathbf{P}_j^+(0) + \mathbf{P}_j^-(0)\right], \qquad j = 1, 2, 3 \tag{7.26}$$

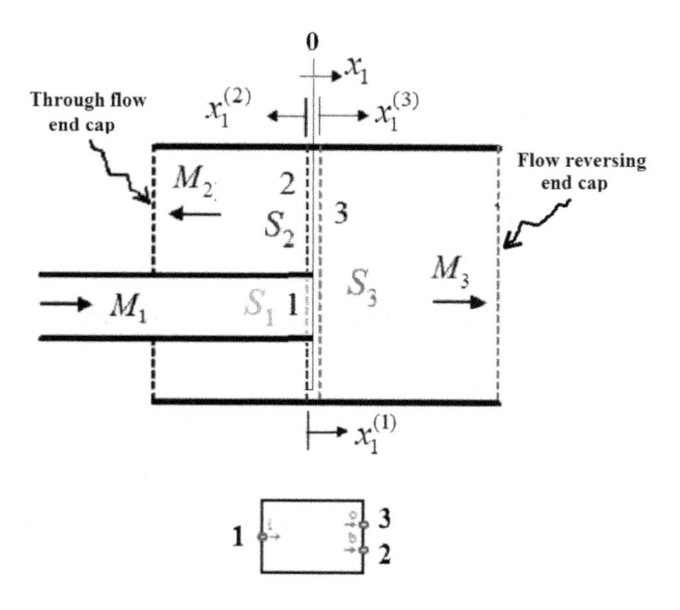

Figure 7.3 Open expansion type area discontinuity and its mode-matching block, with possible closure (end-cap) options.

$$\frac{\partial p_j'}{\partial x_1}\left(0, \bullet^{(j)}\right) = \mathrm{i}k_o \mathbf{\Phi}_j(\bullet^{(j)})\left[\mathbf{K}_j^+ \mathbf{P}_j^+(0) + \mathbf{K}_j^- \mathbf{P}_j^-(0)\right], \qquad j = 1, 2, 3 \tag{7.27}$$

where $\mathbf{P}_j^{\mp}(0)$ denotes the modal pressure wave components vectors for the $+x_1^{(j)}$ and $-x_1^{(j)}$ wave systems at the interface and $\mathbf{K}_j^{\mp}$ denote the corresponding axial propagation constants matrices, which are defined by Equation (6.63). Mean flow velocity in each duct is assumed to be uniform and axial, that is, $\mathbf{v}_o = \mathbf{e}_1^{(j)} v_{o1}^{(j)}$, where $\mathbf{e}_1^{(j)}$ denotes the unit vector in the positive direction of axis $x_1^{(j)}$. Now, apply Equation (7.7) by splitting the surface integral as

$$\sum_{j=1}^{3} \int_{S_j} \left\{ \left(\delta \breve{p}_j'\left(0, \bullet^{(j)}\right)\right) \beta_j \left[\left(1 - M_j^2\right) \frac{\partial p_j'}{\partial x_1}\left(0, \bullet^{(j)}\right) + \mathrm{i}k_o M_j p_j'\left(0, \bullet^{(j)}\right) \right] \right\} \mathrm{d}S = 0. \tag{7.28}$$

Here, $k_o = \omega / c_o$, c_o denotes the speed of sound, $M_j = v_{o1}^{(j)}/c_o$ and S_j denote the mean flow Mach number and the cross-sectional area of duct $j \in 1, 2, 3$, and $\beta_j = \mathbf{e}_1^{(j)} \cdot \mathbf{n}$, that is, for the sign convention of Figure 7.3, $\beta_1, \beta_2, \beta_3 = -1, 1, 1$.

Because the variations $\delta p'$ are arbitrary, it is essential that continuity of acoustic pressure is imposed at the interfaces of the ducts:

$$p_1'\left(0, \bullet^{(1)}\right) = p_3'\left(0, \bullet^{(3)}\right) \qquad \text{on } S_1 \tag{7.29}$$

$$p_2'\left(0, \bullet^{(2)}\right) = p_3'\left(0, \bullet^{(3)}\right) \qquad \text{on } S_2. \tag{7.30}$$

Substituting Equations (7.26) and (7.27) and using the foregoing conditions, Equation (7.28) may be expressed as

$$\sum_{j=1}^{3} \beta_j \mathbf{H}_{3j} \left\{ \left(1 - M_j^2\right) \left(\mathbf{K}_j^+ \mathbf{P}_j^+ + \mathbf{K}_j^- \mathbf{P}_j^-\right) + M_j \left(\mathbf{P}_j^+ + \mathbf{P}_j^-\right) \right\} = 0. \tag{7.31}$$

Here, we put $\mathbf{P}_j^{\mp} = \mathbf{P}_j^{\mp}(0)$ for simplicity of notation and

$$\mathbf{H}_{kj} = \int_{S_j} \breve{\mathbf{\Phi}}_k^{\mathrm{T}}(\bullet^{(k)})\, \mathbf{\Phi}_j(\bullet^{(j)}) \mathrm{d}S, \qquad j, k = 1, 2, 3 \tag{7.32}$$

where the superscript "T" denotes matrix transposition. Introducing the diagonal matrices

$$\mathbf{a}_j = \begin{bmatrix} a_1 & 0 & \cdots \\ 0 & a_2 & \cdots \\ \vdots & \vdots & \ddots \end{bmatrix}_j, \qquad j = 1, 2, 3 \tag{7.33}$$

with a_m, $m = 1, 2, \ldots$, denoting the mode cut-off ratios defined by Equation (6.50), Equation (7.31) simplifies to

$$\sum_{j=1}^{3} \beta_j \mathbf{H}_{3j} \mathbf{a}_j \left(\mathbf{P}_j^+ - \mathbf{P}_j^-\right) = 0. \tag{7.34}$$

On the other hand, upon substituting Equation (7.7), pre-multiplying by $\breve{\boldsymbol{\Phi}}_1^{\mathrm{T}}(\bullet^{(1)})$ and integrating over S_1, Equation (7.29) gives

$$-\mathbf{H}_{11}\left(\mathbf{P}_1^+ + \mathbf{P}_1^-\right) + \mathbf{H}_{13}\left(\mathbf{P}_3^+ + \mathbf{P}_3^-\right) = 0. \tag{7.35}$$

Similarly, from Equation (7.30)

$$-\mathbf{H}_{22}\left(\mathbf{P}_2^+ + \mathbf{P}_2^-\right) + \mathbf{H}_{23}\left(\mathbf{P}_3^+ + \mathbf{P}_3^-\right) = 0. \tag{7.36}$$

Combining Equations (7.34)-(7.36)

$$\begin{bmatrix} \beta_1\mathbf{H}_{31}\mathbf{a}_1 & -\beta_1\mathbf{H}_{31}\mathbf{a}_1 \\ -\mathbf{H}_{11} & -\mathbf{H}_{11} \\ \mathbf{0} & \mathbf{0} \end{bmatrix} \begin{bmatrix} \mathbf{P}_1^+ \\ \mathbf{P}_1^- \end{bmatrix} + \begin{bmatrix} \beta_2\mathbf{H}_{32}\mathbf{a}_2 & -\beta_2\mathbf{H}_{32}\mathbf{a}_2 \\ \mathbf{0} & \mathbf{0} \\ -\mathbf{H}_{22} & -\mathbf{H}_{22} \end{bmatrix} \begin{bmatrix} \mathbf{P}_2^+ \\ \mathbf{P}_2^- \end{bmatrix}$$

$$+ \begin{bmatrix} \beta_3\mathbf{H}_{33}\mathbf{a}_3 & -\beta_3\mathbf{H}_{33}\mathbf{a}_3 \\ \mathbf{H}_{13} & \mathbf{H}_{13} \\ \mathbf{H}_{23} & \mathbf{H}_{23} \end{bmatrix} \begin{bmatrix} \mathbf{P}_3^+ \\ \mathbf{P}_3^- \end{bmatrix} = \mathbf{0}. \tag{7.37}$$

This is the required mode matching condition at the open expansion discontinuity depicted in Figure 7.3. It is in a modal three-port element format and may be represented in a block diagram, by the block shown in Figure 7.3.

This formulation neglects the effect of mean jet formation downstream of the expansion discontinuity. The onset of this phenomenon can be modeled by penetrating the mean flow in the inlet duct downstream of the discontinuity as core flow surrounded by a thin vortex sheet. Although this model is not capable of representing the subsequent rapid non-linear roll-up of this unstable vortex sheet to form a train of vortices, it explains the role of vortex shedding at the edges of the discontinuity as a mechanism for acoustic absorption, which is observed in experiments. Mathematical analyses of this model have been presented by using the Wiener–Hopf method [8–9, 15, 23] and the acoustic analogy based on the vortex sound theory [11].

7.4.1.1 Closed Through-Flow Expansion

In this case duct 2 (the annulus) acts as a side branch and if it terminates with known reflection coefficient matrix (for example, the through-flow end-cap in Figure 7.3), the reflection coefficient matrix of duct 2 at the area discontinuity, $\mathbf{R}_2(0)$, may be determined from Equation (6.69). Then, Equation (7.37) is supplemented by the relation

$$[\mathbf{R}_2 \quad -\mathbf{I}] \begin{bmatrix} \mathbf{P}_2^+ \\ \mathbf{P}_2^- \end{bmatrix} = 0. \tag{7.38}$$

This can be used to eliminate the vectors $\mathbf{P}_2^{\mp}$ to obtain the modal two-port wave transfer equation

$$\begin{bmatrix} \mathbf{P}_1^+ \\ \mathbf{P}_1^- \end{bmatrix} = \mathbf{T}_{13} \begin{bmatrix} \mathbf{P}_3^+ \\ \mathbf{P}_3^- \end{bmatrix}, \tag{7.39}$$

where the modal wave transfer matrix is given by

$$\mathbf{T}_{13} = \begin{bmatrix} \mathbf{H}_{31}\mathbf{a}_1 & -\mathbf{H}_{31}\mathbf{a}_1 \\ \mathbf{H}_{11} & \mathbf{H}_{11} \end{bmatrix}^{-1} \left\{ \begin{bmatrix} \mathbf{H}_{33}\mathbf{a}_3 & -\mathbf{H}_{33}\mathbf{a}_3 \\ \mathbf{H}_{13} & \mathbf{H}_{13} \end{bmatrix} - \begin{bmatrix} \mathbf{H}_{32}\mathbf{a}_2 & -\mathbf{H}_{32}\mathbf{a}_2 \\ \mathbf{0} & \mathbf{0} \end{bmatrix} \mathbf{T}_{23} \right\}. \tag{7.40}$$

Here,

$$\mathbf{T}_{23} = \begin{bmatrix} \mathbf{H}_{22} & \mathbf{H}_{22} \\ \mathbf{R}_2 & -\mathbf{I} \end{bmatrix}^{-1} \begin{bmatrix} \mathbf{H}_{23} & \mathbf{H}_{23} \\ \mathbf{0} & \mathbf{0} \end{bmatrix}. \tag{7.41}$$

7.4.1.2 Closed Flow-Reversing Expansion

In this case, duct 3 acts as the side branch and if the reflection coefficient matrix at its termination (for example, the flow-reversing end-cap in Figure 7.3) is known, the reflection coefficient matrix of duct 3 at the area discontinuity, $\mathbf{R}_3(0)$, may be determined from Equation (6.69). Then, Equation (7.37) is supplemented by

$$\begin{bmatrix} \mathbf{R}_3 & -\mathbf{I} \end{bmatrix} \begin{bmatrix} \mathbf{P}_3^+ \\ \mathbf{P}_3^- \end{bmatrix} = 0. \tag{7.42}$$

This may be used to eliminate the vectors $\mathbf{P}_3^{\mp}$ to obtain the modal two-port equation

$$\begin{bmatrix} \mathbf{P}_1^+ \\ \mathbf{P}_1^- \end{bmatrix} = \mathbf{T}_{12} \begin{bmatrix} \mathbf{P}_2^+ \\ \mathbf{P}_2^- \end{bmatrix}. \tag{7.43}$$

Here, the modal transfer matrix is given by

$$\mathbf{T}_{12} = \left[\begin{bmatrix} -\mathbf{H}_{11} & -\mathbf{H}_{11} \\ \mathbf{0} & \mathbf{0} \end{bmatrix} + \begin{bmatrix} \mathbf{H}_{13} & \mathbf{H}_{13} \\ \mathbf{H}_{23} & \mathbf{H}_{23} \end{bmatrix} \mathbf{B}_{33}\mathbf{B}_{31} \right]^{-1} \left[\begin{bmatrix} \mathbf{0} & \mathbf{0} \\ \mathbf{H}_{22} & \mathbf{H}_{22} \end{bmatrix} + \begin{bmatrix} \mathbf{H}_{13} & \mathbf{H}_{13} \\ \mathbf{H}_{23} & \mathbf{H}_{23} \end{bmatrix} \mathbf{B}_{33}\mathbf{B}_{32} \right], \tag{7.44}$$

where

$$\mathbf{B}_{33} = \begin{bmatrix} \mathbf{H}_{33}\mathbf{a}_3 & -\mathbf{H}_{33}\mathbf{a}_3 \\ \mathbf{R}_3 & -\mathbf{I} \end{bmatrix}^{-1} \tag{7.45}$$

$$\mathbf{B}_{31} = \begin{bmatrix} \mathbf{H}_{31}\mathbf{a}_1 & -\mathbf{H}_{31}\mathbf{a}_1 \\ \mathbf{0} & \mathbf{0} \end{bmatrix} \tag{7.46}$$

$$\mathbf{B}_{32} = \begin{bmatrix} \mathbf{H}_{32}\mathbf{a}_2 & -\mathbf{H}_{32}\mathbf{a}_2 \\ \mathbf{0} & \mathbf{0} \end{bmatrix}. \tag{7.47}$$

The vectors $\mathbf{P}_3^{\mp}$ can be computed from

$$\begin{bmatrix} \mathbf{P}_3^+ \\ \mathbf{P}_3^- \end{bmatrix} = \mathbf{B}_{33}\left\{\mathbf{B}_{31}\begin{bmatrix} \mathbf{P}_1^+ \\ \mathbf{P}_1^- \end{bmatrix} + \mathbf{B}_{32}\begin{bmatrix} \mathbf{P}_2^+ \\ \mathbf{P}_2^- \end{bmatrix}\right\}. \tag{7.48}$$

7.4.2 Sudden Area Contraction

If the inlet, outlet and annulus ducts are numbered as shown in Figure 7.4, the analyses and results of Section 7.4.1 apply also for the corresponding sudden area contractions, provided that the factors β_j in Equation (7.37) are taken as $\beta_1, \beta_2, \beta_3 = 1,\ 1,\ -1$ for the sign convention in Figure 7.4a and as $\beta_1, \beta_2, \beta_3 = 1,\ -1,\ 1$ for the sign convention in Figure 7.4b.

7.4.3 Open Area Change with Multiple Ducts

The foregoing mode-matching analyses of axial expansion and contraction type area changes may be generalized for sudden expansion from or sudden contraction to multiple ducts. As usual, we consider an infinitesimally thin control volume just enclosing the inlet and outlet discontinuities of the sudden contraction and sudden expansion configurations, as shown in Figure 7.5a and b, respectively. In both cases, ducts are numbered as duct 1,2,. . .,N, where duct 1 is always the main duct, and the default positive directions of the local duct axes are taken as indicated. Analyses of the previous sections may be applied for the two configurations by taking into account multiple number of inlet or outlet ducts.

Thus, invoking the condition for the continuity of acoustic pressure at the area discontinuity

$$\mathbf{\Phi}_1(\bullet^{(1)})\{\mathbf{P}_1^+ + \mathbf{P}_1^-\} = \mathbf{\Phi}_j(\bullet^{(j)})\{\mathbf{P}_j^+ + \mathbf{P}_j^-\},\ j = 2, 3, 4, \ldots, n \tag{7.49}$$

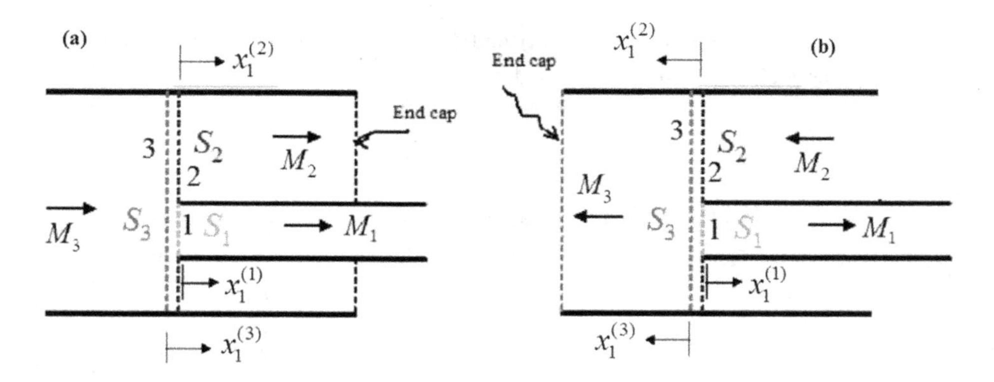

Figure 7.4 Open contraction type area discontinuity with possible closure (end-cap) options. (a) Through-flow closure; (b) Flow-reversing closure.

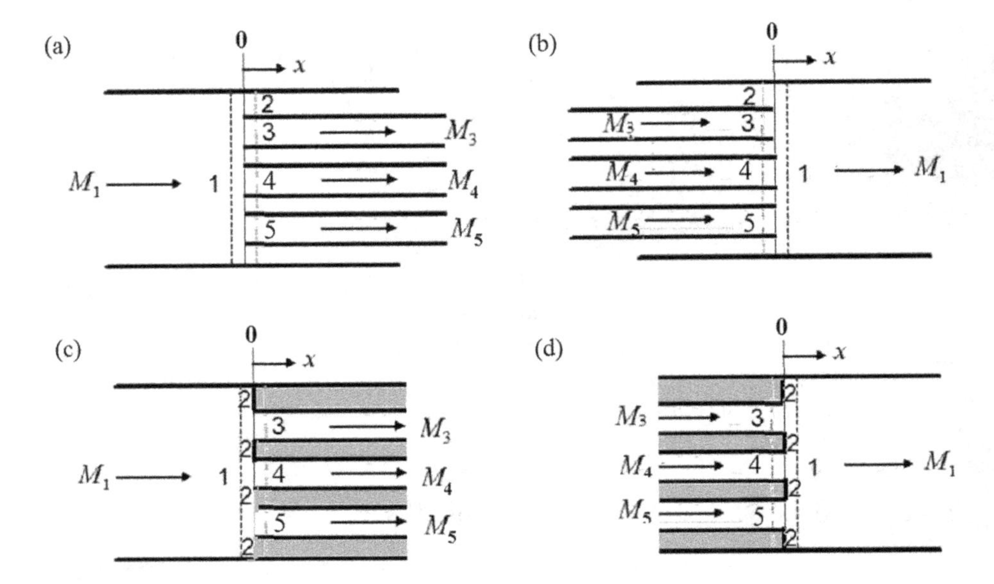

Figure 7.5 (a) Sudden contraction type open area discontinuity with multiple outlet ducts, (b) Sudden expansion type open area discontinuity with multiple inlet ducts, (c) Contraction type duct split, (d) Expansion type duct split.

where $\mathbf{P}_i^{\mp}$ denotes $\mathbf{P}_i^{\mp}(0)$, $i = 1, 2, \ldots, n$. Pre-multiplying these equations by $\breve{\mathbf{\Phi}}_j^{\mathrm{T}}$, respectively, and integrating over the respective duct cross-sectional area S_j gives

$$\mathbf{H}_{j1}\left(\mathbf{P}_1^+ + \mathbf{P}_1^-\right) = \mathbf{H}_{jj}\left(\mathbf{P}_j^+ + \mathbf{P}_j^-\right), j = 2, 3, 4, \ldots, n \tag{7.50}$$

where $\mathbf{H}_{ji}$ are defined as in Equation (7.32). Similarly, Equation (7.7) yields, for the sign convention in Figure 7.5, the compatibility condition

$$-\mathbf{H}_{11}\mathbf{a}_1\left(\mathbf{P}_1^+ - \mathbf{P}_1^-\right) + \sum_{j=2}^{n}\mathbf{H}_{1j}\mathbf{a}_j\left(\mathbf{P}_j^+ - \mathbf{P}_j^-\right) = 0. \tag{7.51}$$

Combining Equations (7.50) and (7.51), the mode-matching condition of the open area-change can be expressed as

$$\begin{bmatrix} \mathbf{H}_{11}\mathbf{a}_1 & -\mathbf{H}_{11}\mathbf{a}_1 & -\mathbf{H}_{12}\mathbf{a}_2 & \mathbf{H}_{12}\mathbf{a}_2 & \ldots & -\mathbf{H}_{1n}\mathbf{a}_n & \mathbf{H}_{1n}\mathbf{a}_n \\ \mathbf{H}_{21} & \mathbf{H}_{21} & -\mathbf{H}_{22} & -\mathbf{H}_{22} & \ldots & \mathbf{0} & \mathbf{0} \\ \vdots & \vdots & \vdots & \vdots & \vdots & \vdots & \vdots \\ \mathbf{H}_{n1} & \mathbf{H}_{n1} & \mathbf{0} & \mathbf{0} & \ldots & -\mathbf{H}_{nn} & -\mathbf{H}_{nn} \end{bmatrix} \begin{bmatrix} \mathbf{P}_1^+ \\ \mathbf{P}_1^- \\ \mathbf{P}_2^+ \\ \mathbf{P}_2^- \\ \vdots \\ \mathbf{P}_n^+ \\ \mathbf{P}_n^- \end{bmatrix} = \mathbf{0}, \tag{7.52}$$

which has the format of a modal n-port model. This result applies for both the contraction and expansion area changes, provided that the numbering schemes shown in Figure 7.5a and b are used.

Sometimes the inner duct inlets in Figure 7.5a (or outlets in Figure 7.5b) are staggered axially by substantial distances. Such area changes can be modeled by using the simple through-flow area expansion or contraction elements described in the previous sections. The method has already been described in Section 4.4.2 (see Figure 4.5).

If duct 2 (always an annulus) is closed by a boundary of known impedance, then Equation (7.52) is complemented by

$$\begin{bmatrix} \mathbf{R}_2 & -\mathbf{E} \end{bmatrix} \begin{bmatrix} \mathbf{P}_2^+ \\ \mathbf{P}_2^- \end{bmatrix} = 0, \tag{7.53}$$

where $\mathbf{R}_2$ denotes the reflection coefficient at interface of duct 2 with the discontinuity plane and $\mathbf{E}$ denotes a unit matrix of conforming size. Combining Equation (7.53) with Equation (7.50) for $j = 2$ gives the relationship

$$\begin{bmatrix} \mathbf{P}_2^+ \\ \mathbf{P}_2^- \end{bmatrix} = \mathbf{A}_{21} \begin{bmatrix} \mathbf{P}_1^+ \\ \mathbf{P}_1^- \end{bmatrix}, \tag{7.54}$$

where

$$\mathbf{A}_{21} = \begin{bmatrix} \mathbf{A}_{21}^{(1)} \\ \mathbf{A}_{21}^{(2)} \end{bmatrix} = \begin{bmatrix} \mathbf{H}_{22} & \mathbf{H}_{22} \\ \mathbf{R}_2 & -\mathbf{E} \end{bmatrix}^{-1} \begin{bmatrix} \mathbf{H}_{21} & \mathbf{H}_{21} \\ \mathbf{0} & \mathbf{0} \end{bmatrix} = \frac{1}{2} \begin{bmatrix} \mathbf{H}_{22}^{-1}\mathbf{H}_{21} & \mathbf{H}_{22}^{-1}\mathbf{H}_{21} \\ \mathbf{H}_{22}^{-1}\mathbf{H}_{21} & \mathbf{H}_{22}^{-1}\mathbf{H}_{21} \end{bmatrix}. \tag{7.55}$$

Here, the third equality holds if $\mathbf{R}_2 = \mathbf{E}$, which holds if duct 2 is a rigid baffle flush with the discontinuity plane, since then $v_1'(0, \bullet) = 0$ on S_2 and, therefore, $\mathbf{R}_2 = \mathbf{E}$. Substituting Equation (7.54) in Equation (7.51) and combining the resulting equation with Equation (7.50) for $j = 3, 4, \ldots, n$, the wave transfer matrix of the closed area change becomes

$$\begin{bmatrix} \mathbf{H}_{11}\mathbf{a}_1 - \mathbf{H}_{12}\mathbf{a}_2\mathbf{A}_{21}^{(1)} & -\mathbf{H}_{11}\mathbf{a}_1 + \mathbf{H}_{12}\mathbf{a}_2\mathbf{A}_{21}^{(2)} & -\mathbf{H}_{13}\mathbf{a}_3 & \mathbf{H}_{13}\mathbf{a}_3 & \ldots & -\mathbf{H}_{1n}\mathbf{a}_n & \mathbf{H}_{1n}\mathbf{a}_n \\ \mathbf{H}_{31} & \mathbf{H}_{31} & -\mathbf{H}_{33} & -\mathbf{H}_{33} & \ldots & \mathbf{0} & \mathbf{0} \\ \vdots & \vdots & \vdots & \vdots & \vdots & \vdots & \vdots \\ \mathbf{H}_{n1} & \mathbf{H}_{n1} & \mathbf{0} & \mathbf{0} & \ldots & -\mathbf{H}_{nn} & -\mathbf{H}_{nn} \end{bmatrix} \begin{bmatrix} \mathbf{P}_1^+ \\ \mathbf{P}_1^- \\ \mathbf{P}_3^+ \\ \mathbf{P}_3^- \\ \vdots \\ \mathbf{P}_n^+ \\ \mathbf{P}_n^- \end{bmatrix} = \mathbf{0}. \tag{7.56}$$

This result applies for both the contraction and expansion area-changes. Note that, if $\mathbf{R}_2 = \mathbf{E}$, then $\mathbf{H}_{12}\mathbf{a}_2\mathbf{A}_{21}^{(1)} = \mathbf{H}_{12}\mathbf{a}_2\mathbf{A}_{21}^{(2)} = 0$. In general, $\mathbf{R}_2$ is determined from the impedance of the boundary that terminates duct 2. Thus, Equation (7.56) also holds for the duct splits shown in Figure 7.5c and d, with planar splitter faces of reflection matrix $\mathbf{R}_2$.

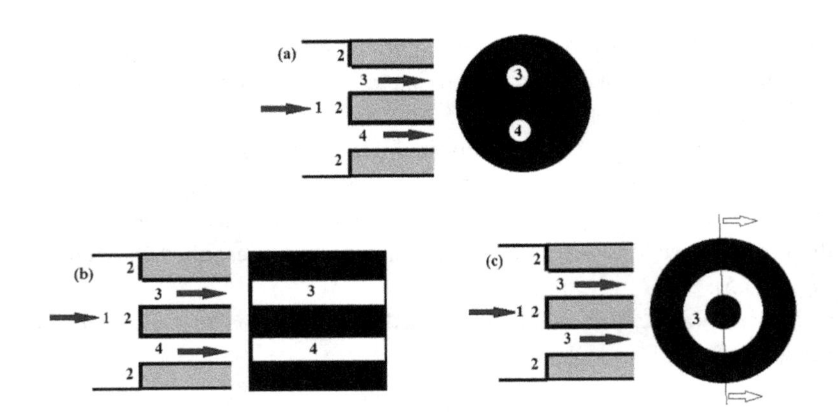

Figure 7.6 (a) A closed sudden area contraction discontinuity with two outlet ducts. (b) A rectangular duct split with two channels. (c) A circular duct split with single annular channel.

The numbering of the ducts is important when a multiple area change or duct-split configuration is defined by side-view figures like Figure 7.5a–c. In the case of a multiple area change, each number must correspond to a duct. Consequently, possible geometries always consist of a number of inner ducts all enclosed by the main duct, as depicted in Figure 7.6a. For a duct split, however, each number except that given to the splitter face must correspond to a duct. Therefore, the splitter geometry depends on the number of inner ducts. For example, in the configuration shown in Figure 7.6b, there are two inner ducts and, hence, three splitters are implied, as shown for a rectangular enclosing duct. But if there is only one inner duct in the same configuration, a single circular splitter is implied, as illustrated in Figure 7.5c.

7.5 Perforated Baffles

Shown in Figure 7.7, is a perforated baffle plate of thickness h, which divides a uniform duct of an arbitrary section of area S into two regions, region L and region R. The plate is assumed to be hard walled except for n apertures of cross-sectional area σ_j, $j = 1, 2, \ldots, n$.

Using the matrix notation of Section 6.4.2 and the sign convention of Figure 7.7, the sound pressure and the axial component of particle velocity in regions L and R can be written as

$$p'_\beta(x_{1_\beta}, \bullet^{(\beta)}) = \mathbf{\Phi}_\beta(\bullet^{(\beta)})\left\{\mathbf{P}^+_\beta\left(x_{1_\beta}\right) + \mathbf{P}^-_\beta\left(x_{1_\beta}\right)\right\} \tag{7.57}$$

$$\rho_\beta c_\beta v'_{1_\beta}(x_{1_\beta}, \bullet^{(\beta)}) = \mathbf{\Phi}_\beta(\bullet^{(\beta)})\left\{\mathbf{A}^+_\beta\mathbf{P}^+_\beta\left(x_{1_\beta}\right) + \mathbf{A}^-_\beta\mathbf{P}^-_\beta\left(x_{1_\beta}\right)\right\}, \tag{7.58}$$

where $\beta \in \mathrm{L}, \mathrm{R}$ and $\mathbf{\Phi}$, $\mathbf{P}^\mp$ and $\mathbf{A}^\mp$ denote, respectively, the modal matrix of the duct transverse modes, the modal pressure wave components and the modal admittance

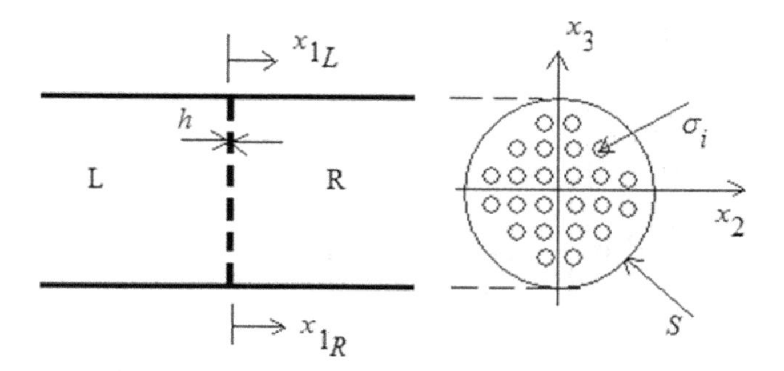

Figure 7.7 A perforated baffle.

matrices. Since the duct is divided by the baffle, $\mathbf{\Phi}_\beta\left(\bullet^{(\beta)}\right) = \mathbf{\Phi}(\bullet)$ and, for simplicity, it is assumed that $\rho_\beta c_\beta = \rho_o c_o$ and $\mathbf{A}_\beta^\mp = \mathbf{A}^\mp$. The fluid motion in the vicinity of the apertures is assumed to be incompressible, which implies that the normal velocity distributions on the two sides of the baffle are identical. Then, the boundary conditions to be satisfied may be expressed as follows:

$$p'_L(0,\bullet) - p'_R(0,\bullet) = \rho_o c_o \zeta(\bullet)\, u(\bullet) \tag{7.59}$$

$$u(\bullet) = v'_{1_\beta}(0,\bullet) \tag{7.60}$$

$$v_{1_\beta}(0,\bullet) = 0, \bullet \text{ on } S \cap \sum_j \sigma_j, \tag{7.61}$$

where $\zeta(\bullet)$ and $u(\bullet)$ denote, respectively, the normalized impedance (impedance divided by $\rho_o c_o$) and normal particle velocity distribution on each surface of the baffle plate. Upon substituting Equation (7.58), Equation (7.60) gives

$$\mathbf{A}^+\mathbf{P}_L^+(0) + \mathbf{A}^-\mathbf{P}_L^-(0) = \mathbf{A}^+\mathbf{P}_R^+(0) + \mathbf{A}^-\mathbf{P}_R^-(0). \tag{7.62}$$

Using Equation (7.57), Equation (7.59) is expressed as

$$\mathbf{\Phi}(\bullet)\left\{\mathbf{P}_L^+(0) - \mathbf{P}_R^+(0) + \mathbf{P}_L^-(0) - \mathbf{P}_R^-(0)\right\} = \rho_o c_o \zeta(\bullet) u(\bullet) \tag{7.63}$$

Multiplying this from left by $\breve{\mathbf{\Phi}}^T(\bullet)$, integrating over the duct cross-sectional area, S, and making note of Equation (7.61),

$$\mathbf{H}\left\{\mathbf{P}_L^+(0) - \mathbf{P}_R^+(0) + \mathbf{P}_L^-(0) - \mathbf{P}_R^-(0)\right\} = \left(\rho_o c_o \sum_{j=1}^{n} \zeta_j \mathbf{H}_j\right)\left\{\mathbf{A}^+\mathbf{P}_L^+(0) + \mathbf{A}^-\mathbf{P}_L^-(0)\right\}, \tag{7.64}$$

where n denotes the number of apertures, ζ_j denotes the normalized impedance of aperture j = 1,2,. . .,n, and

$$\mathbf{H} = \int_S \breve{\mathbf{\Phi}}^T(\bullet)\mathbf{\Phi}(\bullet)\mathrm{d}S \tag{7.65}$$

$$\mathbf{H}_j = \int_{\sigma_j} \breve{\mathbf{\Phi}}^T(\bullet)\mathbf{\Phi}(\bullet)\mathrm{d}S, \qquad j = 1, 2, \ldots, n \tag{7.66}$$

Equations (7.62) and (7.64) may be combined in matrix form as

$$\begin{bmatrix} \mathbf{A}^+ & \mathbf{A}^- \\ \mathbf{H} & \mathbf{H} \end{bmatrix} \begin{bmatrix} \mathbf{P}_L^+(0) \\ \mathbf{P}_L^-(0) \end{bmatrix} = \begin{bmatrix} \mathbf{A}^+ & \mathbf{A}^- \\ \mathbf{H} + \mathbf{Z}\mathbf{A}^+ & \mathbf{H} + \mathbf{Z}\mathbf{A}^- \end{bmatrix} \begin{bmatrix} \mathbf{P}_R^+(0) \\ \mathbf{P}_R^-(0) \end{bmatrix}, \tag{7.67}$$

where

$$\mathbf{Z} = \rho_o c_o \sum_{j=1}^{n} \zeta_j \mathbf{H}_j. \tag{7.68}$$

Thus, the modal two-port wave transfer matrix of the baffle with R–L causality may be expressed as

$$\mathbf{T}_{\mathrm{RL}} = \mathbf{I} + \begin{bmatrix} \mathbf{A}^+ & \mathbf{A}^- \\ \mathbf{H} & \mathbf{H} \end{bmatrix}^{-1} \begin{bmatrix} \mathbf{0} & \mathbf{0} \\ \mathbf{Z}\mathbf{A}^+ & \mathbf{Z}\mathbf{A}^- \end{bmatrix}, \tag{7.69}$$

where $\mathbf{I}$ denotes a unit matrix of conforming size.

7.6 Cavity Coupled with Multiple Ducts

In this section, we consider the problem of sound propagation through a cavity with several duct connections. Figure 7.8 schematically depicts the general geometry of the problem. The ducts are numbered consecutively as duct 1, duct 2, ... The duct–cavity interfaces, S_j, $j = 1, 2, \ldots$, are assumed to be normal to the duct axes and have surface areas equal to the cross-sectional area of respective ducts, the wall thickness of the ducts being assumed to be negligible. The volume occupied by the cavity is denoted by V, and the total surface area enclosing V by Σ. The latter includes the interface areas S_j and, in the case of an extended duct, for example, ducts 2 in Figure 7.8, the outer lateral surface

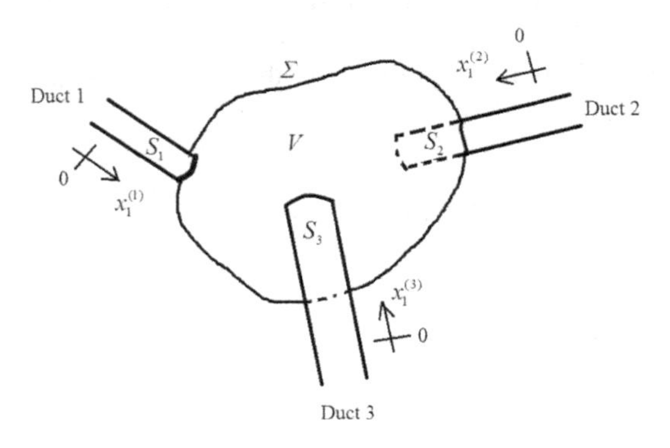

Figure 7.8 A cavity with duct connections.

area of the duct. The cavity surface Σ is assumed to be rigid everywhere except on S_j, where the acoustic field in the cavity must match the acoustic fields in ducts. The default positive acoustic flow directions in the ducts are taken as shown in Figure 7.8. The mean flow field in the cavity is usually complex and difficult to specify with certainty and is neglected in order to render the problem tractable to analysis. Thus, the fluid in the cavity is assumed to be inviscid, homogeneous and quiescent.

7.6.1 Closed Cavity Modes

It is convenient to consider first the cavity with no duct connections. The acoustic field in the cavity is determined, in frequency domain, by the Helmholtz equation

$$(\nabla^2 + k_o^2)p' = 0, \qquad \frac{\partial p'}{\partial n_0} = 0 \quad \text{on } \Sigma_0, \tag{7.70}$$

where $k_o = \omega/c_o$, $p' = p'(\mathbf{x})$ denotes the acoustic pressure at a point $\mathbf{x} \in V_0$ of the cavity, V_0 denotes volume of the cavity, Σ_0 denotes the surface of V_0 (underformed by duct connections) and $\partial/\partial n_0$ ($= \mathbf{n}_0 \cdot \nabla$) denotes differentiation along the unit outward normal vector $\mathbf{n}_0$ of Σ_0. The solution of this eigenvalue problem determines the acoustical modes of the cavity. Cavity modes are characterized by real eigenvalues λ_m^2, and corresponding eigenfunctions, $\psi_m(\mathbf{x})$, $m = 1, 2, \ldots$, which satisfy Equation (7.70) and, therefore, the orthogonality condition $\int_{V_0} \psi_m \psi_m dV = 0, \quad n \neq m$, which can be proved as in Section 6.5.1.

In general, for a cavity of an arbitrary shape, the cavity modes can be determined only numerically, for example, by using the finite element method. Here, we will assume that the cavity geometry can be approximated by a uniform duct of arbitrary cross section with rigid closed ends normal to the duct axis. Let the cavity geometry be defined in a local coordinate system $(x_1, \bullet)$, where the symbol "$\bullet$" is used as usual to denote the transverse coordinates and x_1 denotes the cavity axis, the closed-end surfaces being at $x_1 = 0$ and $x_1 = L$. The general solution of Equation (7.70) can be separated as $p'(\mathbf{x}) = P(x_1)\,\varphi(\bullet)$. The axial function $P(x_1)$ and the transverse function $\varphi(\bullet)$ are given by

$$\frac{\nabla_\perp^2 \varphi(\bullet)}{\varphi(\bullet)} = -\left(\frac{\dfrac{d^2}{dx_1^2} P(x_1)}{P(x_1)} + k_o^2 \right) = -\alpha^2. \tag{7.71}$$

The solution for $\varphi(\bullet)$ are exactly the transverse duct eigenfunctions $\varphi_\mu(\bullet)$ corresponding to the eigenvalues $\alpha_\mu^2, \mu = 1, 2, \ldots$, the determination of which is discussed in depth in Section 6.5. The axial function $P(x_1)$ is given by the solution of the differential equation $d^2P/d^2x_1 + \beta^2 P = 0$, where $k_o^2 = \alpha^2 + \beta^2$ and β denotes an undetermined parameter. The general solution of this equation is $P(x_1) = Be^{i\beta x_1} + Ce^{-i\beta x_1}$, where B and C are integration constants. In view of the boundary condition on $P(x_1)$, that is, $dP/dx_1 = 0$ at $x_1 = 0, L$, solutions exist in the form

$$P_\eta(x_1) = B\cos(\beta_\eta x_1), \qquad \beta_\eta = \frac{\eta\pi}{L}, \quad \eta = 1, 2, \ldots \tag{7.72}$$

Here, β_η^2 and corresponding $P_\eta(x_1)$ are called the axial eigenvalues and eigenfunctions of the cavity. Thus, the cavity modes may be expressed as

$$\psi_{\mu\eta}(\mathbf{x}) = \varphi_\mu(\bullet)\cos(\beta_\eta x_1), \qquad k_o^2 = \alpha_\mu^2 + \beta_\eta^2, \qquad \mu, \eta = 1, 2, \ldots. \tag{7.73}$$

The transverse and axial duct modes are ordered in ascending order of the eigenvalues as $0 = \alpha_1^2 \leq \alpha_2^2 \leq \cdots$ and $\beta_1^2 \leq \beta_2^2 \leq \cdots$. The cavity modes are then ordered also in ascending order of eigenvalues as $\lambda_1^2 \leq \lambda_2^2 \leq \cdots$, where $\lambda_m^2 \in (\alpha_1^2, \alpha_2^2, \ldots) \cup (\beta_1^2 \leq \beta_2^2 \leq \cdots)$, $m = 1, 2, \ldots$, and the mth cavity eigenvalue λ_m^2 corresponds to the eigenfunction $\psi_m(\mathbf{x})$,

7.6.2 The Green Function of the Cavity

With mean flow neglected, the acoustic pressure field in the actual cavity with duct connections can be expressed in the frequency domain by the surface integral formula

$$p'(\mathbf{x}) = \int_\Sigma \mathbf{J}(p', g)\cdot\mathbf{n}\,d\Sigma, \qquad \mathbf{x} \in V \tag{7.74}$$

Here, g denotes the Green function of the cavity, $\mathbf{n}$ denotes the outward unit vector of Σ and $\mathbf{J}(p', g)$ denotes the vector

$$\mathbf{J}(p', g) = g\nabla p' - p'\nabla g. \tag{7.75}$$

By definition, the Green function gives the acoustic pressure field at any point $\mathbf{x} \in V$ due to a point source of unit strength located at point $\mathbf{x}_S \in V$. This is usually indicated by using the syntax $g = g(\mathbf{x}_S|\mathbf{x})$. Mathematically, it is a particular solution of the inhomogeneous Helmholtz equation

$$\left(\nabla^2 + k_o^2\right)g(\mathbf{x}_S|\mathbf{x}) = \delta(\mathbf{x} - \mathbf{x}_S), \tag{7.76}$$

where $\delta(\mathbf{x})$ denotes a spatial Dirac function located at point $\mathbf{x} = 0$. If Equation (7.76) is multiplied by p' and integrated over V, application of the divergence theorem yields Equation (7.74), since $p'\left(\nabla^2 + k_o^2\right)g \equiv \nabla\cdot\mathbf{J}(p', g)$.

The Green function for an arbitrary cavity with arbitrary number of duct connections is difficult to determine. However, it can be represented by an infinite series of orthogonal functions as

$$g(\mathbf{x}_S|\mathbf{x}) = \sum_{m=0}^{\infty} b_m\psi_m(\mathbf{x}), \tag{7.77}$$

where ψ_m denotes the mth eigenfunction of the closed cavity. Substituting this into Equation (7.76), multiplying the result by $\psi_n(\mathbf{x})$ and integrating over the volume of the cavity gives

$$\int_V \psi_n(\mathbf{x}) \sum_{m=1}^{\infty} \left(k_o^2 - \lambda_m^2\right) b_m \psi_m(\mathbf{x}) dV = \int_V \delta(\mathbf{x} - \mathbf{x}_S) \psi_n(\mathbf{x}) dV, \quad n = 1, 2, \ldots \tag{7.78}$$

We assume that V is not substantially changed from V_0 by the connection of the ducts, so that the orthogonality of the closed cavity modes may still be invoked. Then, solving the foregoing equation

$$b_m = \frac{\psi_m(\mathbf{x}_S)}{N_m\left(k_o^2 - \lambda_m^2\right)}, \qquad N_m = \int_V \psi_m^2 dV. \tag{7.79}$$

Hence, a Green function of the cavity is

$$g(\mathbf{x}_S|\mathbf{x}) = \sum_{m=1}^{\infty} \frac{\psi_m(\mathbf{x})\, \psi_m(\mathbf{x}_S)}{N_m\left(k_o^2 - \lambda_m^2\right)}. \tag{7.80}$$

7.6.3 Coupling the Cavity with Ducts

Since the cavity eigenfunctions satisfy the hard-wall condition $\partial\psi_m/\partial n = 0$ and the normal component of the particle velocity vanishes on rigid parts of Σ, Equation (7.74) simplifies to

$$p'(\mathbf{x}) = \sum_{j=1}^{n} \int_{S_j} g(\mathbf{x}_S|\mathbf{x}) \frac{\partial p'_j}{\partial n_j} \mathrm{d}S_j, \tag{7.81}$$

where the subscript $j = 1, 2, \ldots, n$ refers to connecting duct j. The normal derivative of the acoustic pressure at duct interfaces must be compatible with the acoustic pressure fields in the connecting ducts. Therefore, we can use Equation (6.6), the momentum equation for the acoustic field in the ducts, to determine the normal derivative terms in Equation (7.81). Hence, for the sign convention of Figure 7.8, we obtain for duct $j \in 1, 2, \ldots, n$

$$\frac{\partial p'_j}{\partial n_j} = \left[\rho_o c_o \left(-\mathrm{i}k_o v'_1 + M \frac{\partial v'_1}{\partial x_1}\right)\right]_j. \tag{7.82}$$

Using matrix notation (Section 6.4.2), this is expressed as

$$\frac{\partial p'_j}{\partial n_j} = \mathrm{i}k_o \mathbf{\Phi}_j(\bullet^{(j)}) \mathbf{h}_j \mathbf{P}_j. \tag{7.83}$$

Here, $\mathbf{P}_j = \left\{ \mathbf{P}_j^+ \quad \mathbf{P}_j^- \right\}$ denotes the vector of modal pressure wave components at the interface of duct j with the cavity volume, and

$$\mathbf{h}_j = \left[\mathbf{A}_j^+\left(-\mathbf{I} + M_j\mathbf{K}_j^+\right) \quad \mathbf{A}_j^-\left(-\mathbf{I} + M_j\mathbf{K}_j^-\right)\right], \tag{7.84}$$

where **I** denotes a unit square matrix of conforming size. Substituting Equation (7.83) in Equation (7.81),

$$p'(\mathbf{x}) = \mathrm{i}k_o \sum_{j=1}^{n} \left(\int_{S_j} g(\mathbf{x}_S|\mathbf{x}) \mathbf{\Phi}_j(\bullet^{(j)}) \mathrm{d}S_j \right) \mathbf{h}_j \mathbf{P}_j. \tag{7.85}$$

Finally, using Equation (7.80) in Equation (7.85), the sound pressure field in the cavity can be expressed as

$$p'(\mathbf{x}) = \mathrm{i}k_o \sum_{j=1}^{n} \left(\sum_{m=1}^{\infty} \frac{\psi_m(\mathbf{x}) \mathbf{F}_{m,j}}{N_m \left(k_o^2 - \lambda_m^2\right)} \right) \mathbf{h}_j \mathbf{P}_j, \tag{7.86}$$

where

$$\mathbf{F}_{m,j} = \int_{S_j} \psi_m \mathbf{\Phi}_j(\bullet^{(j)}) \mathrm{d}S_j. \tag{7.87}$$

Now, consider a point $\mathbf{x}_i = \left(x_1^{(i)}, \bullet^{(i)}\right) \in S_i, \quad i = 1, 2, \ldots, n$, on the interface of the cavity with duct i. The acoustic pressure at this point can be expressed in matrix notation as

$$p'(\mathbf{x}_i) = \mathbf{\Phi}_i(\bullet^{(i)}) [\mathbf{I} \quad \mathbf{I}] \mathbf{P}_i, \tag{7.88}$$

where $\mathbf{P}_i = \mathbf{P}_i\left(x_1^{(i)}\right)$. Writing Equation (7.86) for $\mathbf{x} = \mathbf{x}_i$, multiplying from the left by $\mathbf{\Phi}_i^{\mathrm{T}}\left(\bullet^{(i)}\right)$, where the superscript "T" denotes matrix transpose, and integrating over the interface area S_i yields n equations of the form

$$[\mathbf{H}_{ii} \quad \mathbf{H}_{ii}] \mathbf{P}_i = \mathrm{i}k_o \sum_{j=1}^{n} \left(\sum_{m=1}^{\infty} \frac{\mathbf{F}_{m,i}^{\mathrm{T}} \mathbf{F}_{m,j}}{N_m \left(k_o^2 - \lambda_m^2\right)} \right) \mathbf{h}_j \mathbf{P}_j, \qquad i = 1, 2, \ldots, n \tag{7.89}$$

where $\mathbf{H}_{ii} = \int_{S_i} \mathbf{\Phi}_i^{\mathrm{T}}\left(\bullet^{(i)}\right) \mathbf{\Phi}_i\left(\bullet^{(i)}\right) \mathrm{d}S$. These equations can be collated as

$$\begin{bmatrix} -\mathbf{L}_{11} + \boldsymbol{\alpha}_{11} & \boldsymbol{\alpha}_{12} & \cdots & \boldsymbol{\alpha}_{1n} \\ \boldsymbol{\alpha}_{21} & -\mathbf{L}_{22} + \boldsymbol{\alpha}_{22} & \cdots & \boldsymbol{\alpha}_{2n} \\ \vdots & \vdots & \ddots & \vdots \\ \boldsymbol{\alpha}_{n1} & \boldsymbol{\alpha}_{n2} & \cdots & -\mathbf{L}_{nn} + \boldsymbol{\alpha}_{nn} \end{bmatrix} \begin{bmatrix} \mathbf{P}_1 \\ \mathbf{P}_2 \\ \vdots \\ \mathbf{P}_n \end{bmatrix} = \mathbf{0} \tag{7.90}$$

where

$$\boldsymbol{\alpha}_{ij} = \mathrm{i}k_o \sum_{m=1}^{\infty} \left(\frac{\mathbf{F}_{m,i}^{T} \mathbf{F}_{m,j}}{N_m \left(k_o^2 - \lambda_m^2\right)} \right) \mathbf{h}_j, \qquad \mathbf{L}_{ii} = [\mathbf{H}_{ii} \quad \mathbf{H}_{ii}]. \tag{7.91}$$

Equation (7.90) is the required wave transfer equation of the cavity and can be used in block diagrams as a modal n-port block. This is a versatile acoustic element. Two

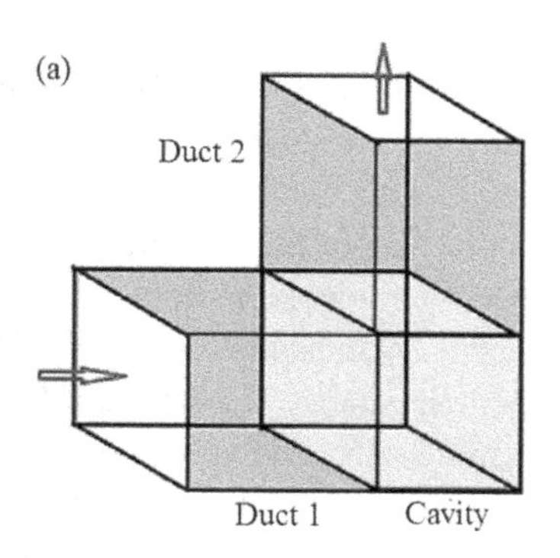

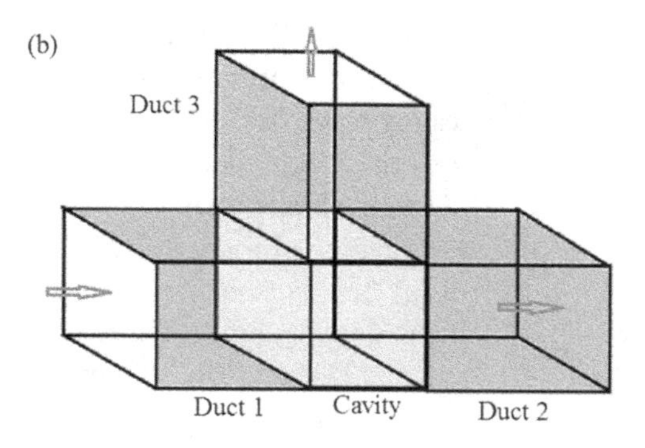

Figure 7.9 Use of the junction cavity element for modeling a 90° elbow and a branch.

typical uses are shown in Figure 7.9 and many different configurations can be generated by changing the shapes and sizes of the ducts and the cavity, and adding ducts to or removing ducts from the surfaces of the cavity. The effect of the soft surfaces of the cavity, which is neglected in the foregoing formulation, may be taken into account approximately by modifying the denominator of Equation (7.91) as follows

$$\boldsymbol{\alpha}_{ij} = \mathrm{i}k_o \sum_{m=1}^{\infty} \left(\frac{\mathbf{F}_{m,i}^T \mathbf{F}_{m,j}}{N_m \left(k_o^2 - \lambda_m^2 + \mathrm{i}2\xi_m k_o \lambda_m\right)} \right) \mathbf{h}_j, \tag{7.92}$$

where ξ_m denotes an overall non-dimensional damping ratio for cavity mode m. An estimate of ξ_m for a lightly damped cavity due to surface impedance is:

$$\xi_m = \frac{\bar{\alpha} A}{8\lambda_m N_m}, \tag{7.93}$$

where $\bar{\alpha}$ is the random incidence absorption coefficient of the cavity surface and A denotes the total treated surface area. On the other hand, including the mean flow in the connecting ducts, but neglecting it in the cavity is of course not consistent, but is convenient for the simplicity of the element. The actual mean flow structures in duct junctions are complex, specific to each geometry and flow entrance conditions and their measurement or computation requires considerable effort. For practical purposes, effect of the mean flow in the cavity is akin to damping and may be taken into account, at least as a first approximation, by adjusting the ξ_m correction for the surface impedance heuristically.

7.7 Coupled Perforated Ducts

Considered in this section is three-dimensional modeling of sound transmission in an arbitrary number of uniform ducts of finite length enclosed in another duct acting as casing, sound fields in the ducts being coupled by compact apertures distributed on their walls. The theory presented is based on the assumption that fluid motion in apertures is incompressible and excited by acoustic pressure fields in the communicating pipes. This is a generalization of the row-wise discrete one-dimensional theory described in Section 4.8 to three dimensions. Although perforations are actually distributed discretely over the duct walls, they are often modeled as continuous distribution of finite wall impedance swamped over duct surfaces (Section 4.7). This is, in fact, the common approach used previously by most authors; however, applications in three dimensions are limited to the simple straight-through muffler configuration [25–34]. The discrete method is more versatile in modeling aperture distributions and is simpler to implement, since it does not require the solution of any eigenvalue problem. It can be applied for any perforation pattern, if the local coordinates of all apertures are specified. Practical perforate patterns are usually regular and the aperture coordinates are generated relatively easily. On the other hand, in the frequency ranges of interest in many practical applications, the effect of perforation pattern on sound transmission usually becomes discernible when the open area porosity is low enough, less than about 5%, say. For relatively larger porosities, adequately accurate results can be obtained by replacing the actual perforation pattern by any convenient uniform pattern of the same porosity. Consequently, when porosity is sufficiently high, it may not be necessary to be concerned about the exact coordinates of the apertures; a suitable algorithm for generating a uniform aperture pattern from a given open area porosity may be used.

The following sections describe the calculation of the wave transfer matrix across a transverse section of single- or double-coupled perforated ducts (Figure 4.8 or Figure 7.10). The procedure for connecting such section matrices with those of solid pipe segments so as to form the wave transfer matrix of a pack of perforated ducts

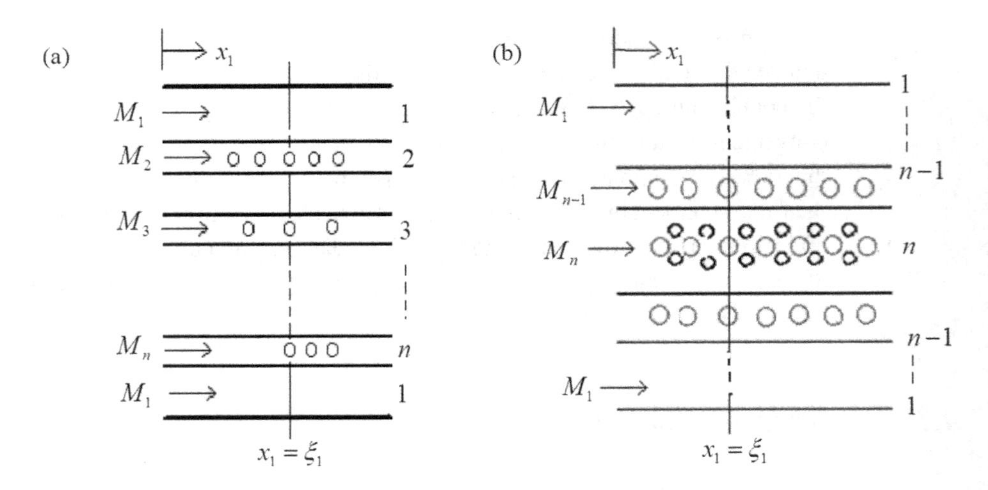

Figure 7.10 Discretely coupled perforates: (a) single-coupled and (b) double-coupled configurations.

communicating over a given length has been described in Section 4.8.4 and need not be repeated here.

7.7.1 Acoustic Field in a Duct with a Single Aperture

We begin with the integral form of the law of conservation of mass, which is given as Equation (A.18). Upon linearization, this equation yields

$$\frac{\partial}{\partial t}\int_V \rho' \, dV + \int_\Sigma (\rho_o \, \mathbf{v}' + \rho' \mathbf{v}_o) \cdot \mathbf{n} dS = \int_V \dot{m}' dV, \tag{7.94}$$

where V denotes a volume of the duct enclosed between two transverse cross sections and the duct walls, $\dot{m}'$ denotes the fluctuating part the rate of mass inflow per unit volume into V, and Σ denotes the surface area of V, $\mathbf{n}$ denotes the unit normal vector of the surface Σ, $\mathbf{v}_o = \mathbf{e}_1 v_{o1}$ denotes the mean flow velocity, which is assumed to be uniform and axial, and $\mathbf{e}_1$ denotes the unit vector in the $+x_1$ direction (the duct axis). The remaining symbols in Equation (7.94) have the usual meanings as defined in Appendix A. Upon application of the divergence theorem and using the state equation $p' = c_o^2 \rho'$, Equation (7.94) yields the acoustic continuity equation in differential form

$$\frac{\partial p'}{\partial t} + v_{o1}\frac{\partial p'}{\partial x_1} + \rho_o c_o^2 \nabla \cdot \mathbf{v}' = \dot{m}' c_o^2. \tag{7.95}$$

Combining this with the acoustic momentum equation, Equation (6.6), results in the inhomogeneous convective wave equation

$$\left[\nabla^2 - \frac{1}{c_o^2}\left(\frac{\partial}{\partial t} + v_{o1}\frac{\partial}{\partial x_1}\right)^2\right] p' = -\left(\frac{\partial}{\partial t} + \mathbf{v}_o \cdot \nabla\right)\dot{m}'. \tag{7.96}$$

In the frequency domain, this can be written as

$$\left(\nabla^2 - M^2 \frac{\partial^2}{\partial x_1^2} + \mathrm{i}2k_o M \frac{\partial}{\partial x_1} + k_o^2\right) p' = -\left(M \frac{\partial}{\partial x_1} - \mathrm{i}k_o\right)\dot{m}' \, c_o, \tag{7.97}$$

where $k_o = \omega / c_o$ and $M = v_{o1}/c_o$ denotes the Mach number of the mean flow velocity.

Let $\mathbf{x}$ denote any point on Σ and let there be single compact aperture centered at point $\boldsymbol{\xi} \in \Sigma$ of strength $\dot{m}' = \rho_o \dot{Q} \delta(\mathbf{x} - \boldsymbol{\xi})$, where $\dot{Q}$ denotes the rate of volume injection from the aperture and $\delta(\mathbf{x})$ denotes a Dirac function located at point $\mathbf{x} = 0$. The solution of Equation (7.97) can be searched in the form

$$p'(\mathbf{x}) = \sum_{m=1}^{\infty} \varphi_m(\bullet) \, q_m(x_1), \tag{7.98}$$

in which $\varphi_m(\bullet)$, $m = 1, 2, \ldots$, denote the complete set of eigenfunctions of the transverse duct modes, "$\bullet$" as usual denotes the transverse coordinates (e.g., $\bullet \equiv (x_2, x_3)$ in rectangular coordinates and $\bullet \equiv (r, \theta)$ in cylindrical polar coordinates) and $q_m(x_1)$ denote some unknown functions of the duct axis, x_1. Upon substituting

Equation (7.98) into Equation (7.97) and integrating the result over the duct cross-sectional area, we find the following differential equation for $q_m(x_1)$:

$$\left(1-M^2\right)\frac{d^2q_m}{dx_1^2}+\mathrm{i}2k_oM\frac{dq_m}{dx_1}+\left(k_o^2-\alpha_m^2\right)q_m=b_{0m}\delta(x_1-\xi_1)+b_{1m}\frac{d}{dx_1}\delta(x_1-\xi_1), \tag{7.99}$$

where α_m^2 are the transverse eigenvalues of the duct cross section and

$$b_{0m}=\frac{\mathrm{i}k_o\rho_oc_o\dot{Q}\varphi_m\left(\bullet_\zeta\right)}{H_m},\qquad b_{1m}=-\frac{Mb_{0m}}{\mathrm{i}k_o} \tag{7.100}$$

$$H_m=\int_S\varphi_m^2(\bullet)\mathrm{d}S, \tag{7.101}$$

where we use the notation $\boldsymbol{\xi}\equiv\left(\xi_1,\bullet_\xi\right)$ to denote the coordinates of point $\boldsymbol{\xi}$. The general solution of Equation (7.99) can be expressed as [35]:

$$q_m(x_1)=C_m^+\mathrm{e}^{\mathrm{i}k_oK_m^+x_1}+C_m^-\mathrm{e}^{\mathrm{i}k_oK_m^-x_1}+G_m(x_1)H(x_1-\xi_1),\qquad m=1,2,\ldots \tag{7.102}$$

where the first two terms on the right represent the solution of the homogeneous form of Equation (7.99), $C_m^\mp$ being the integration constants and $K_m^\mp$ denote the propagation constants in single index notation (see Section 6.5). The third term on the right of Equation (7.102) is the particular integral of Equation (7.99), where $H(x_1)$ denotes a Heaviside function located at $x_1=0$, that is, $H(x_1)=0$ for $x_1<0$ and $H(x_1)=1$ for $x_1>0$, and the Green function $G_m(x_1)$ is given by the solution of

$$\left(1-M^2\right)\frac{d^2G_m}{dx_1^2}+\mathrm{i}2k_oM\frac{dG_m}{dx_1}+\left(k_o^2-\alpha_m^2\right)G_m=0 \tag{7.103}$$

satisfying the boundary conditions

$$\left(1-M^2\right)\frac{dG_m}{dx_1}(\xi_1)+\mathrm{i}2k_oMG_m(\xi_1)=b_{0m},\ \left(1-M^2\right)G_m(\xi_1)=b_{1m}. \tag{7.104}$$

Solution of this boundary value problem for mode m is given by

$$G_m(x_1)=\frac{1}{2}\left(B_m^+\mathrm{e}^{\mathrm{i}k_oK_m^+(x_1-\xi_1)}+B_m^-\mathrm{e}^{\mathrm{i}k_oK_m^-(x_1-\xi_1)}\right)\dot{\mu}_m,\qquad m=1,2,\ldots \tag{7.105}$$

where

$$B_m^\pm=\frac{\pm1-Ma_m}{a_m\left(1-M^2\right)} \tag{7.106}$$

$$\dot{\mu}_m=\frac{\rho_oc_o\dot{Q}\varphi_m\left(\bullet_\xi\right)}{H_m} \tag{7.107}$$

and a_m are defined by Equation (6.50). Thus, Equation (7.98) can be expressed as an infinite sum of pressure wave components:

$$p'(x_1, \bullet) = \sum_{m=1}^{\infty} \{p_m^+(x_1) + p_m^-(x_1)\} \varphi(\bullet) \tag{7.108}$$

where

$$p_m^{\pm}(x_1) = C_m^{\pm} e^{ik_o K_m^{\pm} x_1} + \frac{1}{2}\mu_m B_m^{\pm}\ e^{ik_o K_m^{\pm}(x_1 - \xi_1)} H(x_1 - \xi_1). \tag{7.109}$$

Equation (7.108) is the general solution of the inhomogeneous wave equation (7.97) for a uniform hard-walled duct containing single aperture.

7.7.2 Wave Transfer Across a Row of Apertures

From Equation (7.108), the relationship between the modal pressure wave components just upstream, $x_1 = \xi_{1-}$, and just downstream, $x_1 = \xi_{1+}$, of an aperture may be expressed for mode m as[1]

$$\begin{bmatrix} p_m^+(\xi_{1+}) \\ p_m^-(\xi_{1+}) \end{bmatrix} = \begin{bmatrix} p_m^+(\xi_{1-}) \\ p_m^-(\xi_{1-}) \end{bmatrix} + \frac{1}{2}\dot{\mu}_m \begin{bmatrix} B_m^+ \\ B_m^- \end{bmatrix}, \qquad m = 1, 2, \ldots. \tag{7.110}$$

For all modes, these equations can be collected in matrix notation as

$$\begin{bmatrix} \mathbf{P}^+(\xi_{1+}) \\ \mathbf{P}^-(\xi_{1+}) \end{bmatrix} = \begin{bmatrix} \mathbf{P}^+(\xi_{1-}) \\ \mathbf{P}^-(\xi_{1-}) \end{bmatrix} + \frac{1}{2} \begin{bmatrix} \mathbf{B}^+ \\ \mathbf{B}^- \end{bmatrix} \boldsymbol{\mu} \tag{7.111}$$

where

$$\mathbf{B}^{\pm} = \begin{bmatrix} B_1^{\pm} & 0 & \cdots \\ 0 & B_2^{\pm} & \cdots \\ \vdots & \vdots & \ddots \end{bmatrix} \tag{7.112}$$

$$\mathbf{P}^{\pm}(x_1) = \begin{bmatrix} p_1^{\pm}(x_1) \\ p_2^{\pm}(x_1) \\ \vdots \end{bmatrix}, \qquad \boldsymbol{\mu} = \begin{bmatrix} \dot{\mu}_1 \\ \dot{\mu}_2 \\ \vdots \end{bmatrix}. \tag{7.113}$$

Still more concisely, Equation (7.111) is written as

$$\mathbf{P}(\xi_{1+}) = \mathbf{P}(\xi_{1-}) + \frac{1}{2}\mathbf{B}\boldsymbol{\mu}, \tag{7.114}$$

where

[1] If only the fundamental mode is retained, then $m = 1$, $B_1^{\pm} = \pm 1/(1 \pm M)$, $\dot{\mu}_1 = (\rho_o c_o/S)\sum \dot{Q}$, where S denotes the duct cross-sectional area, and Equation (7.110) is comparable with Equation (4.136). The two equations become identical if the effect of mean through flow is neglected and the slip-flow parameter is taken equal to unity in Equation (4.136).

$$\mathbf{B} = \begin{bmatrix} \mathbf{B}^+ \\ \mathbf{B}^- \end{bmatrix}, \qquad \mathbf{P}(x_1) = \begin{bmatrix} \mathbf{P}^+(x_1) \\ \mathbf{P}^-(x_1) \end{bmatrix}. \tag{7.115}$$

Acoustic wave transfer across a transverse row of apertures is still governed by Equation (7.97), but $\dot{\mu}_m$ must be redefined by invoking the principle of superposition

$$\dot{\mu}_m = \frac{\rho_o c_o}{H_m} \sum_{\text{apertures}} \dot{Q}\varphi_m(\bullet_\xi), \qquad m = 1, 2, \ldots \tag{7.116}$$

where the summation is taken over all apertures in the row considered. The volume strength of each aperture is $\dot{Q} = -Au'$, where A denotes the effective cross-sectional area of an aperture, u' denotes the unsteady normal velocity of the fluid in the aperture, which is taken as positive in the outward direction from the duct surface, and the minus sign accounts for m' being taken as positive if into the duct. Aperture impedance, Z, is defined as $Z = [p']/u'$, where $[p']$ denotes the acoustic pressure difference across the aperture. Then, the volume injection strength of a single aperture can be expressed as

$$\dot{Q} = -\frac{[p']A}{\rho_o c_o \zeta}, \tag{7.117}$$

where $\zeta = Z/\rho_o c_o$ denotes the normalized aperture impedance (see Appendix C).

For systematic application of Equation (7.114) for any arrangement of communicating perforated ducts, it is convenient to number the ducts as duct 1, duct 2, duct 3, . . ., where duct 1 is always the casing (an annular duct). Let two ducts, duct j and duct k, communicate through the apertures on a transverse plane of duct j, all duct axes being assumed to be parallel. Let the center coordinates of an aperture on duct j be denoted by $\boldsymbol{\xi} = \left(\xi_1, \bullet_\xi\right)$ in the local coordinate system of duct j. Then, the volume injection strength of this aperture is, from Equation (7.117),

$$\dot{Q}_j = -\left(\frac{A}{\rho_o c_o \zeta}\right)_j \left\{p_j'(\boldsymbol{\xi}) - p_k'(\boldsymbol{\xi})\right\}, \tag{7.118}$$

where the subscripts refer to the pipe number. Thus, for the case of multiple apertures on the row at plane $x_1 = \xi_1$ of pipe j, Equation (7.116) can be expressed as

$$(\dot{\mu}_m)_j = -\left(\frac{A}{H_m \zeta}\right)_j \left\{ \sum_{\substack{\text{apertures} \\ \text{on duct } j}} \left(p_j'(\boldsymbol{\xi}) - p_k'(\boldsymbol{\xi})\right) \left[\varphi_m\left(\bullet_\xi\right)\right]_j \right\}, \qquad m = 1, 2, \ldots \tag{7.119}$$

where the summation is taken over all apertures in the row considered and, for the simplicity of the presentation, all apertures on the same row are assumed to be identical. In matrix notation of Sec. 6.4.2, the acoustic pressure field in duct j is expressed as

$$p'_j(x_1^{(j)}, \bullet^{(j)}) = \mathbf{\Phi}_j(\bullet^{(j)}) \left(\mathbf{P}_j^+(x_1^{(j)}) + \mathbf{P}_j^-(x_1^{(j)})\right), \tag{7.120}$$

where $\mathbf{\Phi}_j$ is the modal matrix of pipe j. Upon substituting Equation (7.120) in Equation (7.119), and collecting the resulting equations in matrix form, it may be shown that that, for duct j (communicating with duct k)

$$\boldsymbol{\mu}_j = -g_j\mathbf{G}_{j,j}^{(j)}\mathbf{P}_j(\xi_1) + g_j\mathbf{G}_{j,k}^{(j)}\mathbf{P}_k(\xi_1) \tag{7.121}$$

and for duct k (communicating with duct j)

$$\boldsymbol{\mu}_k = -g_j\mathbf{G}_{k,k}^{(j)}\mathbf{P}_k(\xi_1) + g_j\mathbf{G}_{k,j}^{(j)}\mathbf{P}_j(\xi_1). \tag{7.122}$$

Here,

$$g_j = \frac{A_j}{\zeta_j} \tag{7.123}$$

$$\mathbf{G}_{j,k}^{(i)} = \mathbf{H}_j^{-1} \sum_{\substack{\text{apertures} \\ \text{on duct } i}} \mathbf{\Phi}_j^T(\bullet_\xi)\left[\mathbf{\Phi}_k(\bullet_\xi) \quad \mathbf{\Phi}_k(\bullet_\xi)\right], \tag{7.124}$$

where the summation is taken over the row considered.

A modal wave transfer matrix for wave transfer across a transverse plane of any pack of multiple communicating perforated ducts can be derived by using Equation (7.114) together with Equations (7.121) and (7.122). Two common arrangements are the single-coupled and double-coupled configurations shown in Figure 7.10a and b. Here, the duct axes are assumed to be parallel, but not necessarily collinear, and have their origins in the same transverse plane of the pack. It suffices to derive a modal wave transfer matrix across a transverse plane $x_1 = \xi_1$ where all inner ducts are assumed to have a row of apertures. This can be cascaded with the transfer matrix of the adjoining solid pipe sections to reach the next row of holes, and so on until the plane containing the last perforate row. As it will transpire subsequently, although all inner ducts are assumed to have an aperture row at plane $x_1 = \xi_1$, the resulting transfer matrix can also be used when one or more ducts does not have any aperture row in this plane. All apertures in a given row of a duct are assumed to be identical, but apertures on the same plane of different ducts may be different.

Referring to the single-coupled configuration, Figure 7.10a, Equation (7.114) is written for the inner ducts

$$\mathbf{P}_j(\xi_{1+}) = \mathbf{P}_j(\xi_{1-}) + \frac{1}{2}\mathbf{B}_j\boldsymbol{\mu}_j, \qquad j = 2, 3, \ldots, n \tag{7.125}$$

with $\boldsymbol{\mu}_j$ given by Equation (7.121), and for the casing (duct 1)

$$\mathbf{P}_1(\xi_{1+}) = \mathbf{P}_1(\xi_{1-}) + \frac{1}{2}\mathbf{B}_1\sum_{k=2}^{n}\boldsymbol{\mu}_k, \tag{7.126}$$

where $\boldsymbol{\mu}_k$ is given by Equation (7.122) and n denotes the total number of ducts, including the casing. Upon substituting Equations (7.121) and (7.122), these equations may be collated as

$$\begin{bmatrix} \mathbf{P}_1(\xi_{1+}) \\ \mathbf{P}_2(\xi_{1+}) \\ \mathbf{P}_3(\xi_{1+}) \\ \vdots \\ \mathbf{P}_n(\xi_{1+}) \end{bmatrix} = \begin{bmatrix} \mathbf{I}-\frac{1}{2}\mathbf{B}_1\sum_{j=2}^{n} g_j\mathbf{G}_{1,1}^{(j)} & \frac{1}{2}\mathbf{B}_1 g_2\mathbf{G}_{1,2}^{(2)} & \frac{1}{2}\mathbf{B}_1 g_3\mathbf{G}_{1,3}^{(3)} & \cdots & \frac{1}{2}\mathbf{B}_1 g_4\mathbf{G}_{1,4}^{(4)} \\ \frac{1}{2}\mathbf{B}_2 g_2\mathbf{G}_{2,1}^{(2)} & \mathbf{I}-\frac{1}{2}\mathbf{B}_2 g_2\mathbf{G}_{2,2}^{(2)} & \mathbf{0} & \cdots & \mathbf{0} \\ \frac{1}{2}\mathbf{B}_3 g_3\mathbf{G}_{3,1}^{(3)} & \mathbf{0} & \mathbf{I}-\frac{1}{2}\mathbf{B}_3 g_3\mathbf{G}_{3,3}^{(3)} & \cdots & \mathbf{0} \\ \vdots & \vdots & \vdots & \ddots & \vdots \\ \frac{1}{2}\mathbf{B}_n g_n\mathbf{G}_{n,1}^{(n)} & \mathbf{0} & \mathbf{0} & \cdots & \mathbf{I}-\frac{1}{2}\mathbf{B}_n g_n\mathbf{G}_{n,n}^{(n)} \end{bmatrix}$$

$$\times \begin{bmatrix} \mathbf{P}_1(\xi_{1-}) \\ \mathbf{P}_2(\xi_{1-}) \\ \mathbf{P}_3(\xi_{1-}) \\ \vdots \\ \mathbf{P}_n(\xi_{1-}) \end{bmatrix} \tag{7.127}$$

where $\mathbf{I}$ denotes a unit matrix of conforming size ($2m \times 2m$, where m denotes the modality). This is the required modal wave transfer equation across the plane $x_1 = \xi_1$ of single-coupled perforates. Although it is derived by assuming that all inner ducts have an aperture row at plane $x_1 = \xi_1$, it is now clear that, if an inner duct does not have an aperture row in this plane, the corresponding form of Equation (7.127) is obtained simply by making its g_j or, what is the same, the aperture cross-sectional area equal to zero.

In the case of double-coupled perforates (Figure 7.10b), ducts are numbered consequently as duct 1, duct 2, ..., in the order of outermost to the innermost duct. Considered here is a pack with $n - 1$ perforated ducts in a solid casing. Again, it suffices to derive the wave transfer matrix at a generic transverse plane, $x_1 = \xi_1$, where all inner ducts are assumed to have a row of apertures. Equation (7.114) is now written for the innermost duct

$$\mathbf{P}_n(\xi_{1+}) = \mathbf{P}_n(\xi_{1-}) + \frac{1}{2}\mathbf{B}_n\boldsymbol{\mu}_n, \tag{7.128}$$

where $\boldsymbol{\mu}_n$ is given by Equation (7.121) with $j = n$ and $k = n - 1$; for the casing

$$\mathbf{P}_1(\xi_{1+}) = \mathbf{P}_1(\xi_{1-}) + \frac{1}{2}\mathbf{B}_1\boldsymbol{\mu}_1, \tag{7.129}$$

where $\boldsymbol{\mu}_1$ is given by Equation (7.122) with $j = 2$ and $k = 1$, and for the centric ducts between the casing and the innermost duct

$$\mathbf{P}_j(\xi_{1+}) = \mathbf{P}_j(\xi_{1-}) + \frac{1}{2}\mathbf{B}_j\boldsymbol{\mu}_j + \frac{1}{2}\mathbf{B}_k\boldsymbol{\mu}_k, \qquad j = 2, 3, \ldots, n-1, \quad k = j+1 \quad (7.130)$$

where $\boldsymbol{\mu}_j$ and $\boldsymbol{\mu}_k$ are given by Equations (7.121) and (7.122), respectively. Combining Equations (7.128)–(7.130) in matrix form after appropriate substitutions of Equations (7.121) and (7.122), the wave transfer across the generic row of apertures at $x_1 = \xi_1$ is obtained as

$$\begin{bmatrix} \mathbf{P}_1(\xi_{1+}) \\ \mathbf{P}_2(\xi_{1+}) \\ \mathbf{P}_3(\xi_{1+}) \\ \vdots \\ \mathbf{P}_n(\xi_{1+}) \end{bmatrix} = \begin{bmatrix} \mathbf{I} - \frac{1}{2}\mathbf{B}_1 g_2 \mathbf{G}^{(2)}_{1,1} & \frac{1}{2}\mathbf{B}_1 g_2 \mathbf{G}^{(2)}_{1,2} & \mathbf{0} & \cdots & \mathbf{0} \\ \frac{1}{2}\mathbf{B}_2 g_2 \mathbf{G}^{(2)}_{2,1} & \mathbf{I} - \frac{1}{2}\mathbf{B}_2 \sum_{i=2,3} g_i \mathbf{G}^{(i)}_{2,2} & \frac{1}{2}\mathbf{B}_2 g_3 \mathbf{G}^{(3)}_{2,3} & \cdots & \mathbf{0} \\ \mathbf{0} & \frac{1}{2}\mathbf{B}_3 g_3 \mathbf{G}^{(3)}_{3,2} & \mathbf{I} - \frac{1}{2}\mathbf{B}_3 \sum_{i=3,4} g_j \mathbf{G}^{(j)}_{3,3} & \cdots & \mathbf{0} \\ \vdots & \vdots & \vdots & \ddots & \vdots \\ \mathbf{0} & \mathbf{0} & \mathbf{0} & \cdots & \mathbf{I} - \frac{1}{2}\mathbf{B}_n g_n \mathbf{G}^{(n)}_{n,n} \end{bmatrix}$$

$$\times \begin{bmatrix} \mathbf{P}_1(\xi_{1-}) \\ \mathbf{P}_2(\xi_{1-}) \\ \mathbf{P}_3(\xi_{1-}) \\ \vdots \\ \mathbf{P}_n(\xi_{1-}) \end{bmatrix} \quad (7.131)$$

Again, if an inner duct does not have an aperture row in this plane, we may simply set the corresponding g_j equal to zero.

7.7.3 Dissipative Silencers

The modal wave transfer matrices, Equations (7.127) and (7.131) are valid when one or more ducts are packed with sound absorbent material, provided that the bulk acoustic properties of the packing material are used for these ducts in place of the ambient fluid properties. (See Appendix B for an introduction to the bulk acoustic properties of porous materials.) With appropriate application of sound absorbent material to the internal ducts or the casing, the single- and double-coupled perforated discrete n-duct elements can be used for the modeling the various types dissipative silencers used in the industry. In these silencers, the perforates usually serve as protective sheets over sound absorption materials or structures, but their presence must be taken into account in acoustic analyses, as they can have a substantial effect on the intended function of the silencer. This can become a challenging issue for the dissipative silencers used in ventilation systems and gas turbines, not only because acoustic performance is usually required up to an octave band center frequency of 8

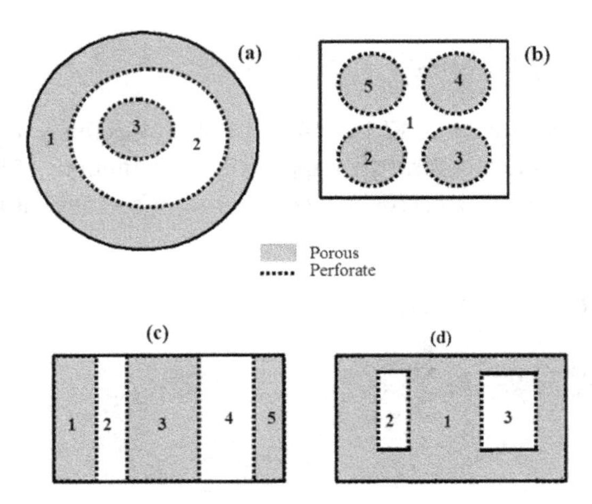

Figure 7.11 Dissipative silencers (front views). (a) Pod silencer. (b) Bar silencer. (c) Splitter silencer with non-connected packing (or parallel baffle silencer). (d) Splitter silencer with simply connected packing.

kHz, but also because these silencers tend to be very large, with section size over 1 m being common. Such sizes and frequencies demand three-dimensional models and the finite element method, the boundary element method and semi-analytical implementation of the finite element method for rectangular or circular ducts have been shown to have potential uses in such studies [29, 30, 36]. In these methods the perforate impedance is modeled as continuous distribution over the treated surfaces and, consequently, solution of an eigenvalue problem is required for the determination of the axial wavenumbers. Modal single- and double-coupled perforated discrete n-duct models offer an alternative approach which does not require solution of an eigenvalue problem.

Shown in Figure 7.11a is a well-known circular dissipative silencer which is usually called a pod silencer (the figures can be imagined as extruded out of the paper for three-dimensional views of the silencers). It consists of a number of circular ducts, which are all perforated except for the outermost one, placed one inside the other, and the annuli thus formed are alternately filled with bulk reacting porous material. As such, therefore, a pod silencer is actually a double-coupled n-duct element and the theory described in the previous section is directly applicable. For example, the pod silencer in Figure 7.11a can be modeled by using a double-coupled three-duct element. In applications, the element can be inserted to a main duct of diameter equal to that of duct 1 or duct 2. In either case, packed annuli are closed at both ends and, if facing the air flow, streamlined so as to reduce the static pressure losses. The theory of Section 7.7.2 applies for ducts of arbitrary section and can be used for designing, for example, a rectangular pod silencer, too.

The dissipative silencer in Figure 7.11b is usually called a bar silencer. It is constructed by mounting in the main duct parallel axial perforated ducts of finite length, usually of circular or rectangular section, which are filled with porous bulk reacting material and are closed at both ends. So, a bar silencer is actually a single-coupled n-duct element and the theory described in Section 7.7.2 is directly applicable for arbitrary bar section shapes. The example shown in Figure 7.11b is a single-coupled five-duct element with all inner ducts filled with porous material.

The silencer shown in Figure 7.11c is called a splitter silencer (also called a parallel baffle silencer). It is commonly used in rectangular ventilation ductworks and consists of a number of parallel splitters, each splitter containing bulk reacting porous material separated from the airway by a thin perforated metal sheet, the faces of the splitters again being closed at both ends by metallic fairings which aid streamlined air flow in the channels between the splitter. Closer examination of the theory described in Section 7.7.2 reveals that it is applicable as is to such rectangular splitter silencers, provided that the channels and splitters are numbered consecutively left-to-right as in Figure 7.11c. Thus, for example, the splitter silencer shown in Figure 7.11c may be modeled as a double-coupled five-duct element. The differences with the corresponding circular configuration pertain only to the transverse duct modes, calculation of duct cross-sectional areas and the coordinates of perforate apertures.

Finally, Figure 7.11d shows a dissipative silencer which can also be called a splitter silencer in the sense that it splits the flow in the main duct into several passages surrounded by sound absorbing faces. In this case, however, the porous filling is coupled with the acoustic fields in all flow passages. Thus, this is really a single-coupled n-duct element and the theory of Section 7.7.2 is directly applicable for this version of splitter silencer configuration and duct shapes are not limited to the rectangular geometry shown in Figure 7.11d.

In calculating the acoustic performance of such dissipative silencers, or their analogues, it is necessary to pay attention to several issues. Perhaps the most important is the acoustic model used for the porous material and the impedance of perforates. Appendices B and C contain introductory information about the common models used in predictive studies on acoustic performance of silencers, but it is important to be aware of the fact that the bulk acoustic properties of most production materials can have substantial variability. Streamlined impervious end pieces used for closing the ends of the ducts filled with porous material can also affect the acoustic performance of a dissipative silencer. These pieces are impervious and form non-uniform hard-walled duct segments at the inlets and outlets of air flow channels. A precise modeling of such non-uniform duct segments in three dimensions is not easy. For this reason, they are usually modeled as flat hard surfaces. The effect of these surfaces on the acoustic performance used to be neglected in the early works on the subject, but more recent work shows that it can be important if multiple modes are incident on the silencer [37].

A single- or double-coupled n-duct model of a dissipative silencer is connected to a ductwork by using the modal duct split element described in Section 7.4.3. This is illustrated in Figure 7.12 for a splitter silencer. Blackened node circles represent closed ends and correspond to the ends of the splitter fairings.

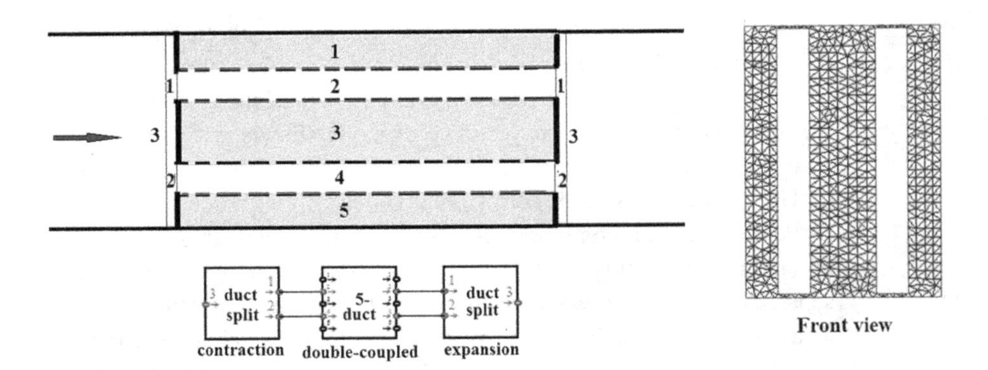

Figure 7.12 A splitter silencer and its block diagram. The front view shows the finite element mesh used for the calculation of the modal characteristics of the inlet and outlet duct splits.

7.7.4 Lined Ducts

The simplest form of dissipative silencer is a two-duct model with porous annulus. This is usually called lined duct. If the porous filling is bulk-reacting, the modal wave transfer across a row of holes of the face sheet can be obtained from the two-duct cases of Equation (7.127), or (7.131), as

$$\begin{bmatrix} \mathbf{P}_1(\xi_{1+}) \\ \mathbf{P}_2(\xi_{1+}) \end{bmatrix} = \begin{bmatrix} \mathbf{I} - \frac{1}{2}\mathbf{B}_1 g_2 \mathbf{G}_{1,1}^{(2)} & \frac{1}{2}\mathbf{B}_1 g_2 \mathbf{G}_{1,2}^{(2)} \\ \frac{1}{2}\mathbf{B}_2 g_2 \mathbf{G}_{2,1}^{(2)} & \mathbf{I} - \frac{1}{2}\mathbf{B}_2 g_2 \mathbf{G}_{2,2}^{(2)} \end{bmatrix} \begin{bmatrix} \mathbf{P}_1(\xi_{1-}) \\ \mathbf{P}_2(\xi_{1-}) \end{bmatrix}, \tag{7.132}$$

where the modal characteristics of duct 1 must be determined by taking into account the presence of the sound absorbent packing.

If the porous filling is locally reacting, its impedance acts in series with the impedance of the apertures. This case is usually treated by modeling the impedance of the inner duct surface as continuously distributed over the internal duct surface (see Section 6.8). It can also be modeled as a discrete distribution of the impedance of the apertures combined in series with the impedance of the porous material backing. Then, the modal wave transfer relation across a row of the perforated facesheet holes may be computed from the one-duct case of Equation (7.131) as

$$\mathbf{P}_1(\xi_{1+}) = \left(\mathbf{I} - \frac{1}{2}\mathbf{B}_1 g_2 \mathbf{G}_{1,1}^{(2)}\right)\mathbf{P}_1(\xi_{1-}), \tag{7.133}$$

7.8 Contracted Models of Silencers

This section presents some applications of the modal elements described in this chapter to some typical chambers and resonators for which three-dimensional solutions have been published in prestigious journals. Since, in the frequency ranges considered in these applications, no higher-order modes propagate in the inlet and outlet ducts, multi-modal assemblies are contracted to modality of unity (modal wave transfer matrix size of 2×2) while still including the higher-order mode effects in system internals.

7.8.1 Expansion Chamber with Offset Inlet and Outlet Ducts

Consider the expansion chamber shown in Figure 7.13a. It has an oval cross section and inlet and outlet ducts are circular. An oval section is defined by four circular arcs, two of radius R_1 and the other two of radius R_2 which are joined tangentially and symmetrically. For a given depth, width and ovality ratio R_2/R_1, two distinct oval sections may exist. A scaled view of the option chosen for a depth of 279.4 mm, width of 114.3 mm and ovality of 8.5 is shown in Figure 7.13a. For this oval chamber section, the calculated cut-on frequency of the smallest transverse higher-order mode is 714.1 Hz. Note that this corresponds to the first non-axisymmetric mode and is

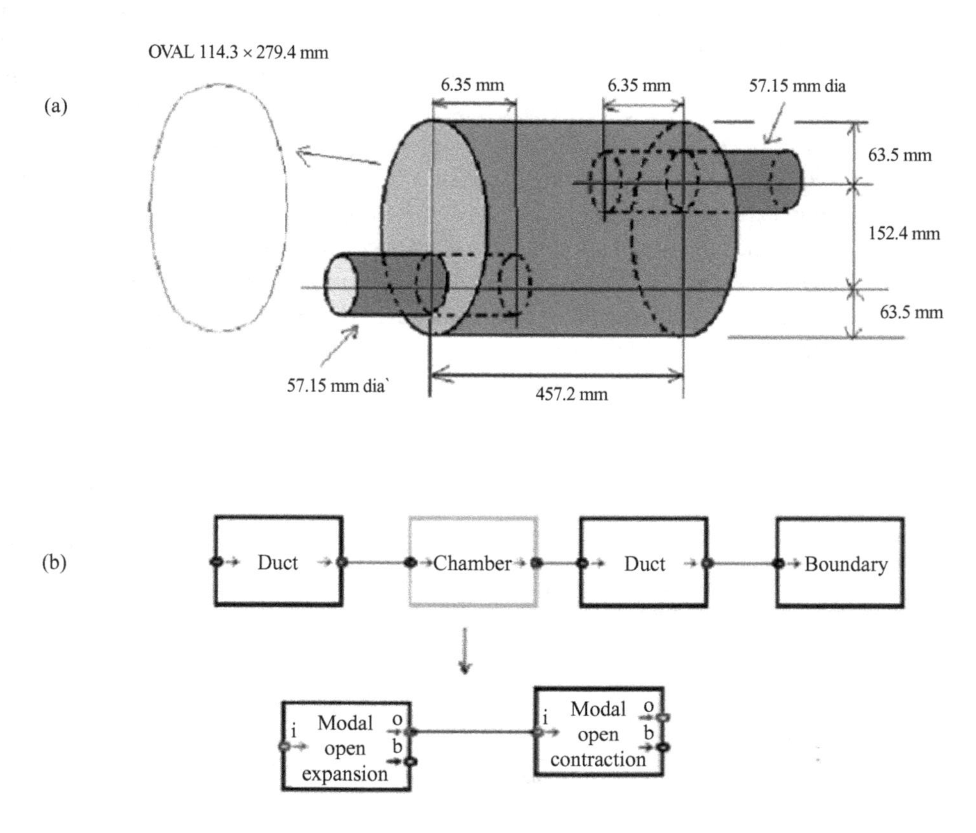

Figure 7.13 (a) An expansion chamber having an oval section and offset inlet and outlet ducts. (b) block diagram model of the expansion chamber. The subassembly "chamber" consists of modal open sudden expansion and sudden contraction elements (with duct extensions) in cascade and is contracted from modality of 4 to modality unity.

appropriate here, since the chamber is not axisymmetrical. So, if we are interested also in higher frequencies than this, it is necessary to construct its acoustic model by using modal elements.

A block diagram model of the chamber is shown in Figure 7.13b. Here the block labeled "chamber" represents a subassembly consisting of two multimodal area-change elements (Section 7.4) connected by a modal duct element (representing the shell), but the latter is not visible in the block diagram in Figure 7.13b, because it is integrated with one of the modal area changes for optimizing the size of the assembled global matrix (see Section 2.4.3). Although the dimensions of the oval shell justifies the use of these modal elements, we observe that the cut-on frequency of the first higher-order mode in the inlet and outlet ducts is about 4000 Hz and if we are interested only in, for instance, the 0–1000 Hz range, they can be modeled accurately by using one-dimensional duct elements. We can take advantage of the situation by contracting the chamber block to modality of unity so that it can be cascaded with one-dimensional duct elements. The resulting acoustic model is not a plane-wave model, because it includes the effects of the higher-order modes in the chamber block.

The solid gray curve in Figure 7.14 shows the transmission loss spectrum of the chamber computed by using Equation (5.2) after the modality of the chamber block is reduced from four to unity. These characteristics are not altered discernibly if the modality of the chamber element is increased, which confirms the convergence of the modal expansion used in this frequency range. The solid black wavy curves are the corresponding measured characteristics (with several different outlet mean flow Mach numbers, including zero mean flow, superimposed).

If the higher-order modes are neglected, the transmission-loss spectrum of the chamber consists of periodic repetition of domes at approximately the closed–closed duct axial resonance frequencies of 376 Hz, 752 Hz and so on (see Section 5.4.3). The first plane-wave transmission-loss dome is observed in both the computed and

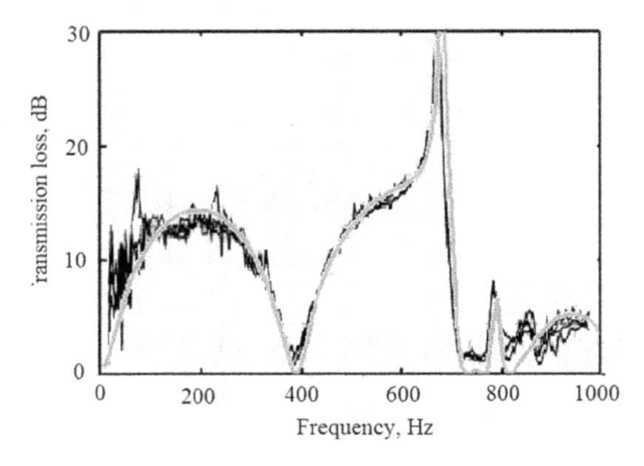

Figure 7.14 Transmission loss of the expansion chamber in Figure 7.13a. (Speed of sound is $c_o = 343.73$ m/s and the mean flow Mach number is 0.1 in the outlet duct.) Solid gray: modal element contracted from modality of 4 to modality of 1; solid dark (wavy): measured (by courtesy of Arvin N.A., Columbus, Indiana, USA).

measured characteristics. This shows that, since the first axial natural frequency of the chamber is quite remote from the cut-on frequency of first higher-order transverse mode, the coupling between these modes is negligible. But the second axial resonance frequency is quite close to the cut-on frequency of the first higher-order mode and, consequently, we observe effects of strong coupling between these modes around 700 Hz. The higher-order mode is entrapped in the chamber, which manifests itself in the transmission-loss spectrum by the appearance of two resonance spikes at about the uncoupled frequencies.

7.8.2 Expansion Chamber with Double Outlet

Shown in Figure 7.15a is a chamber with a single inlet and double outlets. The inlet and outlet pipes are of 50 mm diameter; the chamber is of 150 mm diameter and 225 mm length. The center of the inlet pipe is on the chamber axis; the centers of the two tailpipes are at 12.5 mm radius from the chamber axis and at polar angles of 90° and 330°. Wu et al. computed the transmission loss of this chamber with no mean flow by using ANSYS™ [38], employing 3005 linear elements and 14675 nodes. Denia and Selamet improved the results of the latter transmission loss calculations using SYSNOISE™ with 5389 quadratic finite elements and 8526 nodes [39].

A block diagram model of this muffler is shown in Figure 7.15b. The chamber is modeled by using a subassembly labeled “chamber,” the block diagram of which is shown in Figure 7.15c. Its acoustic model is constructed by using a cylindrical cavity with three duct connections (Section 7.6). This element is represented by the block labeled “cavity” in Figure 7.15c. The modeling strategy consists of terminating one of the outlet ducts of the cavity with appropriate end conditions (anechoic for

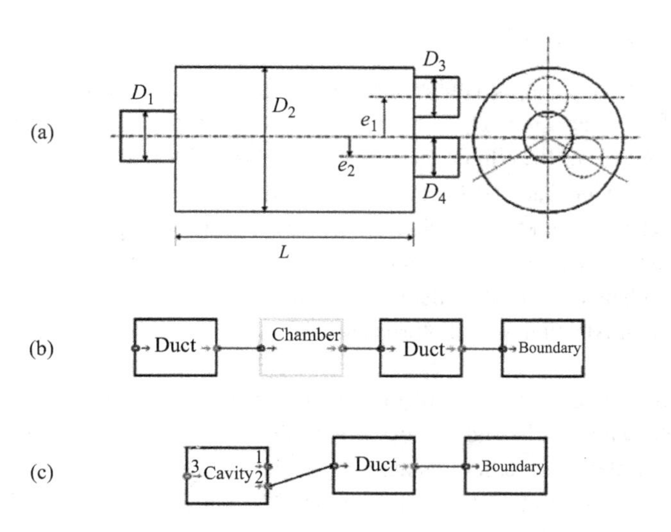

Figure 7.15 Chamber with multiple outlet ducts.

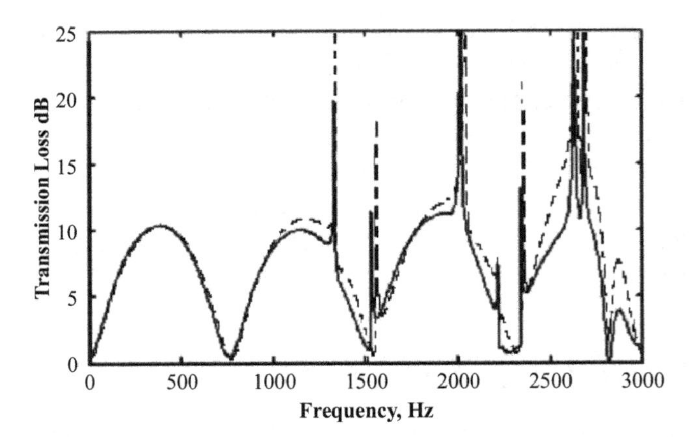

Figure 7.16 Transmission loss of the chamber in Figure 7.15a. Solid: three-port cavity with modality 6 contracted to modality unity; dashed: finite element model, 5389 quadratic elements with 8526 nodes [39].

transmission loss calculation) in the subassembly (Figure 7.15c). The remaining duct forms the outlet port of the subassembly and is terminated in the main system model (Figure 7.15b).

Transmission loss spectrum of the chamber is calculated by using Equation (5.102). The results obtained by using modal elements of modality 6 and assuming a speed of sound of 343.7 m/s are compared in Figure 7.16 with the corresponding finite element solution of Denia and Selamet [39].

7.8.3 Flow-Reversing Chamber

The elliptical flow-reversing chamber shown in Figure 7.17a, has circular inlet and outlet ducts of diameter $D_1 = D_2 = 51$ mm. The chamber length is $L = 400$ mm, the major and minor axes of the elliptic cross section are $2a = 230$ mm and $2b = 130$ mm, respectively, the inlet duct is central ($e_1 = 0$) and the outlet duct is offset by $e_2 = 73$ mm. The inlet and outlet side branch lengths are negligible ($L_1 = L_2 = 0$). The block diagram in Figure 7.13b applies also for this chamber, provided that the expansion area-change element is defined as a flow-reversing one. Alternatively, the chamber may be modeled as an elliptical with two axial duct connections, as indicated in Figure 7.17b.

The transmission loss spectrum of the chamber for $c_o = 344$ m/s and zero mean flow is shown in Figure 7.18 and compared with the corresponding analytical, experimental and finite element method results presented by Denia et al. [40]. Note that the transverse natural frequencies of the chamber are 891, 1505, 1621, 2041, 2334, 2624, 2903, 3039, ... Hz and the closed–closed axial resonance frequencies of the chamber are 215, 645, 1075, 1504, 1934, 2363, 2793, ... Hz. The effects of coupling between these modes manifest themselves on the transmission loss spectrum.

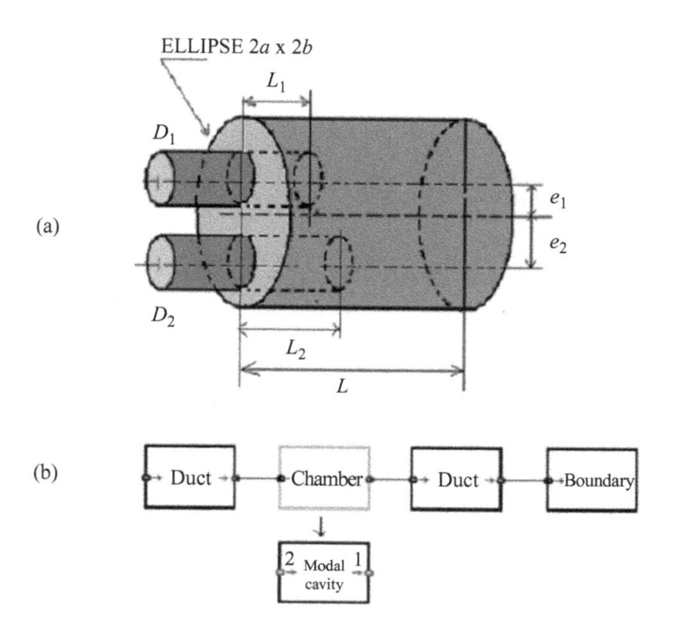

Figure 7.17 A flow-reversing chamber and its block diagram model.

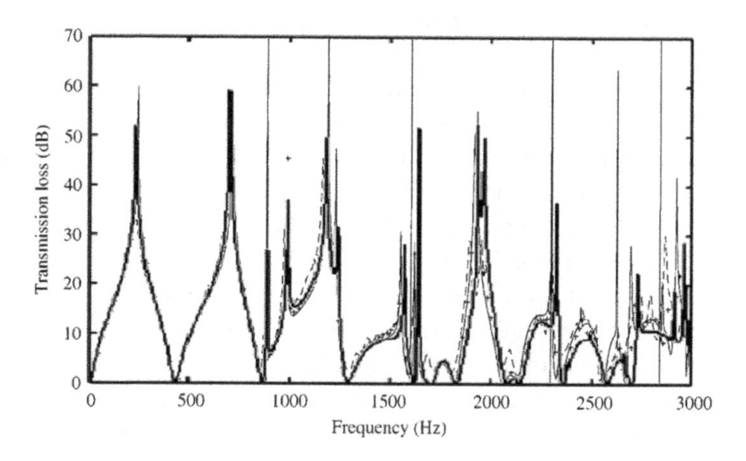

Figure 7.18 Transmission loss of a flow-reversing chamber. Solid thick line: cavity of modality 8 contracted to modality 1; solid thin line: mode-matching [40]; dashed line: experimental [40]; +: finite element method [40].

7.8.4 Through-Flow Resonator and Muffler

In a case of an elliptical through-flow resonator shown in Figure 7.19a, published by Albeda et al. [33], the elliptical major and minor axes are $2a$ = 230 mm and $2b$ = 130 mm and the perforated pipe carries no mean flow. The perforate length is $L = 200$ mm, the internal diameter of the perforated pipe is $D = 50$ mm, the side-branch lengths are L_1 = 80 mm, L_2 = 40 mm and $e = 0$. The apertures are circular of 3 mm diameter and 1 mm thickness and have an open area porosity of 5% over L.

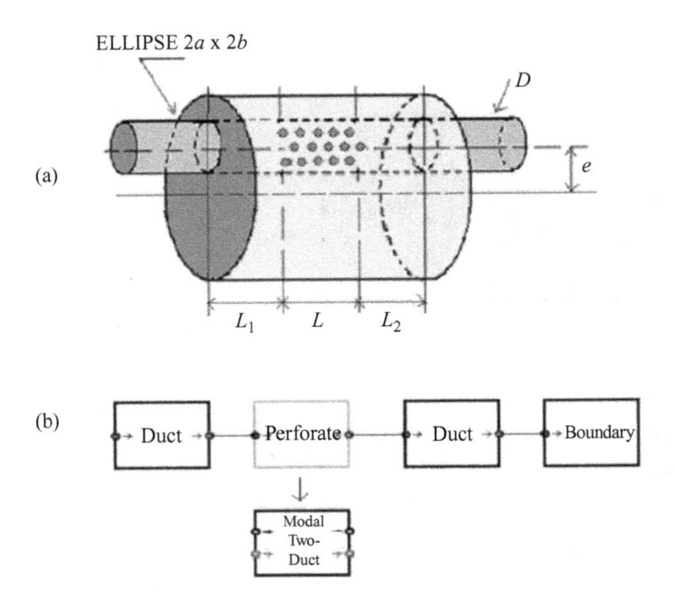

Figure 7.19 A through flow resonator and its block diagram model.

The block diagram model of the resonator is also shown in Figure 7.19b. The block labelled "perforate" models the perforated section of the two-duct unit (Section 7.7), with the side-branches integrated with the unit for reducing the size of the wave transfer matrix. Since this is a discrete perforate model, knowledge of the actual hole pattern is required. In the method described by Albeda et al., this was not needed [33] because the authors used a continuously distributed impedance model of the perforations. For this reason, a rectangular and uniformly distributed hole pattern is assumed in the calculations. In this pattern, the holes are distributed in equally spaced rows, each row containing an equal number of holes of equal circumferential pitch, without any stagger between the rows. For example, using 225 holes in 15 rows over $L = 200$ mm in this way gives an open area porosity of 5%, which is also the value used in Albeda et al. [33].

The transmission-loss spectrum computed using this model with four modes for zero mean flow and the same perforate impedance model as in Albeda et al. is shown in Figure 7.20 and compared with the results given therein [33]. The non-zero transverse natural frequencies of the casing annulus are 774, 1284, 1671, 1985, 2228, 2575, 2902, 3172, ... Hz, and the axial natural frequencies of the side branches are 2148, 6445, ... Hz. It is seen that the first two transverse natural frequencies, for which the eigenfunctions are asymmetrical about the ellipse axes, are not observed. The one-dimensional transmission-loss characteristics may be expected to be valid below first symmetrical transverse natural frequency, that is, 1671 Hz.

The annulus of the through-flow resonator is often packed by sound absorbing material. Figure 7.21 shows the effect of the mean flow in a similar elliptical muffler with $2a = 220$ mm, $2b = 120$ mm, $L_1 = 0$ mm, $L_2 = 0$ mm, $L = 350$ mm, $e = 0$, $D = 74$ mm. The circular perforations in this case are of 3.5 mm diameter and 1 mm thickness, the open area porosity of the perforated length, L, being 26.3%, the annulus

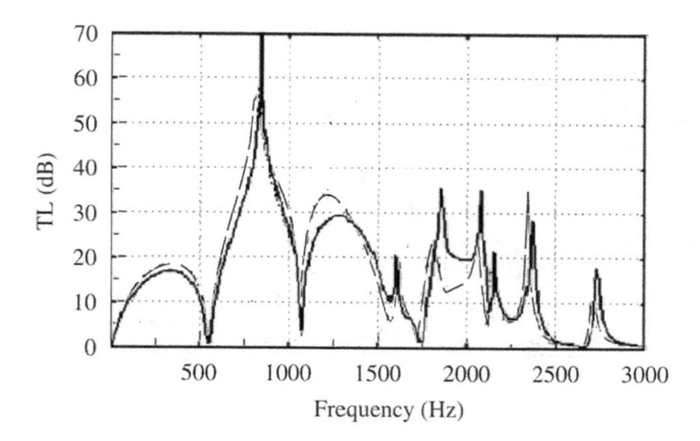

Figure 7.20 Transmission loss of a through-flow resonator. Solid: discrete perforate of modality 8 contracted to modality 1; dashed: mode-matching [33] ($c_o = 343.73$m/s).

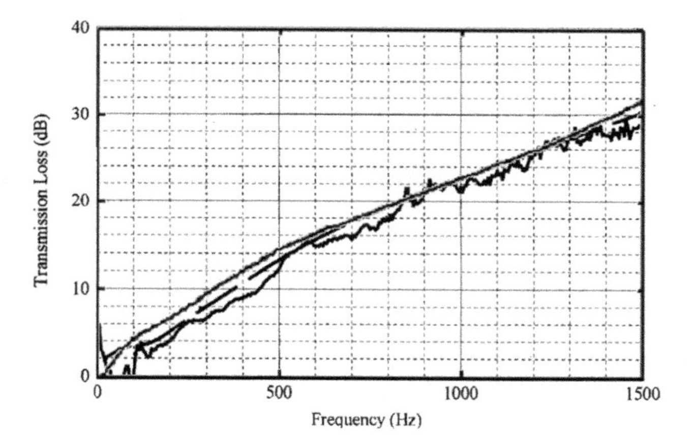

Figure 7.21 Transmission loss of an elliptic dissipative silencer with mean flow. Solid: discrete perforate with modality 6 contracted to modality 1; solid wavy: experimental [30]; dashed: mode-matching [30].

is packed with basalt and the mean flow Mach number is 0.15. This is same as the elliptical silencer considered by Kirby [30], who used a mode-matching method in which the perforate impedance is modeled as continuous distribution. The theoretical and experimental results given by Kirby are also shown Figure 7.21 [30].

7.8.5 Three-Pass Muffler

This muffler consists of three perforated pipes mounted on two 12.7 mm thick solid baffle plates in a solid circular casing (Figure 7.22a). The perforated pipes are all 0.8 mm thick and have an internal diameter of 48.9 mm. The casing diameter is 165.1 mm, and the pipes are arranged at an angular pitch of 120°, around a concentric circle of radius of 39.69 mm. The perforate segments in the center chamber contain 448 circular holes of 2.47 mm diameter,

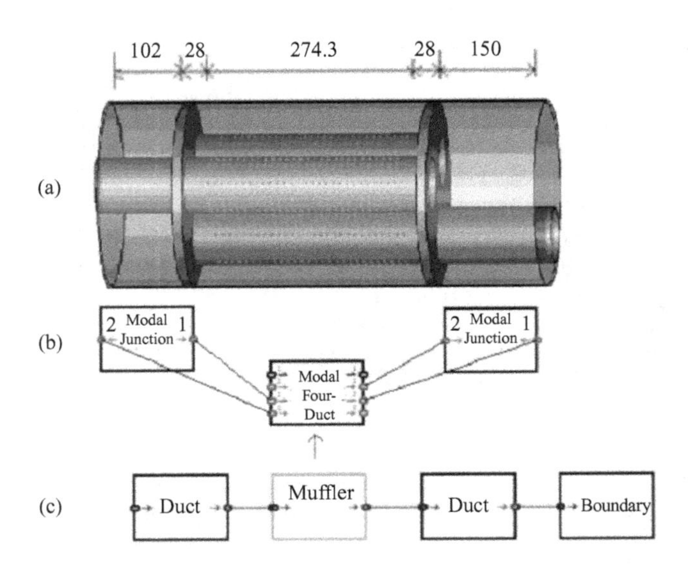

Figure 7.22 A three-pass muffler and its block diagram model

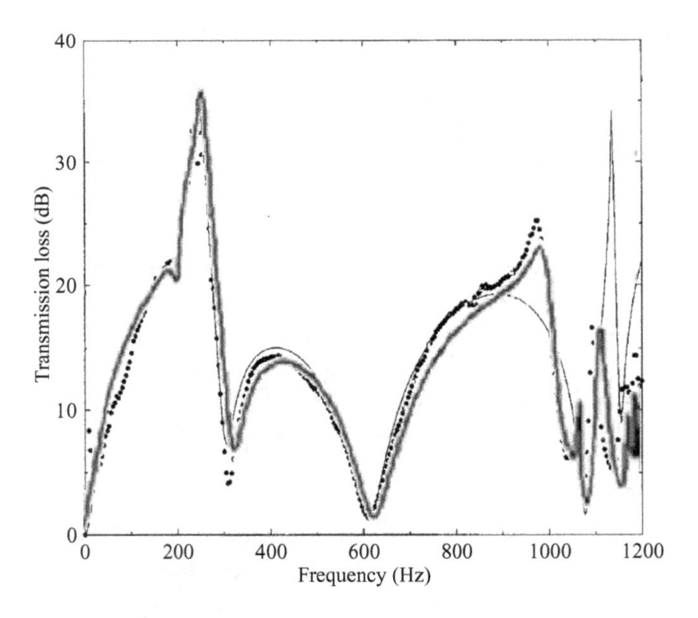

Figure 7.23 Transmission loss of a three-pass muffler. Solid thick: discrete perforate of modality 3 contracted to modality 1; solid thin: one-dimensional [41]; dot: measured [41].

giving an open area porosity of 4.5%. The block diagram of the muffler subassembly is shown in Figure 7.22b. It consists of two junction elements modeling the end chambers and a four-duct perforate unit. The experimentally determined transmission-loss spectrum of this muffler has been presented by Selamet et al. [41], who also calculated the transmission loss by using a one-dimensional model. The transmission-loss spectrum computed by using modal elements of modality 3 (contracted to modality 1) is shown in Figure 7.23 and compared with the results of Selamet et al. [41]. The cut-on frequencies of the first higher order mode in the perforate chamber and the flow-reversing chambers are 920 Hz and 1054 Hz, respectively.

References

[1] G. Gabard and R.J. Astley, A computational mode-matching approach for sound propagation in three-dimensional ducts with flow, *J. Sound. Vib.* **315** (2008), 1103–1124.

[2] S.W. Rienstra, Acoustic scattering at a hard-soft lining transition in a flow duct, *J. Eng. Math.* **59** (2007), 451–475

[3] G.D. Furnell and D.A. Bies, Matrix analysis of acoustic wave propagation within curved duct systems, *J. Sound Vib.* **132** (1989), 245–263.

[4] F.C. Karal, The analogous acoustical impedance for discontinuities and constrictions of circular cross-section, *J. Acoust. Soc. Am.* **78** (1953), 327–334.

[5] H. Hudde and U. Letens, Scattering matrix of a discontinuity with a nonrigid wall in a lossless circular duct, *J. Sound Vib.* **78** (1985), 1826–1837.

[6] J. Kergomard and A. Garcia, Simple discontinuities in acoustic waveguides at low frequencies: critical analysis and formulae, *J. Sound. Vib.* **114** (1987), 465–479.

[7] K.S. Peat, The acoustical impedance at the junction of an extended inlet or outlet duct, *J. Sound. Vib.* **150** (1991), 101–110.

[8] B. Nilsson and O. Brander, The propagation of sound in cylindrical ducts with mean flow and bulk-reacting lining: II. Bifurcated ducts, *J. I. Math. Appl.* **26** (1980), 381–410.

[9] B. Nilsson and O. Brander, The propagation of sound in cylindrical ducts with mean flow and bulk-reacting lining: III. Step discontinuities, *IMA J. Appl. Math.* **27** (1981), 105–131.

[10] R.T. Muehleisen and D.C. Swanson, Modal coupling in acoustic waveguides: planar discontinuities, *Appl. Acoust.* **63** (2002), 1375–1392.

[11] I.D.H. Dupere and A.P. Dowling, The absorption of sound near abrupt axisymmetric area expansion, *J. Sound. Vib.* **239** (2001), 709–730.

[12] A. Cummings, Sound transmission in a folded annular duct, *J. Sound. Vib.* **41** (1975), 375–379

[13] A.E. E-Sharkawy and A.H. Nayfeh, Effect of expansion chamber on the propagation of sound in circular pipes, *J. Acoust. Soc. Am.* **63** (1978), 667–674.

[14] L.J. Eriksson, Higher order mode effects in circular ducts and expansion chambers, *J. Acoust. Soc. Am.* **68** (1980), 545–550.

[15] B. Nilsson and O. Brander, The propagation of sound in cylindrical ducts with mean flow and bulk-reacting lining: IV. Several interacting discontinuities, *IMA J. Appl. Math.* **27** (1981), 263–289.

[16] J.G. Ih and B.H. Lee, Analysis of higher order mode effects in the circular expansion with mean flow, *J. Acoust. Soc. Am.* **77** (1985), 1377–1388.

[17] S.I. Yi and B.H. Lee, Three-dimensional acoustic analysis of circular expansion chambers with side inlet and side outlet, *J. Acoust. Soc. Am.* **79** (1986), 1299–1306.

[18] S.I. Yi and B.H. Lee, Three-dimensional acoustic analysis of circular expansion chambers with side inlet and end outlet, *J. Acoust. Soc. Am.* **81** (1987), 1279–1287.

[19] J.G. Ih and B.H. Lee, Theoretical prediction of the transmission loss of circular reversing chamber mufflers, *J. Sound. Vib.* **112** (1987), 261–272.

[20] J. Kergomard, A. Garcia, G. Tagui and J.P. Dalmont, Analysis of higher order mode effects in an expansion chamber using modal theory and equivalent electrical circuits, *J. Sound. Vib.* **129** (1989), 457–475.

[21] M. Abom, Derivation of four-pole parameters including higher-order mode effects for expansion chamber mufflers with extended inlet and outlet, *J. Sound. Vib.* **137** (1990), 403–418.

[22] F.D. Denia, J. Albelda and F.J. Fuenmayor, Acoustic behaviour of elliptical chamber mufflers, *J. Sound. Vib.* **241** (2001), 401–421.

[23] S. Boij and B. Nilsson, Scattering and absorption of sound at flow duct expansions, *J. Sound. Vib.* **289** (2006), 577–594.

[24] A. Selamet and Z.L. Li, Circular asymmetric Helmholtz resonators, *J. Acoust. Soc. Am.* **107** (2000), 2360–2369.

[25] B. Nilsson and O. Brander, The propagation of sound in cylindrical ducts with mean flow and bulk-reacting lining: I. Modes in an infinite duct, *J. I. Math. Appl.* **26** (1980), 269–298

[26] S.-H. Ko, Theoretical analysis of sound attenuation in acoustically lined flow ducts separated by porous splitters (rectangular, annular and circular ducts), *J. Sound. Vib.* **39** (1975), 471–487.

[27] R.J. Astley, A. Cummings and N. Sormaz, A finite element scheme for acoustic propagation in flexible walled ducts with bulk reacting liners and comparison with experiment, *J. Sound. Vib.* **150** (1991), 119–138.

[28] K.S. Peat and K.L. Rathi, A finite element analysis of the convected acoustic wave motion in dissipative silencers, *J. Sound. Vib.* **184** (1995), 529–545.

[29] R. Glav, The point–point matching method on dissipative silencers of arbitrary cross-section, *J. Sound. Vib.* **189** (1996), 123–135.

[30] R. Kirby, Transmission loss predictions for dissipative silencers of arbitrary cross section in the presence of mean flow, *J. Acoust. Soc. Am.* **114** (2003), 200–209.

[31] A. Selamet, M.B. Xu and I.H. -Lee, Analytical approach for sound attenuation in perforated dissipative silencers with inlet/outlet extensions *J. Acoust. Soc. Am.* **117** (2005), 2078–2089.

[32] J.B. Lawrie and R. Kirby, Mode-matching without root finding: application to a dissipative silencer, *J. Acoust. Soc. Am.* **119** (2006), 2050–2061.

[33] J. Albeda, F.D. Denia, M.I. Torres and F.J. Fuenmayor, A transversal sub-structuring mode matching method applied to the acoustic analysis of dissipative mufflers, *J. Sound. Vib.* **303** (2007), 614–631.

[34] N. Sohei, N. Tsuyoshi and Y. Takashi, Acoustic analysis of elliptical muffler chamber having a perforated pipe, *J. Sound. Vib.* **297** (2006), 761–773.

[35] R.P. Kanwal, *Generalized Functions: Theory and Technique*, (Boston: Birkhäuser, 1998).

[36] R. Kirby and J.B. Lawrie, A point collocation approach to modeling large dissipative silencers, *J. Sound. Vib.* **286** (2005), 313–339.

[37] R. Kirby, The influence of baffle fairings on acoustic performance of rectangular splitter silencers, *J. Acoust. Soc. Am.* **118** (2005), 2302–2312.

[38] C.J. Wu, X.J. Wang and H.B. Tang, Transmission loss prediction on a single-inlet/double-outlet cylindrical expansion chamber muffler using the modal meshing method, *Appl. Acoust.* **69** (2008), 173–178.

[39] F.D. Denia and A. Selamet, "Transmission loss prediction on a single-inlet/double-outlet cylindrical expansion chamber muffler using the modal meshing method" by C.J. Wu, X.J. Wang and H.B. Tang (*Applied Acoustics 69* (2008), 173–178), *Appl. Acoust.* **69** (2008) 280–281.

[40] F.D. Denia, J. Albeda, F.J. Fuenmayor and A.J. Torregrosa, Acoustic behaviour of elliptical chamber mufflers, *J. Sound. Vib.* **243** (2001), 401–421.

[41] A. Selamet, W. Easwaran and A.G. Falkowski, Three pass muffler with uniform perforations, *J. Acoust. Soc. Am.* **105** (1999), 1548–1562.

8 Effects of Viscosity and Thermal Conductivity

8.1 Introduction

So far, we have neglected the effects of the viscosity and thermal conductivity of the fluid on the propagation of sound waves. The neglect of shear viscosity[1] is justified because the major effects of acoustic shear strains are confined to acoustic viscous boundary layers. They are about two orders of magnitude smaller in the bulk of the fluid [1], and, therefore, may become discernible only after propagation over very long distances. The thickness of the viscous boundary layer is given by $\delta_v = \sqrt{2\mu_o/\omega\rho_o}$ [1], where μ_o denotes the shear viscosity of the fluid. For frequencies of interest in practical applications, δ_v remains less than a fraction of one millimeter and is very small compared to the acoustic wavelength $\lambda = c_o/f$, where $f = \omega/2\pi$ is the frequency in hertz. For example, for dry air at 100 kPa, $\delta_v = 0.22/\sqrt{f}$ mm at 15 °C and $\delta_v = 0.50/\sqrt{f}$ mm at 500 °C. A useful metric for the assessment of the effect of the fluid viscosity on sound wave propagation is the shear wavenumber, or the Stokes number, which we denote by s (not to be confused with specific entropy). This is defined as

$$s = \frac{a\sqrt{2}}{\delta_v}, \tag{8.1}$$

where a denotes a characteristic dimension of the duct cross section, for example, the radius, if the duct is circular. In general, the effect of the fluid viscosity on sound wave propagation can be neglected when $s \gg 40$ or so.

Historically, the effect of shear viscosity on sound propagation was considered by Helmholtz, but his results did not explain accurately enough the experimental findings of Kundt about the frequency dependence of propagation constants in narrow circular tubes containing a quiescent perfect gas. (References to the work of Helmholtz and Kundt are given by Rayleigh [2].) Kirchhoff derived, neglecting mean flow but taking into account also the effect of the thermal conductivity of the fluid, a dispersion equation for axisymmetric mode propagation and presented an approximate solution for the fundamental mode, which is now known as the wide pipe solution [3]. This did

[1] The effect of bulk viscosity is usually neglected at audio frequencies because, in view of the relaxation times of most fluids, it becomes discernible at about ultrasonic frequencies.

not agree with Kundt's results in all details, but showed that the effect of heat conduction on the attenuation of sound in a perfect gas is of the same order as that of viscosity. The theory of Kirchhoff is described in detail by Rayleigh [2], who also proposed a solution for the fundamental mode for another extreme case of the problem, that is, very narrow pipes in which the acoustic boundary layer extends over the whole cross section. Since these pioneering works, numerous approximate forms of Kirchhoff's theory have been proposed for evaluation of the fundamental propagation constants for specific frequency ranges and duct sizes. A review of the various approximations has been presented by Tijdeman [4].

As shown by Tijdeman for circular ducts and by Stinson for ducts of arbitrary cross section [5], in the absence of mean flow, the determination of the fundamental mode solution of Kirchhoff's theory simplifies considerably if

$$k_o a << \min\{1, s\}, \tag{8.2}$$

where $k_o = \omega/c_o$. This range covers a large proportion of duct sizes and frequencies of interest in practical systems. The viscothermal duct models derived in this chapter are based on approximations justified by the foregoing condition. Following Tijdeman, this is referred to as the low reduced frequency approximation, but it is also referred to sometimes as the Zwikker–Kosten approximation.

8.2 Convected Wave Equation for a Viscothermal Fluid

Kirchhoff's theory is based on the linearized continuity, momentum and energy equations for unsteady flow of a perfect gas with non-negligible viscosity and thermal conductivity in straight uniform hard-walled circular ducts with no mean flow. Here, we present the theory in a generality encompassing arbitrary duct shapes with finite impedance walls containing a Newtonian fluid with uniform mean flow of velocity $\mathbf{v}_o = \mathbf{e}_1 v_o$, where v_o may be regarded as the average value of the actual mean flow velocity over a duct cross section. The linearized continuity equation is still given by Equation (6.1), which is repeated below for convenience:

$$\frac{\partial \rho'}{\partial t} + \mathbf{v}_o \cdot \nabla \rho' + \rho_o \nabla \cdot \mathbf{v}' = 0. \tag{8.3}$$

The momentum equation for unsteady flow is given by Equation (A.34) in Appendix A4.1. This is linearized by neglecting the fluctuations of the shear viscosity ($\mu \approx \mu_o$) and the bulk viscosity ($\eta \approx \eta_o$) of the fluid. The linearized momentum equation for acoustic fluctuations is then

$$\rho_o \left(\left(\frac{\partial}{\partial t} + \mathbf{v}_o \cdot \nabla \right) \mathbf{v}' + \mathbf{v}' \cdot \nabla \mathbf{v}_o \right) + \nabla p' = \mu_o \nabla^2 \mathbf{v}' + \left(\frac{\mu_o}{3} + \eta_o \right) \nabla \nabla \cdot \mathbf{v}', \tag{8.4}$$

where, as usual, the prime denotes a fluctuating quantity and the subscript "o" denotes a mean value. Several forms of the energy equation for unsteady fluid flow are given in Appendix A.4. These equations are linearized neglecting the fluctuations of the thermal conductivity ($\kappa \approx \kappa_o$). Here, it is convenient to use the linearized form of Equation (A.38), namely,

$$\rho_o T_o \left(\frac{\partial}{\partial t} + \mathbf{v}_o \cdot \nabla \right) s' = \kappa_o \nabla^2 T', \tag{8.5}$$

where s' denotes the specific entropy fluctuations. Equations (8.3)–(8.5) can be manipulated to derive the convected wave equation [6]:

$$\left(\frac{1}{c_o^2} \frac{\mathrm{D}^3}{\mathrm{D}t^3} - \frac{1}{\rho_o c_o^2} \left(\frac{4\mu_o}{3} + \eta_o + \frac{\gamma_o \kappa_o}{c_{po}} \right) \frac{\mathrm{D}^2}{\mathrm{D}t^2} \nabla^2 - \frac{\mathrm{D}}{\mathrm{D}t} \nabla^2 \right.$$
$$\left. + \frac{1}{\rho_o^2 c_o^2} \left(\frac{4\mu_o}{3} + \eta_o \right) \frac{\gamma_o \kappa_o}{c_{po}} \frac{\mathrm{D}}{\mathrm{D}t} \nabla^4 + \frac{\kappa_o}{\rho_o \, c_{po}} \nabla^4 \right) T' = 0 \tag{8.6}$$

where $\mathrm{D}/\mathrm{D}t = \partial/\partial t + \mathbf{v}_o \cdot \nabla$. For zero mean flow, this wave equation reduces to that derived by Bruneau et al. [7].

Solution to Equation (8.6) in the frequency domain is subject to the following boundary conditions on the duct boundary Γ (apart from the condition that it is finite over the duct cross section):

$$\left. \begin{aligned} \mathbf{v}_t' &= 0 \\ \mathbf{v}_n' &= \left(-\mathrm{i}\omega + v_o \frac{\partial}{\partial x} \right) \frac{p'}{-\mathrm{i}\omega Z} \\ T' &= 0 \end{aligned} \right\} \text{ on } \Gamma, \tag{8.7}$$

where $\mathbf{v}_t'$ and $\mathbf{v}_n'$ denote, respectively, the component of the particle velocity tangential to and normal to Γ. The first condition in Equation (8.7) represents the no-slip condition for the acoustic flow and the second condition accounts for the effect of wall impedance Z according to the Ingard–Myers theory (see Section 6.3.3). The third expresses the condition that the temperature fluctuations vanish on Γ.[2] The latter condition is based on the view that small temperature fluctuations in the fluid cannot cause discernible changes in the temperature of the wall because the corresponding heat transfer occurs to and from the wall periodically and, consequently, the boundary temperature remains practically the same as the mean temperature of the fluid [1].

An exact solution of the boundary value problem posed by Equations (8.6) and (8.7) in the frequency domain is given in Reference [6] for circular ducts with soft walls. In many practical applications, it suffices to consider only the fundamental axisymmetric mode of propagation and in most cases low reduced frequency approximation of the fundamental mode gives adequately accurate results for engineering

[2] This condition, which is sometimes referred to as the conducting wall condition, is used in most of the previous studies. It is also adopted in this chapter. If the wall is thermally insulated so that no heat exchange takes place between the gas and the wall, it should be replaced by the condition $\mathbf{n} \cdot \nabla T' = 0$, where $\mathbf{n}$ denotes the outward unit normal vector of the duct surface.

purposes. The subsequent sections of this chapter are devoted to the description and applications of the low reduced frequency theory. The efficacy of this theory with mean flow is confirmed by the modal solutions of Equation (8.6) [6].

8.3 Low Reduced Frequency Theory

It is convenient to write Equation (8.3) as

$$\frac{\partial \rho'}{\partial t} + v_o \frac{\partial \rho'}{\partial x} + \rho_o \left(\frac{\partial v'}{\partial x} + \nabla_\perp \cdot \mathbf{v}'_\perp \right) = 0, \tag{8.8}$$

where v' denotes the axial component of the particle velocity, $\mathbf{v}'_\perp$ denotes the transverse component of the acoustic particle velocity (i.e., $\mathbf{v}' = \mathbf{e}v' + \mathbf{v}'_\perp$, where $\mathbf{e} \perp \mathbf{v}'_\perp$ denotes a unit vector in the positive direction of the duct axis) and $\nabla_\perp$ denotes the gradient operator on the duct cross section. The density fluctuations can be expressed as function of acoustic pressure and acoustic temperature as

$$\rho' = \frac{\gamma_o}{c_o^2} p' - \rho_o \beta_o T'. \tag{8.9}$$

This relationship follows from the linearization of the state equation $\rho = \rho(p, T)$, which can be obtained by eliminating the specific entropy between Equations (A.15) and (A.16), and then using Equation (A.17) for the speed of sound; that is, $\beta_o^2 T_o c_o^2 = (\gamma_o - 1) c_{po}$.

Treatment of the linearized momentum equation, Equation (8.4), involves some judicious steps. Firstly, we note that the axial component of Equation (8.4) is

$$\rho_o \frac{\partial v'}{\partial t} + \rho_o v_o \frac{\partial v'}{\partial x} = -\frac{\partial p'}{\partial x} + \mu_o \left(\nabla^2 v' + \left(\frac{1}{2} + \frac{\eta_o}{\mu_o} \right) \frac{\partial^2 v'}{\partial x^2} \right). \tag{8.10}$$

Implicit in the condition in Equation (8.2) is the assumption that the length scales in axial and transverse directions are dissimilar enough so as the variations of the same order result in similarly dissimilar gradients in the corresponding directions. Then the axial gradients can be neglected as small compared to transverse gradients. Accordingly, dropping the second order axial gradients on its right-hand side, Equation (8.10) simplifies to

$$\rho_o \frac{\partial v'}{\partial t} + \rho_o \, v_o \frac{\partial v'}{\partial x} = -\frac{\partial p'}{\partial x} + \mu_o \nabla_\perp^2 v', \tag{8.11}$$

where $\nabla_\perp^2 = \nabla_\perp \cdot \nabla_\perp$ is the Laplacian on the duct cross section. The components of the momentum equations in the transverse directions are then simplified by assuming further that the axial component of the particle velocity is much larger than its transverse components. This implies that the terms in the transverse components of the momentum equation are much smaller in comparison to those in Equation (8.11) and, since this includes the acoustic pressure gradient in transverse directions, the transverse components of the momentum equation become approximately equivalent to the statement that

$$p' = p'(t, x). \tag{8.12}$$

Thus, to this approximation, the acoustic pressure field in the duct is planar.

For uniform mean flow, Equation (8.5) becomes

$$\rho_o T_o \left(\frac{\partial s'}{\partial t} + v_o \frac{\partial s'}{\partial x} \right) = \kappa_o \nabla_\perp^2 T'. \tag{8.13}$$

Here, the specific entropy fluctuations are related to the temperature and pressure fluctuations as

$$s' = \frac{c_{po}}{T_o} T' - \frac{\beta_o}{\rho_o} p', \tag{8.14}$$

which follows from the linearization of Equation (A.16).

The foregoing equations can be manipulated so that the determination of acoustic state variables p', v', T' is closed under the boundary conditions, Equation (8.7). Analytical solutions of this boundary value problem are given in the following sections for circular and rectangular ducts. It can be solved only numerically for arbitrary section shapes. A finite element formulation based on the Galerkin weighted residual is described by Astley and Cummings [8].

8.3.1 Circular Hollow Ducts

For circular ducts, the governing acoustic equations are written in symmetrical cylindrical co-ordinates (x, r) using the expansions

$$\nabla_\perp^2 = \frac{1}{r} \frac{\partial}{\partial r} \left(r \frac{\partial}{\partial r} \right) \tag{8.15}$$

$$\nabla \cdot \mathbf{v}' = \frac{\partial v'}{\partial x} + \frac{1}{r} \frac{\partial}{\partial r} (v'_r r) \tag{8.16}$$

$$\nabla_\perp \cdot \mathbf{v}'_\perp = \frac{1}{r} \frac{\partial}{\partial r} (v'_r r), \tag{8.17}$$

where r denotes the radial coordinate and v_r denotes the radial component of the particle velocity. The boundary conditions in Equation (8.7) become, respectively,

$$\left. \begin{aligned} & v' = 0 \\ & v'_r = \left(-\mathrm{i}\omega + v_o \frac{\partial}{\partial x} \right) \frac{p'}{-\mathrm{i}\omega Z} \\ & T' = 0 \end{aligned} \right\} \text{ at } r = a. \tag{8.18}$$

Solutions of the governing acoustic equations described in the previous section are sought in the wavenumber domain by the ansatz

$$p' = Pe^{ik_o Kx} \tag{8.19}$$

$$v' = H(r)p' \tag{8.20}$$

$$T' = F(r)p', \tag{8.21}$$

where P denotes a constant, K denotes the propagation constants (to be determined). The density and entropy fluctuations can then be determined from Equations (8.9) and (8.14) as, respectively,

$$\rho' = \left(\frac{\gamma_o}{c_o^2} - \rho_o \beta_o F\right)p' \tag{8.22}$$

$$s' = \left(\frac{c_{po}}{T_o} F - \frac{\beta_o}{\rho_o}\right)p' \tag{8.23}$$

Upon substitution of these equations, Equations (8.11) and (8.13) become, respectively,

$$\frac{1}{r}\frac{\partial}{\partial r}\left(r\frac{\partial H}{\partial r}\right) + B^2 H = \frac{ik_o K}{\mu_o} \tag{8.24}$$

$$\frac{1}{r}\frac{\partial}{\partial r}\left(r\frac{\partial F}{\partial r}\right) + B^2\sigma^2 F = i(1 - KM_o)\frac{\omega\beta_o T_o}{\kappa_o}, \tag{8.25}$$

where a denotes the radius of the duct, $M_o = v_o/c_o$ denotes the average Mach number of the mean flow velocity, $\sigma^2 = \mu_o c_{po}/\kappa_o$ denotes the Prandtl number and

$$B^2 a^2 = i(1 - KM_o)s^2. \tag{8.26}$$

Equation (8.24) is an inhomogeneous Bessel equation of the first kind and its solution, which is finite at $r = 0$ and satisfies the first condition in Equation (8.18), is

$$H(r) = \frac{K}{\rho_o c_o(1 - KM_o)}\left(1 - \frac{J_o(Br)}{J_o(Ba)}\right), \tag{8.27}$$

where, J_m denotes a Bessel function of order m. Equation (8.25) is also an inhomogeneous Bessel equation and its solution satisfying the boundary condition on the temperature fluctuations is

$$F(r) = \frac{\beta_o T_o}{\rho_o c_{po}}\left(1 - \frac{J_o(\sigma Br)}{J_o(\sigma Ba)}\right). \tag{8.28}$$

Next, elimination of the acoustic density between Equations (8.8) and (8.9) gives, in the frequency domain,

$$\frac{1}{r}\frac{\partial}{\partial r}(v_r' r) = \left\{i\omega\left(\frac{\gamma_o}{\rho_o c_o^2} - \beta_o F\right)(1 - M_o K) - ik_o KH\right\}p'. \tag{8.29}$$

After substitution for H and F from Equations (8.27) and (8.38), integration of the resulting equation with respect r gives

$$v'_r = \frac{ik_o}{\rho_o c_o}\left\{\left(\frac{\gamma_o r}{2} - (\gamma_o - 1)\left(\frac{r}{2} - \frac{1}{\sigma B}\frac{J_1(\sigma Br)}{J_o(\sigma Ba)}\right)\right)(1 - M_o K) - \frac{K^2}{(1 - KM_o)}\left(\frac{r}{2} - \frac{1}{B}\frac{J_o(Br)}{J_o(Ba)}\right)\right\}p'. \tag{8.30}$$

Note that the integration constant vanishes, since the radial component of the particle velocity must vanish at the center of the duct cross section. Thus, using the foregoing solution for v'_r in the second boundary condition in Equation (8.18), the following dispersion equation for K may be deduced:

$$\gamma_o + (\gamma_o - 1)\frac{J_2(\sigma Ba)}{J_o(\sigma Ba)} + \left(\frac{K}{1 - KM_o}\right)^2 \frac{J_2(Ba)}{J_o(Ba)} = \frac{2}{ik_o a\zeta}, \tag{8.31}$$

where $\zeta = Z/\rho_o c_o$ denotes the normalized wall impedance, which is assumed to be uniform for simplicity, and $\gamma_o = c_{po}/c_{vo}$ denotes the ratio of the specific heat coefficients. The propagation constants K satisfying this dispersion equation can be obtained numerically. Evolution of these, as frequency and liner characteristics are varied, may be studied in a polar plane similar to that described in Section 6.7.1.2.

8.3.1.1 Hard-Walled Ducts

In this case ($\zeta = \infty$), roots of Equation (8.31) for K correspond to acoustic modes [9], and hydrodynamic modes [10]. The hydrodynamic modes are given approximately by the roots of $J_o(Ba) \approx 0$ and $J_o(\sigma Ba) \approx 0$, which correspond, respectively, to vorticity and entropy modes.[3] These modes are convected with the mean flow and, as such, are not relevant for the acoustic field in the duct.

The propagation constants corresponding to the acoustic modes are denoted by K^+ and K^-. The real and imaginary parts of K^+ and K^- are shown in Figure 8.1 as the function of the shear wavenumber and the mean flow Mach number. The imaginary part of K^+ is always positive, whilst the imaginary part of K^- is always negative. Application of the Briggs criterion shows that K^+ and K^- correspond to decaying waves propagating in the $+x$ and $-x$ directions, respectively.

Hence, the acoustic pressure field in the duct consists of superposition of pressure waves propagating in opposite directions

$$p' = p^+(x) + p^-(x) \tag{8.32}$$

and, in view of Equation (8.19), the wave transfer equation of the duct may be expressed as

$$\begin{bmatrix} p^+(x) \\ p^-(x) \end{bmatrix} = \begin{bmatrix} e^{ik_o K^+ x} & 0 \\ 0 & e^{ik_o K^- x} \end{bmatrix}\begin{bmatrix} p^+(0) \\ p^-(0) \end{bmatrix}. \tag{8.33}$$

[3] Actually, the roots of equations $J_o(Ba) = 0$ and $J_o(\sigma Ba) = 0$ are singularities of Equation (8.31) and, therefore, cannot be associated with the propagation constants corresponding to wave modes in the duct. But the hydrodynamic modes occur extremely close to the roots of these equations, which are $Ba = j_{on}$ and $\sigma Ba = j_{on}$, respectively (see Table 6.1).

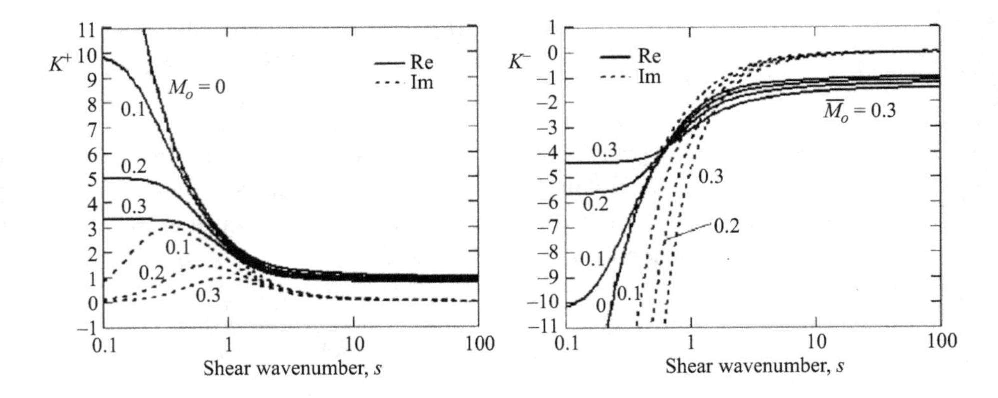

Figure 8.1 Effect of shear wavenumber on acoustic propagation constants ($\sigma^2 = 0.7, \gamma_o = 1.4$). As $s \to \infty$, the propagation constants tend to $K^{\mp} = \mp 1/(1 \mp M_o)$ asymptotically.

Other acoustic properties are related to pressure wave components by Equations (8.20)–(8.23):

$$v' = H^+ p^+ + H^- p^- \tag{8.34}$$

$$T' = F^+ p^+ + F^- p^- \tag{8.35}$$

$$\rho' = G^+ p^+ + G^- p^- \tag{8.36}$$

$$s' = E^+ p^+ + E^- p^-. \tag{8.37}$$

Here, $H^{\mp} = H^{\mp}(r)$ and $F^{\mp} = F^{\mp}(r)$ denote the values of Equations (8.27) and (8.28) evaluated for the propagation constants $K^{\mp}$, respectively, that is,

$$H^{\mp} = \frac{K^{\mp}}{\rho_o c_o (1 - K^{\mp} M_o)} \left(1 - \frac{J_o(B^{\mp} r)}{J_o(B^{\mp} a)}\right) \tag{8.38}$$

$$F^{\mp} = \frac{\beta_o T_o}{\rho_o c_{po}} \left(1 - \frac{J_o(\sigma B^{\mp} r)}{J_o(\sigma B^{\mp} a)}\right), \tag{8.39}$$

where $B^{\mp 2} a^2 = \mathrm{i}(1 - K^{\mp} \bar{M}_o) s^2$. $G^{\mp} = G^{\mp}(r)$ and $E^{\mp} = E^{\mp}(r)$ are similarly evaluated from Equations (8.22) and (8.23):

$$G^{\mp} = \frac{1}{c_o^2} \left(1 + (\gamma_o - 1) \frac{J_o(\sigma B^{\mp} r)}{J_o(\sigma B^{\mp} a)}\right) \tag{8.40}$$

$$E^{\mp} = -\frac{\beta_o}{\rho_o} \frac{J_o(\sigma B^{\mp} r)}{J_o(\sigma B^{\mp} a)}. \tag{8.41}$$

A peculiarity of this duct model is, as seen from Equation (8.32), that the pressure wave components propagate as plane waves, although other acoustic variables ($v', \rho', T', s', \ldots$) are not uniformly distributed over the duct cross section. For example, the radial distribution of the axial component of the particle velocity is

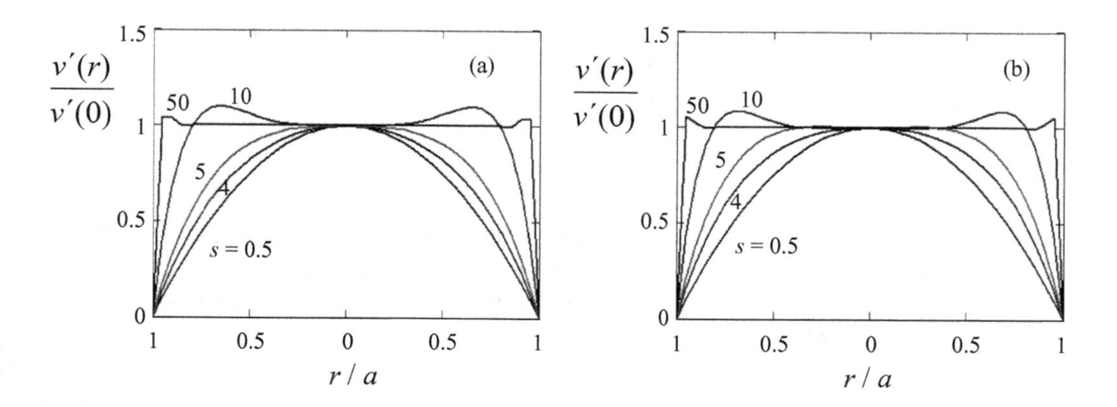

Figure 8.2 Radial profiles of the axial component of the particle velocity with uniform mean flow, (a) K^+ wave, (b) K^- wave ($M_o = 0.1$, $\sigma^2 = 0.7$, $\gamma_o = 1.4$).

shown in Figure 8.2 for $M_o = 0.1$. For this reason, the wave transfer matrix given in Equation (8.30) may not be connected directly to one-dimensional duct elements (see Section 8.4).

8.3.1.2 Wide-Duct Approximation

It can be seen in Figure 8.2 that as the shear wavenumber increases, the axial particle velocity profile tends to become flat, resembling planar propagation. We can take advantage of this situation to simplify the theory for application to ducts with large shear wavenumbers, which are usually referred to as wide ducts.

For hard-walled wide ducts with zero mean flow, Kirchhoff showed that the propagation constants are given by $K^{\mp} = \mp K_o + O(s^{-2})$[3], where

$$K_o = 1 + \frac{1+\mathrm{i}}{s}\left(1 + \frac{\gamma_o - 1}{\sigma}\right)\frac{1}{\sqrt{2}}. \tag{8.42}$$

The corresponding propagation constants for a wide duct with mean flow can be derived from Equation (8.33) by using the asymptotic expansions

$$\frac{\mathrm{J}_2(z)}{\mathrm{J}_o(z)} \approx -1 + \frac{\mathrm{i}2}{z}, \quad \frac{\mathrm{J}_1(z)}{\mathrm{J}_o(z)} \approx \mathrm{i} + \frac{1}{2z}, \tag{8.43}$$

which are valid for $|z| >> 4$. Since the Prandtl number is less than unity, this condition can be translated for Equation (8.31) as $|\sigma Ba| = \left|\sqrt{1 - KM_o}\right|\sigma s >> 4$, or as $s > 30$ or 40, say, since $K \approx \mp 1$ for large enough s (see Figure 8.1). Thus, substituting the first equation in Equation (8.43), the hard-wall case of Equation (8.31) may be expressed in this range of the shear wavenumbers as

$$\frac{K}{1 - KM_o} = \mp\left(1 + \frac{(1+\mathrm{i})}{s\sqrt{1 - KM_o}}\left(1 + \frac{\gamma_o - 1}{\sigma}\right)\frac{1}{\sqrt{2}}\right) + O(s^{-2}) = \mp D(KM_o) + O(s^{-2}). \tag{8.44}$$

Since $D(0) = K_o + O(s^{-2})$, for subsonic sufficiently low Mach numbers, the acoustic propagation constants may be calculated, as a first approximation, by using the formula [9]:

$$K^{\mp} = \frac{\mp K_o}{1 \mp K_o M_o}, \tag{8.45}$$

which follows from Equation (8.44) after putting $D = K_o$ and is attractive because of its simplicity. A better estimate may be obtained by noting that $K^{\mp} \to \mp 1/(1 \mp M_o)$ as $s \to \infty$ (Figure 8.1). Then, approximating the right-hand side of Equation (8.44) by $D(\mp M_o/(1 \mp M_o))$ yields the formula

$$K^{\mp} = \frac{\mp\left(1 + (K_o - 1)\sqrt{1 \mp M_o}\right)}{1 \mp M_o\left(1 + (K_o - 1)\sqrt{1 \mp M_o}\right)}. \tag{8.46}$$

These formulae are compared in Figure 8.3 with the exact dispersion equation, Equation (8.31), of the low reduced frequency theory. It is seen that both formulae produce the real part of the propagation constant accurately in the subsonic low Mach number range even for shear wavenumbers as low as 10. The imaginary part of the propagation constant is more sensitive to the shear wavenumber and satisfactory correlation occurs for about $s \geq 40$ or so. In this case, Equation (8.46) accurately represents the exact dispersion equation. Equation (8.45) is, as expected, is not as satisfactory for Mach numbers greater than about 0.1, but may still be considered adequately accurate for engineering calculations.

In wide ducts, since the radial distribution of the acoustic variables is fairly uniform across the cross section, except in the vicinity of the duct walls (for example,

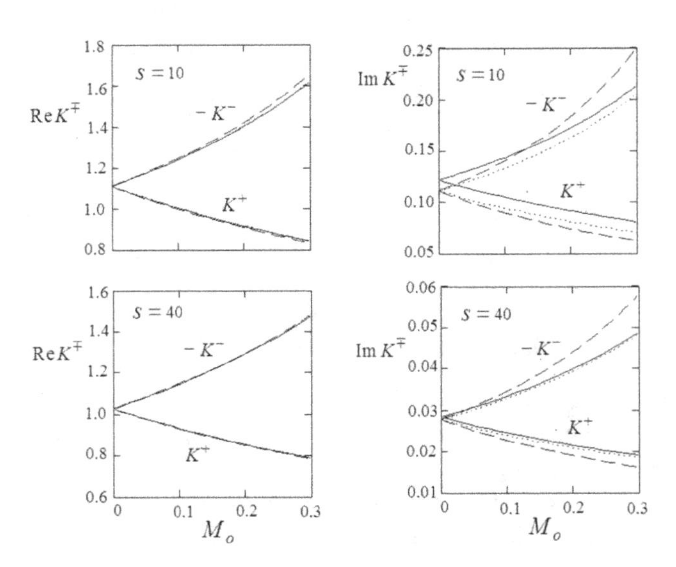

Figure 8.3 Comparison of approximate formulae for wide duct propagation constants with the exact values at subsonic low Mach numbers. Solid: exact, Equation (8.31); dash: Equation (8.45), dot: Equation (8.46). ($\sigma^2 = 0.7$, $\gamma_o = 1.4$).

Figure 8.2), the radial functions defined in Equations (8.34)–(8.37) can be approximated with adequate accuracy by their averages over a duct cross section. The following formulae may be used for calculating these averages:

$$\bar{H} = -\frac{1}{\rho_o c_o}\frac{K}{1 - KM_o}\frac{\mathrm{J}_2(Ba)}{\mathrm{J}_o(Ba)} \tag{8.47}$$

$$\bar{F} = -\frac{\beta_o T_o}{\rho_o c_{po}}\frac{\mathrm{J}_2(\sigma Ba)}{\mathrm{J}_o(\sigma Ba)} \tag{8.48}$$

$$\bar{G} = \frac{1}{c_o^2}\left\{1 + (\gamma_o - 1)\left(1 + \frac{\mathrm{J}_2(\sigma Ba)}{\mathrm{J}_o(\sigma Ba)}\right)\right\} \tag{8.49}$$

$$\bar{E} = -\frac{\beta_o}{\rho_o}\left(1 + \frac{\mathrm{J}_2(\sigma Ba)}{\mathrm{J}_o(\sigma Ba)}\right), \tag{8.50}$$

where an overbar denotes averaging over a duct section. These formulae are valid for any value of the shear wavenumber. In the wide duct case, they may be evaluated by using the asymptotic formulae in Equation (8.43) [11].

The wave transfer matrix of a wide duct is still given by Equation (8.33), but it may not be combined directly with the wave transfer matrix of a one-dimensional inviscid duct, because the cross section averaged particle velocity is now given by

$$\overline{v'} = \bar{H}^+ p^+ + \bar{H}^- p^-, \tag{8.51}$$

which follows upon averaging of Equation (8.34) over a duct section. To match this with its one-dimensional inviscid counterpart, namely, $\overline{v'} = (p^+ - p^-)/\rho_o c_o$, we redefine the pressure wave components in Equation (8.32) as $p^\mp = P^\mp \mathrm{e}^{\mathrm{i}k_o K^\mp x}$ and determine $P^\mp$ so that the particle velocity decomposition in Equation (8.51) transforms to the one-dimensional inviscid form given above. This process yields the transformed wave transfer equation of the duct in the form

$$\begin{bmatrix} p^+(x) \\ p^-(x) \end{bmatrix} = \psi \begin{bmatrix} \mathrm{e}^{\mathrm{i}k_o K^+ x} & 0 \\ 0 & \mathrm{e}^{\mathrm{i}k_o K^- x} \end{bmatrix} \psi^{-1} \begin{bmatrix} p^+(0) \\ p^-(0) \end{bmatrix}, \tag{8.52}$$

where

$$\psi = \begin{bmatrix} 1 + \rho_o c_o \bar{H}^+ & 1 + \rho_o c_o \bar{H}^- \\ 1 - \rho_o c_o \bar{H}^+ & 1 - \rho_o c_o \bar{H}^- \end{bmatrix}. \tag{8.53}$$

The wave transfer matrix in Equation (8.52) may be combined directly with one-dimensional uniform ducts.

8.3.1.3 Effect of Parabolic Mean Flow Velocity Profile

The assumption of uniform mean flow velocity profile may be of concern in smooth narrow pipes with Poiseuille flow. In this case, the mean flow velocity profile is parabolic, but the governing acoustic equations can be solved only approximately or

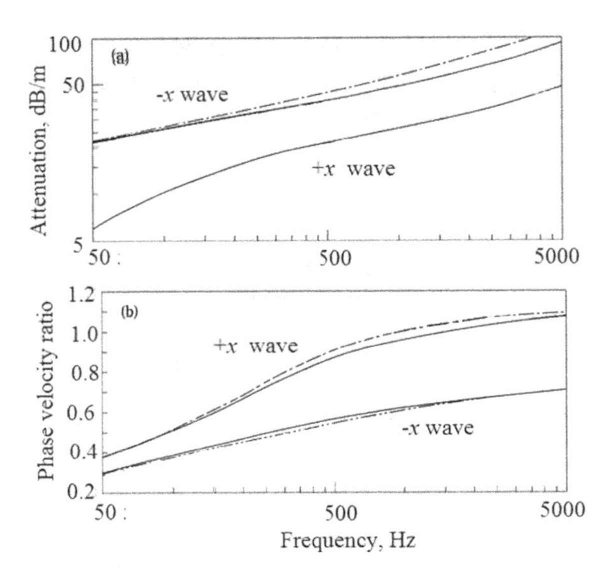

Figure 8.4 Transmission characteristics of sound waves in a circular duct of 1 mm diameter with average mean flow Mach number of 0.2. Solid: uniform mean flow (Equation (8.31)); dash-dot: finite element method with parabolic mean flow velocity profile [8] ($T_o = 1000$ K, $\gamma_o = 1.4$, $\sigma^2 = 0.7$, $\mu_o = 4.15 \times 10^{-5}$ Ns/m^2, $\rho_o = 0.35$ kg/m^3).

numerically. Peat derived an approximate analytical dispersion equation for the wavenumbers of hard-walled ducts using a variational perturbation approach and neglecting the radial component of the acoustic particle velocity [12]. Numerical solutions for the propagation constants are given by using the finite element method [8], and the Runge–Kutta finite-difference integration method [13]. A comparison of the propagation constants computed by using Equation (8.31) with the finite element solutions is shown in Figure 8.4 for a circular duct of 1 mm diameter with an average mean flow Mach number of 0.2, the fluid properties being as given in the figure caption. The real and imaginary parts of the wavenumbers are given as attenuation in dB/m ($= \mp 8.686 k_o \mathrm{Im}\{K^{\mp}\}$) and phase speed ratio ($= \mp 1/\mathrm{Re}\{K^{\mp}\}$) for waves traveling in positive and negative directions of the ducts axis. In general, for subsonic low Mach numbers, the propagation constants are not significantly affected by the shape of the mean flow velocity profile.

8.3.1.4 Effect of Turbulent Boundary Layer

The damping of sound waves becomes stronger when the acoustic boundary layer becomes thicker than the viscous sublayer. A measure of this effect is the normalized viscous acoustic boundary layer thickness, δ_A^+ (Section 3.9.3.4). When δ_A^+ is sufficiently small, the viscous sublayer thickness is much larger than the acoustic boundary layer thickness. Then turbulence has negligible effect and attenuation of sound waves is described adequately accurately by the viscothermal propagation constants. But, when δ_A^+ is sufficiently large, the acoustic boundary layer thickness becomes larger

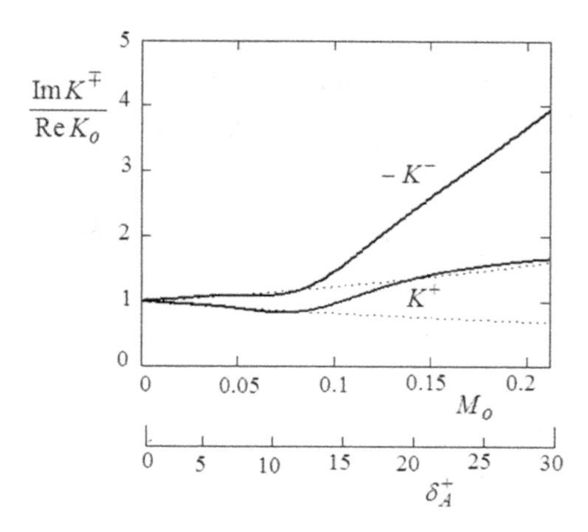

Figure 8.5 Effect of turbulent boundary layer stress on the imaginary part of the propagation constants. Solid: Howe boundary layer admittance; dot: Equation (8.45). ($k_o a = 0.0808$, $s = 179$, $\gamma_o = 1.4$, $\sigma^2 = 0.713$).

than the sublayer thickness and the damping of the acoustic waves is significantly influenced by turbulent stresses. For circular hard-walled ducts, these effects are fairly accurately predicted by the Howe boundary layer admittance, Y (Figure 3.10). This may be applied by using the one-dimensional inviscid uniform duct model described in Section 3.9.2 with wall impedance $Z_W = 1/Y$ and no-slip boundary condition. Figure 8.5 shows the effect of turbulent stresses on the imaginary part of the propagation constants predicted by this model for a case experimentally validated by Allam and Åbom [14]. It is seen that the acoustic boundary layer remains in the viscous sublayer and dominates the attenuation while δ_A^+ is less than about 10 and in this range the attenuation is predicted accurately by Equation (8.45). It suffices to consider the latter, because the shear wavenumber is large ($s = 179$) and the discrepancy observed for larger δ_A^+, which is a manifestation of the acoustic boundary layer entangling the viscous sublayer, will not improve by more accurate calculation of the viscothermal propagation constant.

8.3.2 Circular Annular Ducts

Analysis of an annular circular duct of outer radius a and inner radius b is similar to the hollow duct case (see Section 8.3.1), except that the boundary conditions in Equation (8.18) are now applied at $r = a$ and $r = b$. So, in this case we proceed with the general solutions of Equations (8.24) and (8.25), namely,

$$H(r) = C_1 \mathrm{J}_o(Br) + C_2 \mathrm{Y}_o(Br) + \frac{\mathrm{i} k_o K}{\mu_o B^2} \tag{8.54}$$

$$F(r) = D_1 \mathrm{J}_o(\sigma Br) + D_2 \mathrm{Y}_o(\sigma Br) + \frac{\mathrm{i}(1 - KM_o)\omega\beta_o T_o}{\kappa_o \sigma^2 B^2}, \tag{8.55}$$

where Y_m denotes a Bessel function of the second kind of order m and C_1, C_2, D_1, D_2 denote the integration constants. Upon application of the first and third boundary conditions in Equation (8.18) at $r = a$ and $r = b$, the foregoing equations may be expressed, respectively, as

$$H(r) = \frac{K}{\rho_o c_o(1 - KM_o)}(1 + \chi_1 \mathrm{J}_o(Br) + \chi_2 \mathrm{Y}_o(Br)) \tag{8.56}$$

$$F(r) = \frac{\beta_o T_o}{\rho_o c_{po}}(1 + \xi_1 \mathrm{J}_o(\sigma Br) + \xi_2 \mathrm{Y}_o(\sigma Br)), \tag{8.57}$$

where

$$\chi_1 = -\frac{\mathrm{Y}_o(Ba) - \mathrm{Y}_o(Bb)}{\mathrm{J}_o(Bb)\mathrm{Y}_o(Ba) - \mathrm{J}_o(Ba)Y_o(Bb)} \tag{8.58}$$

$$\chi_2 = \frac{\mathrm{J}_o(Ba) - \mathrm{J}_o(Bb)}{\mathrm{J}_o(Bb)\mathrm{Y}_o(Ba) - \mathrm{J}_o(Ba)Y_o(Bb)} \tag{8.59}$$

$$\xi_1 = -\frac{\mathrm{Y}_o(\sigma Ba) - \mathrm{Y}_o(\sigma Bb)}{\mathrm{J}_o(\sigma Bb)\mathrm{Y}_o(\sigma Ba) - \mathrm{J}_o(\sigma Ba)Y_o(\sigma Bb)} \tag{8.60}$$

$$\xi_2 = \frac{\mathrm{J}_o(\sigma Ba) - \mathrm{J}_o(\sigma Bb)}{\mathrm{J}_o(\sigma Bb)\mathrm{Y}_o(\sigma Ba) - \mathrm{J}_o(\sigma Ba)Y_o(\sigma Bb)}. \tag{8.61}$$

Substituting Equations (8.56) and (8.57) in Equation (8.29) and integrating with respect r we get

$$v_r' = \frac{\mathrm{i}k_o}{\rho_o c_o}(1 - KM_o)R(r)p' + \frac{C}{r}. \tag{8.62}$$

Here, C denotes an integration constant and the function $R(r)$ is

$$R(r) = \frac{r}{2}\left(\gamma_o - (\gamma_o - 1)(1 + \Phi_1(r)) - \left(\frac{K}{1 - KM_o}\right)^2(1 + \Phi_2(r))\right) \tag{8.63}$$

where

$$\Phi_1(r) = 2\frac{\xi_1 \mathrm{J}_1(\sigma Br) + \xi_2 \mathrm{Y}_1(\sigma Br)}{\sigma Br} \tag{8.64}$$

$$\Phi_2(r) = 2\frac{\chi_1 \mathrm{J}_1(Br) + \chi_2 \mathrm{Y}_1(Br)}{Br}. \tag{8.65}$$

The second boundary condition in Equation (8.18) at $r = a$ and $r = b$ gives

$$\frac{\mathrm{i}k_o}{\rho_o c_o}R(a) + \frac{C}{a} = \frac{1}{Z_a} \tag{8.66}$$

$$\frac{\mathrm{i}k_o}{\rho_o c_o}R(b) + \frac{C}{b} = \frac{1}{Z_b}, \tag{8.67}$$

where Z_a and Z_b denote the wall impedance at $r = a$ and $r = b$, respectively. Elimination of the integration constant yields the dispersion equation

$$\gamma_o - (\gamma_o - 1)\left(1 + \frac{\Phi_1(a) - \frac{b^2}{a^2}\Phi_1(b)}{1 - \frac{b^2}{a^2}}\right) - \left(\frac{K}{1 - KM_o}\right)^2\left(1 + \frac{\Phi_2(a) - \frac{b^2}{a^2}\Phi_2(b)}{1 - \frac{b^2}{a^2}}\right)$$

$$= \frac{2}{\mathrm{i}k_o a\zeta_a}\left(1 - \frac{b\zeta_a}{a\zeta_b}\right)\frac{1}{1 - \frac{b^2}{a^2}} \tag{8.68}$$

where $\zeta_a = Z_a/\rho_o c_o$ and $\zeta_b = Z_b/\rho_o c_o$. Acoustic field properties in the duct are still given by Equations (8.32)–(8.37), but with $H^{\mp}$ and $F^{\mp}$ defined by Equations (8.56) and (8.57), and $G^{\mp}$ and $E^{\mp}$ are determined from Equations (8.22) and (8.23). Here, the superscript "∓" refers to values of functions evaluated for the acoustic propagation constants $K^{\mp}$ extracted from Equation (8.68). As can be shown by straightforward integration, the cross section averages of the acoustic property functions can be calculated by using the following formulae:

$$\bar{H} = \frac{1}{\rho_o c_o}\frac{K}{1 - KM_o}\frac{\pi r^2}{S}(1 + \Phi_2(r))\Big|_{r=b}^{r=a} \tag{8.69}$$

$$\bar{F} = \frac{\beta_o T_o}{\rho_o c_{po}}\frac{\pi r^2}{S}(1 + \Phi_1(r))\Big|_{r=b}^{r=a} \tag{8.70}$$

$$\bar{G} = \frac{\pi r^2}{c_o^2 S}(1 + (\gamma_o - 1)\Phi_1(r))\Big|_{r=b}^{r=a} \tag{8.71}$$

$$\bar{E} = \frac{\beta_o}{\rho_o}\frac{\pi r^2}{S}(1 + \Phi_1(r))\Big|_{r=b}^{r=a} \tag{8.72}$$

where $S = \pi(a^2 - b^2)$.

8.3.3 Rectangular Ducts

For rectangular ducts, the governing acoustic equations are written in Cartesian coordinates (x, y, z) by using the expansions:

$$\nabla_\perp^2 = \frac{\partial^2}{\partial y^2} + \frac{\partial^2}{\partial z^2} \tag{8.73}$$

$$\nabla\cdot\mathbf{v}' = \frac{\partial v'}{\partial x} + \frac{\partial v'_y}{\partial y} + \frac{\partial v'_z}{\partial z} \tag{8.74}$$

$$\mathbf{v}'_\perp\cdot\nabla_\perp = v'_y\frac{\partial}{\partial y} + v'_z\frac{\partial}{\partial z}, \tag{8.75}$$

where y and z denote the transverse coordinates, the duct cross section lying in $0 \leq y \leq 2a$, $0 \leq z \leq 2b$, with center at $y = z = 0$, and v'_y and v'_z denote the components of the particle velocity in the y and z directions, respectively. The boundary conditions to be satisfied are:

$$\left.\begin{aligned} v' &= 0 \\ v'_y &= \left(-\mathrm{i}\omega + v_o \frac{\partial}{\partial x}\right) \frac{p'}{-\mathrm{i}\omega Z_y} \\ v'_z &= 0 \\ T' &= 0 \end{aligned}\right\} \text{ at } y = 0 \text{ and } y = 2a \tag{8.76}$$

$$\left.\begin{aligned} v' &= 0 \\ v'_y &= 0 \\ v'_z &= \left(-\mathrm{i}\omega + v_o \frac{\partial}{\partial x}\right) \frac{p'}{-\mathrm{i}\omega Z_z} \\ T' &= 0 \end{aligned}\right\} \text{ at } z = 0 \text{ and } z = 2b \tag{8.77}$$

where Z_y and Z_z denote the impedance of walls $y = 0, 2a$ and $z = 0, 2b$. Solutions of Equations (8.11) and (8.13) are sought as

$$p' = A\mathrm{e}^{\mathrm{i}k_o K x} \tag{8.78}$$

$$v' = H(y, z) p' \tag{8.79}$$

$$T' = F(y, z) p'. \tag{8.80}$$

Upon substitution of these, Equations (8.11) and (8.13) give, respectively,

$$\frac{\partial^2 H}{\partial y^2} + \frac{\partial^2 H}{\partial z^2} + B^2 H = \frac{\mathrm{i}k_o K}{\mu_o} \tag{8.81}$$

$$\frac{\partial^2 F}{\partial y^2} + \frac{\partial^2 F}{\partial z^2} + B^2 \sigma^2 F = \mathrm{i}(1 - KM_o) \frac{\omega \beta_o T_o}{\kappa_o}. \tag{8.82}$$

Solution of Equation (8.82) is searched for in the form of a double Fourier series

$$H(y, z) = \sum_{m,n} a_{mn} \sin\left(\frac{m\pi y}{2a}\right) \sin\left(\frac{m\pi z}{2b}\right), \qquad m, n = 1, 3, 5, \ldots \tag{8.83}$$

which satisfies the first boundary condition in Equations (8.80) and (8.81) identically. The coefficients a_{mn} are determined by substituting Equation (8.83) in Equation (8.81) and averaging the resulting equation over the duct cross-sectional area. This process gives

$$a_{mn} = \mathrm{i} \frac{16 k_o K}{\pi^2 \mu_o} \frac{1}{mnB^2 \alpha_{mn}(Ba)}, \tag{8.84}$$

where the parameter B is as defined by Equation (8.26), but with $2a$ representing the side of the rectangular section, and

$$\alpha_{mn}(\xi) = 1 - \frac{\pi^2}{4\xi^2}\left(m^2 + \frac{a^2}{b^2}n^2\right). \tag{8.85}$$

The solution of Equation (8.82) which satisfies the fourth boundary condition in Equations (8.76) and (8.77) identically can be expressed similarly

$$F(y,z) = \sum_{m,n} b_{mn} \sin\left(\frac{m\pi y}{2a}\right) \sin\left(\frac{m\pi z}{2b}\right), \qquad m,n = 1,3,5,\ \ldots \tag{8.86}$$

where

$$b_{mn} = \mathrm{i}\frac{16\omega(1-KM_o)\beta_o T_o}{\pi^2\kappa_o}\frac{1}{mn\sigma^2B^2\alpha_{mn}(\sigma Ba)}. \tag{8.87}$$

An eigen-equation for K can now be derived as follows: Elimination of the acoustic density between Equations (8.8) and (8.9) gives, in frequency domain,

$$\rho_o\nabla_\perp\cdot\mathbf{v}'_\perp = \left[\mathrm{i}\omega\left(\frac{\gamma_o}{c_o^2} - \rho_o\beta F\right)(1-KM_o) - \mathrm{i}\rho_o k_o KH\right]p'. \tag{8.88}$$

Substituting the foregoing double Fourier series expansions in this equation and integrating the resulting equation over the duct cross-sectional area

$$\rho_o c_o \oint_\ell \mathbf{v}'_\perp\cdot\mathbf{n}d\ell = \mathrm{i}k_o(1-KM_o)\left[\gamma_o - (\gamma_o - 1)\Gamma(\sigma Ba) - \left(\frac{K}{1-KM_o}\right)^2\Gamma(Ba)\right]4abp', \tag{8.89}$$

where the Gauss theorem in two-dimensions is applied for the evaluation of the left-hand side of the integrated form of Equation (8.88), ℓ denotes the perimeter of the duct cross-sectional area, $\mathbf{n}$ denotes the unit outward normal of the duct walls and

$$\Gamma(\xi) = -\frac{64}{\pi^4}\sum_{m,n}\frac{1}{m^2n^2\alpha_{mn}(\xi)}, \qquad m,n = 1,3,5,\ \ldots \tag{8.90}$$

Upon application of the boundary conditions on v'_y and v'_z to the line integral, Equation (8.89) yields

$$\rho_o c_o\left(2bv'_{y=0} + 2bv'_{y=2a} + 2av'_{z=0} + 2av'_{z=2b}\right) = \mathrm{i}k_o(1-K\bar{M}_o)\left[\gamma_o - (\gamma_o - 1)\Gamma(\sigma Ba) - \left(\frac{K}{1-KM_o}\right)^2\Gamma(Ba)\right]4abp'. \tag{8.91}$$

Hence, with substitution of the relevant boundary conditions from Equations (8.76) and (8.77), the dispersion equation for K is found as

$$\gamma_o - (\gamma_o - 1)\Gamma(\sigma Ba) - \left(\frac{K}{1-KM_o}\right)^2\Gamma(Ba) = \frac{1}{\mathrm{i}2k_o a}\left(\frac{1}{\zeta_{y=0}} + \frac{1}{\zeta_{y=2a}} + \frac{a}{b}\left(\frac{1}{\zeta_{z=0}} + \frac{1}{\zeta_{z=2b}}\right)\right), \tag{8.92}$$

where $\zeta = Z/\rho_o c_o$ denotes normalized impedance. Solutions of this dispersion equation for the acoustic propagation constants K^+ and K^- of hard-walled ducts are given in Reference [15] as functions of the shear wavenumber for various values of the aspect ratio, a/b, and subsonic low mean flow Mach numbers. In general, for rectangular sections having an aspect ratio close to unity, the propagation constants may be approximated with adequate accuracy by the propagation constants of a hollow circular section having the same cross-sectional area.

Acoustic field properties in the duct are still given by Equations (8.32)–(8.37), where $H^{\mp}$ and $F^{\mp}$ are now given by Equations (8.83) and (8.86), and $G^{\mp}$ and $E^{\pm}$ can be determined from

$$G(y,z) = \frac{1}{c_o^2}\left(\gamma_o - (\gamma_o - 1)\frac{16}{\pi^2}\sum_{m,n=1,3,5,\ldots}\frac{\sin\dfrac{m\pi y}{2a}\sin\dfrac{n\pi z}{2b}}{mn\alpha_{mn}(\sigma\beta a)}\right) \tag{8.93}$$

$$E(y,z) = \frac{\beta_o}{\rho_o}\left(-1 + \frac{16}{\pi^2}\sum_{m,n=1,3,5,\ldots}\frac{\sin\dfrac{m\pi y}{2a}\sin\dfrac{n\pi z}{2b}}{mn\alpha_{mn}(\sigma\beta a)}\right), \tag{8.94}$$

which follow from Equations (8.22) and (8.23). The cross section averages of these functions can be calculated by using the following formulae:

$$\bar{H} = -\frac{1}{\rho_o c_o}\frac{K}{1-K}\frac{64}{\pi^4}\sum_{m,n=1,3,5,\ldots}\frac{1}{m^2 n^2 \alpha_{mn}(Ba)} \tag{8.95}$$

$$\bar{F} = \frac{\beta_o T_o}{\rho_o c_{po}}\frac{64}{\pi^4}\sum_{m,n=1,3,5,\ldots}\frac{1}{m^2 n^2 \alpha_{mn}(\sigma Ba)} \tag{8.96}$$

$$\bar{G} = \frac{1}{c_o^2}\left(\gamma_o - (\gamma_o - 1)64\sum_{m,n=1,3,5,\ldots}\frac{1}{m^2 n^2 \alpha_{mn}(\sigma\beta a)}\right) \tag{8.97}$$

$$\bar{E} = \frac{\beta_o}{\rho_o}\left(-1 + \frac{64}{\pi^4}\sum_{m,n=1,3,5,\ldots}\frac{1}{m^2 n^2 \alpha_{mn}(\sigma Ba)}\right). \tag{8.98}$$

8.4 Time-Averaged Acoustic Power

The time-averaged acoustic power passing through normal to a duct cross section may be computed in the frequency domain by using Equation (1.48). For the low reduced frequency model considered in this chapter, this gives

$$W = 2\mathrm{Re}\int_S \left(\breve{p}'v' + \rho_o v_o \breve{v}'v' + \frac{v_o}{\rho_o}\breve{p}'\rho' + v_o^2 \breve{v}'\rho'\right)\mathrm{d}S, \tag{8.99}$$

where S denotes the duct cross-sectional area and an inverted over-arc denotes the complex conjugate. The acoustic pressure field in the duct decomposes into the pressure wave components as usual in the one-dimensional theory as $p' = p^+ + p^-$. For the calculation of Equation (8.99), it is convenient to introduce the functions $h^{\mp} = \rho_o c_o H^{\mp}$ and $g^{\mp} = c_o^2 G^{\mp}$, where $H^{\mp}$ and $G^{\mp}$ are given in previous sections for circular and rectangular ducts. Then, the acoustic particle velocity and density are given by

$$v' = \frac{h^+ p^+ + h^- p^-}{\rho_o c_o} \tag{8.100}$$

$$\rho' = \frac{g^+ p^+ + g^- p^-}{c_o^2}, \tag{8.101}$$

respectively. Upon substitution of these decompositions, Equation (8.99) may be expressed as [16]:

$$W = \frac{2S}{\rho_o c_o} \left(\mathrm{Re}\{g_{11}\}|p^+|^2 + \mathrm{Re}\{g_{22}\}|p^-|^2 + \mathrm{Re}\left\{g_{12}\breve{p}^+ p^-\right\} + \mathrm{Re}\left\{g_{21} p^+ \breve{p}^-\right\} \right), \tag{8.102}$$

where

$$g_{11} = \frac{1}{S} \int_S \left(h^+ + M_o\left(\breve{h}^+ h^+ + g^+\right) + M_o^2 \breve{h}^+ g^+ \right) \mathrm{d}S \tag{8.103}$$

$$g_{12} = \frac{1}{S} \int_S \left(h^- + M_o\left(\breve{h}^+ h^- + g^-\right) + M_o^2 \breve{h} + g^- \right) \mathrm{d}S \tag{8.104}$$

$$g_{21} = \frac{1}{S} \int_S \left(h^+ + M_o\left(\breve{h}^- h^+ + g^+\right) + M_o^2 \breve{h}^- g^+ \right) \mathrm{d}S \tag{8.105}$$

$$g_{22} = \frac{1}{S} \int_S \left(h^- + M_o\left(\breve{h}^- h^- + g^-\right) + M_o^2 \breve{h}^- g^- \right) \mathrm{d}S. \tag{8.106}$$

Clearly, if the duct is very long or has an anechoic termination, sound power is transmitted only by the incident pressure wave component and Equation (8.102) reduces to $W = 2S\mathrm{Re}\{g_{11}\}|p^+|^2/\rho_o c_o$. The full form of Equation (8.102) is relevant, however, when the duct has a reflecting boundary and we see that the time-averaged acoustic power transmitted in the duct is then not canonical; that is, it is not given by the algebraic sum of the powers transmitted by the incident and reflected wave components. The presence of the mean flow has a role in the appearance of the non-canonical terms because it may be shown that when the mean flow is neglected, Equation (8.102) simplifies to the canonical form

$$W = \frac{2S}{\rho_o c_o} \left(\mathrm{Re}\{g_{11}\}|p^+|^2 + \mathrm{Re}\{g_{22}\}|p^-|^2 \right), \tag{8.107}$$

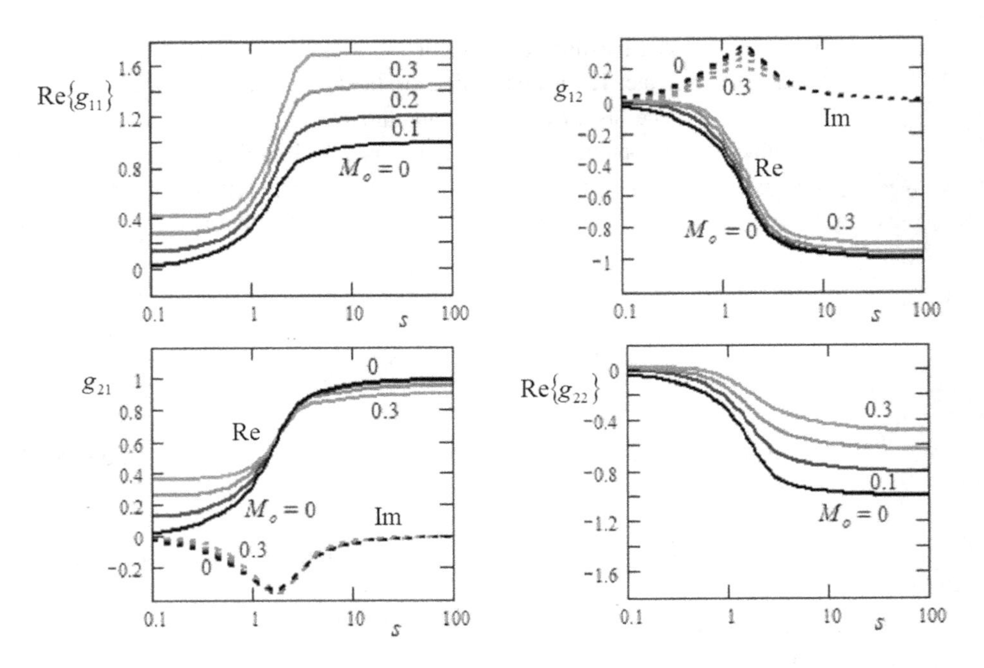

Figure 8.6 Sound power factors for a circular pipe with $\sigma^2 = 0.7$ and $\gamma_o = 1.4$ [16].

with $\text{Re}\{g_{11}\} = -\text{Re}\{g_{22}\}$. The viscosity and thermal conductivity of the fluid also have roles, because the foregoing canonical form still holds, with $g_{11} = (1 + M_o)^2$ and $g_{22} = -(1 - M_o)^2$, in the presence of an inviscid mean flow (see Section 3.4).

The power coefficients g_{ij}, $i, j = 1, 2$, are shown in Figure 8.6 as function of the shear wavenumber and the mean flow Mach number for circular hollow ducts. As can be seen from these characteristics, the values of the factors vary fairly sharply in the range of about $1 < s < 5$ or so, and as the shear wavenumber increases to infinity, the imaginary parts of g_{12} and g_{21} tend to vanish, with $\text{Re}\{g_{11}\} \approx -\text{Re}\{g_{22}\}$ gradually tending to the respective inviscid values given above. In general, the acoustic field may be assumed to be canonical if the shear wavenumber is greater than 40 or so, and this limit tends to be smaller the larger is the Helmholtz number in the range of the low reduced frequency condition, Equation (8.2).

8.5 Sudden Area Changes and Junctions

The wave transfer equations at sudden area changes and compact junctions of ducts may be determined by using the quasi-static conservation equations described in Section 4.2.1. We have seen that for the low reduced frequency model, the acoustic pressure decomposes as $p' = p^+ + p^-$, and the acoustic particle velocity and density decompose, respectively, as in Equations (8.100) and (8.101). In addition, the specific entropy fluctuations also decompose similarly:

$$s' = \frac{\beta_o}{\rho_0}(e^+p^+ + e^-p^-), \tag{8.108}$$

where $e^{\mp} = \rho_o E^{\mp}/\beta_o$ and $E^{\mp}$ are given in previous sections for circular and rectangular ducts. Except for the acoustic pressure, these decompositions are dependent on the transverse coordinates. For this reason, it is necessary to revise the working forms of the quasi-static conservation laws given in Section 4.2. Consider first the quasi-static continuity equation, Equation (4.5). Linearization yields for the acoustic and mean parts

$$\sum_{j=1,2,\ldots} \int_{S_j} \left(\rho_{jo}v'_j + \rho'_j v_{jo}\right) C_j \; \mathrm{d}S_j = 0 \tag{8.109}$$

$$\sum_{j=1,2,\ldots} \rho_{jo}v_{jo}C_jS_j = 0. \tag{8.110}$$

Here, we have assumed, for simplicity, that the surfaces of the control volume are hard, except at the interfaces with the connecting ducts. With substitution of Equations (8.100) and (8.101), the quasi-static acoustic continuity equation becomes

$$\sum_{j=1,2,\ldots} \frac{S_jC_j}{c_{jo}} \begin{bmatrix} \bar{h}_j^+ + M_j\bar{g}_j^+ & \bar{h}_j^- + M_j\bar{g}_j^- \end{bmatrix} \begin{bmatrix} p_j^+ \\ p_j^- \end{bmatrix} = 0, \tag{8.111}$$

where $M_j = v_{jo}/c_{jo}$, the cross section averages $\bar{h}^{\mp}$ and $\bar{g}^{\mp}$ are computed from $\bar{h}^{\mp} = \rho_o c_o \bar{H}^{\mp}$ and $\bar{g}^{\mp} = c_o^2\bar{G}^{\mp}$ by using the formulae given for $\bar{H}^{\mp}$ and $\bar{G}^{\mp}$ in previous sections for circular and rectangular ducts. Similarly, linearization of the quasi-static momentum equation, Equation (4.6), yields for the fluctuations

$$\sum_{j=1,2,\ldots} S_jC_j\mathbf{e}_j \begin{bmatrix} 1 + 2M_j\bar{h}_j^+ + M_j^2\bar{g}_j^+ & 1 + 2M_j\bar{h}_j^- + M_j^2\bar{g}_j^- \end{bmatrix} \begin{bmatrix} p_j^+ \\ p_j^- \end{bmatrix} = 0. \tag{8.112}$$

Finally, consider the quasi-static energy equation, Equation (4.7). With Equations (8.109) and (8.110) holding, it may be shown, as in Section 4.2, that for the linearized form of Equation (4.7) to be satisfied, it is sufficient that

$$\int_{S_j} \left(T_{jo}s'_j + \frac{p'_j}{\rho_{jo}} + v_{jo}v'_j \right) \mathrm{d}S_j = \text{constant}, \quad j \in \{1, 2, \ldots | v_{jo}S_j \neq 0\}. \tag{8.113}$$

Substituting Equations (8.100) and (8.108) and integrating

$$\frac{1}{\rho_{jo}} \begin{bmatrix} \bar{e}_j^+ + 1 + M_j\bar{h}_j^+ & \bar{e}_j^- + 1 + M_j\bar{h}_j^- \end{bmatrix} \begin{bmatrix} p^+ \\ p^- \end{bmatrix} = \text{constant}, \quad j \in \{1, 2, \ldots |M_jS_j \neq 0\} \tag{8.114}$$

where the fluid is assumed to be a perfect gas ($\beta_o T_o = 1$).

Equations (8.111), (8.112) and (8.114) may be used now to derive wave transfer equations at discontinuities formed by coupled ducts. For example, the analysis of

Section 4.6 may be applied similarly for viscothermal ducts meeting at a compact junction. In this case, referring to Figure 4.8, Equation (4.82) is replaced by Equation (8.114), and Equation (4.81) is replaced by Equation (8.111), with the mean density and the speed of sound being assumed to be uniform in the junction cavity. Since the formats of these equations are similar, the transfer of the acoustic pressure wave components at the junction is still governed by Equation (4.83), with the row vectors $\mathbf{a}_j$ and $\mathbf{b}_j$ given now for ducts $j = 1, 2, \ldots$ as

$$\mathbf{a}_j = \begin{bmatrix} 1 + M_j h_j^+ + e_j^+ & 1 + M_j h_j^- + e_j^- \end{bmatrix} \tag{8.115}$$

$$\mathbf{b}_j = S_j \begin{bmatrix} \bar{h}_j^+ + M_j \bar{g}_j^+ & \bar{h}_j^- + M_j \bar{g}_j^- \end{bmatrix}. \tag{8.116}$$

The formulae given in Section 4.6.2 for two-duct junction now apply with these row vectors.

The wave transfer equations at area-changes with multiple inner ducts may also be written directly from the analysis of Section 4.4 when the ducts are low reduced frequency models. For example, referring to the configurations shown in Figure 4.4, Equations (4.62), (4.63) and (4.64) are now replaced by Equations (8.111), (8.114) and (8.112), respectively, and collate similarly to the multi-port wave transfer equation $\mathbf{A}_1\mathbf{P}_1 + \mathbf{A}_2\mathbf{P}_2 + \cdots = \mathbf{0}$, where $\mathbf{P}_j = \{ p_j^+ \; p_j^- \}$. In implementations, usually it is not necessary to model all ducts by including viscothermal effects. If the shear wavenumber for a duct is large enough, it may be modeled as an inviscid duct. As the shear wavenumber increases, $h^{\mp} \to \mp 1$, $g^{\mp} \to 1$ and $e^{\mp} \to 0$ and, hence, Equations (8.111), (8.114) and (8.112) tend to their inviscid forms given in Section 4.4. Reduction to the inviscid duct case is convenient when ducts are not circular or rectangular, because low reduced frequency models of ducts of arbitrary section shapes can be derived only numerically.

An interesting application of this theory is the acoustic modeling of catalytic converters, which are essential components in automotive exhaust systems for controlling emission of noxious gases, and they are also sometimes used as sound absorbing structures in silencers, without the catalytic effect. A catalytic converter typically consists of an inlet expansion from the exhaust duct followed by the monolith block, which may be in one or more pieces tightly enclosed in a sheet metal casing with an elastic wire mesh layer, and an outlet contraction to the exhaust duct (Figure 8.7a). An acoustic block diagram model of a catalytic converter may be constructed as in Figure 8.7b.

For usual dimensions in automotive applications, the inlet and outlet sections of the converter may be modeled adequately accurately by using one-dimensional non-uniform inviscid duct models (Section 3.6). But the pores of the usual automotive monolith block are of about 1 mm^2 in area, which leads, assuming circular pores, to a shear wavenumber of 1.4 at 100 Hz and about 2 at 200 Hz, at an exhaust temperature of 600 °C. These values are in the range of significance of viscothermal effects on the propagation constants (Figure 8.1). Since even lower frequencies may be of importance in the acoustic design of exhaust silencers, sound transmission in monolith pores

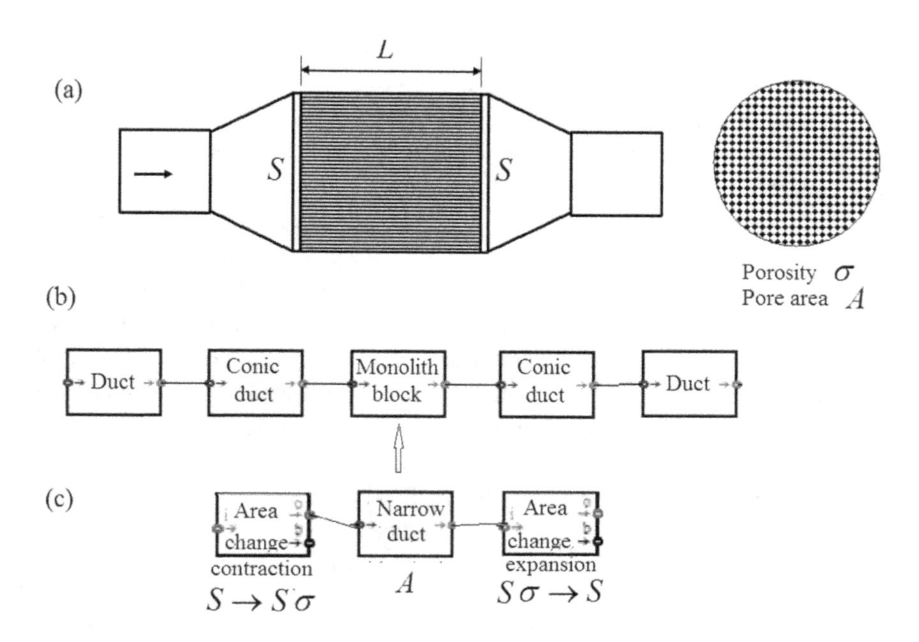

Figure 8.7 (a) Catalytic converter with single monolith block. (b) Acoustic block diagram of the catalytic converter. (c) Acoustic block diagram of the monolith subassembly.

is usually calculated by using a low reduced frequency duct model. Then, it is necessary to take into account the sudden area changes at the ends of a monolith block. These area-change elements represent the above described multi-ports based on the low reduced frequency theory, with their ports connected to distinct pores of the monolith block. But a monolith block usually has a couple of thousand pores and it is not practicable to model them individually. We may simplify this model by taking advantage of the fact that all pores are identical and the gas mass flow at the inlet of the monolith may be assumed to be divided equally into the pores. The procedure is similar to that described in Section 4.3.1, but to see the further restrictions imposed, its useful to review the proof briefly. In this case, Equation (8.114) implies $\mathbf{P}_3 = \mathbf{P}_4 = \cdots$, if all properties of the inner ducts are identical; however, we may not exempt their section areas, because the properties are now dependent on the actual section dimensions. For this reason, the inner ducts may be replaced by a single duct of area equal to the sum of their areas, but with all its other properties being the same as for a single inner duct, provided that all inner ducts are of equal cross-sectional area. The latter condition is satisfied in monolith blocks and, hence, the area changes at the ends of a monolith may be represented by three-ports, as shown in Figure 8.7c. Here, the expansion block corresponds to Figure 4.1a, the contraction block to Figure 4.1c, with inner duct properties of both blocks determined by the above condition, and the narrow duct block represents the low reduced frequency model of a single monolith pore of length L. In a monolith, duct 2 (side-branch) is normally closed flush with the discontinuity plane, at both the expansion and the contraction discontinuities, with a given reflection coefficient, $r_2 = p_2^-/p_2^+$ (usually hard-wall).

Hence, $\mathbf{P}_2$ may be eliminated from the equivalent three-port transfer equation as described in Section 4.3.2, to obtain the two-port subassembly of the monolith used in Figure 8.7b.

A mean temperature increase amounting to about 100 K may occur within the first few centimeters of the monolith. Effects of such ambient gradients are studied by retaining the ambient gradient terms in the low reduced frequency simplification of the acoustic continuity, momentum and energy equations [17–18].

Another application of the theory in automotive technology is charge air coolers, which are used in most turbocharged engines for volumetric efficiency enhancement. In general, a charge air cooler consists of a bundle of a large number of tubes packed in a casing with a gradual inlet expansion and gradual outlet contraction. In this respect, it is akin to a catalytic converter assembly (Figure 8.7a) and can be similarly modeled acoustically, except that the duct sizes in some applications may not require consideration of viscothermal effects. In a study of a car charge air cooler, in which the ducts are created by placing a periodically folded metal sheet between the pairs of plates of a pack of parallel rectangular plates, Knutsson and Åbom report that the ducts have a hydraulic diameter of about 2.5–3 mm and viscothermal effects and boundary layer turbulence diffusion (Section 8.3.1.4) have roles in sound attenuation of the device [19].

8.6 Coupled Narrow Ducts with Porous Walls

Ceramic foams are used to filter the soot particles in diesel engine exhaust gas. These traps essentially consist of a stack of narrow ducts of length L, which communicate through their porous walls. Figure 8.8a shows the cross section of a monolith filter block, which would, in practice, be tightly enclosed in a solid casing. The pores are usually of a square-like cross section and are closed at the other end. The open and closed ends on the inlet face of the monolith are indicated in Figure 8.8b by blank and black squares, respectively.

The filtering action takes place as the inlet gas flow enters through the blank pores into the inlet ducts and then is forced through the porous walls into the neighboring outlet ducts, to be discharged from them at the other end of the monolith. Since the

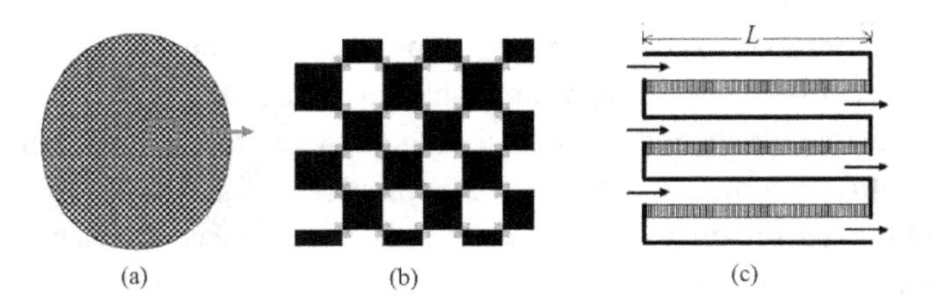

Figure 8.8 Narrow ducts communicating through porous walls.

cross-sectional area of the inlet and outlet ducts are of the order of 1 mm^2, the transmission of sound waves through the trap should be modeled by including viscothermal effects. For this purpose, we may use the low reduced frequency theory, but this now involves an additional approximation because forcing the gas flow through porous walls sets up axial mean pressure and velocity gradients along the monolith ducts, whereas the theory assumes uniform mean flow. However, monoliths are relatively short and it is generally adequately accurate to assume a uniform mean flow having the axially averaged properties of the actual mean flow. Then, following Allam and Åbom [20], we apply Equation (8.91) for the inlet and outlet ducts of a uniform and symmetrical matrix of rectangular pores:

$$\mathrm{i}k_o(1 - KM_o)D(K)p_1' = \left(\frac{w_y'}{a} + \frac{w_z'}{b}\right)_1 \rho_o c_o \tag{8.117}$$

$$\mathrm{i}k_o(1 - KM_o)D(K)p_2' = -\left(\frac{w_y'}{a} + \frac{w_z'}{b}\right)_2 \rho_o c_o. \tag{8.118}$$

Here, the subscripts "1" and "2" refer to the inlet and outlet duct, respectively, $w_y' = v_{y=0}' = v_{y=2a}'$, $w_z' = v_{z=0}' = v_{z=2b}'$ and

$$D(K) = \gamma_o + (\gamma_o - 1)\Gamma(\sigma Ba) + \left(\frac{K}{1 - KM_o}\right)^2 \Gamma(Ba), \tag{8.119}$$

where $\Gamma(\xi)$ is given by Equation (8.90) and K now denotes the propagation constant (yet unknown) for the coupled wave motion in the inlet and outlet ducts. It is convenient to define an average fluctuating wall velocity as

$$w' = \frac{bw_y' + aw_z'}{b + a}. \tag{8.120}$$

Assuming that the fluid motion in the porous wall is incompressible, this average velocity applies at both the inlet and outlet duct wall, and can be related to the acoustic pressures in the ducts as

$$p_1' - p_2' = Z_w w', \tag{8.121}$$

where Z_w denotes the wall impedance. The linearized form of the Ergun equation (see Appendix B) implies the relationship $Z_w = \sigma_R h$, where h denotes the wall thickness and σ_R denotes the flow resistivity of the duct walls, which has to be determined by measurement. For typical filters, the wall thickness is of the order of a few tenths of a millimeter, and the ratio of Z_w to the characteristic acoustic impedance at 20 °C is about 100 and increases with the temperature.

Equations (8.117) and (8.118) may be collected as

$$\begin{bmatrix} (1 - KM_o)D(K) - C & C \\ C & (1 - KM_o)D(K) - C \end{bmatrix} \begin{bmatrix} p_1' \\ p_2' \end{bmatrix} = 0. \tag{8.122}$$

Here,

$$C = \frac{1}{\mathrm{i}k_o\zeta_w}\left(\frac{1}{a} + \frac{1}{b}\right), \tag{8.123}$$

where $\zeta_w(=Z_w/\rho_o c_o)$ denotes the normalized wall impedance. The condition that Equation (8.122) has a non-trivial solution yields the following dispersion equation for the propagation constants

$$[(1 - KM_o)D(K) - 2C](1 - KM_o)D(K) = 0. \tag{8.124}$$

One of the roots of this equation is $K = 1/M_o$, but this may be disregarded because it corresponds to a hydrodynamic mode which is convected with the mean flow. The remaining two groups of roots correspond to acoustic modes. The first group is given by $D(K) = 0$ and consists of, in fact, the propagation constants for a rectangular narrow duct with hard walls. The second group of roots is given by

$$D(K) = \frac{2C}{1 - K\bar{M}_o}. \tag{8.125}$$

This equation also yields two propagation constants which describe acoustic wave motion and are associated, as with the roots of $D(K) = 0$, with waves traveling in forward ($+x$) and backward directions.

Compared in Figure 8.9 are the two groups of the propagation constants for a square duct, for a normalized wall impedance of $\zeta_w = 100$. It is seen that the second group of propagation constants are substantially more damped than the first group. Although the effect of mean flow is not shown in Figure 8.9, this effect is not substantial for mean flow velocities that are typical of diesel particulate filters.

Thus, the acoustic field in the inlet and outlet ducts may be expressed as

$$p_j'(x) = p_j^+(x) + p_j^-(x), \qquad j = 1, 2. \tag{8.126}$$

Here, for the inlet ducts

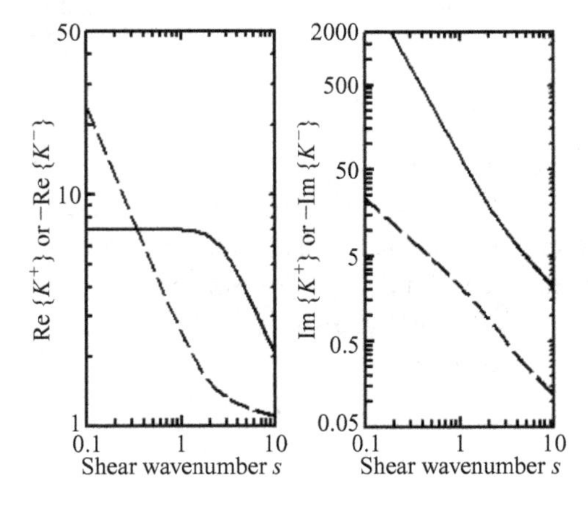

Figure 8.9 Propagation constants of a square narrow duct with zero mean flow. Dashed curve: $D(K) = 0$; solid curve: Equation (8.125). Normalized wall impedance is $\zeta_w = 100$.

$$p_1^{\mp}(x) = A^{\mp}\mathrm{e}^{\mathrm{i}k_oK_1^{\mp}x} + B^{\mp}\mathrm{e}^{\mathrm{i}k_oK_2^{\mp}x}, \tag{8.127}$$

where $K_1^{\mp}$ and $K_2^{\mp}$ denote the first and second groups of the propagation constants, respectively, the superscripts "$\mp$" refer to propagation in the $\mp x$ directions, and $A^{\mp}, B^{\mp}$ are integration constants. The acoustic pressure field in the outlet ducts can be expressed as

$$p_2^{\mp}(x) = A^{\mp}\mathrm{e}^{\mathrm{i}k_oK_1^{\mp}x} - B^{\mp}\mathrm{e}^{\mathrm{i}k_oK_2^{\mp}x}, \tag{8.128}$$

since the acoustic fields in the inlet and outlet ducts are related by Equation (8.122). The two-port wave transfer relation across a filter of length $0 \le x \le L$ is required in the form

$$\begin{bmatrix} p_2^{+}(L) \\ p_2^{-}(L) \end{bmatrix} = \mathbf{T}\begin{bmatrix} p_1^{+}(0) \\ p_1^{-}(0) \end{bmatrix}. \tag{8.129}$$

The foregoing equations can be manipulated as follows to derive this relationship. Firstly, we observe that, the difference field is given by

$$p_1^{\mp}(x) - p_2^{\mp}(x) = 2B^{\mp}\mathrm{e}^{\mathrm{i}k_oK_2^{\mp}x} \tag{8.130}$$

or, solving for the integration constants,

$$p_1^{\mp}(x) - p_2^{\mp}(x) = \left(p_1^{\mp}(0) - p_2^{\mp}(0)\right)\mathrm{e}^{\mathrm{i}k_oK_2^{\mp}x}. \tag{8.131}$$

The sum field, on the other hand, is given by

$$p_1^{\mp}(x) + p_2^{\mp}(x) = 2A^{\mp}\mathrm{e}^{\mathrm{i}k_oK_1^{\mp}x}. \tag{8.132}$$

Similarly solving for the integration constants

$$p_1^{\mp}(x) + p_2^{\mp}(x) = \left(p_1^{\mp}(0) + p_2^{\mp}(0)\right)\mathrm{e}^{\mathrm{i}k_oK_1^{\mp}x}. \tag{8.133}$$

The pressure wave components are determined from Equations (8.131) and (8.133):

$$2p_1^{\mp}(x) = \left(e_2^{\mp}(x) + e_1^{\mp}(x)\right)p_1^{\mp}(0) + \left(e_1^{\mp}(x) - e_2^{\mp}(x)\right)p_2^{\mp}(0) \tag{8.134}$$

$$2p_2^{\mp}(x) = \left(e_1^{\mp}(x) - e_2^{\mp}(x)\right)p_1^{\mp}(0) + \left(e_1^{\mp}(x) + e_2^{\mp}(x)\right)p_2^{\mp}(0), \tag{8.135}$$

where

$$e_1^{\mp}(x) = \mathrm{e}^{\mathrm{i}k_oK_1^{\mp}x}, \qquad e_2^{\mp}(x) = \mathrm{e}^{\mathrm{i}k_oK_2^{\mp}x}. \tag{8.136}$$

The boundary conditions at the closed ends[4] of the inlet and outlet ducts are: $v_1'(L) = 0$ and $v_2'(0) = 0$, where v_j', $j = 1,2$, denote the axial component of the

[4] At a closed end, the boundary conditions are on particle velocity and temperature fluctuations. In the case of diesel particulate filter, the radial component of the particle velocity is allowed at the ends and the velocity condition is expressed only for the axial component of the particle velocity. The condition on fluctuating temperature is ignored. This implies that the heat conductivity of the wall is close to that of the fluid. This is not very far from reality, since the closed ends are normally covered with a soot layer.

particle velocity and the subscripts "1" and "2" refer to the inlet and outlet ducts, respectively. These conditions are applied by averaging over the duct cross-sectional area, that is, $\bar{v}_1(L) = 0$ and $\bar{v}_2(0) = 0$, where

$$\bar{v}_j = \frac{1}{S}\int_S v_j' \mathrm{d}S, \qquad j = 1, 2 \tag{8.137}$$

and S denotes single pore cross-sectional area. The axial component of the particle velocity satisfies the relation $v' = H(y,z)p'$ for each propagation constant. This implies that, the foregoing equations for the acoustic pressure fields of the inlet and outlet ducts can be converted to the corresponding relations for the axial component of the particle velocity by replacing the terms $\mathrm{e}^{\mathrm{i}k_o K_1^{\mp} x}$ and $\mathrm{e}^{\mathrm{i}k_o K_2^{\mp} x}$ by $H_1^{\mp}(y,z)\mathrm{e}^{\mathrm{i}k_o K_1^{\mp} x}$ and $H_2^{\mp}(y,z)\mathrm{e}^{\mathrm{i}k_o K_2^{\mp} x}$, respectively, where $H_1^{\mp}(y,z)$ denotes the function $H(y,z)$ evaluated for $K_1^{\mp}$, and $H_2^{\mp}(y,z)$ for $K_2^{\mp}$. In particular, the counterparts of Equations (8.134) and (8.135) for the axial component of the particle velocity averaged over the pore cross-sectional area are

$$2\rho_o c_o \bar{v}_1^{\mp}(x) = \left(\bar{h}_2^{\mp} e_2^{\mp}(x) + \bar{h}_1^{\mp} e_1^{\mp}(x)\right) p_1^{\mp}(0) + \left(\bar{h}_1^{\mp} e_1^{\mp}(x) - \bar{h}_2^{\mp} e_2^{\mp}(x)\right) p_2^{\mp}(0) \tag{8.138}$$

$$2\rho_o c_o \bar{v}_2^{\mp}(x) = \left(\bar{h}_1^{\mp} e_1^{\mp}(x) - \bar{h}_2^{\mp} e_2^{\mp}(x)\right) p_1^{\mp}(0) + \left(\bar{h}_1^{\mp} e_1^{\mp}(x) + \bar{h}_2^{\mp} e_2^{\mp}(x)\right) p_2^{\mp}(0). \tag{8.139}$$

Here, $\bar{h}_j^{\mp} = \rho_o c_o \bar{H}_j^{\mp}$, $j = 1, 2$, and $\bar{H}_j^{\mp}$ are given by Equation (8.95), evaluated for the propagation constants $K_j^{\mp}$.

Finally, we write Equations (8.130) and (8.132) for $x = L$, apply the conditions $v_1'(L) = 0$, $v_2'(0) = 0$ using Equations (8.138) and (8.139), and collect the resulting six equations in matrix form:

$$\begin{bmatrix} -e_2^+ & 0 & e_2^+ & 0 & 1 & 0 & -1 & 0 \\ 0 & -e_2^- & 0 & e_2^- & 0 & 1 & 0 & -1 \\ -e_1^+ & 0 & -e_1^+ & 0 & 1 & 0 & 1 & 0 \\ 0 & -e_1^- & 0 & -e_1^- & 0 & 1 & 0 & 1 \\ \bar{h}_1^+ - \bar{h}_2^+ & \bar{h}_1^- - \bar{h}_2^- & \bar{h}_1^+ + \bar{h}_2^+ & \bar{h}_1^- + \bar{h}_2^- & 0 & 0 & 0 & 0 \\ \bar{h}_2^+ e_2^+ + \bar{h}_1^+ e_1^+ & \bar{h}_2^- e_2^- + \bar{h}_1^- e_1^- & \bar{h}_1^+ e_1^+ - \bar{h}_2^+ e_2^+ & \bar{h}_1^- e_1^- - \bar{h}_2^- e_2^- & 0 & 0 & 0 & 0 \end{bmatrix} \begin{bmatrix} p_1^+(0) \\ p_1^-(0) \\ p_2^+(0) \\ p_2^-(0) \\ p_1^+(L) \\ p_1^-(L) \\ p_2^+(L) \\ p_2^-(L) \end{bmatrix} = 0 \tag{8.140}$$

where, $e_j^{\mp} = e_j^{\mp}(L)$, $j = 1, 2$. The wave transfer matrix, $\mathbf{T}$ in Equation (8.129), may be derived now by transferring the first two columns of the coefficients matrix in Equation (8.140) to the right of the equality sign and then multiplying the resulting equation by the inverse of the remaining coefficients matrix.

A filter assembly is similar to the catalytic converter assembly shown in Figure 8.7a with the honeycomb monolith replaced by the filter monolith and may be similarly modeled acoustically by using the two-port wave transfer matrix extracted from Equation (8.140). Modeling and transmission loss of a typical filter unit is discussed in Allam and Åbom with experimental validation [20].

The foregoing analysis presumes rectangular monolith pores. If the pore geometry is approximated by a circle of equal area, Equations (8.117) and (8.118) are still valid, but the terms in brackets on the right should be replaced by $2v_r'/a$, where v_r' denotes the radial component of the particle velocity and a denotes the hydraulic radius of the pore. Then, Equation (8.120) becomes simply $w = v_r'$, the function $\Gamma(\xi)$ in Equation (8.119) is given by $\Gamma(\xi) = \mathrm{J}_2(\xi)/\mathrm{J}_o(\xi)$ and the wall thickness h is understood as a notional average quantity. Figure 8.9 still represents the propagation constants for the circular duct case with adequate accuracy. Also, the function $H(x, y)$ should be replaced by $H(r)$, which is given by Equation (8.27), and $\bar{H}_j^{\mp}, j = 1, 2$, are given by Equation (8.47) evaluated for the propagation constants $K_j^{\mp}$. The circular duct approximation may be adequately accurate for rectangular ducts with an aspect ratio close to unity and may be preferred because the propagation constants are quicker to determine.

References

[1] A.D. Pierce, *Acoustics: An Introduction to Its Physical Principles and Applications,* (New York: McGraw-Hill, 1981).

[2] J.W.S. Rayleigh, *Theory of Sound,* Volume II, (New York: Dover Publications, 1945).

[3] G. Kirchhoff, Üeber den Einfluss der Wärmelettung in einem Gase auf dir Schallbewegung, *Annalen der Physik* **134** (1868), 177–193 (English translation in R.B. Lindsay, Ed., *Benchmark Papers in Acoustics: Physical Acoustic*, (Pennsylvania: Dowden, Hutchinson and Ross Inc., 1974).

[4] H. Tijdeman, On the propagation of sound in cylindrical tubes, *J. Sound Vib.* **39** (1975), 1–33.

[5] M.R. Stinson, The propagation of plane sound waves in narrow and wide circular tubes and generalization to uniform tubes of arbitrary cross-sectional shape, *J. Acoust. Soc. Am.* **89** (1991), 550–558.

[6] E. Dokumaci, On the effect of viscosity and thermal conductivity on sound propagation in ducts: a re-visit to the classical theory with extensions for higher order modes and presence of mean flow, *J. Sound Vib.* **333** (2014), 5583–5599.

[7] M. Bruneau, Ph. Herzog, J. Kergomard and J.D. Polack, General formulation of the dispersion equation in bounded viscothermal fluid and applications to some simple geometries, *Wave Motion* **11** (1989), 441–451.

[8] R.J. Astley and A. Cummings, Wave propagation in catalytic converters: formulation of the problem and finite element solution scheme, *J. Sound Vib.* **188** (1995), 635–657.

[9] E. Dokumaci, Sound transmission in narrow pipes with superimposed uniform mean flow and acoustic modeling of automobile catalytic converters, *J. Sound Vib.* **182** (1995), 799–808.

[10] Y. Aurégan, M. Pachebat and V. Pagneux, Hydrodynamic modes in pipes with superimposed mean flow and viscothermal effects, *J. Sound Vib.* **218** (1998), 735–740.

[11] E. Dokumaci, A note on transmission of sound in a wide pipe with mean flow and viscothermal attenuation, *J. Sound Vib.* **208** (1997), 653–655.

[12] K.S. Peat, A first approximation to the effects of mean flow on sound propagation through cylindrical capillary tubes, *J. Sound Vib.* **175** (1994), 475–489.

[13] K.-W. Jeong and J.-G. Ih, A numerical study on the propagation of sound through capillary tubes with mean flow, *J. Sound Vib.* **198** (1996), 67–79.

[14] S. Allam and M. Åbom, Investigation of damping and radiation using full plane wave decomposition in ducts. *J. Sound Vib.* **292** (2006), 519–534.

[15] E. Dokumaci, On transmission of sound in circular and rectangular narrow pipes with superimposed mean flow, *J. Sound Vib.* **210** (1998), 375–389.

[16] E. Dokumaci, On the effect of viscosity and thermal conductivity on sound power transmitted in uniform circular ducts, *J. Sound Vib.* **363** (2016), 560–570.

[17] K.S. Peat, Acoustic wave motion along a narrow duct with a temperature gradient, *Acoustica* **84** (1998), 57–65.

[18] E. Dokumaci, An approximate dispersion equation for sound waves in a pipe with ambient gradients, *J. Sound Vib.* **240** (2001), 607–646.

[19] M. Knutsson and M. Åbom, Acoustic modeling of charge air coolers, *J. Vib. Acoust.* **139** (2017), 1–9.

[20] S. Allam and M. Åbom, Sound propagation in an array of narrow porous channels with application to diesel particulate filters, *J. Sound Vib.* **291** (2006), 882–901.

9 Reflection and Radiation at Open Duct Terminations

9.1 Introduction

Fluid machinery action generates both gross (mean) fluid flow and acoustic waves. The latter are convected with the mean flow and radiated from the open end of the inlet and outlet ducts and can be a noise nuisance for the exterior environment. In duct-borne noise control, acoustic design considerations are largely influenced by the exchange of the fluid conveyed with the exterior environment. Duct-borne acoustic waves may also be radiated to the exterior environment by vibrations of the elastic parts of ductwork walls. This is usually called shell noise and treated separately (see Section 5.8).

Acoustics of open duct terminations are described by the reflection and radiation (transmission to the exterior medium) of incident sound waves. In Sections 9.3 and 9.4, we discuss calculation of the reflection characteristics of flanged and unflanged open ends of circular ducts and present models which may be used as realistic one-port boundary elements in block diagrams. Modeling of the radiation characteristics is considered in Section 9.5.

Mathematical modeling of open duct terminations transpires to be a challenging problem. Notwithstanding the ambiguity and diversity of the actual environments outside open terminals of practical duct systems, coupling of an in-duct acoustic field with the external environment results in mixed boundary value problems which are difficult to solve by using the known analytical or numerical techniques. To simplify the problem, it is generally assumed that only outgoing sound waves exist in the exterior medium. Still, the mixing of the fluid discharged from an open duct termination with the external fluid, or the motion induced in the external fluid by its suction into a duct, fluid–structure interaction at the edges of an open end and the relative motion of the external fluid can have important consequences on reflection and radiation characteristics. A mathematical model which has been fairly successful in accounting for such mean flow effects in unflanged circular ducts is discussed in Section 9.4. When mean flow effects may be neglected, the integral equation formulation of the wave equation provides a general theoretical framework for the determination of the reflection and radiation characteristics at the open end of a duct. This formulation and its applications are discussed in Section 9.3.

9.2 Reflection Matrix and End-Correction

From Equation (2.5), the reflection matrix, $\mathbf{r}$, at the open end of a duct is defined by the relationship $\mathbf{P}^- = \mathbf{r}\mathbf{P}^+$, where $\mathbf{P}^{\mp}$ denote the modal pressure wave components vectors (see Section 6.4.2) at the open end. This is a generalization of the concept of the plane-wave reflection coefficient, which is defined simply as $r = p^-/p^+$ (see Section 3.5.3). In general, we assume that $q(\geq 1)$ modes reach the open end and that these are all cut-on modes. This assumption will be true if the open end is sufficiently far from other discontinuities in the duct. Then $\mathbf{P}^+ = \left\{p_1^+ \; p_2^+ \; \ldots \; p_q^+\right\}$ and, since in theory the incident cut-on modes may generate an infinite number of reflected modes, we should take $\mathbf{P}^- = \{p_1^- \quad p_2^- \quad \cdots\}$, where curly brackets denote a column vector. Obviously, only the first q (≥ 1) reflected modes will be cut-on and the rest evanescent, but the evanescent reflected modes are not dispensable, because a sufficient number of them are necessary for convergence to the correct representation of the acoustic boundary condition at the open end. Thus, theoretically, the reflection matrix is of size $\infty \times q$.

Let an element of reflection matrix $\mathbf{r}$ at row position $\nu \in 1, 2, \ldots, \infty$ and column position $\mu \in 1, 2, \ldots, q$ be denoted by $r_{\nu\mu}$. This is called a modal reflection coefficient. It follows from the definition of $\mathbf{r}$ that $p_\nu^- = r_{\nu 1}p_1^+ + r_{\nu 2}p_2^+ + \cdots + r_{\nu q}p_q^+$; that is, the incident modes $\mu = 1, 2, \ldots, q$ contribute to the reflected modes $\nu \in 1, 2, \ldots, \infty$ in proportion to $r_{\nu\mu}$. If only the plane-wave mode is incident at the open end, then $q = 1$ and the foregoing linear combination reduces to $p_\nu^- = r_{\nu 1}p_1^+$, $\nu = 1, 2, \ldots, \infty$. It is a hypothesis of the one-dimensional theory that all reflected evanescent modes may be neglected. Then, it suffices to consider the relationship $p_1^- = r_{11}p_1^+$, where r_{11} is what we have called previously plane-wave reflection coefficient.

For circular ducts, however, the exact transverse eigenvalues and eigenfunctions emerge from analysis in double-index notation, which we have denoted in Sections 6.5.4.2 and 6.5.4.3 by the circumferential (or azimuthal) and radial mode indices $m, n = 0, 1, 2, \ldots$, respectively. Then it is more convenient to order the modal reflection coefficients also in accordance with this notation, but it will be necessary to do some bookkeeping for tracking the modes in their cut-on order. The commonly used scheme is to denote a modal reflection coefficient as $r_{mn\ell}$, where m $(\in 0, 1, 2, \ldots)$ and n $(\in 0, 1, 2, \ldots)$ denote the circumferential and radial order of the incident cut-on mode, and $\ell = 0, 1, 2,, \ldots$ denotes the radial order of a reflected mode of circumferential order m. Then, if we define a reflection matrix for all modes of circumferential order m as $\mathbf{P}_m^- = \mathbf{r}_m\mathbf{P}_m^+$, where $\mathbf{P}_m^+ = \left\{p_{m0}^+ \quad p_{m1}^+ \quad \cdots\right\}$ and $\mathbf{P}_m^- = \{p_{m0}^- \quad p_{m1}^- \quad \cdots\}$, a reflected pressure wave mode of circumferential index m and radial index ℓ, $p_{m\ell}^-$, is given by the linear combination $p_{m\ell}^- = r_{m0\ell}p_{m0}^+ + r_{m1\ell}p_{m1}^+ + \cdots$. The exact transverse modes of rectangular ducts, and other sections of separable geometry, also come in double index notation and the modal reflection coefficients may be defined by the same scheme.

In this notation, the plane-wave reflection coefficient is denoted as r_{000}. The determination of r_{000} has received the attention of many authors, since one-dimensional elements are widely used in practical acoustic modeling of ductworks. In the usual polar form, it is written as $r_{000} = |r_{000}|e^{i\angle r_{000}}$, where "$\angle$" means "argument of." But the phase is often expressed as end-correction. This is useful insofar as it provides a comparison with the phase of the idealized pressure release condition, which asserts that $r_{000} = -1$ at an open end (see Section 2.3). For plane waves superimposed on a uniform mean flow of Mach number $\bar{M}_o$, the pressure wave components propagate as $p^{\mp}(x) = p^{\mp}(0)e^{\mp ik_o x/(1\mp\bar{M}_o)}$ (see Section 3.4). Then, $r_{000}(x) = r_{000}(0)e^{-i2k_o x/\left(1-\bar{M}_o^2\right)}$ and, therefore, if the waves were to continue traveling outward from the open end, they would have the phase of the pressure release condition, π radians, at a distance

$$\delta = \frac{\pi + \angle r_{000}}{2k_o}\left(1 - \bar{M}_o^2\right) \tag{9.1}$$

from the open end. This formula may be used to calculate the argument of the reflection coefficient when the end-correction is known and vice versa.

9.3 Flanged and Unflanged Open Terminations without Mean Flow

9.3.1 Exterior Surface Helmholtz Equation

Depicted in Figure 9.1 is an open duct termination, of area S, of a duct that is part of a structure occupying a domain D of boundary surface ∂D ($S \subset \partial D$) of unit outward normal vector $\mathbf{n}$, immersed in a free space E. The acoustic pressure field p' in E is governed, in frequency domain, by the Helmholtz equation $\left(\nabla^2 + k_o^2\right)p' = 0$, where $k_o = \omega/c_o$ and c_o denotes the speed of sound of the fluid in E, which is assumed to be homogeneous and initially quiescent. Let $p'(\mathbf{x})$ be a solution of the Helmholtz equation at a point $\mathbf{x}$. The following integral relationship is classical [1–2]:

$$a(\mathbf{x})p'(\mathbf{x}) = -\int_{\partial D} G(R)\mathbf{n}\cdot\nabla p'\mathrm{d}A + \int_{\partial D} p'\mathbf{n}\cdot\nabla G(R)\mathrm{d}A, \quad a(\mathbf{x}) = \begin{cases} 1 & \text{if } \mathbf{x} \in E \\ \dfrac{\Omega(\mathbf{x})}{4\pi} & \text{if } \mathbf{x} \in \partial D \end{cases}. \tag{9.2}$$

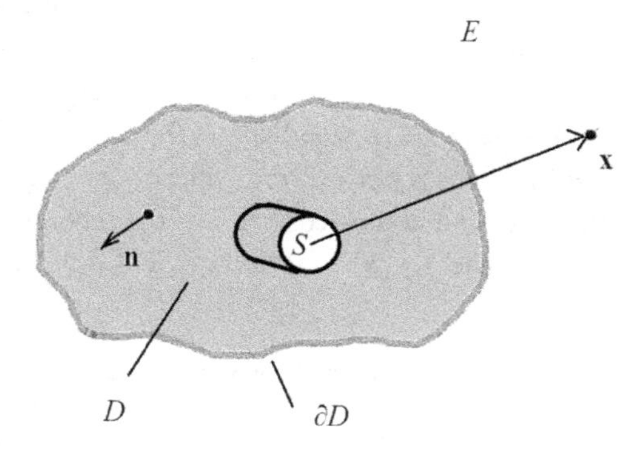

Figure 9.1 Coupling of an open duct termination with exterior free field.

Here, $R = |\mathbf{x} - \mathbf{y}|$ with $\mathbf{y} \in \partial D$, $\mathrm{d}A$ denotes an elementary area on ∂D, $\Omega(\mathbf{x})$denotes the outer solid angle of surface ∂D at point $\mathbf{x} \in \partial D$ and $G = G(R)$denotes the free-space Green function

$$G(R) = \frac{\mathrm{e}^{\mathrm{i}k_o R}}{4\pi R}. \tag{9.3}$$

Equation (9.2) is called the exterior Helmholtz integral formula if $\mathbf{x} \in E$, and it is called the exterior surface integral equation if $\mathbf{x} \in \partial D$.[1]

This classical integral formulation provides a general methodology for the determination of the radiation and reflection characteristics at an open end of a duct: given the normal derivative $(\mathbf{n}\cdot\nabla)$ of the acoustic pressure on ∂D, first the exterior surface Helmholtz integral equation is solved for the acoustic pressure distribution on ∂D and, hence, the reflection matrix at the open end S (Section 9.3.2). This solution is then used in the exterior Helmholtz integral formula to calculate the radiated acoustic pressure at any point $\mathbf{x} \in E$ (Section 9.5.3).

In general, the solution of Equation (9.2) with $\mathbf{x} \in \partial D$ for arbitrary open end geometry requires advanced numerical techniques for geometrical discretization of the boundary ∂D and accurate evaluation of the integrals. Such techniques are described in books on the boundary element method [3]. Several applications of this method are reported in the literature for various forms of open end geometry [4–6], but these are limited to the plane-wave reflection coefficient of hollow circular ducts.

Common forms of open ends of ducts may usually be modeled as unflanged or flanged terminations. The unflanged end model assumes that the duct thickness (or the flange size, if there is one) is much smaller compared to the wavelengths of interest. On the other hand, the flanged end model assumes that the flange size (or the duct thickness, if there is no flange) is large compared to the wavelength. These models are also useful in that the real duct terminations usually lie between these extremes.

9.3.2 Flanged Open End

Shown in Figure 9.2 is the flanged open end configuration. For $\mathbf{x} \in \partial D$, this is obtained from Figure 9.1 by unfolding ∂D to an infinite plane solid surface so that domains D and E become two semi-infinite halves of a three-dimensional free space separated by the plane surface ∂D, which is assumed to be solid everywhere except over S (this unfolding is not permissible if $\mathbf{x} \in E$, see Section 9.5.2).

In view of the identity $\mathbf{n}\cdot\nabla G(R) = \cos\gamma(\mathbf{y},\mathbf{x})\partial G/\partial R$, where $\gamma(\mathbf{y},\mathbf{x})$ denotes the angle between the unit normal vector $\mathbf{n}(\mathbf{y})$ and the vector $\mathbf{x} - \mathbf{y}$, the second integral in Equation (9.2) for $\mathbf{x} \in \partial D$ vanishes since $\gamma(\mathbf{y},\mathbf{x}) = \pi/2$ for $\mathbf{x},\mathbf{y} \in \partial D$. The normal

[1] A derivation of Equation (9.2) when E is a closed domain is given in Section 2.5.2. Take $E = \sum \cap D$, where Σ is a sphere of radius r enclosing D, and let $r \to \infty$. For outgoing waves (the Sommerfeld radiation condition), this process gives Equation (9.2) for $\mathbf{x} \in E$, and converging $\mathbf{x}$ on to ∂D as described in Section 2.5.2 gives its surface form. But $a(\mathbf{x})$ now denotes the outer solid angle, since we converge from the exterior of ∂D.

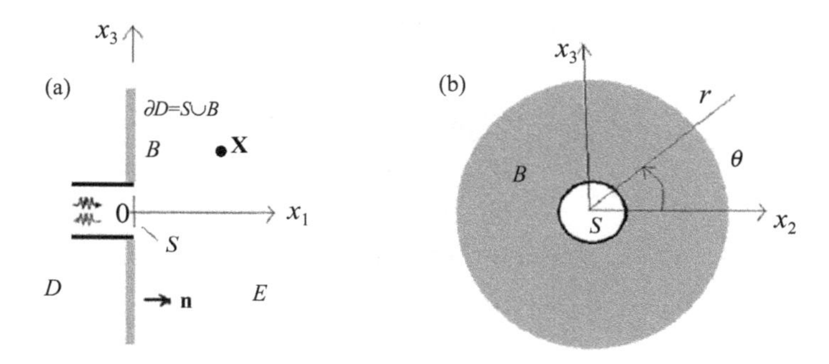

Figure 9.2 Flanged open end.

derivative of the acoustic pressure in the first integral, $\mathbf{n}\cdot\nabla p' = \partial p'/\partial x_1$, also vanishes everywhere on ∂D except over S, because, by the acoustic momentum equation for the exterior we have $-\mathrm{i}\omega\rho_o\mathbf{v}' + \nabla p' = 0$ and $\mathbf{n}\cdot\mathbf{v}' = 0$ over a solid flange surface. Hence, the final form of Equation (9.2) for $\mathbf{x} \in \partial D$ becomes simply

$$p'(\mathbf{x}) = -2\int_S G(R)\frac{\partial p'}{\partial x_1}\mathrm{d}S, \mathbf{x} \in S. \tag{9.4}$$

Now, considering a uniform duct and taking its origin, $x_1 = 0$, at the open end (Figure 9.2), we have from Equations (6.65) and (6.66):

$$p'(0,\bullet) = \mathbf{\Phi}(\bullet)(\mathbf{P}^+(0) + \mathbf{P}^-(0)) \tag{9.5}$$

$$\frac{\partial p'}{\partial x_1}(0,\bullet) = \mathrm{i}k_o\mathbf{\Phi}(\bullet)(\mathbf{K}^+\mathbf{P}^+(0) + \mathbf{K}^-\mathbf{P}^-(0)), \tag{9.6}$$

where $\mathbf{K}^{\mp}$denote the axial propagation constants matrices, and the meanings of the other symbols are the same as defined in Section 6.4.2. In previously published work on the solution of Equation (9.4), it has been traditional to neglect the mean flow in the duct and the temperature and pressure difference in the duct and in the exterior. We will proceed similarly. (The latter assumption can be relaxed by replacing k_o in Equation (9.6) by ω/c_j, where c_j denotes the speed of sound of the fluid in the duct.) Hence, since acoustic pressure at the open end must be continuous, we may substitute Equations (9.5) and (9.6) in Equation (9.4):

$$\mathbf{\Phi}(\bullet_\mathbf{x})(\mathbf{P}^+ + \mathbf{P}^-) = -\mathrm{i}2k_o\left(\int_S G(R)\mathbf{\Phi}(\bullet)\mathrm{d}S\right)(\mathbf{K}^+\mathbf{P}^+ + \mathbf{K}^-\mathbf{P}^-), \mathbf{x} \in S, \tag{9.7}$$

where $(\bullet_\mathbf{x})$ denotes the transverse coordinates at point $\mathbf{x}$ and we have written $\mathbf{P}^{\mp} = \mathbf{P}^{\mp}(0)$ for simplicity of notation. Multiplying both sides of this equation by $\breve{\mathbf{\Phi}}^\mathrm{T}(\bullet_\mathbf{x})$, where the superscript "T" denotes array transposition and an inverted over-arc denotes the complex conjugate, and integrating the resulting equation over the duct cross section

$$\mathbf{H}\ (\mathbf{P}^+ + \mathbf{P}^-) = -\mathbf{B}(\mathbf{K}^+\mathbf{P}^+ + \mathbf{K}^-\mathbf{P}^-), \tag{9.8}$$

where

$$\mathbf{H} = \int_S \breve{\boldsymbol{\Phi}}^{\mathrm{T}}(\bullet)\boldsymbol{\Phi}(\bullet)\ \mathrm{d}S \tag{9.9}$$

$$\mathbf{B} = -\mathrm{i}2k_o \int_S \left(\breve{\boldsymbol{\Phi}}^{\mathrm{T}}(\bullet_{\mathbf{x}}) \int_S G(R)\boldsymbol{\Phi}(\bullet)\ \mathrm{d}S \right) \mathrm{d}S(\bullet_{\mathbf{x}}). \tag{9.10}$$

Rearranging Equation (9.8), we have

$$\mathbf{P}^+ = -(\mathbf{H} + \mathbf{B}\mathbf{K}^+)^{-1}(\mathbf{H} + \mathbf{B}\mathbf{K}^-)\mathbf{P}^-. \tag{9.11}$$

We may partition this equation as follows:

$$\begin{bmatrix} \mathbf{P}_q^+ \\ \mathbf{P}_e^+ \end{bmatrix} = \begin{bmatrix} \mathbf{F}_{qq} & \mathbf{F}_{qe} \\ \mathbf{F}_{eq} & \mathbf{F}_{ee} \end{bmatrix} \begin{bmatrix} \mathbf{P}_q^- \\ \mathbf{P}_e^- \end{bmatrix}. \tag{9.12}$$

Here, subscripts q and e denote the cut-on and evanescent modes incident at the open end. Assuming that the open end is sufficiently far from other discontinuities in the duct, we may take $\mathbf{P}_e^+ = 0$. Then, the reflected evanescent and the reflected cut-on modes are related as $\mathbf{P}_e^- = -\mathbf{F}_{ee}^{-1}\mathbf{F}_{eq}\mathbf{P}_q^-$. Thus, the reflection matrix for the cut-on modes, $\mathbf{r}_q$, is $\mathbf{r}_q = \mathbf{F}_{qq} - \mathbf{F}_{qe}\mathbf{F}_{ee}^{-1}\mathbf{F}_{eq}$ and the complete reflection matrix $\mathbf{r}$ may be expressed as

$$\mathbf{r} = \begin{bmatrix} \mathbf{I} \\ -\mathbf{F}_{ee}^{-1}\mathbf{F}_{eq} \end{bmatrix} \mathbf{r}_q. \tag{9.13}$$

For hard-walled ducts, in view of the orthogonality property of the eigenfunctions, the matrix $\mathbf{H}$ will be a diagonal matrix. Numerical evaluation of $\mathbf{H}$ is quite straightforward, but the evaluation of matrix $\mathbf{B}$ presents a significant computational challenge and the existing solutions are limited to circular ducts.

9.3.2.1 Circular Ducts

In double-index notation for the transverse duct modes, the general solution given in the previous section is applied for a circumferential mode of order m. We assume a hard-walled duct and consider the submatrix of its modal matrix which includes all eigenfunctions of circumferential order m:

$$\boldsymbol{\Phi}_m(r,\theta) = [\,\psi_m(\alpha_{m0}r)\quad \psi_m(\alpha_{m1}r)\quad \psi_m(\alpha_{m2}r)\quad \ldots]\ \mathrm{e}^{\mathrm{i}m\theta}. \tag{9.14}$$

The radial functions ψ_m are defined in Equation (6.98) for annular ducts with $b \le r \le a$ and can be replaced by $\mathrm{J}_m(\alpha_{mn}r)$ for hollow ducts with $r \le a$. We denote the subarrays of $\mathbf{B}$, $\mathbf{H}$, $\mathbf{K}^{\mp}$ and $\mathbf{P}^{\mp}$ corresponding to all modes of circumferential order m similarly as $\mathbf{B}_m$, $\mathbf{H}_m$, $\mathbf{K}_m^{\mp}$ and $\mathbf{P}_m^{\mp}$, respectively. From inspection of Equation (9.10), it can be seen that an element of $\mathbf{B}_m$ in row position n and column position ℓ can be expressed as

$$B_{mn\ell} = -\mathrm{i}2k_o \int_0^{2\pi}\int_b^{a}\left(\int_0^{2\pi}\int_b^{a}\frac{\mathrm{e}^{\mathrm{i}k_oR}}{4\pi R}\psi_m(\alpha_{mn}r')\psi_m(\alpha_{m\ell}r)\mathrm{e}^{\mathrm{i}m(\theta-\theta')}r\mathrm{d}r\mathrm{d}\theta\right)r'\mathrm{d}r'\mathrm{d}\theta', n,\ell=0,1,2,\ldots \tag{9.15}$$

where

$$R = \sqrt{r^2 + r'^2 - 2rr'\cos(\theta-\theta')}. \tag{9.16}$$

Hence, Equation (9.8) expands to the infinite set of equations

$$H_{mnn}\ (p_{mn}^{+} + p_{mn}^{-}) = -\sum_{\ell=0}^{\infty} B_{mn\ell}\left(K_{m\ell}^{+}p_{m\ell}^{+} + K_{m\ell}^{-}p_{m\ell}^{-}\right), n = 0,1,2\ldots \tag{9.17}$$

where $p_{mn}^{\mp}$ denote the pressure wave components for mode (m,n), that is, $\mathbf{P}_m^{\mp} = \{p_{m0}^{\mp}\ p_{m1}^{\mp}\ldots\}$, $K_{m\ell}$ are given by Equation (6.49), and Watson [7] gives

$$H_{mn\ell} = \begin{cases} \frac{1}{2}r^2\left\{\left(1-\frac{m^2}{\alpha_{mn}^2 r^2}\right)\psi_m^2(\alpha_{mn}r) + \psi_m'^2(\alpha_{mn}r)\right\}\Bigg|_{r=b}^{r=a} & \text{if } n=\ell \\ 0 \quad \text{if } n \neq \ell \end{cases} \tag{9.18}$$

Therefore, Equation (9.28) may be recast as

$$\sum_{\ell=0}^{\infty}\left(H_{mn\ell}\delta_{n\ell} + B_{mn\ell}K_{m\ell}^{-}\right)p_{m\ell}^{-}(0) = -\sum_{\ell=0}^{\infty}\left(H_{mn\ell}\delta_{n\ell} + B_{mn\ell}K_{m\ell}^{+}\right)p_{m\ell}^{+}(0), n=0,1,2\ldots \tag{9.19}$$

where $\delta_{n\ell} = 1$ if $n=\ell$ else 0. Usefulness of this infinite set of equations depends on finding integration schemes which can successfully avoid the impact of the singularity of $G(R)$ at $R=0$ and produce converging results for the integrals in Equation (9.15). Norris and Sheng described a solution for the case of $m=0$, which enabled them to compute the plane-wave reflection coefficient r_{000} for hollow ducts [8]. Their results are given in Figure 9.3 as a function of the Helmholtz number ka. Silva et al. give the following curve fit formulae for this solution [9]:

$$|r_{000}| = \frac{1+0.730(k_oa)^2}{1+1.730(k_oa)^2+0.372(k_oa)^4+0.0231(k_oa)^6} \tag{9.20}$$

$$\frac{\delta}{a} = 0.8216\frac{1+0.244(k_oa)^2}{1+0.723(k_oa)^2-0.0198(k_oa)^4+0.00366(k_oa)^6}, \tag{9.21}$$

which are stated to be accurate to within 2% error for $k_oa < 3$. Here, δ denotes the end correction defined by Equation (9.1). A more general solution of Equations (9.19) for arbitrary m and any number of cut-on incident modes is described by Zorumski [10].

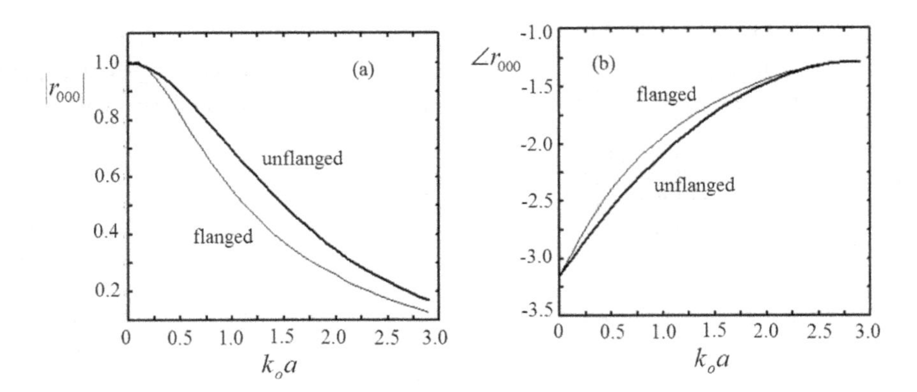

Figure 9.3 Modulus (a) and phase (b) of the plane-wave reflection coefficient of flanged and unflanged open duct terminations.

A disadvantage of these solutions is the slow convergence of the results. Accordingly, a large number of modes are required to obtain adequate accuracy. Several authors have proposed techniques for improving the speed of convergence, but applications are limited to the plane-wave reflection coefficient of hollow circle geometry [11–13]. An interesting approach which seems to give accurate results for the reflection matrix without convergence problems is that proposed by Cho [14]. He envisages a circular hard-walled duct which is smoothly coupled at its open end to a hard-walled hyperboloidal infinite duct and calculates the reflection matrix at the junction of the two parts. This geometry degenerates to the infinite flange case when the flair angle of the hyperboloid is 90°. In this approach, the free-space Green function is not needed, since the Green function of the hyperboloidal extension, expanded in terms of associated eigenfunctions, can be used instead.

The effect of finite flange size on the reflection coefficient r_{000} is investigated by da Silva et al. using the boundary element method and neglecting the effect of the flange thickness [6]. Their results are shown in Figure 9.4. The authors also present formulae which fit these results accurately.

9.3.3 Unflanged Open End

For an unflanged open end, there are two options to consider as boundary ∂D for an integral equation formulation. The boundary shown in Figure 9.5a is appropriate for the determination of the open-end reflection coefficient by solving the exterior surface Helmholtz integral equation. In this case, $\partial D = S \cup S_D$, where S_D denotes the outer surface of the duct. The first integral in Equation (9.2) with $\mathbf{x} \in \partial D$ vanishes only on S_D, but the full equation has to be solved numerically by using the boundary element method.

The boundary in Figure 9.5b, on the other hand, is appropriate for the determination of the open end reflection coefficient by using the exterior Helmholtz integral formula, Equation (9.2) with $\mathbf{x} \in E$. We still have $\partial D = S \cup S_D$, but now S_D represents the

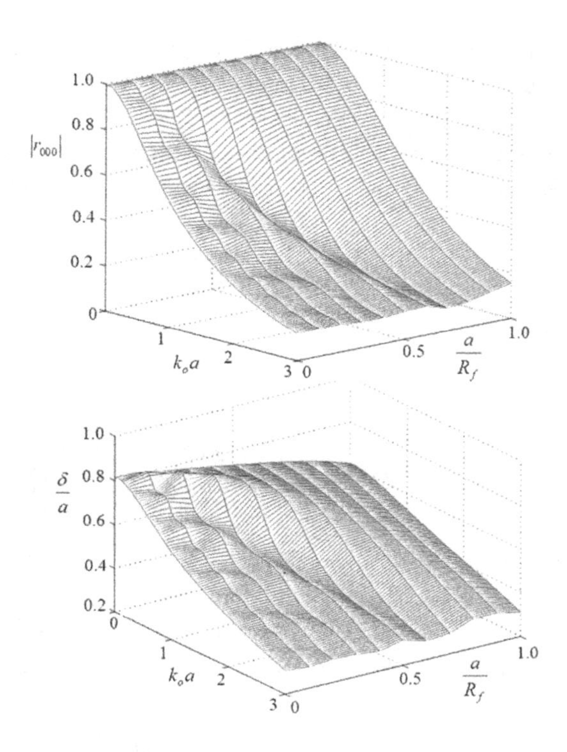

Figure 9.4 Modulus of the reflection coefficient r_{000} and the end-correction (see Equation 9.1) at flanged open end of a circular duct: R_f denotes the flange radius (taken from da Silva et al. [6], CC BY NC-ND 4.0, labels adapted to present notation).

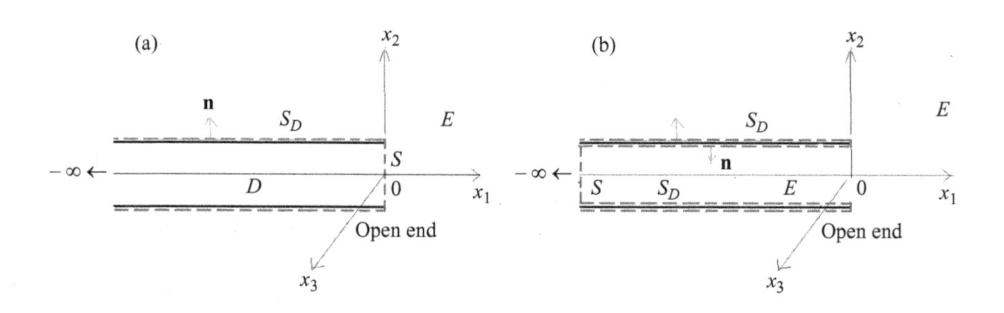

Figure 9.5 Unflanged open end boundary models

union of the inner and outer surfaces of the duct and S denotes the duct section at $x_1 = -\infty$. In this case, the first integral in Equation (9.2) with $\mathbf{x} \in E$ vanishes on S_D. It also vanishes for $\mathbf{x} \in S$, because $R \to \infty$ for points on this surface. Thus, the first integral vanishes completely and Equation (9.2) with $\mathbf{x} \in E$ reduces to

$$p'(\mathbf{x}) = -\int_{S_D} p' \frac{\partial G}{\partial R} \cos \gamma(\mathbf{y}, \mathbf{x}) \mathrm{d}A, \mathbf{x} \in \text{open end}, \tag{9.22}$$

where $\gamma(\mathbf{y}, \mathbf{x})$ denotes the angle between the unit normal vector $\mathbf{n}(\mathbf{y})$ and the vector $\mathbf{x} - \mathbf{y}$. If the duct is circular, this integral equation is tractable to an analytical solution for the calculation of the plane-wave reflection coefficient. In their 1948 paper, Levine and Schwinger, assuming no mean flow and no mismatch between the temperature in the duct and in the exterior, give the following explicit solution for r_{000} of hard-walled ducts [15]:

$$\log|r_{000}| = -\frac{2k_o a}{\pi}\int_0^{k_o a} \frac{\tan^{-1}\left\{-\frac{\mathrm{J}_1(x)}{\mathrm{Y}_1(x)}\right\}}{x\sqrt{k_o^2a^2 - x^2}}\,\mathrm{d}x \tag{9.23}$$

$$\frac{\delta}{a} = \frac{1}{\pi}\int_0^{ka} \frac{\log\left\{\pi \mathrm{J}_1(x)\sqrt{\mathrm{J}_1^2(x) - \mathrm{Y}_1^2(x)}\right\}}{x\sqrt{k_o^2a^2 - x^2}}\,\mathrm{d}x - \frac{1}{\pi}\int_0^{\infty} \frac{\log\{2\mathrm{I}_1(x)\mathrm{K}_1(x)\}}{x\sqrt{k_o^2a^2 + x^2}}\,\mathrm{d}x. \tag{9.24}$$

Here, J_1 and Y_1 denote first order Bessel functions of the first and second kind, I_1 and K_1 denote first order modified Bessel functions of the first and second kind, and δ is the end-correction defined in Equation (9.1). The variation of the modulus and phase of r_{000} are shown in Figure 9.3 by thick lines. The following curve-fits to foregoing equations are given by da Silva et al. [6]:

$$|r_{000}| = \frac{1 + 0.800(k_o a)^2}{1 + 1.300(k_o a)^2 + 0.266(k_o a)^4 + 0.0263(k_o a)^6} \tag{9.25}$$

$$\frac{\delta}{a} = 0.6133\frac{1 + 0.0599(k_o a)^2}{1 + 0.238(k_o a)^2 - 0.0153(k_o a)^4 + 0.00150(k_o a)^6}. \tag{9.26}$$

These are stated to be accurate with less than 2% error for $k_o a < 3$.

Solution of the same problem is also discussed in Morse and Feshbach [1] and in Noble [16]. Rawlins extended the solution of Levine and Schwinger for uniformly lined ducts [17]. His results showed that a liner of a few wavelengths long can reduce the sound radiated considerably. Peake presented an extension for oblique slices to the duct axis under the low frequency and low slice angle restriction $\phi k_o a << 1$, where ϕ denotes the angle of the slice from a plane perpendicular to the duct axis in radians [18]. Such compact open-end asymmetry, although it does not have discernible effect on the total sound power radiated, can alter the directivity pattern of the radiation in that less power is radiated to the closed side of the slice.

9.4 Reflection Matrix at an Unflanged Open End with Mean Flow

9.4.1 The Exhaust Problem

The geometry of the exhaust problem is depicted in Figure 9.6. This consists of a hard-walled semi-infinite uniform duct carrying an axial uniform mean flow of

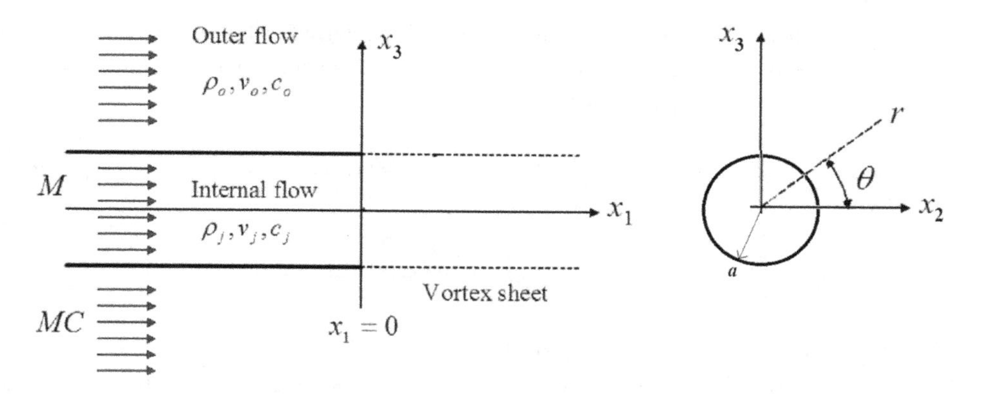

Figure 9.6 A duct with unflanged outlet discharging to a free space.

density ρ_j, velocity v_j and speed of sound c_j, from which issues a uniform mean flow (subsequently also called a jet) of the same cross-sectional area, the whole being immersed in a fluid of density ρ_o, speed of sound c_o and moving axially with velocity v_o relative to the duct. It is convenient to define the following non-dimensional parameters:

$$\gamma = \frac{\rho_o}{\rho_j}, M = \frac{v_j}{c_j}, C = \frac{c_j}{c_o}, \alpha = \frac{v_o}{v_j}. \tag{9.27}$$

The duct thickness is assumed to be negligibly small, and the vortex sheet formed between the jet and the outer flow is taken as the semi-infinite extension of the duct wall in the region $x_1 > 0$, where x_1 denotes the duct axis. For the exhaust problem $M > 0$, and the possibility that the vortex sheet may be separating two fluids of different mean density and speed of sound is included in the formulation. The outer and internal mean flows are restricted by the condition $0 \leq \alpha \leq 1$.

The acoustic fields inside the duct and the vortex sheet and in the outer flow are governed by the convected wave equation, Equation (6.47). Here, it is convenient to formulate these fields in terms of the acoustic velocity potential ϑ', which is defined as $\mathbf{v}' = \nabla\vartheta'$, where $\mathbf{v}'$ denotes the acoustic particle velocity. We have used this transformation in Section 6.9 and shown that in the frequency domain, the acoustic velocity potential satisfies the convected wave equation, Equation (6.198), and that it is related to the acoustic pressure by Equation (6.197). Hence, putting $\mathbf{M}_o = \mathbf{e}_1 M$ in Equation (6.197) and (6.198), we may write the following equations for acoustic fields inside the duct and inside the vortex sheet:

$$p'(x_1, \bullet) = -\rho_j c_j \left(-ik + M\frac{\partial}{\partial x_1}\right)\vartheta'(x_1, \bullet), \bullet \in S, \; -\infty \leq x_1 \leq \infty \tag{9.28}$$

$$\left[\nabla^2 - \left(-ik + M\frac{\partial}{\partial x_1}\right)^2\right]\vartheta'(x_1, \bullet) = 0, \bullet \in S, \; -\infty \leq x_1 \leq \infty, \tag{9.29}$$

where $k = \omega/c_\mathrm{j}$ denotes the wavenumber referred to the speed of sound in the jet, $(\bullet)$ denotes the transverse coordinates and S denotes the duct cross-sectional area. Similarly, in the outer flow

$$p'(x_1, \bullet) = -\gamma\rho_\mathrm{j}c_\mathrm{j}\left(-\mathrm{i}k + \alpha M\frac{\partial}{\partial x_1}\right)\vartheta'(x_1, \bullet), \bullet \in S \cup P, \ -\infty \leq x_1 \leq \infty \tag{9.30}$$

$$\left[\nabla^2 - C^2\left(-\mathrm{i}k + M\alpha\frac{\partial}{\partial x_1}\right)^2\right]\vartheta'(x_1, \bullet) = 0, \bullet \in S \cup P, \ -\infty \leq x_1 \leq \infty, \tag{9.31}$$

where P denotes perimeter of S.

Solutions of the wave equations must satisfy the following boundary conditions:

(i) Normal component of the acoustic particle velocity vanishes on the duct walls:

$$\left.\frac{\partial\vartheta'}{\partial n}\right|_{x_1 \leq 0, \bullet\in P} = \left.\frac{\partial\vartheta'}{\partial n}\right|_{x_1 \leq 0, \bullet\in P_+} = 0, \tag{9.32}$$

where P_+ and P_- denote, respectively, just outer and inner sides of the boundary P (perimeter) of the duct cross section, and $\partial/\partial n$ denotes differentiation along the unit normal vector of P.

(ii) Acoustic pressure is continuous across the vortex sheet:

$$\left.\left(-\mathrm{i}k\vartheta' + M\frac{\partial\vartheta'}{\partial x_1}\right)\right|_{x_1>0, \bullet\in P_-} = \gamma\left.\left(-\mathrm{i}k\vartheta' + \alpha M\frac{\partial\vartheta'}{\partial x_1}\right)\right|_{x_1>0, \bullet\in P_-}. \tag{9.33}$$

(iii) Acoustic particle displacement is continuous across the vortex sheet:

$$\left.\left(-\mathrm{i}k\eta + M\frac{\partial\eta}{\partial x_1}\right)\right|_{x_1>0, \bullet\in P} = \left.\frac{\partial\vartheta'}{\partial n}\right|_{x_1>0, \bullet\in P_-} \tag{9.34}$$

$$\left.\left(-\mathrm{i}k\eta + M\alpha\frac{\partial\eta}{\partial x_1}\right)\right|_{x_1>0, \bullet\in P} = \left.\frac{\partial\vartheta'}{\partial n}\right|_{x_1>0, \bullet\in P_+}, \tag{9.35}$$

where η denotes the displacement of the vortex sheet from its mean position (extension of the duct wall). Note that the normal component of the particle velocity is determined by the material derivative of the vortex sheet displacement, since the fluid is in uniform mean motion on the two sides of the vortex sheet.

(iv) Waves are radiated away from the open end and only outgoing waves exist in the exterior environment.

(v) Appropriate edge condition is satisfied at the duct lip. This is controlled by the amount of the vorticity shed at the edge. The shedding of all available vorticity from the lip can be ensured by applying the full Kutta condition (smooth separation from the trailing edge).

At the present time, solution of the above formulated mixed boundary value problem is known only for hollow circular hard-walled ducts in cylindrical polar coordinates; that is, $\bullet \equiv r, \theta$. It is given by Munt [19–20], using the Wiener–Hopf method [16]. Solutions are also known for the associated problems of a circular annulus with an infinite center body [21–22] and co-axial circular ducts [23–24], which have more relevance for aircraft turbofan engines. Since the hollow duct case is more common, here we describe the salient steps of Munt's solution.

9.4.2 Circular Duct

Recalling the solution of the convected wave equation for the acoustic pressure (Section 6.5.2), Equation (9.28) implies that the acoustic velocity potential inside the duct and the vortex sheet may be decomposed into sum of modes propagating in opposite directions as $\vartheta' = \vartheta^+ + \vartheta^-$, where the superscripts "∓" denote, as usual, the wave modes propagating in the $\mp x_1$ directions. Also, since the acoustic velocity potential satisfies Equation (9.29), the convected wave equation, the circular transverse eigenfunctions in Equation (6.92) apply also for the velocity potential of the acoustic field inside the duct and the vortex sheet. Following Munt, we consider a single cut-on mode m, n of velocity potential

$$\vartheta^+_{mn}(r, \theta, x_1) = \left(\frac{B}{\mathrm{J}_m(\alpha_{mn}a)}\right)\mathrm{J}_m(\alpha_{mn}r)\mathrm{e}^{\mathrm{i}\left(kK^+_{mn}x_1 + m\theta\right)}, \qquad r < a \tag{9.36}$$

incident at the open end of the duct. Here, $B = 1$ m/s^2 denotes a scale factor, a denotes the duct radius, r and θ denote the radial and angular coordinates in cylindrical coordinates, α^2_{mn} denotes the transverse eigenvalue and K^+_{mn} denotes the axial propagation constant

$$K^+_{mn} = \frac{-M + \sqrt{1 - \frac{\alpha^2_{mn}}{k^2}\left(1 - M^2\right)}}{1 - M^2}. \tag{9.37}$$

The condition $k^2 > \left(1 - M^2\right)\alpha^2_{mn}$ holds, since the incident mode is cut-on.

The incident wave is partly reflected back into the duct and partly radiated out to the exterior environment. The acoustic velocity potential field ϑ' consists of the sum of the incident wave, ϑ^+_{mn}, plus a diffracted field which satisfies Equations (9.29) and (9.31) subject to the same boundary conditions stated above and has the same angular dependence as the incident wave. Then, the total acoustic velocity field in the duct may be expressed as

$$\vartheta'(r, \theta, x_1) = \vartheta^+_{mn} + \sum_{\ell=0}^{\infty} Bb_{mn\ell}\mathrm{J}_m(\alpha_{m\ell}r)\mathrm{e}^{\mathrm{i}\left(kK^-_{m\ell}x_1 + m\theta\right)}, \qquad r < a \tag{9.38}$$

where b_{mnl} are related to the modal reflection coefficient as explained below and $K^-_{m\ell}$ denote the axial propagation constants of the reflected waves

$$K^-_{m\ell} = \frac{-M - \sqrt{1 - \frac{\alpha^2_{mn}}{k^2}\left(1 - M^2\right)}}{1 - M^2}. \tag{9.39}$$

Upon substituting Equation (9.38) in Equation (9.28), we get for the acoustic pressure in the duct

$$p'(x_1,r,\theta) = \rho_j c_j\Bigg(ik\left(1-MK^+_{mn}\right)\frac{B}{J_m(\alpha_{mn}a)}J_m(\alpha_{mn}r)e^{i\left(kK^+_{mn}x_1+m\theta\right)} + \sum_{\ell=0}^{\infty} ik\left(1-K^-_{m\ell}M\right)Bb_{mn\ell}J_m(\alpha_{m\ell}r)e^{i\left(kK^-_{m\ell}x_1+m\theta\right)}\Bigg), r<a \tag{9.40}$$

For interpretation of b_{mnl}, we observe, recalling Equations (6.65) and (6.66), that the foregoing equation may be written in matrix notation as $p'(x_1,r,\theta) = \mathbf{\Phi}_m(r,\theta)\left[\mathbf{P}^+_m(x_1) \quad \mathbf{P}^-_m(x_1)\right]$, where $\mathbf{\Phi}_m$ denote the modal matrix encompassing all modes of circumferential order m, $\mathbf{P}^+_m(x_1) = \{0 \quad \cdots \quad p^+_{mn}(x_1) \quad 0 \quad \ldots\}$ and $\mathbf{P}^-_m(x_1) = \{p^-_{m0}(x_1) \quad p^-_{m1}(x_1) \quad \ldots \quad p^-_{m\ell}(x_1) \quad \ldots\}$. With the duct origin $x_1 = 0$ at the open end, the modal pressure reflection coefficients $r_{mn\ell}$ are defined as $p^-_{m\ell}(0) = r_{mn\ell}p^+_{mn}(0)$ and inspection of Equation (9.40) shows that, they are related to $b_{mn\ell}$ as

$$r_{mn\ell} = \frac{1-K^-_{m\ell}M}{1-MK^+_{mn}}b_{mn\ell}J_m(\alpha_{mn}a). \tag{9.41}$$

The determination of the coefficients $b_{mn\ell}$ amounts to solving the mixed boundary value problem defined in Section 9.4.1. After taking the spatial Fourier transforms of Equations (9.28)–(9.35) and applying the full Kutta condition, Munt proved the following formula for $b_{mn\ell}$ through a Wiener–Hopf formulation [20]:

$$b_{mn\ell} = \operatorname*{Lim}_{u\to K^-_{m\ell}}\left\{\frac{\left(u-K^-_{m\ell}\right)(1-Mu)F_+(u)}{v(u)J'_m(kav)}\right\}. \tag{9.42}$$

Here, u denotes the spatial Fourier transform variable defined as $\kappa = ku$, where κ denotes the axial wavenumber (see Section 1.3.1) and

$$F_+(u) = \frac{v_-\left(K^+_{mn}\right)w_-\left(K^+_{mn}\right)\chi_-\left(K^+_{mn}\right)v_+(u)w_+(u)\left(K^+_{mn}-u_o\right)(1-Mu)}{\chi_+(u)\left(u-K^+_{mn}\right)(u-u_o)}, \tag{9.43}$$

with $w^2 = w_+w_-$, $v^2 = v_+v_-$, $w_\mp = \sqrt{C\pm(1\mp M\alpha C)u}$, $v_\mp = \sqrt{1\pm(1\mp M)u}$ and

$$\chi(u) = \frac{\chi_+(u)}{\chi_-(u)} = \frac{\gamma(1-M\alpha u)^2 v(u)H^{(1)}_m kaw)}{H'^{(1)}_m(kaw)} - \frac{(1-Mu)^2 w(u)J_m(kav)}{J'_m(kav)}. \tag{9.44}$$

Here, $H^{(1)}_m(z)$ denotes a Hankel function of order m, and $H'^{(1)}_m(z)$ denotes its first derivative with respect to z. The function $\chi_+(u)$ must be analytic in the lower half of the complex plane, and $\chi_-(u)$ must be analytic and non-zero in the upper half of the plane. In Equation (9.43), u_o denotes the zero of Equation (9.44) in the lower half of the complex plane with a positive real part. This can be determined by using the Newton–Raphson method. Alternatively, Equation (9.44) may be recast in the form $u = f(u)$ for simple Newton iteration, where

$$f(u) = \frac{1}{M}\frac{B(u) \pm 1}{\alpha B(u) \pm 1}, \tag{9.45}$$

with

$$B(u) = \frac{\gamma v J'_m(kav) H'^{(2)}_m(kaw)}{w J_m(kav) H^{(2)}_m(kaw)}. \tag{9.46}$$

In particular, if $\alpha = 0$, $u = (1 \pm B(u))/M$, which converges to the required root quickly when $ka > 1$. Note that $u_o = 1/M$ if $\alpha = 1$.

We omit the details of derivation of Equation (9.42), as this requires mathematical digressions on subtle issues such as the instability of the vortex sheet, causality of the acoustic field, application of the Kutta condition and splitting of the dispersion equation of the problem into functions which are analytic in the upper and lower halves of the complex plane. Here, we may mention that the limit in Equation (9.42) exists, because zeroes of $J'_m(kav)$ are given by K^-_{ml}. In fact, if $n, m = 0$, then K^-_{ml} is a zero of both $v(u)$ and $J'_m(kav)$. Thus, if $n, m \neq 0$, Equation (9.42) may be written as

$$b_{mn\ell} = \frac{(1 - MK^-_{m\ell})F_+(K^-_{m\ell})}{v(K^-_{m\ell})} \lim_{u \to K^-_{m\ell}} \left\{ \frac{u - K^-_{m\ell}}{J'_m(kav)} \right\}. \tag{9.47}$$

Applying l'Hôpital's rule gives

$$b_{mn\ell} = \frac{(1 - MK^-_{m\ell})F_+(K^-_{m\ell})}{v(K^-_{m\ell})} \lim_{u \to K^-_{m\ell}} \left\{ \frac{v(u)}{ka(M + (1 - M^2)u)J''_m(kav)} \right\}. \tag{9.48}$$

Hence,

$$b_{mn\ell} = \frac{(1 - MK^-_{m\ell})F_+(K^-_{m\ell})\alpha^2_{m\ell}a^2}{ka(M + (1 - M^2)K^-_{m\ell})\ (m^2 - \alpha^2_{m\ell}a^2)J_m(\alpha_{m\ell}a)} \tag{9.49}$$

since $kav(K^-_{m\ell}) = \alpha_{ml}a$, $J'_m(\alpha_{m\ell}a) = 0$ and $J''_m(\alpha_{m\ell}a) = (m^2/\alpha^2_{m\ell}a^2 - 1)J_m(\alpha_{m\ell}a)$. This result applies also for the case of $n, m = 0$, since $v(K^-_{m\ell})$ cancels if it is included in the denominator before applying l'Hôpital's rule. With substitution of the foregoing equation in Equation (9.41), the modal reflection coefficients are found as

$$r_{mn\ell} = \frac{(1 - MK^-_{m\ell})^2 F_+(K^-_{m\ell})\alpha^2_{m\ell}a^2}{ka(1 - MK^+_{mn})(M + (1 - M^2)K^-_{m\ell})\ (m^2 - \alpha^2_{m\ell}a^2)}. \tag{9.50}$$

Numerical evaluation of this formula is not trivial, because $\chi_-(K^+_{mn})$ and $\chi_+(K^-_{m\ell})$ are needed for the calculation of $F_+(K^-_{m\ell})$. The splitting of $\chi(u)$ into $\chi_\mp(u)$ may be achieved only numerically by evaluation of the integral forms

$$\ln(\chi_\mp(y)) = \frac{1}{i2\pi} \overset{*}{\int_{C_\mp}} \frac{\ln(\chi(u))}{u - y} du, \tag{9.51}$$

where the asterisk on the integral sign denotes the Cauchy principal value, $C_\mp$ denote integral paths from $-\infty$ to $+\infty$ near the real axis and y denotes a point between the integration paths. We have to omit description of the integration, as it requires a detailed study of the topology of $\chi(u)$. The reader may refer to Panhuis [25], who describes an optimized integration procedure and gives a MATLAB code.

Munt's solution assumes full Kutta condition at the duct lip. It can be generalized for the simulation of partial shedding of vorticity by modifying Equation (9.43) as [22, 26]:

$$F_+(u) = \left\{1 + (\sigma - 1)\frac{u - K^+_{mn}}{u_o - K^+_{mn}}\right\} \times \frac{v_-(K^+_{mn})w_-(K^+_{mn})\chi_-(K^+_{mn})v_+(u)w_+(u)(K^+_{mn} - u_o)(1 - Mu)}{\chi_+(u)(u - K^+_{mn})(u - u_o)}, \tag{9.52}$$

where σ is the control parameter for the amount of shed vorticity at the duct lip: $\sigma = 1$ corresponds to the application of the full Kutta condition (shedding all available vorticity). The Kutta condition is not applied if $\sigma = 0$. Then no vorticity is shed.

9.4.2.1 Plane-Wave Reflection Coefficient

In this case $m, n, \ell = 0$, $K^{\mp}_{00} = \mp 1/(1 \mp M)$ and Equation (9.50) reduces to $r_{000} = (1 + M)F_+(-1/(1 - M))/(1 - M)^2 ka$. Munt computed $|r_{000}|$ for $\alpha = 0$, $C = \gamma = 1$ and $M = 0.01$ to 0.6 and found his results to be in good agreement with the existing experimental data [20]. This conclusion is confirmed by more recent accurate measurements undertaken by Allam and Åbom [27], English [28] and Rammal and Lavrentjev [29]. Some typical results showing the variations of r_{000} with the Helmholtz number ka are shown in the following discussion.

Figure 9.7 shows the magnitude and phase of r_{000} as function of the Helmholtz number ka and the mean flow Mach number, for the cold jet and no co-flow case. An interesting feature of $|r_{000}|$ is that it can have a maximum value greater than unity, but always tends to unity as $ka \to 0$, which is the maximum for $M = 0$. The maxima occur, approximately, at the Strouhal number St $\approx \pi$, where $St = ka/M$[30]. The

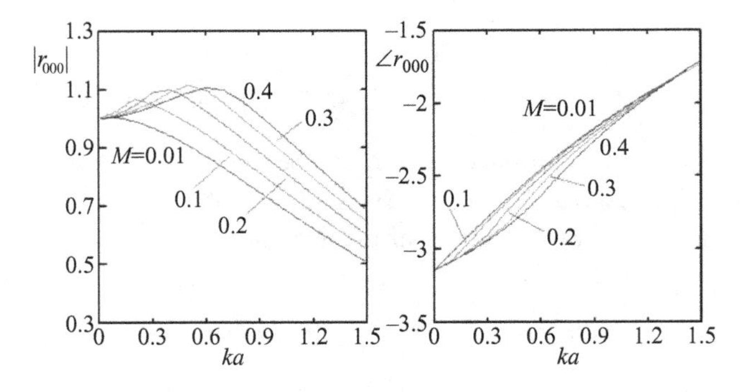

Figure 9.7 The magnitude $|r_{000}|$ and phase $\angle r_{000}$ (radian) of the reflection coefficient r_{000} for cold jets, $\alpha = 0$, $C = \gamma = 1$.

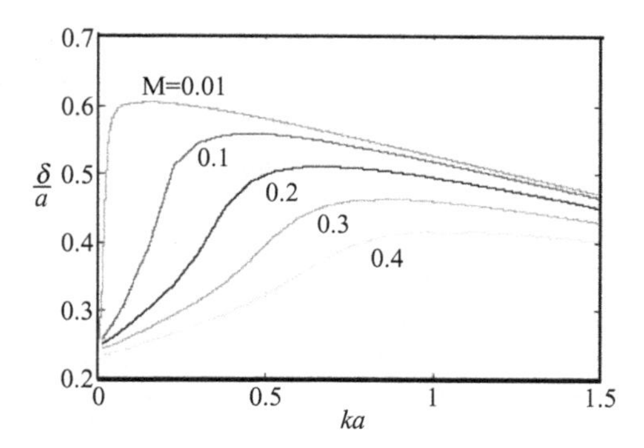

Figure 9.8 The end-correction for the cold jet conditions of Figure 9.6

behavior of Munt's solution for $ka << 1$ and for small Strouhal numbers, St < 1, say, is well investigated [30–31].

The phase characteristics are replotted in Figure 9.8 as end-correction defined in Equation (9.32). A feature of these characteristics is that, as $ka \to 0$ and $ka < M$, $\delta/a \to 0.2554\sqrt{1-M^2}$[30]. It is interesting to note that this limiting behavior applies for $M > 0$. For $M = 0$, $\delta/a \to 0.6133$ as $ka \to 0$ (see Equation (9.26)).

Munt's solution corresponds to $\sigma = 1$; however, $0 < |\sigma| \leq 1$ may be considered to simulate partial shedding of vorticity. English studied the effect of vortex shedding at the edge of duct lip by assuming that the difference in the measured and predicted reflection coefficient $|r_{000}|$ is the consequence of the extent of the vorticity shed [28]. His results indicate that the full Kutta condition ($\sigma = 1$) is not valid for all frequencies and flow rates. Rather, the full Kutta condition is only valid when $\omega\delta_A^+/M \to 0$, where δ_A^+ denotes the normalized acoustic boundary layer thickness (see Section 3.9.3.4). As the frequency and flow rate increases, the value of σ decreases until after the peak value of $|r_{000}|$ is reached (see Figure 9.7), at which point it begins to increase toward the full Kutta condition. The turning point lies between $1000 \leq \omega\delta_A^+/M \leq 2000$, approximately. In addition, σ seems to increase with M and decrease with the thickness of the duct wall.

Figure 9.9 shows the effect of the jet temperature on r_{000} for $M = 0.3$ and no co-flow. The range of C considered corresponds to jet temperatures of up to about 500 °C, the outer air being assumed to be at 25 °C.

Vortex shedding at an open exhaust is accompanied by the absorption of some part of the incident sound wave energy, which goes into perturbing the shear layer. This was explained for subsonic low Mach numbers by Howe using the theory of vortex sound [32], and Cargill [33] showed that this effect is included in Munt's model. Hirshberg and Rienstra point out that the absorption takes place in the first half of an acoustic oscillation period [34], after which the vortex will generate sound owing to the reversal of the direction of the particle velocity, although this will be weaker than the initial absorption, because both the magnitude of the particle velocity and the angle between the vortex path and acoustical streamlines decrease as the vortex moves away from the edge. However, if the initial absorption is minimized by using a duct with rounded edges so that the flow separation is eliminated, a net sound production at the mean flow velocity given by the empirical relationship $kR/M \approx 1.25$, where R denotes the radius of curvature of the outlet, can be obtained.

9.4.2.2 Reflection of Higher-Order Incident Modes

As frequency increases, more and more higher-order modes cut on and the number of reflected modes progressively increase. The evolution of modal reflection coefficients as higher-order modes cut on is portrayed in Table 9.1 for hollow circular ducts. The moduli of some of the modal reflection coefficients are shown as functions of the Helmholtz number in Figure 9.10. The evanescent branches are not shown because they are assumed to have died before reaching the open end. In general, the modal reflections are most significant at cut-on and diminish rather quickly as frequency increases.

Table 9.1 Hollow circular duct modal reflection coefficients scheme

μ	Incident (m, n)	Cut-on ka	Reflected (m, ℓ)	$r_{nm\ell}$
1	(0,0)	0	(0,0), (0,1),...	r_{000},r_{001},...
2	(1,0)	1.841	(1,0), (1,1),...	r_{100},r_{101},...
3	(2,0)	3.054	(2,0), (2,1),...	r_{200},r_{201},...
4	(0,1)	3.832	(0,1), (0,2),...	r_{011} ,r_{012},...
5	(3,0)	4.201	(3,0), (3,1),...	r_{300},r_{301},...
6	(4,0)	5.318	(4,0), (4,1),...	r_{400} r_{401},...
7	(1,1)	5.331	(1,0), (1,1),...	r_{110} r_{111},...
8	(5,0)	6.415	(5,0), (5,1),...	r_{500},r_{501},...

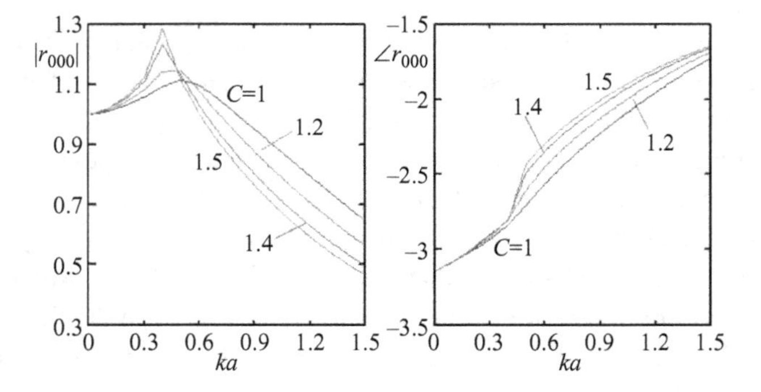

Figure 9.9 The magnitude $|r_{000}|$ and phase $\angle r_{000}$ (rad) of the reflection coefficient r_{000} for hot jets, $\alpha = 0$, $M = 0.3$, $\gamma = C^2$.

9.4.3 Reflection at Flow Intakes

The entrance geometry of duct intakes varies from sharp edged to bell mouthed designs. A sharp-edged intake (Figure 9.11), commonly known as a re-entrant or Borda intake, can be modeled by using the model of the previous section, but with the mean flow direction reversed, that is, $M < 0$, with $\alpha = 1$, $C = \gamma = 1$ and no vorticity shedding at the duct lip. The assumption of uniform mean flow, $\alpha = 1$, is, of course, an approximation, but it is convenient mathematically, as it eliminates the instability issues of the vortex sheet. Accordingly, Equation (9.52) is applied with $\sigma = 0$. This approximates the case of a sharp-edged intake, for which any vorticity shedding just downstream of the duct lip due to flow separation can be neglected. In principle, the presence of vortex shedding at the lip can be taken into account by assuming $0 < |\sigma| \leq 1$; however, the flow separation is dependent on the geometry and

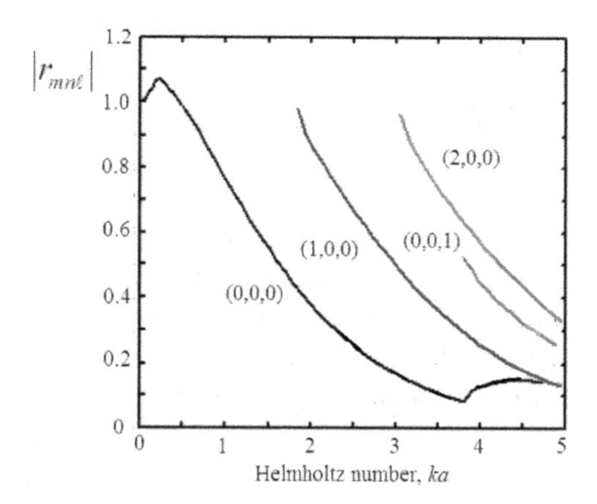

Figure 9.10 Variation of modal reflection coefficients with the Helmholtz number as incident modes progressively cut-on at the open end. $M = 0.1$, $\sigma = 1$, $\gamma = 1$, $C = 1$, $\alpha = 0$.

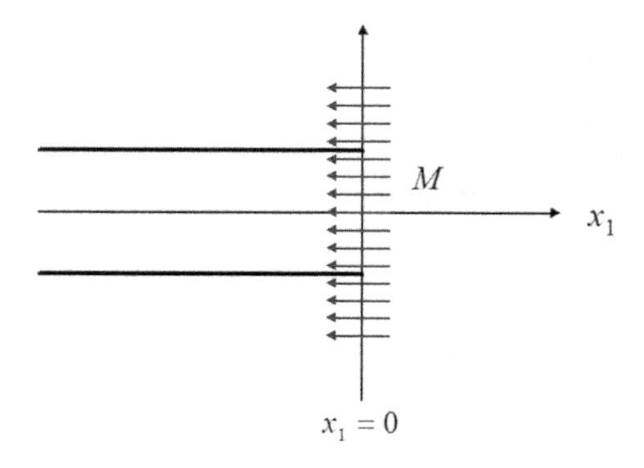

Figure 9.11 Sharp-edged intake.

sharpness of the intake, which are difficult to translate into a value for σ, which can be a complex number.

9.4.3.1 Plane-Wave Reflection Coefficient

Figure 9.12 shows the magnitude and phase of r_{000} as function of the Helmholtz number ka and the mean flow Mach number. These characteristics are in agreement with the small Strouhal number solution of Rienstra [30], which predicts that

$$|r_{000}| \sim \frac{1+M-2\sigma M}{1-M}, \qquad ka \to 0. \tag{9.53}$$

A sharp-edged intake has the disadvantage of giving significant mean pressure loss due to the formation of vena contracta. This is usually counteracted by slight rounding of the entrance. Davies has investigated the effect of the entrance geometry on the modulus of plane-wave reflection coefficient experimentally for circular intakes and found that a good fit to the measurements is given by [35]:

$$|r_{000}| = \left(\frac{1-\left(1+\sqrt{0.4K_p}\right)|M|}{1+\left(1+\sqrt{0.4K_p}\right)|M|}\right)^{0.9} \left|r_{000}^{(0)}\right|, \tag{9.54}$$

where K_p denotes the mean pressure loss coefficient of the entrance, and $r_{000}^{(0)}$ denotes the reflection coefficient of an unflanged open circular duct in the absence of mean flow ($M=0$). Typical values for the pressure loss coefficient will lie between $K_p \approx 1$ for a re-entrant intake and $K_p \approx 0$ for a streamline bell mouthed intake. Davies states that the observed end-correction corresponding to Equation (9.54) is $\delta = \left(1-M^2\right) \delta^{(0)}$[35]. And $r_{000}^{(0)}$ can be calculated relatively easily by using Equations (9.25) and (9.26). However, for larger mean flow Mach numbers, the effects of the exponent 0.9 and of the value of K_p become discernible and, for the re-entrant intake, Equation

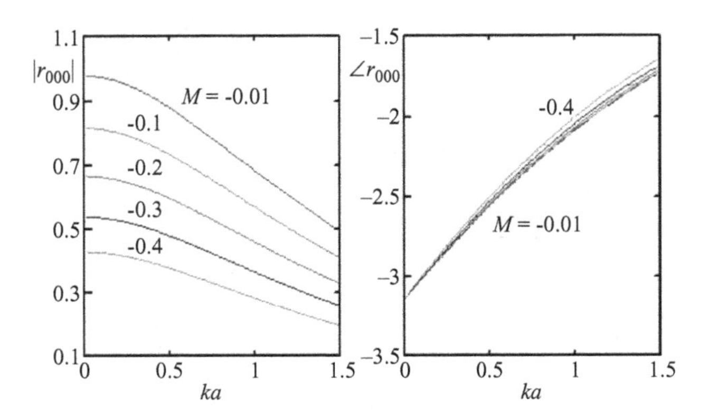

Figure 9.12 The magnitude $|r_{000}|$ and phase $\angle r_{000}$ (rad) of the reflection coefficient r_{000} for a sharp edged intake, $\sigma = 0, \alpha = 1$, $C = \gamma = 1$.

(9.54) gives slightly smaller values for the modulus of the reflection coefficient than the Wiener–Hopf solution shown in Figure 9.12.

9.5 Acoustic Radiation from Open Ends of Ducts

9.5.1 Modal Radiation Transfer Function

Referring to Figure 9.1, let $p'(\mathbf{x})$ denote the acoustic pressure at point $\mathbf{x} \in E$. This may be conceived as sum of modal contributions $p'(\mathbf{x}) = p'_1(\mathbf{x}) + p'_2(\mathbf{x}) + \ldots + p'_n(\mathbf{x})$, where $p'_q(\mathbf{x})$ denote the contributions of the propagating modes $q = 1, 2, \ldots, N$ incident at the open end, only N modes being assumed cut-on at the frequency of interest. The open-end radiation transfer function for mode $q \in 1, 2, \ldots, N$ is defined by the quotient

$$K^{(q)} = \frac{p_q^+(0)}{p'_q(\mathbf{x})}, \tag{9.55}$$

where $p_q^+(0)$ denotes the pressure wave component for cut-on mode q incident at the open end. Thus, the radiated acoustic pressure can be calculated, knowing the radiation transfer function and the incident pressure wave components. Some models which may be used for the determination of the modal radiation transfer functions are described in this section.

9.5.2 Radiated Acoustic Power

The time-averaged acoustic power $\delta \Pi$ crossing an elementary surface area δA in the acoustic free field radiated from the open end of a duct is $\delta \Pi = \langle \mathbf{N}'' \rangle \cdot \mathbf{n} \delta A$, where $\mathbf{n}$ is the outward unit normal vector of surface A, $\mathbf{N}''$ denotes the acoustic intensity vector and the angled brackets denote time averaging. Acoustic intensity is defined as work done by the acoustic pressure per unit time per unit volume and for a homogeneous and quiescent medium, it can be expressed as $\mathbf{N}'' = p'\mathbf{v}'$(see Section 1.4).

For the calculation of radiated acoustic power from an open end, it is convenient to introduce the cone parameters R_0, ϕ and φ. These are defined with respect to a Cartesian frame (x_1, x_2, x_3) as shown in Figure 9.13, with the x_1-axis as duct axis and the open-end section lying on $x_1 = 0$ plane. Thus, the time-averaged acoustic power passing through a surface at distance R_0 seeing an elementary solid angle $\delta\Omega$ in direction ϕ may be expressed as $\delta \Pi = \langle p'\mathbf{v}' \cdot \mathbf{e}_{R_0} \rangle R_0^2 \delta \Omega$, where $\mathbf{e}_{R_0}$denotes the outward unit vector in direction ϕ. To compute the time average, we note that the acoustic momentum equation in the time domain, Equation (6.2), here simplifies to $\rho_o \partial \mathbf{v}' / \partial t + \nabla p' = 0$. But, in the far field, p' scales with R_0 only by way of the free-space Green function, which is, in the frequency domain, $\mathrm{e}^{\mathrm{i}k_o R_0} / 4\pi R_0$. Then, the Fourier transform of the component of the foregoing momentum equation in the $\mathbf{e}_{R_0}$ direction gives

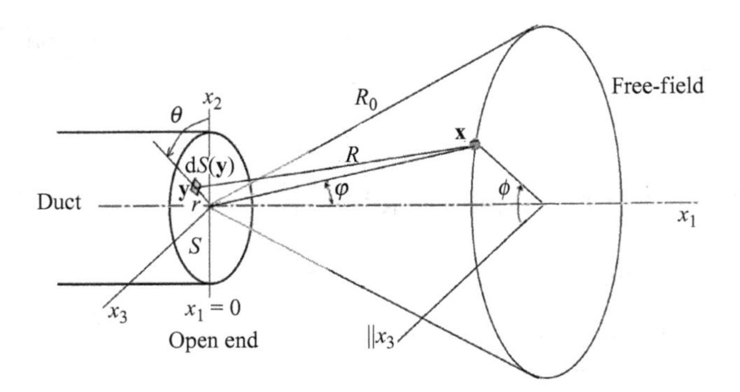

Figure 9.13 Radiation cone. The end may be flanged or unflanged. Flange is not shown. $x_1 = R_0 \sin(\phi)$, $x_2 = R_0 \sin(\phi)\sin(\varphi)$, $x_3 = R_0 \sin(\phi)\cos(\varphi)$. $\mathbf{y} = \mathbf{e}_2 r\cos(\theta) + \mathbf{e}_3 \sin(\theta)$ and $\mathbf{x} = \mathbf{e}_1 x_1 + \mathbf{e}_2 x_2 + \mathbf{e}_3 x_3$. Therefore, $\mathbf{x}\cdot\mathbf{y} = R_0 r \sin(\phi)(\sin(\varphi)\cos(\theta) + \cos(\varphi)\sin(\theta))$.

$$\mathbf{v}'\cdot\mathbf{e}_{R_0} = \frac{1}{\mathrm{i}\omega\rho_o}\frac{\partial p'}{\partial R_0} = \left(\mathrm{i}k_o + \frac{1}{R_0}\right)\frac{p'}{\mathrm{i}\omega\rho_o}. \tag{9.56}$$

This implies that $\mathbf{v}'\cdot\mathbf{e}_{R_0} = p'/\rho_o c_o$, if $k_o R_0 \gg 1$, which is in fact the mathematical condition for the far-field approximation. Hence, since $\mathbf{v}'\cdot\mathbf{e}_{R_0} = p'/\rho_o c_o$ holds also in the time domain, $\langle p'\mathbf{v}'\cdot\mathbf{e}_{R_0}\rangle = \langle p'p'\rangle/\rho_o c_o$ and the radiated time-averaged acoustic power per solid angle can be expressed as

$$\frac{\mathrm{d}\Pi}{\mathrm{d}\Omega} = \frac{2|p'|^2}{\rho_o c_o}R_0^2. \tag{9.57}$$

The total radiated acoustic power is then

$$\Pi = \frac{2}{\rho_o c_o}\int_{\phi=0}^{\widehat{\phi}}\int_{\varphi=0}^{2\pi}|p'|^2 R_0^2 \sin(\phi)\mathrm{d}\phi\mathrm{d}\varphi, \tag{9.58}$$

since $\mathrm{d}\Omega = \sin(\phi)\mathrm{d}\phi\mathrm{d}\varphi$. The angle $\widehat{\phi}$ depends on the solid angle seen by the radiated space, for example, $\widehat{\phi} = \pi/2$ and $\widehat{\phi} = \pi$ for flanged and unflanged open ends, respectively.

Radiated acoustic power may be conceived as the sum of contributions of the incident wave modes at the open end. So we may write $\Pi = \Pi_1 + \Pi_2 + \cdots$, where Π_q denotes the contribution of mode $q \in 1, 2, \ldots$ Then, given the modal radiation transfer function defined by Equation (9.55), we can calculate Π_q when the modulus of the corresponding incident modal wave component, $\left|p_q^+(0)\right|$, is known.

The acoustic power from an open end is sometimes expressed as power gain function. For an incident mode $q \in 1, 2, \ldots$, this is defined as

$$G_q = \left(\frac{\mathrm{d}\Pi_q}{\mathrm{d}\Omega}\right)\left(\frac{\widehat{\Omega}}{W_q}\right), \tag{9.59}$$

where W_q denotes the net time-averaged in-duct acoustic power of mode q at the open end (see Section 6.5.5) and $\hat{\Omega}$ denotes the radiated solid angle, for example, $\hat{\Omega} = 4\pi$ and $\hat{\Omega} = 2\pi$, respectively, for unflanged and flanged open ends. Clearly, the reciprocal of the second factor in Equation (9.59) gives the time-averaged acoustic power that would be radiated per solid angle, if the total power available at the open end radiated into the free-space isotropically. The integral of the power gain function over the solid angle $\hat{\Omega}$ should be equal to $\hat{\Omega}$, if acoustic power losses at the duct lip are neglected.

9.5.3 Flanged Open End without Mean Flow

In sequel to the analysis of Section 9.3.2, we begin with the Helmholtz integral formula, Equation (9.2) with $\mathbf{x} \in E$. However, it is necessary to change the free-space Green function, because Equation (9.3) does not satisfy the hard-wall condition $\mathbf{n}\cdot\nabla G = 0$ on ∂D (plane $x_1 = 0$ in Figure 9.2). This was not an issue in Section 9.3.2, because Equation (9.3) happens to satisfy this condition for points $\mathbf{x} \in \partial D$. The proper free-space Green function to use when points $\mathbf{x} \in E$ are considered is $G^*(\mathbf{y}, \mathbf{x}) = \mathrm{e}^{ik_oR}/4\pi R + \mathrm{e}^{ik_oR^*}/4\pi R^*$, where $\mathbf{y} \in \partial D$, $R = |\mathbf{x} - \mathbf{y}|$ as before, but $R^* = |\mathbf{x}^* - \mathbf{y}|$, where point $\mathbf{x}^*$ denotes the mirror symmetry of point $\mathbf{x}$ with respect to the flange plane. G^* is known as the half-space Green function [1]. It satisfies the condition $\mathbf{n}\cdot\nabla G^* = 0$ for $\mathbf{x} \in E$ and, therefore, it causes the second integral vanish in Equation (9.2) for $\mathbf{x} \in E$. Furthermore, in the first integral we can write $G^* = 2G = 2\mathrm{e}^{ik_oR}/4\pi R$, where G is the free-space Green function in Equation (9.3). Thus, upon substituting G^* for G, Equation (9.2) for $\mathbf{x} \in E$ becomes simply

$$p'(\mathbf{x}) = -2\int_S G(R)\frac{\partial p'}{\partial x_1}\mathrm{d}S, \mathbf{x} \in E. \tag{9.60}$$

This is commonly known as the Rayleigh integral. It is the same as the integral in Equation (9.4), except for the location of the observation point $\mathbf{x}$.

As in Section 9.3.2, we assume that the temperature and pressure in the duct and in the exterior are the same and substitute Equation (9.5) in the Rayleigh integral. This gives

$$p'(\mathrm{x}) = -\mathrm{i}2k_o\left(\int_S G(R)\mathbf{\Phi}(\bullet)\mathrm{d}S\right)(\mathbf{K}^+ + \mathbf{K}^-\mathbf{R}(0))\mathbf{P}^+(0), \mathbf{x} \in E, \tag{9.61}$$

where $\mathbf{P}^+(0) = \{p_1^+(0) \quad p_2^+(0) \quad \ldots \quad p_n^+(0)\}$, $p_j^+(0)$, $j \in 1, 2, \ldots, n$, denote the incident modal pressure wave components at the open end and $\mathbf{R}(0)$ denotes the reflection matrix at the open end. The modal radiation transfer functions $K^{(j)}$ may be calculated when the coefficients array in Equation (9.61) is evaluated. It is convenient to write Equation (9.61) in the form

$$p'(\mathbf{x}) = -\mathrm{i}2k_o[F_1 \quad F_2 \quad \ldots \quad F_n](\mathbf{K}^+ + \mathbf{K}^-\mathbf{R}(0))\mathbf{P}^+(0), \tag{9.62}$$

with

$$F_q = \int_S \varphi_q(\bullet) \frac{e^{ik_o R}}{4\pi R} dS = SD_q \frac{e^{ik_o R_0}}{4\pi R_0}. \tag{9.63}$$

Here, $\varphi_q(\bullet)$ denotes the eigenfunction of the transverse duct mode $q = 1, 2, \ldots, n$ and D_q is called directivity factor of mode q and, as it may be shown by inspection of Figure 9.13:

$$R^2 = r^2 + R_0^2 - 2rR_0 \sin(\phi) \sin(\varphi + \theta). \tag{9.64}$$

In practice, the observation point $\mathbf{x}$ is usually taken far away from the center of the open end. If R_0 is larger than 3–4 times the characteristic dimension of the duct section, it is sufficiently accurate to write

$$R \approx R_0 - r \sin(\phi) \sin(\varphi + \theta). \tag{9.65}$$

This is called the far-field approximation. The directivity factor D_q for mode q may then be expressed as

$$D_q(\phi, \varphi) = \frac{1}{S} \int_S \varphi_q(\bullet) e^{-ik_o r \sin(\phi) \sin(\varphi+\theta)} dS. \tag{9.66}$$

We use the identity $|p'|^2 = p' \left(\breve{p}'\right)^{\mathrm{T}}$, where an inverted over-arc denotes complex conjugate and the superscript "T" denotes matrix transpose, to express the modulus of the acoustic pressure from Equation (9.62) in the form

$$|p'|^2 = 4\left(\frac{Sk_o}{4\pi R_0}\right)^2 \breve{\mathbf{P}}^+(0)^{\mathrm{T}} \left(\breve{\mathbf{K}}^+ + \breve{\mathbf{K}}^- \breve{\mathbf{R}}(0)\right)^{\mathrm{T}} [\breve{D}_1 \quad \breve{D}_2 \quad \ldots \quad \breve{D}_n]^{\mathrm{T}} \times$$
$$[D_1 \quad D_2 \quad \ldots \quad D_n](\mathbf{K}^+ + \mathbf{K}^- \mathbf{R}(0))\mathbf{P}^+(0) \tag{9.67}$$

This result is used in Equation (9.58) for the calculation of the radiated acoustic power.

9.5.3.1 Circular Ducts

The foregoing equations can be applied as in Section 9.3.2.1 for circular ducts by using the cylindrical coordinates $\bullet \equiv r, \theta$ and the transverse eigenfunctions defined in double index notation in Equation (9.14). Then, for circumferential modes of order m, Equation (9.62) takes the form

$$p'_m(\mathbf{x}) = -i2k_o [F_{m0} \quad F_{m1} \quad \cdots] \left(\mathbf{K}_m^+ + \mathbf{K}_m^- \mathbf{R}_m(0)\right) \mathbf{P}_m^+(0), \tag{9.68}$$

with

$$F_{mn} = \int_S \psi_m(\alpha_{mn} r) e^{im\theta} \frac{e^{ik_o R}}{4\pi R} dS = SD_{mn} \frac{e^{ik_o R_0}}{4\pi R_0}, n = 0, 1, 2, \ldots \tag{9.69}$$

where, for an annular duct $b \le r \le a$, the directivity factor is given by

$$D_{mn}(\phi,\varphi) = \frac{1}{S}\int_{b}^{a}\left(\psi_m(\alpha_{mn}r)\left(\int_{0}^{2\pi} e^{im\theta}e^{-ik_o r\sin(\phi)\sin(\varphi+\theta)}d\theta\right)\right)rdr. \tag{9.70}$$

This integral can be evaluated exactly. First, we use Hansen's integral [7]:

$$J_n(z) = \frac{1}{2\pi}\int_{\alpha}^{2\pi+\alpha} e^{i(n\theta - z\sin\theta)}d\theta \tag{9.71}$$

to put it into the form

$$D_{mn}(\phi,\varphi) = \frac{1}{S}2\pi e^{-im\varphi}\int_{b}^{a}\psi_m(\alpha_{mn}r)J_m(k_o r\sin\phi)rdr. \tag{9.72}$$

Next, the formula for the integral of products two cylinder functions, of the same order but different arguments [36] is used to get the final result:

$$D_{mn}(\phi,\varphi) = \frac{2e^{-im\varphi}}{1-\frac{b^2}{a^2}}\frac{\left[\alpha_{mn}r\psi_{m+1}(\alpha_{mn}r)J_m(k_o r\sin\phi) - k_o r\sin(\phi)\psi_m(\alpha_{mn}r)J_{m+1}(k_o r\sin\phi)\right]_b^a}{\alpha_{mn}^2a^2 - k_o^2a^2\sin^2(\phi)} \tag{9.73}$$

This equation applies for both hollow and annular ducts. For the hollow duct case, using the recurrence relation $J_{m+1}(z) = (m/z)J_m(z) - J'_m(z)$ and noting that $J'_m(\alpha_{mn}a) = 0$, it simplifies to

$$D_{mn}(\phi,\varphi) = 2e^{-im\varphi}k_o a\sin(\phi)\frac{J_m(\alpha_{mn}a)J'_m(k_o a\sin\phi)}{\alpha_{mn}^2a^2 - k_o^2a^2\sin^2(\phi)}. \tag{9.74}$$

Foregoing equations suggest that some sort of maximum occurs near

$$k_o a\sin(\phi) = \alpha_{mn}a. \tag{9.75}$$

In fact, both Equation (9.73) and Equation (9.74) are indeterminate when this condition occurs, but application of l'Hôpital's rule immediately shows that the limits are finite. Then Equation (9.75) provides a quick way of predicting the strongest direction of sound radiation. Since mean flow is neglected, the axial propagation constants of cut-on modes in the duct are given by $K_{mn}^2 = 1 - \alpha_{mn}^2/k_o^2$ with $k_o^2 > \alpha_{mn}^2$. Then, if the cone angle that satisfies Equation (9.75) is $\phi_o = \phi_o(k_o, \alpha_{mn}|k_o > \alpha_{mn})$, say, then $K_{mn}^2 = \cos^2(\phi_o)$. Accordingly, close to its cut-on frequency, a cut-on mode radiates most strongly at $\phi_o \approx 90^\circ$, but a well cut-on mode tends to radiate approximately along the duct axis. This is illustrated for $k_o a = 10$ in Figure 9.14, which shows the variation of the modulus of the modal directivity factor with the cone angle. It is seen that $|D_{mn}(\phi)|$ has a lobe structure. The principal lobe of mode (2,2), which is just cut-on at $k_o a = 10$, is nearly perpendicular ($\phi_o \approx 85.5^\circ$) to the duct axis and is quite wide, which is typical near cut-off for all modes. For modes (2,1) and (2,2), which are well

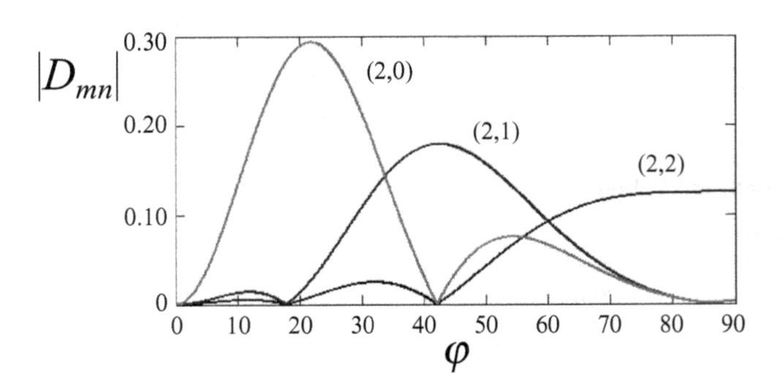

Figure 9.14 Modal directivity factors for a flanged open end at $k_o a = 10$. The eigenvalues for the displayed modes are $\alpha_{20}a = 3.054,\ \alpha_{21}a = 6.706,\ \alpha_{22}a = 9.969$.

cut-on, the principal lobes shift toward the duct axis and become narrower. It can also be concluded from Equation (9.74) that $D_{mn} = 0$ for $\phi = 0$ and also for $k_o a \sin(\phi) = \alpha_{mj}a,\ j < n$. Thus, we can locate the minor lobes at approximately midway between the zeroes. Knowledge of the lobes can be important in the control of sound radiated from open ends of ducts for diverting the direction of principal radiation away from sensitive areas.

Measurements carried out by Torregrosa et al. on sound radiated from a flanged tailpipe of an IC engine in the plane-wave range [37], show that D_{00} for hollow ducts (Equation (9.74) gives acceptable predictions at medium and high engine speeds, but it overestimates the radiated sound pressure at low engine speeds, by about 5 dB and for some harmonics by 10 dB. The authors argue that this is attributable to the neglect of the sound absorption in shear layers at the open end.

The power gain function $G_1 = G_1(\phi)$ for a hollow circular duct was computed by Norris and Sheng [8]. Their results are shown in Figure 9.15a as function of the Helmholtz number and the cone angle. In this case, since mean flow in the duct is neglected,

$$W_1 = \frac{2S}{\rho_0 c_0}\left(1 - |r_{000}|^2\right)|p_1^+(0)|^2 \tag{9.76}$$

and it can be shown that $G_1(0) = k_o^2 a^2 |1 - r_{000}|^2 / 2\left(1 - |r_{000}|^2\right)$.

9.5.3.2 Rectangular Ducts

Referring to Figure 6.3, we first move the origin of the Cartesian frame x_1, x_2, x_3 to the duct center so that the duct section lies in $-b/2 \le x_2 \le b/2$ and $-d/2 \le x_3 \le d/2$ and the radiation cone in Figure 9.13 is centered at the geometric center of the duct section. Accordingly, the coordinates in transverse eigenfunctions in Equation (6.86) are translated by $b/2$ and $d/2$ in x_2 and x_3 directions, respectively. With this modification, Equation (9.66) can be expressed in central x_1, x_2, x_3 frame as

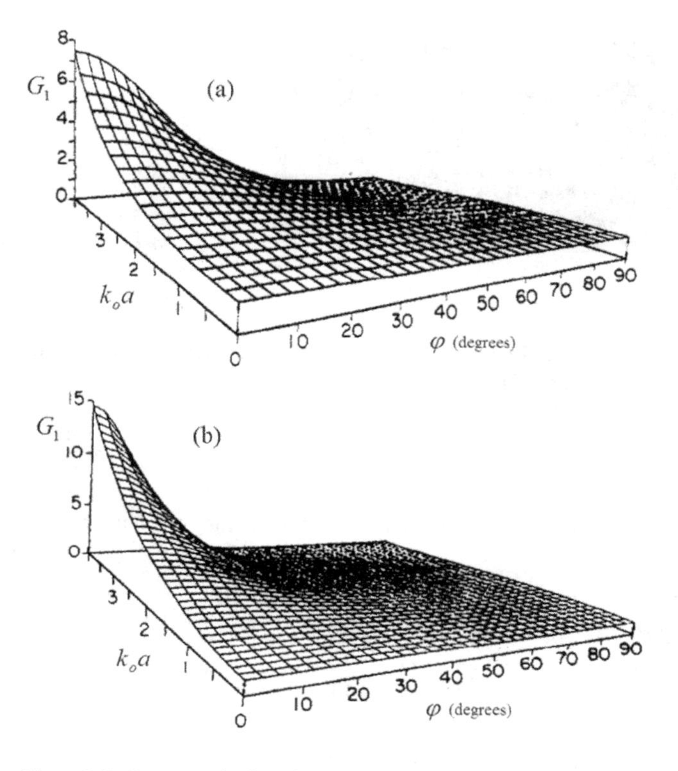

Figure 9.15 Power gain function, (a) Flanged open end, (b) Unflanged open end (Norris and Sheng [8], by permission from Elsevier).

$$D_{mn}(\phi,\varphi)=\frac{1}{S}\left(\int_{-d/2}^{d/2}\cos\left(\frac{n\pi(x_3+d/2)}{d}\right)\mathrm{e}^{-\mathrm{i}k_o x_3\sin(\phi)\cos(\varphi)}\mathrm{d}x_3\right)\times\left(\int_{-b/2}^{b/2}\cos\left(\frac{m\pi(x_2+b/2)}{b}\right)\mathrm{e}^{-\mathrm{i}k_o x_2\sin(\phi)\sin(\varphi)}\mathrm{d}x_2\right), \tag{9.77}$$

since $\cos\theta = x_2/r$ and $\sin\theta = x_3/r$. For integration, we substitute $x_2 = x_2' - b/2$ and $x_3 = x_3' - d/2$, and drop the primes in the resulting equation. Then partial integration gives, for example, the second integral in Equation (9.77) as

$$\frac{2k_0 b^2\sin(\phi)\sin(\varphi)}{k_o^2 b^2\sin^2(\phi)\sin^2(\varphi)-m^2\pi^2}\times\begin{cases}\sin\left(\frac{1}{2}k_o b\sin(\phi)\sin(\varphi)\right) & \text{if } m=0,2,4\ldots\\ -\mathrm{i}\cos\left(\frac{1}{2}k_o b\sin(\phi)\sin(\varphi)\right) & \text{if } m=1,3,\ldots\end{cases} \tag{9.78}$$

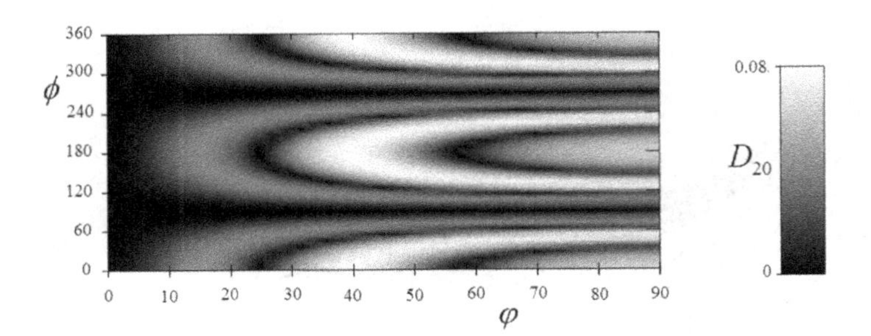

Figure 9.16 Directivity index of mode (2,0) of a flanged rectangular duct of side ratio $d/b = 2$ at $k_o d = 15$. Cut-on frequency of this mode occurs at $k_o d = 12.566$.

and a similar result with b replaced by d for the first integral. Hence,

$$D_{mn}(\phi,\varphi) = \frac{4k_o^2 bd \sin^2(\phi)\sin^2(\varphi)}{\left(k_o^2 b^2 \sin^2(\phi)\sin^2(\varphi) - m^2\pi^2\right)\left(k_o^2 d^2 \sin^2(\phi)\sin^2(\varphi) - n^2\pi^2\right)}$$
$$\times\begin{cases} \sin\left(\frac{1}{2}k_o b\sin(\phi)\sin(\varphi)\right)\sin\left(\frac{1}{2}k_o d\sin(\phi)\sin(\varphi)\right) & \text{if } m,n = 0,2,4...\\ -\cos\left(\frac{1}{2}k_o b\sin(\phi)\sin(\varphi)\right)\cos\left(\frac{1}{2}k_o d\sin(\phi)\sin(\varphi)\right) & \text{if } m,n = 1,3,...\\ -i\sin\left(\frac{1}{2}k_o b\sin(\phi)\sin(\varphi)\right)\cos\left(\frac{1}{2}k_o d\sin(\phi)\sin(\varphi)\right) & \text{if } m = 0,2,..,n = 1,3,...\\ -i\cos\left(\frac{1}{2}k_o b\sin(\phi)\sin(\varphi)\right)\sin\left(\frac{1}{2}k_o d\sin(\phi)\sin(\varphi)\right) & \text{if } m = 1,3,..,n = 2,4,... \end{cases} \tag{9.79}$$

It is seen that, in this case the modal directivity factor is dependent on the polar angle φ and the ratio of the section sides, as well as the cone angle ϕ. Again, some sort of maximum is indicated at the zeroes of the denominator and, it can be shown that, D_{mn} has finite limits at these zeroes. However, the principal lobe now occurs at a substantially different angle than that indicated by the factors in the denominator.

Variation of the directivity factor with ϕ and φ is shown in Figure 9.16 for mode (2,0) of a rectangular duct of side ratio $d/b = 2$ at Helmholtz number $k_o d = 15$. This is a well cut-on mode at this frequency and it can be observed that there are three lobs near $\varphi = 0^o$ and 180°, but there is no radiation near $\varphi = 90^o$ and 270°.

9.5.4 Unflanged Circular Open End without Mean Flow

The modulus of the directivity factor of unflanged circular ducts with no mean flow is given by Homicz and Lordi in the following form [38]:

$$|D_{mn}(\phi)| = 4e^{-im\varphi}\left|\frac{J_m(\alpha_{mn}a)a_{mn}}{\cos(\phi) - a_{mn}}\right|\sqrt{\frac{\alpha_{mn}^2 a^2 - m^2}{\pi\alpha_{mn}^2}\sin(\varpi(k_o a\sin(\phi)))}\sqrt{\prod_{\substack{j=j_1\\ j\neq n\\ a_{00}=1}}\left(\frac{a_{mj}+a_{mn}}{a_{mj}-a_{mn}}\right)}$$

$$\times\sqrt{\prod_{j=j_1}\left(\frac{a_{mj}-\cos(\phi)}{a_{mj}+\cos(\phi)}\right)}e^{\frac{1}{\pi}\int_{-1}^{1}\frac{\varpi\left(k_o a\sqrt{1-w^2}\right)dw}{w-a_{mn}}}e^{\frac{-1}{\pi}\int_{-1}^{1}\frac{\varpi\left(k_o a\sqrt{1-w^2}\right)dw}{w-\cos(\phi)}} \tag{9.80}$$

with

$$\varpi(\xi) = \tan^{-1}\left(\frac{Y'_m(\xi)}{J'_m(\xi)} \mp \frac{\pi}{2}\right), \tag{9.81}$$

where $j_1 = 0$ if $n = 0$, else 1, the + sign is used if $n = 0$, the Cauchy principal values of the integrals are taken and upper limit of products is truncated at the largest cut-on mode and a_{mn} is given by Equation (6.51) with $\bar{M}_o = 0$. Applications of Equation (9.80) are discussed in Dowling and Ffowcs Williams [39]. Here, it suffices to note that Equation (9.80) suggests that principal radiation modes occur at cone angles $\cos(\phi) = a_{mn}$.. With use of Equation (6.51), it may readily be shown that this condition is the same as Equation (9.75). The location of the zeroes can be determined by noting that, $\varpi(\alpha_{mn}a)$ is an integral multiple of π and, therefore, $|D_{mn}|$ vanishes whenever $k_o a\sin(\phi) = \alpha_{mj}a$, $j < n$. This is also the same as the condition derived in Section 9.5.3.1.

The power gain function for the fundamental mode of an unflanged circular open end without mean flow is given explicitly by Levine and Schwinger as [15]:

$$G_1(\phi) = \frac{4}{\pi\sin^2\phi}\left[\frac{J_1(k_o a\sin\phi)e^{\frac{2k_o a\cos\phi}{\pi}\int_0^{k_o a}\frac{x\tan^{-1}\left(-\frac{J_1(x)}{Y_1(x)}\right)}{(x^2-(k_o a\sin\phi)^2)\sqrt{x^2+(k_o a)^2}}dx}}{\sqrt{J_1^2(k_o a\sin\phi)+Y_1^2(k_o a\sin\phi)}}\right]\frac{|r_{000}|}{1-|r_{000}|^2}, \tag{9.82}$$

where J_1 and Y_1 denote Bessel functions of the first and second kind of order one, respectively, and Equation (9.74) applies. This function is depicted in Figure 9.14. In this case, $G_1(0) = k_o^2a^2/\left(1-|r_{000}|^2\right)$ and $G_1(\pi) = |r_{000}|^2G_1(0)$ [15]. It should be noted that, the principal value of the integral in Equation (9.82) is relevant, since it is singular for $x = |k_o a\sin\phi|$.

9.5.5 Radiation from Unflanged Circular Open End with Mean Flow

In this section, we describe, in sequel to Section 9.4.2, Munt's solution for the far-field radiation from an unflanged open end of a circular duct with mean flow [19].

Upon taking spatial Fourier transform (see Section 1.3.1) of Equation (9.29) with respect to the axial coordinate, x_1, the radiated acoustic pressure field in the outer flow region $r > a$ can be expressed as

$$p'(u) = \mathrm{i}\omega\rho_o(1 - M\alpha u)\vartheta'(u), \qquad r > a \tag{9.83}$$

where $\kappa = ku$ denotes the spatial Fourier variable. After having solved the Wiener–Hopf problem described in Section 9.4.2 for $\vartheta'(u)$, the radiated pressure field is determined by taking the inverse spatial Fourier transform of Equation (9.83) as

$$p'(x_1, r) = \frac{k}{2\pi}\int_{-\infty}^{\infty} p'(u)e^{ikux_1}\,du. \tag{9.84}$$

Since the incident wave, Equation (9.36), at the open end has $e^{im\theta}$ dependence on the angular coordinate, the radiated field also has the same angular dependence; however, this dependency is suppressed in the foregoing equation for the simplicity of notation. Munt shows that Equation (9.84) may be integrated as [19]:

$$p'(x_1, r) = \underset{u \to u_o}{\text{Residue}}\left[A(u)H_n^{(2)}(kwr)e^{ikux_1}\right] + \int_{C_1} A(u)H_n^{(2)}(kwr)e^{ikux_1}\,du, r > a \tag{9.85}$$

where the integration contour C_1 goes from $-\infty e^{-\varepsilon}$ to $+\infty e^{-\varepsilon}$ through the origin, with $\varepsilon \to 0$, and $A(u)$ is given by

$$A(u) = \frac{\omega\rho_o}{2\pi}\frac{(1 - M\alpha u)^2 F_+(u)}{w(u)H'^{(2)}_n(kwa)}. \tag{9.86}$$

Evaluation of Equation (9.85) involves some conceptual and mathematical difficulties. The residue term, which is required for the satisfaction of the boundary conditions of the problem [19], grows exponentially with x_1 in $x_1 > 0$. This phenomenon is associated with the Helmholtz instability of vortex sheet [38]. The amplitude diminishes as r increases and is essentially confined to the cone $r/x_1 \leq \mathrm{Im}(u_o)/\mathrm{Im}(w_o)$ [22]. However, whether or not this term should be observable in experiments is a subject of debate, since, in practice, the downstream growth of the instability of the vortex sheet is quickly broken by non-linear effects not included in the linear theory. Munt argues that the shear layer in the vicinity of the duct lip should still be thin enough for the linear theory to be valid [20], and that it is the behavior of this which determines the radiation of sound out of the duct. It can be shown that the residue term is proportional to the vorticity shedding factor σ defined in Equation (9.51) [26]. Accordingly, the instability considerations are not relevant for the intake problem depicted in Figure 9.11.

On the other hand, general numerical evaluation of Equation (9.85) is prone to convergence problems and is time consuming, since the split functions $\chi_\mp(u)$ need to be computed accurately for a large number of points on the integration contour. Specific numerical procedures are described by Gabard and Astley [22], who present near-field acoustic pressure contour maps to demonstrate the efficacy of these procedures.

In many cases, however, the far-field approximation is acceptable, because the radiated sound pressure is monitored at several diameters away from the radiating outlet (or inlet). Using the method of stationary phase, Munt derived, from Equation (9.85), the following formula for the far-field sound pressure [19]:

$$p'(R_o, \phi, \varphi) = \frac{2A(u')\mathrm{e}^{\frac{\mathrm{i}kR_o C}{1-(M\alpha C)^2}\left\{M\alpha C\cos\phi - \sqrt{1-(M\alpha C\sin\phi)^2}\right\} + \frac{\mathrm{i}}{2}(1+m)\pi}\mathrm{e}^{\mathrm{i}m\theta}}{kR_o\sqrt{1-(M\alpha C\sin\phi)^2}}, \tag{9.87}$$

where ϕ denotes the cone angle (Figure 9.12) and

$$u' = \frac{C\cos\phi\sqrt{1-(M\alpha C\sin\phi)^2} - M\alpha C^2}{1-(M\alpha C)^2} \tag{9.88}$$

$$w(u') = C\sin\theta\sqrt{1-(M\alpha C\sin\theta)^2}. \tag{9.89}$$

In getting this solution, the instability wave is disregarded on the premise that the vortex sheet is irrelevant for the far field.

Munt's formula presumes that the incident wave at the open end has the velocity potential defined by Equation (9.36) and, therefore, has the same polar dependence as the incident wave. From Equation (9.28), the incident acoustic pressure at $x_1 = 0$ may be expressed as

$$p^+_{mn}(r, \theta, \varphi) = \mathrm{i}B\omega\rho_\mathrm{j}\left(1 - MK^+_{mn}\right)\frac{\mathrm{J}_m(\alpha_{mn}r)}{\mathrm{J}_m(\alpha_{mn}a)}\mathrm{e}^{-\mathrm{i}m\varphi}. \tag{9.90}$$

This should be used as scaling factor in evaluating Equation (9.87) for given m and n. If the incident wave is a plane wave, Equation (9.90) reduces to $p^+ = \mathrm{i}B\omega\rho_\mathrm{j}/(1+M)$. This far-field solution still requires numerical calculation of the split functions. For the calculation of the radiation transfer functions $K^{(q)}$ defined in Equation (9.55), it will be necessary to convert Equations (9.87) and (9.90) to the single index notation.

Approximations to Equation (9.90) for small Strouhal numbers are given by Rienstra [30].

9.5.6 Simple-Source Approximation

So far in this chapter, the acoustics of an open end is treated as a problem of matching the in-duct and exterior acoustic fields. A less rigorous, but simpler, approach for the calculation of the radiated far field is to model the open end of a duct as a source region for the exterior acoustic field. The most obvious sound source mechanism with which an open duct end may be associated is the injection of fluctuating mass into the exterior free field (see Section 10.1). Accordingly, we model the exterior field as being free except for a source region and scale the rate of the injected mass with the momentum density fluctuations at the open end of the duct. The exterior acoustic field is then governed by the wave equation

$$(\nabla^2 + k_o^2)p' = \mathrm{i}\omega\dot{\mu}'(\mathbf{x}), \tag{9.91}$$

where $\dot{\mu}'(\mathbf{x})$ denotes the rate of fluctuating density injection at source point $\mathbf{x}$. The derivation of this equation proceeds as in Section 7.7.1, except that the control volume is now envisaged to enclose the source region. It can be obtained from Equation (7.97) by neglecting the mean flow (and a change of notation for the rate of fluctuating mass injection per unit volume).

Now, the free space Green function $G(R) = \mathrm{e}^{\mathrm{i}k_oR}/4\pi R$, where $R = |\mathbf{x} - \mathbf{y}|$, is an outgoing wave solution of the wave equation

$$(\nabla^2 + k_o^2)G(R) = -\delta(\mathbf{y} - \mathbf{x}) \tag{9.92}$$

in a three-dimensional homogeneous space. Here, $\delta(\mathbf{x})$ denotes a three-dimensional Dirac function placed at point $\mathbf{x} = 0$. This is a generalized function and is defined by the identity

$$q(\mathbf{x}) = \int_V q(\mathbf{y})\delta(\mathbf{x} - \mathbf{y})\mathrm{d}V(\mathbf{y}), \tag{9.93}$$

where q is any scalar point function and $\mathrm{d}V$ denotes a volume element of volume V where q is defined. Thus, invoking the principle of superposition, solution of Equation (9.91) may be written as

$$p'(\mathbf{x}) = -\mathrm{i}\omega\int_{V_S} \dot{\mu}'(\mathbf{y})G(R)\mathrm{d}V(\mathbf{y}), \tag{9.94}$$

where V_S denotes the source region. Here, a plausible choice for $\dot{\mu}'$ is the distribution $\dot{\mu}' = \dot{m}'\delta(x_1)$, where $\delta(x_1)$ denotes a one-dimensional Dirac function located at the duct origin (the open end) $x_1 = 0$ and $\dot{m}'$ is the rate of mass injected per unit area over the duct section S, and zero elsewhere. And $\dot{m}'$ is tantamount to the axial momentum density fluctuations in the duct and may be expressed, including the effect of convection, as $\dot{m}' = \rho_j v_1' + v_j\rho'$, where the subscript "j" denotes the properties in the duct (see Figure 9.6). Hence, using the relationship $p' = c_\mathrm{j}^2\rho'$, Equation (9.94) can be expressed as

$$p'(\mathbf{x}) = -\mathrm{i}k_\mathrm{j}\int_S \Big(\rho_\mathrm{j}c_\mathrm{j}v_1'(0,\bullet) + M_\mathrm{j}p'(0,\bullet)\Big)G(R)\mathrm{d}S, \tag{9.95}$$

where "•" denotes the transverse duct coordinates, as usual. Upon substitution of Equations (6.66) and (6.67), this expands to the form

$$p'(\mathrm{x}) = \mathrm{i}k_j\left(\int_S \boldsymbol{\Phi}(\bullet)G(R)\mathrm{d}S\right)\left(\mathrm{A}^+ + M_\mathrm{j}\mathrm{I} + (\mathrm{A}^- + M_\mathrm{j}\mathrm{I})\mathbf{R}(0)\mathbf{P}^+(0)\right), \tag{9.96}$$

where $\mathbf{I}$ denotes a unit matrix of conforming size and we may use the functions F_q defined in Equation (9.63) to recast this equation as

$$p'(\mathbf{x}) = -\mathrm{i}k_{\mathrm{j}}[F_1 \quad F_2 \quad \ldots \quad F_n]\big(\mathbf{A}^+ + M_{\mathrm{j}}\mathbf{I} + \big(\mathbf{A}^- + M_{\mathrm{j}}\mathbf{I}\big)\mathbf{R}(0)\big)\mathbf{P}^+(0), \quad (9.97)$$

where the results given in Section 9.5.3 for the directivity factor D_q, $q = 1, 2, \ldots$, are also valid. The modal radiation transfer functions $K^{(j)}$ can be calculated now once the coefficients array of this equation is determined. For example, for the plane-wave mode

$$\frac{1}{K^{(1)}} = -\mathrm{i}k_{\mathrm{j}}SD_1\frac{\mathrm{e}^{\mathrm{i}k_o R_0}}{4\pi R_0}(1 + M - (1 - M)r_1). \quad (9.98)$$

This may be compared with Equation (9.68). For $M = 0$, it reduces to the half of Equation (9.68), which is as expected because the formulation tacitly assumes an unflanged open end radiating over a solid angle of 4π.

This simple radiation model neglects the presence of duct surfaces and the jet flow in the radiation field. However, it has some practical advantages due to its simplicity. It is not restricted to any specific open-end geometry and can be used relatively easily when the duct is lined. Although the foregoing formulation tacitly assumes an unflanged open end, the flanged case can be included by modifying the free-space Green function as described in Section 9.5.3 or by using the image source method (Section 9.5.8). The effect convection in the duct is taken into account and the model may be extended for the effect of the vorticity shedding at the duct lip as will be discussed next.

A general formulation in which the diffractive and refractive effects in a non-uniform jet flow are also taken into account is given in Mungur et al. [40]. The authors use a semi-numerical method to obtain the far-field directivity pattern of the radiated noise from the inhomogeneous wave equation in spherical coordinates.

9.5.6.1 Effect of Vorticity

According to the vortex theory of aerodynamically generated sound [32], the effect of vorticity generated by mean flow at the edge of a rigid pipe open end may be included in Equation (9.91) as

$$\left(\nabla^2 + k_o^2\right)p'(\mathrm{x}) = \mathrm{i}\omega\dot{\mu}(\mathrm{x}) - \rho_o\nabla\cdot(\boldsymbol{\omega}\times\mathbf{v})', \quad (9.99)$$

where $\boldsymbol{\omega}$ $(= \nabla \times \mathbf{v})$ denotes the vorticity vector and $\mathbf{v}$ denotes the particle velocity of the outlet flow and a prime denotes fluctuations, as usual. Then, by the principle of superposition, the contribution of the vorticity term, p'_ω, say, to the radiated acoustic pressure field may be expressed as

$$p'_\omega(\mathbf{x}) = \rho_o\int_V(\boldsymbol{\omega}\times\mathbf{v})'\cdot\nabla G(R)\mathrm{d}V(\mathbf{y}). \quad (9.100)$$

An approximate far-field and low Mach number evaluation of this integral for an incident plane wave at a compact circular outlet is given by Howe as [32]:

$$\frac{p'_\omega(R,\theta)}{p^+(0)} = -\mathrm{i}k_o SM(1 - r_{000})\left(2\frac{c_j}{c_o}\cos\theta - 1\right)\frac{\rho_o c_o}{\rho_j c_j}\frac{\mathrm{e}^{\mathrm{i}k_o R_o}}{4\pi R_0}, \quad (9.101)$$

where M denotes the Mach number of the outlet mean flow velocity. Upon addition of this contribution, the plane-wave radiation transfer function in Equation (9.98) becomes

$$\frac{1}{K^{(000)}} = -ik_oS(1 - r_{000})\frac{c_o}{c_j}\left\{D_{000}\left(1 + M\frac{1 + r_{000}}{1 - r_{000}}\right) + M\left(2\frac{c_j}{c_o}\cos\theta - 1\right)\frac{\rho_o}{\rho_{jj}}\right\}\frac{e^{ik_oR_0}}{4\pi R_0} \tag{9.102}$$

This result shows that the magnitude of the radiated sound pressure with the vorticity contribution included is smaller than when it is excluded if the measurement angle θ is greater than $\cos^{-1}(c_o/2c_j)$.

9.5.7 Power Source Model

In many practical problems, the total time-averaged power, W, at an open end is the only parameter that is given or may be estimated from the information available about the source (see Section 11.6). This precludes the possibility of calculation of the modal radiation transfer functions and directivity factors. However, we can estimate the radiated sound pressure in the far field by assuming that the power gain function for all cut-on modes at the open end is approximately equal to unity. Then, Equation (9.57) gives

$$|p'|^2 = \frac{1}{2}\rho_o c_o \frac{W}{\hat{\Omega}R_0^2}. \tag{9.103}$$

This formula may be modified by replacing W by $W - W_D$, where W_D denotes a heuristic correction for the dissipated time-averaged power due to flow-acoustic interactions at the open end.

A radiation transfer function for this model may be derived if only plane waves are cut-on at the open end. In this case, the time averaged net acoustic power at the open end is given by (see Section 3.4):

$$W = \frac{2S}{\rho_j c_j}\left(1 + 2M + (1 + \beta)M^2 - (1 - 2M + (1 + \beta)M^2)|r_{000}|^2\right)|p^+(0)|^2 \tag{9.104}$$

and according to the axisymmetric cylindrical vortex sheet model of Howe [32]:

$$W_D = \frac{4SM}{\rho_j c_j}|1 - r_{000}|^2|p^+(0)|^2, \tag{9.105}$$

for an unflanged nozzle. Hence, the plane-wave radiation transfer function may be expressed as

$$\frac{1}{|K(R,\theta)|^2} = \frac{1}{2}\frac{\rho_o c_o}{\rho_j c_j}\frac{1}{\hat{\Omega}R_o^2}\left(1 + 2M + (1 + \beta)M^2\right.$$

$$\left. - (1 - 2M + (1 + \beta)M^2)|r_{000}|^2 - 2M|1 - r_{000}|^2\right) \tag{9.106}$$

Recall that, $\hat{\Omega} = 4\pi$ if radiation is into the full space, $\hat{\Omega} = 2\pi$ if into a half space (flanged open end) and $\hat{\Omega} = \pi/2$ if into a quarter space (open end at the corner of three plane walls).

9.5.8 Effect of Reflecting Surfaces

The free-field radiation is simulated in the laboratory by using rectangular (usually cube-like) anechoic rooms. The acoustic response of an anechoic room is frequency dependent and the free-field conditions will exist in a central volume and above the cut-off frequency of the room. As a rule of thumb, for the free-field condition to exist in an anechoic room at a certain frequency, the room should have at least one dimension of the same order as the wavelength at that frequency.

Fully anechoic chambers are treated with sound absorbing wedges on all surfaces to reduce the reflection coefficient to an adequately small value, usually, typically less than 0.1 at the cut-off frequency. Rooms are also designed with all surfaces but the floor, or two or three mutually perpendicular walls, treated with wedges. The untreated surface(s) is constructed to be hard enough to be assumed as a perfectly reflecting surface(s).

If only the floor is hard, the room simulates the sound radiation in a half space with a perfectly reflecting delimiting infinite plane and is called a hemi-anechoic room. Similarly, when two (three) mutually perpendicular walls are hard, the sound radiation may be assumed to be into a quarter space (one-eight space) delimited by mutually perpendicular two (three) infinite reflecting planes. In such fractional free-field rooms, the sound pressure at a point in the room is determined by sound waves arriving there directly from the source and the waves which arrive after reflection from the hard surfaces of the room. Therefore, the contribution of the reflected waves should be taken into account in the calculation of sound pressure radiated from the outlet of a duct into a fractional free-field room. This is usually done by using the image source method. For example, the image source for a hemi-anechoic room is shown in Figure 9.17a. Since all surfaces except the floor are treated with sound absorbent wedges, it can be assumed that only the waves reflected from the floor can have discernible contribution to the sound pressure at a point in the room. Then, as indicated by the ray paths in Figure 9.17a, a reflected wave may be envisaged as a wave directly arriving at the field point from an exact replica of the actual source located at the indicated image source point. Hence, the contribution of a reflected wave may be calculated by using the cone angle and slant length for the image source in the radiation models described in previous sections. For example, for the hemi-anechoic case of Figure 9.17a, the cone slant length, R', polar angle, φ', and the cone angle, θ', for the image source are given by

$$R' = \sqrt{R^2 + 4h^2 + 4hR\sin\theta\sin\varphi} \tag{9.107}$$

$$\varphi' = \tan^{-1}\frac{2h + r\sin\varphi}{r\cos\varphi} \tag{9.108}$$

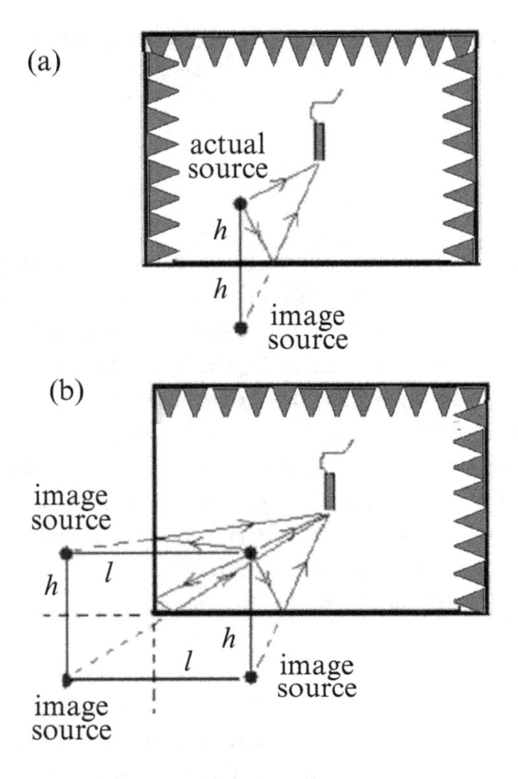

Figure 9.17 Fractional free-field rooms: (a) hemi-anechoic (b) quarter-anechoic.

$$\theta' = \sin^{-1}\frac{R\cos\varphi}{R'\cos\varphi'}, \tag{9.109}$$

where R, φ and θ denote the cone slant length, the polar angle and the cone angle of the actual source. The sound pressure at a point in the room may then be determined in the time domain by superposition of the sound pressures of the actual source and the image source with due account taken of the temporal phase of the arriving waves. The radiation model based on sound power (Section 9.5.7) may not be used to predict the effect of the hard wall reflections by the image source method, since it does not contain the phase information.

Application of the image source concept to a fractional free-field room with two hard walls is illustrated in Figure 9.17b. In this case, three image sources are required to simulate the hard-wall reflections. It can be extended similarly for a fractional free-field room with three hard surfaces, using seven image sources.

The classical reverberant field theory of large rooms is often invoked to compute the sound radiation into large non-free-field rooms. In this theory, it is assumed that, an acoustic field of uniform time-averaged local energy density is set up in the major portion of the room by multiple reflections from the walls. The diffused energy density is proportional to the time-averaged acoustic power, W, incident at the open end and is given by $4W/c_oC_R$, where c_o is the speed of sound and C_R is a frequency dependent parameter called the room constant, which is computed by using the formula

$$C_R = \frac{\bar{\alpha}S}{1-\bar{\alpha}}. \tag{9.110}$$

Here, S is the total surface area of the room and

$$\bar{\alpha} = \frac{\sum_i \alpha_i A_i}{\sum_i A_i}, \tag{9.111}$$

where α_i denotes the sound absorption coefficient of the room surface (including objects in the room) having area A_i, $i = 1,2,\ldots$. The room constant can be derived also from a reverberation-time measurement [41].

Since the time-averaged energy density is equal to the mean-squared acoustic pressure, p_{rms}^2, divided by $\rho_o c_o^2$, the diffuse field can be determined from

$$\left(p_{rms}^2\right)_{\text{diffuse}} = \frac{4\rho_o c_o W}{C_R}, \tag{9.112}$$

where ρ_o denotes the mean density. Then, the total time-averaged acoustic energy density at a point in the room is given by the sum of the free-field and reverberant field contributions and the mean square acoustic pressure may be computed as

$$p_{rms}^2 = \left(p_{rms}^2\right)_{f-f} + \left(p_{rms}^2\right)_{\text{diffuse}} \tag{9.113}$$

where the first term on the right-hand side denotes the free-field contribution, which can be determined by using one of the free-field radiation models described in the previous section. If multiple modes are propagating at the radiating outlet, both terms in Equation (9.113) are determined by summing the contributions of all propagating modes.

References

[1] P.M. Morse and H. Feshbach, *Methods of Theoretical Physics,* (New York: McGraw-Hill, 1953).

[2] B.B. Baker and E.T. Copson, *The Mathematical Theory of Huygen's Principle*, (Oxford: Clarendon Press, 1950).

[3] S.M. Kirkup, *The Boundary Element Method in Acoustics*, Integrated Sound Software, available online at www.boundary-element-method.com, (2019[1998/2007]).

[4] A. Selamet, Z.L. Li and R.A. Kach, Wave reflections from duct terminations, *J. Acoust. Soc. Am.* **109** (2001), 1304–1311.

[5] J.-P. Dalmont, C.J. Nederveen and N. Joly, Radiation impedance of tubes with different flanges: numerical and experimental investigations, *J. Sound Vib.* **244** (2001), 505–534.

[6] A.R. Da Silva, P.H. Mareze and A. Lenzi, Approximate expressions for the reflection coefficient of ducts terminated by circular flanges, *J. Braz. Soc. Mech. Sci. Eng.* **34** (2012), 219–224.

[7] G.N. Watson, *A Treatise on the Theory of Bessel Functions*, (Cambridge: Cambridge University Press, 1966).

[8] A.N. Norris and I.C. Sheng, Acoustic radiation from a circular pipe with an infinite flange, *J. Sound Vib.* **135** (1989), 85–93.

[9] F. Silva, Ph. Guillerman, J. Kergomard, B. Mallorini and A.N. Norris, Approximation formulae for the acoustic radiation impedance of a cylindrical pipe, *J. Sound Vib.* **322** (2009), 255–263.

[10] W.E. Zorumski, Generalized radiation impedances and reflection coefficients of circular and annular ducts, *J. Acoust. Soc. Am.* **54** (1973), 1667–1673.

[11] J.C. Wendolowski, F.R. Fricke and R.C. McPhedran, Boundary conditions of cylindrical flanged pipe, *J. Sound Vib.* **162** (1993), 89–96.

[12] H.J. Bom and I.J. Park, A series solution for acoustic radiation from a flanged circular pipe, *Acustica* **80** (1994), 315–316.

[13] N. Amir and H. Matzner, Acoustics of flanged circular pipe using singular basis functions, *J. Acoust. Soc. Am.* **107** (2000), 714–724.

[14] Y.C. Cho, Rigorous solutions for sound radiation from circular ducts with hyperbolic horns or infinite plane baffle, *J. Sound Vib.* **69** (1980), 405–425.

[15] H. Levine and J. Schwinger, On the radiation of sound from an unflanged circular pipe, *Physical Review* **73** (1948), 383–406.

[16] B. Noble, *The Wiener–Hopf Technique*, (London: Pergamon Press, 1958).

[17] A.D. Rawlins, Radiation of sound from an unflanged rigid cylindrical duct with an acoustically absorbent internal surface, *Proc. Royal Soc. Lond. A.* **361** (1978), 65–91.

[18] N. Peake, On radiation properties of an asymmetric cylinder, *Wave Motion* **22** (1995), 371–385.

[19] R. Munt, The interaction of sound with a subsonic jet issuing from a semi-infinite cylindrical pipe, *J. Fluid Mech.* **83** (1977), 609–640.

[20] R. Munt, Acoustic transmission properties of a jet pipe with subsonic jet flow, *J. Sound Vib.* **142** (1990), 413–436.

[21] S.W. Rienstra, Acoustic radiation from a semi-infinite annular duct in a uniform subsonic mean flow, *J. Sound Vib.* **94** (1984), 267–288.

[22] G. Gabard and R.J. Astley, Theoretical model for sound radiation from annular jet pipes: far- and near-field solutions, *J. Fluid Mech.* **549** (2006), 315–341.

[23] M.V. Taylor, D.G. Crighton and A.M. Cargill, The low frequency aeroacoustics of buried nozzle systems, *J. Sound Vib.* **163** (1993), 493–526.

[24] B. Veitch and N. Peake, Acoustic propagation and scattering in the exhaust flow from coaxial cylinders, *J. Fluid Mech.* **613** (2008), 275–307.

[25] P. in't Panhuis, Calculation of the sound pressure reflection coefficient and the acoustic end correction of a semi-infinite circular pipe issuing a subsonic cold or hot jet with co-flow, Internal report, MWL, Aeronautical and Vehicle Engineering, KTH, Stockholm, (2003).

[26] S. Rienstra, On the acoustical implications of vortex shedding from an exhaust pipe, *Trans. AMSE.* **103** (1981), 378–384.

[27] S. Allam and M. Åbom, Investigation of damping and radiation using full plane wave decomposition in ducts, *J. Sound Vib.* **292** (2006), 519–534.

[28] E.J. English, A measurement based study of the acoustics of pipe systems with flow, PhD thesis, ISVR, University of Southampton, UK, (2010) (https://bit.ly/3loWU6L).

[29] H. Rammal and J. Lavrentjev, Sound reflection at an open end of a circular duct exhausting hot gas, *Noise Control Eng. J.* **56** (2008), 107–114.

[30] S.W. Rienstra, Small Strouhal number analysis for acoustic wave-jet flow-pipe interactions, *J. Sound Vib.* **86** (1983), 539–556.
[31] A. Cargill, Low frequency acoustic radiation from a jet pipe-a second order theory, *J. Sound Vib.* **83** (1982), 339–354.
[32] M.S. Howe, *Acoustics of Fluid-Structure Interaction*, (Cambridge: Cambridge University Press, 1998).
[33] A. Cargill, Low-frequency sound radiation and generation due to the interaction of unsteady flow with a jet pipe, *J. Fluid Mech.* **121** (1982), 59–105.
[34] A. Hirschberg and S.W. Rienstra, *An Introduction to Aeroacoustics,* (The Netherlands: Eindhoven University of Technology, 2004).
[35] P.O.A.L. Davies, Plane wave reflection at flow intakes, *J. Sound Vib.* 115 (1987), 560–564.
[36] M. Abromovitch, I.A. Stegun, *Handbook of Mathematical Functions*, (Washington D.C: National Bureau of Standards, Applied Mathematics Series 55, 1972).
[37] A.J. Toregrosa, A. Broatch, V. Bermúdez and I. Andrés, Experimental assessment of emission models used for IC engine exhaust noise prediction, *Experimental Thermal and Fluid Science* **32** (2005), 97–107.
[38] G.F. Homicz and J.A. Lordi, A note on the radiative directivity patterns of duct acoustic modes, *J. Sound Vib.* **41** (1975), 283–290.
[39] A.P. Dowling and J.E. Ffowcs Williams, *Sound and Sources of Sound*, (Chichester: Ellis Horwood Ltd., 1983).
[40] P. Mungur, H.E. Plumblee and P.E. Doak, Analysis of acoustic radiation in a jet flow environment, *J. Sound Vib.* **36** (1974), 21–52.
[41] F.A. Bies and C.H. Hansen, *Engineering Noise Control*, (London: E and FN Spon, 2009).

10 Modeling of Ducted Acoustic Sources

10.1 Introduction

Without mathematical models of acoustic sources, we can predict acoustic transfer functions or wave transmission matrices in ductworks, but not the actual values of the acoustic variables involved in these relations. So whenever the actual values of the acoustic variables are required, it is inevitable to get involved with modeling of the sources in acoustic analyses (see Chapter 11).

In general, the basic sound generation mechanisms are associated with effects that apply a non-steady compressive strain to a fluid. Such effects can be produced by external agents such as moving solid surfaces or by stationary solid surfaces obstructing the path of fluid flow, by non-steady injection of a mass of fluid into or suction from the duct, or by non-steady heating of the fluid in the duct. Also, the fluid flow in the duct itself can generate sound waves by mechanisms which can be traced to the flow momentum fluctuations due to turbulent mixing and vortex generation in boundary layers and free shear layers. Such source mechanisms can be explained mathematically by including the source terms in the basic conservation laws of fluid dynamics [1–2].

The continuity, momentum and energy equations of the unsteady motion of inviscid fluids are given in Appendix A as Equation (A.21), (A.22) and (A.24), respectively, and are repeated here for convenience:

$$\frac{\partial \rho}{\partial t} + \mathbf{v}\cdot\nabla\rho + \rho\nabla\cdot\mathbf{v} = \dot{\mu} \tag{10.1}$$

$$\frac{\partial}{\partial t}(\rho\mathbf{v}) + \mathbf{v}\nabla\cdot\rho\mathbf{v} + \rho\mathbf{v}\cdot\nabla\mathbf{v} + \nabla p = \mathbf{f} \tag{10.2}$$

$$\frac{\partial}{\partial t}(e\rho) + \nabla\cdot(\rho\, h^o\mathbf{v}) = \dot{U}, \tag{10.3}$$

where $\dot{\mu}$ denotes the local mass per unit volume created per unit time, $\mathbf{f}$ denotes the volume force per unit volume applied on the fluid, $\dot{U}$ denotes the local internal energy created per unit volume per unit time and the other symbols have the usual meanings. For a perfect gas ($\gamma p = \rho c^2$) the continuity and momentum equations may be combined as

$$\gamma p\left(\frac{\partial \mathbf{v}}{\partial t} + \mathbf{v}\nabla\cdot\mathbf{v}\right) + c^2\nabla p = c^2(\mathbf{f} - \dot{\mu}\mathbf{v}). \tag{10.4}$$

Equation (10.3), on the other hand, may also be expressed in terms of $p, \mathbf{v}$ by first using the continuity equation to derive Equation (A.26), and then using the state equation for the specific entropy, Equation (A.15a). The resulting equation for a perfect gas ($\beta T = 1$) is

$$\frac{\partial p}{\partial t} + \mathbf{v}\cdot\nabla p + \gamma p \nabla\cdot\mathbf{v} = \dot{\mu}(\gamma - 1)\left(\frac{\dot{U}}{\dot{\mu}} - \frac{1}{2}v^2 - \mathbf{f}\cdot\mathbf{v}\right). \tag{10.5}$$

In these equations, $\dot{\mu}$, $\mathbf{f}$ and $\dot{U}$ are called source terms, which may also be decomposed into mean (time independent) fluctuating parts, which we denote as usual as $q = q_o + q'$, $q \in \dot{\mu}, \mathbf{f}, \dot{U}, \ldots$ The mean parts represent the gross flow of the fluid imparted to the fluid by the source process and the fluctuating parts are hypothesized to be the sources of the acoustic wave motion generated in the fluid by the same process. Thus, the basic acoustic source processes are: (i) injection of fluctuating mass, (ii) application of fluctuating force and (iii) addition of fluctuating heat. These processes are readily identified when they occur by direct action of fluid machinery. For example, the primary source of exhaust noise in reciprocating fluid machinery is the sudden discharge of compressed gas through the valves; forces applied on the fluid by rotating blades are the primary noise source in turbomachinery; and fluctuating heat addition is the primary source of noise generated in combustion chambers. Subtler, however, is the sound generated by the motion of the fluid itself. According to Lighthill's theory of aerodynamic sound [3], the source of sound generated by turbulence is the Reynolds stress distribution in the fluid. At low Mach numbers, the dominant component of this source comes from vorticity fluctuations, which is equivalent to fluctuating body force distribution [4–5].

Formulation of working mathematical models for these basic source mechanisms is perhaps the least developed link in duct acoustics. The state-of-the-art methods mostly rely on fitting measured acoustic data to generic equations, for example Equation (2.10), that are postulated to be equations of generic source processes. Such measurements methods are described in Section 12.4. In the present chapter, we derive, from first principles, simple mathematical models for primary source mechanisms in fluid machinery. These idealized models are, of course, not as decisive as measured models, but they are useful for insight into the physics underlying the processes and as first approximations in engineering calculations when no better models are available.

One of the difficulties in the formulation of mathematical models of the basic source mechanisms is the ambiguity of the boundaries of the actual source region. A plausible resolution of this problem is to assume that the actual source region is confined to an axially compact slice of a duct (Figure 10.1). This is usually called an actuator disk and is characterized by the jumps it creates in the mean and acoustic properties across its transverse planes. If the sound wave motion generated is constrained to propagate in only one direction along the duct, the model is called a one-port source (Figure 10.1a). If the generated sound waves are free to propagate to both sides of the duct, it is called a two-port source (Figure 10.1b). These are considered in the subsequent subsections with typical applications.

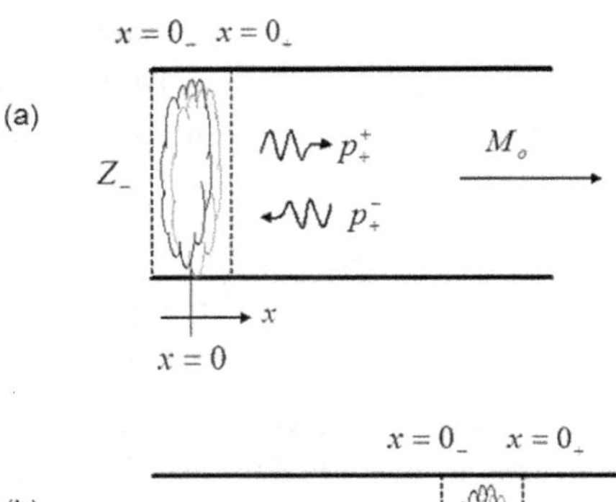

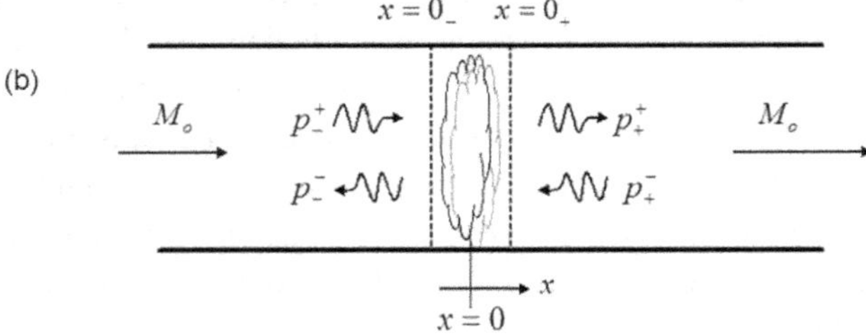

Figure 10.1 Actuator-disk models: (a) one-port source, (b) two-port source.

10.2 One-Port Sources Characterized by Unsteady Mass Injection

Consider a uniform duct of cross-sectional area S, with origin at $x = 0$, where x denotes the duct axis (Figure 10.1). The duct geometry is defined in a right-handed Cartesian frame of axes x_1, x_2, x_3, where $x_1 \equiv x$. The source region is defined at the origin $x = 0$, as an elementary slice of volume $\delta V = S\delta x$ of surface $\sigma = \sigma_1 \cup \sigma_2 \cup \sigma_3$, where σ_1 and σ_2 denote surfaces of δV normal to the duct axis (of area S) and σ_3 denotes the interfacial surface between δV and the inner surface of the duct walls (assumed to be hard). Let non-steady addition (injection) or removal (suction) of fluid mass occur in the source region at the rate of $\dot{\mu}$ per unit volume per unit time. We denote the specific energy carried by this unsteady mass creation by H^o and assume that the fluctuating forces applied by the process on the fluid are negligible. The source region is assumed to be compact in the axial direction so that the source distribution may be represented by the generalized function $\dot{\mu} = \dot{m}\delta(x)/S$, where $\delta(x)$ denotes a Dirac function at $x = 0$ and $\dot{m} = \dot{m}(t)$ denotes the rate of mass addition or removal. The following Heaviside function is used throughout this chapter in defining the jumps in the values of fluid properties across axially compact source regions:

$$H_{1/2}(x) = \begin{cases} 0 & \text{if} \quad x < 0 \\ \dfrac{1}{2} & \text{if} \quad x = 0\,. \\ 1 & \text{if} \quad x > 0 \end{cases} \tag{10.6}$$

Then a fluid property q which undergoes a jump $[q]$ across the source region is expressed as $q = \tilde{q} + [q]H_{1/2}(x)$, where the square brackets denote a jump of the enclosed variable and an over-tilde denotes the same variable with the jump subtracted [6]. Therefore, even though q is not differentiable at the origin due to the jump, $\tilde{q}$ is continuous there and is differentiable in the usual sense. The source process may cause jumps in axial components and axial derivatives of some fluid properties across the source region, but we assume that the transverse components and the transverse derivatives of fluid properties remain continuous.

Upon substitution of $\mathbf{f} = 0$ and $\dot{\mu} = \dot{m}(t)\delta(x)/S$, integration of Equation (10.5) and the axial component of Equation (10.4) across the source region yield the following jump equations:

$$\langle \gamma p \rangle \left[\frac{1}{2} v^2 \right] + \langle c^2 \rangle [p] = -\frac{1}{S} \dot{m} \langle v \rangle \langle c^2 \rangle \tag{10.7}$$

$$\langle v \rangle [p] + \langle \gamma p \rangle [v] = \frac{1}{S} (\langle \gamma \rangle - 1) \dot{m} \left(H^o - \frac{1}{2} \langle v^2 \rangle \right). \tag{10.8}$$

Here, v denotes the axial component of $\mathbf{v}$ (that is, $v = \mathbf{e}_1 \cdot \mathbf{v}$) and angled brackets $\langle \; \rangle$ denote arithmetic mean. The following examples should clarify definitions of the jump and the arithmetic-mean brackets:

$$\langle p \rangle \equiv \frac{1}{2} (p_+ + p_-), \quad [p] \equiv p_+ - p_-, \quad \langle v \rangle \equiv \frac{1}{2} (v_+ + v_-), \quad [v] \equiv v_+ - v_-, \tag{10.9}$$

where the subscripts minus "−" and plus "+" refer to the just negative ($x = 0_-$) and just positive ($x = 0_+$) sides of $x = 0$. For an example of integration across the source region, consider the second term, $\mathbf{v} \cdot \nabla p$, on the left-hand side of Equation (10.5). Firstly, this is expanded as $\mathbf{v} \cdot \nabla p = v \partial p / \partial x + \mathbf{v}_\perp \cdot \nabla_\perp p$, where the subscript "$\perp$" denotes the transverse (in-plane) component of a vector. Since only the axial component and the axial derivatives of fluid properties undergo jumps across the source region, it remains to integrate $v \partial p / \partial x$ and the following line explains this operation:

$$\int_{x=0_-}^{x=0_+} v \frac{\partial p}{\partial x} \mathrm{d}x = \int_{x=0_-}^{x=0_+} \left(\tilde{v} + H_{1/2}(x)[v] \right) \left(\frac{\partial \tilde{p}}{\partial x} + \delta(x)[p] \right) \mathrm{d}x = [p] v_- + \frac{1}{2} [v][p] = \langle v \rangle [p] \tag{10.10}$$

where we have used the relationship $\mathrm{d}H_{1/2}(x)/\mathrm{d}x = \delta(x)$ in expanding $\partial p / \partial x$. In general, the terms with time derivatives in Equations (10.4) and (10.5) have no contribution to the jumps.

Equations (10.7) and (10.8) may be linearized by substituting the usual decomposition $q = q_o + q', q \in p, v, \dot{m}, H^o \ldots$. We assume that the fluctuations c' and γ' are negligible with c_o and γ_o constant in the source region. On substitution of the foregoing decompositions, Equations (10.7) and (10.8) yield the following acoustic jump equations[1]

[1] The continuity equation, Equation (10.1), may be integrated similarly across the source region to get the jump equations $\langle v_o \rangle [\rho_o] + \langle \rho_o \rangle [v_o] = \dot{m}_o / S$ and $v_o \rho'_+ + \langle v' \rangle [\rho_o] + \langle \rho_o \rangle [v'] = \dot{m}' / S$.

$$S\left(\gamma_o \langle p_o\rangle [v_o v'] + \gamma_o \left[\frac{1}{2}v_o^2\right] \langle p'\rangle + c_o^2[p']\right) = -c_o^2 \dot{m}_o \langle v'\rangle - c_o^2 \langle v_o\rangle \dot{m}' \tag{10.11}$$

$$S(\langle v_o\rangle [p'] + \langle v'\rangle [p_o] + \gamma_o \langle p'\rangle [v_o] + \gamma_o \langle p_o\rangle [v']) = (\gamma_o - 1)\left((\dot{m}H^o)' - \dot{m}_o \langle v_o v'\rangle - \frac{1}{2}\dot{m}' \langle v_o^2\rangle\right) \tag{10.12}$$

Since the mean pressure p_o is continuous across the source region and the mean flow is induced only in the $x > 0$ side of the source region (Figure 10.1a), we have

$$[p_o] = 0, \quad \langle p_o\rangle = p_o, \quad [v_o] = v_o, \quad \langle v_o\rangle = \frac{1}{2}v_o. \tag{10.13}$$

Consequently, Equations (10.11) and (10.12) may be expressed as

$$S\left(\gamma_o p_o v_o v'_+ + \frac{1}{2}\gamma_o \langle p'\rangle v_o^2 + c_o^2[p']\right) = -c_o^2 \dot{m}_o \langle v'\rangle - \frac{1}{2}c_o^2 v_o \dot{m}' \tag{10.14}$$

$$S\left(\frac{1}{2}v_o[p'] + \gamma_o v_o \langle p'\rangle + \gamma_o p_o [v']\right) = (\gamma_o - 1)\left((\dot{m}H^o)' - \frac{1}{2}\dot{m}_o v_o v'_+ - \frac{1}{4}\dot{m}' v_o^2\right) \tag{10.15}$$

respectively. Upon expanding the jump and arithmetic mean brackets, the foregoing two equations may be combined as a single equation of the form [7]:

$$p_S = Z_S v'_+ + p'_+, \tag{10.16}$$

where

$$p_S = \frac{1}{B}\left(\frac{2(\dot{m}H^o)'}{S v_o} - \frac{\dot{m}' v_o}{2S} - A\frac{2c_o^2 v_o \dot{m}'}{S}\right) \tag{10.17}$$

$$Z_S = \frac{1}{B}\left(\frac{2\gamma_o p_o}{v_o(\gamma_o - 1)} + \frac{\dot{m}_o}{S} + A\left(4\gamma_o p_o v_o - \frac{2c_o^2 \dot{m}_o}{S}\right)\right) \tag{10.18}$$

$$A = \left(1 - \frac{2\gamma_o p_o}{v_o(\gamma_o - 1)Z_-}\right)\left(4c_o^2 - \gamma_o v_o^2 - \frac{2c_o^2 \dot{m}_o}{S Z_-}\right)^{-1} \tag{10.19}$$

$$B = \frac{\gamma_o + 1}{\gamma_o - 1} + A\left(4c_o^2 + \gamma_o v_o^2\right) \tag{10.20}$$

and $Z_- = p'_-/v'_-$. One-port sources are characterized by $Z_- = \infty$, because, for the generated sound wave motion to be directed in one direction, the source region should be closed on the other side in some way. For $Z_- = \infty$, Equation (10.19) simplifies to $A = 1/\left(4c_o^2 - \gamma_o v_o^2\right)$ and, assuming subsonic low Mach numbers $M_o^2 = v_o^2/c_o^2 << 1$, Equations (10.17) and (10.18) may be expressed as

$$p_S = \frac{\gamma_o - 1}{\gamma_o}\left(\frac{(\dot{m}H^o)'}{S v_o} - \frac{\dot{m}' v_o}{2S}\right) \tag{10.21}$$

$$Z_S = \frac{p_o}{v_o} = \rho_o c_o \left(\frac{1}{\gamma_o M_o} \right). \tag{10.22}$$

It is notable that Z_S is determined by the mean flow only. Accordingly, the one-port source equation, Equation (10.16), applies both in the time domain and the frequency domain. In the latter case, the primed quantities are understood as the Fourier transforms of the respective physical quantities in the time domain or, if the fluctuations are periodic, the complex Fourier coefficients, over the frequency spectrum of the source. We will generally assume the frequency domain representation, unless stated otherwise, and refer to the parameters p_S and Z_S as the source pressure strength and source impedance, respectively. It is also notable that the sign of the source impedance depends on the sign convention used for v_o. According to the sign convention used in this book, v_o is positive if in the positive direction of the duct axis and the direction of the axis is taken positive away from the source region. Consequently, source impedance will be a positive (negative) number for an exhaust (intake) process. If the source process generates no mean flow, this one-port source model degenerates to a pure volume velocity source, which is characterized by an infinite source impedance.[2]

Linearization of Equations (10.7) and (10.8) is carried out on the understanding that a single duct mode is being considered. If multiple cut-on modes are generated at the source plane and interactions between these modes are negligible, each mode may be assumed to be governed by an equation in the form of Equation (10.16). The latter may be expressed, by using Equations (6.55) and (6.58), in terms of modal pressure wave components as

$$p_S^{(\mu)} = \left(1 + \zeta_S A_\mu^+\right) p_\mu^+(0_+) + \left(1 + \zeta_S A_\mu^-\right) p_\mu^-(0_+), \quad \mu = 1, 2, \ldots, q \tag{10.23}$$

with $\varphi_\mu(\bullet)$ dependence on the transverse coordinates suppressed. Here, q denotes the number of cut-on modes and $\zeta_S = Z_S/\rho_o c_o$ denotes the normalized source impedance, which is the same for all modes, as it depends only on the mean flow.

Unsteady fluid mass injection or removal processes typically occur in exhaust and intake ports of reciprocating fluid machinery. The working gas is injected every working cycle from the cylinders into the ports (exhaust) or from the ports into the cylinders (intake) during the open periods of the respective valves. The source region may be assumed to be just on the port side of the valves and the infinite backup impedance assumption is adequately accurate, because the maximum valve displacements are usually much small compared to the wavelengths of interest. In the exhaust process, we may take $\dot{m}H^o \approx \dot{m}h_{\mathrm{cyl}}$, where h_{cyl} denotes the specific enthalpy of the gas in the cylinder, since the stagnation enthalpy is conserved across the valve throat and the kinetic energy of the gas is negligible in the cylinder. Then, Equation (10.21) implies $p_S \approx (1 - 1/\gamma_o)\left(\dot{m}h_{\mathrm{cyl}}\right)'/Sv_o$, which shows that the source pressure strength

[2] When $Z_S = \infty$, Equation (10.16) may be written as $v'_+ = \lim_{v_o \to 0} (p_S/Z_S)$. This limit is finite, which implies that volume velocity Sv'_+ is impressed at the source plane.

of an exhaust process is determined by the fluctuations of the rate of cylinder enthalpy. Consequently, it would be determined by the fluctuations of the rate of gas mass injection, if the exhaust process is cold, that is, the temperature in the cylinder remains fairly constant during the exhaust stroke. For the intake process, a similar argument gives $H^o = h_{\text{in}} + v_{\text{in}}^2/2$, where the subscript "in" denotes the properties of the fresh air drawn from the intake duct, but the kinetic energy term may not be dismissed now. So the source pressure strength is dominated by the fluctuations of the rate of gas mass removal from the intake duct, since the temperature is practically constant during the intake stroke, and also by the fluctuations of the rate of kinetic energy of the flow.

In general, the data that are necessary for the evaluation of the source pressure strength can be determined by simulation of a working cycle of the fluid machine at a given operation point. For example, Figure 10.2a shows the $\dot{m}H^o$ injection at an exhaust port of a 4-stroke spark-ignition engine running at 3000 rpm, as determined by thermodynamic cycle simulation. The engine stroke and bore are both 0.1 m, and the compression ratio is 8. The fuel is C_4H_{10}, used at the equivalence ratio of unity and the burned residual gas content is about 4%. Exhaust valve opening and closing angles are, respectively, 145° and 375° after the top dead centre. The Fourier magnitude spectrum of the $\dot{m}H^o$ time-history of Figure 10.2a is shown in Figure 10.2b for the first 50 harmonics of the firing cycle frequency (FCF), which is equal to the half of the engine crankshaft frequency, since the engine is a four-stroke one. The port Mach number and time-averaged port temperatures are 0.1 and 850 °C, respectively. For the fuel used in the cycle simulation, this corresponds to a speed of sound of about

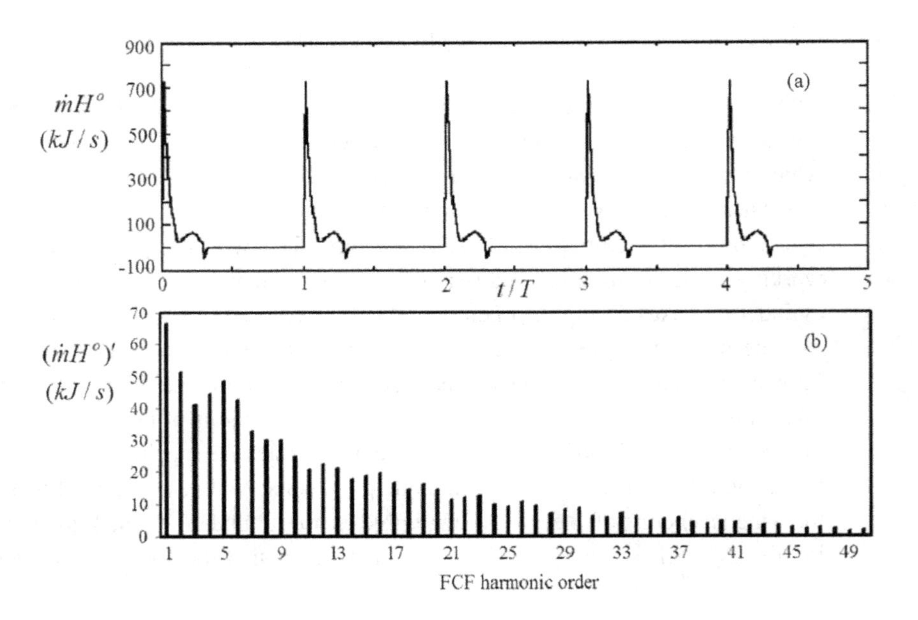

Figure 10.2 (a) Exhaust gas $\dot{m}H^o$ injection from a cylinder of a four-stroke spark-ignition engine and (b) its Fourier spectrum (FCF = 25 Hz).

650 m/s, a ratio of specific heat coefficients of 1.286 and, from Equation (10.22), normalized source impedance of $Z_S/\rho_o c_o = 7.776$.

The source strength calculations simplify considerably if $\dot{m}'$ and $(\dot{m}H^o)'$ characteristics during the intake or exhaust processes can be estimated without cycle simulation. This is more feasible for the intake side, because rate of fresh air flow through the intake valve, $\dot{m} = \dot{m}_o + \dot{m}'$, scales approximately with the intake valve lift profile [8]. Thus, knowing the volumetric efficiency of the engine and the intake conditions, the source pressure strength at each cylinder port can be estimated as function of the crank angle by Fourier analysis.

The situation at a hot exhaust is more complicated due to the blow-down phase of the exhaust process. In Figure 10.2a, the first peak of the $\dot{m}H^o$ characteristics corresponds to the blow-down period, during which hot gas at high pressure is suddenly discharged from the cylinder into the port as the exhaust valve opens. In the example considered, the blow-down occurs during about 1/3 of the full valve-lift period and in the second phase of the exhaust process, as the valve is closing, the piston just pushes the remaining gas in the cylinder into the port, with some backflow before the valve closes.

Exhaust blow-down is difficult to model in a realistic manner without cycle simulation. As a rough approximation, discharge into an exhaust port may be represented by replacing the actual exhaust pulse by a pulse of known discrete Fourier spectrum, such as a rectangular or a triangular pulse, both of which have discrete decaying Fourier spectra akin to that in Figure 10.2b. Laville and Soedel describe a practical approach for scaling such ideal pulses so that the average exhaust injection over one cycle may be matched approximately [9]. They show that the average values over one cycle of the exhaust temperature (T_e) and the velocity of the exhaust gas (v_e) through an exhaust port of area S may be estimated by the formulae

$$T_e = \beta T_i \tag{10.24}$$

$$v_e = \frac{V_d f_{\text{cycle}}}{S}\left(\frac{\beta r - 1}{r - 1}\right). \tag{10.25}$$

Here, T_i denotes the absolute intake temperature, V_d denotes the displacement volume, r denotes the compression ratio, f_{cycle} denotes the firing frequency and

$$\beta = \left(1 + \frac{P_{\text{bhp}}}{V_d p_i \eta_T \eta_M f_{\text{cycle}}}\left(\frac{r}{r-1}\right)^{-1}\left(\frac{r^{n-1} - 1}{n - 1}\right)^{-1}\right)^{\frac{1}{\gamma_e}}, \tag{10.26}$$

where P_{bhp} denotes the brake horse-power of the engine per cylinder, η_T and η_M denote, respectively, the indicated thermal efficiency and the mechanical efficiency of the engine, n denotes the polytropic exponent for the compression and expansion strokes, γ_e denotes the ratio of specific heat coefficients at the exhaust temperature and p_i denotes the intake pressure. These formulae assume that the ratio of intake pressure to the exhaust pressures, p_e, is approximately equal to unity, which is adequately accurate, considering the rather approximate nature of the calculations. Hence, cycle averages of quantities relevant to source calculations, such as $\dot{m}$, h and H^o at an exhaust port may be estimated by using the perfect gas relations, for example, $\dot{m}_e = \rho_e S v_e$, $h_e = c_e^2/(\gamma_e - 1)$, $(H^o)_e = \dot{m}_e\left(h_e + v_e^2/2\right) \approx \dot{m}_e h_e$.

10.3 Moving the Active Plane of One-Port Sources

Silencers are usually mounted on ductworks quite far from the primary acoustic source(s) driving the system. The intermediate ductwork may contain several fixed components with different primary functions. The computational overhead of the acoustic modeling of the silencer with sources can then be simplified considerably if the actual source(s) are replaced by their effect at the inlet of the silencer. This is also convenient when the source characteristics are to be determined by measurement (see Section 12.4), because the conditions at the actual source region may not be suitable for instrumentation.

For example, consider Figure 10.3a, which shows schematically an exhaust manifold of a four-cylinder in-line engine. Here, the actual active one-port source planes – planes 1, 2, 3 and 4 – are located at the exhaust ports. For simplicity, we assume one valve per cylinder; however, the analysis is valid also if there are multiple valves per cylinder. Let the source planes be numbered as $j = 1, 2, \ldots, \sigma$, where σ denotes the number of sources ($\sigma = 4$ in Figure 10.3a), and denote the impedance and strength of source $j \in 1, 2, \ldots, \sigma$, by Z_{Sj} and p_{Sj}, respectively. Then, from Equation (10.16):

$$p_{Sj} = Z_{Sj} v_j' + p_j', \tag{10.27}$$

where p_j' and v_j' denote the acoustic pressure and particle velocity just downstream of source plane j. Now, consider moving these active source planes to plane n, so that a fictive one-port source of equation

$$p_{Sn} = Z_{Sn} v_n' + p_n', \tag{10.28}$$

where p_n' and v_n' denote the acoustic pressure and particle velocity just downstream of source plane n, generates, downstream of plane n, the same in-duct acoustic field as that generated by the actual sources. This fictive source is called an equivalent source and the activated plane n is called an equivalent source plane. If the pressure strength and impedance of this equivalent source, p_{Sn} and Z_{Sn}, respectively, are known for given operation conditions, then the acoustic performance of the complete system may be studied without concern about the components in the part of the system upstream of the equivalent source plane. We will show that Equation (10.28) is feasible and that the equivalent source characteristics may be computed in terms of the characteristics of the actual sources upstream of the equivalent source plane and the acoustic path in between. An important implication of this in the design of silencers is that we need to inquire only about the characteristics of an equivalent source close to its inlet, or measure them at thereabouts, if to be determined by measurement.

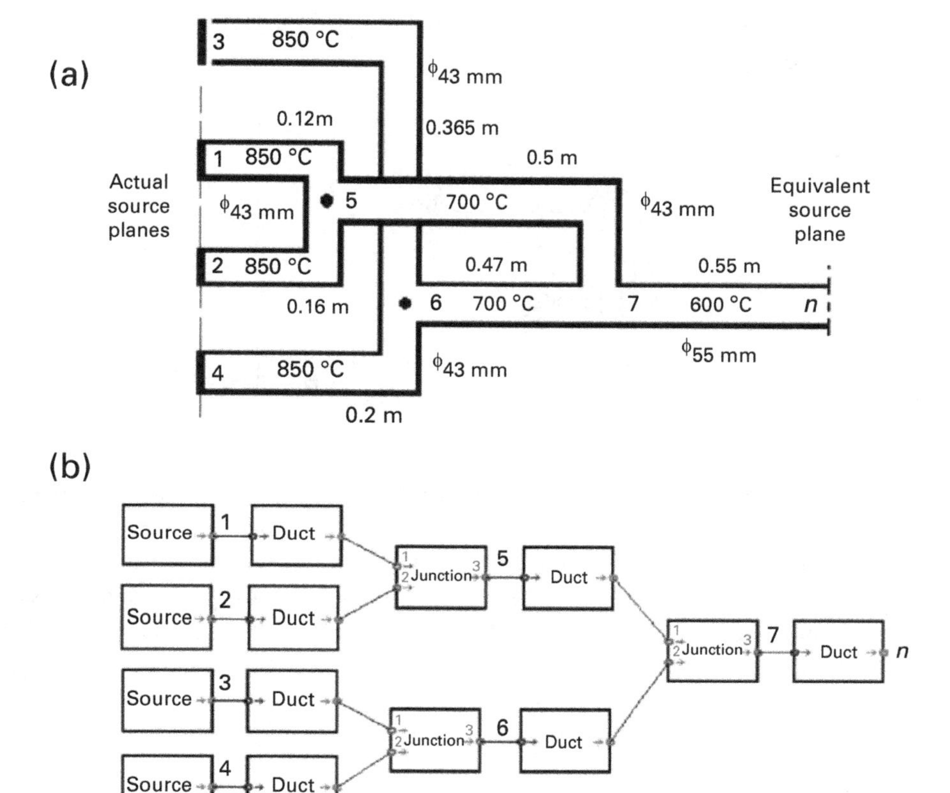

Figure 10.3 (a) Exhaust manifold of a four-cylinder engine with runners, firing order is 3-2-4-1 (actual ducts are continuous, 90° bends are only for the clarity of the geometry), (b) acoustic block diagram model of the manifold.

The equivalent source at plane n in Figure 10.2 may be determined by moving the actual active source planes across the elements along the path to the equivalent source plane [7, 10]. Here we describe the operations involved by considering a block diagram of the system. For example, Figure 10.3b shows a block diagram of the manifold in Figure 10.3a. For simplicity of the discussion, we assume one-dimensional blocks, as modal blocks may be treated similarly. Since we generally use the pressure wave components as acoustic variables in the formulation of the blocks (see Chapters 3–4), first we re-cast Equation (10.27) as

$$p_{Sj} = \begin{bmatrix} 1+\zeta_{Sj} & 1-\zeta_{Sj} \end{bmatrix} \mathbf{P}_j, \tag{10.29}$$

where $\mathbf{P}_j = \begin{Bmatrix} p_j^+ & p_j^- \end{Bmatrix}$ denotes the pressure wave components at plane j and $\zeta_{Sj} = Z_{Sj}/\rho_j c_j$ denotes the normalized impedance of source j. The pressure wave components are defined by the usual plane wave transformations $p'_j = p_j^+ + p_j^-$ and $\rho_j c_j v'_j = p_j^+ - p_j^-$, where ρ_j and c_j denote the mean fluid density and the speed of sound at plane j. The global matrix (see Section 2.4.2) of the block diagram may be expressed as

$$\begin{bmatrix} \begin{bmatrix} \text{source} & \text{equations} \\ & \vdots \end{bmatrix} \\ \begin{bmatrix} \text{model} & \text{equations} \\ & \vdots \end{bmatrix} \end{bmatrix} \begin{bmatrix} \mathbf{P}_1 \\ \mathbf{P}_2 \\ \vdots \\ \mathbf{P}_\sigma \\ \mathbf{P}_{\sigma+1} \\ \vdots \\ \mathbf{P}_{n-1} \\ \mathbf{P}_n \end{bmatrix} = \begin{bmatrix} p_{S1} \\ p_{S2} \\ \vdots \\ p_{S\sigma} \\ 0 \\ \vdots \\ 0 \end{bmatrix}, \tag{10.30}$$

where the nodes are numbered so that the equivalent source plane number, n, is the largest node number of the acoustic model. Let $\mathbf{C}_1$ denote the coefficients matrix in Equation (10.30). The number of columns of this matrix is one greater than the number of rows. Let the column of the coefficients matrix corresponding to the pressure wave component p_n^- be denoted by $-\mathbf{b}_1$. Then, Equation (10.30) can be written in the form

$$\mathbf{C} \begin{bmatrix} \mathbf{P}_1 \\ \mathbf{P}_2 \\ \vdots \\ \mathbf{P}_\sigma \\ \mathbf{P}_{\sigma+1} \\ \vdots \\ \mathbf{P}_{n-1} \\ p_n^+ \end{bmatrix} = \mathbf{a}_1 + \mathbf{b}_1 p_n^-, \tag{10.31}$$

where $\mathbf{a}_1$ denotes the vector on the right-hand side of Equation (10.30) and $\mathbf{C}$ denotes the square matrix which is obtained from matrix $\mathbf{C}_1$ by deleting its column corresponding to p_n^-. Upon pre-multiplying Equation (10.31) by $\mathbf{C}^{-1}$,

$$\begin{bmatrix} \mathbf{P}_1 \\ \mathbf{P}_2 \\ \vdots \\ \mathbf{P}_\sigma \\ \mathbf{P}_{\sigma+1} \\ \vdots \\ \mathbf{P}_{n-1} \\ p_n^+ \end{bmatrix} = \mathbf{C}^{-1}\left(\mathbf{a}_1 + \mathbf{b}_1 p_n^-\right) = \mathbf{a} + \mathbf{b} p_n^-. \tag{10.32}$$

Hence,

$$p_n^+ = \alpha + \beta \; p_n^-, \tag{10.33}$$

where α and β denote the elements of vectors **a** and **b** corresponding to p_n^+. Since the plane-wave decompositions apply also at the equivalent source plane with $j = n$, Equation (10.33) may be expressed in the form Equation (10.28) with

$$p_{Sn} = \frac{2\alpha}{1-\beta} \tag{10.34}$$

$$Z_{Sn} = \rho_n c_n \left(\frac{1+\beta}{1-\beta} \right), \tag{10.35}$$

which are, respectively, the required equivalent source pressure strength and impedance at plane n. This proves the equivalent one-port source concept, but it should be noted that the analysis tacitly assumes that the actual source characteristics are not dependent on the part of the system downstream of the equivalent source plane (for example, a silencer and tailpipe). In the case of a reciprocating engine exhaust, or intake, our source model (Section 10.2) predicts that the source impedance depends on the mean flow and the source strength is determined on the fluctuating enthalpy injection at the valves. These may be modified discernibly, if the silencer has high pressure loss and changes the phase of the waves at the back of valves just before they open (the latter action is well-known in engine tuning). Therefore, the possibility that the source characteristics depend on the silencer may not be ruled out. The effect depends on the type of fluid machinery and the operational conditions, but, in most cases, it may be neglected with adequate accuracy in engineering calculations.

The equivalent one-port source concept is usually introduced in the literature in analogy with the electrical circuit shown Figure 10.4. This circuit is known as the Helmholtz–Thévenin equivalent of linear electrical networks consisting of arbitrary passive electrical components and independent electrical sources. Helmholtz–Thévenin theorem states that any linear two-terminal electrical network is equivalent to a voltage source in series with the network in which all sources are set to zero [11]. Acoustic analogy may then be established, for example, by using the acoustic pressure for voltage difference and the particle velocity for electrical current. Indeed, interpreting p_S and Z_S as the source pressure strength and impedance at the source plane, the electric circuit in

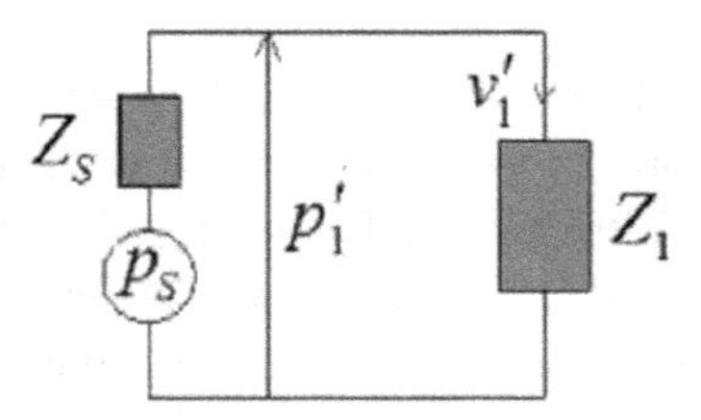

Figure 10.4 Electro-acoustic analogy for a ducted one-port equivalent source (which is assumed to be at a section numbered as plane 1).

Figure 10.3 yields Equation (10.16) or (10.28). But, the principle of reciprocity, which is a fundamental feature of passive linear electrical components, does not apply for linear acoustic models of duct systems with mean flow [12] (see also Section 5.5). Furthermore, the intrinsic characteristics of ducted fluid machinery sources are not analogous to the electrical sources envisaged in the circuit of Figure 10.4. As we have noted in Section 10.2.1, acoustic sources are characterized by processes such as unsteady mass or heat injection or unsteady force application, which are in general dependent on the operational point of the associated fluid machinery, which may in turn be dependent to some extent on the acoustic load presented by the duct system downstream of the source plane, as pointed out in the previous paragraph. Such features are missed completely if the electro-acoustic analogy is taken as basis for the definition of equivalent one-port acoustic sources.

The electro-acoustic analogy of Figure 10.3 is not power consistent when there is mean flow, because the product of acoustic pressure and particle velocity is not equal to the instantaneous acoustic power (see Section 3.4), whereas the product of the analogous electrical variables gives the instantaneous electrical power. For this reason, the electro-acoustic analogy is defined by some authors by using variables, the product of which give the acoustic power of plane waves superimposed on uniform mean flow, for example, $p' + \rho_o c_o M_o v'$ and $v' + M_o p'/\rho_o c_o$ [10]. Such modification is not essential in modeling linear acoustic networks, because the continuity of the pressure wave components is necessary and sufficient for the continuity of acoustic power flow in ducts in one-dimensional as well as in three-dimensional networks.

As an example for the calculation of the equivalent source characteristics, consider the exhaust manifold shown in Figure 10.3a and assume that the source characteristics for each cylinder are as those shown in Figure 10.2, that is, $\zeta_S = 1/\gamma_o M_o = 7.776$ and $p_S \approx (1 - 1/\gamma_o)(\dot{m}h_{\text{cyl}})'/Sv_o$ with $(\dot{m}h_{\text{cyl}})' \approx (\dot{m}H^o)'$, but phased according to the firing order of 3-2-4-1 at equal intervals without overlap (that is, only one exhaust valve is open at any time). The computed equivalent source pressure strength magnitude and the real and imaginary parts of the equivalent source impedance are shown in Figure 10.5. The dominant harmonics of the latter correspond to the first few multiples of the fundamental firing frequency, which is equal to four (number of in-line cylinders) times the firing cycle frequency (FCF). The same exhaust manifold is considered by Åbom et al. who computed the equivalent source impedance at plane n by assuming several values for the actual source impedance at the exhaust valves [13]. The characteristics given in Figure 10.5b,c for ζ_S (=7.776) are in fairly good agreement with their results obtained by assuming that the normalized source impedance at the valves is equal to 10.

As observed from Figures 10.2b and 10.5a, most of the FCF harmonics that dominate strength of the sources at the ports are practically absent at the equivalent source plane. This is due the interferences between sound waves meeting at junctions of the manifold and the runners (also called down pipes). The resulting reduction in the transmitted sound power is not usually substantial, but the modified frequency content may be more convenient for its control downstream. The interference effects depend on the cylinder configuration, the firing order and the mean flow conditions and, in general, can be predicted from the acoustic block diagram of the system by an

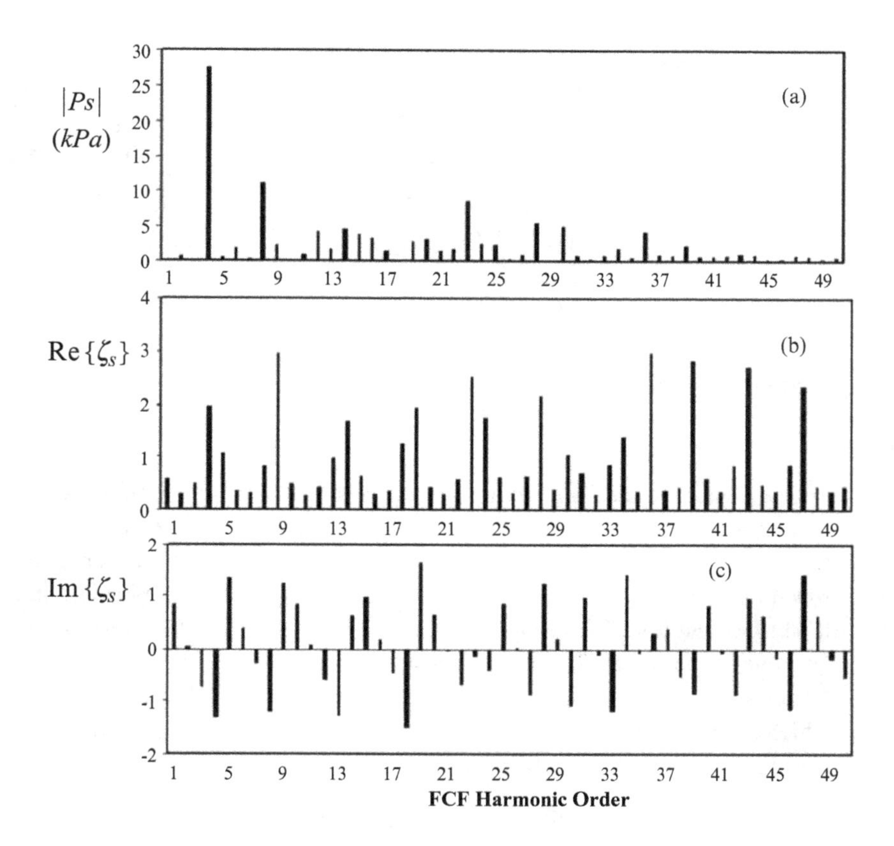

Figure 10.5 (a) Fourier magnitude spectrum of the source pressure, (b) real part of normalized source impedance and (c) imaginary part of normalized source impedance of a four-cylinder engine manifold having the configuration and dimensions of Figure 10.3a.

equivalent source analysis similar to that described above, including the effects of valve overlap and the presence of additional components, such as a turbocharger, an after-treatment device, a muffler or even a small chamber on a runner, upstream of the equivalent source plane.

10.4 Two-Port Sources Characterized by Fluctuating Force Application

A two-port ducted source model (Figure 10.1b) is appropriate when the sound wave motion generated by the source process can propagate both downstream and upstream of the source region. This is typical of ducted rotors in axial turbomachines. Aeroacoustics of ducted rotors has been the subject of extensive research in view of its importance in turbomachinery applications and still presents many challenges [14]. Here, we consider the simplest configuration, which is a single rotor installed in a duct as shown schematically in Figure 10.6.

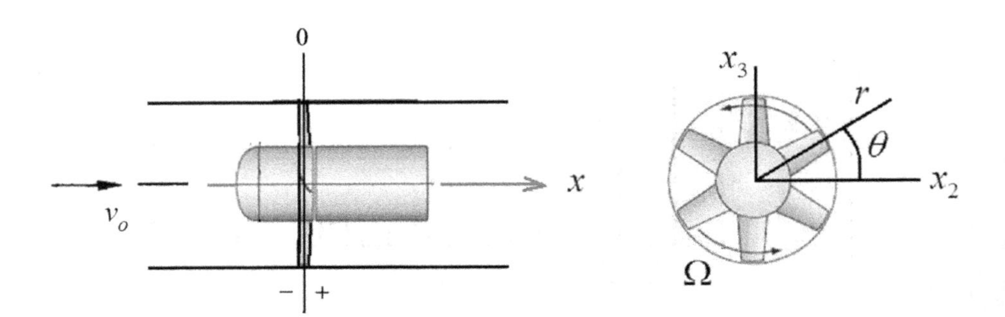

Figure 10.6 An axial flow fan.

The blades fixed to the hub impart energy to the fluid flow (compressors and fans) or take energy from the fluid flow (turbines). The sound generation mechanisms associated with rotating blades are multifarious [15], however, for subsonic low Mach number mean flows, the major mechanism is the pressure loading imposed on the blades and both the steady and unsteady components of the pressure loading can generate sound waves. To see this, consider a blade rotating in the rotor plane x_2x_3 in a Cartesian frame (x_1, x_2, x_3), where $x_1 = x$ is the rotor axis (also the axis of the duct in which the rotor operates). In a fixed observer's time, the force applied by length dr of the blade on the fluid may be expressed as

$$\mathrm{d}\mathbf{F}(r,t) = -\delta_{2\pi}(\theta - \theta_0 - \Omega t)\mathbf{f}_L \mathrm{d}r. \tag{10.36}$$

Here, $\mathbf{f}_L = \mathbf{f}_L(r,t)$ denotes the aerodynamic lift on the blade per unit length at radius r from the rotor axis (drag is not included, because its effect on sound generation is usually negligible), Ω denotes the angular velocity of the rotor, which is assumed to be constant, θ denotes the polar angle in the cylindrical frame (x_1, r, θ), θ_0 denotes the reference angle of the blade and the function $\delta_{2\pi}(\theta)$ is defined as $\delta_{2\pi}(\theta) = 1$ if $\theta = 2h\pi$, $h = 0, 1, 2, \ldots$, else $\delta_{2\pi}(\theta) = 0$. Hence, upon using the Fourier series expansion of $\delta_{2\pi}(\theta)$, we obtain

$$\mathrm{d}\mathbf{F}(r,t) = -\left(\frac{1}{2\pi} + \frac{1}{\pi}\sum_{h=1}^{\infty} \cos\left(h(\theta - \theta_0) - h\Omega t\right)\right)\mathbf{f}_L \mathrm{d}r. \tag{10.37}$$

If there are B equally spaced identical blades on the rotor, which is normally the case in turbomachines, the contributions of each blade to the force applied on the fluid are summed and the expression in the summation sign in Equation (10.37) becomes

$$\sum_{b=1}^{B} \cos\left(h\theta - \frac{2h(b-1)\pi}{B} - h\Omega t\right),$$

since the angular pitch of the blades is $2\pi/B$ and one of the blades may be taken as reference. It may be shown that the foregoing summation vanishes unless h is an integer multiple of B, for which case Equation (10.37) may be written as

$$-\sum_{b=1}^{B}\mathrm{d}\mathbf{F}(r,t) = \left(\frac{B}{2\pi}+\frac{B}{\pi}\sum_{h=1}^{\infty}\cos\left(hB(\theta-\theta_0)-hB\Omega t\right)\right)\mathbf{f}_L\mathrm{d}r$$
$$= \left(\frac{B}{2\pi}\sum_{h=-\infty}^{\infty}\mathrm{e}^{\mathrm{i}(hB(\theta-\theta_0)-hB\Omega t)}\right)\mathbf{f}_L\mathrm{d}r \qquad (10.38)$$

where the second equality gives the complex Fourier series representation. This shows that when the pressure loading is steady ($\mathbf{f}_L$ is time-independent), rotating blades can still generate periodic sound of fundamental frequency $B\Omega$, which is called blade passing frequency. However, for a ducted rotor only the spinning modes (Section 6.5.3.4) with circumferential index $m = hB$ are excited, because the generated modes have to match the rotating blades. The lowest order spinning modes can propagate if $k_o > \alpha_{m0}\sqrt{1-M_o^2}$ or, in terms of the blade passing frequency, if $ahB|\Omega|/c_o\sqrt{1-M_o^2} > \alpha_{m0}a$, where a denotes the duct radius (assumed to be approximately the blade radius). For subsonic low Mach number flows, this condition implies $a|\Omega|/c_o > 1$, since $\alpha_{m0}a > m$ (Section 6.5.3.2). Consequently, for any spinning mode to cut-on under steady pressure loads, it is necessary that the blades rotate with supersonic tip speeds. The inequality $\alpha_{m0}a > m$, which has cardinal role in this conclusion, can be proved rigorously for hollow circular ducts [16], however, as may be inspected from Table 6.2, it is also valid for annular ducts.

A rotating blade can also generate sound by the mechanism of unsteady mass injection, since its rotation is accompanied by continuous displacement and replenishment of the fluid. The volume displaced at a given instant by an elementary strip $\mathrm{d}r$ of the blade at radius r is $S\mathrm{d}r$, where $S = S(r)$ denotes the cross-sectional area of the blade. The corresponding displaced fluid mass is $\mathrm{d}m = \rho_o S(r)\mathrm{d}r$ and a fixed observer observes this as $\mathrm{d}m = \rho_o S\delta_{2\pi}(\theta-\theta_0-\Omega t)\mathrm{d}r$, or as the rate of mass injection

$$\mathrm{d}\dot{m}(r,t) = -\frac{\rho_o S\Omega}{\pi}\sum_{h=1}^{\infty}h\sin\left(h(\theta-\theta_0)-h\Omega t\right)\mathrm{d}r. \qquad (10.39)$$

The extension of this expression for the case of a rotor with equally spaced identical B blades is obvious from considerations which led to Equation (10.38). This result shows that steady rotation of rotor blades is associated with non-steady rate of mass injection at the harmonics of the blade passing frequency. The noise generated by this mechanism is called thickness noise. However, as in the case of steady pressure loading, thickness noise can be neglected if the blades rotate with subsonic tip speeds.

Unsteady loading ($\mathbf{f}_L$ is time-dependent) arises when the aerodynamic lift and drag are caused to fluctuate whilst they rotate. These fluctuations may be caused by many external agents that distort the steady flow. The generated sound may have both discrete and broadband frequency components depending on the characteristics of the incoming flow distortions and the excitation of axisymmetric modes, as well as the spinning modes, may become possible. To see this, let us assume that the blades cut a flow distortion, resulting in a periodic loading over every blade of fundamental period

$2\pi/\Omega$. Then, the harmonics of the lift in Equation (10.38) will have $e^{\mp i\upsilon\Omega t}$ time dependence, where $\upsilon = 1, 2, \ldots$ denotes the harmonic order. Consequently, the spectrum of the generated sound will consist of discrete frequencies $(hB\mp\upsilon)\Omega$ and spinning modes of radial order m may still be excited, but now the necessary condition for this is $m = hB\mp\upsilon$. This condition is known as the Tyler and Sofrin criterion and has important ramifications for noise control in turbomachinery [17]. Proceeding as in the above discussion for the case of steady pressure loading, the condition $ahB\Omega/c_o\sqrt{1 - M_o^2} > m$ is still required for the duct modes to cut on, but now it can be satisfied for integer values of υ that satisfy the condition $1\mp\upsilon/hB < a\Omega/c_o\sqrt{1 - M_o^2}$. Here, we may dismiss the condition with the plus sign, because it requires that the tip speeds are supersonic. So, we see now that supersonic tip speeds are not required for spinning modes to cut-on, if the spectrum of the pressure loading has an appropriate harmonic content. For example, let the periodic unsteady pressure loading be due to the wakes of equally spaced identical inlet guide vanes. If the number of guide vanes is V, every rotor blade will cut V wakes and, therefore, the spectrum of the unsteady loading will consist of the harmonics $\upsilon = sV$, $s = 1, 2, \ldots$ For this case the Tyler and Sofrin criterion gives $m = hB - sV$. If we take $B = V$, then $m = 0$ if $s = h$, which implies that axisymmetric modes, including the plane wave mode, may be excited. But the same is true also if $B \neq V$, but $s/B = h/V$. In theory, this condition is not avoidable, but, in subsonic applications, we may choose B and V such that when this condition is satisfied, it is obliterated for important frequencies by the requirement of supersonic blade tip speed.

It should be noted that, if there is a stator downstream of the rotor, this can also generate sound by the same mechanism. In this case, although the stator blades are stationary, they are subject to unsteady pressure loading caused by the impingement of the upstream blade wakes which rotate with the speed of the rotor. Accordingly, the Tyler and Sofrin criterion $m = hB - sV$ applies for this case too, provided that V is interpreted as the number of blades on the stator.

The duct modes generated by rotating blades (or fixed blades subject rotating pressure loading) can be related to the distribution of the pressure loading on the blades in the general framework of the Ffowcs Williams and Hawkings equation [18], but the solution of the resulting problem is computation intensive and requires specialized computer codes [19]. A simple actuator-disk model of the process assumes that the source region in Figure 10.1b encloses a blade row, and that the blades may be replaced with a volume force distribution, $\mathbf{f}$, which represents the action of the blades on the fluid. This is, of course, only a rough model, as it neglects the presence of the rotor hardware, however, it includes the essential feature of the source process. The axial component of $\mathbf{f}$, denoted by f_x, is represented by the distribution $f_x(t, x) = F_x\delta(x)/S$, where $F_x = F_x(t)$ denotes the total axial force applied on the fluid by the blades and may be computed by integrating the axial component of Equation (10.38) along the blades. The jump equations across the source region are determined by integrating Equation (10.5) and the axial component of Equation (10.4) as described in

Section 10.1. Assuming that no mass or energy is created in the source region ($\dot{\mu} = \dot{U} = 0$) and that $\gamma \approx \gamma_o$ and $c \approx c_o$ are approximately constant in the source region, the jump equations may be expressed as

$$S\left(\gamma_o\langle p\rangle\left[\frac{1}{2}v^2\right] + c_o^2[p]\right) = c_o^2 F_x \tag{10.40}$$

$$\langle v\rangle[p] + \gamma_o\langle p\rangle[v] = 0. \tag{10.41}$$

Upon linearization of the foregoing equations, we obtain

$$S\left(\gamma_o\langle p_o\rangle[v_o v'] + \gamma_o\left[\frac{1}{2}v_o^2\right]\langle p'\rangle + c_o^2[p']\right) = c_o^2 F'_x \tag{10.42}$$

$$\langle v_o\rangle[p'] + \langle v'\rangle[p_o] + \gamma_o\langle p'\rangle[v_o] + \gamma_o\langle p_o\rangle[v'] = 0. \tag{10.43}$$

We discard the jump equation associated with the steady flow part of Equation (10.43) because the shaft work is not included in the parent energy equation, Equation (10.3). Assuming that the axial mean flow velocity does not undergo a jump across the source plane, $[v_o] = 0$ and $\langle v_o\rangle = v_o$. Hence, on expanding the jump and arithmetic mean brackets, Equations (10.42) and (10.43) may be expressed in matrix form

$$\mathbf{p}_S = \mathbf{Z}_S\begin{bmatrix} p'_+ \\ v'_+ \end{bmatrix} + \begin{bmatrix} p'_- \\ v'_- \end{bmatrix}. \tag{10.44}$$

Here, $M_o = v_o/c_o$ and with $\Delta = -[p_o]/2 + \gamma_o\langle p_o\rangle(1 - M_o^2)$,

$$\mathbf{p}_S = \begin{bmatrix} \frac{1}{2}[p_o] + \gamma_o\langle p_o\rangle \\ v_o \end{bmatrix}\frac{F'_x}{S\Delta} \approx \begin{bmatrix} 1 \\ \dfrac{v_o}{\gamma_o\langle p_o\rangle} \end{bmatrix}\frac{F'_x}{S} \tag{10.45}$$

$$\mathbf{Z}_S = \frac{1}{\Delta}\begin{bmatrix} \frac{1}{2}[p_o] - (1 - M_o^2)\gamma_o\langle p_o\rangle & \dfrac{\gamma_o}{c_o}[p_o]\langle p_o\rangle M_o \\ 0 & -\frac{1}{2}[p_o] - (1 - M_o^2)\gamma_o\langle p_o\rangle \end{bmatrix} \approx \begin{bmatrix} -1 & \dfrac{[p_o]M_o}{c_o} \\ 0 & -1 \end{bmatrix} \tag{10.46}$$

where the approximate equalities indicate simplified forms for $\langle p_o\rangle >> [p_o]$ and $M_o^2 << 1$ and Equation (10.44) holds both in the time domain and the frequency domain. In the frequency domain, it may be transformed to the form of the generic two-port source equation defined in Section 2.4.1, Equation (2.29), by decomposing the acoustic pressure and the axial component of the particle velocity into the modal pressure wave components by using Equations (6.55) and (6.58). Suppressing the dependence of the variables on the transverse duct coordinates, Equation (10.44) may be expressed for a single mode $\mu = 1, 2, \ldots$ as

$$\mathbf{p}_S^{(\mu)} = \mathbf{T}_S^{(\mu)}\begin{bmatrix} p_\mu^+(0_+) \\ p_\mu^-(0_+) \end{bmatrix} + \begin{bmatrix} p_\mu^+(0_-) \\ p_\mu^-(0_-) \end{bmatrix}, \tag{10.47}$$

with

$$\mathbf{p}_S^{(\mu)} = \frac{1}{A_\mu^- - A_\mu^+}\begin{bmatrix} A_\mu^- & -\rho_o c_o \\ -A_\mu^+ & \rho_o c_o \end{bmatrix}\mathbf{p}_S \tag{10.48}$$

$$\mathbf{T}_S^{(\mu)} = \frac{1}{\rho_o c_o\left(A_\mu^- - A_\mu^+\right)}\begin{bmatrix} A_\mu^- & -\rho_o c_o \\ -A_\mu^+ & \rho_o c_o \end{bmatrix}\mathbf{Z}_S\begin{bmatrix} \rho_o c_o & \rho_o c_o \\ A_\mu^+ & A_\mu^- \end{bmatrix}. \tag{10.49}$$

For plane wave propagation, $A_1^{\pm} = \mp 1$ and the foregoing equations reduce to

$$\mathbf{p}_S^{(1)} \approx \frac{1}{2}\begin{bmatrix} 1+M_o \\ 1-M_o \end{bmatrix}\frac{F_x'}{S}, \mathbf{T}_S^{(1)} \approx \begin{bmatrix} -1 & -\dfrac{[p_o]M_o}{2\rho_o c_o^2} \\ \dfrac{[p_o]M_o}{2\rho_o c_o^2} & -1 \end{bmatrix}. \tag{10.50}$$

Another actuator-disk formulation, proposed by Morfey [20], assumes that the mean flow is the same on both sides of the source region, but takes into account the jump in vorticity fluctuations produced by the non-uniformities in the force distribution. Then, Equations (10.42) and (10.43) reduce to

$$S\left(\rho_o c_o M_o\left[v_a' + v_v'\right] + [p']\right) = F_x' \tag{10.51}$$

$$M_o[p'] + \rho_o c_o\left[v_a' + v_v'\right] = 0, \tag{10.52}$$

where $v' = v_a' + v_v'$ and the subscripts "a" and "v" denote the acoustic and vorticity component of the fluctuations, respectively. For uniform mean flow of Mach number M_o, Morfey gives

$$\left[v_v'\right] = \frac{1-a_\mu^2}{1-a_\mu^2 M_o}\frac{M_o F_x'}{S\rho_o c_o} \tag{10.53}$$

for mode μ. Upon substitution of this, Equations (10.51) and (10.52) may be cast in the format of Equation (10.44) with $\mathbf{Z}_S = -\mathbf{I}$, where $\mathbf{I}$ denotes a 2×2 unit matrix, and

$$\mathbf{p}_S = \begin{bmatrix} -1 \\ \dfrac{M_o\left(2-M_o^2-a_\mu^2\right)}{\rho_o c_o\left(1-a_\mu^2 M_o\right)} \end{bmatrix}\frac{F_x'}{S\left(1-M_o^2\right)}. \tag{10.54}$$

The determination of the lift force on the blades is covered in the literature extensively, but the calculations depend on blade geometry and its environment. Here, for simplicity, we assume that the axial component of the lift force is uniformly distributed over the span of a blade. Then, if the pressure loading is steady, the axial component of the total force applied by the blades on the fluid may be expressed from Equation (10.38) in frequency domain as $F_x = (Bf_{Lx}/2\pi)\sum_{h=-\infty}^{\infty} e^{ihB(\theta-\theta_0)}\delta(\omega - hB\Omega)$, where f_{Lx} denotes the axial component of the lift on a

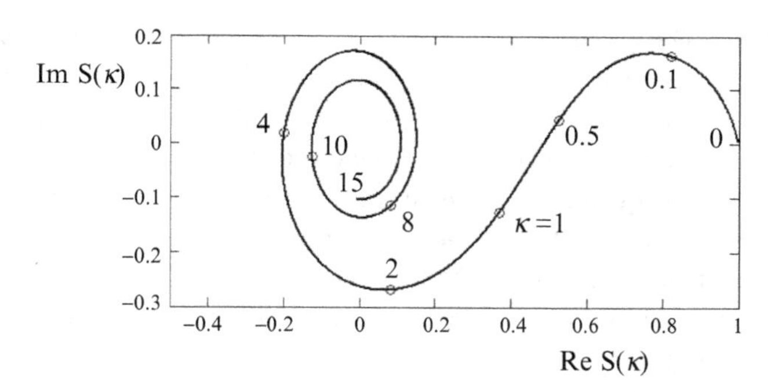

Figure 10.7 Sears' function.

blade (note that the sense of F_x is opposite to the total lift, as the latter is, by definition, force applied by the fluid on a blade) and $\delta(\omega)$ denotes unit impulse function at $\omega = 0$. F_x' is given by the unsteady part of F_x. This confirms the previous discussion on the excitation of the spinning modes at blade passing frequencies.

If pressure loading is unsteady, the axial component of the lift on a blade is time dependent, that is, $f_{Lx} = f_{Lx}(t)$. Then, we similarly deduce from Equation (10.38) $F_x = (B/2\pi)\sum_{h=-\infty}^{\infty} \mathrm{e}^{\mathrm{i}hB(\theta-\theta_0)} f_{Lx}(\omega - hB\Omega)$ in the frequency domain. The lift function f_L is usually modeled by using Sears' incompressible thin airfoil theory [21]. According to this theory, the axial component of the lift produced by an unsteady flow (usually called a gust) consisting of an upwash wave convected with constant velocity U toward a thin flat plate airfoil of cord $2b$ at zero angle of attack may be expressed in frequency domain (here, the lift and stagger angle being assumed to be uniform) by $f_{Lx} = \pi\rho_o(2b)U\ell S(k_o b)v_g(\omega)$. Here ℓ denotes the span of the blade, v_g denotes velocity of the upwash wave, $k_o = \omega/c_o$ and

$$S(\kappa) = \frac{\mathrm{J}_0(\kappa)\mathrm{K}_1(-\mathrm{i}\kappa) - \mathrm{i}\mathrm{J}_1(\kappa)\mathrm{K}_0(-\mathrm{i}\kappa)}{\mathrm{K}_0(-\mathrm{i}\kappa) + \mathrm{K}_1(-\mathrm{i}\kappa)}, \tag{10.55}$$

where J_n and K_n denote Bessel functions of the first kind and modified Bessel function of the second kind, respectively, of order n. The function $S(\kappa)$, which is known as Sears' function, is shown in Figure (10.7). The effect of the non-zero angle of attack may be taken into account approximately by rotating F_x anti-clockwise by the angle of attack. Sears' theory neglects the effects of the thickness and camber of the blades, the angle of the impinging gust, the presence of neighboring blades and the decay of gust velocity with convection; however, all these and other issues have been investigated in later studies [15].

10.4.1 Flow Noise

Another area of application of the above described two-port actuator-disk model is the modeling of the flow noise produced when fluid flow in a duct is constricted by a hard surface, such as an orifice, a damper or a vane. According to the theory of Nelson and

Morfey [5], the noise produced by such a flow spoiler is primarily caused by fluctuating drag forces arising from the flow turbulence in the vicinity of the constriction. Assuming that the drag force distribution is confined to an axially compact control volume, Equation (10.47) may be applied for this source process; however, it will be necessary to quantify F'_x, which, in this case, is determined by the axial component of the fluctuating part of the drag force (the lift force being assumed to have no significant fluctuating part). Nelson and Morfey resolved this problem by hypothesizing that, in subsonic low Mach number flows, the root-mean-square (rms) value of fluctuating part of the drag force is proportional to its mean part, that is, $(F'_x)_{\text{rms}} = K(F_x)_o$, where the usual decomposition $F_x = (F_x)_o + F'_x$ is implied, K denotes a proportionality constant, the root-mean-square (rms) value is evaluated in suitable octave bands and the mean force component is computed from $(F_x)_o = S[p_o]$. They show that, in a rectangular duct, the time-averaged power of the noise generated by the flow to downstream of rectangular flow spoilers normal to the flow may be determined from $W_{\text{flow}} = (F_x)_o^2 K^2 / 2S\rho_o c_o$ for plane waves and by about $(2\pi f_o / c_o)^2 S / 6\pi$ times this when higher-order modes are cut-on, where f_o denotes the center frequency of the octave band considered. On measuring $(F_x)_o$ and the flow noise power, W_{flow}, of rectangular plate orifices of different sizes, Nelson and Morfey found that the measured data collapse very satisfactorily, if K is scaled with a Strouhal number, St, say. Later work on flow noise in subsonic low Mach number flows are based on the Nelson and Morfey theory with slight adjustments about the definition of the Strouhal number. For example, Kårekull et al. propose $\text{St} = \sqrt{4S\sigma/\pi} f_o \sigma / v_o$, where $\sigma = 1/(1+\sqrt{C_L})$ and $C_L \left(= [p_o]/0.5\rho_o v_o^2\right)$ denotes the usual static pressure loss coefficient of the constriction [22]. This seems to collate the results of previous studies on rectangular and circular orifices, bends and dampers with adequate accuracy to the best-fit formula $K = 10^{-1.4 \log_{10}(\text{St})} \times 10^{-2.6}$ in 1/3 octave bands for about $\text{St} > 3$. For constrictions terminating air flow, they propose to determine the mean force by the momentum of the mean flow at vena contracta, which is tantamount to multiplying the Nelson and Morfey definition of the mean force by $2/\sqrt{C_L}$, but seems to give better correlation with their experimental results on air terminal units.

The flow noise power prediction of the Nelson and Morfey theory presumes waves propagating in one direction. This is usually considered to be of acceptable accuracy in flow generated noise calculations in air handling and ventilation systems (see Section 11.6.1). But when a duct system is modeled by wave transfer equations with reflections included, the two-port source model in Equation (10.47), with F'_x based on an appropriate scaling law, should be used for the modeling of the source process. A problem arises when using this model in sound pressure level (SPL) calculations (see Chapter 11), because it does not contain phase information with respect to the primary sound source. Perhaps the simplest approach to this problem is to compute the SPL with only one source present and evaluate the results by ignoring the phase effects. But if the primary source is given in a Fourier spectrum, we may like to keep its frequency resolution. One way to achieve this is to apply the Nelson and Morfey hypothesis in the frequency domain as $F'_x = G(F_x)_o$, where G is complex now. Then the Fourier components of F'_x can be determined by assuming that they are randomly phased and that $|G|$ may be interpolated from the published scaling laws over the Fourier spectrum, that is, $|G| \approx K(St)$. An advantage of this approach is that it can be used with both the primary source and the flow noise source present in the acoustic model of the system.

10.5 Two-Port Sources Characterized by Ducted Combustion

The heat release resulting from the burning of fuel in combustion chambers of gas turbines, industrial burners, domestic boilers, etc., is the major sources of noise in these systems. Figure 10.8 depicts schematically the burning zone of a modern combustor. The premixed reactants ignite as they enter the hot chamber and heat energy is released over the surface, called the flame front, which separates the burned and unburned gas mixtures. (In diffusion or non-premixed type combustors, the fuel and air enter the chamber in separate streams.) The combustion process is very complicated and involves multiple physics. The reader is referred to books on the subject for fundamentals of combustion and combustors [23]. Under normal conditions, the thickness of the flame front is much smaller than its length ($\delta_f << L_f$) and when the flame is compact, that is, the length of the flame is much small compared to the acoustic wavelength, $\lambda >> L_f$, the flame front may be assumed to be infinitesimally thin. Then, the distribution of heat release becomes unimportant and the actuator-disk model in Figure 10.1b may be used for the modeling of the combustion process as an acoustic two-port source element. This simplified model is very convenient for gaining physical insight into the effect of chamber acoustics on the combustion process because in acoustic calculations we need to consider only the states of the fluid in the chamber before and after combustion.

So we begin again with Equations (10.4) and (10.5), but now integrate these equations across the source region (flame front thickness) by neglecting any mass or momentum creation in this region. This yields the jump equations

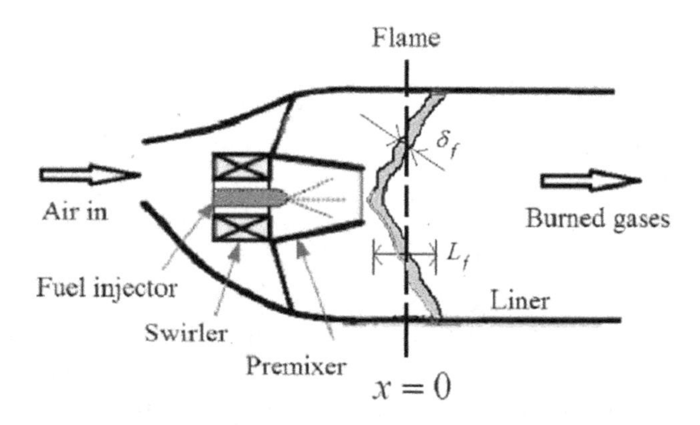

Figure 10.8 Primary zone of a modern combustion chamber.

$$\gamma_o\langle p\rangle\left[\frac{1}{2}v^2\right]+\langle c_o^2\rangle[p]=0 \tag{10.56}$$

$$S(\langle v\rangle[p]+\gamma_o\langle p\rangle[v])=(\gamma_o-1)\dot{H}, \tag{10.57}$$

where it is assumed that $c\approx c_o$, $\gamma\approx\gamma_o=$ constant. The latter is adequately accurate since γ is a very slowly varying function temperature for perfect gases. The heat release rate per unit volume is represented by the distribution $\dot{U}(t,x)=\dot{H}\delta(x)/S$, where $\dot{H}=\dot{H}(t)$ denotes the heat release rate of the combustion process. Equations (10.56) and (10.57) are linearized as usual and yield the acoustic jump equations

$$\gamma_o\langle p'\rangle\left[\frac{1}{2}v_o^2\right]+\gamma_o\langle p_o\rangle[v_ov']+\langle c_o^2\rangle[p']=0 \tag{10.58}$$

$$S(\langle v_o\rangle[p']+\langle v'\rangle[p_o]+\gamma_o\langle p_o\rangle[v']+\gamma_o\langle p'\rangle[v_o])=(\gamma_o-1)\dot{H}'. \tag{10.59}$$

This applies for a single duct mode, but here it suffices to consider one-dimensional acoustics, as combustion generated noise is dominated by discrete components in the low-frequency range. Also, since the pressure drop in combustion chambers is usually small, we may take $[p_o]\approx 0$, which is tantamount, with $\gamma_o=$ const, to $[\rho_oc_o^2]=0$, although $[c_o^2]\neq 0$ and $[\rho_o]\neq 0$. We retain a mean flow velocity jump in the analysis, because the mean flow velocity is usually reduced before ignition for improving the burning efficiency of the reactants and it is determined by the expansion of hot gases after the combustion. Equations (10.58) and (10.59) can be combined in the two-port source format of Equation (10.44) with

$$\mathbf{p}_S=\begin{bmatrix}\dfrac{v_{o-}}{\langle c_o^2\rangle}\\ \dfrac{[v_o^2]}{4\langle c_o^2\rangle p_o}-\dfrac{1}{\gamma_op_o}\end{bmatrix}(\gamma_o-1)\frac{\dot{H}'}{S\Delta} \tag{10.60}$$

$$\mathbf{Z}_S=\frac{1}{\Delta}\begin{bmatrix}-1-\gamma_o\dfrac{\left[\frac{1}{2}v_o\right]^2}{\langle c_o^2\rangle}+\dfrac{v_{o-}\langle v_o\rangle}{\langle c_o^2\rangle} & -\dfrac{\gamma_op_o[v_o]}{\langle c_o^2\rangle}\\ \dfrac{1}{p_o}\left(\dfrac{\left[\frac{1}{2}v_o^2\right]}{\langle c_o^2\rangle}\langle v_o\rangle-[v_o]\right) & -1-\gamma_o\dfrac{\left[\frac{1}{2}v_o\right]^2}{\langle c_o^2\rangle}+\dfrac{v_{o+}\langle v_o\rangle}{\langle c_o^2\rangle}\end{bmatrix}, \tag{10.61}$$

where

$$\Delta=1-\gamma_o\frac{\left[\frac{1}{2}v_o\right]^2}{\langle c_o^2\rangle}-\frac{v_{o-}\langle v_o\rangle}{\langle c_o^2\rangle} \tag{10.62}$$

and formulae like $\langle y\rangle=2[y^2]/[y]$ and $2y_-=2\langle y\rangle-[y]$, $y\in c_o,v_o,\ldots$, may be used to determine the parameters when the jumps are known. For subsonic low Mach number mean flows, we may assume $v_{o\mp}^2/c_{o\mp}^2<<1$ to simplify Equations (10.60) and (10.61):

$$\mathbf{p}_S = \begin{bmatrix} \dfrac{v_{o-}}{\langle c_o^2 \rangle} \\ -\dfrac{1}{4\gamma_o p_o} \end{bmatrix} (\gamma_o - 1) \frac{\dot{H}'}{S} \tag{10.63}$$

$$\mathbf{Z}_S = - \begin{bmatrix} 1 & \dfrac{\gamma_o p_o [v_o]}{\langle c_o^2 \rangle} \\ \dfrac{[v_o]}{p_o} & 1 \end{bmatrix}. \tag{10.64}$$

This impedance form of the two-port source equation can be transformed to the generic form of Equation (2.29) involving the pressure wave components as described in Section 10.4 for Equation (10.44).

This two-port source model can be used given $\dot{H}'$, but the determination of $\dot{H}'$ is not trivial. Obviously, the simplest heat release model is $\dot{H}' \approx 0$. This corresponds to the ideal situation where the fuel is burned at a constant rate. Then the source pressure strength vanishes and no sound is generated by the combustion process. In general, however, the magnitude of heat release is given in the time domain by $|\dot{H}| = y_f h_f \int_A |\dot{m}| \mathrm{d}A$, where h_f denotes the specific enthalpy of reaction of the fuel, A denotes area of the flame front, $\dot{m}$ denotes the instantaneous mass flow rate of the reactants per unit flame area and y_f denotes the mass fraction of the fuel, which is assumed to be distributed uniformly over the flame front area. Since the fuel mass fraction is small compared to that of the mixture, $\dot{m} \approx \rho v$ and the fluctuating heat release may be expressed as $|\dot{H}| \approx y_f h_f \int_A |\rho v| \mathrm{d}A$. This implies that the magnitude of heat release fluctuations scales as $|H'| \approx |\rho'_-/\rho_{o-} + v'_-/v_{o-}||\dot{H}_o|$. Here, the fluctuations are assumed to be small to the first order, but they may not necessarily be of acoustic origin. For example, heat release fluctuations may be caused by variations in the reactant composition or turbulence in the mean flow and if they can be quantified, the generated acoustic field can be calculated. But, once acoustic waves are generated, they also interact with the flame front, forming a feedback loop with the acoustic source process. The acoustic components of the fluctuations are of particular interest in the design of combustion chambers, because this feedback loop can force unstable standing waves in the chamber.

For the analysis of such self-excited standing waves, it suffices to consider only the acoustic fluctuations and link them with the two-port source model derived above. However, the magnitude of the heat release fluctuations is not sufficient for this purpose; it is also necessary to know the phase lag of the heat release response with respect to impinging acoustic waves. For this reason, it is common practice to stipulate the linear interaction between the heat release and acoustic fluctuations in frequency domain in the form

$$\dot{H}' = F_f v'_-, \tag{10.65}$$

where $F_f = F_f(\omega)$ is called the flame frequency response function (flame FRF). This global form is convenient, because F_f can be determined experimentally by subjecting the flame to the acoustic field of a loudspeaker and measuring the resulting velocity fluctuations (normal to the flame front) by laser Doppler velocimetry (LDV) and the heat release fluctuations by collecting chemiluminescence emission from the flame [24]. And F_f can also be determined by numerical simulations [25], and analysis of idealized flame geometry [26]. Clearly, the flame FRF may also be defined with reference to the acoustic pressure or density; however, the form of Equation (10.65), where F_f has the dimension of force, is more common in the literature.

Thus, we substitute Equation (10.65) in Equation (10.60) and write the resulting source pressure strength vector as:

$$\mathbf{p}_S = \begin{bmatrix} a_S \\ b_S \end{bmatrix} F_f v'_-, \tag{10.66}$$

where a_S and b_S follow from Equation (10.60). Consequently, the two-port source equation, Equation (10.44), now transforms to the two-port flame impedance transfer equation

$$\begin{bmatrix} p'_- \\ v'_- \end{bmatrix} = -\begin{bmatrix} 1 & \dfrac{a_S F_f}{1 - b_S F_f} \\ 0 & \dfrac{1}{1 - b_S F_f} \end{bmatrix} \mathbf{Z}_S \begin{bmatrix} p'_+ \\ v'_+ \end{bmatrix}, \tag{10.67}$$

where $\mathbf{Z}_S$ is given by Equation (10.61). Equation (10.67) may be expressed in terms of the plane pressure wave components $p_\mp^\mp$ as

$$\begin{bmatrix} p_-^+ \\ p_-^- \end{bmatrix} = \mathbf{T}_f \begin{bmatrix} p_+^+ \\ p_+^- \end{bmatrix}, \tag{10.68}$$

with the flame wave transfer matrix $\mathbf{T}_f$ is given by

$$\mathbf{T}_f = \frac{1}{A^+ - A^-} \begin{bmatrix} A^- & -\rho_{o-} c_{o-} \\ -A^+ & \rho_{o-} c_{o-} \end{bmatrix} \begin{bmatrix} 1 & \dfrac{a_S F_f}{1 - b_S F_f} \\ 0 & \dfrac{1}{1 - b_S F_f} \end{bmatrix} \frac{\mathbf{Z}_S}{\rho_{o+} c_{o+}} \begin{bmatrix} \rho_{o+} c_{o+} & \rho_{o+} c_{o+} \\ A^+ & A^- \end{bmatrix} \tag{10.69}$$

where $A^\mp = \mp 1$. This result may be generalized to any single mode $\mu \in 1, 2, \ldots, q$, by understanding the pressure wave components as the modal pressure wave components $p_\mu^\mp$, which are given by Equations (6.55) and (6.58), and $A^\mp$ as the modal admittances $A_\mu^\mp$.

The flame wave transfer matrix can be cascaded with the passive duct elements at upstream and downstream of the flame as usual (see Section 2.4.1). Here, however, attention should be paid to the fact that, although it may be sufficiently accurate to neglect the entropy wave component in the upstream of the flame front, this may not be accurate for the downstream of the flame, where strong temperature gradients and non-homentropic flow conditions are usually present. The effect of the entropy wave component ε' may be taken into account by using the inhomogeneous duct element described in Section 3.9, if the value of ε' at the duct origin (in this case, $x = 0_+$ in Figure 10.8) is known. To determine this, we first note from the continuity Equation (10.1) that $[\rho v] = 0$ and that linearization of this gives

$$[v_o\rho'] + [\rho_o v'] = 0. \tag{10.70}$$

Also, assuming that the upstream wave propagation is isentropic, $\rho'_- = p'_-/c^2_{o-}$ and $[\varepsilon'] = \varepsilon'_+$. Hence, expanding the foregoing equation and substituting $\rho'_+ = (p'_+ + \varepsilon'_+)/c^2_{o+}$, we find

$$\varepsilon'_+ = -\frac{c^2_{o+}}{v_{o+}}\left(\left[\frac{v_o}{c^2_o}p'\right] + [\rho_o v']\right). \tag{10.71}$$

This may be expressed in terms of the source characteristics as

$$\varepsilon'_+ = -\frac{c^2_{o+}}{v_{o+}}\left(\left[\frac{v_{o+}}{c^2_{o+}} \quad \rho_{o+}\right] + \left[\frac{v_{o-}}{c^2_{o-}} \quad \rho_{o-}\right]\mathbf{Z}_S\right)\begin{bmatrix} p'_+ \\ v'_+ \end{bmatrix} - \left[\frac{v_{o-}}{c^2_{o-}} \quad \rho_{o-}\right]\mathbf{p}_S \tag{10.72}$$

The speed of sound varies substantially across the flame and also a relatively large mean temperature gradient may be present downstream of the flame. Thus, the prediction of the flame temperature is an important prerequisite for acoustic calculations. Li and Morgan discuss calculation of the flame temperature when dissociation and large number species are involved [27].

10.5.1 Combustion Oscillations and Instability

Shown in Figure 10.9a is an idealized model of a premixed combustor. The mixing region is modeled by a duct, with the flame front at just downstream of the area discontinuity. In order to make the calculations as transparent as possible, we will

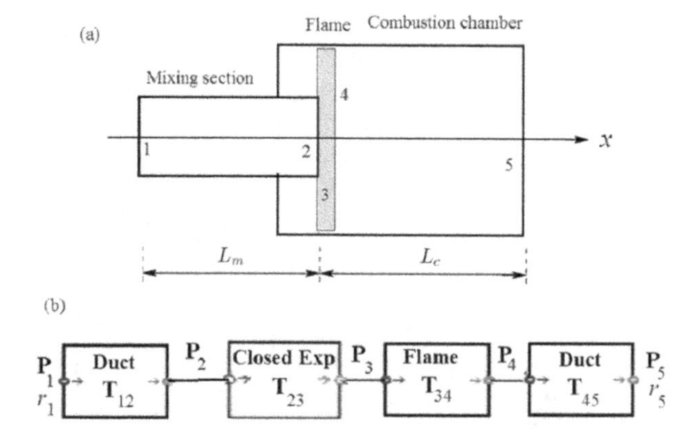

Figure10.9 (a) A simplified model of a premixed combustion chamber, (b) acoustic block diagram model of the chamber.

neglect the mean flow and assume that the mean temperatures in the mixing section and the combustion chamber are uniform. Hence, the entropy wave component may also be neglected.

An acoustic block diagram model of the combustor is shown in Figure 10.9b. The wave transfer matrices of the elements are summarized below assuming plane waves and hard-walled uniform ducts:

Duct 1-2: $\mathbf{P}_1 = \mathbf{T}_{12}\mathbf{P}_2$, where

$$\mathbf{T}_{12} = \begin{bmatrix} e^{-i\frac{\omega L_m}{c_{o-}}} & 0 \\ 0 & e^{i\frac{\omega L_m}{c_{o-}}} \end{bmatrix}. \tag{10.73}$$

Closed-expansion 2-3: $\mathbf{P}_2 = \mathbf{T}_{23}\mathbf{P}_3$, from Equation (4.41):

$$\mathbf{T}_{23} = \frac{1}{2}\begin{bmatrix} \beta+1 & -\beta+1 \\ -\beta+1 & \beta+1 \end{bmatrix}, \tag{10.74}$$

where $\beta = S_4/S_2$ and S_2 and S_4 denote the cross-sectional areas of the mixing section and the combustion chamber, respectively. Note that we have neglected the side-branch for simplicity of the calculations. This is justified if the side-branch is short enough so that its resonances occur outside the frequency range of interest.

Flame 3-4: $\mathbf{P}_3 = \mathbf{T}_{34}\mathbf{P}_4$, from Equation (10.69):

$$\mathbf{T}_{34} = \frac{1}{2}\begin{bmatrix} 1+\frac{\varsigma}{1+\chi} & 1-\frac{\varsigma}{1+\chi} \\ 1-\frac{\varsigma}{1+\chi} & 1+\frac{\varsigma}{1+\chi} \end{bmatrix}, \tag{10.75}$$

where $\varsigma = \rho_{o-}c_{o-}/\rho_{o+}c_{o+}$ and $\chi = (\gamma_o - 1)F_f/4\gamma_o p_o S_4$.

Duct 4-5: $\mathbf{P}_4 = \mathbf{T}_{45}\mathbf{P}_5$, where

$$\mathbf{T}_{45} = \begin{bmatrix} e^{-i\frac{\omega L_c}{c_{o+}}} & 0 \\ 0 & e^{i\frac{\omega L_c}{c_{o+}}} \end{bmatrix}. \tag{10.76}$$

The wave transfer equation of the assembled system is

$$\begin{bmatrix} p_1^+ \\ p_1^- \end{bmatrix} = \mathbf{T}_{15}\begin{bmatrix} p_5^+ \\ p_5^- \end{bmatrix}, \tag{10.77}$$

where $\mathbf{T}_{15} = \mathbf{T}_{12}\mathbf{T}_{23}\mathbf{T}_{34}\mathbf{T}_{45}$ or, as may be shown by carrying out the indicated matrix multiplications,

$$\mathbf{T}_{15} = \frac{1}{2}\begin{bmatrix} \left(1+\frac{\beta\varsigma}{1+\chi}\right)e^{-i\frac{\omega L_m}{c_{o-}}}e^{-i\frac{\omega L_c}{c_{o+}}} & \left(1-\frac{\beta\varsigma}{1+\chi}\right)e^{-i\frac{\omega L_m}{c_{o-}}}e^{i\frac{\omega L_c}{c_{o+}}} \\ \left(1-\frac{\beta\varsigma}{1+\chi}\right)e^{i\frac{\omega L_m}{c_{o-}}}e^{-i\frac{\omega L_c}{c_{o+}}} & \left(1+\frac{\beta\varsigma}{1+\chi}\right)e^{i\frac{\omega L_m}{c_{o-}}}e^{i\frac{\omega L_c}{c_{o+}}} \end{bmatrix}. \tag{10.78}$$

It is important to note that Equation (10.77) represents an active two-port element which can transmit acoustic power from its inlet and outlet nodes. Therefore, it is constrained by the acoustic boundary conditions $p_1^-/p_1^+ = r_1$ and $p_5^-/p_5^+ = r_5$, where r_1 and r_5 denote the reflection coefficient at nodes 1 and 5 (Figure 10.9b). Upon combining these conditions with Equation (10.77), we obtain a homogeneous set of equations for the pressure wave components:

$$\begin{bmatrix} -1 & 0 & \frac{1}{2}\left(1+\frac{\beta\varsigma}{1+\chi}\right)e^{-i\frac{\omega L_m}{c_{o-}}}e^{-i\frac{\omega L_c}{c_{o+}}} & \frac{1}{2}\left(1-\frac{\beta\varsigma}{1+\chi}\right)e^{-i\frac{\omega L_m}{c_{o-}}}e^{i\frac{\omega L_c}{c_{o+}}} \\ 0 & -1 & \frac{1}{2}\left(1-\frac{\beta\varsigma}{1+\chi}\right)e^{i\frac{\omega L_m}{c_{o-}}}e^{-i\frac{\omega L_c}{c_{o+}}} & \frac{1}{2}\left(1+\frac{\beta\varsigma}{1+\chi}\right)e^{i\frac{\omega L_m}{c_{o-}}}e^{i\frac{\omega L_c}{c_{o+}}} \\ r_1 & -1 & 0 & 0 \\ 0 & 0 & r_5 & -1 \end{bmatrix} \begin{bmatrix} p_1^+ \\ p_1^- \\ p_5^+ \\ p_5^- \end{bmatrix} = 0 \tag{10.79}$$

To determine the frequencies of the possible wave modes in the combustion chamber, we set the determinant of the coefficients matrix equal to zero and on expanding the determinant, the following dispersion equation is obtained:

$$\left(1-\frac{\beta\varsigma}{1+\chi}\right)\left(r_1 r_5 e^{i\frac{2\omega L_c}{c_{o+}}} - e^{i\frac{2\omega L_m}{c_{o-}}}\right) + \left(1+\frac{\beta\varsigma}{1+\chi}\right)\left(r_1 - r_5 e^{i\frac{2\omega L_m}{c_{o-}}}e^{i\frac{2\omega L_c}{c_{o+}}}\right) = 0. \tag{10.80}$$

It is convenient to recast this as

$$\tan\left(A + \frac{\omega L_m}{c_{o-}}\right)\cot\left(B - \frac{\omega L_c}{c_{o+}}\right) = \frac{\beta\varsigma}{1+\chi}, \tag{10.81}$$

where the angles A and B are defined by

$$\tan A = i\left(\frac{r_1 - 1}{r_1 + 1}\right) \tag{10.82}$$

$$\tan B = i\left(\frac{r_5 - 1}{r_5 + 1}\right). \tag{10.83}$$

The roots of Equation (10.81) are in general complex and, in view of the $e^{-i\omega t}$ time dependence of the pressure wave components, the roots with negative imaginary part correspond to standing wave modes which decay with time and those roots with positive imaginary part correspond to standing wave modes whose amplitudes grow indefinitely in time. Occurrence of the latter type of modes is called combustion instability.

In general, the roots of Equation (10.81) may be determined only numerically, since the reflection coefficients and the flame FRF are complex numbers and dependent on frequency. It is, however, instructive to examine some simple cases which are tractable to further analysis. For this purpose, we consider a simple flame, the FRF of which is given by the classical model of Crocco and Cheng [28]:

$$F_f = \left(\frac{4S_4\gamma_o p_o}{\gamma_o - 1}\right)n e^{i\omega\tau}. \tag{10.84}$$

Here, n denotes a non-dimensional constant, called interaction index and the scaling factor in brackets is introduced for convenience. For insight into the physics of this flame FRF, we may use the shifting property $h(t-t_o) \leftrightarrow h(\omega)e^{i\omega t_o}$ of the Fourier transform, Equation (1.16). Then, taking the inverse Fourier transform, we see that the flame FRF of Equation (10.84) corresponds to temporal heat release fluctuations of the form $\dot{H}'(t) \propto nv'_{-}(t-\tau)$, where the proportionally factor is equal to the parameter in brackets in Equation (10.84). Accordingly, Equation (10.84) simply hypothesizes that the heat release fluctuations are proportional to the particle velocity of the impinging acoustic waves with a temporal phase lag of τ. Both n and τ are heuristic parameters. The former is typically a number between 1 and 10 and the temporal phase lag is determined by using a convenient datum and is practically equal to the convection time of the fuel to the flame front, since the speed of sound is generally much larger than the fuel convection speed. Upon substituting this flame FRF, Equation (10.81) becomes

$$\tan\left(A + \frac{\omega L_m}{c_{o-}}\right)\cot\left(B - \frac{\omega L_c}{c_{o+}}\right) = \frac{\beta\varsigma}{1 + ne^{i\omega\tau}} \tag{10.85}$$

This may be simplified further by approximating the acoustic conditions at the ends of the chamber by the ideal acoustic boundary conditions (Section 2.3). Thus, a nearly closed end is modeled as a rigidly closed end and a nearly open end is modeled as a pressure release boundary. For example, referring to Figure 10.9a, for a nearly closed fuel supply side ($r_1 = 1$, $A = 0$) and nearly open exhaust side ($r_5 = -1$, $B = \pi/2$), Equation (10.85) becomes

$$\tan\left(\frac{\omega L_m}{c_{o-}}\right)\tan\left(\frac{\omega L_c}{c_{o+}}\right) = \frac{\beta\varsigma}{1 + ne^{i\omega\tau}}. \tag{10.86}$$

We will consider first the special case of $n = 0$, which corresponds to the case $\dot{H}' = 0$, that is no fluctuating heat release. This may seem to be an over-simplification; however, when the rate of combustion is proportional to the rate of fuel injection (as in ideal non-premixed flames), it can be realized approximately by injecting the fuel at a constant rate. In this case, putting $F_f = 0$, Equation (10.86) can be expressed for a perfect gas as

$$\tan\left(\frac{\omega L_m}{c_{o-}}\right)\tan\left(\frac{\omega L_c}{c_{o-}}\sqrt{\frac{T_{o-}}{T_{o+}}}\right) = \beta\sqrt{\frac{T_{o+}}{T_{o-}}}, \tag{10.87}$$

where we have assumed that the gas constant is not substantially altered by the combustion. Roots of this equation are real and represent the resonance frequencies (see Section 5.3.1) of the combustion chamber (with specified boundary conditions), when the temperature distribution in the combustor is as determined by a steady heat release rate flame. The smallest resonance frequency parameter $\omega_1 L/\pi c_{o-}$ is shown in Figure 10.10a as function of the area ratio β and the absolute temperature ratio T_{o+}/T_{o-} for flame position at $L_m = L/2$, where $L = L_m + L_c$. The $\beta = 1$ curve corresponds to a uniform duct of length L. The effect of changing the flame position

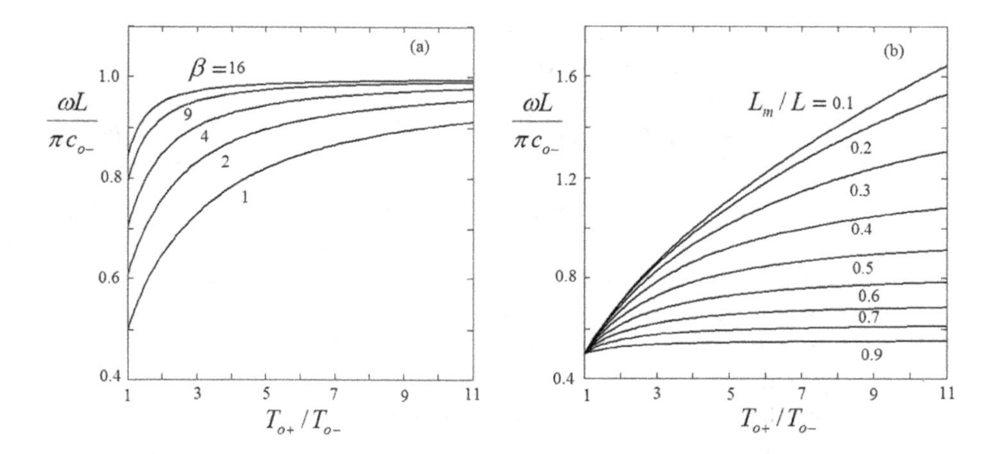

Figure 10.10 Variation of the resonance frequencies of the combustor in Figure 10.9a with the temperature ratio ($n = 0$ in Equation (10.101)). (a) $L_m/L = 0.5(L = L_m + L_c)$, (b) $\beta = 1.0$.

on the resonance frequency parameter is shown for this case in Figure 10.10b. It can be seen that as the flame is moved towards the burner, the resonance frequency parameter $\omega_1 L/\pi c_{o-}$ increases monotonically and the opposite happens as the flame is moved towards the open end of the chamber. It should be noted that if there were no flame (and, hence, no mean temperature jump), the first resonance frequency would be $\omega_1 L/\pi c_{o-} = 0.5$ (see Section 3.5.3). The corresponding values of the first resonance frequency parameter for $\beta > 1$ can be inspected from Figure 10.10a. Similar characteristics can be derived from Equation (10.81) for different end conditions.

Next, consider the case of $\tau \neq 0$. (If $\tau = 0$, then only the standing waves with resonance frequencies depicted in Figure 10.10 are sustained in the combustion chamber, but with β replaced by $\beta/(1+n)$.) In this case, the roots of Equation (10.86) may be complex. For this reason, it is convenient to express the tangent function in terms of its exponential representation. It may thus be shown that Equation (10.86) may be expressed for a perfect gas as

$$z^{2L_c/\kappa L_m+2+\eta} + z^{\eta} - z^{2L_c/\kappa L_m+\eta} - z^{2+\eta} + \left(\frac{\beta\kappa+1}{n}\right)z^{2L_c/\kappa L_m+2} + \left(\frac{\beta\kappa-1}{n}\right)z^{2L_c/\kappa L_m} + \left(\frac{\beta\kappa-1}{n}\right)z^2 + \frac{\beta\kappa+1}{n} = 0, \tag{10.88}$$

where $z = e^{i\omega L_m/c_{o-}}$, $\kappa^2 = T_{o+}/T_{o-}$ and $\eta = c_{o-}\tau/L_m$. Solution of this equation simplifies tremendously if it can be reduced to a polynomial in z. This occurs when η is a positive integer and $L_c/\kappa L_m$ is an integer multiple of 0.5. And $\eta = 1$ may be considered as typical, since the combustion lag is usually of the of the order of few milliseconds. Also, with the flame position at $L_m = L/y$, $y > 1$, the condition $L_c/\kappa L_m = N/2$, $N = 1, 2, \ldots$, corresponds to temperature ratios $T_{o+}/T_{o-} = 4(y-1)^2/N^2$. Thus, using a value of N that makes $T_{o+}/T_{o-} \geq 1$, we may simulate several practically meaningful cases by reducing Equation (10.88) to a polynomial. Here, it suffices to consider the case of $y = 2$ with $T_{o+}/T_{o-} = 4$ $(N = 1)$, for which Equation (10.88) reduces to

$$z^5 + \left(\frac{2\beta + 1}{n}\right)z^4 - 2z^3 + 2\left(\frac{2\beta - 1}{n}\right)z^2 + z + \frac{2\beta + 1}{n} = 0. \qquad (10.89)$$

This equation has one negative real root and two pairs of complex conjugate roots. The principal value of the real part of the frequency parameter $\omega_1 L/\pi c_{o-}$ corresponding to the real root is equal to 2 and this is independent of the values of β and n. Its imaginary part is negative and represents a quickly decaying mode. The real and imaginary parts of the frequency parameter $\omega_1 L/\pi c_{o-}$ corresponding to the principal values of the complex roots of Equation (10.89) with positive real part are shown in Figure 10.11 as function of β for $n = 1$, similar characteristics being valid for different values of n. It is seen that one of the frequency parameters has a negative imaginary part and represents damped oscillations. But the other root has a positive imaginary part and corresponds to temporally amplifying oscillations which characterize combustion instabilities.

Although the theory predicts that the amplitudes of unstable oscillations increase forever with time, this is not observed in practice. The reason for this is the intervention of non-linear effects after the amplitudes of the fluctuations exceed the limits of the linear theory. However, there is continuity between the linear and non-linear descriptions of the dynamics of combustion processes and, therefore, the linear theory is useful in that it gives information about the onset of combustion instabilities, which is generally what is needed in practical design problems.

The foregoing examples show how the two-port combustion source element developed in this section can be used with a block diagram model of a combustor for the analysis combustion instabilities and provide some physical insight on the role of the combustion time lag on the occurrence of combustion instability. More realistic block diagram models including the effects of mean flow, the entropy wave component and geometrical irregularities of actual combustors can be constructed and used

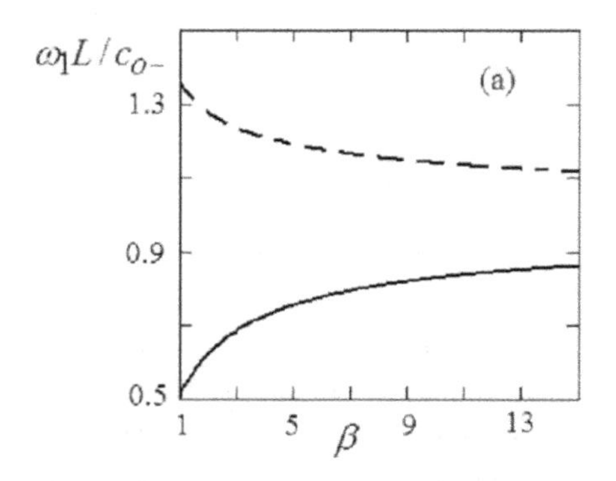

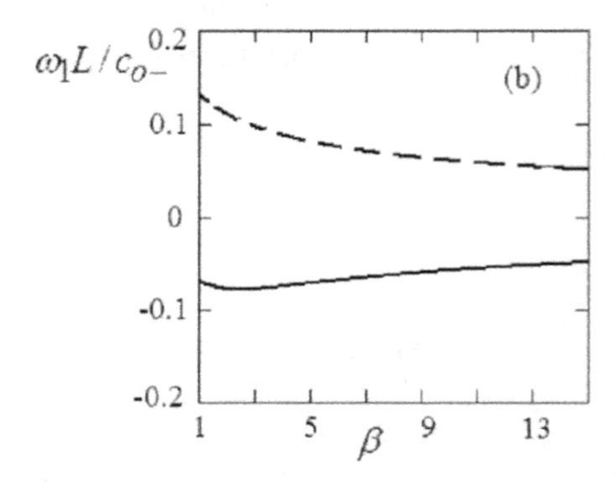

Figure 10.11 Variation of the frequency parameter corresponding to the principal values of the complex roots of Equation (10.89) with the area ratio of the combustor model in Figure 10.8a. (a) Modulus of the real parts, (b) imaginary parts ($T_{o+}/T_{o-} = 4$, $L_m/L = 0.5$, $c_{o-}\tau/L_m = 1$, $n = 1$).

similarly. Furthermore, flames of finite thickness can be modeled by dividing the flame thickness into a cascade of sufficiently large number compact two-port combustion elements [29]; however, the processes governing the flame FRF in real combustor systems are complicated and accurate prediction of combustion instabilities still stands as a highly active research field. The literature is very rich with publications on various aspects of the subject [30].

10.6 Moving Source Planes of Two-Port Sources

Source planes of a two-port source may be moved similarly as the source plane of a one-port source (Section 10.3) and this process may be justified by similar reasons. Consider the two-port source in Figure 10.1b. We wish to move the active source plane $x = 0_+$ to a downstream equivalent source plane 1_a and the plane $x = 0_-$ to an upstream equivalent source plane 1_b, as shown in Figure 10.12. The transfer paths may be arbitrary, but are represented in Figure 10.12 by uniform ducts for simplicity. Let the acoustic path between $x = 0_+$ and section 1_a of the downstream part of the ductwork, which is assumed to be passive, be defined by the impedance transfer equation

$$\begin{bmatrix} p'_+ \\ v'_+ \end{bmatrix} = \mathbf{Z}_{1a} \begin{bmatrix} p'_{1a} \\ v'_{1b} \end{bmatrix}. \tag{10.90}$$

Similarly, the acoustic path between $x = 0_-$ and section 1_b of the upstream part of the duct work may be defined by the impedance transfer matrix

$$\begin{bmatrix} p'_- \\ v'_- \end{bmatrix} = \mathbf{Z}_{1b} \begin{bmatrix} p'_{1_a} \\ v'_{1_b} \end{bmatrix}. \tag{10.91}$$

Then the original two-port source equation, which is generally in the form of Equation (10.44), may be expressed as

$$\mathbf{p}_{Seq} = \mathbf{Z}_{Seq} \begin{bmatrix} p'_{1_a} \\ v'_{1_a} \end{bmatrix} + \begin{bmatrix} p'_{1_b} \\ v'_{1_b} \end{bmatrix}, \tag{10.92}$$

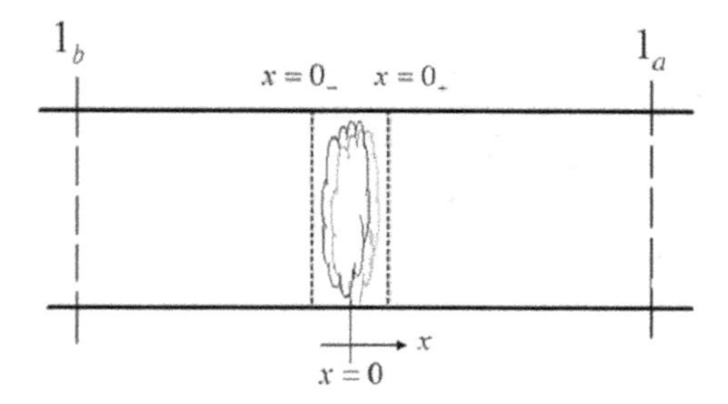

Figure 10.12 Equivalent two-port source with equivalent source planes 1_a and 1_b.

where the equivalent source characteristics, denoted by the subscript "*eq*," are given by

$$\mathbf{p}_{Seq} = \mathbf{Z}_{1_b}^{-1}\mathbf{p}_S \tag{10.93}$$

$$\mathbf{Z}_{Seq} = \mathbf{Z}_{1_b}^{-1}\mathbf{Z}_S\mathbf{Z}_{1_a} \tag{10.94}$$

10.7 Ducted Loudspeakers

Electro-mechanical devices such as loudspeakers and pressure drivers are the primary duct-borne sound sources used in the laboratory for studying acoustic response of ductworks. Figure 10.13 shows the two commonly employed methods of mounting these devices on ducts. Mean flow has to be introduced into the tested ducts separately and this usually dictates the mounting method shown in Figure 10.13b. This method also has the advantage of allowing use of multiple pressure drivers, which may be necessary for generating sufficiently strong in-duct acoustic fields, with flexibility for the selection of the excitation points. Laboratory tests may be simulated by using a block diagram model of the test setup driven by acoustic models of loudspeakers. In this section, we derive a one-port source model of for a loudspeaker.

Essentially, a loudspeaker uses an electrical coil to create a force on a diaphragm proportional to the applied current (Figure 10.14). The Thiele and Small theory [31, 32] treats the diaphragm assembly as a single-degree-of-freedom system with an effective mass m, damping coefficient c and spring stiffness k, which are determined from frequency response tests on the loudspeaker. Application of Newton's second law for the diaphragm assembly gives

$$m\frac{\mathrm{d}^2x}{\mathrm{d}t^2} + c\frac{\mathrm{d}x}{\mathrm{d}t} + kx(t) + S\big(p'_{1+}(t) - p'_{1-}(t)\big) = B\ell i(t), \tag{10.95}$$

where, t denotes the time, x the displacement of the diaphragm from its mean position, S the projected area of the diaphragm, $B\ell$ is the product of magnetic flux density, B, and the length of the coil ℓ, $i(t)$ denotes the current in the coil; and $p'_{1\mp}$ denote the acoustic pressure on the front and the back (which are denoted by subscripts "+" and

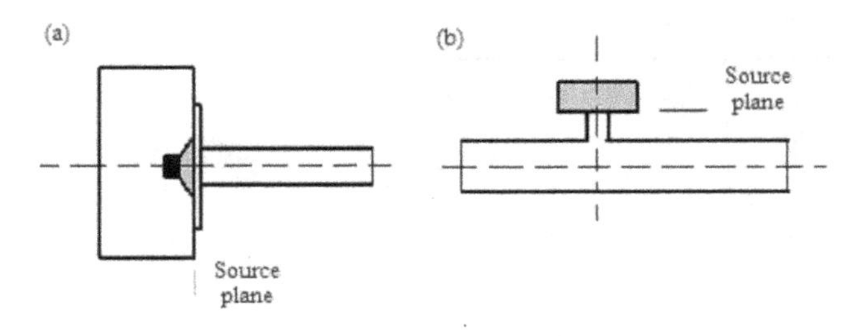

Figure 10.13 Duct-mounted loudspeakers.

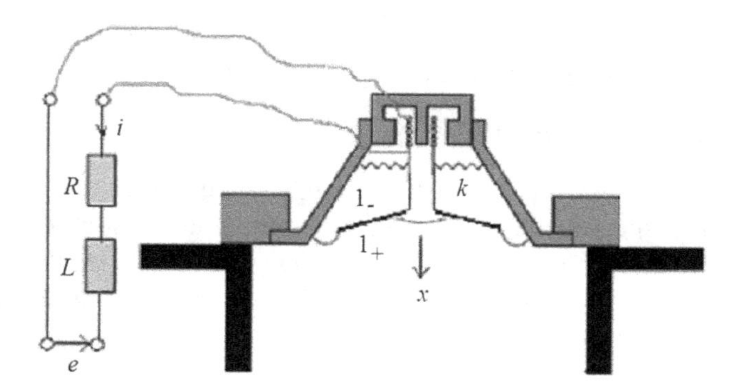

Figure 10.14 Electrical and mechanical components of a loudspeaker.

"−," respectively) of the diaphragm. The term on the right-hand side of Equation (10.95) gives the Lorentz force on the coil. Taking into account the voltage induced in the coil due to its motion, the loop equation of the electrical circuit gives

$$Ri(t) + L\frac{\mathrm{d}(i(t))}{\mathrm{d}t} + B\ell\frac{\mathrm{d}x}{\mathrm{d}t} = e(t), \tag{10.96}$$

where R and L denote, respectively, the coil resistance and inductance and $e(t)$ denotes the open circuit voltage of the loudspeaker amplifier. Upon taking the Fourier transform of the foregoing equation, the current in the coil can be expressed in frequency domain as

$$i(\omega) = \frac{e(\omega) - B\ell v_1'(\omega)}{Z_{ec}}, \tag{10.97}$$

where $Z_{ec} = R + (-\mathrm{i}\omega)L$ denotes the electrical impedance of the coil and $v_{1\mp}'(t) = \mathrm{d}x/\mathrm{d}t$ denote the particle velocity at the front and back of the diaphragm. Assuming that only plane waves propagate at the front and back of the diaphragm, $p_{1\mp}' = p_{1\mp}^{+} + p_{1\mp}^{-}$ and $\rho_o c_o v_1' = p_{1\mp}^{+} - p_{1\mp}^{-}$, where $p^{\mp}$ denote the pressure wave components in the $\mp x$ directions. Taking the Fourier transform of Equation (10.95) and substituting Equation (10.97) and the foregoing pressure wave components

$$Z_S v_{1+}'(\omega) + \mathbf{E}\ (\mathbf{P}_{1+} - \mathbf{P}_{1-}) = p_S, \tag{10.98}$$

where $\mathbf{E} = [1 \quad 1]$, $\mathbf{P}_{1\mp} = \{p_{1\mp}^{+} \quad p_{1\mp}^{-}\}$

$$Z_S = \frac{B\ell^2}{S}\left(\frac{1}{Z_{ec}} + \frac{1}{Z_{em}}\right) \tag{10.99}$$

$$p_S = \frac{B\ell}{S}\frac{e(\omega)}{Z_{ec}} \tag{10.100}$$

and

$$Z_{em} = \frac{-\mathrm{i}\omega B^2\ell^2}{-\omega^2 m - \mathrm{i}\omega c + k} \tag{10.101}$$

represents the electrical impedance of the mechanical parts moving with the coil. The foregoing equations may be manipulated to obtain the one-port source equation:

$$p_S = (Z_S - Z_{1-})v_1' + p_{1+}', \tag{10.102}$$

where $Z_{1-} = \rho_o c_o(1 - r)/(1 + r_{1-})$ with $r_{1-} = p_{1-}^-/p_{1-}^+$ being the reflection coefficient at the back of the diaphragm. This is determined by the acoustic model of the enclosure side of the loudspeaker. For frequencies sufficiently low that only the compliant acoustic mode of the enclosure is significant, an approximate expression for the back impedance Z_{1-} is [33]:

$$Z_{1-} = \frac{SR_{\text{box}}}{1 - \mathrm{i}\omega R_{\text{box}} C_{\text{box}}}, \tag{10.103}$$

where $C_{\text{box}} = V_{\text{box}}/\rho_o c_o^2$, $R_{\text{box}} \approx 8\rho_o c_o/\bar{\alpha} S_{\text{box}}$, V_{box} denotes the enclosure volume and R_{box}, S_{box} and $\bar{\alpha}$ denote, respectively, the net acoustic resistance, the area and the mean random incidence absorption coefficient of the enclosure walls.

References

[1] A.P. Dowling and J.E. Ffowcs Williams, *Sound and Sources of Sound* (Chichester, UK: Ellis Horwood Ltd., 1983).

[2] C.L. Morfey, Sound transmission and generation in ducts with flow, *J. Sound Vib.* **14** (1971), 37–55.

[3] M.J. Lighthill, On sound generated aerodynamically. Part I: General theory, *Proc. Roy. Soc. London* **A222** (1952), 1–32.

[4] M.S. Howe, *Acoustics of Fluid-Structure Interaction*, (Cambridge: Cambridge University Press, 1998).

[5] P.A. Nelson and C.L. Morfey, Aerodynamic sound production in low speed flow ducts, *J. Sound Vib.* **79** (1981), 263–289.

[6] R.P.Kanwal, *Generalized Functions: Theory and Technique*, (Boston, MA: Birkhäuser, 1998).

[7] E. Dokumaci, Prediction of source characteristics of engine exhaust manifolds, *J. Sound Vib.* **280** (2005), 925–943.

[8] M.F. Harrison and P.T. Stanev, A linear acoustic model for intake wave dynamics of IC engines, *J. Sound. Vib.* **269** (2004), 361–387.

[9] F. Laville and W. Soedel, Some new scaling rules for use in muffler design, *J. Sound Vib.* **60** (1978), 273–288.

[10] M.L. Munjal and A.G. Doige, On uniqueness, transfer and combination of acoustic sources in one dimensional systems, *J. Sound Vib.* **12** (1988), 25–35.

[11] D.H. Johnson, Origins of the equivalent source concept: the current-source equivalent, *Proc. IEEE* **91** (2003), 817–821

[12] P.O.A.L. Davies, Transmission matrix representation of exhaust system acoustic characteristics, *J. Sound Vib.* **151** (1991), 333–338.

[13] M. Åbom, H. Bodén and H. Hirvonen, A simple linear model for the source impedance of an internal combustion engine and manifold. *Proc. Inter Noise '88,* (1988), 1273–1276.

[14] N. Peake and A.B. Parry, Modern challenges facing turbomachinery aeroacoustics, *Annu. Rev. Fluid Mech.* **44** (2012), 227–248.

[15] S. Glegg and W. Davenport, *Aeroacoustics of Low Mach Number Flows* (London: Academic Press, 2017).

[16] G.N. Watson, *A Treaties on the Theory of Bessel Functions* (Cambridge: Cambridge University Press, 1966).

[17] J.M. Tyler and T.G. Sofrin, Axial flow compressor noise studies, *SAE Trans.* **70** (1962), 309–332.

[18] J. E. Ffowcs Williams and D. L. Hawkings, Sound generation by turbulence and surfaces in arbitrary motion, *Proc. Roy. Soc. London*, **A264** (1969), 321–342.

[19] C.L. Ramsey, R.T. Biedron, F. Farassat and P.L. Spence, Ducted-fan engine acoustic predictions using a Navier-Stokes code, *J. Sound Vib.* **213** (1998), 643–664.

[20] C.L. Morfey, Rotating pressure patterns in ducts: their generation and transmission, *J. Sound. Vib.* **1** (1964), 60–87.

[21] W.R. Sears, Some aspects of non-stationary airfoil theory and its practical applications, *J. Aeronaut. Sci.* **8** (3), 104–108.

[22] O. Kårekull, G. Efraimsson and M. Åbom, Prediction model of flow duct constriction noise, *Appl. Acoust.* **82** (2014), 45–52.

[23] T. Poinsot and T. Veynante, *Theoretical and Numerical Combustion*, (Philadelphia, PA: R.T.Edwards Inc., 2005).

[24] A. Cuquel, D. Durox and T. Shuller, Experimental determination of flame transfer function using random velocity perturbations, Proceedings of ASME Turbo Expo: Power for Land, Sea, and Air, Volume 2: Combustion, Fuels and Emissions, Parts A and B, 793–802. doi:10.1115/GT2011-45881.

[25] H.J. Krediet, C.H. Beck, W. Krebs, et al. Identification of the flame describing function of a premixed swirl flame from LES, *Combust. Sci. Technol.* **184** (2012), 888–900.

[26] T. Shuller and D. Durox, A unified model for the prediction of laminar flame transfer functions: comparisons between laminar and V-flame dynamics, *Combust. Flame* **134** (2003), 21–34.

[27] J. Li and A.S. Morgans, Simplified models for the thermodynamic properties along a combustor and their effect on thermoacoustic instability prediction, *Fuel* **184** (2016), 735–748.

[28] L. Crocco and S. Cheng, *Theory of Combustion Instability in Liquid Propellant Rocket Motors*, (London: Butterworth, 1956).

[29] K.T. Kim, J.G. Lee, B.D. Quay and D.A. Santavicca, Spatially distributed flame transfer functions for predicting combustion dynamics in lean premixed turbine combustors, *Combust. Flame* **157** (2010), 1718–1730.

[30] T.C. Lieuven, *Unsteady Combustion Physics*, (Cambridge: Cambridge University Press, 2012).

[31] A.N. Thiele, Loudspeakers in vented boxes, *J. Audio Eng. Soc.* **19** (1971), 382–392 (reprint from *Proc. IRE* **22** (1961), 487–508).

[32] R.H. Small, Direct-radiator loudspeakers systems analysis, *IEEE Trans. Audio Electroacoust.* **19** (1971), 269–281.

[33] P.A. Nelson and S.J. Elliot, *Active Control of Sound*, (London: Academic Press, 1992).

11 Radiated Sound Pressure Prediction

11.1 Introduction

This chapter is mainly devoted to the calculation of sound pressure at a target point in the acoustic field of an in-duct acoustic source. This important topic is taken up late in this book because it requires appreciation of the source characteristics, as well as an accurate acoustic model of the path between the source and the target point. As pointed out in the previous chapter, the weakest link here is the source, as its acoustic characteristics can usually be determined decisively only by measurements on the actual source (Section 12.4). Since this is difficult and not always feasible to make at the design stage, sound pressure predictions are often hindered in practice by lack of information about the source characteristics. For this reason, the ability to assess the effects of design changes on the sound pressure at a target point without actually calculating it is also important. Some parameters which can be computed independently of the source may be useful for this purpose under certain conditions and are also discussed (Section 11.4).

The need for prediction of sound pressure usually arises when the silencing requirement of a ductwork is specified directly by the sound pressure at a target point. Such specifications are usually based on the experience of fluid machinery manufacturers and may stem from general environmental noise regulations and customer requirements.

11.2 Calculation of Sound Pressure Field of Ducted Sources

We consider the two basic forms of ducted sources, namely, the one-port source and the two-port source configurations shown in Figure 11.1. In Figure 11.1a, it is assumed that the ductwork is driven by a one-port source from plane 1. This plane may be the actual source plane, but, for the generality of the subsequent discussion, it will be understood as an equivalent source plane (see Section 10.3). The system radiates from plane 2, which is typically an open end of a duct, to an exterior environment, which we may model as a free field or fractional free field or a reverberant room (see Section 9.5.8). The acoustic path between plane 1 and plane 2 is assumed to be passive and although this path is represented in Figure 11.1a by a single straight duct, it can be any passive ductwork of known wave transfer equation $\mathbf{P}_1 = \bar{\mathbf{T}}_{12}\mathbf{P}_2$, where $\bar{\mathbf{T}}_{12}$ denotes the wave transfer matrix of the acoustic path.

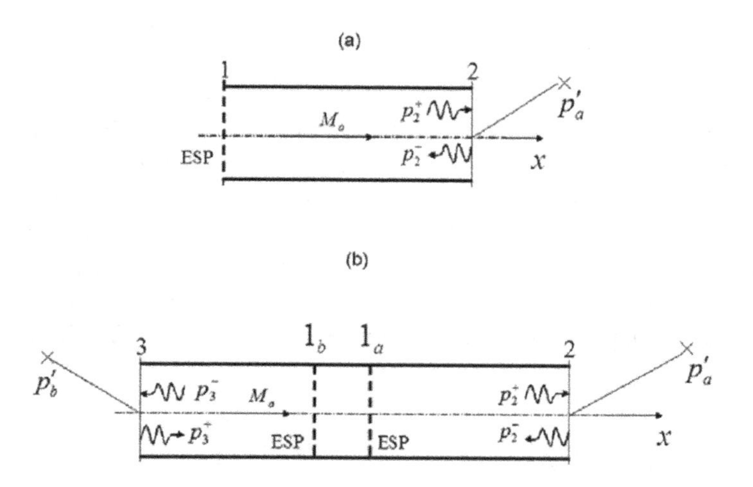

Figure 11.1 Ducted one-port and two-port sources (ESP: equivalent source plane).

The system considered in Figure 11.1b depicts a ductwork driven by a two-port source, the planes 1_a and 1_b denoting the equivalent source planes (see Section 10.7). The system may radiate from plane 2 or plane 3, or both, to an exterior acoustic field. In the latter case, we assume that the radiation from one end has no contribution to the sound pressure field created by radiation from the other end. The acoustic paths between planes 1_a and 2 and 1_b and 3 are assumed to be passive and defined by wave transfer equations $\mathbf{P}_{1_a} = \mathbf{T}_{1_a,2}\mathbf{P}_2$ and $\mathbf{P}_{1_b} = \mathbf{T}_{1_b,3}\mathbf{P}_3$, respectively. Again, although these paths are shown in Figure 11.4b as straight ducts, they may be any assembly of passive components having wave transfer matrices $\mathbf{T}_{1_a,2}$ and $\mathbf{T}_{1_b,3}$.

In both ducted source configurations shown in Figure 11.1, the acoustic path between plane 2 and an exterior target point at distance a from the acoustic center of plane 2 in a given orientation is determined by modal radiation transfer functions (see Section 9.5.1):

$$K_2^{(j)} = \frac{p_2^{+(j)}}{p'^{(j)}_a}, \tag{11.1}$$

where p'_a denotes the acoustic pressure at the target point, p_2^+ denotes the incident sound pressure wave component at plane 2 and the superscript "(j)" denotes the contribution of propagating mode $j = 1, 2, \ldots, n$. We will drop the superscript notation when $n = 1$; that is, only the fundamental mode is cut-on at plane 2. In the case of a two-port source, the radiation transfer function for the acoustic path between plane 3 and an exterior target point is similarly defined.

Although Figure 11.1 implies that the target point is located in the exterior environment, the subsequent analysis is also applicable for interior target points, provided that the radiation transfer function is defined appropriately. For example, assuming fundamental mode propagation for simplicity, if the target is plane 2, then $p'_a = p'_2$, which implies that $K_2 = 1/(1 + r_2)$, where

$$r_k = \frac{p_k^-}{p_k^+} \tag{11.2}$$

denotes the plane-wave reflection coefficient at plane $k \in 1, 2, \ldots$ If the target point is any internal point N, say, this result may be generalized as $K_2 = p_2^+/(1 + r_N)p_N^+$, where the transfer function p_2^+/p_N^+ is determined from the wave transfer equation $\mathbf{P}_N = \mathbf{T}_{N2}\mathbf{P}_2$. Since the exterior target points are usually of more interest due to environmental noise regulations, we will not consider internal target points explicitly in the rest of this chapter; however, it is clear that all results can be applied for internal target points by appropriate redefinition of the radiation transfer functions.

11.2.1 Ducted One-Port Sources

Referring to Figure 11.1a, we consider n modes at planes 1 and 2, not necessarily all cut-on. In general, acoustic wave transfer along this path is represented by the modal wave transfer equation $\mathbf{P}_1 = \bar{\mathbf{T}}_{12}\mathbf{P}_2$. It is convenient to write this in partitioned form

$$\begin{bmatrix} \mathbf{P}_{1n}^+ \\ \mathbf{P}_{1n}^- \end{bmatrix} = \begin{bmatrix} \mathbf{T}_{11} & \mathbf{T}_{12} \\ \mathbf{T}_{21} & \mathbf{T}_{22} \end{bmatrix} \begin{bmatrix} \mathbf{P}_{2n}^+ \\ \mathbf{P}_{2n}^- \end{bmatrix}, \tag{11.3}$$

where the subscript tag "n" is used to indicate the modality (the number of modes), the submatrices $\mathbf{T}_{ij}$, $i,j = 1, 2$, are of size $n \times n$ and

$$\mathbf{P}_{kn}^\pm = \begin{bmatrix} p_k^{(1)\pm} \\ p_k^{(2)\pm} \\ \vdots \\ p_k^{(n)\pm} \end{bmatrix}. \tag{11.4}$$

Here, the subscript $k = 1, 2$ denotes planes 1 and 2, respectively, $p_k^{(j)\pm}$ denote the pressure wave components for mode $j = 1, 2, \ldots, n$ and the superscripts "$\mp$" denote the direction of propagation of the wave components.

Let the number of cut-on modes at a given frequency at planes 1 and 2 be denoted by $q(\leq n)$ and $m(\leq n)$, respectively. Equation (11.3) may be contracted to modality b, where b denotes the larger of q and m, as described in Section 2.4.2. Let the contracted wave transfer matrix be written as

$$\begin{bmatrix} \mathbf{P}_{1b}^+ \\ \mathbf{P}_{1b}^- \end{bmatrix} = \begin{bmatrix} \mathbf{S}_{11} & \mathbf{S}_{12} \\ \mathbf{S}_{21} & \mathbf{S}_{22} \end{bmatrix} \begin{bmatrix} \mathbf{P}_{2b}^+ \\ \mathbf{P}_{2b}^- \end{bmatrix}, \tag{11.5}$$

where $\mathbf{P}_{kb}^\pm$ are defined as in Equation (11.4) and the submatrices $\mathbf{S}_{ij}$, $i,j = 1, 2$, are of size $b \times b$. In general, the vectors $\mathbf{P}_{2b}^-$ and $\mathbf{P}_{2b}^+$ are related as $\mathbf{P}_{2b}^- = \mathbf{R}_{2b}\mathbf{P}_{2b}^+$, where $\mathbf{R}_{2b}$ denotes the reflection coefficient matrix at plane 2 comprising the first b modes. On substituting this relationship, Equation (11.5) may be split as

$$\mathbf{P}_{1b}^{+} = (\mathbf{S}_{11} + \mathbf{S}_{12}\mathbf{R}_{2b})\mathbf{P}_{2b}^{+} = \mathbf{A}\mathbf{P}_{2b}^{+} \tag{11.6}$$

$$\mathbf{P}_{1b}^{-} = (\mathbf{S}_{21} + \mathbf{S}_{22}\mathbf{R}_{2b})\mathbf{P}_{2b}^{+} = \mathbf{C}\mathbf{P}_{2b}^{+} \tag{11.7}$$

where the matrices **A** and **C** are defined by the second equalities.

A multi-modal one-port source imposes constraints on the pressure wave components at the equivalent source plane, which may be expressed generally in the form of Equation (2.10). Here, we express this equation in the present notation in partitioned form

$$\mathbf{P}_S = \mathbf{G}_S\mathbf{P}_{1q} = \begin{bmatrix}\mathbf{G}_S^{+} & \mathbf{G}_S^{-}\end{bmatrix}\begin{bmatrix}\mathbf{P}_{1q}^{+} \\ \mathbf{P}_{1q}^{-}\end{bmatrix}, \tag{11.8}$$

where $\mathbf{G}_S^{\pm}$ denote the appropriate blocks of size $q \times q$ of the source matrix $\mathbf{G}_S$. It is convenient to re-scale the source pressure strength vector $\mathbf{P}_S$ as

$$\mathbf{p}_S = \left(\mathbf{A}_1^{+} - \mathbf{A}_1^{-}\right)\left(-\mathbf{G}_S^{+}\mathbf{A}_1^{-} + \mathbf{G}_S^{-}\mathbf{A}_1^{+}\right)^{-1}\mathbf{P}_S, \tag{11.9}$$

where $\mathbf{A}^{\mp}$ denote the modal admittance matrices defined in Equation (6.64). This may be expressed in the form

$$\mathbf{p}_S = \begin{bmatrix} p_S^{(1)} \\ p_S^{(2)} \\ \vdots \\ p_S^{(q)} \end{bmatrix} = \begin{bmatrix}\mathbf{I} + \zeta_S\mathbf{A}_1^{+} & \mathbf{I} + \zeta_S\mathbf{A}_1^{-}\end{bmatrix}\begin{bmatrix}\mathbf{P}_{1q}^{+} \\ \mathbf{P}_{1q}^{-}\end{bmatrix}, \tag{11.10}$$

where **I** denotes a unit matrix of size $q \times q$ and

$$\zeta_S = \left(\mathbf{A}_1^{+} - \mathbf{A}_1^{-}\right)\left(-\mathbf{G}_S^{+}\mathbf{A}_1^{-} + \mathbf{G}_S^{-}\mathbf{A}_1^{+}\right)^{-1}\left(\mathbf{G}_S^{+} - \mathbf{G}_S^{-}\right)\left(\mathbf{A}_1^{+} - \mathbf{A}_1^{-}\right)^{-1} \tag{11.11}$$

denotes the normalized source impedance matrix (impedances divided by $\rho_o c_o$ at the equivalent source plane) and $p_S^{(j)}$ denotes the equivalent source pressure strength for mode $j = 1, 2, \ldots, q$. Thus, the equivalent sources are characterized by $\mathbf{p}_S$ and ζ_S.

Depending on values of q and m, the following two cases need to be distinguished:

(i) $q = b \geq m$: in this case, Equations (11.5) and (11.7) yield directly

$$\mathbf{P}_{2b}^{+} = \left(\left(\mathbf{I} + \zeta_S\mathbf{A}_1^{+}\right)\mathbf{A} + \left(\mathbf{I} + \zeta_S\mathbf{A}_1^{-}\right)\mathbf{C}\right)^{-1}\mathbf{p}_S,\ b = q. \tag{11.12}$$

If $q > m$, the last $q - m$ elements of vector $\mathbf{P}_{2q}^{+}$ will correspond to non-propagating (evanescent) modes.

(ii) $q < b = m$: in this case, Equation (11.6) may be partitioned as

$$\begin{bmatrix}\mathbf{P}_{1q}^{+} \\ \mathbf{0}\end{bmatrix} = \begin{bmatrix}\mathbf{A}_{11} & \mathbf{A}_{12} \\ \mathbf{A}_{21} & \mathbf{A}_{22}\end{bmatrix}\begin{bmatrix}\mathbf{P}_{2q}^{+} \\ \mathbf{P}_{2(b-q)}^{+}\end{bmatrix}. \tag{11.13}$$

Here, the subscript "$b - q$" denotes a subvector which contains all but first q terms of the main vector of size b. Equation (11.7) is partitioned similarly:

$$\begin{bmatrix} \mathbf{P}_{1q}^{-} \\ \mathbf{P}_{1(b-q)}^{-} \end{bmatrix} = \begin{bmatrix} \mathbf{C}_{11} & \mathbf{C}_{12} \\ \mathbf{C}_{21} & \mathbf{C}_{22} \end{bmatrix} \begin{bmatrix} \mathbf{P}_{2q}^{+} \\ \mathbf{P}_{2(b-q)}^{+} \end{bmatrix}. \tag{11.14}$$

Equations (11.13) and (11.14) may be manipulated to get

$$\begin{bmatrix} \mathbf{P}_{1q}^{+} \\ \mathbf{P}_{1q}^{-} \end{bmatrix} = \mathbf{F}\mathbf{P}_{2q}^{+} \tag{11.15}$$

$$\mathbf{P}_{2(b-q)}^{+} = -\mathbf{A}_{22}^{-1}\mathbf{A}_{21}\mathbf{P}_{2q}^{+}, \tag{11.16}$$

where

$$\mathbf{F} = \begin{bmatrix} \mathbf{A}_{11} - \mathbf{A}_{12}\mathbf{A}_{22}^{-1}\mathbf{A}_{21} \\ \mathbf{C}_{11} - \mathbf{C}_{12}\mathbf{A}_{22}^{-1}\mathbf{A}_{21} \end{bmatrix}. \tag{11.17}$$

Thus, from Equation (11.10) and the foregoing equations, we obtain

$$\mathbf{P}_{2b}^{+} = \begin{bmatrix} \mathbf{I} \\ -\mathbf{A}_{22}^{-1}\mathbf{A}_{21} \end{bmatrix} \left(\begin{bmatrix} \mathbf{I} + \zeta_S \mathbf{A}_1^{+} & \mathbf{I} + \zeta_S \mathbf{A}_1^{-} \end{bmatrix} \mathbf{F} \right)^{-1} \mathbf{p}_S, \; b = m \tag{11.18}$$

The results derived above for $\mathbf{P}_{2b}^{+}$ for the two cases considered may be used now in multi-modal radiation models (see Section 9.5) to calculate the sound pressure at a target point. The calculations has to be carried out numerically, but the main difficulty lies in the fact that $\mathbf{p}_S$ and ζ_S involve $q(1+q)$ complex scalar unknowns for q cut-on modes at the source plane, which have to be known to carry out the sound pressure calculations. This data is not usually practicable to get, because measurement is generally the only decisive approach and the complexity of measurements increase rapidly with the number of cut-on modes. For this reason, multi-modal acoustic sources are handled more conveniently for engineering purposes by resorting to acoustic power considerations (Sections 11.6). However, only two scalar unknowns, p_S and ζ_S, are involved when $q = 1$ and there are well-investigated general purpose experimental methods for the measurement of these parameters (see Section 12.4). This case, with which we continue in the next subsection, is of some practical importance, because it may also be regarded as an approximate representation of a multi-modal source, with an equivalent single mode source of equal power. This approximation is localized to the source plane, as multi-modal propagation is allowed elsewhere in the system. Modal distribution of sound power is usually taken into account by making similar conjectures or assuming specific ideal equivalent source types, since it is not known a priori and difficult to measure with adequate accuracy in most applications [1].

11.2.1.1 One-Dimensional Sources

For $q = 1$, Equation (11.10) simplifies to

$$p_S = \begin{bmatrix} 1 + \zeta_S & 1 - \zeta_S \end{bmatrix} \begin{bmatrix} p_1^{+} \\ p_1^{-} \end{bmatrix} \tag{11.19}$$

where we have dropped the use of superscript in brackets for the simplicity of notation. At plane 2, one or more modes may be propagating. If only the fundamental mode is cut-on, then $m = 1$ also and contraction is not relevant. The matrices **A** and **C** in Equations (11.6) and (11.7) then reduce to scalars A and C, respectively, and Equation (11.12) simplifies to

$$p_2^+ = \frac{p_S}{(1+\zeta_S)A + (1-\zeta_S)\,C}. \tag{11.20}$$

Since the submatrices $\mathbf{T}_{ij}$ in Equation (11.3) also reduce to scalars T_{ij}, and A and C are given by $A = T_{11} + T_{12}r_2 = p_1^+/p_2^+$ and $C = T_{21} + T_{22}r_2 = p_1^-/p_2^+$, then, Equation (11.20) may be expressed as

$$p_2^+ = \frac{p_S}{\alpha((1+\zeta_S) + (1-\zeta_S)\,r_1)}, \tag{11.21}$$

where the transfer function

$$\alpha = \frac{p_1^+}{p_2^+} \tag{11.22}$$

is called fundamental forward wave transmission coefficient between planes 1 and 2. The acoustic pressure at a target point may be computed from

$$p'_a = \frac{p_S}{K_2\alpha((1+\zeta_S) + (1-\zeta_S)\,r_1)}, \tag{11.23}$$

where K_2 denotes the fundamental mode radiation transfer function defined by Equation (11.1).

On the other hand, if $m > 1$, inspection of Equation (11.15) shows that the matrix **F** simplifies to

$$\mathbf{F} = \alpha\begin{bmatrix} 1 \\ r_1 \end{bmatrix}. \tag{11.24}$$

Consequently, Equation (11.18) gives

$$\mathbf{P}_2^+ = \frac{p_S}{\alpha(1+\zeta_S + (1-\zeta_S)\;r_1)}\begin{bmatrix} 1 \\ -\mathbf{A}_{22}^{-1}\mathbf{A}_{21} \end{bmatrix}, \tag{11.25}$$

where the matrix product $\mathbf{A}_{22}^{-1}\mathbf{A}_{21}$ reduces to a column vector. Thus, introducing the radiation transfer functions defined by Equation (11.1), we obtain

$$\begin{bmatrix} p'^{(1)}_a \\ p'^{(2)}_a \\ \vdots \\ p'^{(m)}_a \end{bmatrix} = \frac{p_S}{\alpha(1+\zeta_S + (1-\zeta_S)\;r_1)}\mathbf{K}_2^{-1}\begin{bmatrix} 1 \\ -\mathbf{A}_{22}^{-1}\mathbf{A}_{21} \end{bmatrix}, \tag{11.26}$$

where

$$\mathbf{K}_2 = \begin{bmatrix} K_2^{(1)} & 0 & 0 & 0 \\ 0 & K_2^{(2)} & 0 & 0 \\ 0 & 0 & \ddots & 0 \\ 0 & 0 & 0 & K_2^{(m)} \end{bmatrix}. \tag{11.27}$$

Since the matrix product $\mathbf{A}_{22}^{-1}\mathbf{A}_{21}$ is a column vector, Equations (11.15) and (11.16) combine as

$$\begin{bmatrix} p_2^{(1)+} \\ p_2^{(2)+} \\ \vdots \\ p_2^{(m)+} \end{bmatrix} = \begin{bmatrix} 1 \\ -\mathbf{A}_{22}^{-1}\mathbf{A}_{21} \end{bmatrix} p_2^{(1)+} = \begin{bmatrix} \beta_2^{(1)} \\ \beta_2^{(2)} \\ \vdots \\ \beta_2^{(m)} \end{bmatrix} p_2^{(1)+}, \tag{11.28}$$

where $\beta_2^{(j)} = p_2^{(j)+}/p_2^{(1)+}$, $j = 1, 2, \ldots, m$, are called the participation factors of the forward wave modes at plane 2. Hence, Equation (11.26) yields

$$p'^{(j)}_a = \frac{\beta_2^{(j)} p_S}{K_2^{(j)} \alpha (1 + \zeta_S + (1 - \zeta_S) r_1)} \tag{11.29}$$

for the contribution of propagating modes $j = 1, 2, \ldots, m$ to the acoustic pressure at the target point. The total acoustic pressure may then be found from $p'_a = p'^{(1)}_a + p'^{(2)}_a + \cdots + p'^{(m)}_a$. For $j = 1$, Equation (11.29) reduces to Equation (11.21).

11.2.2 Ducted Two-Port Sources

Referring to Figure 11.1b, first consider radiation from plane 2. The acoustic path between planes 1_a and 2 is defined by a multi-modal wave transfer relation of modality n as

$$\begin{bmatrix} \mathbf{P}_{1_a n}^{+} \\ \mathbf{P}_{1_a n}^{-} \end{bmatrix} = \mathbf{T}_{1_a,2} \begin{bmatrix} \mathbf{P}_{2n}^{+} \\ \mathbf{P}_{2n}^{-} \end{bmatrix}, \tag{11.30}$$

where the notation is similar to that used in Equation (11.3). Again, let the number of propagating (cut-on) modes at a given frequency at planes 1_a and 2 be denoted by $q(\leq n)$ and $m(\leq n)$, respectively, and contract Equation (11.30) to modality b, where b denotes the larger of q and m. The contracted wave transfer Equation will be exactly in the form of Equation (11.5), but with the subscript "1" replaced by "1_a". Then, Equations (11.6) and (11.7) also apply with the same subscript change. Now, assuming that q modes are propagating at the equivalent source planes 1_a and 1_b, the multi-modal two-port source equation, Equation (2.29), may be written in partitioned form as

$$\begin{bmatrix} \mathbf{a}_S \\ \mathbf{b}_S \end{bmatrix} = \begin{bmatrix} \mathbf{A}_S & \mathbf{C}_S \\ \mathbf{B}_S & \mathbf{D}_S \end{bmatrix} \begin{bmatrix} \mathbf{P}^+_{1_a q} \\ \mathbf{P}^-_{1_a q} \end{bmatrix} + \begin{bmatrix} \mathbf{P}^+_{1_b q} \\ \mathbf{P}^-_{1_b q} \end{bmatrix}. \tag{11.31}$$

Let $\mathbf{R}_{1_b}$ denote the reflection coefficient matrix for q modes at plane 1_b, that is,

$$\mathbf{P}^-_{1_b q} = \mathbf{R}_{1_b} \mathbf{P}^+_{1_b q}. \tag{11.32}$$

Equation (11.31) may be written in the one-port source form of Equation (11.7) as

$$\mathbf{p}_{S_a} = \begin{bmatrix} p^{(1)}_{S_a} \\ p^{(2)}_{S_a} \\ \vdots \\ p^{(q)}_{S_a} \end{bmatrix} = \begin{bmatrix} \mathbf{G}^+_{S_a} & \mathbf{G}^-_{S_a} \end{bmatrix} \begin{bmatrix} \mathbf{P}^+_{1_a q} \\ \mathbf{P}^-_{1_a q} \end{bmatrix}, \tag{11.33}$$

where

$$\mathbf{p}_{S_a} = \mathbf{b}_S - \mathbf{R}_{1_b} \mathbf{a}_S \tag{11.34}$$

$$\mathbf{G}^+_{S_a} = \mathbf{B}_S - \mathbf{R}_{1_b} \mathbf{A}_S \tag{11.35}$$

$$\mathbf{G}^-_{S_a} = \mathbf{D}_S - \mathbf{R}_{1_b} \mathbf{C}_S. \tag{11.36}$$

This result reduces the problem of determination of the target sound pressure p'_a due to the two-port source to that of an equivalent one-port source defined by Equation (11.33). Consequently, the results of Section 11.2.1 can be applied directly by using the foregoing characteristics of the equivalent two-port source. In particular, if propagation at the equivalent source planes 1_a and 1_b is one-dimensional ($q = 1$), Equation (11.31) reduces to

$$\begin{bmatrix} a_S \\ b_S \end{bmatrix} = \begin{bmatrix} A_S & B_S \\ C_S & D_S \end{bmatrix} \begin{bmatrix} p^+_{1_a} \\ p^-_{1_a} \end{bmatrix} + \begin{bmatrix} p^+_{1_b} \\ p^-_{1_b} \end{bmatrix} \tag{11.37}$$

and Equation (11.33) becomes

$$b_S - r_{1_b} a_S = \begin{bmatrix} B_S - r_{1_b} A_S & D_S - r_{1_b} C_S \end{bmatrix} \begin{bmatrix} p^+_{1_a} \\ p^-_{1_a} \end{bmatrix}, \tag{11.38}$$

since the reflection coefficients matrix $\mathbf{R}_{1_b}$ then reduces to the fundamental reflection coefficient r_{1_b}. This is formally the same as Equation (11.19) and, therefore, Equation (11.29) applies now as

$$p'^{(j)}_a = \frac{\beta^{(j)}_2 (b_S - r_{1_b} a_S)}{K^{(j)}_2 \alpha (B_S - r_{1_b} A_S + (D_S - r_{1_b} C_S) r_{1a})}, \tag{11.39}$$

where $\beta_2^{(j)}$ are defined as in Equation (11.28), $K_2^{(j)}$ denotes the radiation transfer function at plane 2 for mode $j = 1, 2, \ldots, m$, and $\alpha = p_{1_a}^+/p_2^+$ denotes the fundamental mode attenuation of path $1_a - 2$.

The radiation from plane 3 (Figure 11.1b) is calculated similarly. In this case we use $\mathbf{P}_{1_aq}^- = \mathbf{R}_{1_a}\mathbf{P}_{1_aq}^+$ to eliminate $\mathbf{P}_{1_aq}^+$ in Equation (11.31) and the equivalent one-port source equation may be shown to be

$$\mathbf{b}_S - \mathbf{S}\mathbf{a}_S = \begin{bmatrix} -\mathbf{S} & \mathbf{I} \end{bmatrix} \begin{bmatrix} \mathbf{P}_{1_bq}^+ \\ \mathbf{P}_{1_bq}^- \end{bmatrix}, \tag{11.40}$$

where

$$\mathbf{S} = (\mathbf{B}_S + \mathbf{D}_S\mathbf{R}_{1_a})(\mathbf{A}_S + \mathbf{C}_S\mathbf{R}_{1_a})^{-1}. \tag{11.41}$$

Alternatively, we may re-define Equation (11.31) with plane 1_a interchanged with plane 1_b, in which case the equivalent one-port source equation comes out formally in the same form as Equation (11.33).

11.2.3 Multiple Radiating Outlets

The foregoing analyses may be extended for equivalent one-port and two-port sources radiating from multiple outlets. Figure 11.2 shows a generalization of Figure 11.1a for multiple outlets. Assuming sound propagation is one-dimensional at the equivalent source plane, plane 1, the contribution of each radiating outlet to the sound pressure is given by Equation (11.29):

$$p'_{a_i} = \frac{p_S}{\alpha_i(1 + \zeta_S + (1 - \zeta_S)r_1)} \sum_{j=1}^{m} \left(\frac{\beta_i^{(j)}}{K_i^{(j)}} \right), \tag{11.42}$$

where $\alpha_i = p_1^+/p_i^+$, subscript $i = 2, 3, 4, \ldots$ denotes the radiating planes, subscript "1" denotes plane 1. The sound pressure at the target point is given by the sum of contributions of all outlets; that is, $p'_a = p'_{a_2} + p'_{a_3} + \cdots$, that is

$$p'_a = \frac{p_S}{1 + \zeta_S + (1 - \zeta_S)r_1} \sum_{i=2,3,\ldots} \left(\frac{1}{\alpha_i} \sum_{j=1}^{m} \frac{\beta_i^{(j)}}{K_i^{(j)}} \right). \tag{11.43}$$

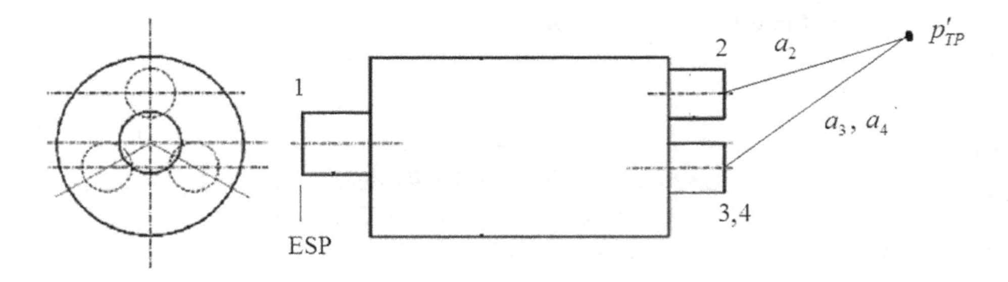

Figure 11.2 One-port source radiating from three outlets (abbreviations: ESP: equivalent source plane; TP: target point).

A two-port source with multiple outlets can be analyzed similarly by reducing it to one-port source systems as described in Section 11.2.2.

11.3 Analysis of Sound Pressure

Assuming a one-dimensional source, the sound pressure at a target point may be expressed, in frequency domain, in the generic form

$$p'_a(f) = H(f)p_S(f), \tag{11.44}$$

where $p_S(f)$ denotes the source pressure strength and f denotes the frequency. Explicit forms of the transfer function $H(f)$ for cut-on modes may be observed in Equation (11.29), or (11.39). Then $p'_a(f)$ is calculated from the foregoing transfer equation and is usually expressed in decibel units by using the formula

$$L_{p'_a}(f) = 20\log_{10}\left(\frac{(p'_a(f))_{\text{rms}}}{p_{\text{ref}}}\right) \quad \text{dB(L)}, \tag{11.45}$$

where the subscript "rms" denotes root-mean-square value (see Section 1.2.4), the reference pressure $p_{\text{ref}} = 0.00002$ Pa is approximately equal to the lowest sound pressure audible to the normal human ear at 1000 Hz. And $L_{p'_a}$ is called sound pressure level (SPL). In the abbreviation for the decibel unit, the "(L)" tag stands for "linear," which means that the dB level is computed directly from $p'_a(f)$. Equation (11.44) should be understood with right-to-left causality. This means that $p'_a(f)$ is determined over the same frequency spectrum in which $p_S(f)$ is given. For example, the fast Fourier transform (FFT) is standard in measurement of $p_S(f)$. Then, $p'_a(f)$ is determined from Equation (11.44) over the same FFT spectrum and $p'_{\text{rms}}(f) = \sqrt{2}|p'_a(f)|$.

The use of the dB unit transcends Weber's law, which states that human sound sensation is proportional to the logarithmic change in the stimulus. Ability to specify sound pressure in a unit which scales with human sound sensation is important, because sound pressure calculations are usually undertaken for the assessment of the likely effects of noise on people at or near the target point during the day or night. In fact, environmental noise criteria are often given in the dB(A) unit, as this unit is found to scale better with human hearing sensation than the dB(L). These units are related as $\text{dB(A)} = \text{dB(L)} + C_A$, where the correction term C_A is shown in Figure 11.3. Similar corrections which give the SPL in units designated as dB(B) and dB(C) are also standard, but their use is not as widespread as the dB(A) unit.

Human perception of sound is frequency selective. Our ears can distinguish changes in frequency better at lower frequencies than at higher frequencies. (for example, Figure 11.3 approximates the response of normal human ear at relatively low frequencies.) In fact, our ears have hearing bands, the widths of which increase with frequency. The bandwidths of the first four hearing bands are constant at 100 Hz, then increase to 160 Hz at the ninth hearing band, the center frequency of which is 1000 Hz. For engineering purposes, these hearing bands are usually approximated by

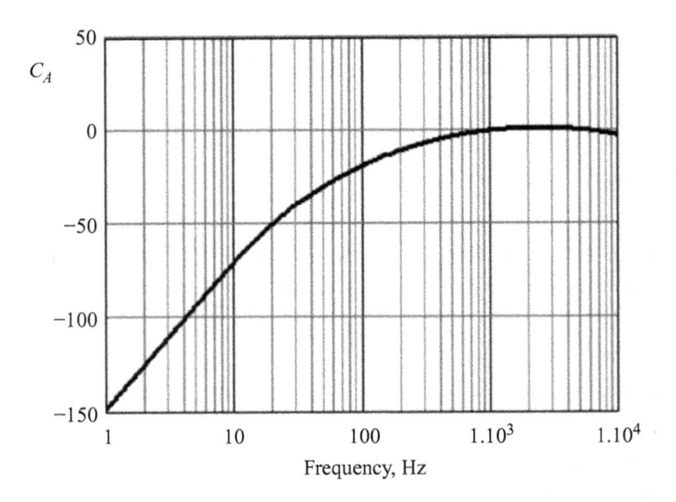

Figure 11.3 The dB(A) correction.

1/3 octave bands (see Section 1.2.4). Accordingly, SPL is often measured in 1/3 octave bands.

Sound pressure level spectra of ducted acoustic sources due to cyclic action in fluid machinery are usually displayed in the so-called Campbell diagram form or by vertical and horizontal cuts of a Campbell diagram. Figure 11.4a shows a typical Campbell diagram, which shows SPL as function of the cycle frequency (the vertical scale) and sound frequency (the horizontal scale), color intensity being used to scale SPL in dB units. Vertical high intensity lines on the diagram signify dominant resonances of the acoustic path, $H(f)$. High intensity inclined lines, on the other hand, indicate the dominant regimes of the acoustic source. This is because the discrete spectrum components of cyclic source processes occur at integer multiples (harmonics) of the cycle frequency. Horizontal cuts on a Campbell diagram may be taken to show more clearly the variation of SPL with frequency at fixed operation points. Similarly, cuts along inclined lines may be taken to see the variation of SPL with harmonics of the cycle frequency in more detail. Such inclined cuts are called order plots.

Actually, the Campbell diagram in Figure 11.4a gives the predicted SPL at 0.141 m from the intake outlet of a four-cylinder four-stroke engine in a free field. This is in fairly good agreement with the measured SPL, which is shown in Figure 11.4b, also in the Campbell diagram form. The system is shown schematically in Figure 11.5a. The speed range of the engine is 1000–6000 rpm and, at full load, the fresh air mass flow rate varies approximately linearly from about 100 kg/h at 1000 rpm to about 500 kg/h at 6000 rpm at 40 °C. So the speed of sound is constant at about 355 m/s and the maximum Mach number is less than about 0.08. The frequencies of interest in this application are in the range 0–1000 Hz. The source characteristics used in the calculations were measured in this frequency range at the equivalent source plane in the speed range of 1000–6000 rpm at full load by using the two-load method (see Section 12.4.1). The magnitude of the measured source pressure strength, $|p_S|$, at

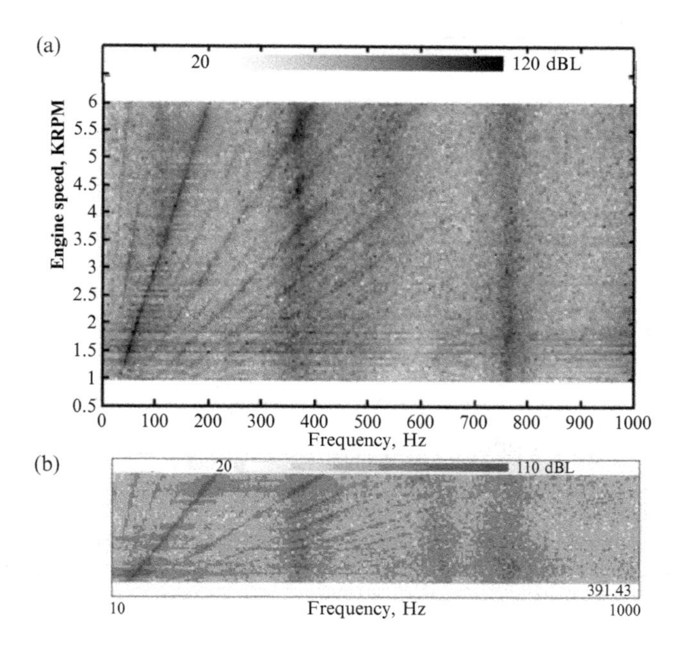

Figure 11.4 Campbell diagrams. (a) The predicted SPL at 0.141 m from the open end of an intake duct shown in Figure 11.5a, (b) the corresponding measured SPL (data by courtesy of Mark IV Systèmes Moteurs, France).

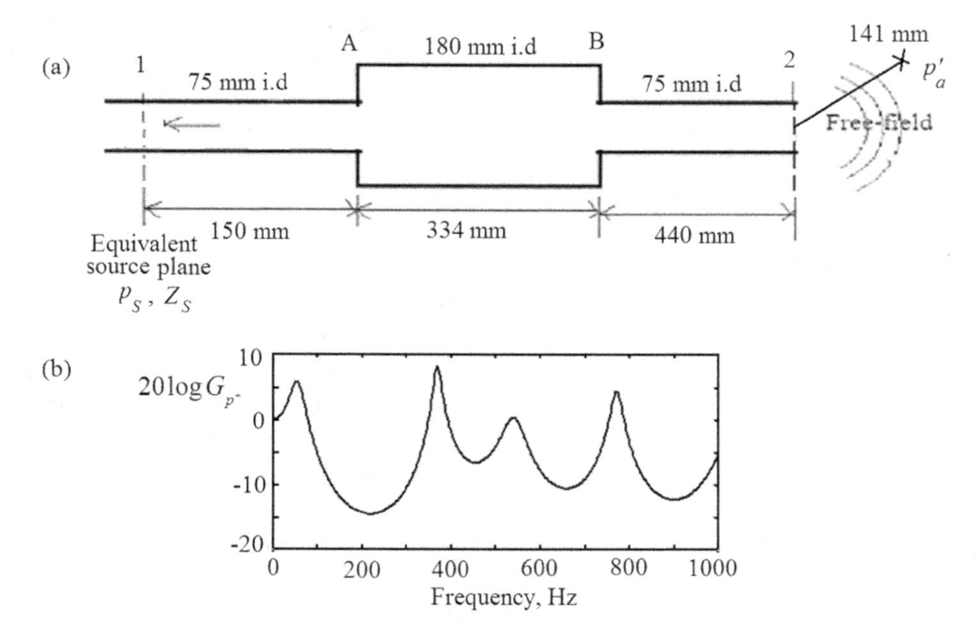

Figure 11.5 (a) An expansion chamber driven from an equivalent source plane, (b) system frequency response function $G_{p^+} = p_2^+/p_1^+$.

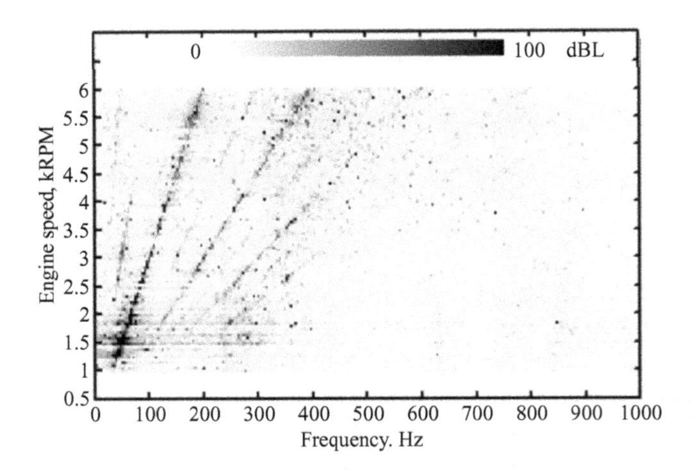

Figure 11.6 Campbell plot of the magnitude of the measured source pressure strength at an equivalent source plane of a four-cylinder four-stroke engine intake. Slopes of the inclined lines are $120/n$ rpm/Hz, $n = 1, 2, \ldots$ (data by courtesy of Mark IV Systèmes Moteurs, France).

various engine speeds is shown in Figure 11.6 in the Campbell diagram form. Since the engine is a four-stroke one, the discrete frequency components at integer multiples of half of the crankshaft frequency manifest as immersed in a background of random noise. The inclined lines in the Campbell diagram occur with slopes $120/n$ rpm/Hz, where $n = 1, 2, \ldots$ denotes the harmonic order. The orders n =1,2,4,6,8,10 are markedly visible in the speed range considered, $n = 4$ being the strongest. The predicted sound pressure levels at the target point (Figure 11.4a) are also dominant at these orders. Thus, the equivalent source provides a direct explanation of the order lines (the inclined lines) in the predicted Campbell diagram.

The system in Figure 11.5a is a special case of Figure 11.1a and results of Section 11.2.1 for SPL calculations are applicable. The wave transfer matrix in Equation (11.3) is computed by cascading the wave transfer matrices of duct 1-A, sudden area expansion at A, duct A-B, sudden area contraction at B and duct B-2. We use the model described in Section 9.4.3 to calculate the reflection coefficient matrix at plane 2 and, for example, the model described in Section 9.5.6 for the radiation transfer functions $K_1^{(j)}, j = 1, 2, \ldots, m$. Then the matrix $\mathbf{A}$ in Equation 11.6 is computed and, assuming the equivalent source is one-dimensional ($q = 1$), matrix $\mathbf{A}$ is partitioned as in Equation (11.13) and the transfer function $H^{(j)}(f)$ in Equation (11.44) is computed from Equation (11.29) for the cut-on modes $j = 1, 2, \ldots, m$. Finally, the acoustic pressure at the target point follows from $p'_a = \left(H^{(1)} + H^{(2)} + \cdots + H^{(m)}\right)p_S$. The assumption of a one-dimensional source is plausible because the cut-on frequency of the first non-axisymmetric higher order mode at the equivalent source plane is about 2500 Hz. This also applies for the intake duct, which implies that the SPL calculations are confined to the case $m = 1$ in the frequency range of interest. This considerably simplifies the computations, because plane wave models may be used for the reflection coefficient r_2, and the radiation transfer function, K_2.

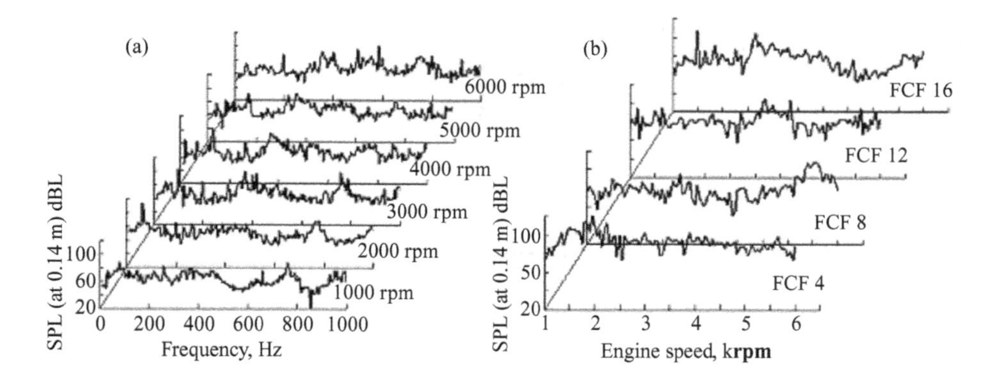

Figure 11.7 Cuts on the Campbell diagram in Figure 11.4a. (a) Horizontal cuts at various engine speeds. (b) Order cuts for various harmonics of the firing cycle frequency (FCF).

The predicted resonance lines in Figure 11.4a are not explainable by single duct resonances (Section 5.3), because the acoustic fields in the ducts are coupled strongly by wave reflections at the discontinuities. A suitable frequency response function to see the system resonances is $G_{p^+} = p_2^+/p_1^+$. This is shown in Figure 11.5b in dB units as function of frequency. Inspection of Figure 11.4a shows that, of the four resonances predicted in Figure 11.5b, the second and fourth are observed markedly throughout the speed range of the engine and are more intensive than the first and third, which tend be discernible in the background noise (impressed by the measured source) at relatively higher engine speeds.

Horizontal cuts of Figure 11.4a are shown in Figure 11.7a for several engine speeds. For fluid machinery having only few discrete operating points, it suffices to give SPL in this format. Order cuts of Figure 11.4a are shown Figure 11.7b for the major harmonic orders of the firing cycle frequency, $n = 4, 8, 12, 16$. When it is known a priori that some engine harmonics are annihilated in engine manifolds by interference, SPL order plots for the undestroyed harmonics in the frequency range of interest are usually sufficient for acoustic design purposes.

11.4 Insertion Loss

Insertion loss is defined as the change in the sound pressure level (SPL) at a target point when a passive component (usually a silencer or its internals) of a ductwork is modified or replaced by a new one, all other conditions independent of that component remaining the same. This concept, which is popular in the acoustic design of silencers, is illustrated schematically in Figure 11.8. Here, the component to be modified or replaced is referred to as the reference component and the insertion loss of the new component is given by

$$L_{IL} = L_{\text{ref}} - L_{\text{new}} \tag{11.46}$$

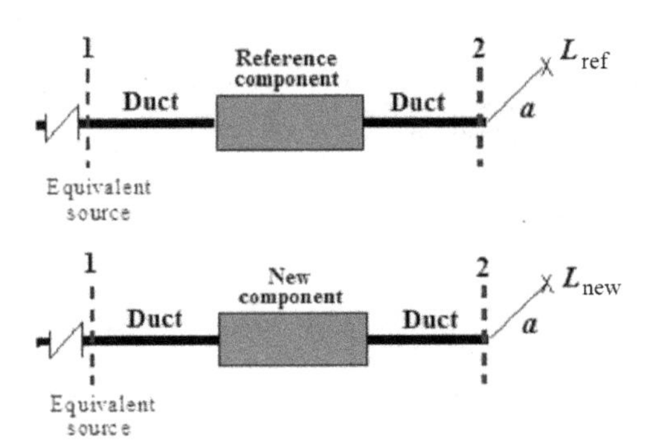

Figure 11.8 Definition of insertion loss: $L_{IL} = L_{ref} - L_{new}$.

where L_{ref} and L_{new} denote the SPL at the target point with the reference component and the new component mounted on the system, respectively, and L_{IL} denotes the insertion loss (of the new component with respect to the reference component). Thus, if a system prototype is available, the measurement of insertion loss is very simple. Its calculation is more difficult, because it is necessary to calculate the SPL at the target point twice. Furthermore, if the source process is markedly dependent on the load (the component to be modified), the source characteristics will have to be determined separately for each case, too.

Under certain conditions, however, insertion loss may be estimated approximately without knowledge of the source characteristics. This possibility is discussed in the next two subsections for systems driven by an equivalent one-port source. The two-port source cases may be analyzed similarly by reducing it to two one-ports as described in Section 11.2.2. We assume one-dimensional propagation at the equivalent source plane.

In Figure 11.9a, the wavy curve shows the computed insertion loss spectrum of the chamber in Figure 11.5a with respect to a reference system, which is the same of Figure 11.5a, except for the chamber which is replaced by a straight hard-walled uniform duct of length 334 mm and of 75 mm internal diameter. The calculations assume that the source pressure strength and impedance are the same for both systems. The waviness of the spectrum is due to the noise in the measured source impedance. The real and imaginary parts of the latter at 3000 rpm are shown in Figure 11.9b and c, respectively, as normalized (divided by) by the characteristic impedance at the source plane.

11.4.1 Noise Reduction

Referring to Figure 11.1a, for a one-dimensional source ($q = 1$), the following identity holds

$$p'_a = \frac{1}{\tau} \frac{p_S}{1 + \dfrac{1 - r_1}{1 + r_1} \zeta_S}, \tag{11.47}$$

where p'_a denotes the sound pressure at the target point, r_1 denotes the reflection coefficient at plane 1, the equivalent source plane, and τ denotes the transfer function

$$\tau = \frac{p'_1}{p'_a}. \tag{11.48}$$

Then, SPL at the target point may be expressed in dB(L) units as

$$L_{p'_a} = L_{p_S} - L_{NR} - 20\log_{10}\left|\frac{1+\zeta_S+(1-\zeta_S)r_1}{1+r_1}\right|, \tag{11.49}$$

where the source pressure level is given by

$$L_{p_S} = 20\log_{10}\left(\frac{\sqrt{2}|p_S|}{p_{\text{ref}}}\right) \tag{11.50}$$

and the source independent parameter

$$L_{NR} = 20\log_{10}|\tau| \tag{11.51}$$

is called noise reduction. The sound pressure levels calculated from Equation (11.49) may be converted to dB(A), dB(B) and dB(C) units by applying the appropriate corrections, for example, Figure 11.3 for dB(A).

Referring to Figure 11.8 and using Equation (11.49) in Equation (11.46), insertion loss of the new component may be expressed as

$$L_{IL} = L_{NR\text{new}} - L_{NR\text{ref}} + 20\log_{10}\left(\frac{|1+r_{1\text{ref}}|}{|1+r_{1\text{new}}|}\right) + C \tag{11.52}$$

where

$$C = 20\log_{10}\left(\frac{|p_S|_{\text{ref}}}{|p_S|_{\text{new}}}\right) + 20\log_{10}\left(\frac{|1+\zeta_S+(1-\zeta_S)r_1|_{\text{new}}}{|1+\zeta_S+(1-\zeta_S)r_1|_{\text{ref}}}\right). \tag{11.53}$$

In practical calculations this correction term is usually evaluated by assuming that the source impedance and strength are not substantially altered by the introduction of the new component. Then, Equation (11.53) simplifies to

$$C = 20\log_{10}\left(\frac{|1+\zeta_S+(1-\zeta_S)r_{1\text{new}}|}{|1+\zeta_S+(1-\zeta_S)r_{1\text{ref}}|}\right). \tag{11.54}$$

This shows that only the source impedance is needed for the calculation of insertion loss. If, furthermore, the change in the reflection coefficient r_1 does not result in a substantial change of C and the quotient in Equation (11.52) remains practically equal to unity, insertion loss of the new component may be estimated from

$$L_{IL} \approx L_{NR\text{new}} - L_{NR\text{ref}}. \tag{11.55}$$

However, two noise reduction runs are required.

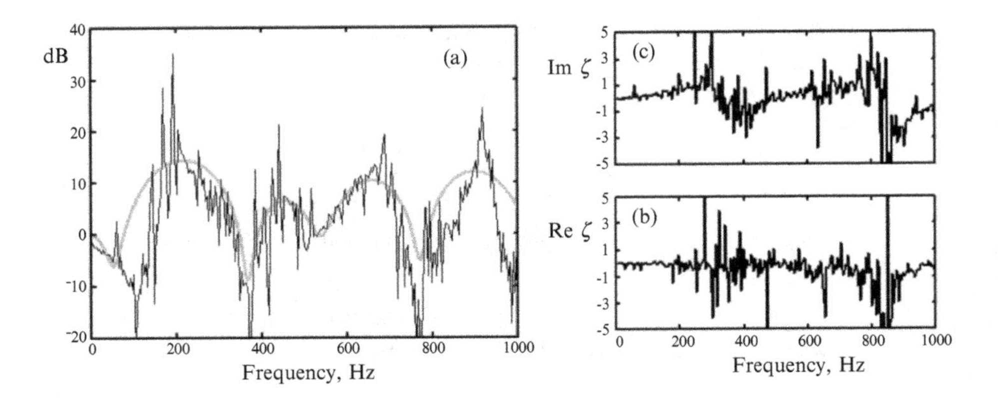

Figure 11.9 (a) A comparison of insertion loss and attenuation of the expansion chamber in Figure 11.5a at 3000 rpm. Smooth curve in gray: attenuation, wavy curve: insertion loss of the chamber relative to a straight lossless duct of length 334 mm and 75 mm i.d. (b,c) Real and imaginary parts of the measured source impedance at 3000 rpm.

11.4.2 Attenuation

Referring again to Figure 11.8, let p'_a denote the acoustic pressure at the target point. Contribution of single mode to p'_a is given by Equation (11.29). If m modes are propagating at plane 2, the sound pressure at the target point is given by the sum $p'_a = p'^{(1)}_a + p'^{(2)}_a + \cdots + p'^{(m)}_a$, that is,

$$p'_a = \frac{p_S}{\alpha(1 + \zeta_S + (1 - \zeta_S) r_1)} \sum_{j=1}^{m} \left(\frac{\beta_2^{(j)}}{K_2^{(j)}} \right) \tag{11.56}$$

or, in the dB(L) unit,

$$L_{p'_a} = L_{p_S} - A + 20 \log_{10} \left| \sum_{j=1}^{m} \left(\frac{\beta_2^{(j)}}{K_2^{(j)}} \right) \right| - 20 \log_{10} |1 + \zeta_S + (1 - \zeta_S) r_1|. \tag{11.57}$$

Here,

$$A = 20 \log_{10} |\alpha| \tag{11.58}$$

is called attenuation of acoustic path 1-2. Note that, for calculation of A, acoustic path 1-2 need not necessarily be modeled by using only one-dimensional elements; sound propagation has to be one-dimensional only at planes 1 and 2. Consequently, attenuation can be calculated to various degrees of accuracy, depending on the accuracy of the mathematical model used for path 1-2.

Using Equation (11.57) in Equation (11.46), insertion loss may be expressed as

$$L_{IL} = A_{\text{new}} - A_{\text{ref}} + 20 \log_{10} \left(\frac{\left| \sum_{j=1}^{m} \left(\frac{\beta_2^{(j)}}{K_2^{(j)}} \right) \right|_{\text{ref}}}{\left| \sum_{j=1}^{m} \left(\frac{\beta_2^{(j)}}{K_2^{(j)}} \right) \right|_{\text{new}}} \right) + C, \tag{11.59}$$

where the correction term C is given by Equation (11.53) or Equation (11.54), and may be neglected on the premise that the conditions at the source are not substantially altered by the introduction of the new component. Then Equation (11.59) may expressed approximately as

$$L_{IL} \approx A_{\text{new}} - A_{\text{ref}} \tag{11.60}$$

and if we choose the reference component to be a lossless straight duct, then this becomes simply

$$L_{IL} \approx A_{\text{new}} \tag{11.61}$$

In Figure 11.9a, the attenuation spectrum of the chamber in Figure 11.5a at 3000 rpm (the smooth curve) is compared with the measured insertion loss of the chamber. Considering the relative simplicity of calculation of attenuation, the correlation may be considered quite useful.

11.5 Multi-Modal Transmission Loss Calculations

The definition of transmission loss given in Equation (5.2) is still valid for multi-modal propagation in the inlet and outlet ducts. The time-averaged acoustic power crossing the inlet or outlet planes, which here are called planes 1 and 2, respectively, may be expressed from Equation (6.116) as

$$W_k = \sum_{j=1} \left(W_k^{(j)+} + W_k^{(j)-} \right) = \frac{2S_k}{c_k \rho_k} \sum_{j=1} \left(\Lambda_k^{(j)+} \left| p_k^{(j)+} \right|^2 + \Lambda_k^{(j)-} \left| p_k^{(j)-} \right|^2 \right). \tag{11.62}$$

Here, subscript $k = 1, 2$ denotes planes 1 and 2, respectively, superscript "(j)" denotes the cut-on mode number and

$$\Lambda_k^{(j)\mp} = \left(1 + M_k^2\right) A_k^{(j)\mp} + M_k + M_k A_k^{(j)\mp} A_k^{(j)\mp}, \tag{11.63}$$

where $A_1^{(j)\mp}$ denote mode j admittance coefficients defined in Equation (6.57) and M_k denotes the Mach number of the mean-flow velocity at plane k. For cut-on modes, the modal admittance coefficients and, hence, $\Lambda_k^{(j)\mp}$ are real (ducts being assumed to be hard-walled). Thus, if q modes are propagating at plane 1 and m modes are propagating at plane 2, Equation (5.2) may be expressed as

$$TL(f) = 10 \log_{10} \left(\frac{\sum_{j=1}^{q} \Lambda_1^{(j)+} \left| \beta_1^{(j)} \right|^2}{\sum_{j=1}^{m} \Lambda_2^{(j)+} \left| \beta_2^{(j)} \right|^2} |\alpha|^2 \right)_{r_2^{(j)}=0} + 10 \log_{10} \left(\frac{S_1 c_2 \rho_2}{S_2 c_1 \rho_1} \right), \tag{11.64}$$

where the subscript "$r_2^{(j)} = 0$" means that the quantity in brackets is to be evaluated by assuming that the outlet plane is anechoic to all propagating modes $j = 1, 2, \ldots, m$. Mode participation factors $\beta_k^{(j)}$ and the reflection coefficients $r_k^{(j)}$ are defined, respectively, as

$$\beta_k^{(j)} = \frac{p_k^{(j)+}}{p_k^{(1)+}}, r_k^{(j)} = \frac{p_k^{(j)-}}{p_k^{(j)+}}, k = 1, 2. \tag{11.65}$$

In view of Equation (11.6), mode participation factors at planes 1 and 2 are related as

$$\begin{bmatrix} 1 \\ \beta_2^{(2)} \\ \vdots \\ \beta_2^{(b)} \end{bmatrix} = \mathbf{A}^{-1} \begin{bmatrix} 1 \\ \beta_1^{(2)} \\ \vdots \\ \beta_1^{(b)} \end{bmatrix} \alpha, \tag{11.66}$$

where, by definition, $\beta_1^{(q+1)} = \cdots = \beta_1^{(m)} = 0$ if $b = m > q$ or $\beta_2^{(m+1)} = \cdots = \beta_2^{(q)} = 0$ if $b = q \geq m$. Note that, if $q = 1$, then the mode participation factors $\beta_2^{(j)}$ are given by Equation (11.28). As can be deduced from Equation (11.6), the condition $r_2^{(j)} = 0$, $j = 1, 2, \ldots, m$, implies that $\mathbf{A} = \mathbf{S}_{11}$ in Equation (11.66). The mode participation factors $\beta_1^{(j)}, j = 1, 2, \ldots, q$, depend on the source. Thus, we see that transmission loss is in general not a source-independent parameter, although statements describing it to be so are ubiquitous in the literature. An exception is the case of $q = 1$, since then $\beta_1^{(1)} = 1$ and Equation (11.34) yields $\beta_2^{(j)} = \left(\mathbf{S}_{11}^{-1}\right)_{j1}\alpha, j = 1, 2, \ldots, m$.

For $q > 1$, transmission loss is usually computed by introducing the hypothesis that the incident cut-on modes at plane 1 are uncorrelated and transmit equal time-averaged acoustic power. This hypothesis, which is a general feature of propagating random acoustic fields, imposes the conditions $W_1^{(1)+} = W_1^{(j)+}, j = 2, 3, \ldots, q$. Hence, the mode participation factors at plane 1 are fixed as:[1]

$$\beta_1^{(j)} = \sqrt{\frac{\Lambda_1^{(1)+}}{\Lambda_1^{(j)+}}}, \tag{11.67}$$

since $\Lambda_1^{(j)+} > 0$ for cut-on modes for subsonic mean flow Mach numbers. Hence, Equation (11.64) may be expressed as

$$TL(f) = 10\log_{10}\left(\frac{q|\alpha|^2}{1 + \sum_{j=2}^{m} \frac{\Lambda_2^{(j)+}}{\Lambda_2^{(1)+}} \left|\beta_2^{(j)}\right|^2} \right)_{r_2^{(j)}=0} + C_{TL}. \tag{11.68}$$

This formula, which is exact when $q = 1$, may be tractable to practical calculations up to several cut-on higher order modes at plane 2, but, as m increases, the modality of the system model must also be increased. Since the number of cut-on modes increases at a rate proportional to about frequency squared (see the next section), evaluation of Equation (11.68) is likely to be hindered at relatively high frequencies due to the high

[1] For the implementation of the hypothesis, only the incident modes are relevant, because the reflected modes are correlated to the incident modes by deterministic local properties of the system.

density of cut-on modes and the resulting large complex matrices which are to be inverted. Here, a plausible strategy is to use Equation (11.68) for up to a threshold number of cut-on modes, m_1, say, for which it can be evaluated without computational problems and, for cut-on modes of orders higher than this threshold, to invoke the hypothesis of equipartition of time-averaged acoustic energy. Then $W_2^{(j)+} = W_2^{(m_1)+}$, $j > m_1$, and this fixes the mode participation factors at plane 2 as $\left|\beta_2^{(j)+}\right|^2 = \left(\Lambda_2^{(m_1)+}/\Lambda_2^{(j)+}\right)\left|\beta_2^{(m_1)}\right|^2, j > m_1$. Thus, Equation (11.68) may be evaluated approximately as

$$TL(f) = 10\log_{10}\left(\frac{q|\alpha|^2}{\sum_{j=1}^{m_1}\frac{\Lambda_2^{(j)+}}{\Lambda_2^{(1)+}}\left|\beta_2^{(j)}\right|^2 + (m-m_1)\frac{\Lambda_2^{(m_1)+}}{\Lambda_2^{(1)+}}\left|\beta_2^{(m_1)}\right|^2}\right)_{r_2^{(j)}=0} + C_{TL}. \tag{11.69}$$

It should be noted that, this approximation presumes that the mode participation factors $\beta_2^{(j)}$ are uncorrelated with $\beta_1^{(j)}$ for modes of order higher than threshold order m_1, since it renders Equation (11.66) redundant for these modes.

The overall transmission loss (see Section 5.2.2) in a frequency interval $f_1 \leq f \leq f_2$ is computed from

$$\bar{TL}(f_1,f_2) = 10\log_{10}\left(\frac{\int_{f_1}^{f_2} W_1^+ \mathrm{d}f}{\int_{f_1}^{f_2} W_2^+ \mathrm{d}f}\right)_{r_2^{(j)}=0} \tag{11.70}$$

Again, assuming q equal time-averaged power modes are propagating at the inlet plane and m modes are propagating at the outlet plane, substitution for the incident modal acoustic powers from Equation (11.62) yields

$$\bar{TL}(f_1,f_2) = 10\log_{10}\left(\frac{q\int_{f_1}^{f_2} S_{p_1^{(1)+}}(f)\mathrm{d}f}{\sum_{j=1}^{m}\frac{\Lambda_2^{(j)+}}{\Lambda_2^{(1)+}}\int_{f_1}^{f_2}\left|\beta_2^{(j)}\right|^2 S_{p_2^{(1)+}}(f)\mathrm{d}f}\right)_{r_2^{(j)}=0} + C_{TL} \tag{11.71}$$

since $S_{p_2^{(j)+}}(f) = \left|\beta_2^{(j)}\right|^2 S_{p_2^{(1)+}}(f)$. This may be expressed as

$$\bar{TL}(f_1,f_2) = 10\log_{10}\left(\frac{\int_{f_1}^{f_2} S_{p_1^{(1)+}}(f)\mathrm{d}f}{\int_{f_1}^{f_2} 10^{-\frac{TL(f)}{10}} S_{p_1^{(1)+}}(f)\mathrm{d}f}\right) \tag{11.72}$$

where $TL(f)$ is given by Equation (11.68) or (11.69). In practical calculations, it is convenient to assume that the incident power spectral density at the inlet section is constant in the frequency range of interest. Then, Equation (11.72) simplifies to

$$\bar{TL}(f_1,f_2) = 10\log_{10}(f_2 - f_1) - 10\log_{10}\left(\int_{f_1}^{f_2} 10^{-\frac{TL(f)}{10}}\mathrm{d}f\right). \tag{11.73}$$

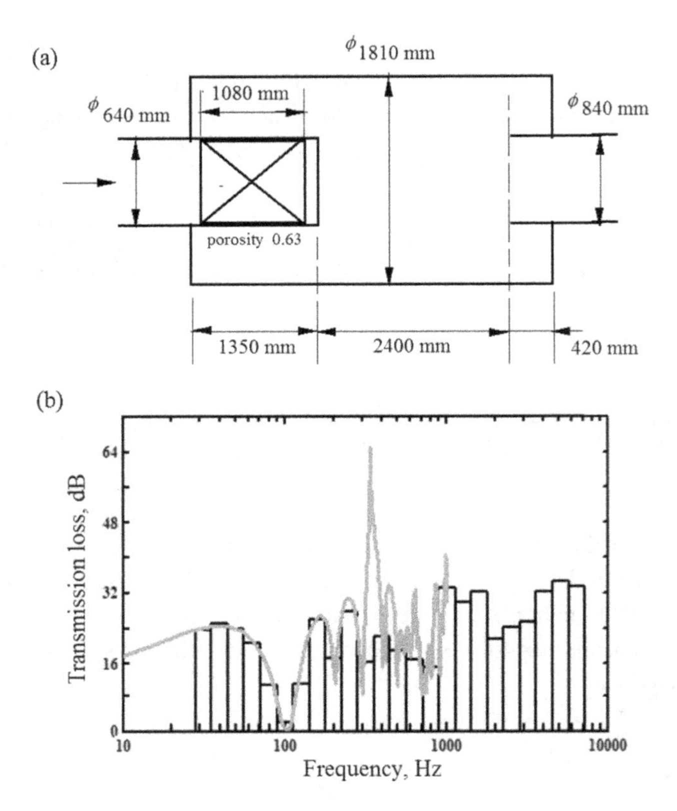

Figure 11.10 (a) Sketch of a silencer. (b) Transmission loss of the silencer. Bars: overall levels in 1/3 octave bands; curve: discrete frequency levels in the modality range of the outlet duct.

This can be shown to be formally the same as Equation (5.28), except that $TL(f)$ is now given by Equation (11.68) or (11.69).

The foregoing formula may be used to compute transmission loss in octave bands. An example is shown in Figure 11.10 for a rather large expansion chamber with a diffuser (which is typical of industrial vent silencers) receiving fluid at 250 °C and 0.1 Mach number. The threshold modality in the outlet duct is set arbitrarily at four and transmission loss is computed in 1/3 octave bands, beginning with center frequency of 31.5 Hz. The cut-on frequencies of the higher order modes in the outlet duct are, in Hz units, 395, 655, 821, 900, 1140,... Also shown in Figure 11.10b is the transmission loss of the silencer with 5 Hz resolution in the 1000 Hz range. The number of cut-on modes increases rapidly with frequency (about 324 modes cut on in the outlet duct at 1000 Hz) and the average levels become more meaningful for engineering purposes.

11.6 In-Duct Sources Characterized by Acoustic Power

The measurement of equivalent source characteristics, ζ_S and $\mathbf{p}_S$, defined in Equation (11.10) may not be practicable for all $q > 1$. To see this, let us assume that the equivalent source plane lies in a uniform hard-walled circular duct. The number of cut-on modes in one direction at a given frequency f can then be estimated by using the Rice formula [2]:

$$N = \mathrm{int}\left\{1 + \left(1 + \frac{\pi f D}{2c_o\sqrt{1 - M_o^2}}\right)\left(\frac{\pi f D}{2c_o\sqrt{1 - M_o^2}}\right)\right\}, \tag{11.74}$$

where D denotes the duct diameter, M_o denotes the mean flow Mach number, c_o denotes the speed of sound and "int" means the integer part. For example, if $c_o = 500$ m/s and $M_o = 0.1$, then, for example, $N = 1$ for $f < 3500$ Hz , if $D = 0.05$ m, which is typical of ducts in passenger car exhaust systems. The frequencies of interest for acoustic design of such systems are usually lower than this and it may suffice to consider only one-dimensional propagation ($q = 1$) at the equivalent source plane.[2] But in many industrial installations, duct diameters can be larger and $D = 0.5$ m is not an uncommon value. For this duct size, assuming the same values for c_o and M_o, the number of cut-on modes transpires to be $N = 8$ at 1000 Hz, which involves 72 unknowns in the equivalent source equation, Equation (11.10). Furthermore, if the frequency range is required to be extended to 8 kHz, say, Equation (11.74) predicts $q = 173$!

On the other hand, standardized procedures are widely practiced in industry for the measurement of sound power in unsilenced fluid machinery inlet and outlet ducts. This data is not in general sufficient by itself for the prediction of the characteristics of the in-duct sound source. But, in many practical silencer design problems, it may be the only information available about the source. So, it is expedient to look at the extent to which it can be useful in sound pressure level predictions.

Let $W_S(f)$ be the given unsilenced time-averaged in-duct acoustic power of the source, which is usually determined by using a test duct. It is assumed that this remains substantially the same when a silencer is connected to the system (usually through the test duct). So, referring to Figure 11.1, we define plane 1 with $W_1 = W_S(f)$. Then, if q modes are propagating at plane 1 and m modes are propagating at plane 2 (radiation plane of the silenced system), the time-averaged acoustic power at plane 2 can be written, using Equation (11.62), as

$$W_2 = \frac{c_1\rho_1 S_2}{c_2\rho_2 S_1}\left(\frac{\sum_{j=1}^{m}\left(\Lambda_2^{(j)+} + \Lambda_2^{(j)-}\left|r_2^{(j)}\right|^2\right)\left|\beta_2^{(j)}\right|^2}{\sum_{j=1}^{q}\left(\Lambda_1^{(j)+} + \Lambda_1^{(j)-}\left|r_1^{(j)}\right|^2\right)\left|\beta_1^{(j)}\right|^2}\right)\frac{W_S}{|\alpha|^2}, \tag{11.75}$$

[2] This does not necessarily mean that sound propagation is to be one-dimensional everywhere in the system. Higher-order modes can influence acoustic transmission at the discontinuities even if they are evanescent and may cut-on at parts of the system having larger dimensions. However, the wave transfer matrix of a multi-modal acoustic path can be contracted to one dimension, if sound propagation at the inlet and outlet of the path is one-dimensional at the frequencies of interest (see, for example, Section 7.8).

where the mode participation factors $\beta_k^{(j)}$ and the reflection coefficients $r_k^{(j)}$ are as defined in Equation (11.65) and the mode participation factors are related by Equation (11.66). Implementation of Equation (11.75) may be simplified as described in the previous section, by invoking the principle of equipartition of time-averaged acoustic energy after a threshold for the number of correlated cut-on modes at plane 2 is exceeded. Then Equation (11.67) holds and Equation (11.75) may be expressed approximately as

$$W_2 = \frac{c_1\rho_1 S_2}{c_2\rho_2 S_1}\left(\frac{\sum_{j=1}^{m_1}\left(\Lambda_2^{(j)+}+\Lambda_2^{(j)-}\left|r_2^{(j)}\right|^2\right)\left|\beta_2^{(j)}\right|^2+(m-m_1)\left(\Lambda_2^{(m_1)+}+\Lambda_2^{(m_1)-}\left|r_2^{(m_1)}\right|^2\right)\left|\beta_2^{(m_1)}\right|^2}{|\alpha|^2\sum_{j=1}^{q}\left(1+\frac{\Lambda_1^{(j)-}}{\Lambda_1^{(j)+}}\left|r_1^{(j)}\right|^2\right)\Lambda_1^{(1)+}}\right)W_S \tag{11.76}$$

This is similar to Equation (11.69), except for the terms involving reflection coefficients. Neglecting the effects of wave reflections at the source plane gives an estimate of W_2 as

$$W_2 = 10^{-\frac{C_{TL}}{10}}\left(\frac{\sum_{j=1}^{m_1}\frac{\Lambda_2^{(j)+}}{\Lambda_2^{(1)-}}\left|\beta_2^{(j)}\right|^2+(m-m_1)\frac{\Lambda_2^{(m_1)+}}{\Lambda_2^{(1)-}}\left|\beta_2^{(m_1)}\right|^2}{q|\alpha|^2}\right)W_S+\eta_2 W_S, \tag{11.77}$$

where

$$\eta_2 = \frac{10^{-\frac{C_{TL}}{10}}}{q|\alpha|^2}\left(\sum_{j=1}^{m_1}\frac{\Lambda_2^{(j)-}}{\Lambda_1^{(1)+}}\left|r_2^{(j)}\right|^2\left|\beta_2^{(j)}\right|^2+(m-m_1)\frac{\Lambda_2^{(m_1)-}}{\Lambda_1^{(1)+}}\left|r_2^{(m_1)}\right|^2\left|\beta_2^{(m_1)}\right|^2\right) \tag{11.78}$$

denotes the contribution of the reflection coefficients at the outlet plane to the total time-averaged acoustic power there as fraction of the source power. The expression in the large brackets in Equation (11.77) is the reciprocal of the expression in the large brackets in Equation (11.69), but the mode participation factors $\beta_2^{(j)}$ are now determined by Equation (11.66), since the boundary condition at the outlet plane may not be anechoic. Nevertheless, it may be approximated by the transmission loss of the system, since the effects of reflection at plane 2 are deferred to η_2. Then, Equation (11.77) may be expressed as

$$W_2 = \left(10^{-\frac{TL(f)}{10}}+\eta_2\right)W_S, \tag{11.79}$$

where $TL(f)$ is given by Equation (11.69). Time-averaged sound power is usually expressed in dB level units by the formula

$$L_W = 10\log_{10}\left(\frac{W}{W_{\mathrm{ref}}}\right), \tag{11.80}$$

where $W_{ref} = 10^{-12}$ watt and L_W is the sound power level. Thus, Equation (11.79) may be expressed as

$$L_{W_2} = L_{W_S} + 10\log_{10}\left(10^{-\frac{TL(f)}{10}} + \eta_2\right). \tag{11.81}$$

The sound pressure level at the target point, $L_{p'_a}$, may now be calculated by using Equation (9.103). On introducing the definitions of sound pressure level and sound power level given in Equations (11.45) and (11.80), respectively, Equation (9.103) may be written in the present notation as

$$L_{p'_a} = L_{W_2} - 10\log_{10}\left(\hat{\Omega}R_0^2\right) + 10\log_{10}\left(\rho_o c_o \frac{W_{ref}}{p_{ref}^2}\right). \tag{11.82}$$

The last term on the right-hand side is usually neglected on the premise that it is equal to $\rho_o c_o/400$ and the characteristic impedance of dry air is 405 Rayl at normal atmospheric conditions (20 °C and 1 bar). In-duct power levels of unsilenced fluid machinery sound sources may be obtained from the manufacturers or, as a first approximation alternative to manufacturers' data, from correlations published in the literature. They are usually given in 1/1 octave bands, which means that the level terms in Equations (11.81) and (11.82) are to be understood as overall averages in 1/1 octave bands.

In industrial engine exhaust applications, it is customary to specify the full-load octave sound pressure level spectrum at a specified distance from the outlet of a uniform test duct (usually the exhaust duct) connected to the engine, on the understanding that the far-field radiation is approximately isotropic and the test duct may be assumed to be lossless and anechoic. This data can be converted to L_{W_S} by using Equation (11.82), since the second term on the right-hand side of Equation (11.81) vanishes for the test duct. Once L_{W_S} is computed, Equations (10.81) and (10.82) may be used with a silencer added to the exhaust line. For example, for a 1750 kVA engine operating steadily at 1500 rpm at a full load, the measured sound level spectrum at 1.5 m from a test duct outlet, $\left(L_{p'_a}\right)_{test}$, say, is provided as 110, 120, 115, 106, 105, 108, 109, 105 dB(L) at 1/1 octave center frequencies of 63, 125, 250, 500, 1000, 2000, 4000, 8000 Hz, respectively. Hence, assuming normal atmospheric conditions and free-field radiation at the measurement point, the source sound power level is found as 125, 135, 130, 121, 120, 123, 124, 120 dB in the same octave bands. Let the target point after a silencer is mounted on the exhaust duct be again at 1.5 m from the tailpipe outlet in a free field. Neglecting reflections at the tailpipe outlet, the target sound pressure level may be expressed from Equation (11.81) as $L_{p'_a} = \left(L_{p'_a}\right)_{test} - TL_{silencer}$, where $TL_{silencer}$ denotes the transmission loss of the silencer in the above octave bands. An advantage of this approach is that the effect tailpipe outlet reflections is separated in Equation (10.81) and may be studied by computing η_2, if required.

11.6.1 The ASHRAE Method

An important corollary to Equation (11.81), which may be confirmed by inspection of the analysis leading to it, is that it remains valid for any two-port acoustic element, provided that we replace W_S by W_{in} and W_2 by W_{out}, where W_{in} and W_{out} denote the time-averaged acoustic powers at the inlet and outlet ports of the element. Thus, for an acoustic two-port, Equation (11.81) may written in the form

$$L_{W_{\text{out}}} = L_{W_{\text{in}}} - \Delta L_W \tag{11.83}$$

where

$$\Delta L_W = -10 \log_{10}\left(10^{-\frac{TL(f)}{10}} + \eta_{\text{out}}\right) \tag{11.84}$$

Equation (11.83) forms the basis of a method of acoustic analysis initially developed for air distribution and ventilation systems. This method, which is known as the ASHRAE method [3] (also called the power method), consists representing specific two-port ductwork components by their ΔL_W characteristics, usually in 1/1 octave bands. Junctions of two-ports are modeled by assuming that the input power divides into branches in proportion to the ratio of single branch cross-sectional area to the total cross-sectional area of the branches. Equation (11.84) shows that ΔL_W accounts for the transmission loss of the component, as well as the effects of reflections at its outlet, adequately accurately. But these parameters are strictly determined by the wave solutions described in previous sections of this book. For this reason, implementation of the ASHRAE method is subject to a lot of heuristics. The user has to rely on tables provided for the determination of ΔL_W. These are based on results of theoretical and experimental research over the years, but are specific to particular conditions of the borrowed studies and generally neglect the effects of wave reflections.

In applications to air distribution and ventilation systems, Equation (11.83) is usually modified as

$$L_{W_{out}} = L_{W_{in}} - \Delta L_W + L_{W_{\text{flow}}}, \tag{11.85}$$

where $L_{W_{\text{flow}}}$ denotes the air-flow noise power level, which is also determined from tables provided for specific components. This modification is justified in fan operated systems, because the acoustic efficiency of pressure loading on rotating blades, which is the primary source mechanism of fan noise, is of $O(M^3)$, which is about the same order as the efficiency of vortex generated noise in air flows, where M denotes the mean flow Mach number (see Section 1.1.2).

For an overview of the working of the ASHRAE method, consider the air distribution system shown in Figure 11.11, where a fan supplies air to a room and the sound pressure level at the target point is required to comply with environmental noise regulations. We assume that there is no noise entering the room or the ducts through their walls and vice versa, and that there is no air return outlet in the room. So, as usual, we look for the sound pressure level at the target point due to the primary source, the fan, only. The contributions of other noise components may be calculated by using the appropriate tables given in the ASHRAE Handbook [3].

Let us write Equation (11.85) for the relevant components:

Duct 1-2: $L_{W_2} = L_{W_S} - (\Delta L)_{1-2}$
Elbow: $L_{W_3} = L_{W_2} - (\Delta L)_{2-3} + (L_{W_{\text{flow}}})_{2-3}$

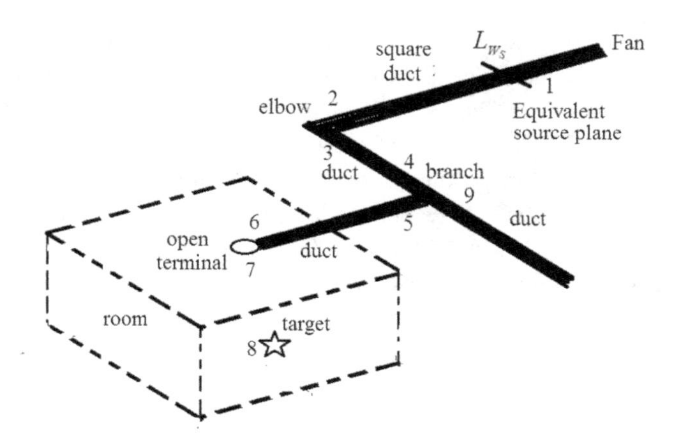

Figure 11.11 A simple air distribution system.

Duct 3-4: $L_{W_4} = L_{W_3} - (\Delta L)_{3-4}$
Branch 4-5: $L_{W_5} = L_{W_4} - (\Delta L)_{4-5} + (L_{W_{\text{flow}}})_{4-5}$
Duct 5-6: $L_{W_6} = L_{W_5} - (\Delta L_{5-6})$
Terminal 6-7: $L_{W_7} = L_{W_6} - (\Delta L)_{6-7}$

Finally, we calculate the sound pressure level at the target point from an equation which is basically in the form of Equation (11.82), but is corrected for the acoustic properties of the room, that is:

Room 7-8: $L_{p'_a} = L_{W_7} - (L_{\text{rad}})_{7-8}$.

Summing all these equations gives the required target sound pressure level as

$$L_{p'_a} = L_{W_S} - \left\{(\Delta L)_{1-2} + (\Delta L)_{2-3} + (\Delta L)_{3-4} + (\Delta L)_{4-5} + (\Delta L_{5-6}) + (\Delta L)_{6-7}\right\} + \left\{(L_{W_{\text{flow}}})_{2-3} + (L_{W_{\text{flow}}})_{4-5}\right\} - (L_{\text{rad}})_{7-8}, \tag{11.86}$$

where all terms on the right-hand side are determined from the tables given in the ASHRAE Handbook [3]. The need for a silencer will depend on how $L_{p'_a}$ compares with the corresponding environmental criterion at the target point and the difference will determine the type and size of the silencer to be used. It is usually sufficient to use dissipative silencers and lined ducts. Obviously, the in-duct sound power level of the fan plays a major role in the silencing requirements, and the selection of a sufficiently quiet one may be cost-effective. The parameters governing the in-duct sound power of fans seem to coalesce as in the following formula

$$L_{W_S} = L_{\text{fan}} + 10\log_{10}(\dot{Q}_o) + 20\log_{10}(\Delta p_o) + C_{\text{bpf}} + C_\eta. \tag{11.87}$$

Here, $\dot{Q}_o$ denotes the volume flow rate (m^3/s), Δp_o denotes the static pressure of the fan (kPa), C_η denotes a correction for the static efficiency ($\eta = \dot{Q}_o \Delta p_o / 1002\times$ brake kW) of the fan, which is zero for $\eta \geq 90\%$ and is increased by 3 dB for the first 5% and next 10% decreases in the efficiency; C_{bpf} denotes a correction to be added to level of the frequency band including the blade passing frequency (bpf = fan rotational speed (Hz)$\times$ number of blades), for example, for vaneaxial type fans, $C_{bpf} = 8$ and bpf $= 125$ Hz, if blade information not available; and L_{fan} denotes 1/1 octave base power levels for specific fan design, for example, for vaneaxial fans with hub ratio in the range 0.4–0.6, the recommended values are 98, 88, 81, 88, 86, 81, 75, 73 at center frequencies 63, 125, 250, 500, 1000, 2000, 4000, 8000 Hz, respectively. Correlations for other fan types are also given in the Handbook [3].

References

[1] P. Joseph, C.L. Morfey and C.R. Lowis, Multi-mode sound transmission in duct with flow, *J. Sound. Vib.* 264 (2003) 523–544.

[2] E.J. Rice, Modal density function and the number of propagating modes in ducts, NASA TM X-73539, (1976).

[3] ASHRAE, *Application Handbook*, Chapter 46: Sound and vibration control, (Atlanta GA: American Society of Heating, Refrigerating and Air Conditioning Engineers Inc., 1999).

12 Measurement Methods

12.1 Introduction

Acoustic measurements in ducts are usually undertaken for validation or refinement of mathematical models or for formulation of empirical models of passive and active flow duct components. In this chapter we present some of the contemporary methods used in duct acoustics measurements. Our treatment assumes familiarity with principles of acquisition of acoustic signals by microphones and instrumentation requirements, which are usually given in technical publications and catalogues of the manufacturers. Fast Fourier transform (see Section 1.2.3) is normally used for the calculation of Fourier transform or the power spectral density (see Section 1.2.4) of the measured signals.

Measurements may be carried out with the actual source, when available, or a test source driving the system. The usual choice for a test source is the loudspeaker (or pressure driver). The actual source is needed, however, for the measurement of equivalent characteristics of ducted sources. Measurements taken on steady running fluid machinery can be transformed directly to the frequency domain by Fourier analysis. Acoustic tests on fluid machinery having a range of operational speeds, such as automobile engines, are often undertaken by gradually accelerating (run-up) or decelerating (run-down) the engine stepwise through the speed range. In such tests, the sound pressure signals are taken simultaneously with crankshaft rotation pulses, so that the engine speeds corresponding to the segments of the time records can be determined and analyzed in the frequency domain.

On the other hand, if a ducted loudspeaker is used as a source, the duct can be excited by using several signal forms. The popular excitation techniques utilize steady random signals, but a sine-sweep signal or pseudo-random bit sequence can also be used effectively. The acoustic fields set up by such signals can be analyzed in the frequency domain in a unified manner again by using the Fourier transform technique.

12.2 Measurement of In-Duct Acoustic Field

The sound pressure at a particular point in the acoustic field of a ducted source can, in principle, be measured by placing a microphone at the target point, provided that the presence of the microphone does not modify the acoustic field to be measured. This condition is normally satisfied with modern measuring microphones if the target point is in a free field. In-duct measurements are commonly made by mounting microphones or microphone probes on the duct walls.

Thus, insertion loss can be measured by using one microphone and taking two sound pressure measurements at the target point, and noise reduction can be measured by taking simultaneous measurements from two microphones located at the two reference points in the acoustic field.

But parameters such as transmission loss and attenuation cannot be measured directly from acoustic pressure measurements because they require information about the acoustic pressure wave components incident at two reference planes. Similarly, if we wanted to measure the wave transfer matrix of a passive flow duct component across its inlet and outlet sections, we cannot do this by only measuring the acoustic pressure at these sections. In such situations, it is necessary to also measure the acoustic particle velocity at the reference points. Acoustic particle velocity can be measured by using methods such as particle image velocimetry or laser Doppler velocimetry. However, these techniques and their several derivatives are used at present in fundamental research. It transpires that techniques based on making sound pressure measurements by using multiple number of wall-mounted microphones are more convenient for engineering purposes.

12.2.1 Multi-Modal Wave Field Decomposition

From Equation (6.66), the acoustic pressure at any point in a uniform straight duct carrying a uniform mean flow is given in frequency domain by

$$p'(\bullet, x) = \mathbf{\Phi}(\bullet) \left(\mathrm{e}^{ik_o\mathbf{K}^+x}\mathbf{P}^+(0) + \mathrm{e}^{ik_o\mathbf{K}^-x}\mathbf{P}^-(0) \right), \tag{12.1}$$

where the notation used is same as in Section 6.4.2, except that the subscript "1" is dropped from x_1 for simplicity of the notation. We assume that n modes are propagating at the frequency of interest. Then,

$$\mathbf{P}^{\mp}(x) = \begin{bmatrix} p_1^{\mp}(x) \\ p_2^{\mp}(x) \\ \vdots \\ p_n^{\mp}(x) \end{bmatrix}, \tag{12.2}$$

where $p_j^{\mp}$ denote the pressure wave components for mode $j = 1, 2, \ldots, n$, which are considered as unknowns on the right-hand side of Equation (12.1). The basic idea underlying the multiple wall-mounted microphone methods is this: if we measure the sound pressure at $2n$ axial locations $x = x_1, x_2, \ldots, x_{2n}$ in a duct, then we should, in principle, be able to solve $\mathbf{P}^{\mp}(0)$ from the set of $2n$ equations

$$\begin{bmatrix} p'(\bullet_1,x_1) \\ p'(\bullet_2,x_2) \\ \vdots \\ p'(\bullet_{2n},x_{2n}) \end{bmatrix} = \begin{bmatrix} \mathbf{\Phi}(\bullet_1)\mathrm{e}^{\mathrm{i}k_o\mathbf{K}^+x_1} & \mathbf{\Phi}(\bullet_1)\mathrm{e}^{\mathrm{i}k_o\mathbf{K}^-x_1} \\ \mathbf{\Phi}(\bullet_2)\mathrm{e}^{\mathrm{i}k_o\mathbf{K}^+x_2} & \mathbf{\Phi}(\bullet_2)\mathrm{e}^{\mathrm{i}k_o\mathbf{K}^-x_2} \\ \vdots & \vdots \\ \mathbf{\Phi}(\bullet_{2n})\mathrm{e}^{\mathrm{i}k_o\mathbf{K}^+x_{2n}} & \mathbf{\Phi}(\bullet_{2n})\mathrm{e}^{\mathrm{i}k_o\mathbf{K}^-x_{2n}} \end{bmatrix} \begin{bmatrix} \mathbf{P}^+(0) \\ \mathbf{P}^-(0) \end{bmatrix}, \tag{12.3}$$

which are obtained by writing Equation (12.1) for the points in question. Obviously, the success of this approach largely depends on the ability to choose the $2n$ microphone positions so that the $2n \times 2n$ coefficient matrix in Equation (12.3) is not ill-conditioned. This requirement can be quite critical, because ill-conditioning problems arising from improper microphone positions are further aggravated by inevitable uncontrollable measurement errors. The problem of selecting independent measurement points does not have a known general solution, but can be handled in some particular cases [1]. Here we consider the simplest, but most widely implemented, form of the method, which is known as the two-microphone method, which is now standard in measurement of one-dimensional in-duct acoustic fields. Since only two microphones are used, the frequency range of the measurements is necessarily limited by the cut-on frequency of the first higher-order mode in the duct.

12.2.2 The Two-Microphone Method

The two-microphone measurement setup is shown schematically in Figure 12.1. Microphones can be mounted on the duct walls in several ways, but mounting them in a sealed holder flush with the inner duct surface is the method usually recommended. In Chapter 3 we saw that one-dimensional wave propagation in a duct is governed, in the frequency domain, by the two-port wave transfer equation

$$\mathbf{P}_0 = \mathbf{T}_{0x}\mathbf{P}_x. \tag{12.4}$$

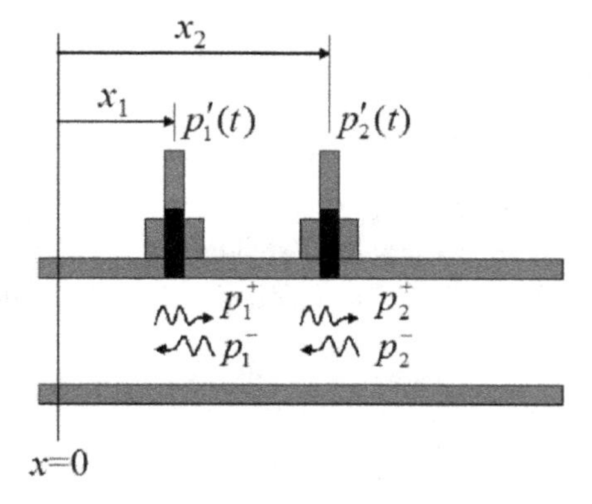

Figure 12.1 The two-microphone setup.

Here

$$\mathbf{P}_x = \begin{bmatrix} p^+(x) \\ p^-(x) \end{bmatrix}, \tag{12.5}$$

where x denotes the duct axis and $\mathbf{T}_{x0}$ denotes the wave transfer matrix between the origin, $x = 0$, and any duct plane x. Let the sound pressure signals measured by the two microphones be $p_1'(t)$ and $p_2'(t)$, and apply Equation (12.4) for the segment of the duct between the two microphone planes:

$$\mathbf{P}_{x_1} = \mathbf{T}_{x_1x_2}\mathbf{P}_{x_2}. \tag{12.6}$$

Multiplying this equation from the left by $\mathbf{E} = [1 \;\; 1]$ and combining the resulting equation with $p_2'(f) = p^+(x_2) + p^-(x_2)$, where $p_2'(f)$ denotes the Fourier transform of $p_2'(t)$, we find

$$\begin{bmatrix} p_1'(f) \\ p_2'(f) \end{bmatrix} = \begin{bmatrix} \mathbf{E}\mathbf{T}_{x_1x_2} \\ \mathbf{E} \end{bmatrix} \mathbf{P}_{x_2}, \tag{12.7}$$

which is a generalization of Equation (12.3) for the case of $n = 1$. Hence, the pressure wave components at microphone 2 are given by

$$\mathbf{P}_{x_2} = \begin{bmatrix} \mathbf{E}\mathbf{T}_{x_1x_2} \\ \mathbf{E} \end{bmatrix}^{-1} \begin{bmatrix} p_1'(f) \\ p_2'(f) \end{bmatrix} \tag{12.8}$$

and the pressure wave components at microphone 1 are determined by using this result in Equation (12.6). Let

$$\mathbf{T}_{x_1x_2} = \begin{bmatrix} T_{11} & T_{12} \\ T_{21} & T_{22} \end{bmatrix}. \tag{12.9}$$

Then, Equation (12.8) can be expressed explicitly as

$$\mathbf{P}_{x_2} = \frac{1}{T_{11} + T_{21} - T_{12} - T_{22}} \begin{bmatrix} 1 & -T_{12} - T_{22} \\ -1 & T_{11} + T_{21} \end{bmatrix} \begin{bmatrix} p_1'(f) \\ p_2'(f) \end{bmatrix}. \tag{12.10}$$

Therefore, the microphone spacing must be such that

$$T_{11} + T_{21} \neq T_{12} + T_{22}, \tag{12.11}$$

which is the condition for the two microphone signals to be linearly independent.

If the duct is uniform, as is usually specified in standard tests, then the wave transfer matrix $\mathbf{T}_{x_1x_2}$ is given by Equation (3.54) and Equation (12.10) gives

$$\mathbf{P}_{x_2} = \frac{1}{\mathrm{e}^{\mathrm{i}k_oK^+s} - \mathrm{e}^{\mathrm{i}k_oK^-s}} \begin{bmatrix} \alpha & -\left(\dfrac{\alpha-1}{2}\right)\mathrm{e}^{\mathrm{i}k_oK^+s} - \left(\dfrac{\alpha+1}{2}\right)\mathrm{e}^{\mathrm{i}k_oK^-s} \\ -\alpha & \left(\dfrac{\alpha+1}{2}\right)\mathrm{e}^{\mathrm{i}k_oK^+s} + \left(\dfrac{\alpha-1}{2}\right)\mathrm{e}^{\mathrm{i}k_oK^-s} \end{bmatrix} \begin{bmatrix} p_1'(f) \\ p_2'(f) \end{bmatrix} \tag{12.12}$$

where $s = x_1 - x_2$ denotes the microphone spacing and $K^{\mp}$ denote the propagation constants for forward (+) and backward propagation. These are given by Equation (3.62) for a uniform hard-walled duct. Viscothermal effects may be taken into account by using the theory described in Section 8.3.1.2 when the viscous sublayer thickness of the mean flow is much larger than the acoustic boundary layer thickness; and Equation (3.150) may be used as described in Section 3.9.3.4 in order to account for the effect of turbulent boundary layer stresses on sound propagation (see also Section 8.3.1.4).

In the case of a uniform hard-walled duct with propagation constants given by Equation (3.62), the condition for microphone spacing, Equation (12.11), can be expressed as

$$k_o(K^+ - K^-)\,|s| \neq 0, 2\pi, 4\pi, \ldots \tag{12.13}$$

which implies that, for a given microphone spacing, the measurement frequencies must satisfy the range condition

$$2n\pi < k_o(K^+ - K^-)\,|s| < 2(n+1)\pi, \qquad n = 0, 1, 2, \ldots \tag{12.14}$$

Substituting from Equation (3.55) for the propagation constants, we can express explicitly as

$$n\pi\left(\alpha^2 - \bar{M}_o^2\right) < k_o\alpha\,|s| < (n+1)\pi\left(\alpha^2 - \bar{M}_o^2\right). \tag{12.15}$$

Since the frequency has to be less than the cut-on frequency of the first higher order mode, this range is practicable for $n = 0$. This implies that microphone spacing can be arbitrarily small, but in practice it has to be greater than a certain minimum due to finite size of the measuring microphone probes. In general, the upper limit of the frequency range is limited by the duct size and the upper limit in Equation (12.15) determines the upper limit for the microphone separation.

The results of the two-microphone method are subject to errors arising from the effects of the mean flow and finite microphone size, as well as the usual measurement errors in the calculated quantities and input data [2–3]. Accurate implementation of the method requires careful calibration of the gain and phase factors of the entire measurement system, and a signal enhancement technique to be employed to improve the signal-to-noise ratio in the calculations. Ramifications of the two-microphone method in which multiple microphones are used to reduce the measurement errors in the least-squares sense have also been proposed [4].

12.2.2.1 Calibration of Microphones

A simple and robust method for the calibration of the two microphones for transfer function evaluation is the microphone switching technique first described by Chung and Blazer [5]. Let $p_1'(t)$ and $p_2'(t)$ denote the true instantaneous acoustic pressures to be measured at microphone positions 1 and 2, respectively. The transfer function H_{12} is defined by

$$p'_1(f) = H_{12} p'_2(f). \tag{12.16}$$

Let $p_1^M(t)$ and $p_2^M(t)$ be the actually measured signals by microphones 1 and 2, respectively, and define the transfer functions

$$p_1^M(f) = H_{11} p'_1(f) \tag{12.17}$$

$$p_2^M(f) = H_{22} p'_2(f) \tag{12.18}$$

$$p_1^M(f) = H_{12}^M p'_2(f), \tag{12.19}$$

where the superscript "*M*" refers to the measured signals. Now, switch the microphones and repeat the measurement. Let $p_1^S(t)$ and $p_2^S(t)$ denote the new measured signals at original microphone positions 1 and 2. For this case, the transfer functions corresponding to Equations (12.17)–(12.19) can be expressed as, respectively,

$$p_1^S(f) = H_{22} p'_1(f) \tag{12.20}$$

$$p_2^S(f) = H_{11} p'_2(f) \tag{12.21}$$

$$p_1^S(f) = H_{12}^S p'_2(f), \tag{12.22}$$

where the superscript "*S*" refers to signals measured when the microphones are switched. With the use of these equations, it may be shown that

$$H_{12} = \sqrt{H_{12}^M H_{12}^S}. \tag{12.23}$$

An alternative calibration technique is to expose the microphones to the same sound field simultaneously at a single section of the duct and measure the mutual sound pressure transfer functions between the microphone signals. The measured transfer functions in this calibration test are then used as calibration factors to correct the measured signals in the actual two microphone measurements. Let $G_{1,n}$ denote the sound pressure transfer function between microphone 1 and $n\,(= 1, 2, 3, \ldots)$, that is, $p'^{(C)}_1(f) = G_{1,n} p'^{(C)}_n(f)$, where the superscript "(*C*)" denotes the calibration setup and $G_{1,1} = 1$. To determine $G_{1,n}$, multiply both sides of this relationship by $\breve{p}'^{(C)}_1(f)$, where an inverted over-arc denotes the complex conjugate. Hence,

$$G_{1,n}(f) = \frac{S_{1,1}^{(C)}(f)}{S_{1,n}^{(C)}(f)}, \tag{12.24}$$

where $S_{1,n}^{(C)}(f) = \breve{p}'^{(C)}_1(f) p'^{(C)}_n(f)$, $n = 1, 2, \ldots$ Now, let $p'^{(M)}_n(f)$ denote the outputs of microphones $n = 1, 2, \ldots$ in an actual two-microphone measurement. The corresponding calibrated signals, $p'_n(f)$, can be determined from

$$p'_n(f) = G_{1,n}(f) p'^{(M)}_n(f). \tag{12.25}$$

Multiple microphones can be calibrated at one time by using this single plane method.

12.2.2.2 Signal Enhancement

The measurements will be erroneous in frequency regions where the signal strength is low, due to the interference of noise from the flow or unavoidable external sources. The signal-to-noise ratio of the transfer function evaluations can be improved by using a coherence function technique. It may be shown that the transfer function H_{12} without noise contamination is given by [5]:

$$H_{12} = \left(\frac{\gamma_{23}}{\gamma_{12}\gamma_{31}}\right) H_{12}^{c}, \tag{12.26}$$

where, H_{12}^{c} denotes the measured transfer function with noise contamination and the coherence function γ_{ij} between measured signals i and j is given by

$$\gamma_{ij} = \frac{\left|\overline{p_i'(f)\, \breve{p}_j'(f)}\right|_c}{\left|\overline{p_i'(f)}\right|_c \left|\overline{p_j'(f)}\right|_c}, \qquad i,j = 1,2,3, \qquad i \neq j. \tag{12.27}$$

Here, an inverted over-arc denotes complex conjugate, an over-bar denotes averaging of the spectral densities over a number of records and the subscript "c" denotes contaminated signals. Thus, this method requires a third microphone signal. This should also be in the coherent sound field and as close to the main microphones as possible to avoid effects of non-linearity, if any. It should be noted that, for a single record, the coherence function will always be unity. Therefore, the method presumes that a sufficient number of samples have been measured.

A discussion of the alternative signal-to-noise ratio enhancement methods and their comparison is presented by Allam and Bodén [6].

12.2.3 Measurement of the Plane-Wave Reflection Coefficient

From Equation (12.4), the pressure wave components at $x = 0$ can be expressed as

$$\mathbf{P}_0 = \mathbf{T}_{0,x_2} \begin{bmatrix} \mathbf{ET}_{x_1 x_2} \\ \mathbf{E} \end{bmatrix}^{-1} \begin{bmatrix} p_1'(f) \\ p_2'(f) \end{bmatrix} = \begin{bmatrix} A_{11} & A_{12} \\ A_{21} & A_{22} \end{bmatrix} \begin{bmatrix} p_1'(f) \\ p_2'(f) \end{bmatrix}. \tag{12.28}$$

Hence, the reflection coefficient at $x = 0$ can be computed from

$$r(0) = \frac{p^-(0)}{p^+(0)} = \frac{A_{21}H_{12} + A_{22}}{A_{11}H_{12} + A_{12}}, \tag{12.29}$$

where the transfer function H_{12} is defined as

$$p_1'(f) = H_{12} p_2'(f) \tag{12.30}$$

and H_{12} should strictly be calibrated and signal enhanced as described in Section 12.2.2.

12.2.4 Measurement of Wavenumbers

An apparatus for the measurement of the axial wavenumbers (or propagation constants) in a uniform duct is shown in Figure 12.2. It is usually designed so that an acoustic source or an anechoic termination can be installed at either end of the duct. Thus, by switching the source locations, wavenumbers for propagation in the direction of a mean flow and in the opposite direction can be measured. Acoustic pressure is measured along the axis of the duct by N equidistant identical (calibrated) wall-mounted microphones which are mounted at the same position with respect to duct sections (e.g., if the duct is rectangular, along the center line of one side).

In the configuration of the apparatus with source 1, Equation (12.3) reduces to

$$\begin{bmatrix} p'(\bullet_1, x_1) \\ p'(\bullet_2, x_2) \\ \vdots \\ p'(\bullet_N, x_N) \end{bmatrix} = \begin{bmatrix} \mathbf{\Phi}(\bullet_1)\mathrm{e}^{\mathrm{i}k_o\mathbf{K}^+x_1} \\ \mathbf{\Phi}(\bullet_2)\mathrm{e}^{\mathrm{i}k_o\mathbf{K}^+x_2} \\ \vdots \\ \mathbf{\Phi}(\bullet_N)\mathrm{e}^{\mathrm{i}k_o\mathbf{K}^+x_N} \end{bmatrix} \mathbf{P}^+(0), \tag{12.31}$$

where $x_j, j = 1, 2, \ldots, N$ denote the axial positions of the microphones measured from the first one at $x_1 = 0$. Equation (12.31) gives N equations of the form

$$p'(\bullet_j, x_j) = \sum_{\mu=1}^{\infty} \varphi_\mu(\bullet_j)\mathrm{e}^{\mathrm{i}k_o K_\mu^+ \Delta x \xi_j} p_\mu^+(0), \qquad j = 1, 2, \ldots, N \tag{12.32}$$

where $\xi_j = x_j/\Delta x$ and φ_μ denotes the transverse eigenfunction for mode μ. Note that the only unknowns in these equations are the modal pressure wave components $p_\mu^+(0)$ and the propagation constants K_μ^+ (or the wavenumbers $\kappa_\mu^+ = k_o K_\mu^+$), which are to be determined. To this end, it is convenient to write Equation (12.32) as

$$f_j = \sum_{\mu=1}^{\infty} C_\mu \eta_\mu^{\xi_j}, \quad j = 1, 2, \ldots, N \tag{12.33}$$

where $\eta_\mu = \mathrm{e}^{\mathrm{i}k_o K_\mu^+ \Delta x}$ and $C_\mu = \varphi_\mu(\bullet_j)p_\mu^+(0)$ is a constant for each mode, since the microphones are located so that $\varphi_\mu(\bullet_j) = \varphi_\mu(\bullet_1)$. Then, it remains to determine η_μ and C_μ from Equation (12.33). But, there are only N equations in this set and, therefore, the summation in Equation (12.33) has to be truncated. The number of modes to be retained should be equal to the number of cut-on modes plus a number of evanescent

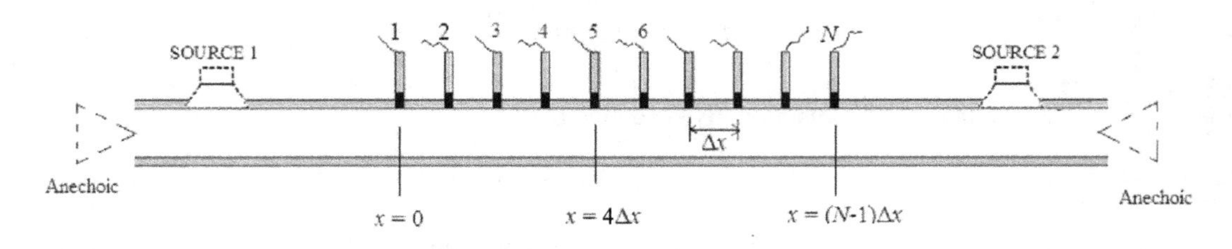

Figure 12.2 Apparatus for axial wavenumber measurement.

modes. The latter are necessary because of the impedance discontinuity at interfaces of the test duct with the end modules with which it is connected.

Truncating the sum at n terms (the selection of n is specified subsequently), Equation (12.33) yields N non-linear equations, which are written in matrix notation as

$$\begin{bmatrix} f_0 \\ f_1 \\ \vdots \\ f_{N-1} \end{bmatrix} = \begin{bmatrix} 1 & 1 & \cdots & 1 \\ \eta_1 & \eta_2 & \cdots & \eta_n \\ \vdots & \vdots & \vdots & \vdots \\ \eta_1^{N-1} & \eta_2^{N-1} & \cdots & \eta_n^{N-1} \end{bmatrix} \begin{bmatrix} C_1 \\ C_2 \\ \vdots \\ C_n \end{bmatrix}. \tag{12.34}$$

Solution of this system of equations was considered in a 1795 article by Prony [7]. His ideas are still used for the extraction of η_μ and C_μ, $\mu = 1, 2, \ldots, n$, from this set of equations. The crucial point of the solution is the construction of the algebraic equation

$$\eta^n + \alpha_1 \eta^{n-1} + \alpha_2 \eta^{n-2} + \cdots + \alpha_{n-1} \eta + \alpha_n = 0 \tag{12.35}$$

the n roots of which are equal to η_μ, $\mu = 1, 2, \ldots, n$. The coefficients of this polynomial can be determined from the following system of linear equations:

$$\begin{bmatrix} f_{n-1} & f_{n-2} & \cdots & f_0 \\ f_n & f_{n-1} & \cdots & f_1 \\ \vdots & \vdots & \cdots & \vdots \\ f_{N-2} & f_{N-3} & \cdots & f_{N-n-1} \end{bmatrix} \begin{bmatrix} \alpha_1 \\ \alpha_2 \\ \vdots \\ \alpha_n \end{bmatrix} = - \begin{bmatrix} f_n \\ f_{n+1} \\ \vdots \\ f_{N-1} \end{bmatrix}, \tag{12.36}$$

which is derived by multiplying ith row of Equation (12.34) by α_{1-i+n} and adding the rows thus derived for $i = 1, 2, \ldots, n$, and repeating this process by multiplying the $i + j$-th row of Equation (12.34) by $\alpha_{1-i-j+n}$, $j = 1, \ldots, n - 1$ [8]. Equation (12.36) can be solved for α_μ, $\mu = 1, 2, \ldots, n$, directly if $N = 2n$, or, if $N > 2n$, it can be solved in the sense of least squares. Thus, the method requires that the number of microphones should be equal to twice the number of modes retained in the modal expansion of the acoustic field in the duct.

After coefficients α_μ are determined, η_μ, $\mu = 1, 2, \ldots, n$, are found from the roots of Equation (12.35) and the propagation constants K_μ^+ are determined from $\eta_\mu = e^{ik_o K_\mu^+ \Delta x}$. These can be used for validation of a mathematical model of the duct or to educe an unknown parameter of which K_μ^+ is a function. For example, impedance of locally reacting liners with grazing mean flow are often determined in this way and the salient details of the test rigs used for this purpose around world are described by Bodén et al. [9]. Since the propagation constants of several modes are measured, the selection of the one to be used for parameter eduction can be an issue. Normally, the method yields the propagation constant of the least attenuated mode most accurately.

The accuracy of the Prony method is apt to be affected by numerical ill-conditioning problems and, in order to smooth out the effects of errors in the measurement data, it is usually employed by using over-determined systems of linear equations. The coefficients C_μ, $\mu = 1, 2, \ldots, n$, or the pressure wave components $p_\mu^+(0)$ at the origin, are generally of less significance. With η_μ known, they can be determined relatively easily from the first n of Equations (12.34) or an over-determined extension of this set.

The configuration of the apparatus with source 2 is analyzed in the same way, except that in this case we begin with the form of Equation (12.31) for the wave components in the opposite direction.

12.3 Measurement of Passive Acoustic Two-Ports

12.3.1 Basics of the Four Microphone Method

A passive one-dimensional acoustic two-port element with outlet to inlet causality between the pressure wave components $p_i^\mp$ and $p_o^\mp$ at its inlet and outlet planes, respectively, is characterized by the wave transfer equation (see Section 2.3.2):

$$\begin{bmatrix} p_i^+ \\ p_i^- \end{bmatrix} = \begin{bmatrix} T_{11} & T_{12} \\ T_{21} & T_{22} \end{bmatrix} \begin{bmatrix} p_o^+ \\ p_o^- \end{bmatrix} \tag{12.37}$$

or, briefly, $\mathbf{P}_i = \mathbf{T}_{io}\mathbf{P}_o$. The pressure wave components at the inlet and outlet sections of the element can be measured by using the two-microphone method in the four-microphone setup shown in Figure 12.3. Applying Equation (12.28) for the downstream section gives

$$\begin{bmatrix} p_o^+(f) \\ p_o^-(f) \end{bmatrix} = \begin{bmatrix} B_{11} & B_{12} \\ B_{21} & B_{22} \end{bmatrix} \begin{bmatrix} p_3'(f) \\ p_4'(f) \end{bmatrix}, \tag{12.38}$$

where

$$\begin{bmatrix} B_{11} & B_{12} \\ B_{21} & B_{22} \end{bmatrix} = \mathbf{T}_{o,4} \begin{bmatrix} \mathbf{ET}_{34} \\ \mathbf{E} \end{bmatrix}^{-1}. \tag{12.39}$$

Similarly, for the upstream microphones

$$\begin{bmatrix} p_i^+(f) \\ p_i^-(f) \end{bmatrix} = \begin{bmatrix} A_{11} & A_{12} \\ A_{21} & A_{22} \end{bmatrix} \begin{bmatrix} p_1'(f) \\ p_2'(f) \end{bmatrix}, \tag{12.40}$$

where

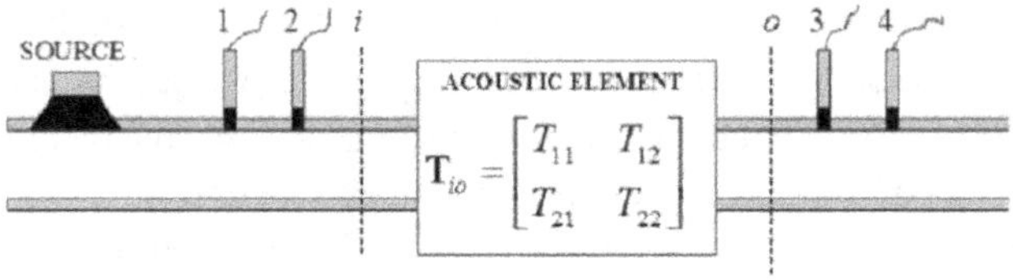

Figure 12.3 Basic two-port measurement set-up with four microphones.

$$\begin{bmatrix} A_{11} & A_{12} \\ A_{21} & A_{22} \end{bmatrix} = \mathbf{T}_{1,i}^{-1} \begin{bmatrix} \mathbf{E} \\ \mathbf{E}\mathbf{T}_{12}^{-1} \end{bmatrix}^{-1}. \tag{12.41}$$

The four microphones may be calibrated by using the single plane method (Section 12.2.1) and cross-spectrum based frequency domain averaging (using a reference signal) may be used for signal-to-noise ratio enhancement. Let $g = g(f)$ denote the reference signal. Expanding Equations (12.40) and (12.38) and multiplying through by the complex conjugate of the reference signal, we obtain the following cross-spectrum relations:

$$S_{g,i}^{+}(f) = A_{11}S_{g,1}(f) + A_{12}S_{g,2}(f) \tag{12.42}$$

$$S_{g,i}^{-}(f) = A_{21}S_{g,1}(f) + A_{22}S_{g,2}(f) \tag{12.43}$$

$$S_{g,o}^{+}(f) = B_{11}S_{g,3}(f) + B_{12}S_{g,4}(f) \tag{12.44}$$

$$S_{g,o}^{-}(f) = B_{21}S_{g,3}(f) + B_{22}S_{g,4}(f), \tag{12.45}$$

where

$$S_{g,x}^{\mp}(f) = \breve{g}\ (f)p_x^{\mp}(f), x \in i, o, 1, 2, 3, 4 \tag{12.46}$$

$$S_{g,o}^{\mp}(f) = \breve{g}\ (f)p_o^{\mp}(f) \tag{12.47}$$

$$S_{g,n}(f) = \breve{g}\ (f)p_n'(f), n = 1, 2, 3, 4. \tag{12.48}$$

12.3.2 Measurement of Attenuation

Attenuation is given by Equation (11.58) and can be calculated from the measured cross-spectra as

$$A(f) = 10\log_{10}\left|\frac{S_{g,i}^{+}}{S_{g,o}^{+}}\right|. \tag{12.49}$$

If the flow noise effects are known to be negligible, this can be evaluated by frequency-domain averaging, which is tantamount to taking the reference signal equal to $p_i^{+}(f)$ for the inlet section and to $p_o^{+}(f)$ for the outlet section. Then, Equations (12.42) and (12.44) expand, respectively, as

$$S_{g,i}^{+}(f) = |A_{11}|^2 S_{11}(f) + A_{11}\breve{A}_{12}S_{21}(f) + A_{12}\breve{A}_{11}S_{12}(f) + |A_{12}|^2 S_{22}(f) \tag{12.50}$$

$$S_{g,i}^{+}(f) = |B_{11}|^2 S_{33}(f) + B_{11}\breve{B}_{12}S_{43}(f) + B_{12}\breve{B}_{11}S_{34}(f) + |B_{12}|^2 S_{44}(f), \tag{12.51}$$

where $S_{ij} = \breve{p_i'}\ (f)p_j'(f)$, $i, j = 1, 2$ and $3, 4$.

12.3.3 Measurement of Transmission Loss

Transmission loss is given by Equation (5.2) and can be calculated from the measured cross-spectra as

$$TL(f) = 10\log_{10}\left|\frac{S_{g,i}^{+}}{S_{g,o}^{+}}\right| + 10\log_{10}\left(\frac{\rho_o c_o S_i (1+M_o)^2}{\rho_i c_i S_o (1+M_i)^2}\right), \tag{12.52}$$

where M denotes the mean flow Mach number, ρc denotes the characteristic impedance and the subscripts "o" and "i" refer, respectively, to the outlet and inlet sections. Since the outlet section should be anechoic, Equation (12.44) reduces to

$$S_{g,o}^{+}(f) = \left(B_{11} - \frac{B_{12}B_{21}}{B_{22}}\right)S_{g,3}(f) \tag{12.53}$$

and, therefore, one of the downstream microphones is actually not required. Note that, the counterpart of Equation (12.51) for Equation (12.53) is simply $S_{g,i}^{+}(f) = |B_{11}|^2 S_{33}(f)$.

The main difficulty in transmission loss (TL) measurements is the experimental simulation of the anechoic (non-reflective) outlet condition. Several techniques, such as using sound absorbing wedges [10], a composite absorber packing [11] and horns [12], have been proposed over the years. The performance of an anechoic termination depends on the frequency and the duct size and, therefore, a design given for a specific size should be scaled to another size with caution. The size and frequency effects can be predicted by constructing a block diagram model of the system as described in previous chapters of this book. In general, however, the anechoic condition cannot be simulated experimentally exactly. For this reason, although only three microphone signals are required for TL measurement, the use of the fourth microphone is recommended, as it can be used for the evaluation of the efficacy of the anechoic terminal design.

12.3.4 Measurement of the Wave Transfer Matrix

Substitute Equations (12.38) and (12.40) in Equation (12.37):

$$\begin{bmatrix} A_{11} & A_{12} \\ A_{21} & A_{22} \end{bmatrix}\begin{bmatrix} p_1'(f) \\ p_2'(f) \end{bmatrix} = \begin{bmatrix} T_{11} & T_{12} \\ T_{21} & T_{22} \end{bmatrix}\begin{bmatrix} B_{11} & B_{12} \\ B_{21} & B_{22} \end{bmatrix}\begin{bmatrix} p_3'(f) \\ p_4'(f) \end{bmatrix} \tag{12.54}$$

or, briefly,

$$\mathbf{A}\begin{bmatrix} p_1'(f) \\ p_2'(f) \end{bmatrix} = \begin{bmatrix} T_{11} & T_{12} \\ T_{21} & T_{22} \end{bmatrix}\mathbf{B}\begin{bmatrix} p_3'(f) \\ p_4'(f) \end{bmatrix}. \tag{12.55}$$

When the pressure wave components are measured, Equation (12.54) gives two equations for the four complex unknowns T_{ij}, $i,j = 1,2$. Therefore, the elements of a two-port transfer matrix can be calculated, if $\mathbf{P}_i$ and $\mathbf{P}_o$ are measured for two

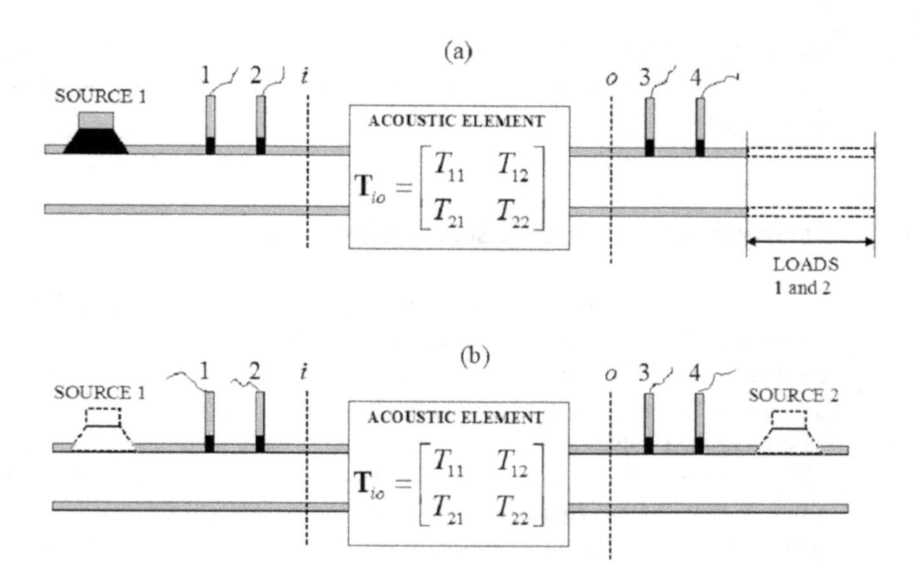

Figure 12.4 Transfer matrix measurement: (a) two-load method, (b) two-source method.

independent acoustic states, states a and b, say, of the component. These two states are usually created by using two loads [13], or two sources [14], as shown schematically in Figure 12.4a and 12.4b, respectively.

Let the pressure wave components measured by using the two-microphone method (Section 12.2.2) be denoted as $\mathbf{P}_{i,a}$, $\mathbf{P}_{o,a}$, $\mathbf{P}_{i,b}$ and $\mathbf{P}_{o,b}$, where the second subscripts refer to states a and b. Since the transfer matrices $\mathbf{A}$ and $\mathbf{B}$ are not dependent on different test states, the form of Equation (12.55) for the two states may be combined as

$$\mathbf{A}\begin{bmatrix} p'_{1,a}(f) & p'_{1,b}(f) \\ p'_{2,a}(f) & p'_{2,b}(f) \end{bmatrix} = \begin{bmatrix} T_{11} & T_{12} \\ T_{21} & T_{22} \end{bmatrix}\mathbf{B}\begin{bmatrix} p'_{3,a}(f) & p'_{3,b}(f) \\ p'_{4,a}(f) & p'_{4,b}(f) \end{bmatrix}. \tag{12.56}$$

After signal enhancement by multiplying both sides of this equation by a reference signal $g(f)$, the elements of the wave transfer matrix can be determined from

$$\begin{bmatrix} T_{11} & T_{12} \\ T_{21} & T_{22} \end{bmatrix} = \mathbf{A}\begin{bmatrix} S^{(a)}_{g,1}(f) & S^{(b)}_{g,1}(f) \\ S^{(a)}_{g,2}(f) & S^{(b)}_{g,2}(f) \end{bmatrix}\begin{bmatrix} S^{(a)}_{g,3}(f) & S^{(b)}_{g,3}(f) \\ S^{(a)}_{g,4}(f) & S^{(b)}_{g,4}(f) \end{bmatrix}^{-1}\mathbf{B}^{-1}, \tag{12.57}$$

where $S^{(k)}_{g,n} = \breve{g}\,(f)p'_{n,k}(f)$, $n \in 1,2,3,4$ and $k = a, b$. Upon expanding the right-hand side, Equation (12.57) can be expressed as

$$\begin{bmatrix} T_{11} & T_{12} \\ T_{21} & T_{22} \end{bmatrix} = \mathbf{A}\begin{bmatrix} \dfrac{H^{(a)}_{14} - H^{(b)}_{14}}{H^{(a)}_{34} - H^{(b)}_{34}} & \dfrac{H^{(a)}_{13} - H^{(b)}_{13}}{H^{(a)}_{43} - H^{(b)}_{43}} \\ \dfrac{H^{(a)}_{24} - H^{(b)}_{24}}{H^{(a)}_{34} - H^{(b)}_{34}} & \dfrac{H^{(a)}_{23} - H^{(b)}_{23}}{H^{(a)}_{43} - H^{(b)}_{43}} \end{bmatrix}\mathbf{B}^{-1}, \tag{12.58}$$

where $H_{nm}^{(k)} = S_{g,m}^{(k)}(f)/S_{g,n}^{(k)}(f)$. As possible reference signals, the incident pressure wave component at the microphone closest to the source, or the electric signal driving the loudspeaker may be used [15].

12.4 Measurement of One-Port Source Characteristics

When q modes are cut-on at the equivalent source plane, the equivalent one-port source equation is of the form given in Equation (11.16) and involves $q(1+q)$ complex scalar unknown terms. For the determination of these unknowns by measurement, it is necessary to create at least as many independent test states without modifying their values and carry out multi-modal wave field decomposition (Section 12.2.1) in these states. No general procedure is available for this at the present for arbitrary multi-modal equivalent source characterization. There are, however, well-studied methods for the measurement of the characteristics of equivalent one-port and two-port sources operating in the plane wave mode ($q = 1$) [16]. In this section, we describe commonly practiced forms of these methods in which the independent test states are created by using external acoustic loads.

In Section 10.3, we saw that a one-dimensional linear and time-invariant one-port source is defined in frequency domain by the relationship

$$p_S(f) = Z_S(f)v_1'(f) + p_1'(f), \tag{12.59}$$

where, p_S and Z_S denote, respectively, the equivalent source pressure strength and impedance, and $p_1'(f)$ and $v_1'(f)$ denote the Fourier transform of the acoustic pressure, $p_1'(t)$, and particle velocity, $v_1'(t)$, at the equivalent source plane, which is always taken as plane 1 in this chapter (see Figure 12.5). Here, time is denoted by t, and the frequency, f, is related to the Fourier variable ω as $\omega = 2\pi f$. Essential to considerations on the use of Equation (12.59) is the relationship

$$p_1'(f) = Z_1(f)v_1'(f), \tag{12.60}$$

where the transfer function Z_1 is called the acoustic load impedance. Let r_1 be the reflection coefficient at plane 1. The load impedance is given by

$$Z_1(f) = \rho_1 c_1 \left(\frac{1 + r_1(f)}{1 - r_1(f)}\right), \tag{12.61}$$

where $\rho_1 c_1$ is the characteristic impedance at the equivalent source plane. Thus, the load impedance can be determined by measuring the reflection coefficient at the equivalent source plane by using the two-microphone method (Figure 12.5a). However, if an adequately accurate mathematical model is available for the calculation of Z_1, it need not be measured. Such a load is usually called calibrated load. But, a calibrated load requires not only an accurate acoustic wave transfer matrix, $\mathbf{T}_{12}$, of the path between planes 1 and 2, but also an accurate model of the reflection coefficient at plane 2, r_2 (Figure 12.5b).

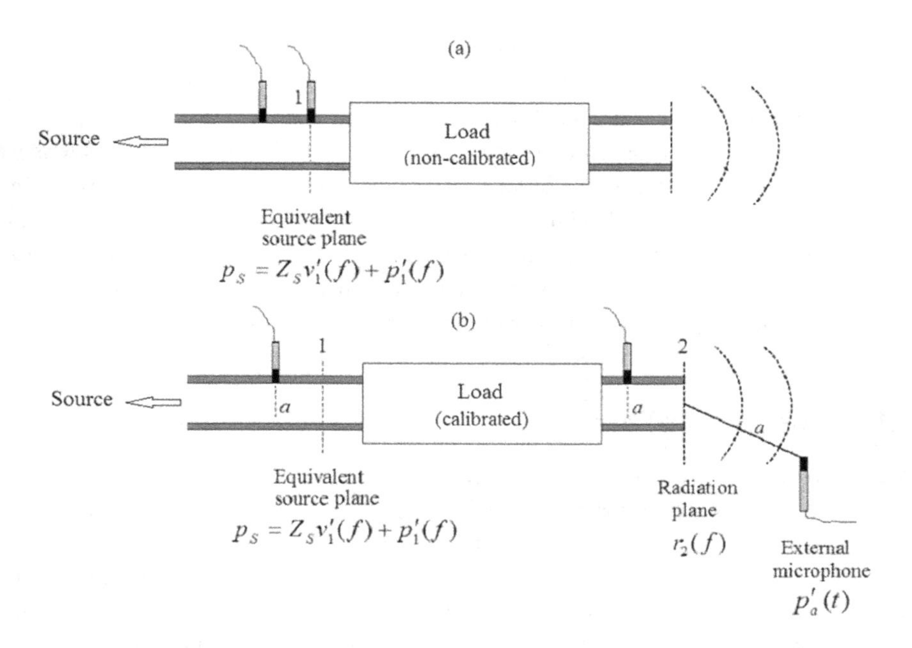

Figure 12.5 Examples of microphone locations for source measurements (a) with a non-calibrated load, Z_1 is measured by using the two-microphone method; (b) with a calibrated load and single microphone, Z_1 is calculated from a mathematical model of the system (figure shows three alternative microphone positions).

In principle, any system with a validated acoustic model can be used as a calibrated load, however, in practice, simple flow duct elements such as straight ducts or side-branches of different lengths are usually used, since their mathematical models are relatively simple and well-proven. Equations (12.59) and (12.60) may be combined as

$$p_S(f) = (Z_S(f) + Z_1(f))v_1'(f) \tag{12.62}$$

or, as

$$p_S(f) = \left(1 + \frac{Z_S(f)}{Z_1(f)}\right)p_1'(f). \tag{12.63}$$

In real form, Equation (12.62) may be expressed as

$$|Z_S(f) + Z_1(f)| = R(f), \tag{12.64}$$

where

$$R(f) = \frac{|p_S(f)|}{|p_1'(f)|}|Z_1(f)|. \tag{12.65}$$

Equation (12.64) is the equation of a circle of radius R and center $Z_1 = X_1 + \mathrm{i}Y_1$ in the complex $Z_S = X_S + \mathrm{i}Y_S$ plane. It is expressed explicitly as

$$(X_S(f) + X_1(f))^2 + (Y_S(f) + Y_1(f))^2 = R(f)^2. \tag{12.66}$$

Thus, for example, if the acoustic pressure at the equivalent source plane, $p'_1(t)$, is measured with a calibrated or non-calibrated load, Equation (12.63) provides one linear complex equation in the two complex unknowns, $p_S(f)$ and $Z_S(f)$, for each frequency in the spectrum of $p'_1(f)$. Then, the equivalent source characteristics can, in principle, be determined by measuring $p'_1(t)$ for two independent loads. Similarly, if $|p'_1(f)|$ is measured with one load, Equation (12.64) provides one non-linear real equation for three real unknowns X_S, Y_S and p_S, which can be determined by measuring $|p'_1(f)|$ with three independent loads. These options are called the two-load method and three-load method, respectively.

12.4.1 The Two-Load Method

12.4.1.1 Implementation with Non-Calibrated Loads

In this case, the setup of Figure 12.5a is used to measure the pressure wave components and, hence, p'_1 and v'_1, at the equivalent source plane 1 for two non-calibrated independent loads $Z_1^{(1)}$ and $Z_1^{(2)}$. Let $p'^{(n)}_1$ and $v'^{(n)}_1$ denote the measured acoustic pressure and the particle velocity with load $Z^{(n)}$, where the superscript (n), $n \in 1, 2$, refers to the load number. For each load, Equation (12.59) gives

$$p_S = p'^{(n)}_1 + Z_S v'^{(n)}_1, \tag{12.67}$$

where we have back-substituted $v'^{(n)}_1 = p'^{(n)}_1 / Z_1^{(n)}$ and dropped the frequency dependence for notational simplicity. In view of the unavoidable measurement and modeling errors, the equality in Equation (12.67) is not exact and should be written as

$$Z_S v'^{(n)}_1 - p_S + p'^{(n)}_1 = e^{(n)}, \tag{12.68}$$

where $e^{(n)}$ accounts for the error involved with load $n \in 1, 2$. Collecting these equations in matrix notation,

$$\begin{bmatrix} v'^{(1)}_1 & -1 \\ v'^{(2)}_1 & -1 \end{bmatrix} \begin{bmatrix} Z_S \\ p_S \end{bmatrix} + \begin{bmatrix} p'^{(1)}_1 \\ p'^{(2)}_1 \end{bmatrix} = \begin{bmatrix} e^{(1)} \\ e^{(2)} \end{bmatrix}. \tag{12.69}$$

The solution of this equation which minimizes the sum of the squared moduli of the error terms can be found by the least-squares method. This is tantamount to multiplying Equation (12.69) from left by the Hermitian transpose of the coefficient matrix and letting the error vector vanish. Thus, we obtain

$$\begin{bmatrix} v'^{(1)}_1 \breve{v}'^{(1)}_1 + \breve{v}'^{(2)}_1 \breve{v}'^{(2)}_1 & -\breve{v}'^{(1)}_1 - \breve{v}'^{(2)}_1 \\ -v'^{(1)}_1 - v'^{(2)}_1 & 2 \end{bmatrix} \begin{bmatrix} Z_S \\ p_S \end{bmatrix} = \begin{bmatrix} -\breve{v}'^{(1)}_1 p'^{(1)}_1 - \breve{v}'^{(2)}_1 p'^{(2)}_1 \\ p'^{(1)}_1 + p'^{(2)}_1 \end{bmatrix} \tag{12.70}$$

where an inverted over-arc denotes the complex conjugate. Hence, the condition for the two loads to be linearly independent can be stated as

$$v'^{(1)}_1 \neq v'^{(2)}_1 \tag{12.71}$$

for each frequency. When this condition is satisfied, a unique solution of Equation (12.70) exists and it is relatively easily found by matrix inversion. For this solution to be physically valid, however, the measured sound pressure signals for the two loads must have equal phase with respect to the source process. For this reason, it is also necessary to measure, simultaneously with the microphone signals, a signal that is triggered by the actual source mechanism. For example, for the measurement of the breathing noise source characteristics of a reciprocating internal combustion engine, the crankshaft rotation pulses can be used as such reference signal, since the source mechanism, sudden mass injection during the valve-open periods, is timed by the crankshaft rotation.

An error analysis for the contemporary implementations of the two-load method to fluid machinery is given by Bodén [17].

12.4.1.2 Implementation with Calibrated Loads

When calibrated loads are used (Figure 12.5b), a single sound pressure measurement per load is sufficient, but this must be taken instantaneously with the phase referencing signal. The sound pressure at the source plane can be computed from the transfer function

$$p'^{(n)}_1(f) = H_{1,a}\, p'^{(n)}_a(f), \tag{12.72}$$

where $p'^{(n)}_a(f)$ denotes the Fourier transform of the measured sound pressure signal, $p'^{(n)}_a(t)$, for load $n = 1, 2, \ldots, N$, and the subscript "a" refers to the location of the microphone. Then, $p'^{(n)}_1$ is determined from Equation (12.72) and the particle velocity $v'^{(n)}_1$ from Equation (12.60).

Various possibilities for the location of microphone "a" are shown schematically in Figure 12.5b. If the sound pressure measurement is taken by using a wall-mounted microphone, the transfer function $H_{1,a}$ can be computed from the knowledge of the transfer matrix, $\mathbf{T}_{1,a}$, of the path between planes 1 and a, and the reflection coefficient at plane a. The sound pressure measurement may also be taken at a point outside the radiating open end of the system [18]. In this case, noting that p_1^+ and p_1' are related as

$$\frac{p_1^+}{p_1'} = \frac{1}{2}\left(1 + \frac{\rho_1 c_1}{Z_1}\right) \tag{12.73}$$

and, using Equation (9.55), it may be shown that

$$H_{1,a} = \frac{2}{1 + \dfrac{\rho_1 c_1}{Z_1}} \frac{\alpha}{K} \tag{12.74}$$

for each load. Here, K denotes the plane wave radiation transfer function and $\alpha = p_1^+/p_2^+$.

12.4.2 Geometrical Interpretation of the Two-Load Method

Equation (12.66) may be recast as

$$(X_S + X_1)^2 + (Y_S + Y_1)^2 = \frac{|Z_1|^2}{|p_1'|^2} |p_S|^2. \tag{12.75}$$

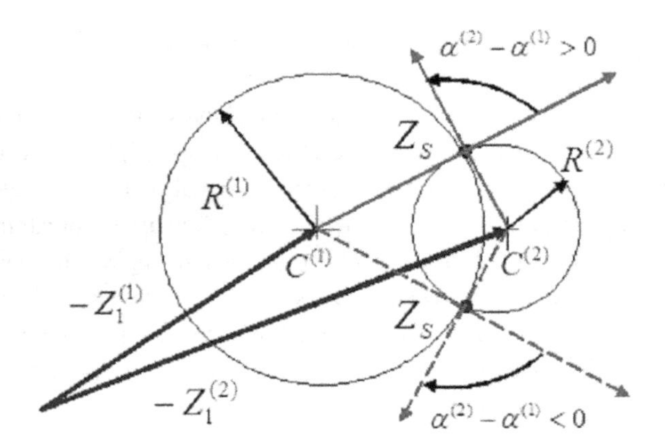

Figure 12.6 Load circle for two loads $Z_1^{(1)}$ and $Z_1^{(2)}$.

This represents a circle, which is subsequently called a load circle, of radius $R = |Z_1||p_S|/|p_1'|$ and center $C \equiv (-X_1, -Y_1)$ in the complex plane (X_S, Y_S). Figure 12.6 shows the load circles of a specific one-port source drawn for two independent loads, $Z_1^{(1)}$ and $Z_1^{(2)}$. Obviously, the impedance of the source is determined by one of the intersection points of the two load circles. The physically correct intersection point can be picked by inspecting the sign of the angle, $\alpha^{(2)} - \alpha^{(1)} = \angle v_1'^{(1)} - \angle v_1'^{(2)}$, where "$\angle$" denotes the argument of a complex quantity and $\alpha^{(n)} = \angle(Z_S + Z_1^{(n)})$, $n \in 1, 2$. As indicated in Figure 12.6, the angle $\alpha^{(2)} - \alpha^{(1)}$ has opposite signs at the two intersection points, although their magnitudes are equal. Therefore, the physically correct intersection point can be picked if we measure also the phase difference $\angle v_1'^{(1)} - \angle v_1'^{(2)}$, since $\alpha^{(2)} - \alpha^{(1)} = \angle v_1'^{(1)} - \angle v_1'^{(2)}$.

This proves geometrically that the phases of the time signals $v_1'^{(1)}(t)$ and $v_1'^{(2)}(t)$ with respect to a source event datum need to be measured when the two-load method is used. Furthermore, the two load circles must not be tangential and must not have equal radii, since, if otherwise, the phase difference $\angle v_1'^{(1)} - \angle v_1'^{(2)}$ vanishes. The conditions that the load circles have different radii and that the angle $\angle v_1'^{(1)} - \angle v_1'^{(2)}$ is not equal to zero are tantamount to the inequality given in Equation (12.71).

12.4.3 The Apollonian Circle of Two Loads

The foregoing geometrical interpretation of the two-load method in terms of load circles is not useful for synthetic determination of the source characteristics, because the load circles can be drawn or calculated only if the source pressure strength is already known. The Apollonian definition of a circle is more useful for synthetic considerations, because it gives a circle which is independent of the source characteristics. Apollonius[1]

[1] Apollonius of Perga was born about 262 BCE in Perga (now in Antalya, Turkey). Eudemuin s of Pergamum taught Apollonius at Ephesus and recommended that he continue his studies at Alexandria. Apollonius went to Alexandria where he studied under the followers of Euclid; he later taught and died there (about 190 BCE), having become known as the Great Geometer of the era.

defined the circle as the locus of points, the ratio of the distances of which to two fixed points (in the same plane) is constant. To use this definition, consider the quotient

$$q_{12} = \frac{R^{(1)}}{R^{(2)}}. \tag{12.76}$$

This is called the characteristic ratio of load 1 with respect to load 2. Assuming that the source characteristics are independent of the load, substitution from Equations (12.63) and (12.66) gives

$$q_{12} = \frac{\left|p'^{(2)}_1\right|\left|Z^{(1)}_1\right|}{\left|p'^{(1)}_1\right|\left|Z^{(2)}_1\right|}. \tag{12.77}$$

According to Apollonius' definition, this quotient defines a circle (for a given frequency), the two fixed points being the points $-Z_1^{(1)}$ and $-Z_1^{(2)}$ in the complex plane. We call this circle the Apollonian circle of the two loads and denote it by Θ for brevity.

In Figure 12.7, two load circles are shown together with their Apollonian circle. Here, Θ is the locus of the source impedance Z_S, since the ratio of the distances of all points on Θ to the centers of the load circles $C^{(1)}$ and $C^{(2)}$ are equal to q_{12}. Furthermore, Θ cuts the line connecting the centers of the two load circles at points M and N (see Figure 12.7), which divide the line segment $\overline{C^{(1)}C^{(2)}}$ internally and externally in the ratio of q_{12}, respectively. It is localized to the center of one of the load circles. The center of the load circle onto which Θ is localized is called the focal point of Θ (for example, point $-Z_1^{(2)}$ in Figure 12.7) and the vector which describes the position of a point on Θ with respect to its focal point is called the focal vector (for

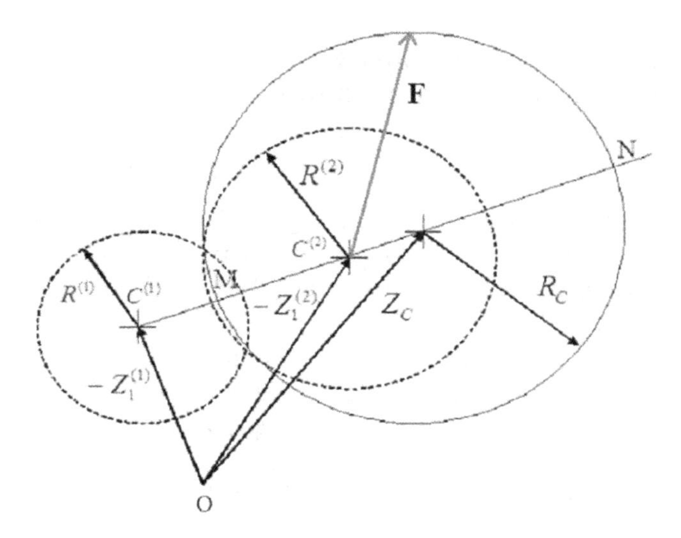

Figure 12.7 Apollonian circle of two loads; dotted circles are the load circles.

example, vector **F** in Figure 12.7). The smallest and largest magnitudes of **F** occur at points M and N, respectively.

The focal point depends on the value of the characteristic ratio. If $q_{12} < 1$, it is the center of the load circle of load 1; else, if $q_{12} > 1$, it is the center of load circle of load 2. For $q_{12} = 1$, Θ degenerates to the perpendicular bisector of $\overline{C^{(1)}C^{(2)}}$. Then the load circles are tangential and have equal radii, which implies that the two loads are not independent, providing yet another interpretation of Equation (12.71).

From the inspection of the geometry of Figure 12.7, it can be seen that the center, Z_C, and the radius, R_C, of Θ are given by

$$Z_C = -\frac{Z_1^{(1)} - q_{12}^2 Z_1^{(2)}}{1 - q_{12}^2} \tag{12.78}$$

$$R_C = \frac{q_{12}\left|Z_1^{(1)} - Z_1^{(2)}\right|}{\left|1 - q_{12}^2\right|}. \tag{12.79}$$

Therefore, Θ is independent of source characteristics and can be drawn, or calculated, from the microphone data acquired for the application of the two-load method. (Phase data is not needed.) Equation (12.79) implies that as $q_{12} \to 0$ or $q_{12} \to \infty$, the radius of Θ gets increasingly smaller, in the limit becoming a point circle. Therefore, very small or very large, values of q_{12} imply $Z_S \approx Z_C$ or, more accurately, $Z_S \approx Z_1^{(n^*)}$, where the superscript $n^* \in 1, 2$ in brackets refers to the load number associated with the focal point of Θ ($n^* = 2$ in Figure 12.7). Variations of the measured Apollonian circle radii with the frequency of an equivalent intake noise source of a four-stroke four-cylinder engine running at 4000 rpm are shown in Figure 12.8a. The real and imaginary parts of the corresponding centers of the Apollonian circles are shown in Figure 12.8b.

12.4.3.1 Upper and Lower Bounds for Source Pressure Strength

Since Θ is the locus of the source impedance, the physically correct intersection point of the load circles lies on it. Let $Z \in \Theta$ be any point on Θ and $\mathbf{F}(Z)$ be the corresponding focal vector and consider the infinite set $B = \left\{ |\mathbf{F}(Z)| \left| v_1'^{(n^*)} \right| : Z \in \Theta \right\}$. In view of Equation (12.64), the interval $P_S = [\inf(B), \sup(B)]$ necessarily contains the value of $|p_S|$. Thus, knowing the Apollonian circle of two loads, upper and lower bounds to the magnitude of the crisp source pressure strength $|p_S|$ can be obtained by calculating a discrete sample of the set B.

An analytical description of this process is possible. Referring again to Figure 12.7, it can be seen that the focal vectors are represented by $Z_S + Z_1^{(n^*)}$ on the complex plane. Therefore, the magnitude of the focal vector $Z_S + Z_1^{(n^*)}$ is at once bracketed as

$$\left|Z_{\mathrm{M}} + Z_1^{(n^*)}\right| \leq \left|Z_S + Z_1^{(n^*)}\right| \leq \left|Z_{\mathrm{N}} + Z_1^{(n^*)}\right| \tag{12.80}$$

and, hence, the magnitude of the source pressure strength is then bracketed as

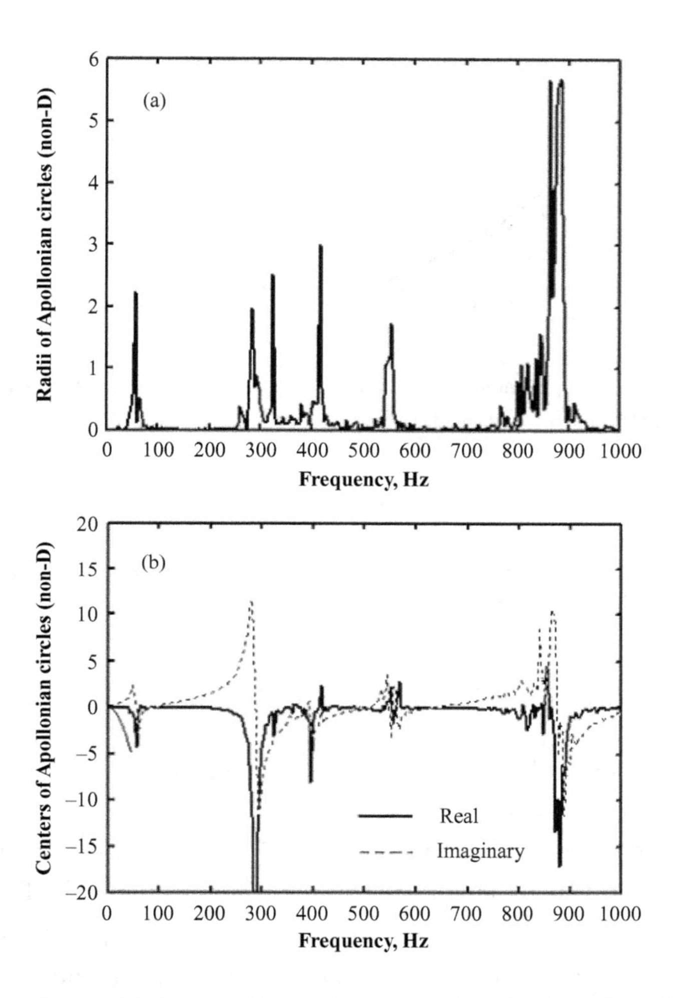

Figure 12.8 Intake breathing noise source radii and centers of Apollonian circles of an engine running at 4000 rpm (normalized by the specific impedance at the equivalent source plane).

$$\left|1+\frac{Z_{\mathrm{M}}}{Z_1^{(n^*)}}\right|\left|p'^{(n^*)}_1\right| \le |p_S| \le \left|1+\frac{Z_{\mathrm{N}}}{Z_1^{(n^*)}}\right|\left|p'^{(n^*)}_1\right|, \tag{12.81}$$

where Z_{M} and Z_{N} designate the points M and N on the complex plane (Figure 12.7). Clearly, Equation (12.81) can be expressed briefly as $|p_{\mathrm{M}}| \le |p_S| \le |p_{\mathrm{N}}|$, where $|p_{\mathrm{M}}|$ and $|p_{\mathrm{N}}|$ denote the lower and upper bounds for the magnitude of the source pressure strength. This proves that the interval P_S is given by

$$P_S = [|p_{\mathrm{M}}|, |p_{\mathrm{N}}|]. \tag{12.82}$$

The interval end-points (bounds) in these equations can be calculated by using the following formulae for the points M and N of Θ:

$$Z_{\mathrm{M}} + Z_1^{(n^*)} = \left(\sigma - \frac{1}{1+q_{12}}\right) Z_1^{(1,\,2)} \tag{12.83}$$

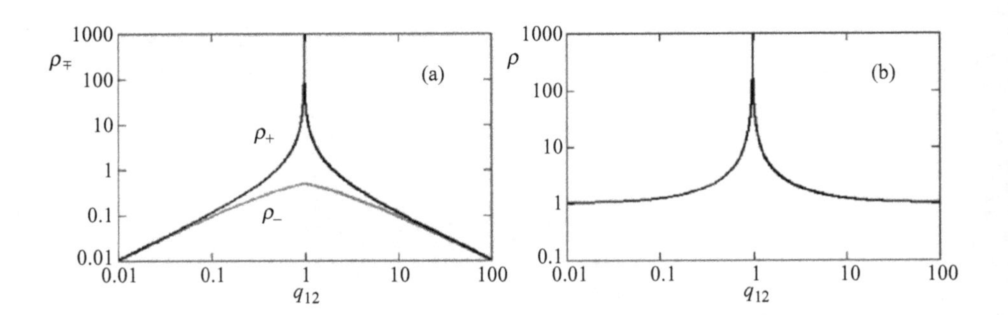

Figure 12.9 (a) Variation of the bounds $\rho_{\mp} = |\sigma - 1/(1 \pm q_{12})|$ in Equation (12.86) with q_{12}; (b) variation of the quotient $\rho = \rho_+/\rho_-$ with q_{12}.

$$Z_{\mathrm{N}} + Z_1^{(n^*)} = \left(\sigma - \frac{1}{1 - q_{12}}\right) Z_1^{(1,2)} \tag{12.84}$$

where $\sigma = 1$ if $q_{12} < 1$ and $\sigma = 0$ if $q_{12} > 1$. Upon substitution of these, Equations (12.80) and (12.81) may be recast as

$$\left|\sigma - \frac{1}{1 + q_{12}}\right| \left|Z_1^{(1,2)}\right| \leq \left|Z_S + Z_1^{(n^*)}\right| \leq \left|\sigma - \frac{1}{1 - q_{12}}\right| \left|Z_1^{(1,2)}\right| \tag{12.85}$$

$$\left|\sigma - \frac{1}{1 + q_{12}}\right| \left|Z_1^{(1,2)}\right| \left|v_1'^{(n^*)}\right| \leq |p_S| \leq \left|\sigma - \frac{1}{1 - q_{12}}\right| \left|Z_1^{(1,2)}\right| \left|v_1'^{(n^*)}\right| \tag{12.86}$$

where $Z_1^{(1,2)} = Z_1^{(2)} - Z_1^{(1)}$ denotes the vector $\overrightarrow{C^{(1)}C^{(2)}}$. The lower and upper bounds in the foregoing equations are proportional to $\rho_{\mp} = |\sigma - 1/(1 \pm q_{12})|$, the proportionality factors being $\left|Z_1^{(1,2)}\right|$ for Equation (12.85) and $\left|Z_1^{(1,2)}\right| \left|v_1'^{(n^*)}\right|$ for Equation (12.86). $\rho_\pm$ are shown in Figure 12.9a as functions of q_{12}. Their ratio, $\rho = \rho_+/\rho_-$, depends only on q_{12} and is shown in Figure 12.9b.

The following features are noteworthy in these figures:

(i) As q_{12} tends to be sufficiently small or large, the bounds get increasingly tighter: if q_{12} is of $O(\varepsilon)$ or $O(\varepsilon^{-1})$, where $\varepsilon << 1$, then the bounds are equal to $O(\varepsilon)$ and, in the limit of $\varepsilon \to 0$, we have $Z_S \to -Z_1^{(n^*)}$.

(ii) As q_{12} approaches unity to $O(\varepsilon)$, $\varepsilon << 1$, the upper bound increases to infinity asymptotically and is $O(\varepsilon^{-1})$ larger than the lower bound.

The capability to bracket the source characteristics in this way enhances the efficacy of the two-load method. Note that only auto-power spectral density measurements with two loads are needed for the calculation of the bounds. Consequently, they can be determined even when there is insufficient data for the application of the two-load method.

12.4.4 Calculation Bounds to Sound Pressure

The calculation of sound pressure at a target point is described in detail in Chapter 11. However, if source characteristics are determined in intervals, as described in the previous section, further considerations are required.

Let a generic load, Z_1, be driven by a measured source from the equivalent source plane (plane 1) and let $p'_a(f)$ denote the acoustic pressure at the target point. From Equation (11.55):

$$p'_a = \frac{p_S}{1+\dfrac{Z_S}{Z_1}}\left(\frac{G_{a,1}}{Z_1}\right), \tag{12.87}$$

where the transfer function $|Z_1|/|G_{a,1}|$ in dB units, that is, $L_{NR} = 20\log_{10}(|Z_1|/|G_{a,1}|)$ is called noise reduction (see Section 11.4.1). Applying Equation (12.87) for points $Z_S \in \Theta$ generates the set $Q_a = \{|G_{a,1}||p|/|Z+Z_1| : Z \in \Theta,\ |p| \in B\}$, where B denotes the set $B = \left\{|\mathbf{F}(Z)|\left|v_1'^{(n^*)}\right| : Z \in \Theta\right\}$ (Section 12.4.3). Then the value of $|p'_a|$ is bounded by the interval $P_a = [\min(Q_a), \max(Q_a)]$.

We can also derive an a posteriori inequality for this interval. Substituting Equation (12.87) into the abridged form of Equation (12.81), that is, $|p_\mathrm{M}| \le |p_S| \le |p_\mathrm{N}|$, we obtain

$$|G_{a,1}|\left(\frac{|p_\mathrm{M}|}{|Z_S+Z_1|}\right) \le |p'_a| \le |G_{a,1}|\left(\frac{|p_\mathrm{N}|}{|Z_S+Z_1|}\right). \tag{12.88}$$

But, as can be seen from the topology of the load circle of Z_1 relative to Θ, the magnitude of the vector $Z_S + Z_1$ can be bracketed as

$$||Z_C+Z_1| - R_C| \le |Z_S+Z_1| \le |Z_C+Z_1| + R_C. \tag{12.89}$$

Hence, Equation (12.88) can be expressed in a slightly more relaxed form as

$$|G_{a,1}|\left(\frac{|p_\mathrm{M}|}{|Z_C+Z_1|+R_C}\right) \le |p'_a| \le |G_{a,1}|\left(\frac{|p_\mathrm{N}|}{||Z_C+Z_1|-R_C|}\right), \tag{12.90}$$

which is the required inequality. It bounds $|p'_a|$ to an interval with upper and lower limits which are determined by the source parameters p_M, p_N, Z_C and R_C, which are in turn determined by Θ.

12.4.5 The Three-Load Method

The physically correct intersection point in Figure 12.6 may also be picked by repeating the measurements by using a third load. As illustrated in Figure 12.10, if the third load is independent of the other two, then, theoretically, the corresponding load circle has to pass through the physically correct intersection point of the first two load circles. The application of this idea is called the three-load method. It can be implemented algebraically by solving equations of the three load circles, namely,

$$\left(X_S + X_1^{(1)}\right)^2 + \left(Y_S + Y_1^{(1)}\right)^2 = \left(R^{(1)}\right)^2 \tag{12.91}$$

$$\left(X_S + X_1^{(2)}\right)^2 + \left(Y_S + Y_1^{(2)}\right)^2 = \left(R^{(2)}\right)^2 \tag{12.92}$$

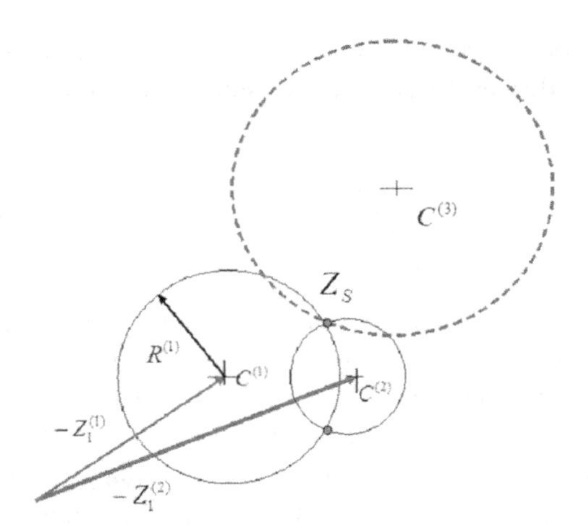

Figure 12.10 Geometric demonstration of the three-load method

$$\left(X_S + X_1^{(3)}\right)^2 + \left(Y_S + Y_1^{(3)}\right)^2 = \left(R^{(3)}\right)^2 \tag{12.93}$$

for the three unknowns X_S, Y_S and $|p_S|$ (recall that $R^{(n)} = \left|Z_1^{(n)}\right| |p_S| / \left|p'^{(n)}\right|$. Alves and Doige suggested eliminating $|p_S|$ between pairs of these equations and then solving the resulting two non-linear equations in X_S and Y_S numerically by using the Newton–Raphson method [19]. The magnitude of the source pressure strength, $|p_S|$, could then be determined from the equation of one of the load circles. A dilemma with this approach is that the two reduced equations are quadratic in both X_S and Y_S, and, in the presence of inevitable measurement errors, they can yield spurious solutions which cannot be distinguished systematically from the two that correspond to the physically correct intersection point of the load circles.

The main advantages of the three-load method are: (i) phase measurement with respect to the source process is not necessary and (ii) only power spectral density of $p'^{(n)}$ is required. With calibrated loads, the method is implemented by using a single microphone, which can be located as shown in Figure 12.5b.

12.4.6 Over-Determined Methods

Experience shows that the two-load method and, in particular, the three-load method are sensitive to measurement errors. The normal measurement errors in data acquisition and processing and acoustic path models interfere with the accuracy of the results and there is usually a non-negligible error in the measured source characteristics, even though extreme care is exercised in measurements to keep these errors under control. On the other hand, it is also known that the primary sound source mechanisms in fluid machinery, such as unsteady fluid mass injection by valve action, are load dependent, time-variant and non-linear to some extent. Possible errors due to these effects, and other measurement errors, can be minimized in the least-squares sense, by using multiple loads. Considerable progress has been achieved in the optimization of load combinations [20], detecting the consistency of the linear source model [21] and testing non-linearity inherent to the measured system [22].

12.4.6.1 Over-Determined Two-Load Method

Load-dependency, in particular, has the potential to jeopardize the reliability of the two-load method, because, even with a linear time-invariant source, different pairs of loads can yield different source characteristics. Since it takes two loads to determine the source characteristics, the method does not lend itself to the determination of the source characteristics as function of the loads (see Section 12.4.8). However, the modeling errors (due to non-linearity, time-invariancy and load-dependency of the source) as well as the measurement errors can be smoothed out by using more than two loads. If measurements are carried out by using $N > 2$ independent loads, writing Equation (12.70) for $n = 1, 2, \ldots, N$ gives an over-determined system of equations of the form

$$\begin{bmatrix} v_1'^{(1)} & -1 \\ v_1'^{(2)} & -1 \\ \vdots & \vdots \\ v_1'^{(N)} & -1 \end{bmatrix} \begin{bmatrix} Z_S \\ p_S \end{bmatrix} + \begin{bmatrix} p_1'^{(1)} \\ p_1'^{(2)} \\ \vdots \\ p_1'^{(N)} \end{bmatrix} = \begin{bmatrix} e_1^{(1)} \\ e_1^{(2)} \\ \vdots \\ e_1^{(N)} \end{bmatrix}. \tag{12.94}$$

The least-squares solution of this equation is

$$\begin{bmatrix} \sum_{n=1}^{N} v_1'^{(n)} \breve{v}_1'^{(n)} & -\sum_{n=1}^{N} \breve{v}_1'^{(n)} \\ -\sum_{n=1}^{N} v_1'^{(n)} & N \end{bmatrix} \begin{bmatrix} Z_S \\ p_S \end{bmatrix} = \begin{bmatrix} -\sum_{n=1}^{N} \breve{v}_1'^{(1)} p_1'^{(1)} \\ \sum_{n=1}^{N} p_1'^{(1)} \end{bmatrix}. \tag{12.95}$$

It is clear that when extra loads are used, the condition of Equation (12.71) is no longer required and a unique solution of Equation (12.94) may exist even if some pairs of loads are linearly dependent in this sense.

Figure 12.11 shows the equivalent characteristics of an intake breathing noise source of a four-stroke four-cylinder engine measured by using the two-load method with non-calibrated loads (Figure 12.5a). The modulus of the source strength and the source impedance are shown as function of the crankshaft speed for the fourth harmonic of the firing cycle frequency. The dashed curves are the characteristics measured by using two loads and the thick solid curves are those measured by using six loads.

12.4.6.2 Over-Determined Three-Load Method

Since the three-load method is very sensitive to measurement errors, several authors suggested the use of a fourth load to improve its efficacy. Interestingly, when

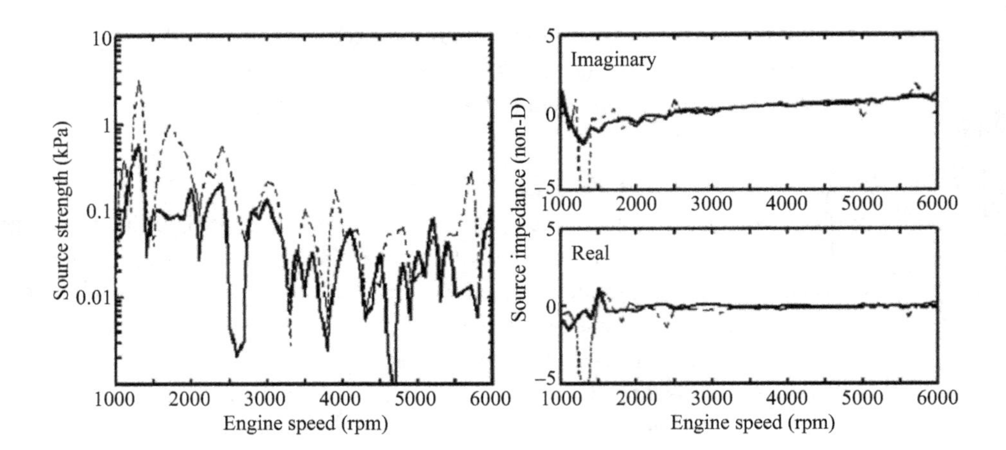

Figure 12.11 Equivalent source characteristics of an internal combustion engine intake noise measured by the two-load method using two loads (dashed curves) and six loads (thick curves).

Equations (12.91)–(12.93) are augmented with a fourth one, the three non-linear equations, which result from the elimination of $|p_S|$ in paired combinations of the four loads, can be manipulated into two linear equations in X_S and Y_S [12, 23, 24]. This reduction is flawed because the solution of the two linear equations is not sufficient for enforcing this solution to be also a solution of the parent non-linear equations [25]. However, procedures suggested later to avoid this flaw have not been particularly effective and progress in more recent applications has been in the direction of using many more than four loads for smoothing the modeling and measurement errors in the least-squares sense [26–27], or by minimizing an error function [28]. These approaches are considerably complex, since the over-determined system now consists of many non-linear equations of the form of Equations (12.91)–(12.93).

12.4.7 The Fuzzy Two-Load Method

A disadvantage of the over-determined methods is that the computed point (or average) estimates for the equivalent source parameters carry no information of how closely the true values of the parameters are estimated. This information is best specified by estimating intervals that include the point estimates. In probability sampling, such intervals are calculated routinely and are known as confidence intervals. But, in over-determined two- and three-load methods, the loads do not constitute a probability sample and, therefore, calculation of such confidence intervals is not feasible. However, in Section 12.4.3 we have seen that intervals containing a unique solution of Equations (12.67) can be determined by using the properties of the Apollonian circle of two loads. This possibility can be extended so that the uncertainty due to modeling and measurement errors are taken into account by fuzzification of the characteristic ratio of the Apollonian circle of two loads [29].

12.4.8 The Explicit *N*-Load Method

In time-domain, a linear time-variant one-port equivalent source that is periodic in time can be defined as

$$p_S(t) = \left[\mathrm{L}\{v_1'(t)\}\right]' + p_1'(t), \tag{12.96]}$$

where prime denotes the fluctuating part of a quantity and L denotes a linear integro-differential operator of the form

$$\mathrm{L} = a\mathrm{D}^{\alpha} + b\mathrm{D}^{\beta} + \cdots, \tag{12.97}$$

where D^{λ} denotes λ-fold differentiation with respect to time for $\lambda > 0$ and integration if $\lambda < 0$; $\alpha, \beta, \ldots$ denote distinct integers and the coefficients a, b,... may be constants or periodic functions of time. Typically, $\mathrm{L} = a + b\mathrm{D}$, which represents a predominantly resistive and inductive source. Upon expansion of Equation (12.96) into complex Fourier series, it may be shown that [30]:

$$p_{S_k} = \sum_{\substack{m=1 \\ m \neq k}}^{M} z_m^{(k-m)} v_{k-m}' + \sum_{m=1}^{M} z_m^{(-k-m)} v_{k+m}' + z_o^{(k)} v_k' + p_k', \qquad k = 1, 2, \ldots, K \tag{12.98}$$

where p_{S_k}, v_k' and p_k' denote the kth complex Fourier coefficient of $p_S(t), v_1'(t)$ and $p_1'(t)$, respectively, and

$$z_m^{(k)} = (ik\omega)^{\alpha} a_m + (ik\omega)^{\beta} b_m + \cdots. \tag{12.99}$$

Here, $a_m, b_m, \ldots$ denotes the mth complex Fourier coefficient of $a(t), b(t), \ldots$, respectively, which are to be determined from the measured $v_1'(t)$ and $p_1'(t)$ signals for a number of acoustic loads. As in the two-load method, the measurements can be carried out by using the two-microphone method, or a single microphone with calibrated loads. The number of loads required depends on the upper limit assumed for the convolution summations in Equation (12.98), M, say. Then, the total number of Fourier coefficients $a_m, b_m, \ldots$ present $J(1 + 2M)$ real unknowns, where J denotes the number of coefficients $a, b, \ldots$ in Equation (12.97). These are to be determined by writing the source equation, Equation (12.98), for first K harmonics in the source spectrum. But each source equation will in turn introduce a new complex unknown, that is, p_{S_k}, the source strength. Therefore, the number of loads required to determine the source spectrum in a frequency range covering the first K harmonics is given by

$$N = 1 + \frac{J(1 + 2M)}{2K}. \tag{12.100}$$

The solution of this equation for an integer N can be expressed as

$$K \mid (1 + 2M) = 0, \qquad \left(\frac{K}{J/2}\right) \mid (1 + 2M) = 0, \qquad J = 2, 4, 6, \ldots \tag{12.101}$$

where $A \mid B$ means the remainder of B divided by A. For a set of integers J,K,M,N satisfying Equation (12.101), the source operator L can be determined from Equation

Table 12.1 Solutions of Equation (12.101)

J	K	M	N	J	K	M	N	J	K	M	N	J	K	M	N
2	7	3	2	2	9	4	2	2	11	5	2	2	13	6	2
4	14	3	2	4	18	4	2	4	22	5	2	4	26	6	2
6	21	3	2	6	27	4	2	6	33	5	2	6	39	6	2
4	7	3	3	4	9	4	3	4	11	5	3	4	13	6	3
8	14	3	3	8	18	4	3	8	22	5	3	8	26	6	3
12	21	3	3	12	27	4	3	12	33	5	3	12	39	6	3

(12.98) to the accuracy of M harmonics in J operators, and the source pressure strength p_{S_k} to the accuracy of K harmonics, by using N loads, provided that the frequency spectrum of the measurements required to determine the particle velocity terms in Equation (12.98) span over at least K+M harmonics. This approach is called explicit N-load method. Some solutions of Equation (12.101) for two and three loads are shown in Table 12.1, which can be generalized by extending its pattern horizontally and vertically.

The basic implementation of the explicit N-load method consists of solving the system of K linear equations resulting from application of Equation (12.98) for $k = 1, 2, \ldots, K$, for the Fourier coefficients $a_m, b_m, \ldots$, which present $J(1 + 2M)$ real unknowns, with J,K,M,N satisfying Equation (12.101). For given J and M, this basic form of the method allows least-squares overdetermination by using more than N loads, or more than K harmonics. Load overdetermination consists of using $N + n$ loads, where n is the number of extra loads. Similarly, the harmonic overdetermination is applied by writing Equation (12.98) for $K + k$ harmonics, where k denotes the extra number of harmonics. In both cases, J and M remaining the same, the extra number of loads and harmonics yield more equations than the number of unknowns. Hence, the latter can be determined in the least-squares sense, as described for the two-load method.

Equation (12.98) encompasses both linear time-invariant and time-variant source formulations in frequency domain. If time-invariance is assumed a priori, it simplifies to the one-port source equation

$$p_{S_k} = z_o^{(k)} v'_k + p'_k, \qquad k = 1, 2, \ldots, K \tag{12.102}$$

since the convolution sums vanish when the coefficients $a, b, \ldots$ are constants. In general, Equation (12.98) can also be expressed in the usual one-port source form as

$$p_{S_k} = z_e^{(k)} v'_k + p'_k, \qquad k = 1, 2, \ldots, K \tag{12.103}$$

where

$$z_e^{(k)} = z_o^{(k)} + \sum_{\substack{m=1 \\ m \neq k}}^{M} z_m^{(k-m)} \frac{v'_{k-m}}{v'_k} + \sum_{m=1}^{M} z_m^{(-k-m)} \frac{v'_{k+m}}{v'_k}, \qquad k = 1, 2, \ldots, K \tag{12.104}$$

is called the effective source impedance. After computation of the source operator coefficients, the effective source impedance is determined for each load used and can be curve fitted to obtain an empirical equation giving the source impedance as function of the load. The simplest form of such an equation is the linear regression

$$Z_S = a_0 + a_1 Z_1, \tag{12.105}$$

where $Z_1(=p'_1/v'_1$, or p'_k/v'_k in the notation of the present section) denotes the load impedance at the source plane and Z_S denotes the effective source impedance defined by Equation (12.104). Hence, the explicit N-load method can be used for the measurement of one-port source characteristics, with the added advantage of yielding the source impedance as function of the load.

Shown in Figure 12.12 are the regression coefficients a_0 and a_1 for the intake noise source impedance of a four-stroke internal combustion engine at 4000 rpm, measured by using the explicit 2-load method. The source operator is assumed to be $\mathrm{L} = a + b\mathrm{D}$ and the operator coefficients are represented by $M = 6$ harmonics. As can be seen from Table 12.1, this combination determines the source spectrum to $K = 13$ harmonics, or, for a crankshaft speed of 4000 rpm of a four-stroke engine, to about 430 Hz. This range can be increased by harmonic overdetermination, which will also have the effect of smoothing the possible measurement and model errors in the least squares sense. The regression coefficients shown in Figure 12.12 are determined by using $K + k = 40$ harmonics in the source spectrum.

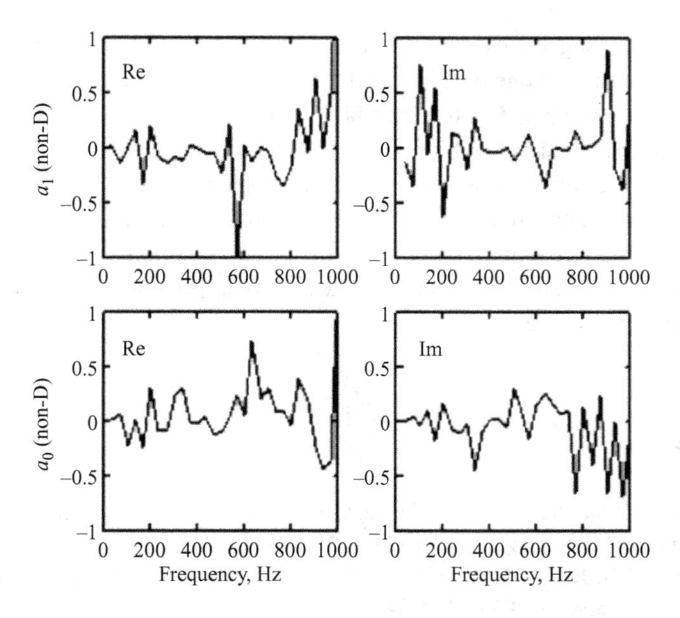

Figure 12.12 Linear regression coefficients modeling load-dependence of the intake noise source impedance of an engine at 4000 rpm.

12.5 Measurement of Two-Port Source Characteristics

Referring to Figure 11.1b, where planes 1_a and 1_b represent the equivalent source planes, a ducted linear time-invariant two-port acoustic source can be represented mathematically by Equation (2.17). Upon applying the usual plane wave decompositions $p'_{1_{a,b}} = p^+_{1_{a,b}} + p^-_{1_{a,b}}$ and $v'_{1_{a,b}} = \left(p^+_{1_{a,b}} - p^-_{1_{a,b}}\right)/(\rho_o c_o)_{1_{a,b}}$, where $p^{\mp}_{1a}$ and $p^{\mp}_{1b}$ denote the traveling pressure wave components at planes 1_a and 1_b, respectively Equation (2.17) can be expressed in the form

$$\mathbf{p}_S = \mathbf{T}_S \begin{bmatrix} p^+_{1a} \\ p^-_{1a} \end{bmatrix} + \begin{bmatrix} p^+_{1b} \\ p^-_{1b} \end{bmatrix}. \tag{12.106}$$

Here, $\mathbf{p}_S$ denotes the source pressure strength vector and $\mathbf{T}_S$ denotes a 2×2 transfer matrix. And $\mathbf{p}_S$ and $\mathbf{T}_S$ represent six complex unknowns and a method involving a number of test states are needed to determine these by measurement. Lavrentjev et al. proposed creating the required test states by using external sound sources [31]. This approach is based on the hypothesis that $\mathbf{T}_S = \mathbf{T}^*_{1_b 1_a}$, where $\mathbf{T}^*_{1_b 1_a}$ denotes the wave transfer matrix between the equivalent source planes when the source is operating. Thus, in order to measure $\mathbf{T}^*_{1_b 1_a}$, it is necessary to generate a secondary acoustic field which is uncorrelated with the acoustic field of the actual source and strong enough to dominate over it, so that $\mathbf{T}^*_{1_b 1_a}$ can be measured (Section 12.3.4). Once $\mathbf{T}_S$ is known, $\mathbf{p}_S$ is determined by repeating the measurements with the external source turned off and using Equation (12.106).

A mathematical proof of the hypothesis underlying this external source approach is not available. The main advantage of the external source method is the possibility of using the signal driving the external source, usually a loudspeaker, as the reference signal for the two-microphone measurements required for the measurement of in-duct acoustic fields. The external source method may also be used similarly for the measurement of the characteristics of one-port sources [16].

The characteristics of two-port acoustic sources may be measured also by creating the required the test states by using independent acoustic loads. This approach is best applied by reducing a two-port source to an equivalent one-port source (see Section 11.2.2) and then using one of the methods described in previous sections for measurement of one-port sources.

References

[1] M. Åbom, Modal decomposition in ducts based on transfer function measurements between microphone pairs, *J. Sound Vib.* **135** (1989) 95–114.

[2] M. Åbom and H. Bodén, Error analysis of two-microphone measurements in ducts with flow, *J. Acoust. Soc. Am.* **83** (1988), 2429–2438.

[3] T. Schultz, M. Sheplak and L.N. Cattafesta III, Uncertainty analysis of the two-microphone method, *J. Sound Vib.* 304 (2007), 91–109.

[4] S.-H. Jang and J.-G. Ih, On the multiple microphone method for measuring in-duct acoustic properties in the presence of mean flow, *J. Acoust. Soc. Am.* **103** (1998), 1520–1526.
[5] J.Y. Chung and D.A. Blaser, Transfer function method of measuring in-duct acoustic properties. I. Theory, II. Experiment, *J. Acoust. Soc. Am.* **68** (1980) 907–921.
[6] S. Allam and H. Bodén, Methods for accurate determination of acoustic two-port data in flow ducts, ICSV12, Lisbon, Portugal, (2005).
[7] R. Prony, Essai experimentale et analytique, *Journal de l'École Polytechnique* **1** (1795), 24–26.
[8] F. B. Hildebrand, *Introduction to Numerical Analysis*, (New York: McGraw-Hill, 1956).
[9] H. Boden, L. Zhou, J. Cordioli, A. Medeiros and A. Spillere, On the effect of flow direction on impedance eduction results, Proceedings of the 22nd AIAA-CEAS Aeroacoustics Conference, No. AIAA-2016-2727, Lyon, France, (2016).
[10] L.L. Beranek, J.L. Reynolds and K.E. Wilson, Apparatus and procedures for predicting ventilation system noise, *J. Acoust. Soc. Am.* **25** (1953), 313–321.
[11] I. Dunn and W. Dawern, Calculation of acoustic impedance of multi-layer absorbers, *Applied Acoustics*, **19** (1986), 321–334.
[12] K. Myers, Design of catenoidal shaped anechoic termination, (2012), Master's thesis, 41, available online at http://scholarworks.wmich.edu/masters_theses/41.
[13] T.Y. Lung and A.G. Doige, A time averaging testing method for acoustic properties of piping systems and mufflers with flow, *J. Acoust. Soc. Am.* **73** (1983), 867–876.
[14] M.L. Munjal and A.G. Doige, Theory of a two-source-location method for direct experimental evaluation of the four-pole parameters of an aeroacoustic element, *J. Sound Vib.* **141** (1990), 323–333.
[15] M. Åbom, Measurement of the scattering-matrix of acoustic two-ports, *Mech. Syst. Signal Process.* **5** (1991), 89–104.
[16] H. Bodén and M. Åbom, Modeling of fluid machines as sources of sound in duct and pipe systems, *Acta Acoustica* **3** (1995), 549–560.
[17] H. Bodén , Error analysis for the two-load method used to measure the source characteristics of fluid machines, *J. Sound Vib.* **126** (1988), 173–177.
[18] L. Desmons, J. Hardy and Y. Auregan, Determination of the acoustical source characteristics of an internal combustion engine by using several calibrated loads, *J. Sound Vib.* **179** (1995), 869–878.
[19] H.S. Alves and A.G. Doige, A three-load method for noise source characterization in ducts, Proc. NOISE-CON, 87, 329–334.
[20] S.F. Zheng, H.T. Liu, J.B. Dan and X.M. Lian, Analysis of the load selection on the error of source characteristics identification for an engine exhaust system, *J. Sound Vib.* **344** (2015) 126–137.
[21] V. Macian, A.J. Torregrosa, A. Broatch, P.C. Niven and S.A. Amphlett, A view on the internal consistency of linear source identification for IC engine exhaust noise prediction, *Math. Comput. Model.* **57** (2013) 1867–1875.
[22] H. Bodén and F. Albertson, Linearity tests for in-duct one-port sources, *J. Sound Vib.* **237** (2000), 45–65.
[23] M.L. Kathuriya and M.L. Munjal, Experimental evaluation of the aero-acoustic characteristics of a source of pulsating gas flow, *J. Acoust. Soc. Am.* **65** (1979), 240–248.
[24] M.G. Prasad, A four load method for evaluation of acoustical source impedance in a duct, *J. Sound Vib.* **114** (1987), 347–355.

[25] B.S. Sridhara and M.J. Crocker, Error analysis for the four-load method used to measure the source impedance in ducts, *J. Acoust. Soc. Am.* **92** (1992), 2924–2931.

[26] L. Desmons and J. Hardy, A least squares method for evaluation of characteristics of acoustical sources, *J. Sound Vib.* **175** (1994), 365–376.

[27] H. Bodén, On multi-load methods for determination of the source data of acoustic one-port sources, *J. Sound Vib.* **180** (1995), 725–743.

[28] S.-H. Jang and J.-G. Ih, Refined multiload method for measuring acoustical source characteristics of an intake and exhaust system, *J. Acoust. Soc. Am.* **107** (2000) 3217–3225.

[29] E. Dokumaci, The fuzzy two-load method for measurement of ducted one-port sources, *J. Sound Vib.* **397** (2017), 31–50.

[30] E. Dokumaci, On one-port characterization of noise sources in ducts by using external loads, *J. Sound Vib.* 260 (2003), 389–402.

[31] L. Lavrentjev, M. Åbom and H. Bodén, A measurement method for determining the source data of acoustic two-port sources, *J. Sound Vib.* **183** (1993), 517–531.

13 System Search and Optimization

13.1 Introduction

Acoustic design of a silencer usually begins with the selection of one or more trial concepts for the system configuration. This stage of design is based on first principles, previous applications and experience. In the next stage, combinations of possible values of system parameters are iterated for concept testing and optimization for a given design target. Some cases can be handled based on heuristic data and straightforward parameter analysis, but, in general, systematic search methods can be of great assistance for quick and efficient progress in this stage.

In general, the search process involves determination of the values of a set of parameters that makes an acoustic performance parameter (called an object function) take a value that satisfies the design objective for the frequencies of interest. The methods which can be used for this purpose may be classified into three broad categories: (1) direct search methods, which use only the function values, (2) gradient methods, which require accurate values of the first derivative of the object function and (3) second-order methods, which require accurate values of both the first and second derivatives of the objective function. These methods are capable of finding the minimum(s) or the maximum(s) of an objective function; however, in practical calculations, as long as the acoustic design objective is satisfied, it is usually not necessary to know whether this occurs very close to a global extremum or not. For this reason, direct search methods are quite suitable for acoustic design iterations, even though they offer no guarantees of convergence to the global optimum. The gradient and second-order methods can yield closer estimations of the optimum with guaranteed convergence. But for practical duct systems, it is generally impractical to obtain analytical expressions for the derivatives of acoustic transfer functions, which are highly oscillatory. Then it is inevitable to resort to the numerical difference approximations, but these are prone to failure due to the difficulty of finding a proper balance between step size and the truncation errors. Section 13.2 of the present chapter describes one class of direct search procedures, known as direct random search, in acoustic design problems of silencers.

On the other hand, there is always some uncertainty on values system parameters used in acoustic performance predictions. This may arise from factors such as geometrical approximations, neglected three-dimensional effects in theoretical predictions, manufacturing and assembly tolerances and even the service conditions.

Acoustic performance of a robust design should be reasonably insensitive under such uncertainty at the frequencies of interest. Numerical interval analysis is an effective method for the estimation of the effect of the variations of a system parameter when uncertainty exists about its precise value and the sensitivity of the acoustic performance to possible variations is of concern. Section 13.3 describes an application of numerical interval analysis.

We have seen in Chapter 11 that the sound pressure at a target point can be calculated, given the equivalent source characteristics and the acoustic path. This may be called the direct method. Whether a silencer prototype satisfies a given target sound pressure is traditionally determined by this direct method. In Section 13.4, we describe an alternative approach, which is called the inverse method, and discuss its practical applications.

13.2 Direct Random Search

Direct search methods for constrained optima may be divided into adaptations of unconstrained direct search methods and random search procedures. With former methods, considerable care must be exercised for accommodating the constraints, but sophisticated algorithms do not usually result in significant improvements. The direct random search techniques, on the other hand, are very easy to implement because of their simple search logic, and are quite reliable in yielding improved, if not optimal, solutions.

A constrained search problem may be stated as optimizing an object function which is expressed generically as

$$F = F(x_1, x_2, \ldots, x_M) \tag{13.1}$$

where the M real variables $x_1, x_2, \ldots, x_M$ are subject to inequality constraints

$$G_j(x_1, x_2, \ldots, x_M) \geq 0, \qquad j = 1, 2, \ldots, m \tag{13.2}$$

$$H_j(x_1, x_2, \ldots, x_M) \leq 0, \qquad j = 1, 2, \ldots, r \tag{13.3}$$

and equality constraints

$$Q_j(x_1, x_2, \ldots, x_M) = 0, \qquad j = 1, 2, \ldots, s \tag{13.4}$$

In order to have a meaningful problem, the number of the equality constraints, s, must be less than the number of variables, M; there is, however, no limit to the number of the inequality constraints. The inequality and equality constraints may be re-cast without loss of generality as

$$x_i = q_i(x_1, x_2, \ldots, x_{i-1}, x_{i+1}, x_{i+2}, \ldots, x_M), \qquad i = i_1, i_2, \ldots, i_s \tag{13.5}$$

$$x_j > g_j\left(x_1, x_2, \ldots, x_{j-1}, x_{j+1}, x_{j+2}, \ldots, x_M\right), \qquad j = j_1, j_2, \ldots, j_m, \qquad j \notin i \tag{13.6}$$

$$x_k < h_k(x_1, x_2, \ldots, x_{k-1}, x_{k+1}, x_{k+2}, \ldots, x_M), \qquad k = k_1, k_2, \ldots, k_r, \quad k \notin i \tag{13.7}$$

The parameters x_i defined by equality constraints are called the equality, or, dependent, variables. The remaining $N = M - s$ variables are called the independent variables. The random search space is defined by the ranges of variation of the independent variables. A convenient scheme for range assignment is to specify an initial value and a search radius for each independent variable.

In the basic form of direct random search, the trial points in the space of the independent variables are generated as

$$x_n = \hat{x}_n + R_n(2\rho_n - 1), \qquad n \in n_1, n_2, \ldots, n_N \tag{13.8}$$

where $\hat{x}_n$ denotes the value of the independent variable which produced the current best value of the objective function, R_n denotes the allowable radius of search range about $\hat{x}_n$, ρ_n denotes a pseudo-random number distributed uniformly between 0 and 1, and x_n denotes the new value of the independent variable to be tried. And ρ_n can be generated by using the random number routines available in most computer programming languages. Each generated trial point is used to calculate the dependent variables and tested by evaluating the inequality constraints, and if it is found to be feasible, the corresponding objective function value is compared to the previous best value corresponding to $\hat{x}_n$. If the current point yields a better value, it is retained as new $\hat{x}_n$; if not, it is rejected. The process continues until a given number of trial points have been generated.

This process is known to be quite inefficient in its use of function evaluations. It may be shown that, in order to attain 90% confidence that the initial range of uncertainty, $2R_n$, is reduced by a factor of ε_n, the number of function evaluations required is of the order of [1]:

$$2.3\prod_n \varepsilon_n^{-1}, \qquad n \in n_1, n_2, \ldots, n_N. \tag{13.9}$$

For example, for $N = 5$ and $\varepsilon_n = 0.01$, the number of function evaluations required will be of the order of 2.3×10^{10}.

Since this large number of trials prohibits the practical use of the basic direct random search, many other pseudo-random search techniques have been proposed to improve the efficiency of function evaluations. In general, these methods work by altering the allowable search region. One strategy is to divide the sampling into a series of blocks. The best point of each block is used to initiate the next one, which is executed with a range reduction. For example, Luus and Jaakola proposed that, after each block consisting of μ function evaluations, the ranges of the independent variables are reduced by a factor of $1 - p$, and the search is terminated after B blocks have been exhausted [2]. The typical values suggested for the search were $p = 0.05, \mu = 100$ and $B = 200$, which correspond to a total range reduction by a factor of 3.5×10^{-5} at the cost of 20 000 function evaluations. This is a significant improvement in the number of function evaluations over the basic direct random search; however, depending on the CPU time for each function evaluation, this may or may not be considered an acceptable computational burden.

Since x_n is random, Equation (13.8) corresponds to random sampling from a uniform probability distribution. The sampling region can also be controlled by causing the probability distribution to become more concentrated around $\hat{x}_n$. This can be achieved by modifying Equation (13.8) as [3]

$$x_n = \hat{x}_n + R_n(2\rho_n - 1)^K, \qquad n \in n_1, n_2, \ldots, n_N, \qquad K = 3, 5, 7, \ldots \tag{13.10}$$

As K increases, the probability distribution becomes more and more concentrated around $\hat{x}_n$. Since K serves as an adaptive parameter whose value regulates the search region, this approach is referred to as adaptive random direct search. Heuckroth et al. suggest to start with $K = 1$ and increase K by 2 after a specified number of improved object function evaluations and decrease by 2 when no improvements occur after a certain number of function evaluations [3]. This periodic increase in the search region seems to help to avoid local optima and speed up convergence. An advantage of this approach is that the number of function evaluations between adjustments is flexible, varying with the progress of the search. As the probability of achieving better function evaluations becomes less as K increases, it is proposed that further range reduction is applied by modifying Equation (13.10) as

$$x_n = \hat{x}_n + \frac{R_n}{K}(2\rho_n - 1)^K, \qquad n \in n_1, n_2, \ldots, n_N, \qquad K = 3, 5, 7, \ldots \tag{13.11}$$

Another adaptive ramification of Equation (13.8), proposed by Martin and Gaddy [4], aims to search randomly for the improvements in the objective function from the surface of an N-dimensional ellipsoid centered at $\hat{x}_n$. Thus, the current values of the independent variables are determined by

$$x_n = \hat{x}_n + \frac{R_n}{\kappa} \frac{2\rho_n - 1}{\sqrt{\sum_n (2\rho_n - 1)^2}}, \qquad n \in n_1, n_2, \ldots, n_N \tag{13.12}$$

where $\kappa = 1, 2, \ldots$ denotes a range reduction coefficient. The axes of the ellipsoid are the ranges R_n/κ. The ranges and, hence, the size of the ellipsoid, are decreased as the search converges to maintain a satisfactory success ratio, that is, the ratio of the number of better function evaluations to the number function evaluations. It is recommended that the search is started with $\kappa = 1$ and increased by one if the success ratio drops below 0.4. The initial range of the variables is important in achieving good function evaluation efficiency. Therefore, it is suggested that the initial ranges be halved if no improvements are found after a number function evaluation. Direct random search based on Equation (13.12) is called randomly directed search by the proposing authors.

Numerous similar strategies which claim to improve the efficiency of function evaluations have also been proposed, most without extensive comparative testing to verify their relative efficiency. Comparison with random search techniques is difficult since the search itself is a stochastic (Monte Carlo) process. Different search histories can be obtained from the same starting point, depending on the sequence of random numbers used. For this reason, random search efficiencies are best compared on the basis of average number of function evaluations. But average efficiencies are reported in only few publications. On the whole, the body of available evidence indicates that the above described simple random search techniques are quite adequate if highly accurate determination of optima is not required.

There are publications in the open literature demonstrating the application of some optimization techniques for the transmission loss of some simple silencers [5–7]. Although such studies imply that the same methods can be extended for practical problems, this is usually not feasible, because it is very time consuming even for the specialist to derive a realistic explicit mathematical model for a given system configuration. For this reason, optimization techniques are best implemented in practice with the support of an acoustic network program. For example, with the block diagram technique described in Chapter 2, after a block diagram model is constructed, the search process can be set up in few simple steps involving the selection of the object function, the type of the target bound, the variant of random direct search method and the parameters which control its strategy. The search variables can be selected from the datasheets of the blocks and assigned with algebraic symbols, so that the equality and inequality constraints and the search ranges of the independent variables are defined in the format of Equations (9.5)–(9.7) in symbolic form.

As an example, consider a muffler location optimization problem. Figure 13.1 depicts schematically the cold end of the exhaust line of a commercial car and its block diagram model. (Block diagrams of the mufflers are not shown.) An upper-bound target is specified for the sound pressure level (SPL) in dB(B) at a given distance and angle from the tailpipe outlet over the speed range of the engine, for frequencies equal to integer multiples of twice the engine crankshaft rotational speed. This target is shown in Figure 13.2 by the dashed curve. The optimization problem is to find locations for the two mufflers on a given wheelbase, L, that meet this target. The lengths of ducts that may be varied are denoted as x_a, x_b and x_c, but these lengths are subject to constraints imposed by the fixed wheelbase and the muffler dimensions. The front muffler is not to be closer than a given distance to the source plane; the rear

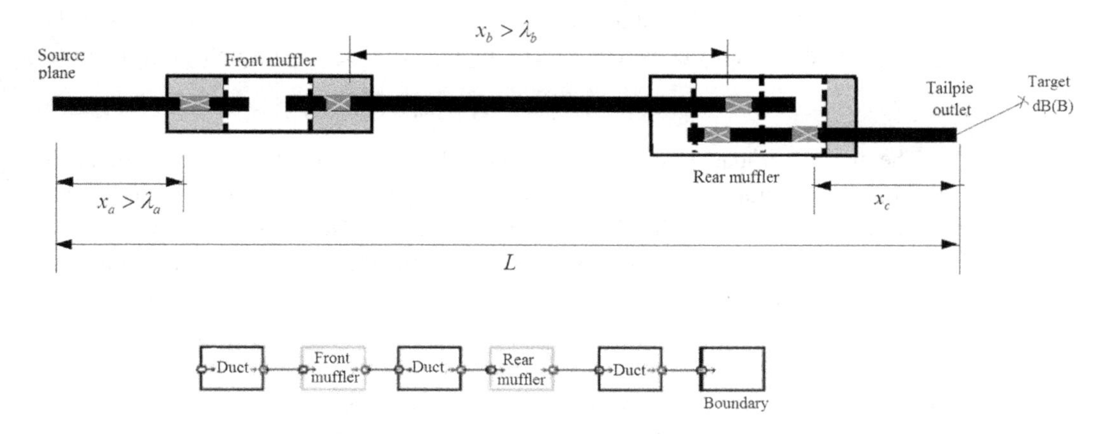

Figure 13.1 The cold end of the exhaust line of a commercial vehicle and its block diagram model. The gray shading indicates sound absorbent packing. Crosses indicate perforated ducts. Dashed lines indicate perforated baffle plates.

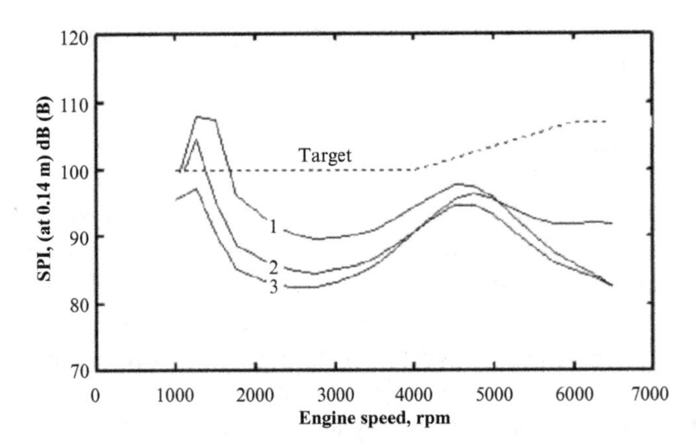

Figure 13.2 Adaptive randomly directed search results for the sound pressure level (SPL) at twice the engine rotational speed: curve 1: $x_a = 0.931$ m, $x_b = 0.959$ m, fitness function $= 90.9\%$; curve 2: $x_a = 1.1649$ m, $x_b = 0.3629$ m, fitness function $= 95.5\%$; curve 3: $x_a = 0.5102$ m, $x_b = 0.6422$ m, fitness function $= 100\%$.

muffler is not to be closer than a given distance to the tailpipe outlet, and the two mufflers must be separated at least by a given distance. These constraints can be expressed mathematically as

$$x_a > \lambda_a \tag{13.13a}$$

$$x_b > \lambda_b \tag{13.13b}$$

$$x_c = \ell - (x_a + x_b), \tag{13.13c}$$

where the minimum values of λ_a, λ_b and ℓ are fixed by the above requirements and the equality constraint follows from the fact that L is fixed. There is one more inequality constraint, that is, $x_c > \lambda_c$, but this may not be given in this form, because x_c is already specified by the equality constraint. Accordingly, we write this inequality constraint as

$$x_b < \ell - \lambda_c - x_a. \tag{13.13d}$$

The progress of the search may be monitored by observing the number of function evaluations, the number of successful function evaluations, the value of the fitness function and the number of retries with range reduction. The fitness function is defined as the ratio of the number of spectrum lines satisfying the target to the total number of spectrum lines. Figure 13.2 shows the progress of a typical search using the adaptive randomly directed search method, Equation (13.5). Curve 1 is the predicted sound pressure level for the input initial values of the independent variables. This gives a fitness function value of 0.909. The search yielded, after a number of unsuccessful function evaluations, the curve 2. Since no improvements seemed to occur for the subsequent function evaluations, a range reduction is applied. Then, curve 3 is obtained with fitness function of unity. Thus, insofar as the meeting of the target is concerned, this solves the problem. However, the search may be continued to see if there are other solutions.

13.3 Interval Analysis

Interval analysis is a simple but effective method to determine the effect of the variations of system parameters in specified intervals on acoustic performance silencers. However, rigorous analytical calculations using interval arithmetic are unwieldy because of the complexity of mathematical operations involved and yield rather loose bounds even in the simplest cases. Numerical interval analysis by repeated simulations is better suited for practical calculations. In this approach, a stochastic (Monte Carlo) sampling technique is employed to obtain approximate bounds for the design objective.

To demonstrate an application of this technique, consider modeling of the branch shown in Figure 13.3 using plane-wave elements. This model is prone to yield a shifted resonance frequency for the branch, since the effects of the finite extent of the discontinuity and the evanescent waves generated there are neglected by the one-dimensional acoustic elements. It is common to try to correct this shift by applying a heuristic end-correction to the length of the branch. This correction depends strongly on the geometry and the correct amount is not known until the test results are available or the problem is solved in three dimensions. This uncertainty inherent to a one-dimensional model can be taken into account by treating the lengths of the ducts as interval parameters. For example, let the nominal lengths of the branch and the outlet ducts be a_1 and a_2 (Figure 13.3) and assume that these are subject to uncertainty within the intervals $(a_1,a_1+\varepsilon_1)$ and $(a_2-\varepsilon_2,a_2+\varepsilon_2)$, respectively. Here, in order to emulate the effect of the end-correction, we have taken the lower bound of the interval for the branch length equal to its nominal length. For the outlet duct length, on the other hand, variations both above and below the nominal value are considered. No

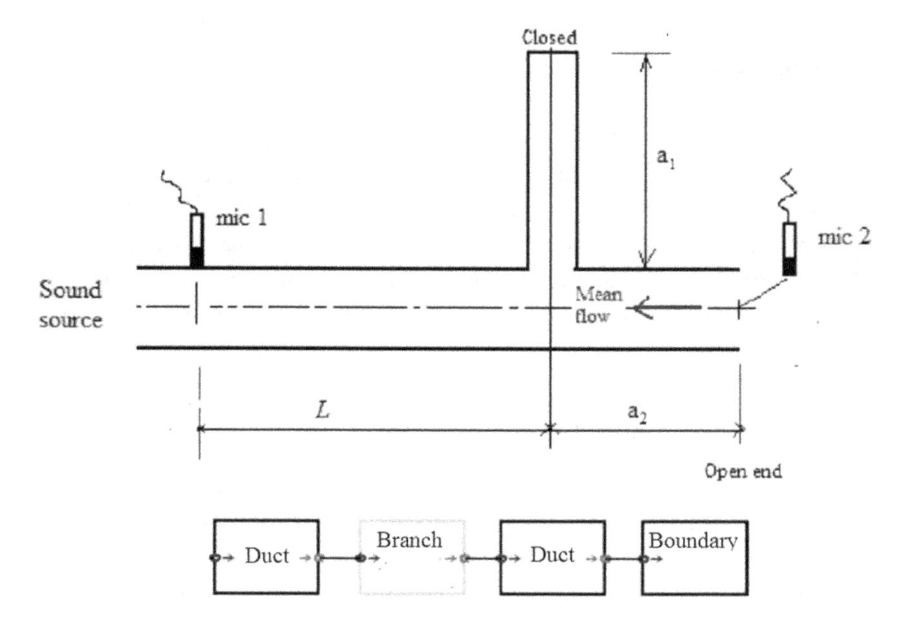

Figure 13.3 A duct with a closed duct branch.

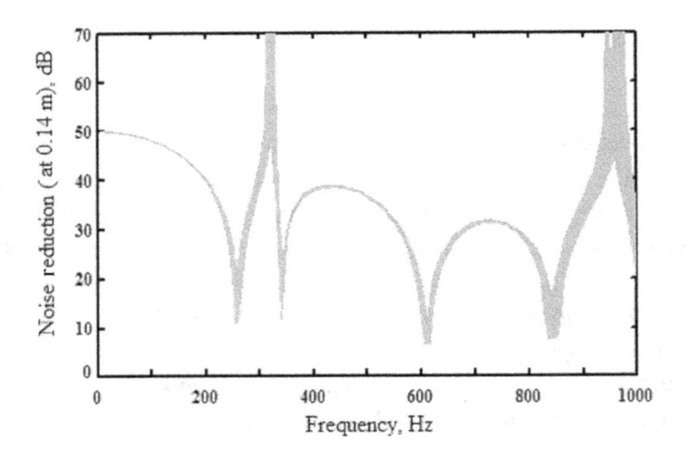

Figure 13.4 Noise reduction between microphone 1 and 2, for branch length and outlet duct length ranges of $(a_1, a_1 + \varepsilon_1)$ and $(a_2 - \varepsilon_2, a_2 + \varepsilon_2)$, respectively: $a_1 = 265$ mm, $a_2 = 65$ mm, $\varepsilon_1 = 10$ mm, $\varepsilon_2 = 5$ mm.

end-correction is applied for the open unflanged end, since an accurate model is used for the reflection coefficient there (see Section 9.4.3).

Shown in Figure 13.4 is the predicted noise reduction, computed for 50 randomly drawn samples of branch and outlet duct lengths. This figure clearly shows that the prediction of acoustic performance in interval form when some system parameters are fuzzy has decided advantage over crisp predictions based on crisp but uncertain heuristic estimates of the duct parameters. Numerical interval calculations not only show the possible variations, but also depict the sensitivity of the system to parameter variations, which is an important asset at the design stage.

13.4 The Inverse Method

The techniques described in the previous sections of this chapter are useful for speeding up the sound pressure calculations by the direct method (Section 11.2). In this section, we describe a different approach which can provide this information without explicit knowledge of the acoustic model of the path between the equivalent source plane and the target point and discuss its applications in acoustic design of silencers. This method is called the inverse method, because it is based on an inverse formulation of the direct calculations. First, we introduce the concept of acoustic path space.

13.4.1 Acoustic Path Space

Consider a duct system radiating sound from its open end (Figure 13.5). Let plane 1 denote the equivalent one-port source plane[1] and plane 2 the radiating open-end

[1] It suffices to consider a one-port source, as sound radiation from ducted two-port sources may be calculated by using its one-port representations (see Section 11.2.2).

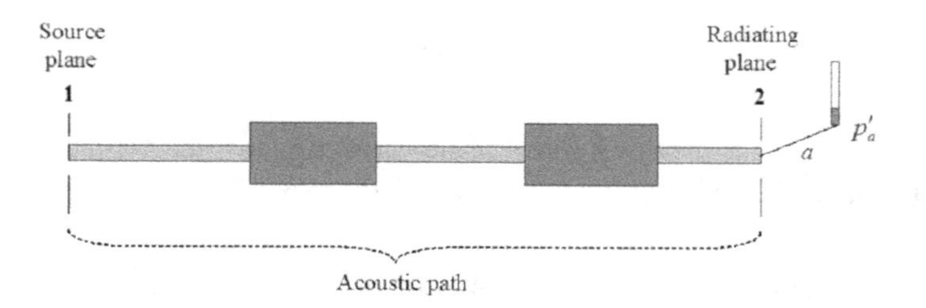

Figure 13.5 Basic elements of an acoustic path.

plane. The part of the system between planes 1 and 2 is called the acoustic path. We assume that the shape and size of duct cross sections at planes 1 and 2, and the equivalent source characteristics are known, and that the mean flow conditions at these sections may be estimated with fair accuracy. The sound pressure, p'_a, at distance a from the acoustic center of the open end is given by the sum $p'_a = p'^{(1)}_a + p'^{(2)}_a + \cdots + p'^{(b)}_a$, where b denotes the modality of the acoustic path (see Section 11.2.1). Using the radiation transfer function definition in Equation (9.55), this can be expressed as

$$p'_a = \frac{p_1^+}{K^{(1)}} + \frac{p_2^+}{K^{(2)}} + \cdots + \frac{p_b^+}{K^{(b)}} = \begin{pmatrix} \frac{1}{K^{(1)}} & \frac{1}{K^{(2)}} & \cdots & \frac{1}{K^{(b)}} \end{pmatrix} \mathbf{P}_{2b}^+, \tag{13.14}$$

where the vector $\mathbf{P}_{2b}^+ = \{ p_1^+ \quad p_2^+ \quad \ldots \quad p_b^+ \}$ is given, in terms of the source and the acoustic path, by Equations (11.12) or (11.18).

An acoustic path which satisfies the target specification

$$|p'_a| \leq p_{\text{target}} \tag{13.15}$$

is called an admissible path. Let q modes be propagating at plane 1 and m modes at plane 2. We can always contract the acoustic path so that $q = b \geq m$ or $q < b = m$ (see Section 11.2.1). In the latter case, inspection of Equation (11.18) shows that, provided that the shape and size of the duct at plane 1 are known, the acoustic path determines the elements of matrices $\mathbf{A}_{22}^{-1}\mathbf{A}_{21}$ and $\mathbf{F}$, which are of size $(m - q) \times q$ and $(2q) \times q$, respectively. On the other hand, if $q = b \geq m$, we may conclude from Equation (11.12) that the acoustic path determines the elements of two matrices of sizes $q \times q$. Thus, for given modality, however complex the acoustic path may be, p'_a depends on a finite number of parameters. Consequently, whether or not Equation (13.15) is satisfied can be studied on a space spanned by these parameters. In the following study of this problem, we focus on the case of $q = 1$, as it is tractable to explicit analysis. In this case, Equation (11.29) applies and, hence, Equation (13.13) can be expressed as

$$p'_a = \frac{p_S}{A^+(1 + \zeta_S + (1 - \zeta_S) r_1)} \sum_{j=1}^{m} \left(\frac{\beta_2^{(j)}}{K_2^{(j)}} \right), \tag{13.16}$$

where $\beta_2^{(1)} = 1$ and we have used the notation

$$A^{+} = \frac{p_1^{+}}{p_2^{+}} \tag{13.17}$$

for the forward or incident ($+x$ wave) plane-wave transmission coefficient α defined by Equation (11.22). Equation (13.15) and (13.16) may be combined as

$$|A^{+}(1+\zeta_S) + (1-\zeta_S)A^{-}r_2| \geq \frac{|p_S|}{p_{target}} \left| \sum_{j=1}^{m} \left(\frac{\beta_2^{(j)}}{K_2^{(j)}} \right) \right|, \tag{13.18}$$

where r_2 denotes the reflection coefficient at plane 2 (which is assumed to be known) and

$$A^{-} = \frac{p_1^{-}}{p_2^{-}} \tag{13.19}$$

is called the reflected plane-wave transmission coefficient. Equation (13.18) contains $m+1$ parameters $(A^{+}, A^{-}, \beta_2^{(2)}, \ldots, \beta_2^{(m)})$ of the acoustic path when $m > 1$, and only 2 parameters (A^{+}, A^{-}) when $m = 1$. The latter case is very convenient, because it allows the topology of the inequality in Equation (13.18) to be studied on a two-dimensional space spanned by A^{+} and A^{-}. To bring also about this simplicity in the case of $m > 1$, we invoke the hypothesis of equipartition of acoustic power at plane 2. This fixes the mode participation factors in Equation (13.18) as

$$\beta_2^{(j)} = \sqrt{\frac{\Lambda_2^{(1)+}}{\Lambda_2^{(j)+}}}, \qquad j = 1, 2, \ldots, m \tag{13.20}$$

where $\Lambda_2^{(j)+}$ is given by Equation (11.63). Hence, Equation (13.18) may be expressed as

$$|\alpha A + \beta B| \geq \Pi, \tag{13.21}$$

where

$$\alpha = 1 + \zeta_S \tag{13.22}$$

$$\beta = (1 - \zeta_S)r_2 \tag{13.23}$$

$$\Pi = \frac{|p_s|}{p_{\text{target}}} \sum_{j=1}^{m} \left(\frac{1}{K_2^{(j)}} \sqrt{\frac{\Lambda_2^{(1)+}}{\Lambda_2^{(j)+}}} \right). \tag{13.24}$$

If $m = 1$, then the latter equation reduces to

$$\Pi = \frac{|p_s|}{\left|K_2^{(1)}\right| p_{\text{target}}}. \tag{13.25}$$

It is convenient to recast Equation (13.21) in the form

$$X^2 + Y^2 + 2eXY \geq \gamma^2, \tag{13.26}$$

with

$$X = \left| \frac{A^+}{\beta} \right| \tag{13.27}$$

$$Y = \left| \frac{A^-}{\alpha} \right| \tag{13.28}$$

$$\gamma = \frac{\Pi}{|\alpha||\beta|} \tag{13.29}$$

$$e = \cos \varphi, \qquad \varphi = \angle Y - \angle X \tag{13.30}$$

where "$\angle$" denotes the argument of a complex quantity.

The parameter e is called the acoustic path index. It can be expressed in a physically more intuitive form as

$$e = \cos(\angle r_1 + \angle r_S), \tag{13.31}$$

with

$$r_S = \frac{1 - \zeta_S}{1 + \zeta_S}, \tag{13.32}$$

where $\zeta_S = Z_S/\rho_1 c_1$ denotes the normalized source impedance. Equation (13.31) shows that, for a given equivalent source impedance, the acoustic path index is determined by the phase of the reflection coefficient at the source plane.

The topology of the inequality in Equation (13.26) on the (X, Y) plane. is called the acoustic path space. The basic geometric features of this space are depicted in Figure 13.6. The curves correspond to Equation (13.26) with the equality sign and represent the boundaries of the domains containing all admissible paths. Any acoustic path which falls in a region enclosed by the boundary curves and the X and Y axes is inadmissible, that is, it does not satisfy the target, Equation (13.15). The filled regions in Figure 13.6 are the domain of admissible paths for $e = -1$. It is seen that, as the path index increases, the boundaries tend to the boundary for $e = 1$. Therefore, since $|e| \leq 1$, all acoustic paths which lie in the filled regions will always satisfy the target, irrespectively of the value of the acoustic path index.

The variation of the domains of admissible paths with the acoustic path index is shown in Figure 13.7. It is seen that the domain of admissible paths is largest for $e = 1$; however, for practical design purposes, this condition can be replaced by $e \geq 0$ without substantial loss of accuracy.

13.4.2 Acoustic Path Space on the Attenuation Plane

In the foregoing section, acoustic path space is described on the X, Y plane. This is convenient to elucidate the role of the acoustic path index. For practical calculations, however, it is more convenient to work in a space spanned by parameters which are

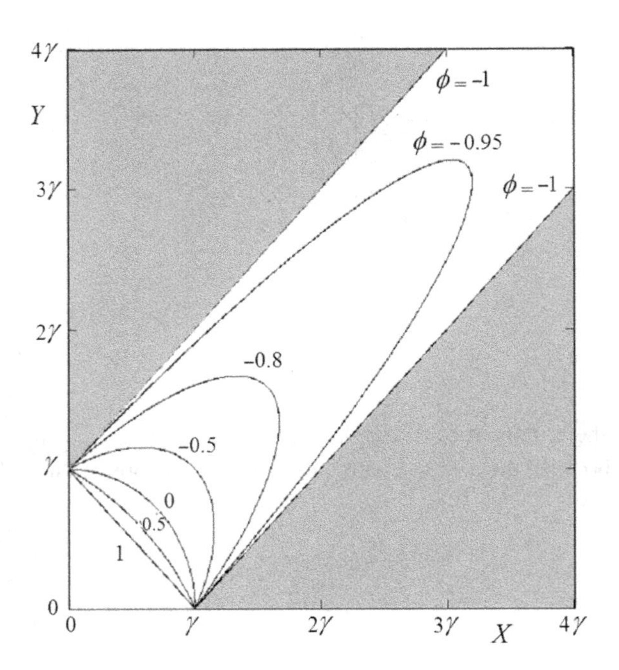

Figure 13.6 Admissible path boundaries in the acoustic design space.

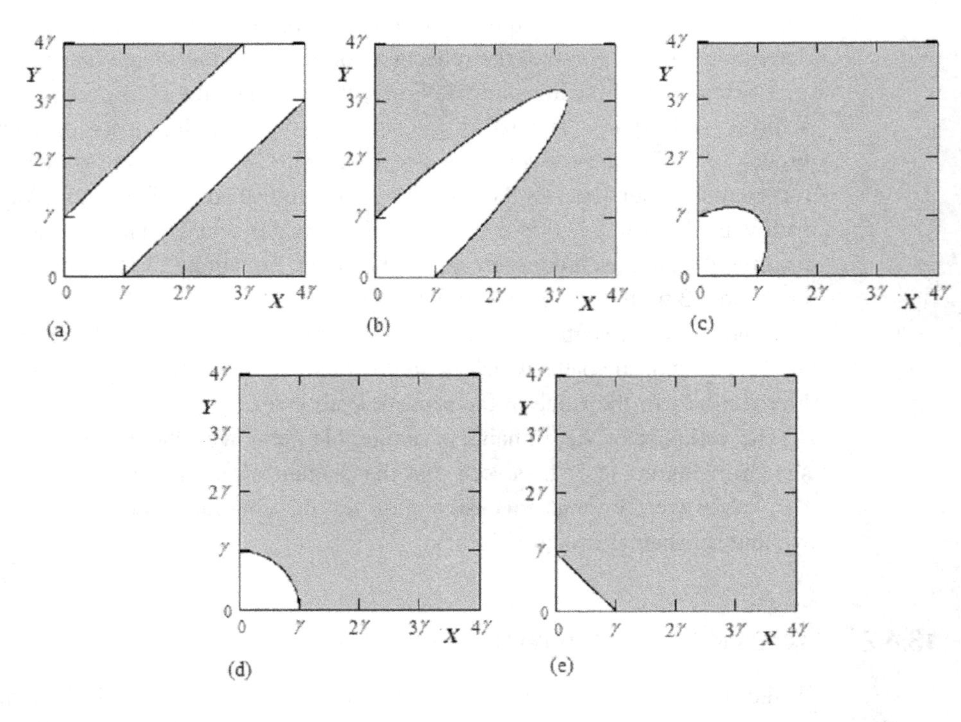

Figure 13.7 Variation of domains of admissible paths with the path index: (a) $e = -1$; (b) $e = -0.95$; (c) $e = -0.5$; (d) $e = 0$; (e) $e = 1$.

related to the performance of the acoustic path. The natural parameters for this purpose are the attenuation A in Equation (11.58), which is defined in the present notation as

$$A = 20\log_{10}|A^{+}| \tag{13.33}$$

and

$$B = 20\log_{10}|A^{-}|, \tag{13.34}$$

which is called, for lack of a better term, the reflection index of the acoustic path. From Equation (13.26), the boundaries of the domains of solutions satisfying the target may be expressed as

$$B = A + 20\log_{10}\left|\frac{\alpha}{\beta}\right| + 20\log_{10}\left(-e\mp\sqrt{e^2 + \frac{\gamma^2|\beta|^2}{10^{\frac{A}{10}}} - 1}\right), \tag{13.35}$$

where the sign(s) which makes the value of the bracket positive is applicable. This equation defines the acoustic path space on the (A, B) plane, which we call the attenuation plane. In this plane, the boundary curves are explicitly dependent on the parameters α and β, as well as γ and e. Typical forms of the boundaries on the attenuation plane are shown in Figure 13.8 for $\alpha = 1.8856$, $\beta = 0.6043$ and $\gamma = 5.41965$. The filled area shows the domain of admissible paths for $e = -0.99$. As the acoustic path index increases, this domain enlarges towards the $e = 1$ boundary. The domain of admissible paths is tightest for $e = -1$. The boundary corresponding to this case is indicated in Figure 13.8 by broken lines. Since this is not a continuous curve, it is usually convenient to indicate it approximated by using a value of the acoustic path index which is just larger than -1.

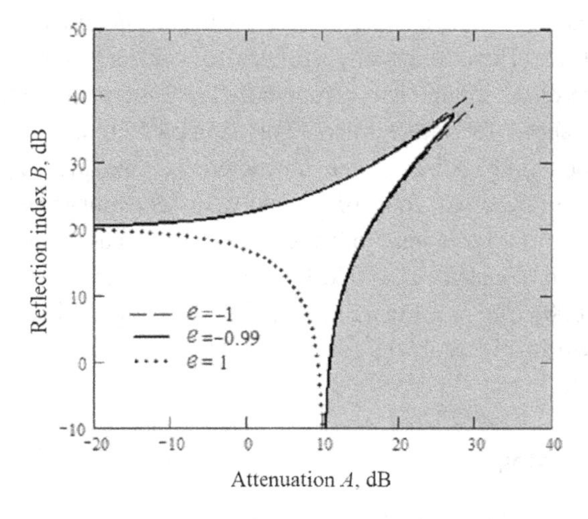

Figure 13.8 Acoustic path space on the attenuation plane. $\alpha = 1.8856$, $\beta = 0.6043$, $\gamma = 5.41965$

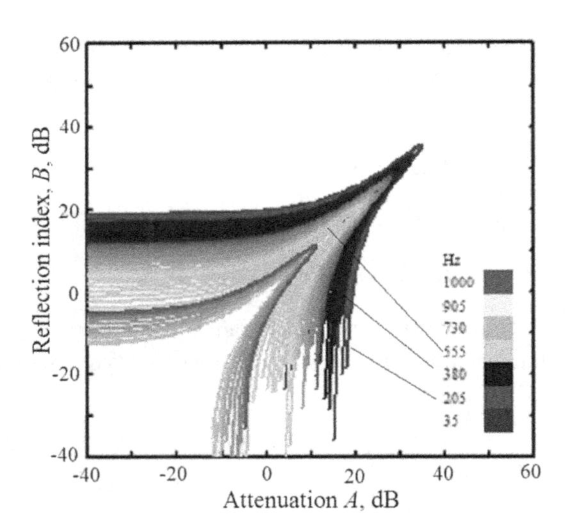

Figure 13.9 The acoustic path design space for $e = -0.99$ for an engine running at 4000 rpm.

The admissible path boundaries shown in Figure 13.8 apply for a given frequency. The boundaries corresponding to different frequencies may be plotted similarly. For example, Figure 13.9 shows the acoustic path space boundaries for the exhaust line of a four-cylinder four-stroke engine running at 4000 rpm. The boundaries are computed for $e = -0.99$ and a constant target of 100 dB(L), for firing cycle frequency harmonics up to 1000 Hz and a color palette is used to display the variation of the boundaries with the frequency. As can be deduced from Figure 13.9, the critical frequency range for the acoustic path is about 200–500 Hz. An attenuation of about 20 dB is required in this range. This attenuation requirement may be relaxed in the final design for frequencies for which the actual acoustic path index is larger than the assumed $e \approx -1$ in Figure 13.9.

In multi-cylinder engines, depending on the cylinder arrangement and manifold design, some of the fundamental cycle frequency orders are annihilated by interference (see Section 10.3). Then, it usually suffices to consider only the major orders in the speed range of the engine and compute the acoustic path boundaries as functions of the engine speed for these orders. For example, Figure 13.10 shows the equivalent intake breathing noise source characteristics for the fundamental cycle frequency harmonic of order four of a four-cylinder, four-stroke internal combustion engine. The sound radiation from the inlet of the intake line is subject to the SPL target which is constant at 100 dB(B) up to 3000 rpm and increases linearly to 120 dB(B) at 6000 rpm (see Figure 13.12). The design space boundaries for this intake line are shown in Figure 13.11 for $e = -0.99$ as function of the engine speed.

13.4.3 Signature of Acoustic Paths

Specific acoustic paths have distinct signatures on the acoustic path space. These signatures are curves, equations of which are given on the attenuation plane by

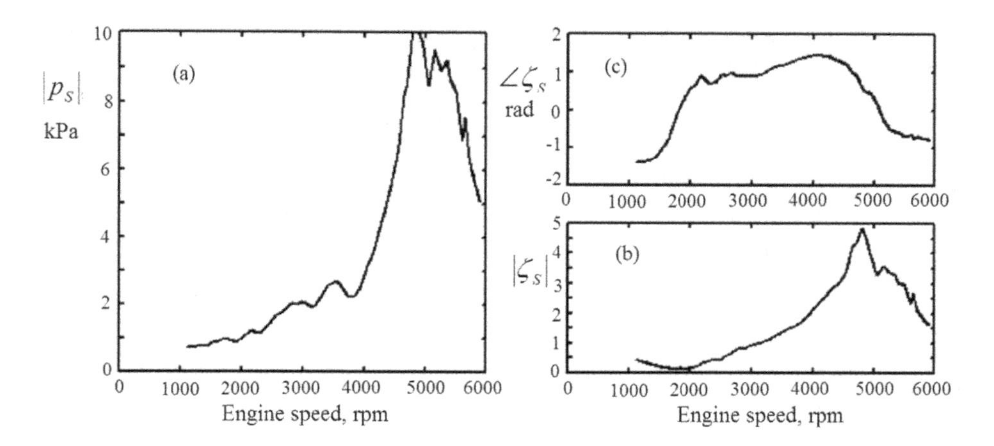

Figure 13.10 Intake breathing noise equivalent source characteristics of a four-cylinder, four-stroke internal combustion engine for the fundamental cycle frequency harmonic of order four. (a) Modulus of source pressure strength; (b) modulus of normalized source impedance, (c) argument of normalized source impedance in radians.

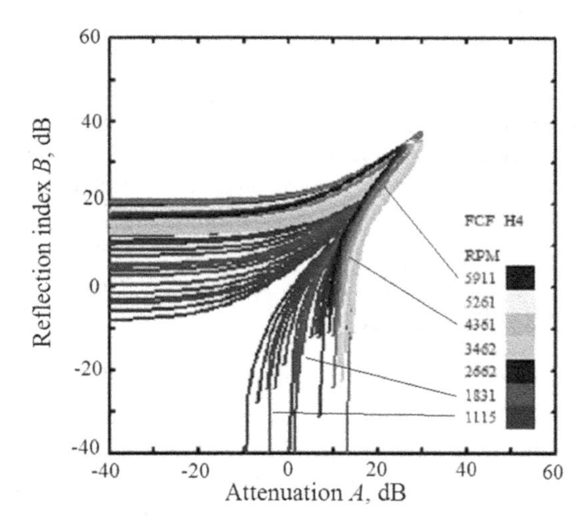

Figure 13.11 Acoustic path space boundaries for $e = -0.99$, for the intake line of a four-cylinder four-stroke engine for the fundamental cycle frequency harmonic of order four. Equivalent source characteristics are as in Figure 13.11 (see Figure 13.12 for the target).

$$B = A + 20\log_{10}\left|\frac{r_1}{r_2}\right|. \tag{13.36}$$

Points on a signature curve correspond one-to-one to values of the acoustic path index. Hence, admissibility of a specific acoustic path at specific frequencies can be examined by superimposing its signature on the attenuation plane with acoustic path space boundaries.

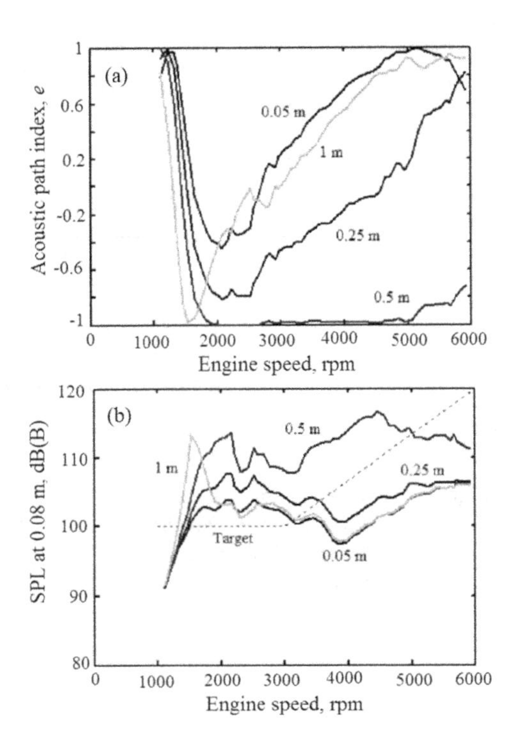

Figure 13.12 Correlation between the acoustic path index and the radiated sound pressure level (SPL) for a simple expansion chamber having various outlet duct lengths: (a) variation of the acoustic path index with the engine speed, (b) radiated sound pressure level as function of the engine speed.

13.4.4 System Search in Acoustic Path Space

Admissible paths can be searched by the path index criterion $e \geq 0$. For a demonstration, consider the engine referred to in Section 13.4.2 (Figures 13.10 and 13.11). For simplicity, let the acoustic path be a simple expansion chamber of diameter 240 mm, with inlet and outlet ducts of 64 mm diameter, the inlet duct being 520 mm long. Four options are considered for the lengths of the outlet duct (the snorkel): 50 mm, 250 mm, 500 mm and 1000 mm. Figure 13.12a shows the acoustic path indices corresponding to these lengths as function of the engine speed (recall that the fourth harmonic of the fundamental cycle frequency is being considered). As can be inferred from these characteristics, the best length option for the outlet duct is 50 mm. This conclusion can be checked by computing the radiated sound pressure level directly. Figure 13.12b shows the predicted results for the four outlet duct lengths considered. These results are seen to be well correlated with the acoustic path index characteristics in Figure 13.12a with regard to the $e \geq 0$ criterion and confirm the above conclusion for the 50 mm option, although this is still above the target by few dB(B) in the 1500–3000 rpm range.

Admissible paths may also be searched by using attenuation criteria deduced from the acoustic path boundaries. The procedure is based on the observation that the

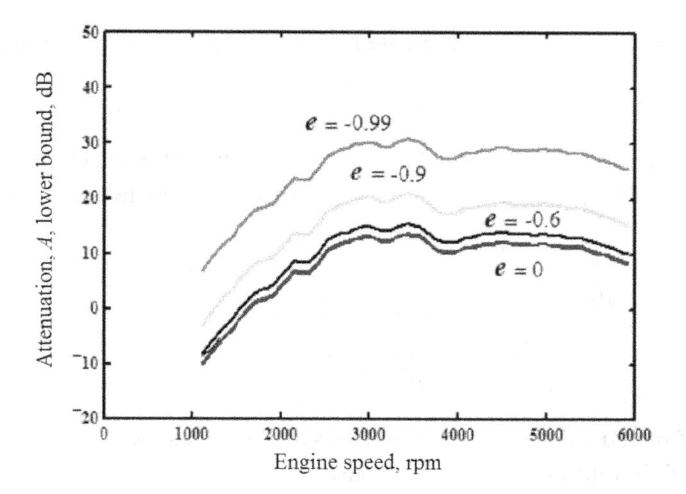

Figure 13.13 Maximum attenuation lower bounds for an internal combustion engine intake source.

admissible path boundaries on the design space are arcs of an ellipse; see Equation (13.26). The calculation of the point of this ellipse farthest away on the $+X$ direction, or the attenuation axis on the attenuation plane, yields the following maximum lower bound for the attenuation of admissible paths:

$$A \geq \begin{cases} 20\log_{10}|\gamma\beta| & e \geq 0 \\ 20\log_{10}\left|\dfrac{\gamma\beta}{\sqrt{1-e^2}}\right| & e < 0 \end{cases} \tag{13.37}$$

where the equality signs correspond to the extreme points of the boundary curves. Thus, the problem of the determination of an admissible silencer reduces to the problem of designing it so that an attenuation criterion established from Equation (13.37) is satisfied. For example, Figure 13.13 shows the maximum attenuation lower bounds determined from Equation (13.37) for typical values of the acoustic path index of the simple expansion chamber considered in the previous section.

13.4.5 Acoustic Path Spaces for Different Targets

This section presents variations of the basic form of the inverse method described in the previous sections for different target parameters.

13.4.5.1 Noise Reduction

Referring to Figure 13.5, noise reduction is defined as (see Section 11.4.1):

$$L_{NR} = 20\log_{10}\left|\frac{p_1'}{p_a'}\right|. \tag{13.38}$$

The target on this parameter will normally be specified as a lower bound, that is,

$$L_{NR} \geq N_T. \tag{13.39}$$

With m modes propagating at plane 2, the sound pressure at the target point is given by $p'_a = p'^{(1)}_a + p'^{(2)}_a + \cdots + p'^{(m)}_a$, where $p'^{(j)}_a$ denotes the contribution of mode j. Hence,

$$\frac{p'_1}{p'_a} = \frac{1 + r_1^{(1)}}{\dfrac{p_2^{+(1)}}{p_1^{+(1)} K_2^{(1)}} + \dfrac{p_2^{+(2)}}{p_1^{+(1)} K_2^{(2)}} + \cdots + \dfrac{p_2^{+(n)}}{p_1^{+(1)} K_2^{(n)}}}. \tag{13.40}$$

Substituting from Equation (11.28) for the ratios of the pressure wave components and using the result in Equation (13.38), we obtain

$$L_{NR} = 20 \log_{10}|A + r_2 B| - 20 \log_{10}\left|\sum_{j=1}^{m} \frac{\beta_2^{(j)}}{K_2^{(j)}}\right|, \tag{13.41}$$

where $\beta_2^{(j)}, j = 1, 2, \ldots, m$, are given by Equation (13.20). Combining this with the noise reduction target gives an inequality in the form of Equation (13.21) with

$$\alpha = 1, \beta = r_2, \Pi = 10^{\frac{N_T}{20}}\left|\sum_{j=1}^{m} \frac{1}{K_2^{(j)}} \sqrt{\frac{\Lambda_2^{(1)+}}{\Lambda_2^{(j)+}}}\right|. \tag{13.42}$$

It may be shown that the acoustic path index is now given by $e = \cos(\angle r_1)$. Previous results about the boundaries and regions on the acoustic path space for sound pressure targets apply now with these modifications.

13.4.5.2 Insertion Loss

Insertion loss is defined in Section 11.4. First consider the acoustic path of the new system. Equality form of Equation (13.21) for this path may be expressed as

$$|\alpha A + \beta B| = \left|\frac{p_S}{p'_a}\right| \left|\sum_{j=1}^{m} \left(\frac{1}{K_2^{(j)}} \sqrt{\frac{\Lambda_2^{(1)+}}{\Lambda_2^{(j)+}}}\right)\right|, \tag{13.43}$$

where p'_a denotes the sound pressure at the target point. Let this acoustic path be replaced by a straight lossless uniform duct of length L, which we take as the reference system. It is assumed that the source characteristics and the modal radiation transfer functions are not affected substantially by this modification. It is also assumed that the source plane and the position of the target point relative to the radiating outlet are not changed. For the reference duct, Equation (13.43) applies with $A = \mathrm{e}^{-\mathrm{i}k_o K^+ L}$ and $B = \mathrm{e}^{-\mathrm{i}k_o K^- L}$. Hence, eliminating the source pressure strength, we obtain

$$|\alpha A + \beta B| = \left|\frac{(p'_a)_{\text{ref}}}{p'_a}\right| \left|\alpha e^{-ik_o K^+ L} + \beta e^{-ik_o K^- L}\right| \tag{13.44}$$

where $(p'_a)_{\text{ref}}$ denotes the acoustic pressure at the target point when the reference duct is mounted, $k_o = \omega/c_o$ denotes the wavenumber, c_o denotes the speed of sound in the reference pipe and $K^{\pm}$ denote the propagation constants of a straight uniform duct. But by definition,

$$\left|\frac{(p'_a)_{\text{ref}}}{p'_a}\right| = 10^{\frac{L_{IL}}{20}}, \tag{13.45}$$

where L_{IL} denotes the insertion loss of the new system with respect to the reference duct. Then, if insertion loss is specified as target lower bounds

$$L_{IL} \geq I_T \tag{13.46}$$

the inequality form of Equation (13.43) takes the form of Equation (13.21) with

$$\Pi = 10^{\frac{I_T}{20}} \left|\alpha e^{-ik_o K^+ L} + \beta e^{-ik_o K^- L}\right|. \tag{13.47}$$

Consequently, all previous results about the boundaries and regions in an acoustic path space (on the attenuation plane) for sound pressure targets also apply, with Π given as above, for acoustic paths satisfying specified insertion loss targets.

References

[1] H.A. Spang, A review of minimization techniques for non-linear function, *SIAM Review* **4** (1962), 343–365.

[2] R. Luus and T.H.I. Jaakola, Optimization by direct search and systematic reduction of the size of search region, *AIChE Journal* **19** (1973), 760–766.

[3] M.W. Heuckroth, J.L. Gaddy and L.D. Gaines, An examination of the adaptive random search technique, *AIChE Journal* **22** (1976), 744–750.

[4] D.L. Martin and J.L. Gaddy, Process optimization with the adaptive randomly directed search, in R.S.H. Mah and G.V. Reklaus (eds.) *Selected Topics in Computer-Aided Process Design and Analysis*, CEP Symposium Series, Volume 14, (New York: American Institute of Chemical Engineers, 1982).

[5] J.S. Lamancusa, Geometric optimization of internal combustion engine induction systems for minimum noise transmission, *J. Sound Vib.* **127** (1988), 303–318.

[6] M.C. Chiu and Y.C. Chang, Shape optimization of multi-chamber cross-flow mufflers by SA optimization, *J. Sound Vib.* **312** (2008), 526–550.

[7] Y.C. Chang, L.J. Yeh and M.C. Chiu, Shape optimization on double-chamber mufflers using a genetic algorithm, *Proc. Inst. Mech. Eng. C. J. Mech. Eng. Sci.* **10** (2005), 31–42.

Appendix A Basic Equations of Fluid Motion

Motion of Newtonian fluids is governed by the classical laws of conservation of mass, momentum and energy, which provide relations between the properties of fluid particles in motion. For example, mass density is a property. Other properties are: velocity, pressure, temperature, internal energy, kinetic energy, potential energy, enthalpy, entropy, momentum. A particle is considered as the smallest volume of the fluid and is assumed to contain enough molecules to adequately describe the fluid by statistical averaging.

The conservation laws are closed by the constitutive (state) equations that apply to specific types of media, and the boundary conditions that apply for the geometry considered. They can be expressed in integral or differential forms. The integral forms are given for a control volume, which is to be understood as a part of a fluid that is separated from the rest of the fluid, but allowing the fluid cross through its boundaries. The differential forms of the conservation laws are derived from the corresponding integral forms by using the divergence theorem, which is stated in three dimensions as

$$\int_S \mathbf{A}\cdot\mathbf{n}\,\mathrm{d}\,S = \int_V \nabla\cdot\mathbf{A}\,\mathrm{d}V, \tag{A.1}$$

where V denotes the control volume, S denotes the surface of the control volume, $\mathbf{n}$ denotes the outward unit normal vector of S, ∇ denotes a vector gradient operator, $\mathbf{A}$ denotes any vector function. This theorem may also be applied in one and two dimensions by appropriate interpretation of the integration domain. Throughout this book, the partial and ordinary derivatives are expressed by using the usual symbols ∂ and d, respectively. For example, in a right-handed three-dimensional orthogonal coordinate system (x_1, x_2, x_3), the gradient operator is $\nabla = \mathbf{e}_1\partial/\partial x_1 + \mathbf{e}_2\partial/\partial x_2 + \mathbf{e}_3\partial/\partial x_3$, where $\mathbf{e}_1, \mathbf{e}_2, \mathbf{e}_3$ denote unit vectors in positive directions of x_1, x_2, x_3 axes, respectively.

A.1 Integral Forms of Conservation Laws

The general conservation laws are given in this section for Newtonian fluids, including the source terms. The following simplifications can be made if the fluid is ideal and the control volume has no sources (or sinks):

- In a source free region: $\dot{m} = 0,\ \mathbf{f} = 0,\ \dot{U} = 0.$
- For an ideal fluid: $\mu = 0,\ \kappa = 0.$

Here, $\dot{m}$ denotes the local mass per unit volume created into the control volume per unit time, $\mathbf{f}$ denotes the volume force per unit volume applied on the fluid, $\dot{U}$ denotes the local internal energy per unit volume created into the control volume per unit time, μ is the shear viscosity coefficient and κ denotes the coefficient of thermal conductivity. An over-dot is used to denote the rate of change with respect to time.

A.1.1 Conservation of Mass

This states that the rate of increase of material in a control volume equals the rate of net flux of material into the control volume plus the rate of creation of new mass into that control volume. Mathematically,

$$\frac{\partial}{\partial t}\int_V \rho\, \mathrm{d}V = -\int_S \rho\ \mathbf{v}\cdot\mathbf{n}\, \mathrm{d}S + \int_V \dot{m}\mathrm{d}V. \tag{A.2}$$

Here, ρ denotes the mass density and the minus sign on the right of Equation (A.2) accounts for the fact that the vector product $\mathbf{v}\cdot\mathbf{n}$ is positive for flow outwards the control volume.

A.1.2 Conservation of Momentum

This states that the sum of the external forces acting on the fluid within a control volume equals the time rate of increase of the momentum within the control volume, plus the net rate of momentum flux out of the control volume:

$$\frac{\partial}{\partial t}\int_V \rho\, \mathbf{v}\, \mathrm{d}V + \int_S \rho\mathbf{v}\ \mathbf{v}\cdot\mathbf{n}\, \mathrm{d}S = -\int_S p\mathbf{n}\, \mathrm{d}S + \int_S \mathbf{T}\mathrm{d}S + \int_V \mathbf{f}\, \mathrm{d}V. \tag{A.3}$$

Here, p denotes the fluid pressure and $\mathbf{T}$ denotes the viscous surface force per unit area. In a right-handed orthogonal coordinate system (x_1, x_2, x_3) $\mathbf{T}$ can be expressed in tensor notation as

$$\mathbf{T} = \mathbf{e}_i\left(\varphi_{ij}\mathbf{e}_j\right)\cdot\mathbf{n}. \tag{A.4}$$

Here, φ_{ij} is the rate-of-shear tensor

$$\varphi_{ij} = \frac{\partial v_i}{\partial x_j} + \frac{\partial v_j}{\partial x_i} - \frac{2}{3}\delta_{ij}\nabla\cdot\mathbf{v}, \tag{A.5}$$

where (v_1, v_2, v_3) denote the components of the velocity vector $\mathbf{v}$ along (x_1, x_2, x_3) axes, respectively, and δ_{ij} denotes the Dirac tensor, which is 1 if $i = j$, else it is zero (it may be noted that, $\varphi_{11} + \varphi_{22} + \varphi_{33} = 0$). Equation (A.3) is often written as

$$\frac{\partial}{\partial t}\int_V \rho\,\mathbf{v}\,\mathrm{d}V + \int_S \rho\mathbf{v}\,\mathbf{v}\cdot\mathbf{n}\,\mathrm{d}S = \int_S \mathbf{e}_i(\sigma_{ij}\mathbf{e}_j)\cdot\mathbf{n}\,\mathrm{d}S + \int_V \mathbf{f}\,\mathrm{d}V, \tag{A.6}$$

where $\sigma_{ij} = -p\delta_{ij} + \mu\varphi_{ij}$.

A.1.3 Conservation of Energy

This states that the time rate of increase of energy in a control volume equals the time rate of influx of energy into the control volume, plus the rate of heat addition, $\dot{Q}$, to the fluid in the control volume, minus the rate of work done by the fluid in the control volume on the surroundings, $\dot{W}$, plus the rate of creation of internal energy per unit volume into that control volume, $\dot{U}$. Mathematically,

$$\frac{\partial}{\partial t}\int_V \rho\, e\,\mathrm{d}V = -\int_S \rho\, e\,\mathbf{v}\cdot\mathbf{n}\,\mathrm{d}S + \dot{Q} - \dot{W} + \int_V \dot{U}\mathrm{d}V. \tag{A.7}$$

Here, T denotes the absolute temperature and e denotes the total stored energy per unit mass

$$e = u + \frac{1}{2}\mathbf{v}\cdot\mathbf{v} + e_P, \tag{A.8}$$

where u is the specific internal energy of the fluid per unit mass, e_P denotes the specific potential energy due to an external field per unit mass, and the second term on the right represents the kinetic energy per unit mass.

The net work done by the fluid contained in the control volume on the surroundings consists of the net work done on surroundings outside the control volume as a result of stresses at those positions on the control surface, S, across which there is a flow of fluid and any other form of work done on the surroundings outside the control volume. For brevity, the latter is referred to as shaft work, $\dot{W}_{\text{shaft}}$, and the former as flow work, $\dot{W}_{\text{flow}}$. The instantaneous power due to flow work is given by

$$\dot{W}_{\text{flow}} = \int_S \mathbf{v}\cdot\mathbf{e}_i(\sigma_{ij}\mathbf{e}_j)\cdot\mathbf{n}\mathrm{d}S = -\int_S p\mathbf{v}\cdot\mathbf{n}\mathrm{d}S + \int_S \mu v_i\varphi_{ij}\mathbf{e}_j\cdot\mathbf{n}\mathrm{d}S. \tag{A.9}$$

The net heat added to the fluid in the control volume, Q, is the sum of heat added to the control volume from its surroundings by conduction and any other form of heat added from the surroundings. Assuming that the Fourier law is valid, the former is given by

$$\dot{Q}_{\text{conduction}} = \int_S \kappa(\nabla T)\cdot\mathbf{n}\mathrm{d}S. \tag{A.10}$$

With the foregoing definitions, neglecting the shaft work and considering heat addition by thermal conduction only, the energy conservation, Equation (A.7), can be expressed as

$$\frac{\partial}{\partial t}\int\limits_V \rho\, e\, \mathrm{d}V + \int\limits_S (e\rho + p)\ \mathbf{v}\cdot\mathbf{n}\,\mathrm{d}S = \int\limits_S \mathbf{v}\cdot\mathbf{e}_i\left(\varphi_{ij}\mathbf{e}_j\right)\cdot\mathbf{n}\mathrm{d}S + \int\limits_S (\kappa\nabla T)\cdot\mathbf{n}\mathrm{d}S + \int\limits_V \dot{U}\mathrm{d}V. \tag{A.11}$$

This is often expressed in the form,

$$\frac{\partial}{\partial t}\int\limits_V \rho\, e\, \mathrm{d}V + \int\limits_S \rho\, h^o\ \mathbf{v}\cdot\mathbf{n}\ \ \mathrm{d}S = \int\limits_S \mu v_i\,\mathbf{e}_j\varphi_{ij}\cdot\mathbf{n}\mathrm{d}S + \int\limits_S \kappa\,(\nabla T)\cdot\mathbf{n}\,\mathrm{d}S + \int\limits_V \dot{U}\mathrm{d}V, \tag{A.11a}$$

where

$$h^o = u + \frac{p}{\rho} + \frac{1}{2}\mathbf{v}\cdot\mathbf{v} \tag{A.12}$$

is called stagnation enthalpy per unit mass.

A.2 State Equations and the Speed of Sound

The thermodynamic state postulate states that any thermodynamic state of a simple compressible fluid is completely determined by the specification of any two independent properties. Lists of state equations can be found in texts on thermodynamics. The state equations which are often useful in acoustics are given below in their total differential forms:

$$u = u(\rho, s): \quad \mathrm{d}u = T\mathrm{d}s + \frac{p}{\rho^2}\mathrm{d}\rho \tag{A.13}$$

$$u = u(p,\rho): \quad \rho\,\mathrm{d}u = \frac{\beta T}{\gamma - 1}\mathrm{d}p + \left(\frac{p}{\rho} - \frac{c_p}{\beta}\right)\mathrm{d}\rho \tag{A.14}$$

$$s = s(p,\rho): \quad \rho\mathrm{d}s = \frac{\beta}{\gamma - 1}\left(\mathrm{d}p - c^2\mathrm{d}\rho\right) \tag{A.15a}$$

$$s = s(p,T): \quad \mathrm{d}s = \frac{c_p}{T}\mathrm{d}T - \frac{\beta}{\rho}\mathrm{d}p \tag{A.15b}$$

$$h = h(p,s): \quad \mathrm{d}h = T\mathrm{d}s + \frac{1}{\rho}\mathrm{d}p \tag{A.16}$$

Here, s denotes the entropy per unit mass, $\beta = -(\partial\rho/\partial T)_p/\rho$ denotes the isobaric compressibility, $c_p = (\partial h/\partial T)_p$ denotes the specific heat coefficient at constant pressure, $\gamma = c_p/c_v$ denotes the ratio of specific heat coefficient at constant pressure to the specific heat coefficient at constant volume, $c_v = (\partial u/\partial T)_{1/\rho}$ denotes the specific heat coefficient at constant volume, and c is the speed of sound

$$c^2 = \frac{c_p\,(\gamma - 1)}{\beta^2 T} = \frac{1}{\rho\alpha}, \tag{A.17}$$

where $\alpha = -(\partial \rho/\partial p)_s/\rho$ is the isentropic (or, adiabatic) compressibility. The reciprocal $1/\alpha$ is called the adiabatic bulk modulus. The isothermal compressibility is defined as $-(\partial \rho/\partial p)_T/\rho = \gamma\alpha$. The inverse of the isothermal compressibility is called the isothermal bulk modulus. For a perfect gas, $\beta T = 1$, $p = \rho RT$ and $c^2 = \gamma RT = \gamma p/\rho$, where $R = c_p - c_v$ is the gas constant. This is determined from $R = R_o/M$, where $R_o = 8314.4$ J/(kg K) is the universal gas constant and M denotes the molecular weight of the gas. Thus, given one of the specific heat coefficients c_p or c_v, which are functions of only temperature for perfect gases, the other can be determined.

Properties of mixtures of perfect gases are given by the weighed sum of the corresponding properties of the constituents, the weighting factors being the mass fractions of the constituents in the mixture. Dry air and exhaust gas may be assumed to behave as a perfect gas. No simple generic correlation is available for the properties of real fluids, e.g., refrigerants in vapor state and liquids, or for multi-phase fluids. Databases such as REFPROP may be consulted for data on properties of real fluids [1].

It should be noted that, the state equations are invoked as relations between the total rates of change $\mathrm{d}/\mathrm{d}t = \partial/\partial t + \mathbf{v}\cdot\nabla$ of the fluid properties with respect to time, since, in the Eulerian view of real fluid motion adopted here, fluid properties are tied to the spatial positions of fluid particles and time.

A.3 Equations of Motion of Ideal Fluids

For an ideal fluid, the integral forms of the conservation laws reduce to the following:

Conservation of mass:

$$\frac{\partial}{\partial t}\int_V \rho \, \mathrm{d}V + \int_S \rho\mathbf{v}\cdot\mathbf{n} \, \mathrm{d}S = \int_V \dot{m}\mathrm{d}V. \tag{A.18}$$

Conservation of momentum:

$$\frac{\partial}{\partial t}\int_V \rho \, \mathbf{v} \, \mathrm{d}V + \int_S \rho\mathbf{v} \, \mathbf{v}\cdot\mathbf{n} \, \mathrm{d}S + \int_S p\mathbf{n} \, \mathrm{d}S = \int_V \mathbf{f}\mathrm{d}V. \tag{A.19}$$

Conservation of energy:

$$\frac{\partial}{\partial t}\int_V \rho \, e \, \mathrm{d}V + \int_S \rho \, h^o \, \mathbf{v}\cdot\mathbf{n}\mathrm{d}S = \int_V \dot{U}\mathrm{d}V. \tag{A.20}$$

The corresponding differential forms are derived by applying the divergence theorem, Equation (A.1).

A.3.1 Continuity Equation

Upon applying the divergence theorem to the surface integral in Equation (A.18) and assuming a control volume with a fixed boundary, the integral form is converted to the differential form

$$\frac{\partial \rho}{\partial t} + \mathbf{v}\cdot\nabla\rho + \rho\nabla\cdot\mathbf{v} = \dot{m}. \tag{A.21}$$

An equation of this form is generally called a continuity equation for the property of interest (in this case the mass density).

A.3.2 Momentum Equation

Applying the divergence theorem to the surface integral in Equation (A.19), its differential form is obtained as

$$\frac{\partial}{\partial t}(\rho\mathbf{v}) + \mathbf{v}\nabla\cdot\rho\mathbf{v} + \rho\mathbf{v}\cdot\nabla\mathbf{v} + \nabla p = \mathbf{f} \tag{A.22}$$

since $\mathbf{e}_i\nabla\cdot v_i\rho\mathbf{v} = \mathbf{v}\nabla\cdot\rho\mathbf{v} + \rho\mathbf{v}\cdot\nabla\mathbf{v}$. In a region with $\dot{m} = 0$, upon expanding the time derivative and using Equation (A.21), this reduces to

$$\rho\left(\frac{\partial\mathbf{v}}{\partial t} + \mathbf{v}\cdot\nabla\mathbf{v}\right) + \nabla p = \mathbf{f}. \tag{A.23}$$

If $\dot{m} \neq 0$ in the region considered, it may be shown that $-\dot{m}\mathbf{v}$ should be added to the right-hand side of this equation. This term accounts for the momentum carried with the mass creation in the source region. Equation (A.23) is called the Euler equation.

A.3.3 Energy Equation

Upon invoking the divergence theorem for the surface integral, the differential form of Equation (20) becomes

$$\frac{\partial}{\partial t}(e\rho) + \nabla\cdot(\rho\, h^o\mathbf{v}) = \dot{U}. \tag{A.24}$$

In a region of the fluid with $\dot{m} = 0$ and $\mathbf{f} = 0$, using the corresponding forms of Equations (A.21) and (A.23), the foregoing equation yields

$$\rho\left(\frac{\partial u}{\partial t} + \mathbf{v}\cdot\nabla u\right) - \frac{p}{\rho}\left(\frac{\partial\rho}{\partial t} + \mathbf{v}\cdot\nabla\rho\right) = \dot{U}, \tag{A.25}$$

where the e_P term in Equation (A.8) is also neglected. In view of Equation (A.13), this is tantamount to

$$\rho T\left(\frac{\partial s}{\partial t} + \mathbf{v}\cdot\nabla s\right) = \dot{U}, \tag{A.26}$$

which shows that in a region free of sources ($\dot{U} = 0$ also) the flow of an ideal fluid is isentropic. This means that the specific entropy of a fluid particle remains constant as it is convected with the flow velocity, but that constant can be different for different particles.

If $\dot{m}, \mathbf{f} \neq 0$ in the region considered, then $-\dot{m}h^o - \mathbf{f}\cdot\mathbf{v}$ should be added to the right-hand sides of Equations (A.25) and (A.26). These terms account for the power carried by the mass creation and force application processes in the region, respectively.

A.4 Equation of Motion of Newtonian Fluids

When the shear viscosity and thermal conductivity of the fluid are taken into account, the differential form of the law of conservation of mass is still given by Equation (A.21).

The integral forms of the conservation laws for momentum and energy are given by Equations (A.6) and (A.11) which are written here for a source-free region:

Conservation of momentum:

$$\frac{\partial}{\partial t}\int_V \rho\,\mathbf{v}\,\mathrm{d}V + \int_S \rho\mathbf{v}\,\mathbf{v}\cdot\mathbf{n}\,\mathrm{d}S = \int_S \mathbf{e}_i\left(\sigma_{ij}\mathbf{e}_j\right)\cdot\mathbf{n}\mathrm{d}S. \tag{A.27}$$

Conservation of energy:

$$\frac{\partial}{\partial t}\int_V \rho\,e\,\mathrm{d}V + \int_S e\rho\,\mathbf{v}\cdot\mathbf{n}\,\mathrm{d}S = \int_S \mathbf{v}\cdot\mathbf{e}_i\left(\sigma_{ij}\mathbf{e}_j\right)\cdot\mathbf{n}\mathrm{d}S + \int_S (\kappa\nabla T)\cdot\mathbf{n}\mathrm{d}S. \tag{A.28}$$

The corresponding differential forms are derived in the following subsections. The volume or body source terms are omitted for simplicity; they can be introduced into the equations as shown in the previous section.

A.4.1 Momentum Equation

Upon expressing velocities by their Cartesian components, applying the divergence theorem, and using the tensor notation for repeating indices, Equation (A.27) is expressed as

$$\mathbf{e}_i\int_V \frac{\partial}{\partial t}(\rho v_i)\mathrm{d}V + \mathbf{e}_i\int_V \frac{\partial}{\partial x_i}(\rho v_i v_j)\mathrm{d}V = \mathbf{e}_i\int_V \frac{\partial \sigma_{ij}}{\partial x_j}\mathrm{d}V, \tag{A.29}$$

which separates into three scalar equations

$$\frac{\partial}{\partial t}\rho v_i + \frac{\partial}{\partial x_j}\rho v_i v_j = \frac{\partial \sigma_{ij}}{\partial x_j}, \qquad i = 1, 2, 3 \tag{A.30}$$

Equation (A.21) can be written similarly

$$\frac{\partial \rho}{\partial t} + \frac{\partial}{\partial x_j}\rho v_j = 0. \tag{A.31a}$$

From this and Equation (A.30)

$$\rho\left(\frac{\partial}{\partial t}+\mathbf{v}\cdot\nabla\right)v_i=\frac{\partial\sigma_{ij}}{\partial x_j},\qquad i=1,2,3 \tag{A.31b}$$

From Equation (A.6)

$$\frac{\partial\varphi_{ij}}{\partial x_j}=\nabla^2 v_i+\frac{\partial}{\partial x_i}\frac{\partial v_j}{\partial x_j}-\frac{2}{3}\frac{\partial}{\partial x_i}\nabla\cdot\mathbf{v}. \tag{A.32}$$

Using this in the partial space derivatives of Equation (A.5), and substituting the result in Equation (A.31b)

$$\rho\left(\frac{\partial}{\partial t}+\mathbf{v}\cdot\nabla\right)v_i=-\frac{\partial p}{\partial x_i}+\mu\left(\nabla^2 v_i+\frac{\partial^2 v_j}{\partial x_i\partial x_j}-\frac{2}{3}\frac{\partial}{\partial x_i}\nabla\cdot\mathbf{v}\right)+\frac{\partial\mu}{\partial x_j}\varphi_{ij}\mathbf{e}_i,\qquad i=1,2,3 \tag{A.33}$$

which can be combined into a single vector equation as

$$\rho\left(\frac{\partial}{\partial t}+\mathbf{v}\cdot\nabla\right)\mathbf{v}+\nabla p=\mu\nabla^2\mathbf{v}+\frac{\mu}{3}\nabla\nabla\cdot\mathbf{v}+\frac{\partial\mu}{\partial x_j}\varphi_{ij}\mathbf{e}_i. \tag{A.34a}$$

Using the vector identity $\nabla^2\mathbf{v}+\nabla\nabla\cdot\mathbf{v}/3\equiv 4\nabla\nabla\cdot\mathbf{v}/3-\nabla\times(\nabla\times\mathbf{v})$, this can be expressed as

$$\rho\left(\frac{\partial}{\partial t}+\mathbf{v}\cdot\nabla\right)\mathbf{v}+\nabla p=\frac{4\mu}{3}\nabla\nabla\cdot\mathbf{v}-\mu\nabla\times(\nabla\times\mathbf{v})+\frac{\partial\mu}{\partial x_j}\varphi_{ij}\mathbf{e}_i. \tag{A.34b}$$

The bulk viscosity is neglected in these equations. It can be included by replacing the factor $\mu/3$ by $\mu/3+\eta$ in Equation (A.34a) and $4\mu/3$ by $4\mu/3+\eta$ in Equation (A.34b), where η denotes the bulk viscosity. Equation (A.34a), or (A.34b), is called the Navier–Stokes equation.

A.4.2 Energy Equation

By applying the divergence theorem to the surface integrals, Equation (A.28) yields the Fourier–Kirchhoff–Neuman equation [2]:

$$\frac{\partial}{\partial t}e\rho+\frac{\partial}{\partial x_j}e\rho v_j=\frac{\partial}{\partial x_j}v_i\sigma_{ij}+\nabla\cdot\kappa\nabla T. \tag{A.35}$$

Rearranging the left-hand side, invoking the continuity equation and neglecting the potential energy term in Equation (A.8):

$$\rho\left(\frac{\partial}{\partial t}+\mathbf{v}\cdot\nabla\right)e=\rho\left(\frac{\partial}{\partial t}+\mathbf{v}\cdot\nabla\right)\left(u+\frac{1}{2}v^2\right)=\frac{\partial}{\partial x_j}v_i\sigma_{ij}+\nabla\cdot\kappa\nabla T. \tag{A.36}$$

From this and Equation (A.31),

$$\rho\left(\frac{\partial}{\partial t}+\mathbf{v}\cdot\nabla\right)u=-p\nabla\cdot\mathbf{v}+\nabla\cdot\kappa\nabla T+\frac{1}{2}\mu\sum_{i,j=1}^{3}\varphi_{ij}^2. \tag{A.37}$$

Using Equation (A.21) for the divergence of particle velocity and Equation (A.13) for the material derivative of the specific internal energy

$$\rho T\left(\frac{\partial}{\partial t}+\mathbf{v}\cdot\nabla\right)s=\nabla\cdot\kappa\nabla T+\frac{1}{2}\mu\sum_{i,j=1}^{3}\varphi_{ij}^{2} \tag{A.38}$$

and using Equation (A.16) for the material derivative of the specific entropy

$$\rho c_p\left(\frac{\partial}{\partial t}+\mathbf{v}\cdot\nabla\right)T=\beta T\left(\frac{\partial}{\partial t}+\mathbf{v}\cdot\nabla\right)p+\nabla\cdot\kappa\nabla T+\frac{1}{2}\mu\sum_{i,j=1}^{3}\varphi_{ij}^{2}. \tag{A.39a}$$

Another form of the energy equation which is often useful follows from Equations (A.14, A.37) by elimination of the specific internal energy:

$$\frac{\beta T}{\gamma-1}\left(\frac{\partial}{\partial t}+\mathbf{v}\cdot\nabla\right)p=-\frac{\rho c_p}{\beta}\nabla\cdot\mathbf{v}+\nabla\cdot\kappa\nabla T+\frac{1}{2}\mu\sum_{i,j=1}^{3}\varphi_{ij}^{2}. \tag{A.39b}$$

References

[1] NIST Reference Fluid Thermodynamic and Transport Properties Database (REFPROP): Version 10, National Institute of Standards and Technology, US Department of Commerce, (2019) (www.nist.gov/srd/refprop).

[2] A.D. Pierce, *Acoustics: An Introduction to Its Physical Principles and Applications*, (New York: McGraw-Hill, 1981).

Appendix B Acoustic Properties of Rigid-Frame Fibrous Materials

A wealth of literature exists on many aspects of the acoustics of porous media [1]. This appendix describes the equivalent fluid characterization of the bulk acoustic properties of rigid-frame fibrous materials which are widely used in ducts for noise control purposes. The rigid frame assumption means that the flexibility of the fibers has indiscernible effect on the bulk acoustic characteristics of the porous medium. This condition can be stated approximately as $d > 10^{-5}\lambda$, or so, where λ denotes the wavelength and d denotes the fiber diameter.

Rock wool, glass wool or basalt wool are the most common fibrous materials used as sound wave energy absorbents in ducts. The description of sound wave propagation in these materials is quite difficult due to the random distribution of pore size and shape, frame orientation and pore interconnections. In practice, however, only the average values on a macroscopic scale much smaller than the wavelength are of interest. Then, under the rigid frame assumption, the fluid inside the porous medium can be replaced by an equivalent fluid. Envisaging, for simplicity, a homogeneous and isotropic equivalent fluid which is at rest initially, acoustic perturbations can be modeled in the frequency domain by the continuity and momentum equations [2]:

$$-\mathrm{i}k_e\tilde{p}' + \rho_e c_e \nabla \cdot \tilde{\mathbf{v}}' = 0 \tag{B.1}$$

$$-\mathrm{i}k_e\tilde{\mathbf{v}}' + \frac{1}{\rho_e c_e}\nabla\tilde{p}' = 0 \tag{B.2}$$

respectively. Here, $\tilde{p}'$ and $\tilde{\mathbf{v}}'$ denote the average acoustic pressure and the average particle velocity, ρ_e and c_e denote, respectively, the density and the speed of sound of the equivalent fluid. The foregoing equations may be deduced ad hoc by first writing the continuity and momentum equations, Equations (1.5) and (1.8), for an inviscid homogeneous fluid at rest and then replacing the density and the speed of sound by corresponding the equivalent fluid properties. However, in view of the inevitable viscothermal losses in a porous medium, the equivalent density and the speed of sound must be conceived as complex quantities in frequency domain. Thus, eliminating the particle velocity between Equations (B.1) and (B.2), the wave equation governing sound wave propagation in a porous material transpires in the familiar Helmholtz equation

$$(\nabla^2 + k_e^2)p' = 0, \tag{B.3}$$

where $k_e = \omega / c_e$ denotes the equivalent wavenumber. Evidently, the utility of this wave equation depends on the availability of working values for the equivalent fluid properties for porous materials. Determination of these properties has been the subject of extensive research and the methods developed for this purpose can be considered in two broad groups, namely, empirical curve fitting and microstructure modeling.

For rigid frame fibrous materials, empirical curve fitting is generally implemented by using the following power law regression equations first suggested by Delany and Bazley [3]:

$$\frac{k_e}{k_o} = 1 + C_3\left(10^3 \times \frac{f}{\sigma}\right)^{C_4} + \mathrm{i}C_1\left(10^3 \times \frac{f}{\sigma}\right)^{C_2} \tag{B.4}$$

$$\frac{\rho_e c_e}{\rho_o c_o} = 1 + C_5\left(10^3 \times \frac{f}{\sigma}\right)^{C_6} + \mathrm{i}C_7\left(10^3 \times \frac{f}{\sigma}\right)^{C_8} \tag{B.5}$$

where ρ_o and c_o denote the density and the speed of sound of the fluid in the pores of the material, $C_j, j = 1, 2, \ldots, 8$, denote the regression coefficients which are to be determined by measurement, f denotes the frequency and σ is called viscous flow resistivity of the material. The latter is determined experimentally by passing a steady air flow through a plug of the porous specimen and measuring the pressure drop across the plug as well as the volume flow through it. Then fitting the measured values for several pressure differentials into the equation $-\mathrm{d}p/\mathrm{d}x = \sigma \bar{v}$, where $\bar{v}$ denotes the area averaged steady mean-flow velocity through the porous medium under a pressure gradient, yields the required flow resistivity. It has the units of $\mathrm{kg/m^3 s}$ (or rayl/m) and, for a wide range of fibrous materials, is closely predicted by the empirical relationship [4]:

$$\frac{\sigma d^2}{\rho_m^{1.53}} = K, \tag{B.6}$$

where the constant K has the value of 3.18×10^{-9} when the fiber diameter, d, is in meters and ρ_m denotes the bulk density of the material. The latter is related to the fiber density, ρ_f, by the relationship

$$\rho_m = \Omega \rho_o + (1 - \Omega)\rho_f, \tag{B.7}$$

where Ω denotes volume porosity of the packing, which is defined as total volume of void divided by the total volume occupied by the material.

Coefficients of Equations (B.4) and (B.5) are determined from the measured values of k_e and $\rho_e c_e$ for fibrous material samples. Tests are carried out in a plane wave standing wave tube and essentially involve measurement of the wave transfer matrix of a specimen (see Section 12.3.4). Then the equivalent fluid parameters can be educed from the theoretical wave transfer matrix determined from the solution of Equation (B.3) for plane wave propagation. This wave transfer matrix is in fact given as Equation (3.112). It is seen that this is a reciprocal two-port element (see Section 5.5) and the elements of the wave transfer matrix are functions of k_e and $\rho_e c_e$. Hence, the latter can be educed from the measured diagonal and off-diagonal elements. Wave transfer matrix measurement can be avoided by carrying out the test with samples of the same material having different thickness. If a specimen of thickness L is backed with a rigid plane surface at $x = L$, then $r(L) = 1$ and Equation (3.112) gives the reflection coefficient on surface ($x = 0$) of the specimen as

$$r(0) = \frac{\tan(k_e L) - \mathrm{i}\dfrac{\rho_e c_e}{\rho_o c_o}}{\tan(k_e L) + \mathrm{i}\dfrac{\rho_e c_e}{\rho_o c_o}} \tag{B.8}$$

Thus, measuring $r(0)$ for $L = L_1, L_2$ yield two such equations, which can be solved for k_e and $\rho_e c_e$ (see Section 12.2.3 for the measurement of the plane-wave reflection coefficient).

By making measurements with fibrous materials with flow resistivity in the range $2000 \le \sigma \le 80\ 000$ mks rayl/m and porosity approximately equal to unity, Delany and Bazley [3] found out that the measurement data coalesced adequately accurately on the power law relations in Equations (B.4) and (B.5) in the frequency range $0.01\ \mathrm{m}^3/\mathrm{kg} \le f/\sigma \le 1\ \mathrm{m}^3/\mathrm{kg}$ with the regression coefficients C_j, $j = 1, 2, \ldots, 8$, given in Table B.1. Errors due to limitations of the measurement technique are inevitable, but have little practical significance in view of the variability of the properties of most production materials. In fact, the use of the regression coefficients with accurately determined flow resistance is likely to give better predictive accuracy than the use of data for a specific material [3].

Extrapolation of the Delany and Bazley formulae beyond the regression range is not recommended, as it can yield unrealistic results, for example, negative values for the real part of the equivalent density. Miki argues that the regression model of Delany and Bazley does not satisfy the positive real function requirement [5], which is an intrinsic property of passive impedances. If this condition is also imposed, the regression model of Delany and Bazley yields different regression coefficients for the same measurement data, which are also shown in Table B.1. Miki claims that this set of coefficients tends to lower the low-frequency threshold of the Delany and Bazley coefficients; however, the low-frequency threshold stems from the limited accuracy of the standing wave tube measurements at low frequencies and prevalence of unrealistic results at low frequencies is not likely to be avoided by such modifications.

The above discussed empirical curve fitting model is based on measurements carried out at laboratory conditions. The question of the effect of the test temperature

Table B.1 Coefficients of curve-fitting formulae

	C_1 (a_1)	C_2	C_3 (a_3)	C_4	C_5 (a_5)	C_6	C_7 (a_7)	C_8
Delany and Bazley [3]	10.3 (0.1948)	−0.59	10.8 (0.0975)	−0.70	9.08 (0.0585)	−0.75	11.9 (0.0878)	−0.73
Miki [5]	11.41 (0.1787)	−0.618	7.81 (0.1223)	−0.618	5.5 (0.0784)	−0.632	8.43 (0.1202)	−0.632

on the regression coefficients has been considered by Christie [6]. He found out that the regression coefficients determined at the room temperature remain valid at temperatures up to about 600 °C, provided that Equations (B.4) and (B.5) are expressed as functions of the non-dimensional frequency parameter

$$\xi = \frac{\rho_o f}{\sigma}, \tag{B.9}$$

where ρ_o denotes the density of the fluid. Assuming that the regression coefficients given in Table B.1 apply for dry air at normal temperature and pressure conditions (20°C and 101 325 Pa) $\rho_o = 1.20\ \mathrm{kg/m^3}$, Equations (B.3) and (B.4) can be recast in the form

$$\frac{k_e}{k_o} = 1 + a_3 \xi^{C_4} + \mathrm{i} a_1 \xi^{C_2} \tag{B.10}$$

$$\frac{\rho_e c_e}{\rho_o c_o} = 1 + a_5 \xi^{C_6} + \mathrm{i} a_7 \xi^{C_8}. \tag{B.11}$$

The coefficients $a_j, j = 1, 3, 5, 7$ are shown in Table B.1 in brackets. In most publications, the Delany and Bazley formulae are quoted in these forms.

Microstructure models can yield physically reasonable results for the bulk acoustic characteristics of fibrous materials over a wide frequency range, whilst retaining some flexibility of injecting empirically determined physical parameters into the model. General theories for porous materials are described by Biot [7] and Attenborough [8]. Taking into account the inertial and viscous couplings between the fluid and the fibrous frame, Johnson et al. proposed the following microstructure expression for the equivalent density of an air-saturated rigid frame porous material [9]:

$$\frac{\rho_e}{\rho_o} = \alpha_\infty - \frac{\Omega}{i 2\pi\xi} \sqrt{1 - \frac{i\pi\xi\alpha_\infty}{\Omega s^2}}, \tag{B.12}$$

where s is an empirical parameter that lies in $0.1 \le s^2 \le 10$ for most fibrous materials, the value of $s = 1$ being usually satisfactory as an average value. The parameter α_∞ is a measurable parameter, which is defined as $\lim_{f\to\infty} \rho_e = \alpha_\infty \rho_o$ and is called tortuosity. Also proposed by the same authors, is the following expression for the equivalent bulk modulus of a fibrous material [9]:

$$\frac{\rho_o c_o^2}{\rho_e c_e^2} = \gamma_o - \frac{\gamma_o - 1}{1 - \dfrac{\Omega}{\mathrm{i}2\pi\xi s'^2 \alpha_\infty \mathrm{Pr}} \sqrt{1 - \dfrac{\mathrm{i}\pi\xi s'^2 \alpha_\infty \mathrm{Pr}}{\Omega}}}, \tag{B.13}$$

where Pr denotes the Prandtl number of the fluid in the pores and again s' denotes an empirical constant. Allard and Champoux show that for fibrous materials, one may take $s' = 2s$ and $s \approx 1$ [10]. With these values, the model predicts the Delany and Bazley equations reasonably closely, with the added advantage of yielding physically reasonable values at low frequencies. There will be some degree of mean fluid flow within a porous material in a flow duct treated with a porous liner. Although the convective effect of this is usually negligible, its inertial effect can be significant even at relatively low fluid velocities. Based on a parallel cylindrical fiber microstructure model, Cummings and Chang argue that the inertial effect of internal mean flow can be taken into account by modifying the zero mean-flow equivalent fluid properties of rigid framed porous materials as [11]:

$$k_e = \lambda k_{e0} \tag{B.14}$$

$$\rho_e c_e = i\lambda(\rho_e c_e)_0, \tag{B.15}$$

where the subscript "0" denotes the zero mean-flow values and

$$\lambda = \sqrt{\frac{\mathrm{i}q^2\rho_o}{\Omega\rho_{eo}}\left(1 + \mathrm{i}\Omega\frac{\sigma F_\mu + 2\eta \mid \bar{M}_o \mid}{\rho_o q^2 \omega}\right)}. \tag{B.16}$$

Here, $\bar{M}_o$ denotes the average internal mean-flow velocity Mach number, η is the inertial flow resistivity and

$$F_\mu = \frac{\omega}{\sigma}\left(\mathrm{Re}\{\rho_{e0}\} + \mathrm{i}\left(\mathrm{Im}\{\rho_{e0}\} + \frac{q^2\sigma\rho_o}{\Omega}\right)\right), \tag{B.17}$$

where Re and Im denote the real and imaginary parts of a complex quantity. The inertial flow resistivity η is determined from the empirical Ergun equation $-\mathrm{d}p/\mathrm{d}x = \sigma\bar{v} + \eta\bar{v}^2$, which relates the area averaged steady mean-flow velocity, $\bar{v}$, in a porous medium under a pressure gradient of $\mathrm{d}p/\mathrm{d}x$. The experimental determination consists of passing a steady air flow through a plug of the porous specimen and measuring the pressure drop across the plug as well as the volume flow through it. Then, fitting the measured values for several pressure differentials into the Ergun equation yields the required flow resistivities.

References

[1] J.F. Allard and N. Atalla, *Propagation of Sound in Porous Media: Modeling of Sound Absorbing Materials*, (Chichester, UK: Wiley, 2009).

[2] P.M. Morse and K.U. Ingard, *Theoretical Acoustics,* (New York: McGraw-Hill, 1968).

[3] M.A. Delany and E.N. Bazley, Acoustic properties of fibrous absorbent materials, *Appl. Acoust.* **3** (1970), 105–116.

[4] F.A. Bies and C.G. Hansen, Flow resistance information for acoustical design, *Appl. Acoust.* **13** (1980), 357–371.

[5] Y. Miki, Acoustical properties of porous materials-modifications of Delany-Bazley models, *J. Acoust. Soc. Jpn.* **11** (1990), 19–24.

[6] D.R.A. Christie, Measurement of the acoustic properties of a sound absorbing material at high temperatures, *J. Sound Vib.* **46** (1976), 347–355.

[7] M.A.A. Biot, Theory of propagation of elastic waves in a fluid saturated porous solid, *J. Acoust. Soc. Am.* **28** (1956), 168–191.

[8] K. Attenborough, Acoustical characteristics of porous materials, *Phys. Rep.* **82** (1982), 179–227.

[9] D.L. Johnson, J. Kopik and R. Dashen, Theory of dynamic permeability and tortuosity in fluid saturated media, *J. Fluid Mech.* **176** (1987), 379–402.

[10] J.F. Allard and Y. Champoux, New empirical equations for sound propagation in rigid frame fibrous materials, *J. Acoust. Soc. Am.* **91** (1992), 3346–3353.

[11] A. Cummings and I.-J. Chang, Acoustic propagation in porous media with internal mean flow, *J. Sound Vib.* **114** (1987), 565–581.

Appendix C Impedance of Compact Apertures

C.1 Empirical and Semi-Empirical Models

Most of the existing aperture impedance models [1–11] are empirical or semi-empirical single-degree-of-freedom models, usually derived from the response of circular holes drilled on flat plates with grazing mean flow over them or with through mean flow. In these models, the fluid in and around the aperture is tacitly assumed to be incompressible and its motion is formulated as a single-degree-of-freedom system driven by an average fluctuating pressure load $[p'] = (1/A)\int_A (\Delta p')\mathrm{d}A$ with average fluctuating velocity $\bar{u}' = (1/A)\int_A u'\mathrm{d}A$, where A denotes the aperture area, u' denotes the component of particle velocity normal to the aperture area and $\Delta p'$ denotes the difference of fluctuating pressures on the two sides of an aperture. Then, in the frequency domain, the motion of the system may be described by the frequency response function

$$Z = \frac{[p']}{\bar{u}'} = \rho_o c_o \zeta, \tag{C.1}$$

which is called the lumped parameter impedance of the aperture. It is often expressed as a non-dimensional (normalized) parameter ζ, as defined by the second equality, where ρ_o denotes the density and c_o the speed of sound of the fluid in the vicinity of the aperture. The (real) and reactive (imaginary) parts of ζ are denoted as $\zeta = \theta - \mathrm{i}\chi$, with $\mathrm{e}^{-\mathrm{i}\omega t}$ time dependence.

Realistic aperture impedance should include the effects of fluid inertia in the aperture, the presence of mean flow, viscothermal losses, the level of acoustic excitation and fluid-structure interactions. Some parameters governing ζ may be stated as

$$\zeta = \zeta(k_o, d, \mu_o, v_o, h, a, \sigma, u_o), \tag{C.2}$$

where $k_o = \omega/c_o$ denotes the wavenumber, d denotes a typical length scale of the aperture (for example, the diameter for a circular hole), μ_o denotes the shear viscosity coefficient, v_o denotes the grazing mean-flow velocity, u_o denotes the through flow mean-flow velocity (positive if out of the duct), h denotes the thickness of the aperture, σ denotes the open area porosity of the duct wall and a denotes the duct radius.

For sound pressure levels less than about 130 dB(L), aperture impedance is linear; that is, it is independent of the level of acoustic excitation. This has been experimentally observed and reported in several publications for subsonic low mean-flow velocity Mach numbers. Perforates used in most practical applications are subject to excitation levels in this range and, accordingly, linear aperture impedance models are widely used. Several aperture impedance models which are also function of the excitation level are also available [12], but they can be used only when the actual sound pressure field in perforates can be calculated, which is not straightforward, since an iterative solution procedure is necessary to predict the level of acoustic excitation. In the presence of grazing mean flow, non-linear aperture response is linked to Kelvin–Helmholtz-type instability inherent to fluid–structure interactions in the vicinity of an aperture. The onset of such instability occurs at excitation levels in the linear range and can be predicted if an appropriate model of the aperture impedance is available (see Section C.2).

The presence of mean flow affects the aperture impedance significantly. Usually grazing mean flow is predominant, since forcing the mean flow through perforations causes large flow losses. However, through flow may also be important in some applications. The through flow Mach number, $M_{thr} = u_o/c_o$, can be calculated from the continuity equation for one-dimensional inviscid mean flow

$$\frac{\mathrm{d}}{\mathrm{d}x}(\rho_o v_o S) + \rho_o \sigma P u_o = 0, \tag{C.3}$$

where S denotes the cross-sectional area of the perforated duct and P denotes the duct perimeter. For a uniform duct this gives

$$M_{thr} = \frac{u_o}{c_o} = -\frac{S}{\sigma P}\frac{\mathrm{d}M_o}{\mathrm{d}x}, \tag{C.4}$$

where $M_o = v_o/c_o$ denotes the grazing mean-flow velocity Mach number.

The effect of evanescent waves generated at the edges of an aperture is accounted for in the reactive part of a lumped parameter model as end-correction applied to the aperture thickness on both sides. Further considerations may be necessary, however, when the apertures are backed with a sound absorbent material. If the perforations are in the form of stamped louvers, or if a protecting layer of steel wool is used, the effect of the absorbent on the end-correction may be neglected. But if the porous backing is tightly closing on the holes, its bulk acoustic properties should be taken into account when applying the end-correction on the absorbent side. The effect of the absorbent packing on the perforate impedance is not clearly understood at the present, although its presence is known to increase the perforate impedance. It is suggested by Kirby and Cummings that this effect can be approximately included by adding a correction term to the lumped parameter model of aperture impedance without absorbent backing, ζ, as [8]:

$$\zeta_{\text{corrected}} = \zeta + 0.425d\left(-\mathrm{i}k_o + k_e\frac{\rho_e c_e}{\rho_o c_o}\right), \tag{C.5}$$

where ρ_e, c_e, k_e are the equivalent density, speed of sound and wave number of the absorptive backing (see Appendix B).

Some of the more commonly known semi-empirical lumped parameter impedance models for circular holes are described in the following.

For the case of negligible grazing and through mean flow, Sullivan and Crocker proposed [1]:

$$\zeta = 0.006 - \mathrm{i}k_o(h + 0.75d), \tag{C.6}$$

where the resistive part is stated to be valid for $|u'| < 3$ m/s, and the reactive part is valid for $|u'| < 0.000102f^{1.541}$ m/s.

For the effect of turbulent grazing mean flow, Cummings gives [3]:

$$\zeta = \theta_V + \theta_T - \mathrm{i}k_o(h + \delta), \tag{C.7}$$

where θ_V and θ_T denote the resistive parts due to shear viscosity and turbulence, respectively, and δ denotes an equivalent thickness. The resistive part due to viscosity is determined from

$$\theta_V = 2\left(1 + \frac{h}{d}\right)\sqrt{\frac{2k_o d}{A_{\mathrm{Re}}}}, \tag{C.8}$$

where $A_{\mathrm{Re}} = \rho_o c_o d/\mu_o$. The resistive part due to turbulent grazing flow is

$$\theta_T = \begin{cases} C\dfrac{fd}{c_o} & \text{if} \quad C \geq 0 \\ 0 & \text{if} \quad C < 0 \end{cases} \tag{C.9}$$

Here, f denotes the frequency and the parameter C is given by

$$C = \left(12.52\left(\frac{h}{d}\right)^{-0.32} - 2.44\right)\frac{v_*}{fd} - 3.2, \qquad v_* = v_o\sqrt{\frac{\psi}{2}} \tag{C.10}$$

where ψ denotes the Prandtl empirical coefficient which is computed from

$$2\sqrt{\psi} = \frac{1}{2\log_{10}\left(2\sqrt{\psi}M_o A_{\mathrm{Re}}\right) - 1.6}. \tag{C.11}$$

If there is fluid on both sides of the aperture, the end-correction, δ, in the reactive part is given by

$$\delta = \begin{cases} 0.425(1 + \varepsilon)d \\ \varepsilon = \left(1 + 0.6\dfrac{h}{d}\right)\mathrm{e}^{\frac{\beta}{0.25 + \frac{h}{d}}} - 0.6\dfrac{h}{d} \\ \beta = \begin{cases} 0 & \text{if} \quad B \geq 0 \\ B & \text{if} \quad B < 0 \end{cases} \\ B = 0.12\dfrac{d}{h} - \dfrac{v_*}{fh} \end{cases}. \tag{C.12}$$

If the side of the aperture with no mean flow is packed with sound absorbent material of equivalent density ρ_e, the following expression is suggested for the end-correction:

$$\delta = 0.425\left(\varepsilon + \frac{\rho_e}{\rho_o}\right). \tag{C.13}$$

Kirby and Cummings later carried out further measurements with and without porous backing [8]. The new equations obtained are similar to the equations described above, but some curve fitting constants have different values, for example, compare Equation (C.13) with Equation (C.5).

Coelho proposed for the grazing mean-flow case [4]:

$$\theta = \begin{cases} 0.6\left(\dfrac{1-\sigma^2}{\sigma}\right)(M_o - 0.025) + \theta_o & \text{if} \quad M_o \leq 0.05 \\ 0.3\left(\dfrac{1-\sigma^2}{\sigma}\right)M_o + \theta_o & \text{if} \quad M_o > 0.05 \end{cases}, \tag{C.14}$$

where

$$\theta_o = 2\left(1 + \frac{h}{d}\right)\sqrt{\frac{2k_o d}{A_{\mathrm{Re}}}} + \frac{1}{8}(k_o d)^2. \tag{C.15}$$

In this model, the reactive part of the impedance is given by

$$\chi = k_o d\left(\sqrt{\frac{8}{k_o d A_{\mathrm{Re}}}}\left(1 + \frac{h}{d}\right) + \frac{h}{d} + \frac{\delta_M}{d}\right), \tag{C.16}$$

where

$$\delta_M = \frac{8d}{3\pi}\frac{1 - 0.7\sqrt{\sigma}}{1 + 305M_o 2}, \qquad \sigma < 0.126. \tag{C.17}$$

Based on curve fitting to measured data, Lee and Ih proposed the following equation to include the effect of grazing mean flow [12]:

$$\begin{aligned}\zeta = {} & 3.94 \times 10^{-4}\left(1 + 7.84 \times 10^{-3}|f - f_o|\right)(1 + 14.9M_o)(1 + 296d)(1 - 127h) - \\ & - \mathrm{i}\ 6 \times 10^{-3}(1 + 194d)(1 + 432h)(1 - 1.72M_o)\left(1 - 6.62 \times 10^{-3}f\right)\end{aligned} \tag{C.18}$$

where

$$f_o = 412\left(\frac{1 + 104M_o}{1 + 274d}\right). \tag{C.19}$$

This equation is stated to be valid for $60 \leq f \leq 4000$ Hz, $0 \leq M_o \leq 0.2$, $2 \leq d \leq 9$ mm, $1\,\mathrm{mm} \leq h \leq 5\mathrm{mm}$, $0.0279 \leq \sigma \leq 0.223$.

Sullivan suggested the following simple model to account for the effect of mean through flow [7]:

$$\theta = 2.056M_{thr} - 0.95k_o(h + 0.75d) \tag{C.20}$$

Bauer suggested the following equation accounting for the effects of both mean grazing and through flows [6]:

$$\theta = \sqrt{\frac{8k_o d}{A_{\mathrm{Re}}}}\left(1 + \frac{h}{d}\right) + 0.3M_o + 1.15M_{thr} - \mathrm{i}k_o(h + 0.25d), \tag{C.21}$$

which is stated to be valid for $M_o < 0.7, \quad 0.02 < \sigma < 0.2, \quad 0 < h/d < 1$.

C.2 Theoretical Models

Theoretical models of forced fluid motion in compact apertures are usually formulated, in the spirit of the classical work of Lord Rayleigh, in the form

$$[p'] = -\mathrm{i}\omega\rho_o \frac{Q'}{K_a}, \tag{C.22}$$

where K_a is called the Rayleigh conductivity of the aperture, $Q' = \int_A u' \mathrm{d}A = \bar{u}' A$ denotes the rate of fluctuating volume flow through an aperture. Comparing Equations (C.22) and (C.1), the Rayleigh conductivity and the aperture impedance may be related as

$$\zeta = -\mathrm{i}k_o \frac{A}{K_a}. \tag{C.23}$$

The vortex sheet method [13–20] is known to yield adequately accurate results for the Rayleigh conductivity of compact apertures under subsonic low Mach number grazing mean flow. In this method, the scheme shown in Figure C.1 is used to derive the Rayleigh conductivity of an aperture. This consists of a uniform rigid thin plate containing an aperture. The plate separates the infinite space into two halves, with incompressible uniform parallel grazing mean flows of velocities U_+ and U_- existing in the upper and lower sides of the plate, respectively. The fluid is assumed to be inviscid and the interface between the fluids on the two sides of the aperture area is modeled by a thin vortex sheet which is subject to a fluctuating pressure load $[p'] = p'_+ - p'_-$. An approximate analytical solution for this scheme is available for a slot, that is, a rectangular aperture with $b/L \gg 1$, the side L being in the direction of

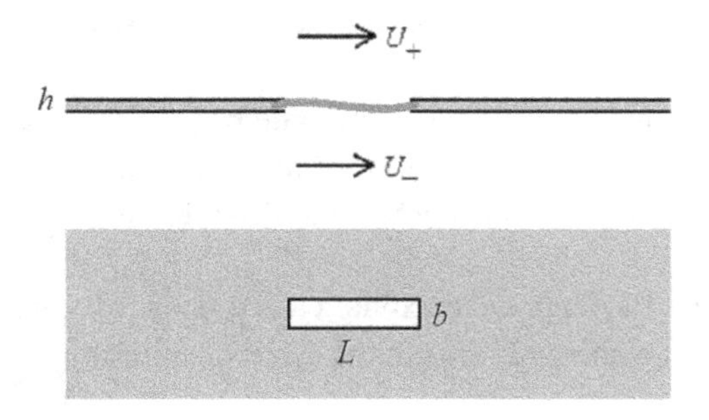

Figure C.1 Compact aperture on a flat plate with two-sided mean flow.

the mean flow (Figure C.1). Rayleigh conductivity of a slot can be expressed as [13–14]:

$$K_a = \frac{1}{2} \frac{\pi b}{F(\sigma_1, \sigma_2) + \ln\left(\frac{8b}{\mathrm{e}L}\right)}. \tag{C.24}$$

Here, e is the exponential number, ln denotes natural logarithm and

$$F(\sigma_1, \sigma_2) = \frac{-\sigma_1 \mathrm{J}_0(\sigma_2)(\mathrm{J}_0(\sigma_1) - 2W(\sigma_1)) + \sigma_2 \mathrm{J}_0(\sigma_1)(\mathrm{J}_0(\sigma_2) - 2W(\sigma_2))}{\sigma_1 W(\sigma_2)(\mathrm{J}_0(\sigma_1) - 2W(\sigma_1)) - \sigma_2 W(\sigma_1)(\mathrm{J}_0(\sigma_2) - 2W(\sigma_2))} \tag{C.25}$$

where J_n denotes a Bessel function of order n,

$$W(x) = \mathrm{i}x(\mathrm{J}_0(x) - \mathrm{i}\mathrm{J}_1(x)) \tag{C.26}$$

and

$$\sigma_1 = \frac{1}{2} \frac{\omega L(1+\mathrm{i})}{U_+ + \mathrm{i}U_-}, \qquad \sigma_2 = \frac{1}{2} \frac{\omega L(1-\mathrm{i})}{U_+ - \mathrm{i}U_-} \tag{C.27}$$

are called Kelvin–Helmholtz wavenumbers. For perforated ducts, usually the one-sided grazing flow case ($U_+ = v_o$, $U_- = 0$) is relevant. With no grazing mean flow, Equation (C.24) applies with $F = 0$.

For other aperture geometry with grazing mean flow, Rayleigh conductivity can be computed only numerically. Solutions are available for circular [13–14], rectangular [13, 15–16], trapezoidal [17], and some irregular [18] apertures. Also, the theory has been extended to include the effect of the finite thickness of the plate [19] and a thin cylindrical gap between two circular ducts [20].

Shown in Figure C.2 is the variation of the real and imaginary parts of the Rayleigh conductivity of thin rectangular apertures with the Strouhal number $\omega L/v_o$ for several aspect ratios $b/L > 1$ under one-sided mean flow $U_+ = v_o$, $U_- = 0$. The characteristics for $b/L = 10$ fall in the category of slots, for which Equation (C.25) holds.

Rayleigh conductivity of a slit, a rectangular aperture with aspect ratio $L/b >> 1$, is shown in Figure C.3 for one-sided mean flow as function of the Strouhal number, $\omega L/U$, for several width-to-length ratios and thickness-to-length ratios. In these figures, the Rayleigh conductivity is defined as $K_a = 2R_{eq}(\Gamma + \mathrm{i}\Delta)$, where $R_{eq} = \sqrt{bL/\pi}$. It is seen that, the effect of the wall thickness is small, but not insignificant.

Shown in Figure C.4, as function of the Strouhal number, $\omega a/U$, the Rayleigh conductivity of a circular hole of radius a for one sided mean grazing flow for several thickness to radius ratios. The real and imaginary parts of the Rayleigh conductivity are defined as $K_a = 2a(\Gamma + \mathrm{i}\Delta)$.

An important feature of the Rayleigh conductivity can be deduced from the time-averaged power flow through an aperture. This may be computed from

$$W = \int_A \langle u'[p'] \rangle \mathrm{d}A, \tag{C.28}$$

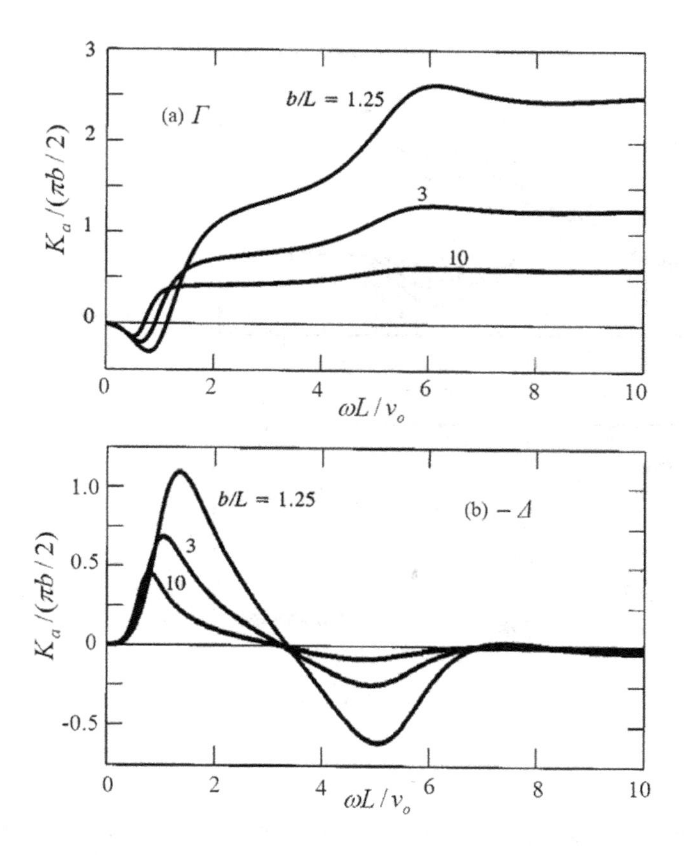

Figure C.2 (a) Normalized Rayleigh conductivity $K_a = \pi b(\Gamma + i\Delta)/2$ of rectangular apertures of different aspect ratio with one-sided grazing mean flow [16].

since only mean grazing flow is present and the average pressure load may be assumed to be uniformly distributed, as the aperture is assumed to be compact. The angled brackets denote time averaging, which implies that the time domain forms of the fluctuations are relevant in the integrand. A straightforward approach for the evaluation of Equation (C.28) is to invoke the $e^{-i\omega t}$ time dependence for the fluctuations and, after the time integration over a period $2\pi/\omega$, take the real part of the result. Thus, it may be shown that $W = ([\breve{p}']Q' + [p']\breve{Q}')/4$, where an inverted over-arc denotes complex conjugate. Then, substitution for Q' from Equation (C.22) yields

$$W = (-\mathrm{Im}\{K_a\})\frac{|[p']|^2}{2\rho_o\omega}. \tag{C.29}$$

This shows that, if the imaginary part of the Rayleigh conductivity is negative, then the applied pressure $[p']$ does positive work, which means that the vortex sheet dissipates energy by stable damped motion. But, if the imaginary part of the Rayleigh conductivity is positive, a negative work is done by the applied pressure, which means that the vortex sheet supplies energy to the applied pressure (by

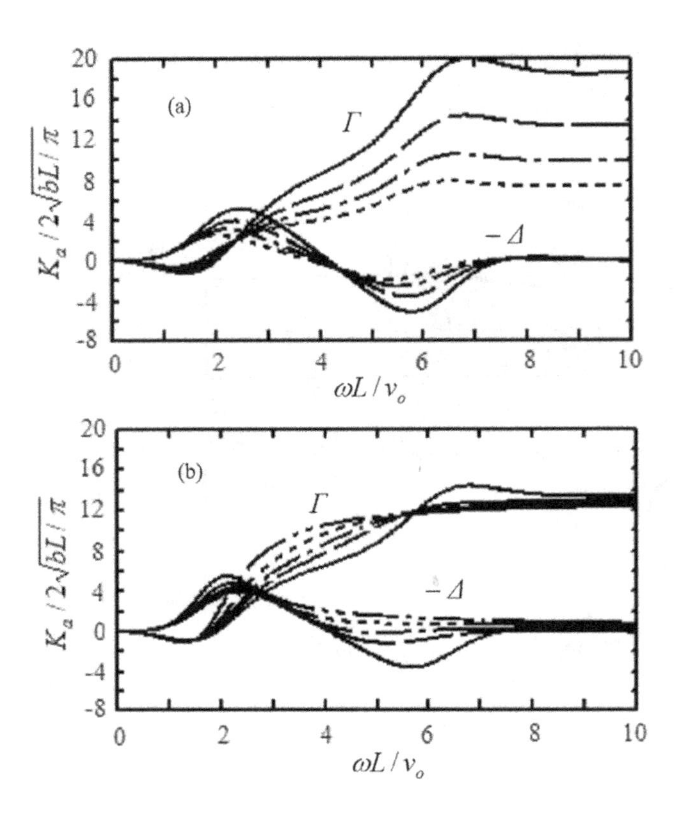

Figure C.3 Normalized Rayleigh conductivity of rectangular narrow slits with one-sided grazing mean flow (computed by the present author). (a) $h/L = 0$: solid: $b/L = 0.025$; long dash: $b/L = 0.05$; dash-dot: $b/L = 0.1$; short dash: $b/L = 0.2$. (b) $b/L = 0.025$: solid: $h/L = 0$; long dash: $h/L = 0.025$; dash-dot: $h/L = 0.05$; short dash: $h/L = 0.1$; dash-dot-dot: $h/L = 0.2$.

extracting energy from the mean flow). Then the sheet sustains a growing motion, which may be shown to be an absolute instability in the case of one-sided mean flow [13]. In practice, however, the growth of motion is curtailed by non-linear mechanisms which are ignored in the linear theory. The vortex sheet quickly breaks down and the shed vortices convect towards the trailing edge of the aperture, generating sound pulses upon interaction with the trailing edge. These pulses propagate upstream where, upon impinging at the leading edge of the aperture, they either reinforce or inhibit vortex shedding, depending on the vortex convection velocity. When the returning sound pulses reach the leading edge with the right frequency to reinforce periodic vortex shedding, they manifest as tonal sound known as vortex sound. The frequency of vortex sound does not seem to depend appreciably on the growth of motion in the non-linear regime and, consequently, may be estimated, as a first approximation, from the instability regions predicted by the Rayleigh conductivity. For example, referring to Figure C.4, it is seen that the aperture motion is absolutely unstable in the Strouhal number range $1.9 < (\omega a/v_o) < 3.9$ for $h/a = 0$ and both the lower and upper limits increase with the aperture thickness. Peat et al. argue that the

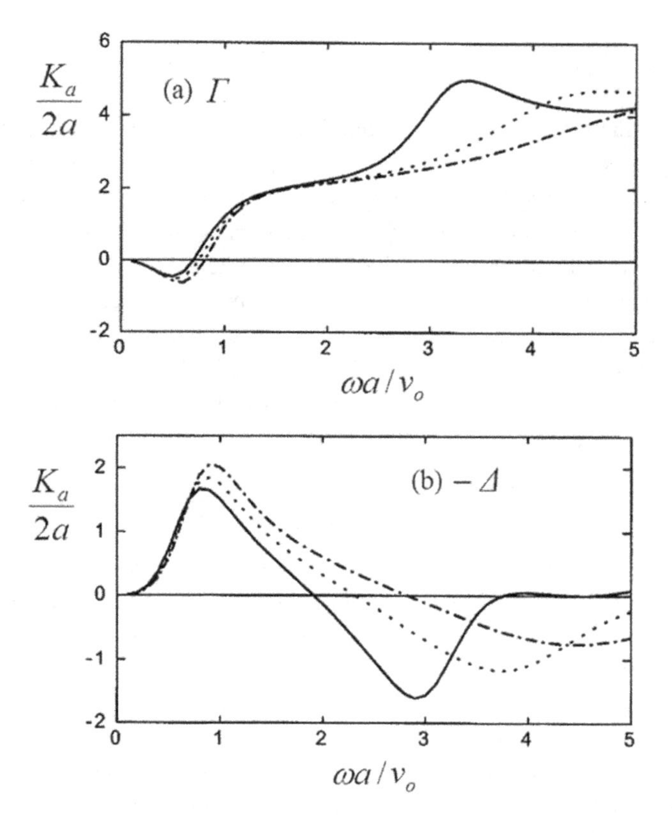

Figure C.4 Normalized conductivity of a circular aperture of radius a in a thin wall for one-sided grazing mean flow [14]. Solid: $h/a = 0$, dashed: $h/a = 0.25$, dash-dot: $h/a = 0.5$.

correlation of these results with experiments improve if the Strouhal number is based on the convection velocity of the vortices [14]. The time it takes for the sound pulse to reach the leading edge after the associated vortex is shed from there is approximately equal to $T = 2a/U_c$, where U_c denotes the convection velocity of the vortex, since $U_c << c_o$ normally. Thus, the critical Strouhal numbers may be expected to occur about $\omega a/U_c = n\pi, n = 1, 2, \ldots$ The first instability region ($n = 1$) in Figure C.4b for $h/a = 0$ becomes compatible with this estimate if $v_o = U_c$. This seems to be in line with the experiments which indicate that the vortex convection velocity is about half of the mean-flow velocity, v_o.

In view of such non-linear phenomenon at frequencies corresponding to the instability regions, predictions of linear acoustic calculations based on the Rayleigh conductivity of apertures should be assessed with caution at these frequencies.

The vortex sheet method is also used for the calculation of the Rayleigh conductivity of apertures with mean through flow. A well-established formula for a compact circular aperture, which was derived by Howe by considering a cylindrical vortex sheet downstream of the aperture, is the following [21]:

$$\frac{K_a}{2a} = 1 + \frac{\pi I_1(\kappa a)e^{-\kappa r} - i2\mathrm{K}_1(\kappa a)\sinh(\kappa a)}{\kappa a(\pi I_1(\kappa a)e^{-\kappa r} + i2\mathrm{K}_1(\kappa a)\cosh(\kappa a))}, \tag{C.30}$$

where a denotes the radius, $\kappa = \omega/U$, I_n and K_n are the modified Bessel functions of order n, U denotes the mean convection velocity of the vortices generated at the aperture edges. U can be approximated by the mean through flow velocity at the slit plane [21]. Figure C.5 shows the real (Γ) and imaginary (Δ) parts of the right-hand side of Equation (C.30) as function of the Strouhal number κa.

Dowling and Hughes calculated the acoustic absorption by a screen having a large number of parallel two-dimensional rectangular slits of compact width, b, assuming two planar vortex sheets for each slit [22]. Although they did not consider the Rayleigh conductivity of a single slit, this can be derived readily from their fluctuating volume flux analysis as

$$\frac{K_a}{\pi L} = \frac{\kappa s(\pi e^{-\kappa s}\mathrm{I}_1(\kappa s) - \mathrm{i}2\sinh(\kappa s)\mathrm{K}_1(\kappa s))}{\pi e^{-\kappa s}(\mathrm{I}_1(\kappa s)\kappa s\ln 2 + \mathrm{I}_0(\kappa s)) + \mathrm{i}2\sinh(\kappa s)(\mathrm{K}_0(\kappa s) - K_1(\kappa s)\kappa s\ln 2)}, \tag{C.31}$$

where $s = b/2$, $\kappa = \omega/U$, U denotes the mean convection velocity of the vortices generated at slit edges and ln denotes natural logarithm. U can be approximated, in the absence of a better estimate, by the average mean through flow velocity at the slit plane. The real (Γ) and imaginary (Δ) parts of the right-hand side of Equation (C.31) are shown in Figure C.6 as function of the Strouhal number κs.

Both in Equation (C.30) and Equation (C.31), the effect of the aperture thickness is neglected. This can be taken into account, approximately, by regarding the pressure difference across the aperture as sum of the contributions to vortex shedding and the inertia of the fluid in the aperture. This gives the thickness corrected Rayleigh conductivity of an aperture as

$$(K_a)_{\text{corrected}} = \frac{1}{\dfrac{1}{K_a} + \dfrac{h}{A}} \tag{C.32}$$

where h denotes the thickness of the slit, which may itself include an end-correction.

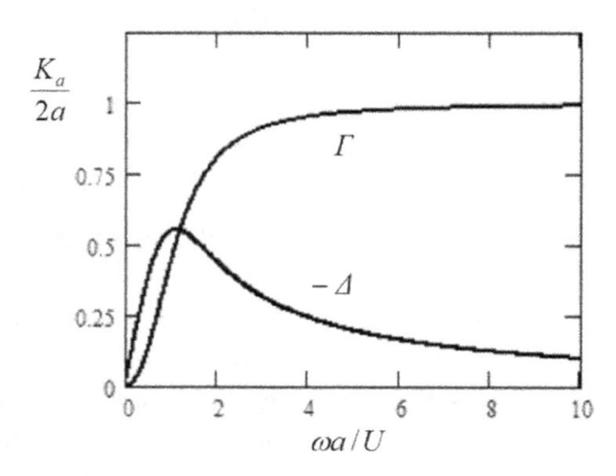

Figure C.5 Rayleigh conductivity, $K_a = 2a(\Gamma + \mathrm{i}\Delta)$, of a circular aperture under subsonic low Mach number mean through flow.

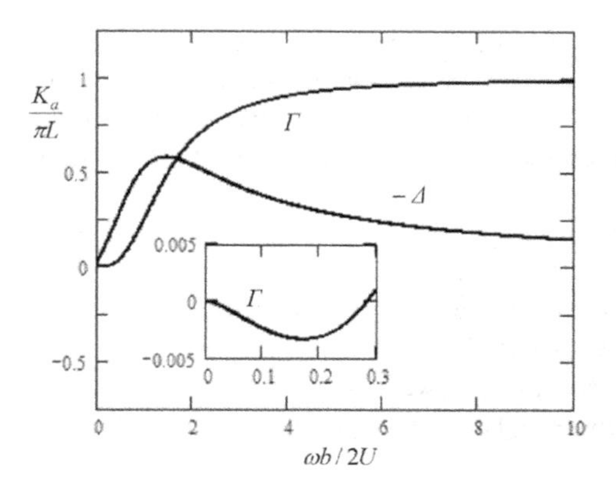

Figure C.6 Rayleigh conductivity $K_a = \pi L(\Gamma + i\Delta)$ of a narrow rectangular slit under subsonic low Mach number mean through flow.

References

[1] J.W. Sullivan and M.J. Crocker, Analysis of concentric tube resonators having unpartitioned cavities, *J. Acoust. Soc. Am.* **64** (1978), 207–215.

[2] K.N. Rao and M.L. Munjal, Experimental evaluation of impedance of perforates with grazing mean flow, *J. Sound Vib.* **108** (1986), 283–293.

[3] A. Cummings, The effects of grazing turbulent pipe flow on the impedance of an orifice, *Acustica* **61** (1986), 233–242.

[4] J.L.B. Coelho, Acoustic characteristics of perforate liners n expansion chambers, PhD Thesis, University of Southampton, UK, (1983).

[5] T. Morel, J. Morel and D.A. Blaser, Fluid dynamic and acoustic modeling of concentric tube resonators/silencers, SAE Technical Paper 910072, (1991).

[6] A.B. Bauer, Impedance theory and measurement on porous acoustic liners, *Journal of Aircraft* **14** (1977), 720–728.

[7] J.W. Sullivan, A method for modeling perforated tube muffler components II: applications, *J. Acoust. Soc. Am.* **66** (1979), 779–788.

[8] R. Kirby and A. Cummings, The impedance of perforated plates subjected to grazing gas flow and backed by porous media, *J. Sound Vib.* **217** (1998), 619–636.

[9] M.S. Dickey, A. Selamet and M.S. Çıray, An experimental study of the impedance of perforated plates with grazing mean flow, *J. Acoust. Soc. Am.* **110** (2001), 2360–2370.

[10] D. Ronneberger, The acoustical impedance of holes in the wall of flow ducts, *J. Sound Vib.* **24** (1972), 133–150.

[11] S.H. Lee and J.G. Ih, Empirical model of the acoustic impedance of a circular orifice in grazing mean flow, *J. Acoust. Soc. Am.* **114** (2003), 98–108.

[12] A. Cummings, Transient and multiple frequency sound transmission through perforated plates at high amplitude, *J. Acoust. Soc. Am.* **79** (1986), 942–951.

[13] M.S. Howe, M.I. Scott and S.R. Sipcic, The influence of tangential mean flow on the Rayleigh conductivity of a aperture, *Proc. R. Soc. Lond.* **A452** (1996), 2303–2317 (Also in: M.S. Howe, *Acoustics of Fluid-Structure Interactions*, (Cambridge: Cambridge University Press, 1998)).

[14] K.S. Peat, J-G. Ih and S-H. Lee, Acoustic impedance of a circular orifice in grazing mean flow: comparison with theory, *J. Acoust. Soc. Am.* **114** (2003), 3076–3086.

[15] M.S. Howe, The influence of mean shear on unsteady aperture flow, with application to acoustical diffraction and self-sustained cavity oscillations, *J. Fluid Mech.* **109** (1981), 125–146.

[16] M.S. Howe, Low Strouhal number instabilities of flow over apertures and wall cavities, *J. Acoust. Soc. Am.* **103** (1997), 772–780.

[17] M.S. Howe, Influence of cross-sectional shape on conductivity of a wall aperture in mean flow, *J. Sound Vib.* **207** (1997), 601–616.

[18] S.M. Grace, K.P. Horan and M.S. Howe, The influence of shape on the Rayleigh conductivity of a wall aperture in the presence grazing flow, *J. Fluids Struct.* **12** (1998) 335–351.

[19] M.S. Howe, Influence of wall thickness on Rayleigh conductivity and flow-induced aperture tones, *J. Fluids Struct.* **11** (1997), 351–366.

[20] S.M. Grace, Prediction of low-frequency tones produced by flow through a duct with a gap, *J. Sound Vib.* **229** (2000), 859–878.

[21] M.S. Howe, On the theory of unsteady high Reynolds number flow through a circular aperture, *Proc. R. Soc. Lond. A,* **366** (1979), 205–223.

[22] A.P. Dowling and I.J. Hughes, Sound absorption by screen with a regular array of slits, *J. Sound Vib.* **156** (1992), 387–405.

Index

CPSIA information can be obtained
at www.ICGtesting.com
Printed in the USA
LVHW051926030821
694430LV00004B/331